ROUTLEDGE LIBRARY EDITIONS:
ECOLOGY

Volume 

THE COMMERCIAL USE OF BIODIVERSITY

THE COMMERCIAL USE OF BIODIVERSITY

Access to Genetic Resources and Benefit-Sharing

KERRY TEN KATE AND SARAH A LAIRD

LONDON AND NEW YORK

First published in 1999 by Earthscan Publications Ltd

This edition first published in 2020
by Routledge
2 Park Square, Milton Park, Abingdon, Oxon OX14 4RN

and by Routledge
52 Vanderbilt Avenue, New York, NY 10017

Routledge is an imprint of the Taylor & Francis Group, an informa business

British Library Cataloguing in Publication Data
A catalogue record for this book is available from the British Library

ISBN: 978-0-367-36640-7 (Set)
ISBN: 978-0-429-35088-7 (Set) (ebk)
ISBN: 978-0-367-35754-2 (Volume 15) (hbk)
ISBN: 978-0-367-35764-1 (Volume 15) (pbk)
ISBN: 978-0-429-34154-0 (Volume 15) (ebk)

Publisher's Note
The publisher has gone to great lengths to ensure the quality of this reprint but points out that some imperfections in the original copies may be apparent.

Disclaimer
The publisher has made every effort to trace copyright holders and would welcome correspondence from those they have been unable to trace.

The Commercial Use of Biodiversity

This book is dedicated to
the memory of Laura Touche.
Her brilliant mind and high
standards, her commitment and
friendship were inspirational, and
made the project not only
possible, but great fun.

The Commercial Use of Biodiversity

Access to Genetic Resources and Benefit-Sharing

Kerry ten Kate and Sarah A Laird

Earthscan Publications Ltd, London

This report was prepared for the
European Commission by
Royal Botanic Gardens, Kew (UK)

First published in the UK in 1999 by
Earthscan Publications Ltd

Reprinted 2000

A catalogue record for this book is available
from the British Library
ISBN: 1 85383 334 7

Edited, designed and typeset by
BDP – Book Development and Production,
Penzance, Cornwall
Printed and bound by Thanet Press,
Margate, Kent
Cover design by Declan Buckley

For a full list of publications please contact:
Earthscan Publications Ltd
120 Pentonville Road
London, N1 9JN, UK
Tel: +44 (0)20 7278 0433
Fax: +44 (0)20 7278 1142
Email: earthinfo@earthscan.co.uk
http://www.earthscan.co.uk

Earthscan is an editorially independent subsidiary
of Kogan Page Ltd and publishes in association with
WWF-UK and the International Institute for
Environment and Development

This book is printed on elemental chlorine-free paper

Contents

List of Boxes, Figures and Tables

Boxes

Figures

Tables

Acronyms and Abbreviations

ABPI Association of the British Pharmaceutical Industry
ACP African, Caribbean and Pacific
ACRE Advisory Committee on Releases into the Environment
ACTS African Centre for Technology Studies (Nairobi)
ADI acceptable daily intake
AH&H Adams Harkness and Hill, Inc
AHA alpha hydroxy acid
AHP American Home Products Corporation
AIPH International Association of Horticultural Producers
ANGAP Association Nationale pour la Gestion des Aires Protegées
AOS American Orchid Society
AOSCA Association of Official Seed Certifying Agencies
APHIS Animal and Plant Health Inspection Service
ARC Agricultural Research Council (South Africa)
ARS Agricultural Research Service
ASEAN Association of South East Asian Nations
ASTA American Seed Trade Association
ATCC American Type Culture Collection
BAC bacterial artificial chromosome
BCCM Belgian Coordinated Collections of Microorganisms
BCP Biological Control Products
BGVS Bedrijf Geneesmiddelen Voorziening Suriname
BHA beta hydroxy acid
BLM Bureau of Land Management (USA)
BMS Bristol Myers Squibb
bn billion
BSPB British Society of Plant Breeders
CABI Commonwealth Agricultural Bureau International
CALM Department of Conservation and Land Management (Western Australia)
CBD Convention on Biological Diversity
CBG Chicago Botanic Gardens
CERES Coalition for Environmentally Responsible Economies
CGIAR Consultative Group on International Agricultural Research
CI Conservation International
CIG Conservation International – Guatemala
CIA Chemicals Industry Association
CIAT Centro Internacional de Agricultura (Colombia)
CIDA Canadian International Development Agency
CILSS Comité Inter-Etats pour la Lutte contre la Secheresse dans le Sahel
CIMMYT Centro Internacional de Mejoramiento de Maiz y Trigo (Mexico)
CIP Centro Internacional de la Papa (Peru)
CITES Convention on International Trade in Endangered Species of Wild Fauna and Flora
CNA Competent National Authority
CONABIA National Advisory Commission for Agriculture Biotechnology (Argentina)
COP Conference of the Parties
CPP crop protection product
CPRO-DLO Centre for Plant Breeding and Reproduction Research, Agricultural Research Department, Wageningen, the Netherlands
CRADA Cooperative Research and Development Agreement
CRO contract research organisation
CTFA Cosmetic, Toiletry, and Fragrance Association
CUP Cambridge University Press
CXL Codex Maximum Limit
DANI Department of Agriculture for Northern Ireland
DFID Department for International Development (UK) (formerly ODA)
DHHS Department of Health and Human Services
DNA deoxyribonucleic acid
DOE Department of the Environment (UK)
DSHEA Dietary Supplement Health and Education Act
DSMZ German Collection of Microorganisms and Cell Cultures (Germany)
DTI Department of Trade and Industry (UK)
DUS distinctness, uniformity and stability
EAPC European American Phytomedicines Coalition
EC European Commission
EFPIA European Federation of Pharmaceutical Industries' Associations
EITO European Information Technology Observatory
EMEA European Agency for the Evaluation of Medicinal Products
EO Executive Order
EPA Environmental Protection Agency (USA)
EPC European Patent Convention
EPPO European Plant Protection Organisation
ESCOP European Scientific Cooperative for Phytotherapy
EU European Union
EUP Experimental Use Permit
FAO Food and Agriculture Organisation
FDA Food and Drug Administration (USA)
FTC Federal Trade Commission
GATT General Agreement on Tariffs and Trade
GDP gross domestic product
GFP Association for the Promotion of German Private Plant Breeding
GIFAP International Group of National Associations of Manufacturers of Agrochemical Products
GMO genetically modified organism
GMP good manufacturing practice
GMTF Green Muscle Trust Fund
GPS global positioning systems
GRAI Genetic Resources Action International
GRAS Generally Recognised As Safe
GRIN Germplasm Resources Information Network
GRRF Genetic Resources Recognition Fund
GTZ Gesellschaft für Technische Zusammenarbeit (Germany)
HDC Horticultural Development Council
HPLC high-performance liquid chromatography
HSE Health and Safety Executive (UK)
IACBR Inter-Agency Committee on Biological and Genetic Resources (Philippines)
IARC International Agricultural Research Centre
ICBG International Cooperative Biodiversity Group
ICH International Conference on Harmonisation of Technical Requirements for Registration of Pharmaceuticals for Human Use
ICRISAT International Crops Research Institute for the Semi-Arid Tropics
IDA Institute fur Demoskopie Allensbach
IDRC International Development Research Centre
IDRI International Development Research Institute
IITA International Institute of Tropical Agriculture
ILO International Labour Organisation

IMI International Mycological Institute
IMS Institute for Medical Statistics
INA Institute for Nutriceutical Advancement
INASE Instituto Nacional de Semillas (Argentina)
INBio National Institute of Biodiversity (Costa Rica)
IND/NDA Investigational New Drug/New Drug Application
INGER International Network for Genetic Evaluation of Rice
INPECO Industria Petenera de Corozo
INTA Centro de Investigaciones de Recursos Naturales del Instituto Nacional de Tecnologia Agropecuaria (Argentina)
InterCEDD International Centre for Ethnomedicine and Drug Development
IOCU International Organisation of Consumers' Unions
IP indigenous peoples
IPGRI International Plant Genetic Resources Institute
IPM integrated pest management
IPPC International Plant Protection Convention
IPR intellectual property right
IRAOH International Registration Authority for Orchid Hybrids
IRRI International Rice Research Institute (Philippines)
ISTA The International Seed Testing Association
ITAL/CAMPINAS Instituto de Téchnologia de Alimentos de Campinas
IU International Undertaking on Plant Genetic Resources for Food and Agriculture
IUCN International Union for Conservation of Nature and Natural Resources
IUPAC International Union of Pure and Applied Chemistry
JAS Japan Agricultural Standard
JMPR Joint Meeting on Pesticide Residues
JPMA Japan Pharmaceutical Manufacturers Association
LAMP Latin American Maize Project
LGU local government unit
LPG liquefied petroleum gas
LRTAP Long-Range Transboundary Air Pollution Convention
LUBILOSA Lutte Biologique contre les Locustes et les Sauteriaux
m million
MAFF Ministry of Agriculture, Fisheries and Food (UK)
MAFF Ministry of Agriculture, Forestry and Fisheries (Japan)
MBG Missouri Botanical Garden
MCBI Marine Conservation Biology Institute
MCP Mount Cameroon Project
MGR microbial genetic resources
MHW Ministry of Health and Welfare (Japan)
MINAE Ministry of Environment and Energy (Costa Rica)
MINEF Ministry of Environment and Forest (Cameroon)
MOSAICC Microorganisms Sustainable Use and Access Regulation International Code of Conduct
MRA Mutual Recognition Agreement
MRL maximum residue level
MSDS material safety data sheet
MTA material transfer agreement
NARS national agricultural research system
NABC National Agricultural Biotechnology Council
NBGSS Nanjing Botanical Garden Service Station
NBPGR National Bureau of Plant Genetic Resources (India)
NC National Collection
NCCPG National Council for the Conservation of Plants and Gardens
NCEs new chemical entities
NCI National Cancer Institute (USA)
NCIMB National Collection of Industrial and Marine Bacteria
NDA new drug application
NEB New England Biolabs
NEDA Netherlands Development Agency
NEJM *New England Journal of Medicine*
NGO non-governmental organisation
NIAR National Institute of Agrobiological Resources (Japan)
NIH National Institutes of Health (USA)
NLEA Nutrition Labeling Education Act (USA)
NPGS National Plant Germplasm System
NYBG New York Botanical Garden
OAEYRG Organizacao dos Agricultores e Extractivistas Yawanawa do Rio Gregorio (Brazil)
OAU Organisation of African Unity
ODA Overseas Development Administration (UK) (now DFID)
ODI Overseas Development Institute
OECD Organisation for Economic Cooperation and Development
OGA Ornamental Growers Association
OTA Office of Technology Assessments
OTC over-the-counter
OUP Oxford University Press
PAMB Protected Area Management Board
PBRs plant breeders' rights
PCR polymerase chain reaction
PCT Patent Cooperation Treaty
PGRFA plant genetic resources for food and agriculture
PhRMA Pharmaceutical Research and Manufacturers of America (USA)
PI plant introduction
PIC prior informed consent
PMA Pharmaceutical Manufacturers Association (USA)
PNGOPRA Papuan New Guinea Oil Palm Research Association
POP persistent organic pollutant
PPRI Plant Protection Research Institute (South Africa)
PVPO Plant Variety Protection Office
PVRO Plant Variety Rights Office
R&D research and development
RAFI Rural Advancement Foundation International
RBG Royal Botanic Gardens
rDNA recombinant DNA
RECA Reflorestamento economico adensado
RECIEL *Review of European Community and International Environmental Law*
RHS Royal Horticultural Society
RNA ribonucleic acid
RRR Renewable Rainforest Resources™
SADC Southern African Development Community
SAES state and territorial agriculture research station (USA)
SASA Scottish Agricultural Science Agency
SBSTTA Subsidiary Body on Scientific, Technical and Technological Advice
SDC Swiss Development Corporation
SEI Sustainable Ecosystems Institute
SENASA National Service for Health and AgriFood Quality
SIDR Strathclyde Institute for Drug Research
SKB SmithKline Beecham
SMP Sarawak Medichem Pharmaceuticals Incorporated
SPC summary of product characteristics

SPREP	South Pacific Regional Environment Programme
SPS	sanitary and phytosanitary
STAFF	Society for Techno-Innovation of Agriculture, Forestry and Fisheries (Japan)
SY	science person years
TBGRI	Tropical Botanic Garden and Research Institute
TCM	Traditional Chinese Medicine
TMEC	Traditional Medicines Evaluation Committee
TRIPS	Trade Related Aspects of Intellectual Property Rights
TSCA	Toxic Substances Control Act
UG	University of Georgia
UIC	University of Illinois at Chicago
ULV	ultra-low volume
UNAM	Universidad Nacional Autonoma de Mexico
UNCED	United Nations Conference on Environment and Development
UNCLOS	United Nations Conference on the Law of the Sea
UNDP	United Nations Development Programme
UN/ECE	United Nations Economic Commission for Europe
UNEP	United Nations Economic Programme
UNIDO	United Nations Industrial Development Organisation
UNP	Universidad Nacional de la Patagonia
UPOV	Union for the Protection of New Varieties of Plants
USAID	United States Agency for International Development
USDA	US Department of Agriculture
USP	United States Pharmacopeia
VCMB	Vanuatu Commodities Marketing Board
VCU	value for cultivation and use
WABEL	Western Australian Biotic Extract Library
WDCM	World Data Centre for Microorganisms
WFCC	World Federation of Culture Collections
WHO	World Health Organisation
WIPO	World Intellectual Property Organisation
WRI	World Resources Institute
WTO	World Trade Organisation
WWF	World Wildlife Fund
ZINATHA	Zimbabwe National Traditional Healers Association

Foreword

The Convention on Biodiversity (CBD) has brought a long overdue international focus and control on the access to, and use of, genetic resources. As the details and consequences of the Convention are worked out by many different bodies, there has been much confusion. For anyone wanting information about the implications of the Convention for their business, research or livelihood, the data are scattered in a confusing array of sources. A book that brings this subject together in such a thorough way will be welcomed by many people. The authors of this book are to be congratulated on their detailed analysis of how genetic resources are used, of the scientific, technological and regulatory trends, and on their analysis of the different markets for many kinds of product based on genetic resources. It is full of examples from which we can learn.

We are sure that this will be an invaluable resource to policymakers, organisations studying and creating uses of genetic resources and the companies that wish to develop them. Anyone who has spent time living in or working with people from the developing world and studying the uses of plants by indigenous peoples will be acutely aware of their interests, but also of the need to protect intellectual property rights in order that uses may be developed which enable these peoples to share in the resulting benefits.

We are also glad to see the extremely broad coverage of the different sectors of industry that use biodiversity. We know of no other analysis of this sort of the horticulture, biotechnology, crop protection and personal care and cosmetics industries, all of which are deeply involved in using biodiversity. The chapters on each sector are presented in a balanced way because of the extensive consultations that the authors held with representatives of each industry discussed. It is our hope that this book will be widely used to help governments to draw up more practical laws and to help industrial companies to ensure the participation of and reasonable compensation for those countries who possess and provide access to so much biodiversity, while at the same time helping industries to avoid legal risks. However, this book will also be invaluable to universities, research institutes and botanic gardens as we work out the new role that has been placed upon us since the CBD was ratified, that of acting as intermediaries between biodiverse countries and would-be users of their genetic resources.

Professor Sir Ghillean Prance, FRS, VMH
Director, Royal Botanic Gardens, Kew

Sir Richard Sykes, FRS
Chairman, Glaxo Wellcome plc

Acknowledgements

Some 300 people were kind enough to provide a wealth of information during the interviews that were the basis for this book, and a team of other dedicated people conducted interviews and background research, worked with us to prepare case studies, and reviewed draft chapters.

Laura Touche helped design the project and guided us in our research, writing and editing, as well as helping to coordinate the administration of a complex research project. Harry Barton, Steve Brewer, Amanda Collis, Mike Griffiths, Dagmar Lange, Ben Lyte, Fiona Mucklow, Kristina Plenderleith, Markus Radscheit, Helene Weitzner, Adrian Wells and China Williams provided much of the data on which the book is based, and helped with the analysis. Their roles in the project are described in the section on research methodology and team (p384). Lucia Liscio volunteered help on quantitative analysis of the results of interviews. Invaluable administrative support was provided by Helen Armitage, Stuart Huntley and Amanda Collis, and the contact details for related institutions were compiled by Stas Burgiel and Craig Metrick.

The following people were very generous with their time and ideas when they reviewed drafts of the chapters and contributed information: **Introduction**: Andrea Bagri and Jeff McNeely of IUCN, and Dominic Moran of the Scottish Agricultural College; **The CBD, national law on access and benefit-sharing and material transfer agreements**: David Downes, CIEL; Colin Pearson, Allen and Overy; and Manolo Ruiz, SPDA; **Pharmaceuticals**: Gordon Cragg, NCI; Todd Capson, Smithsonian Tropical Research Institute; Brad Carte, Ancile Pharmaceuticals; Dwight Baker, Thetagen (formerly MDS Panlabs); David Corley, Monsanto; L H Huang, Pfizer; Murray Tait, AMRAD; Kodzo Gbewonyo, Bioresources International; Peter Hylands, Oxford Natural Products; Hanne Gürtler, Novo Nordisk; Maurice Iwu, Bioresources Development and Conservation Program; Steven King, Shaman Pharmaceuticals; David Newman, NCI; and Josh Rosenthal, NIH; **Botanical medicines**: Dagmar Lange; Steven Foster; Trish Flaster, Botanical Liaisons; Jim Duke; Klaus Duerbeck; Qun Yi Zheng, Pure World Botanicals; Marian Burningham, Nature's Way; Chris Kilham; Trish Stanley; Christopher Hedley and Non Shaw; and staff at *Nutrition Business Journal* and PhytoPharm Consulting; **Major crops**: Henry Shands, USDA; Mike Ambrose, John Innes Centre; Stephen Smith, Pioneer Hybrid; Patrick Heffer, FIS; Patrick Mulvany, ITDG; Richard Hoddinot, Jim Goodford and Anthony Watts, MAFF; Jonathan Davey, SASA; B J Moore, BSPB; and Michael Schechtman, FDA; **Horticulture**: Rob Griesbach, USDA; Bob Bowman, Goldsmith Seeds; Ned Nash, American Orchid Society; Rob Bogers, RIKILT-DLO; Marvin Miller, Ball; Anthony Watts, MAFF; **Crop protection**: David Hunt; Len Copping; Stephen Duke, USDA; Rocky Rowe and Laura Whatley, American Cyanamid; David Dent, CABI; Wolf Maier, European Commission DGVI; Michael Whitaker and Michael Legg, Zeneca Agrochemicals; **Biotechnology (in fields other than healthcare and agriculture)**: Brian and Barbara Kirsop; Lisbeth Anker, Novo Nordisk; Brian Jones, Genencor; Eric Mathur, Diversa; Mike Griffiths; Vanderlei Canhos and Gilson Manfio, Fundaçao Tropical Andre Tosello; Peter Kearns, OECD; James Maryanski, FDA; Peter Hinchcliffe, DETR; Anne Glover, Aberdeen University; David Hales, Ernst and Young; Gert Groot, Gist-Brocades; Seizo Sumida, Japan BioIndustry Association; **Personal care and cosmetics**: Jennifer Morris, Conservation International; May Waddington, Aveda Corporation; Peter Matravers, Aveda Corporation; Helene Weitzner; Paul Pierog and Rob Granader, Kalorama; Kim Holting; **Opinions on the CBD**: Lyle Glowka, IUCN-ELC; and Matthias Jørgensen, European Commission-DG XI; **Conclusions and recommendations**: Matthias Jørgensen, European Commission-DG XI; and Walt Reid, World Resources Institute.

Close collaboration with the individuals involved in individual access and benefit-sharing partnerships is needed in order to prepare case studies. We are extremely grateful to the following for working with us in this way: **Pharmaceuticals**: David J Newman, NP Drugs; **Botanical medicines**: Marian Burningham, Nature's Way; **Major crops**: Gurdev Khush and Michael Jackson, IRRI; **Biotechnology (in fields other than healthcare and agriculture)**: Werner Nader and Nicolas Mateo, INBio; Eric Mathur, Diversa Corporation and Terry Bruggeman (formerly CEO of Diversa); Donald Comb and colleagues, NEB; **Crop protection**: David Dent, CABI; Barbara Timmermann, Arizona University; **Horticulture**: He Shan An, Nanjing Botanic Garden; Pierre Piroche, Piroche Plants; and Peter Hunt, International Orchid Registrar; **Personal care and cosmetics**: May Waddington, Aveda Corporation; and Jennifer Morris, Conservation International.

We would like to thank each and every one of the people who answered questions and provided information, opinions and advice, and directed us to valuable information, who are too many to name. We have been touched and impressed by the kindness of all these people in making extra time, on top of their demanding jobs, to work with us, often to very tight deadlines.

The research leading to this book was supported by the Institute for Prospective Technical Studies of the Joint Research Centre of the European Commission, and coordinated by the Royal Botanic Gardens, Kew. It builds upon earlier research, including interviews with 35 companies, conducted by the same authors with the sponsorship of the World Resources Institute, in collaboration with the Royal Botanic Gardens, Kew. The preliminary findings of this research were presented during workshops at the third meeting of the Subsidiary Body on Scientific, Technical and Technological Advice (SBSTTA) of the Convention on Biological Diversity in Montreal in September 1997 (see ten Kate and Laird, 1997).

We are tremendously grateful to Claude Tahir and Matthias Jørgensen of the European Commission for supporting the research that has culminated in this book; to Ghillean Prance, Simon Owens and Noel McGough of the Royal Botanic Gardens, Kew, for enabling the project to go ahead, and to Kenton Miller, Chip Barber and Walt Reid of WRI for enabling us to start our research and allowing us to take it forward to this stage.

We have benefited enormously from the contributions of all these individuals and institutions, who have made the project possible and this book more accurate and informative. Its remaining shortcomings are our responsibility alone.

KERRY TEN KATE
LONDON

SARAH A LAIRD
NEW YORK

Kerry ten Kate and Sarah A Laird

Chapter 1
Introduction

1.1 The market for genetic resources

This book examines the demand for access to genetic resources for commercial development, the partnerships between scientists, companies, governments and communities involved, and how the benefits that arise from access to genetic resources are shared. It aims to provide governments and other organisations with information on corporate practices and perspectives on access and benefit-sharing. It offers companies information on best practice within each industry sector, and some background information on their legal obligations under the Convention on Biological Diversity (CBD) and the laws and policies introduced in various countries to implement it.

> *'Genetic resources' may be defined as biological materials of actual or potential value, containing functional units of heredity.*[1]

'Genetic resources' are biological materials of animal, plant, microbial, or other origin that contain the hereditary information necessary for life and are responsible for their useful properties and ability to replicate.

For millennia, people have managed genetic resources, selecting crops, brewing beer and harvesting medicinal plants. Genetic resources remain the basis for the improvement of agricultural crops; for medicines for 75 per cent of the world's population that relies upon traditional, largely plant-based treatment for its primary healthcare; and for a myriad of other products such as pharmaceuticals, crop protection products and perfumes. The biotechnology industry has harnessed living organisms to degrade refuse and clean up polluted sites, to provide energy, help make paper, detergents, and to separate valuable minerals from ore. More controversially, genetically modified organisms (GMOs) are now commonly used in healthcare and in agriculture. Molecular biologists developing transgenic organisms require access to genetic resources, but so too do amateur gardeners raising plants, students studying life sciences and companies screening samples from natural sources for research and development. We hope this book will be useful to them all.

The industry sectors we cover are pharmaceuticals, botanical medicines, major crops, horticulture, crop protection products, applications of biotechnology in fields other than healthcare and agriculture, and cosmetics and personal care products. From our study, a crude estimate of combined annual global markets for the products derived from genetic resources in these sectors lies between US$500 billion and US$800 billion (Table 1.1). By comparison, annual global sales of petrochemicals are some US$500 billion[2] and the world computer market (including software, hardware and information technology services) in 1997 was US$800 billion.[3] The broad range in the estimate of the global market for genetic resources results from the challenge of calculating total global figures for highly complex and varied markets, such as agriculture and biotechnology, and of identifying the extent of the use of genetic resources in other markets, such as pharmaceuticals and crop protection. It is important to note that this estimate of certain commercial markets for genetic resources is both rough and speculative, and does not cover the use of genetic resources in other sectors, nor the vast local markets and subsistence-level use of genetic resources that are not captured in national or industry statistics. Additionally, as this book will explore, not only genetic resources but human ingenuity and considerable investment in research and development are needed to develop the products concerned. Sales of final products therefore offer little indication of what companies will pay to gain access to the genetic resources that form the starting point for research.

Table 1.1 *'Ballpark' Estimates for Annual Markets for Various Categories of Product Derived from Genetic Resources*

Sector	Market (US$ bn) LOW	Market (US$ bn) HIGH	Notes/Chapter where estimates developed
Pharmaceuticals	75	150	Some products derived from genetic resources. Low estimate: natural products form 25 per cent of global market. High estimate: 50 per cent. Ch 3.
Botanical medicines	20	40	All products derived from genetic resources. Low estimate for global botanical medicine markets; high estimate includes botanical medicines, minerals, and vitamins. Ch 4.
Agricultural produce (commercial sales of agricultural seed)	300+ (30)	450+ (30)	All products derived from genetic resources. Low estimate: final value of produce reaching consumer 10x commercial sales of seed to farmers. High estimate: 15x commercial sales of seed to farmers. Ch 5.
Ornamental horticultural products	16	19	All products derived from genetic resources. Low estimate: based on available data. High estimate: allows for unreported sales and products. Ch 6.
Crop protection products	0.6	3	Some products derived from genetic resources. High estimate includes wholly synthesised analogues, as well as semi-synthesised products. Ch 7.
Biotechnologies in fields other than healthcare and agriculture	60	120	Some products derived from genetic resources. Low and high estimates based on assessments of environmental biotechnologies. Ch 8.
Personal care and cosmetics products	2.8	2.8	Some products derived from genetic resources. Reflects 'natural' component of the market. Ch 9.
Rounded total	500	800	

At the level of individual products, genetic resources and their derivatives fetch prices that range from just a few cents to tens of millions of dollars per kg, and often command prices far higher than standard indicators of value such as gold (Table 1.2).

Products that result not only from access to genetic resources but also from lengthy programmes of research and development inevitably command far higher prices than basic raw materials. Thus clippings from the yew tree fetch some US$0.75 per kg, but the anti-cancer drug taxol, which is made from these clippings, costs US$12 million per kg.

Sophisticated research programmes are not always required to add value to genetic resources. Illegal trade in a breeding pair of the highly endangered Lear's Macaw can fetch US$260,000 (Burrell, 1998) and tiger bones for traditional Chinese medicine (TCM) are worth US$3,000 per kg (personal communication Crawford Allen, TRAFFIC, 5 February 1999). These prices are one factor which may contribute to the decline of these species, without a raft of properly enforced legal and policy measures.

Table 1.2 *Prices for Selected Genetic Resources and Derivatives*

Commodity (in *italics* if not derived from genetic resources)	Retail price[4] per kg or litre (US$)
Human growth hormone	20,000,000
Taxotere/docetaxol	12,000,000
Vincristine sulphate	11,900,000
Cocaine	150,000
Camptothecin	85,000
Lear's Macaw	24,000
Gold	*10,000*
Dry bear gall bladders	7,000
HIV protease inhibitor	*5,000*
Saffron	6,500
Tiger bones	3,000
Italian truffle	650
Shark fin (personal care)	550
Coffee	10
Cotton	1.5
Petrol	*1.0*

1.2 The economic value of biological diversity

Biological diversity, or 'biodiversity', provides the basis for life on earth. The variability among living organisms and among the ecological complexes of which they are part not only forms the basis of many commercial products, but also underpins our very existence, by providing essential ecosystem services such as the purification of water, prevention of soil erosion and floods and regulation of the climate.

Despite its value in economic terms and to the cultural and spiritual well-being of people around the world, many experts agree that biological diversity is declining rapidly (Box 1.1). In order to catch the attention of policy-makers and to support the conservation of biodiversity, environmental economists have described a number of values of biodiversity (such as consumptive and productive use values, non-consumptive use values, and non-use option and existence values), and a range of techniques to measure them (see, for example, WRI, 1999; Pearce and Puroshothaman, 1993). In 1997, Robert Costanza, from the University of Maryland, and several of his colleagues estimated that the economic value of ecosystem services for the entire biosphere was in the range of US$16–54 trillion ($10^{12}$) per year, with an average of US$33 trillion a year, compared with the annual global gross national product of US$18 trillion (Costanza et al, 1997). Other recent studies have put this exercise into perspective, examining some of its critical assumptions. One such study is by Pearce et al (1999). They evaluate the economic evidence emerging from an increasing body of economic models on pharmaceutical prospecting that seek to mimic the search process and untangle the likely value of the base resources that underly a 'hit'. A wider view of the economics of biodiversity is also provided by Moran and Pearce (1997).

Although such valuation exercises can provide only a crude estimate of the value of biodiversity and do not adequately capture its extensive cultural, religious and aesthetic values, they do demonstrate a central critical point: biological diversity makes an important contribution to the economy, and influences the prospects for sustainable economic development and for conservation.

Box 1.1 *Hierarchical Categories of Biodiversity*

Genetic diversity refers to the variation of genes within species, and might cover distinct populations of the same species or genetic variation within a population. Different combinations of genes within organisms, or the existence of different variants of the same basic gene are the basis of evolution.

Species diversity refers to the variety of species within a region. Current estimates of global species diversity range between 8 million and 100 million species, with 10–13 million being considered a 'best estimate', although only 1.4 million species have been scientifically named. Species extinctions have received the most attention in debates on biodiversity conservation. Conservative estimates are that current extinction rates for well-documented groups of vertebrates and vascular plants are at least 50 to 100 times larger than the expected natural background. There are expectations that extinction rates, particularly in the high-biodiversity tropics, could rise to as much as 10,000 times the natural level.

Ecosystem diversity is more difficult to measure than species or genetic diversity because the associations of species and ecosystems are elusive, but criteria have been used to define communities and ecosystems, primarily at a national and sub-national level. In addition to the components of diversity – species, genes, and ecosystems – it is important to look at diversity in ecosystem structure and function, such as relative abundance of species, the age structure of populations, the pattern of communities in a region, change in community composition and structure over time, and ecological processes.

Cultural diversity refers to the diversity in cultures and cultural practices which have grown from biological diversity, and which in turn have impacted the diversity we see today. The majority of the world's biodiversity is closely tied to traditional management, resource harvesting, and livelihood practices, and many 'natural' areas bear the mark of the interconnection between cultural and biological diversity.

Sources: WRI, 1999; UNEP *Global Biodiversity Assessment*, 1996; Janetos, 1997.

1.3 The CBD and national laws on access and benefit-sharing

During the 1980s, concern about the loss of biodiversity mounted. In developed countries, public attention focused on the vanishing 'medicinal riches' of the rainforest, bolstering an argument for the conservation of biodiversity, while at the same time conservation 'professionals' were grappling with ways to conserve something as complex. Existing environmental legislation and policy, and most conservation programmes, did not address the complex range of approaches necessary to conserve the genetic resources, species, habitats, and ecosystems of biodiversity.

As late as November 1983, the members of the Food and Agriculture Organisation (FAO) (at that time, some 150 governments) adopted the International Undertaking (IU) on Plant Genetic Resources. The objective of this voluntary agreement was to 'ensure that plant genetic resources of economic and/or social interest, particularly for agriculture, will be explored, preserved, evaluated and made available for plant breeding and scientific purposes'. Article 5 of the Undertaking referred to the 'universally accepted principle that plant genetic resources are a heritage of mankind and consequently should be available without restriction'. Within just a few years, that principle turned out to be far from universally accepted.

Following a US proposal for an umbrella treaty on the conservation of biological diversity at the Governing Council of the United Nations Economic Programme (UNEP) in June 1987, developing countries began to ask who would pay for the costly conservation programmes that such a treaty might require. At the same time, they sought to correct the historical inequities in the trade in genetic resources. By the subsequent meeting of the UNEP Governing Council in May 1989, it was plain that developing countries would support the initiative for a biodiversity treaty only if it were founded on national sovereignty over genetic resources, and promoted a more equitable sharing of the benefits arising from the commercial use of genetic resources (McConnell, 1996). Developing countries argued that, historically, they had received next to nothing for their genetic resources, while much of the economic advantage of the colonial powers had been gained through free access to genetic resources (Juma, 1989). Since the pharmaceutical and biotechnology industries used raw material originating in developing countries, these countries felt that any treaty on the conservation of biological diversity should also address 'the use of genetic resources in biotechnology development' (McConnell, 1996). The result was much more than an environmental treaty – it was a hybrid treaty addressing the environment, trade, development, and intellectual property rights issues (Gollin, 1993).

After intense negotiations that started in August 1990, the CBD was opened for signature at the United Nations Conference on Environment and Development held in Rio de Janeiro in 1992. The CBD has subsequently been ratified by 174 countries and the European Union (EU). The objectives of the CBD are the conservation of biological diversity, the sustainable use of its components and the fair and equitable sharing of the benefits arising from the use of genetic resources. The Convention can be seen as an instrument to promote the equitable exchange, on mutually agreed terms, of access to genetic resources and associated knowledge in return for finance, technology and the opportunity to participate in research. Some authors have referred to this exchange, which lies at the heart of the Convention, as the 'grand bargain' (Gollin, 1993).

Provisions of the Convention and national laws that implement them set the scene for any company or individual seeking access to samples of plant, animal, fungal or microbial origin for scientific research or as the starting point for commercial development. At the last count, laws and other policy measures aimed at securing fair business partnerships had been introduced or were under development by the governments of Argentina, Australia (at the Commonwealth level and in the states of Western Australia and Queensland), Belize, Bolivia, Brazil, Cameroon, Colombia, Costa Rica, Ecuador, Ethiopia, Eritrea, Fiji, The Gambia, Ghana, Guatemala, India, Indonesia, Kenya, Laos PDR, Lesotho, Malawi, Malaysia (including the State of Sarawak), Mexico, Mozambique, Namibia, Nigeria, Papua New Guinea, Peru, Philippines, the Republic of Korea, Samoa, Seychelles, Solomon Islands, South Africa, Tanzania, Thailand, Turkey, United States of America, Venezuela, Vietnam, Yemen and Zimbabwe. As well as the regional law introduced by the five member states of the Andean Commission (Bolivia, Colombia, Ecuador, Peru and Venezuela), the Association of South East Asian Nations (ASEAN), the Organisation of African Unity (OAU) and the South Pacific Regional Environment Programme (SPREP) are also exploring the possibility of regional access measures (personal communication, Lyle Glowka, February 1999). Other countries are likely to join this list in the coming years.

As we shall see in the sector Chapters 3–9, the benefits in which providers of genetic resources wish to share can take many forms: financial, scientific, social and environmental. Monetary benefits might include fees per sample, grants to cover agreed research programmes, profit-sharing, stakes in equity, joint ventures, royalties and the prospect of local employment opportunities. Countries supplying samples to industry are increasingly familiar with the costs, risks and delays inherent in product development. Rather than pinning their hopes on the slim chance of a royalty payment twenty years or more down the line, they are increasingly prioritising 'non-monetary' benefits such as the sharing of research results, participation in research, technology transfer, training and capacity building. Some partnerships offer help in kind, such as medical assistance and investment in local infrastructure. Others support conservation projects in the field.

1.4 The actors

The CBD was negotiated by and is now legally binding upon states, but a range of individuals and organisations are involved in the mechanics of access to genetic resources and the development of commercial products from them. The key players are the private sector, the public sector, intermediary organisations (which may be private or public), and local and indigenous communities.

1.4.1 The private sector

Private industry plays the dominant role in the commercial use of biodiversity through the discovery, development and marketing of products based on genetic resources. While the international and national legal and policy framework has evolved rapidly over the last decade, so too have science and technology. Advances in molecular biology, genomics, combinatorial chemistry, robotics, microelectronics and informatics have radically changed the private sector's approach to product discovery and development in medicine, agriculture and environmental management for the 21st century.

The structure of companies and markets is also changing. While small, research-based companies, many with an orientation towards genetic resources, are proliferating, the number of mergers and acquisitions is also on the rise in some sectors. One result is a number of giant 'life science' companies such as Monsanto, Novartis, Aventis, Zeneca and American Home Products, formed through high-profile mergers between multinational companies in the pharmaceutical, biotechnology, seed and crop protection sectors. These companies move genetic resources and their derivatives around the world, and have combined turnovers greater than the gross domestic product (GDP) of several high biodiversity countries.

Over the last 20 years, policy-makers have come to view industry not only as the cause of environmental damage, but as a key player in any plans to develop environmentally sustainable economies (Schmidheiny, 1992), and the approach to regulating industry's impact on the environment has evolved from one of pure command and control to a mixture of different approaches, including covenants and voluntary instruments (ten Kate, 1994).

Motivation on the part of business to comply with the CBD will be essential if the objectives of the treaty are to be fulfilled. To begin with, as the chapters that follow will show, the private sector is a major user of genetic resources. Furthermore, participation in scientific research conducted by industry, and the transfer of technology developed in the private sector are among the most effective ways to share benefits. The views of the private sector on the CBD and the nature of commercial partnerships will inevitably inform the manner in which benefits are shared in practice. They will also influence the extent to which biological resources are used sustainably, and whether or not this use will create incentives for conservation.

The monitoring and enforcement of access legislation and agreements is difficult, since it involves identifying the source and date of collection of specimens and tracking a product's movement through the discovery and development pipeline. Law and procedure on access are unclear in the vast majority of countries, and user countries show little inclination to introduce laws to support enforcement of access agreements in the countries where companies conduct their research and development. For these reasons, voluntary compliance by industry will be essential, if the principles of prior informed consent and the fair and equitable sharing of benefits are to be realised.

1.4.2 The public sector

Government bodies play a number of different roles related to the commercial use of biodiversity. To begin with, many governments fund and maintain large national collections of genetic resources, and these are much used as a source of materials by the private sector. Also, despite the gradual trend towards the privatisation of research institutes around the world, public-sector organisations still play an important role in the research and development, and even the distribution and sale, of products derived from genetic resources. In many countries, for example, there is little private sector involvement in the development and sale of agricultural seed, and public sector organisations such as the US National Institute of Health (NIH) are actively involved in collecting genetic resources as the basis for their pharmaceutical discovery and development programmes.

A completely different role of government is to establish the legal and policy framework within which individuals and organisations may commercialise biodiversity. Relevant laws include not only the regulation of access to genetic resources, phytosanitary permits and trade in endangered species, but also trade and investment, and real and intellectual property law.

1.4.3 Intermediaries

Companies generally gain access to the majority of the samples they acquire through intermediaries such as botanic gardens, universities, research institutions, culture collections, genebanks, and for-profit brokers. In addition to providing collection and scientific services, some intermediaries broker access and benefit-sharing relationships with source countries, sometimes as agents of companies, and sometimes independently. Several intermediaries may be involved between the initial collection stage and the ultimate commercialisation. Consequently, intermediaries play a key role in obtaining access to genetic resources and in determining benefit-sharing relationships, and their role in the commercialisation of biodiversity has excited considerable interest. As *Diversity* magazine puts it, 'Many refer to these intermediary institutions as "leaky", since they allow access to genetic resources without dealing directly with source countries, which might require prior informed consent or the sharing of benefits for access to their genetic resources' (*Diversity*, 1998).

1.4.4 Communities

Local and indigenous communities are involved in biodiversity management, conservation, and trade in genetic resources. Living in close proximity to biodiversity, local communities have developed traditional knowledge and management systems that are closely linked with the natural resources of which they have long been stewards and managers. Academic and commercial researchers seek access to resources on communities' lands, and to traditional knowledge relating to species and ecosystem management as well as uses, such as the many pharmaceutical drugs developed from traditional medical systems. Access and benefit-sharing legislation in some countries, such as the Philippines and the member countries of the Andean Community, require researchers to seek communities' prior informed consent for access to their knowledge or resources.

The commercialisation of genetic resources involves and affects all these actors, since it requires the conservation of biodiversity as the source of genetic resources, access to resources and information about them, research and development, customers for the resulting products, and a regulatory environment that makes the complex relationships between the various actors possible. Where the legal, economic and social environment functions well, there can be a thriving economy based on genetic resources and contributing to sustainable development and the conservation of biological diversity. Where the regulatory framework is inappropriate, when the economic climate favours alternative methods of product development to the use of genetic resources, and where the different actors cannot work together, these opportunities do not arise. One of the main problems is that there is currently little effective dialogue between the different groups involved in and affected by the commercialisation of biodiversity.

1.5 The different perspectives

The lack of effective dialogue on access to genetic resources can be attributed partly to entrenched positions, as well as to limited opportunity for interaction. Although there is a wide range of different perspectives, in this section we have simplified the arguments and will distinguish between two main groups: 'source countries' and 'companies'. 'Source countries' might include governments, community groups, non-governmental organisations (NGOs), and others working on behalf of source country interests. 'Companies' represent the private sector organisations we interviewed as part of this study. Our presentation of positions here is necessarily a caricature, and is intended to introduce the polarised positions that we believe are stalling the more effective dialogue needed to implement the CBD (Box 1.2).

1.5.1 The companies' argument

Investment in research and development and patents

Companies argue that it is the many years of research and development, and the tens or hundreds of millions of dollars invested in them, that convert 'valueless' genetic resources into final products. They point to the financial risk that a company must bear when it develops a new product, and state that patents over products and processes are the only way to recoup the investment needed for product development.

Competition from other approaches to new product discovery and development

Companies believe that it is impossible to pay more than they currently do for the raw materials and remain competitive; if natural product research becomes too expensive it will be abandoned for other more profitable approaches (as several natural product companies did in the 1990s). New technological developments in synthetic and combinatorial chemistry, genomics and bioinformatics offer alternative approaches to natural products, and investment in these approaches in the future is likely to be at the expense of research into natural products. Demand for access to genetic resources has waxed and waned over the last thirty years, but recent advances in biotechnology may have broken the pattern once and for all, so that there will be little demand for access to genetic resources in the future. If nothing else, demand for natural products will be altered to such an extent that the model of resource exchange source countries from the past (with large-scale, 'random' collections) is now largely irrelevant, as more focused, targeted collections of smaller numbers of samples become the norm.

It is business, after all

While companies are profitable, the current economic climate is not favourable, and their shareholders will not countenance lower profit margins or higher prices for raw materials. Companies believe that source

Box 1.2 *Countries and Companies: A Sketch of the Different Perspectives*

Source countries and communities think ...	*Companies think ...*
■ Companies make millions from genetic resources sourced from countries and can afford to help the countries pay to conserve them. ■ Companies don't acknowledge the contribution of the genetic resource to the final product. ■ For hundreds of years, the North has been profiting at the expense of the South, and the time has come to correct inequitable practices.	■ Third World countries are trying to make money from something they do not contribute to. ■ You don't hear them talking about sharing the risk, only the benefits. ■ Greater skills and capacity will always fuel economic development. ■ It is not the role of the private sector to conserve genetic resources or to correct historical inequities.
■ A naturally-derived product could never be discovered if it were not for the genetic resource on which it is based. ■ Genetic resources are hardly 'valueless', but rather hold the genetic and biochemical key to product development. It's fair to ask for benefits in return for granting access to them.	■ It is the ten or more years of research and development, considerable financial risk and the several hundred millions of dollars invested in it that convert a 'valueless' genetic resource into a final product. ■ Source countries value their raw genetic resources too highly and ask too much for them.
■ Patents create monopolies at the expense of those who provided the original material and much of the development. Patents give an unfair advantage to those who took the last step along the road of development. This step often barely satisfies any reasonable test of 'inventiveness' or 'novelty'. ■ Patents result in final products sold back at vastly inflated prices to countries where the material came from.	■ Patents on products and processes are the only way to recoup the investment needed for product development. ■ Patents don't influence access to the original material, and Plant Breeders' Rights don't stop others from using protected varieties for breeding.
■ Companies send teams of collectors to scour the world in search of new drugs, etc. They do not reach proper agreements and are 'ripping off' countries. ■ Companies make massive profits from exploiting genetic reources, witness the 30 per cent profit margin of the pharmaceutical industry. ■ Companies' demand for samples is so high that they will pay high access fees, such as the US$1m paid by Merck to INBio in Costa Rica.	■ Companies have plenty of genetic resources within their own collections or available for free from *ex situ* collections in their own countries, and many approaches to product discovery other than using genetic resources. ■ It is impossible for companies to pay more than they currently do for the raw materials, or natural product research will become uncompetitive. ■ Companies cannot and will not pay 'access fees' of any great magnitude for raw samples. Merck paid INBio to conduct an agreed work programme.

countries value their raw genetic resources too highly and ask too much for them, thus pricing themselves out of the market. Companies are prepared to offer a fair price for materials, but they are highly reluctant to engage in joint research that reveals proprietary technology to potential competitors, particularly during the sensitive discovery stage before a new product is safely protected with intellectual property rights. In any case, source countries frequently cannot provide the quality, volume, efficiency and timeliness of service to make them attractive to industry as partners for collaborative, value-added research. The loss of biological diversity is not a business issue, since companies are confident of access to adequate collections of genetic resources to fuel research for another few decades. It is not economically viable for companies to concern themselves with longer-term considerations. It is up to governments to set the policy framework in such a way as to support conservation and long-term sustainability.

Access legislation

The access legislation introduced in some countries in order to ensure prior informed consent and benefit-sharing is unclear, bureaucratic, time consuming and expensive to comply with. The result is that companies will avoid these countries and work elsewhere. They have plenty of genetic resources within their own collections or available free from *ex situ* collections in their own countries. In these circumstances, they do not need access to genetic resources very often, and so have little motivation to negotiate with source countries.

Benefits

Countries expect unreasonable benefits, all of which are predicated on the profits and research budgets in high-value sectors such as pharmaceuticals. These expectations do not take into consideration the costs and risks of research and development in the pharmaceutical sector, let alone more modest research budgets and profit margins in other industry sectors. Countries need to tailor their expectations for benefit-sharing to the industry sector concerned and also to the circumstances of the individual case.

1.5.2 The source countries' argument

Companies need genetic resources

The source countries argue that a naturally-derived product could never be discovered if it were not for the genetic resource on which it is based, which is hardly 'valueless', but rather holds the genetic and biochemical key to product development. Approaches to product development that use existing collections or synthetic chemistry may provide a greater share of products in the future and may indeed be cheaper than work with genetic resources. However, natural products, or products derived from genetic resources, continue to make an important contribution to the range of new products, and the novelty and diversity of naturally occurring compounds and the attractive traits of genes found in nature have yet to be rivalled by man's ingenuity. Source countries acknowledge ruefully that companies already possess large collections sourced from their territory in years gone by, but point out that while large, *ex situ* collections are already freely available to companies, most genetic resources are not represented in these collections; only 1.4 million species have been scientifically named, of an estimated 10–13 million. Furthermore, many of the same samples are replicated in collections around the world, and specimens held outside their natural habitat do not always exhibit the same desirable properties as they do when *in situ*.

Benefit-sharing is reasonable

Countries are aware of the risks involved in product discovery and development, but are only asking for a 'fair and equitable' share in benefits that arise. It is reasonable to ask for a share of royalties at the stage when the risk has paid off and a product is profitable. Companies would pay royalties on a synthetic lead compound, so why not one derived from nature? Ultimately, companies are profitable or they would not still be in business. The average profit margin of pharmaceutical companies in 1996 was 30 per cent. However, money is not the prime concern of source countries. They would prefer to engage in joint research with companies, in order to receive technology and to build their capacity to add value to their own genetic resources. They will share the financial risk, where appropriate, provided they obtain a fair share of the ultimate reward.

Access legislation

Access legislation is there to protect rights, but is intended to be flexible enough to allow for negotiation and agreement on 'mutually agreed terms'. Companies are accustomed to an 'open access' regime developed when genetic resources were considered the 'common heritage of mankind'. It may take time for academic and commercial researchers to adapt to the new paradigm of national sovereignty, prior informed consent and benefit-sharing, but the approach is more fair and equitable, and will, in the long run, conserve the very resources which the researchers wish to use. It may also take some time for source countries to iron out the problems involved in so difficult a challenge as regulating access to genetic resources, but in time the 'grand bargain' can be effectively achieved.

These perspectives have been articulated in a number of different publications and intergovernmental negotiations. Both sets of arguments are legitimate, and there is room to find solutions that satisfy both sides. However, there is often inadequate factual information available for each side to gauge the merits of the other side's arguments, and insufficient understanding and trust on each side to find solutions. The lack of information about the nature and scope of demand for access or the partnerships that providers and users of genetic resources can form has hampered international negotiations, national policy-making processes, and individual negotiations between institutions. Without such information, it is difficult to create national or corporate strategies, or to enter into partnerships, without running the risk of breaking the law or ending up with a poor bargain.

Countries developing access laws need information on markets and partnerships with business so that they can decide what they hope to gain strategically from granting access to their genetic resources. Such information will help them to draft legislation that is cost-effective and feasible to administer, as well as streamlined and practical to follow. Without it, policy-makers may frame laws that neither encourage beneficial partnerships nor promote conservation. National institutions and local communities collecting and managing genetic resources need to understand the nature of commercial demand for their resources, in order to better control access, protect their rights, and effectively design research collaborations and benefit-sharing packages.

At the same time, companies continue to invest hundreds of millions of dollars in product development based on genetic resources to which they may not be able to guarantee legal title. Few of the companies engaged in natural product research and development (or their industry association representatives) attend the international negotiations that set the scene for access legislation, nor do they lobby individual countries to develop laws that facilitate access. Companies need to understand their legal obligations, and should be actively involved in drafting national access laws that promote the research and collaboration upon which benefit-sharing depends.

1.6 The sectors

This book studies the pharmaceutical industry, the commercial use of biotechnology in fields other than healthcare and agriculture, the crop protection industry, seed companies developing major crops, the horticulture industry, companies developing botanical medicines, and those producing cosmetics and personal care products. There is enormous variety within and between these sectors in market size and growth; expenditure and strategy for research and development; the cost, time, and probabilities of success involved in developing commercial products from natural product samples; and the geographic base for the majority of the research and development activities concerned.

Where Table 1.1 gave estimates of the often substantial markets for various products derived from genetic resources, Table 1.3 gives the other side of the story: the duration of research and development programmes and levels of investment required to bring a product to the market. These factors, coupled with the different probabilities that candidate products will not reach the market at all, offer some explanation for the fact that global markets in products derived from genetic resources do not translate into companies' willingness to pay for access to the genetic resources that they regard as raw materials and as just the starting point of the research and development that produces products.

1.6.1 Market size and growth

By far the largest markets represented by these sectors are those for agricultural produce and the pharmaceutical industry. Global sales of pharmaceuticals are some US$300 billion a year (*The Economist*, 1998), of which the component derived from genetic resources accounts for between US$75 billion and US$150 billion (Chapter 3). In contrast, personal care and cosmetic products are part of an industry with annual 1997 sales of $55 billion, the 'natural' component of which is estimated at US$2.8 billion (*NBJ*, January 1998) (Chapter 9). Similarly, the global market for crop protection products is some US$30 billion, of which products derived from genetic resources account for between US$0.6 billion and US$3 billion, depending on the definitions used (Chapter 7).

It would be a life-time's project, and even then a challenge, to calculate global sales of finished agricultural products. Global sales of commercial seed amount to some US$30 billion a year (Chapter 5). Seed companies sell the seed to farmers, who multiply it and sell it to their customers, which include distributors and industrial and food processing companies. These companies either sell produce (such as potatoes) directly to the consumer, or they manufacture products (such as bread) which are then sold to the consumer, often via other retail outlets. The complex chain from the seed companies to the final consumer makes any estimation of the global market for the products arising from global agricultural and horticultural vegetable crops extremely complex, but it is safe to assume that it is well in excess of the US$300 billion global sales of pharmaceuticals.

Average projected growth varies across sectors. The pharmaceutical industry is projected to grow at a steady 6 per cent per year for the next few years (*The Economist*, 1998). The botanical medicine industry is projected to expand by 10–20 per cent in most countries (*Whole Foods*, 1998; Gruenwald, 1997). The natural component of the personal care and cosmetic industry is estimated to grow between 10 and 20 per cent per year (*NBJ*, January, 1998). A huge range of biotechnological products and processes are used in many industry sectors, so it is difficult to generalise about trends in this

Table 1.3 *Comparison of Duration and Cost of Typical Research and Development Programmes in Different Industry Sectors*

Sector	Years to develop	Cost (US$ m)
Pharmaceutical	10–15+	231–500
Botanical medicines	<2–5	0.15–7
Commercial agricultural seed	8–12	1–2.5
Transgene	4+	35–75
Ornamental horticulture	1–20+	0.05–5
Crop protection	(biocontrol agent) 2–5	1–5
	(chemical pesticide) 8–14	40–100
Industrial enzymes	2–5	2–20
Personal care and cosmetic	<2–5	0.15–7

field, but, to give one example, the bioremediation sector is growing at roughly 10 per cent per year.

1.6.2 Research

Research expenditure, including that on genetic resources, varies dramatically by sector. As a result, while the pharmaceutical industry might generate the greatest revenue, it also spends the most on research and product development. The pharmaceutical industry is the most research-intensive industry in the world, spending on average 20 per cent of sales on research and development in 1998, with a total expenditure of US$21.1 billion (PhRMA, 1998). In contrast, the botanical medicine industry spends very little researching leads for new product development, and instead conducts research primarily in the area of quality control, safety, efficacy, and formulation of known plants. The combined research budgets of the top ten multinationals which account for the breeding of some 90 per cent of ornamental horticultural varieties is probably around US$200 million.

This study concentrates on research and development in Western Europe and the USA, but also includes interviews with companies and organisations in Canada, Japan, Australia and New Zealand and several countries in Central and South America, Africa and Asia. To some degree, this emphasis reflects the concentration of research and development activities in these regions. Europe, the USA and Japan dominate research and development and markets in most of the sectors, including pharmaceuticals, biotechnology, botanical medicines, and personal care and cosmetics.

1.6.3 Cost and time to develop a product

There is enormous variation between industry sectors in the cost and time it takes to develop a product from an original natural product sample, and similar differences in the probabilities that any one sample will succeed as a commercial product. In the pharmaceutical industry, average drug discovery and development times are between 10 and 15 years. The cost of this process varies widely, but estimates generally lie between US$231 million (Di Masi et al, 1991) and US$500 million (PhRMA, 1998). The odds of a successful commercial product arising from a single natural product sample are 1 in 5,000 to 1 in 10,000 (McChesney, 1996; PhRMA, 1998). In the botanical medicine industry, the process of new product development that includes standardisation of active compounds, quality control, and standards for safety and efficacy, can take less than two years, and the cost, which varies dramatically according to the nature of the research, processing and locations where these take place, is often less than US$1 million.

Within each sector, product research and development costs vary enormously. For example, it takes from 8 to 15 years to develop and release a new, modern variety of a major food crop. It costs in the range of US$1–2.5 million for a traditionally-bred variety, and US$35–75 million to genetically engineer and gain regulatory approval for a transgene for use in many different varieties (Chapter 5). In crop protection, a biocontrol agent may take from two to five years to develop and cost US$1–5 million, whereas a new chemical pesticide can take from 8 to 14 years to develop and cost US$40–100 million (Chapter 7).

Despite large global markets in products derived from genetic resources, companies' willingness to pay for access to such resources is balanced against the role of resources in product development, and the costs and risks borne by the company.

1.6.4 Social, environmental and ethical issues

A suite of environmental and social issues is associated with the discovery, development, manufacturing and marketing of products in each of the industry sectors covered in this book. These issues include pricing and distribution policies in the pharmaceutical industry; impacts on the environment and human and animal health associated with crop protection; 'terminator technologies' and genetically modified organisms (GMOs) in the seed and crop protection industries; the use of animals in testing in the cosmetics industry; and the patentability of life forms in all industry sectors. They tend to attract more attention in the press than do questions of access and benefit-sharing. In this book we have chosen to concentrate on access and benefit-sharing. This is not to downplay the importance of the other issues mentioned, nor to ignore the fact that they are all interconnected. However, our focus was necessary in order to maintain a manageable scope and size for a book aiming to cover such a breadth of commercial activity. It has provided the opportunity to look in depth at a range of issues which deserve thorough treatment in their own right.

1.7 The commercial use of biodiversity: about this book

The rapid development of law, policy, science and technology associated with access and benefit-sharing underlines the need for information, but also means that information can become out of date very quickly. While much of the factual information in this book, such as market rates for particular categories of genetic resources, the ranking of companies and the number of countries with access legislation, will inevitably change within the coming months, many of the approaches to product development, opinions of the interviewees and principles discussed are likely to remain the same, at least for the foreseeable future.

Through the evidence gathered in our research, this book will provide some answers to the following questions:

- What is the nature and scope of industry's demand for genetic resources?
- Will industry's demand for access to genetic resources increase or decrease in the future?
- How do access and benefit-sharing arrangements compare in the pharmaceutical, phytomedicine, personal care, biotechnology, seed, crop protection and horticulture industries?
- What kind of contribution can bioprospecting make to:
  - science and product development;
  - countries' development paths;
  - the protection of individuals' and communities' rights and interests; and
  - conservation?

The research for this book was conducted through a survey of literature and 193 interviews, most roughly an hour in length, with representatives from industry, the public sector and intermediary institutions such as government agencies, genebanks, universities and botanic gardens engaged in discovery and development, who answered a set of broadly similar questions (Research Methodology, p384). In addition, some 100 other individuals were interviewed on specific issues or for their general opinion.

Questions addressed in the course of the interviews included:

- What is the scale and nature of demand for access on the part of companies and other organisations in the industry sector concerned?
- Which scientific, technological, and regulatory developments influence demand for genetic resources/natural products? How is demand likely to change in the coming years?
- What are the different types of material and information to which companies seek access (including products derived from genetic resources such as extracts, compounds, genes and cultivars, and associated information, such as traditional knowledge), and what kinds of partnerships (such as 'outsourcing' research, joint ventures, relationships with subsidiaries) do companies rely upon to obtain these?
- What kinds of benefits arise from the research and development process? With whom are these benefits shared?

1.7.1 Content of the book

This book is organised into five parts. The first consists of this introductory chapter and a chapter explaining the provisions of the CBD on access and benefit-sharing, national laws to implement them, and material transfer agreements.

The second part of the book comprises seven sectoral chapters, each of which takes as its theme a major industry sector which requires access to genetic resources for research and development, namely pharmaceuticals; botanical medicines; the development and improvement of major crops; horticulture; crop protection; biotechnology (other than pharmaceutical and agricultural biotechnology), and personal care and cosmetic products.

Each of these sectoral chapters contains an overview of the industry sector concerned, the size of the market and cost and time involved in the research and development process, and information on some important technological, scientific and regulatory trends which influence research and development. A section on access reviews the role of the different actors seeking access to genetic resources in the relevant sector. It explores the nature and extent of the demand for access to genetic resources in the sector, which actors are engaged in collecting activities, and from what other sources they obtain their materials. The next section, on benefit-sharing, reviews the range of benefits that result from the commercialisation of genetic resources in the sector under consideration, and examines the ways in which they arise at different stages in the research and commercialisation process. This section provides the best available information on indicative royalty ranges, fees per sample, and the other, non-monetary benefits that can be shared through commercial partnerships. A further section of each chapter presents a case study describing the role of the different actors involved in the development of a particular product, and examines the type, time-frame, and distribution of benefits in that specific case.

The third part of the book reviews the reactions to the CBD of the representatives from companies and other organisations we interviewed. It reviews the policies developed by companies and other organisations on the acquisition of natural products and other aspects of their business relevant to the CBD and discusses their predictions as to the likely impacts of the CBD on commercial activities.

Part four draws conclusions about industry's demand for access to genetic resources and practices in benefit-sharing, based on the information and analysis in the earlier sections of the book; on companies' awareness, views and opinions of the CBD; and on the impact that different approaches to the regulation of access are likely to have on natural products research by the private sector. It summarises the kind of partners and the development of the legal, policy and administrative environments that companies seek. A final chapter directs a number of recommendations to governments regulating access to genetic resources, on the role of intermediary organisations, and to companies seeking access to genetic resources. It presents a set of possible indicators for the 'fair and equitable sharing of benefits' and sets out a number of elements which companies and other organisations may wish to consider when developing an institutional response to the CBD.

In the final part of the book are some useful contacts in the field of access and benefit-sharing, a glossary of commonly-used terms, and a description of the methodology used for the research and the role of each member of the team involved in this project.

1.8 Conclusion

Successful access and benefit-sharing partnerships can lead to new medicines, crops to feed the growing world population and other useful products for humankind. They can also help to promote legal and policy regimes that protect the rights of countries, individuals, communities and corporations, and sustainable development and the conservation of biological diversity. When laws and policies go wrong, these opportunities can be missed, to the detriment of people and the environment. It is therefore very important to try to find fair and practical solutions to the challenge posed by the 175 parties who have ratified the CBD.

We have written this book because we believe that the information presented here is generally hard to find, and yet is essential for the development of fair and workable regulations on access to genetic resources and traditional knowledge, for realistic expectations relating to benefit-sharing and for a better informed and more collegial dialogue on these issues.

The book does not set out to judge what is 'fair and equitable'. It will not comment on the merit of the prestigious 'Captain Hook Awards for Biopiracy' instituted by the Coalition Against Biopiracy, nor endorse companies' methods of acquiring genetic resources and the quality of their benefit-sharing arrangements. Rather, the intention is to present information and analysis that we believe is badly needed, to serve as a 'bridge' between the different languages and experiences of business, conservation, and development. We acknowledge the inevitable limitations in our efforts to articulate these different perspectives, and know that, despite its size, this book is not comprehensive. We hope that, despite these shortcomings, it will be useful. We leave it to the reader to form an opinion as to whether or not current practice fulfills the 'grand bargain' envisaged by the CBD.

Notes

1 The language in Article 2 of the CBD defines 'genetic resources' as 'genetic material of actual or potential value', and 'genetic material' as 'any material of plant, animal, microbial or other origin containing functional units of heredity'. 'Functional units of heredity' are considered in Chapter 2.

2 This figure includes naptha, LPG and larger organic chemicals, together with any resulting polymers. Personal communication, Nick Sturgeon, Chemicals Industry Association (CIA), 15 February 1999.

3 Source: European Information Technology Observatory (EITO), Frankfurt.

4 These examples of retail prices are provided to give an indication of order of magnitude. They are sourced from the interviews conducted for this study, the price of goods on supermarket shelves, news reports broadcast in 1998 on the street value of cocaine, as well as *The Economist* (1998), Burrell (1998), and Spencer (1998).

Kerry ten Kate

Chapter 2

Regulating Access to Genetic Resources and Benefit-Sharing: the Legal Aspects

2.1 Introduction to the CBD and national law related to access to genetic resources

The CBD and the laws that have followed it mark a watershed in the regulation of access to genetic resources and benefit-sharing. They set the substantive and procedural requirements for any person, organisation or company wishing to obtain genetic, and in some cases, biological, resources, their derivatives and traditional knowledge concerning them for research and commercial development.

This chapter describes the access and benefit-sharing provisions of the CBD and national and regional law to implement it; material transfer agreements; and the implications of all these legal instruments for both providers and users of genetic resources. It starts with a brief introduction to the CBD as a treaty and an intergovernmental process, and to national and regional legal and administrative measures on access to genetic resources and benefit-sharing ('access laws'). It then describes various aspects of the scope of the CBD and access laws, such as the definition of genetic resources; the question of human genetic materials; derivatives; *ex situ* collections; traditional knowledge and the geographical reach of the laws.

A section on 'mutually agreed terms', a phrase that occurs several times in the CBD, describes typical terms involved in agreements to transfer biological material, and sets out some of the legal and other reasons why an individual or organisation acquiring material may wish to do so under an agreement that clearly defines its rights and responsibilities. The subsequent section on 'prior informed consent' considers some of the issues and challenges at stake when obtaining the agreement of government authorities and local-level stakeholders for access to their genetic resources. Finally, a concluding section takes stock of the current position on the regulation of access to genetic resources, and identifies some key points both for regulators and for those seeking access.

2.1.1 The CBD: an overview

The CBD is both an international treaty – and thus a source of international law – and an institutional framework for the continual development of legal, policy and scientific initiatives on biological diversity. Whereas earlier treaties dealt with specific aspects of biodiversity, such as trade, particular ecosystems, geographic areas or species, the CBD is comprehensive in its approach. Its scope is global, covering all components of biological diversity, from ecosystems and habitats, species and communities to genomes and genes, and it deals not only with the conservation of biological diversity *in situ* and *ex situ*, but with its sustainable use and benefit-sharing.

The objectives of the CBD are described in Article 1 as follows:

> *the conservation of biological diversity, the sustainable use of its components and the fair and equitable sharing of the benefits arising out of the utilisation of genetic resources, including by appropriate access to genetic resources and by appropriate transfer of relevant technologies, taking into account all rights over those resources and to technologies, and by appropriate funding.*

The Convention was opened for signature on 5 June 1992, during the Rio Earth Summit (the United Nations Conference on Environment and Development). It entered into force on 29 December 1993, after the necessary 30 countries ratified it. As of February 1999, 175

Parties (174 governments and the European Union) have ratified the Convention.

The breadth of the issues covered by the CBD can be seen in Box 2.1, which sets out the titles of its substantive articles. The CBD obliges governments to take a number of measures in order to fulfil its objectives of conserving biodiversity and using it sustainably. These include:

- monitoring and identification of biodiversity;
- environmental impact assessments;
- national strategies, plans or programmes to conserve and use the components of biological diversity sustainably; and
- the integration of biodiversity policy into relevant sectoral or cross-sectoral plans, programmes and policies.

The CBD establishes rights and responsibilities among its Contracting Parties, which are sovereign states, and not among private individuals. Arguably, institutions that are governmental bodies in countries that are parties to the CBD are bound by the provisions of the CBD. Additionally, public institutions such as universities, genebanks and botanic gardens may also be bound by the CBD, depending on provisions of national law, and on the extent to which they rely on public funding and grants that may require adherence to public law and policy. These bodies are obliged to comply with the objectives of the CBD, whether or not any implementing laws or regulations exist.

Box 2.1 *Some Issues Covered by the CBD*

Preamble	
Article 1	Objectives
Article 2	Use of Terms
Article 3	Principle
Article 4	Jurisdictional Scope
Article 5	Cooperation
Article 6	General Measures for Conservation and Sustainable Use
Article 7	Identification and Monitoring
Article 8	*In situ* Conservation
Article 9	*Ex situ* Conservation
Article 10	Sustainable Use of Components of Biological Diversity
Article 11	Incentive Measures
Article 12	Research and Training
Article 13	Public Education and Awareness
Article 14	Impact Assessment and Minimising Adverse Impacts
Article 15	Access to Genetic Resources
Article 16	Access to and Transfer of Technology
Article 17	Exchange of Information
Article 18	Technical and Scientific Cooperation
Article 19	Handling of Biotechnology and Distribution of its Benefits
Article 20	Financial Resources
Article 21	Financial Mechanism

Yet unless national law has introduced rights and responsibilities for private actors (including individuals and other legal persons such as companies and associations), individual citizens and private organisations are not obliged to comply with the objectives of the CBD. However, even private organisations are obliged to adhere to the national access legislation of other countries when they collect genetic resources there, even if their own government is not a party to the CBD, or has not introduced any implementing legislation.

As this chapter will explain (2.2 below), in addition to compliance with the legal requirements introduced by the CBD and national laws on access, the desire to ensure legal title to genetic resources, to avoid the possibility of legal actions, to maintain a good reputation and positive public relations and to leave open avenues for the future supply of genetic resources are compelling business arguments for obtaining specimens under material acquisition agreements.

2.1.2 The CBD process

The CBD is not a static treaty, but rather a process by which its Parties agree to take certain actions at the national level. For example, the Parties to the Convention, at their fourth meeting in Bratislava in May 1998, decided to establish an expert panel on access and benefit-sharing. The panel will comprise 'a regionally balanced panel of experts appointed by governments, composed of representatives from the private and public sectors as well as representatives of indigenous and local communities'. The mandate of the panel will be to help develop a common understanding of basic concepts involved in access and benefit-sharing and to explore all options for access and benefit-sharing on mutually agreed terms, including guiding principles, guidelines and codes of best practice for access and benefit-sharing arrangements (Decision IV/8). The panel is currently slated to meet for the first time in October 1999. Its work will inform future decisions of the Parties on access and benefit-sharing, and may thus influence developments in national law and policy, and even individual access and benefit-sharing partnerships.

The work of the Parties to the CBD is conducted at meetings of the Conference of the Parties (COP), the Subsidiary Body on Scientific, Technical and Technological Advice (SBSTTA), and other subsidiary bodies such as the Ad Hoc Open Ended Working Group that met six times between July 1996 and February 1999 in an unsuccessful attempt to negotiate a Protocol to the

CBD on Biosafety. Other work of the CBD is carried out by the interim Financial Mechanism (the Global Environment Facility), which provides financial resources to developing country Parties to implement the Convention, and the Clearinghouse Mechanism intended to promote and facilitate technical and scientific cooperation. A Secretariat, based in Montreal, organises the meetings and prepares the background papers for the Parties to use when they meet to take decisions. The function of these bodies is described in the Glossary, and a more detailed description of the CBD process and individual meetings may be found in reports by the Earth Negotiations Bulletin (http://www. iisd.ca) and Glowka et al, 1994.

2.1.3 Access to genetic resources and benefit-sharing in the CBD

Many of the activities touched upon in the Convention, from *in situ* and *ex situ* conservation, through monitoring and assessment of the components of biological diversity, to research and training, require access to genetic resources. Thus, although the provisions on access to genetic resources, traditional knowledge and benefit-sharing (primarily Articles 8(j), 15, 16 and 19) refer to a specific set of activities, they are closely linked to the rest of the Convention. Among other interpretations, the Convention can be seen as an instrument to promote the equitable exchange, on mutually agreed terms, of access to genetic resources and associated knowledge in return for finance, technology and the opportunity to participate in research.

The Convention endorses the sovereign right of states over their biological resources and the consequent authority of national governments to determine access to genetic resources. It also obliges Parties to facilitate access to genetic resources. According to the Convention, such access shall be subject to Parties' prior informed consent, and on mutually agreed terms that promote the fair and equitable sharing of benefits. The Convention strikes a balance between a state's authority to regulate access to genetic resources, and its obligation to facilitate access to genetic resources for environmentally sound uses by other Parties, and not to impose restrictions that run counter to the objectives of the Convention.

The key provisions of the Convention on access to genetic resources and benefit-sharing are summarised in Box 2.2.

As illustrated in Box 2.2, the obligation to share benefits arises in the context of *(a)* access to genetic resources, where it is triggered by the need to obtain prior informed consent (Article 15(5)), and *(b)* access to the knowledge, innovations and practices of indigenous

Box 2.2 *Summary of Provisions of the CBD on Access to Genetic Resources, on the Knowledge, Practices and Innovations of Local and Indigenous Communities and on Benefit-Sharing*

Article 8(j)	Promote the wider application of the knowledge, innovations and practices of indigenous and local communities with their approval and involvement and encourage the equitable sharing of the benefits arising from the utilisation of the knowledge, innovations and practices of indigenous and local communities.
Article 15.1	Sovereign rights of States over their natural resources; the authority of national governments to determine access to genetic resources.
Article 15.2	Endeavour to create conditions to facilitate access to genetic resources for environmentally sound uses by other Contracting Parties and not to impose restrictions that run counter to the objectives of the CBD.
Article 15.3	Articles 15, 16 and 19 apply only to genetic resources acquired 'in accordance with this Convention': eg they do not apply to those obtained prior to its entry into force or from non-Parties.
Article 15.4	Access, where granted, to be on mutually agreed terms and subject to the provisions of Article 15.
Article 15.5	Access to genetic resources to be subject to prior informed consent of the Contracting Party providing such resources, unless otherwise determined by that Party.
Article 15.6	Endeavour to develop and carry out scientific research based on genetic resources provided by other Contracting Parties with the full participation of, and where possible in, such Contracting Parties.
Article 15.7	Take legislative, administrative or policy measures, as appropriate . . . with the aim of sharing in a fair and equitable way the results of research and development and the benefits arising from the commercial and other utilisation of genetic resources with the Contracting Party providing such resources. Such sharing to be upon mutually agreed terms.
Article 16.3	Access to and transfer of technology using genetic resources to countries providing the genetic resources.
Article 19.1	Effective participation by providers of genetic resources in biotechnological research on the genetic resources they provide.
Article 19.2	Priority access on a fair and equitable basis by countries (especially developing countries) providing genetic resources to the results and benefits arising from biotechnologies based on them. Such access to be on mutually agreed terms.

and local communities, for which the approval of the holders of that knowledge is required (Article 8(j)).

The meaning and significance of each of the paragraphs summarised in Box 2.2 are clearly explained in Glowka, 1994. Some aspects of these provisions, together with corresponding measures in national laws, are considered below.

2.1.4 National and regional regulation of access to genetic resources and benefit-sharing

Article 15(1) makes it clear that, in recognising the sovereign rights of states over their natural resources, the authority to determine access to genetic resources rests with national governments and is subject to national legislation. Consequently, countries have a great deal of discretion to decide how to regulate access. In practice, the number of countries developing national laws and policies on this subject has grown fast over the last five years, and now numbers over 40.

To date, access and benefit-sharing measures have been concluded or are under development in Argentina, Australia (at the Commonwealth level and in the states of Western Australia and Queensland), Belize, Bolivia, Brazil, Cameroon, Colombia, Costa Rica, Ecuador, Eritrea, Ethiopia, Fiji, The Gambia, Ghana, Guatemala, India, Indonesia, Kenya, Laos PDR, Lesotho, Malawi, Malaysia (including the State of Sarawak), Mexico, Mozambique, Namibia, Nigeria, Papua New Guinea, Peru, Philippines, the Republic of Korea, Samoa, Seychelles, Solomon Islands, South Africa, Tanzania, Thailand, Turkey, USA, Venezuela, Vietnam, Yemen and Zimbabwe (personal communication, Lyle Glowka, February 1999). Other countries are likely to join this list in the coming years.

On 17 July 1996, the Cartagena Accord, commonly known as the Andean Pact, more properly the Andean Community, published Decision 391 in its Official Gazette (Comisión del Acuerdo de Cartagena, 1996). Decision 391 introduced the Common Regime on Access to Genetic Resources ('the Common Regime'), the first subregional agreement of its kind, legally binding in Bolivia, Colombia, Ecuador, Peru and Venezuela. ASEAN, the OAU and the SPREP are also exploring the possibility of regional access measures.

Glowka (1998) believes that existing and draft access legislation can be categorised into five groups of legislation:

1 *Environmental framework laws* such as those in The Gambia, Kenya, Malawi, the Republic of Korea and Uganda tend to be enabling in nature, and simply charge a competent national authority to provide more specific guidelines or regulations on the export of genetic resources, benefit-sharing and access fees.
2 *Sustainable development, nature conservation or biodiversity laws* such as those in Costa Rica, Eritrea, Fiji, India, Mexico and Peru cover the conservation and sustainable use of biodiversity generally, or implement a number of provisions of the CBD. These laws tend to include more detailed provisions on access than those in the first group, and establish the principles of prior informed consent and mutually agreed terms.
3 *Dedicated or 'stand-alone' national laws and decrees on access to genetic resources*, such as those in the Philippines and the draft laws in Brazil (which has already passed laws in the states of Amapá and Acre, and has draft laws at the Federal level and in the State of São Paolo).
4 *Modification of existing laws and regulations*, for instance modification of the national parks law in Nigeria to establish requirements for prior informed consent for access to genetic resources, and the proposal in the USA to revise the Code of Federal Regulations which deals with the removal of research specimens from national parks. At the sub-national level, examples include the amendment of the Wildlife Conservation and Conservation and Land Management Acts in Western Australia and the amendment of the Forest Ordinance of the State of Sarawak, Malaysia, to prevent the removal of tree parts for pharmaceutical and medicinal research and development.
5 *Regional measures* such as the common access regime introduced by the five members of the Andean Community (Bolivia, Colombia, Ecuador, Peru and Venezuela), and the continuing discussions by South East Asian countries and proposed for the 53 member countries of the OAU.

To varying degrees, the laws introduced or being developed in these countries will govern the terms and procedure under which foreign and national scientists and companies can obtain access to genetic resources and the kind of benefit-sharing expected in return. The laws generally specify the role of the state in authorising access to genetic resources; define (with more or less clarity) the scope of the resources and activities regulated and for which the prior informed consent of the state and other individuals or organisations is needed; describe the procedures which an applicant for access must follow; and nominate existing institutions or create new ones to administer and determine access applications. Some define minimum terms on which access to genetic resources is to be granted. The next section examines some of the key aspects of the regulation of access.

2.2 The scope and requirements of national and regional law on access to genetic resources and benefit-sharing

2.2.1 Scope

Perhaps the first question to ask concerning national laws on access and benefit-sharing is what they cover. The scope of the resources whose access is to be regulated is outlined briefly in the CBD, but since countries have national sovereignty over their natural and other resources, individual states may choose to regulate access to a broader or narrower range of resources than those encompassed by the term 'genetic resources', as defined in the CBD. This section will describe the basic provisions of the CBD regarding scope, and also provide a few examples of the scope chosen by states through the definitions used in their national and regional regulations. The issues covered in this section are genetic resources; human genetic materials; derivatives of genetic resources; traditional knowledge concerning genetic resources; *ex situ* collections of genetic resources; and marine genetic resources.

Genetic resources

The obligations under the CBD on access and benefit-sharing refer to 'genetic resources', defined in Article 2 as 'genetic material of actual or potential value'. Genetic material is defined as 'any material of plant, animal, microbial, or other origin containing functional units of heredity'. There has, as yet, been no cause (such as an international dispute) for courts to interpret this term. Glowka et al (1994) consider that 'functional units of heredity' would include whole organisms, parts of organisms or biochemical extracts from tissue samples that contain deoxyribonucleic acid (DNA) or, in some cases, ribonucleic acid (RNA). A more detailed consideration of the meaning and examples of 'functional units of heredity' is contained in Box 2.3.

Although the language of the CBD does not say so in so many words, it is now widely accepted (see, for example, Glowka et al, 1994, p76) that the scope of Article 15 is confined to the use of genetic resources for their genetic purposes. If this is the case, then it is only necessary to obtain prior informed consent, reach mutually agreed terms and share benefits where the use is made of the material's genetic attributes, for example, exploring the biological activity of compounds extracted from the material or using plants in a breeding programme. Under this interpretation, it would not be necessary to share benefits in return for access for 'non-genetic' uses of genetic resources, such as access to trees for timber (rather than for chemical analysis) or access to a forest for ecotourism. Most national access legislation has not resolved this ambiguity,[1] a problem which is caught up in broader questions concerning the lack of clarity of the exact activities and resources which are regulated (see especially 'derivatives', below).

Although the language of the CBD does not distinguish between human and other categories of genetic resources, the Conference of the Parties, at its second meeting in 1996, 'reaffirmed that human genetic resources were not included within the framework of the Convention (Decision II/11)'. Most national access law and draft laws to date have also specifically excluded human genetic resources from the definition of genetic resources within the scope of their laws. Access to samples of human genetic material for research and indeed for commercialisation are thus largely unregulated (although subject to common law and some statutory provisions in some countries). Since this category of research is common and raises profound ethical issues, many governments will presumably wish to consider developing law or policy on this kind of access outside the ambit of the CBD.

Derivatives

The CBD itself makes no mention of 'derivatives', although the very act of regulating access to genetic resources provides a trigger for negotiations on mutually agreed terms that could (and often do) involve a share of the benefits arising from derivatives of genetic resources, such as sales of semi-synthesised or totally synthesised compounds based on structures discovered from studying genetic resources, or sales of hybrid plants that result from access to two parents.

Recognising that, in sectors such as pharmaceuticals and crop protection, only a few commercial products actually contain unmodified genetic resources, and that the large majority contain products derived from, patterned on or incorporating manipulated compounds and genes found in nature, many access agreements (including many agreed before the CBD) contain benefit-sharing obligations attached to sales or other use of derivatives of the genetic resources themselves. As Chapters 4–11 will show, partnerships can involve benefits such as joint research between the provider of genetic resources and the user on derivatives (see, for example, NCI/Calanolide). Agreements for the supply of genetic resources often involve commitments to pay royalties not only for the commercial use of the genetic resources supplied, but for any derivatives. These terms are agreed at the time when the recipient is granted access to genetic resources themselves.

Box 2.3 *What are 'Functional Units of Heredity'?*

There is no single biological entity that fits the definition of being a unit that functions to convey hereditary information under all circumstances. To answer the question 'What is a functional unit of heredity?', it is necessary to consider which biological entities may be identified as 'units of heredity' and under what circumstances these entities may be considered 'functional' in the context of the CBD. Four candidate entities – intact living cells, whole chromosomes, genes and DNA fragments smaller than genes – can each be considered 'functional units of heredity' under some circumstances.

Living cells can be 'functional units of heredity' because they carry all of the hereditary, or genetic, information necessary for life. Living cells can reproduce and make use of smaller units of genetic information and can also be made to process genetic information from other cells. Functional genetic information, in the form of DNA, can be extracted from living cells and used in a variety of ways.

DNA (and sometimes its close counterpart RNA) carries the hereditary information to build structural and functional proteins (and a few other chemicals) and to control many cellular processes. This hereditary information is often envisioned as discrete 'packets' called 'genes'. In higher organisms ('eukaryotes' – multicellular plants, animals, fungi and some advanced single-celled organisms) DNA is packaged in chromosomes containing hundreds or thousands of genes. Chromosomes also often carry DNA information that does not participate in the biology of the cell, and thus not all DNA is organised as genes. Important genes are also found in mitochondria, sub-cellular structures that process energy in eukaryotes, and in plastids, sub-cellular structures that may be involved in light energy. Most bacteria have a single, large circular DNA molecule. However, many bacteria also contain small DNA circles called plasmids, some of which carry genes conferring antibiotic resistance. Genes in viruses are also usually DNA, but some are RNA.

Genes are important as units of heredity, but the information in DNA is divisible below the level of genes. Both in nature and the laboratory, DNA fragments smaller than genes can be exchanged among organisms with relative ease. Therefore, it is possible to consider chromosomes, genes and DNA fragments smaller than genes as all being functional units of heredity under some circumstances. Relatively small DNA fragments can be removed from one species and inserted into the germ-line of another. Restricting the definition of 'functional units' to larger fragments – such as intact chromosomes or individual genes – may therefore be considered artificially restrictive.

The ability of genetic material to be used as a resource is an important criterion by which 'function' could be judged. To retain hereditary information that can be used to produce biochemical products, an organism does not necessarily have to be 'living' or even 'intact' relative to its natural condition, because its DNA retains its biological information whether in a living cell or in a laboratory solution. However, there are plausible arguments that 'functional' would mean that a unit is able to be expressed and produce the same product that it produced in its natural form. It is equally plausible to argue that any biological specimen that contains intact DNA that can be extracted and manipulated or that can be passed to a living offspring could be considered as containing 'functional units of heredity'.

The functionality of a 'unit of heredity' in the context of a genetic resource is highly dependent upon the evolving sophistication of genetic engineering. In the 1960s, DNA extracted from biological material had little commercial value because at that time the technology to use such material did not exist. Prior to the advent of genetic engineering, 'functional units of heredity' were therefore restricted to living organisms or tissues bearing identifiable characteristics. In the late 1990s, many methods can put extracted DNA to practical use, including transplanting functional genes from one species to another. Future technologies cannot be predicted with certainty, but, in time, it may become possible to make practical use of small DNA samples or even degraded DNA such as may be found in fragmentary, dried or preserved biological material.

What is 'functional' is clearly a question of interpretation; it is based on current science and technology, which is likely to change as technologies advance. The terms 'functional unit' and technology are thus inextricably linked. Any carrier of hereditary information that can be passed from one organism to another could arguably be described as a 'functional unit of heredity', including, among other things, living cells, intact chromosomes or other large packages of genetic information, single genes, fragments of DNA smaller than genes, and RNA samples capable of being retrotranslated into DNA.

Sources: Dr David Galbraith, Royal Botanic Gardens, Hamilton, Canada, and personal communication with Dr Mark Chase, Royal Botanic Gardens, Kew.

A number of national laws include definitions of 'derivative', 'by-product', and 'synthesised product' (see Box 2.4 for a few selected examples). In many cases, the substantive provisions make no further reference to these definitions, so it is unclear which activities with respect to them are regulated, and to what effect. Specifically, it is unclear whether the laws require prior informed consent from the state (and possibly other stakeholders) prior to access to derivatives, or only regulate access to genetic resources themselves, but encourage the sharing of benefits arising from the commercial (and other) use of derivatives of genetic resources. For example, Biodiversity Law 7788 of Costa Rica defines genetic elements, biochemical elements and associated knowledge, but Article 63, the substantive provision concerning the regulation of access, simply refers to 'the requirements for access', without clarifying access to what. Similarly, the Executive Order and Implementing Regulations in the Philippines regulate the prospecting of biological and genetic resources. 'Derivatives' are defined, but not mentioned in the substantive provisions concerning access, except for an allusion to the role of the Department of Health in evaluating proposals for research and development 'including the utilisation of extracts, products and by-products and derivatives for commercial and academic purposes' (Section 10.3.4.). 'By-products' are defined but not mentioned at all in the substantive provisions. By contrast, Decision 391 of the Andean Pact is clear that its objective is to regulate access to genetic resources and their derivatives (Article 2).

Regulators wishing to design laws that will capture a share of the benefits arising from sales of derived products such as pharmaceuticals are faced with a problem. On the one hand, if access to genetic resources alone is regulated, there is the danger that people will make derivatives within the country concerned and export them without ensuring a sharing of benefits, whether by selling traditional medicines or plant cultivars in the market place, or by preparing refined extracts of natural product specimens in a pharmaceutical company, and shipping the resulting samples to a subsidiary company elsewhere. On the other hand, if regulators attempt to control access to derivatives, a broad

Box 2.4 *The Scope of Access Laws, as Described by Definitions*

Law concerned, activity and resources regulated	Definitions
Philippines Executive Order (PEO) and Implementing Regulations (PIRR): **BIOPROSPECTING OF BIOLOGICAL AND GENETIC RESOURCES**	**Bioprospecting or prospecting:** the research, collection and utilisation of biological and genetic resources, for purposes of applying the knowledge derived therefrom for scientific and/or commercial purposes. **Biological resources:** include genetic resources, organisms, or parts thereof, populations or any other biotic components of ecosystems with actual or potential use or value for humanity such as plants, seeds, tissues, and other propagation materials, animals, microorganisms, live or preserved, whether whole or in part thereof. **By-product:** any part taken from biological and genetic resources such as hides, antlers, feathers, fur, internal organs, roots, trunks, branches, leaves, stems, flowers and the like, including compounds indirectly produced in a biochemical process or cycle. (But by-products are only mentioned in the title and definitions, not in the substantive provisions.) **Derivative:** something extracted from biological and genetic resources such as blood, oils, resins, genes, seeds, spores, pollen and the like, taken from or modified from a product.
Andean Community Decision 391: **ACCESS TO GENETIC RESOURCES, DERIVATIVES AND INTANGIBLE COMPONENTS**	**Access:** the acquisition and use of genetic resources conserved in *ex situ* and *in situ* conditions, and of their derivatives or, as applicable, intangible components, for purposes of research, biological prospecting, conservation, industrial application or commercial use, among others. **Derivative:** a molecule or combination or mixture of natural molecules, including raw extracts of living or dead organisms of biological origin, derived from the metabolism of living organisms. **Synthesised product:** a substance obtained by means of an artificial process, using genetic information or other biological molecules. This includes semi-processed extracts and substances obtained through treatment of a derivative using an artificial process (semisynthesis). (NB: Synthesised products are only addressed in the complementary provisions and in the context of IPRs.) **Intangible component:** any knowledge, innovation or individual or collective practice of actual or potential value associated with the genetic resource, its derivatives or the biological resource containing them, whether or not it is protected by intellectual property systems.
Costa Rica Law 7788 **ACCESS TO GENETIC ELEMENTS AND BIOCHEMICALS, AND PROTECTION OF ASSOCIATED KNOWLEDGE**	**Access to biochemical and genetic elements:** the action of obtaining samples of the elements of biodiversity existing in the wild or domesticated, in *ex situ* or *in situ* conditions, and the obtaining of associated knowledge, with the aim of fundamental research, bioprospection and economic exploitation. **Bioprospection:** the systematic search, classification and investigation for commercial purposes of novel sources of chemical compounds, genes, proteins, microorganisms and other products with actual or potential economic value, found in biodiversity. **Biochemical element:** any material derived from plants, animals, fungi or microorganisms which contain specific characteristics, special molecules or leads (lit, 'clues') to design them. **Genetic elements:** any plant, animal, fungi or microorganisms which contain functional units of heredity. **Knowledge:** dynamic product generated by society by different mechanisms in the course of time, including that which occurs in a traditional form, as well as that generated by scientific practice.

definition is needed to capture the full range of products that may be developed (Box 2.4 p19). However, derivatives of genetic resources are ubiquitous in daily life, so it is questionable how governments could effectively regulate access to them.

One way around this problem for legislators, depending on the law governing rights over genetic and biological resources in each country, may be to regulate access only to genetic resources, but to require benefit-sharing for the use of derivatives. Current models of access laws will probably have this effect, but more by accident than by clear design.

Traditional knowledge

Article 8(j) of the CBD provides that each Contracting Party shall:

> *Subject to its national legislation, respect, preserve and maintain knowledge, innovations and practices of indigenous and local communities embodying traditional lifestyles relevant for the conservation and sustainable use of biological diversity and promote their wider application with the approval and involvement of the holders of such knowledge, innovations and practices and encourage the equitable sharing of the benefits arising from the utilisation of such knowledge, innovations and practices.*

There is a growing body of literature on this provision (eg, Posey, 1996). An intersessional Workshop on Traditional Knowledge and Biological Diversity, held under the auspices of the CBD in Madrid in November 1997, focused on Article 8(j) (UNEP/CBD/TBK/1/3), and a meeting of the Ad Hoc Working Group on Article 8(j) is planned for January 2000.

Several national laws on access to genetic resources define terms such as 'knowledge', 'traditional knowledge', 'intangible component' and various categories of peoples such as 'local communities' and 'indigenous cultural communities'. As with other questions of scope, such as derivatives and *ex situ* collections, the exact circumstances in which the 'prior approval' of local and indigenous communities is needed – and for what – vary from country to country and are sometimes unclear. Some access laws require prior informed consent from local and indigenous people for access to genetic resources on their lands but do not address access to their traditional knowledge concerning those resources, while others explicitly require prior approval from local and indigenous peoples for access to their traditional knowledge, but not to genetic resources or their land. This issue is addressed in the section on prior informed consent, on page 27 below.

Ex situ collections

The effect of Article 15(3) of the CBD is to exclude from the remit of its provisions on access and benefit-sharing those genetic resources acquired before the entry into force of the Convention. *Ex situ* collections of genetic resources acquired before the Convention thus fall outside these provisions. Notwithstanding, it is open to governments to consider introducing access requirements and benefit-sharing for these resources, although there are important legal considerations related to the retrospective nature of any such obligations, as well as the fact that the resources may be privately owned (ten Kate, 1995; FAO, 1987).

At the international level, access to pre-CBD *ex situ* collections of plant genetic resources for food and agriculture forms an important part of the negotiations under the auspices of the FAO Commission on Genetic Resources for Food and Agriculture to revise the International Undertaking on Plant Genetic Resources for Food and Agriculture (Chapter 5). It was also one of the most controversial issues discussed during the fourth meeting of the Conference of the Parties (COP), which requested the Executive Secretary of the Convention to invite information from Parties and relevant organisations in respect of those *ex situ* collections which were acquired prior to the entry into force of the CBD and (in order to avoid duplication of the work under the auspices of the FAO) which are not addressed by the Commission on Genetic Resources for Food and Agriculture (Decision IV/8).

At the national level, the majority of countries do not have explicit laws on access and benefit-sharing that apply to *ex situ* collections, particularly to those acquired prior to the entry into force of the Convention. However, it is important to remember that many governments have used collecting permits for decades, or even for hundreds of years, so it is likely that a large proportion of the materials in pre-CBD *ex situ* collections were acquired under this form of agreement. Prior to granting access to its accessions, an *ex situ* collection must thus establish whether it is contractually bound by the terms of specific collecting permits to treat certain accessions in certain ways.

The effect on pre-CBD collections of the laws in those countries that have introduced them is often ambiguous. For example, Section 3 of the Philippine Implementing Regulations does not explicitly state whether, or which, *ex situ* collections established prior to the Convention are in the 'public domain', and are thus covered by the Regulations. The Andean Pact Common System on Access to Genetic Resources governs those genetic resources of which the Member State is the country of origin, defined as the country which possesses those genetic resources in *in situ* conditions (much as in the CBD itself), but also genetic resources that, having been established in *in situ* conditions, are encountered in *ex situ* conditions. Since no date is mentioned, the effect of the Decision is arguably to regulate access to *ex situ* collections of national genetic resources, whether collected prior to or after the entry into force of the CBD. Furthermore, the first Transitory Provision states that:

> *All those who, on the date on which this decision enters into force, are in illegal possession, for purposes of access, of genetic resources for which the Member States are countries of origin, their derivatives or related intangible components, must negotiate such access with the Competent National Authority, subject to the terms of this decision. The Competent National Authorities shall establish deadlines for such negotiation, not to exceed twenty-four months from the entrance into force of this decision.*
>
> *In the event of non-compliance with this requirement, the Member States may disqualify such persons and the bodies which they represent or on whose behalf they are acting from applying for further access to genetic resources or their derivatives in the subregion, without prejudice to application of the corresponding penalties once the period referred to in the previous paragraph has elapsed.*

As well as being clearly retrospective in effect, this clause could be interpreted as being extra-territorial (ie applying to pre-CBD collections of Andean material held outside the jurisdiction of the five member states). The basis for such scope is questionable in international law and may well not have been the intention of the countries, which could use other articles in the Decision to resolve the question in the future, particularly since two major international collections – Centro Internacional de Agricultura (CIAT) in Colombia and Centro Internacional de la Papa (CIP) in Peru – and several national genebanks and botanic gardens are held within their territory. Nonetheless, the provision demonstrates that treatment of its existing *ex situ* collections is likely to affect an organisation's chances of securing access to genetic resources in the future.

Geographical scope

As well as the nature of the genetic resources covered by legislation, another important aspect of defining the scope of access regulations is the location of the genetic resources concerned. As the CBD reaffirms, all countries have sovereign rights over genetic resources within their jurisdiction, and the vast majority require phytosanitary permits for import and export of genetic resources and may well control other activities such as trade in endangered species. However, the approach of different countries to regulating the collection of samples from *in situ* conditions varies. The following examples are simply generalisations to illustrate some major differences in approach. It is necessary to study the law of each individual country in order to understand where access to genetic resources is regulated.

Private land

In some countries (for example, many in Latin America), genetic resources are the 'national patrimony' of the state, so that access to genetic resources is regulated by the state on private as well as public land. In such countries, if one wishes to access plants or microorganisms in someone's private backyard, one will need not only the landowner's permission, but also prior informed consent from the state. In contrast, in Organisation for Economic Cooperation and Development (OECD) countries, the state generally does not regulate access to genetic resources on private land (but there are notable exceptions, for example certain activities involving access to particular species are regulated by domestic laws on wildlife, conservation, hunting, fishing and shooting that apply to private land). In many countries, if one wishes to obtain a plant from private land for research, the permission of the landowner to enter the land and remove the specimen is probably sufficient. No prior informed consent from the government is needed. While ownership of genes is a grey area, ownership of the specimen itself, and authority to dispose of it, probably rests solely with the landowner in these circumstances.

Public land

Regulation of access to genetic resources on public land also varies from country to country. The scope of some countries' access regulations is sometimes defined by reference to particular categories of public land, such as protected areas. In the USA, for example, the Department of the Interior regulates access to genetic resources within the Federal National Parks through the enabling act for the National Park Service (the National Parks Organic Act of 1916), and the regulations under the US Code of Federal Regulations (36 CFR 1) (ten Kate, Touche and Collis, 1998). A number of other departments of the US government regulate access to other Federal lands (ELI, 1996). In Madagascar (where changes in the regulation of access are currently being considered), permission from one body – the Association Nationale pour la Gestion des Aires Protegées (ANGAP) – is needed for access to protected areas, and permission from another (the Direction Générale de la Ministère des Eaux et Forêts) for access to other public lands.

Marine areas

Since several research institutes and companies are known to focus on marine genetic resources, it is perhaps surprising that access laws are often ambiguous in their application to the marine environment. National jurisdiction over marine living resources extends to inland waters, territorial waters and exclusive economic zones. National access law has not, to date, clarified whether countries will distinguish between their territorial waters and the exclusive economic zone, where the jurisdictional powers of the coastal state are understood to be less expansive. Some countries specifically include coastal and inland waters in the geographical scope of their access legislation; the laws in others are ambiguous; and in yet others it is quite clear that access to marine organisms is not covered, since the scope of the law clearly applies only to specific terrestrial areas, such as national parks. As with other issues discussed above, this is a matter that must be checked against the specific law of the country concerned. Countries considering regulating access may wish to consider that, to date, access procedures have been designed with terrestrial prospecting in mind, and may need adaptation for a marine environment.

Although few can afford to work there, researchers and companies are extremely interested in the marine environment outside national jurisdiction in the high seas, since unusual organisms and extremophiles can be found on the deep sea bed and in thermal vents. By virtue of the United Nations Conference on the Law of the Sea (UNCLOS), the seabed, ocean floor and subsoil thereof beyond the limits of national jurisdiction, 'the Area', are designated as 'the common heritage of mankind' (UNCLOS articles 133a and 136). An International Seabed Authority exists to ensure the fair and equitable utilization of the Area's mineral resources (Glowka, 1995).

2.2.2 Mutually agreed terms

According to the CBD, access to genetic resources, where granted, shall be on mutually agreed terms. Typical terms include:

- legal acquisition;
- permitted use of genetic resources;
- restrictions on supply;
- benefit-sharing; and
- other terms.

Since mutually agreed terms for access to genetic resources involve, by definition, two or more parties, at least one party will be supplying the genetic resources and another acquiring them. The terms under which a recipient acquires the genetic resources are likely to affect its freedom to supply the same resources, or their derivatives, to a third party. Thus an important consideration for an 'intermediary' institution, be it a genebank, university department, botanic garden or company, is to ensure that any commitments it makes in agreements in which it acquires genetic resources, are honoured in agreements under which it supplies them. As Figure 2.1 below shows, when resources are initially collected, prior informed consent from government is likely to be required, and this may be granted to the collector subject to certain mutually agreed terms (Material Transfer Agreement 1 (MTA_1)). Before supplying the resources (or their progeny or derivatives) to a third party, this collector will need to check the terms under which they were acquired. If there are restrictions, such as ensuring that the resources are not commercialised without the consent of the government, or passing benefit-sharing obligations on to any subsequent recipient, the collector will need to ensure that the materials are only supplied to any future recipient under an agreement (MTA_2). The terms of MTA_2, mutually agreed with the recipient (in this case, a company), will need to be consistent with those of MTA_1. For its part, the company receiving materials from the collector may also wish the collector to guarantee in MTA_2 that it obtained the resources lawfully and can pass title to them to the company (see Legal acquisition, below).

Legal acquisition

If the organisation acquiring genetic resources is collecting them itself from *in situ* conditions or other situations covered by access legislation, it will need to obtain prior informed consent from the appropriate government authorities, and from others whose consent is needed, such as private landowners, protected area management boards, and local and indigenous communities (see Whose consent is required?, page 27). Some access laws introduce fairly heavy penalties for breach of these requirements, such as fines and imprisonment. For example, the proposed Biological Diversity Act in India would regulate access to biological resources for research or commercial utilisation, collecting or bioprospecting, and even proposes to prohibit the transfer of research results. It states that any contravention of the provisions of the Act will be punishable with imprisonment for a term which may extend up to five years, or a fine which may extend up to Rs 10 lakhs (approximately US$25,000), or both.

If, however, the organisation is not collecting genetic resources itself, but rather acquiring them (or their derivatives) from another organisation, such as a botanic garden or research institute in the source country or elsewhere, this activity may well not be covered by access legislation itself (although that depends on the law in the country concerned. See *Ex situ* collections, page 20). While a recipient is not, *prima facie*, bound by access law if it simply receives genetic resources from a supplier, engaging in such activity without ensuring that the supplier has title to the materials exposes the recipient to a number of legal and other risks. The fundamental danger for a recipient in these circumstances is that

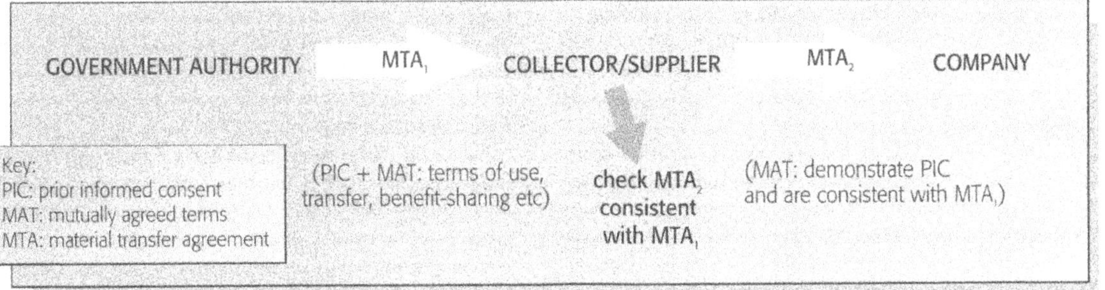

Figure 2.1 *Transmission of the Terms of Material Transfer Agreements*

the genetic resources may not be the supplier's to provide. If the supplier (whether a public or private institution from the country of origin or from elsewhere in the world) does not lawfully own the resources itself, or have the authority to dispose of them to a third party, the recipient's right to commercialise the resources or any derivative from them is thrown into question. Depending on the circumstances and degree of knowledge of the recipient at the time it acquired the materials, it could also itself be guilty of an offence.

There are a number of possible bases for legal actions on the part of the rightful owners or right-holders (who might be a state with sovereign rights, private citizens, or a community) against a recipient of genetic resources. The possibilities depend on the legal provisions in the country of origin, in the country where the collector/supplier was based and in the country where the ultimate recipient company is based. There are two main scenarios. In the first, the supplier has collected or acquired the material legally, under a permit or some other material transfer agreement, but has then supplied the material to the recipient in breach of the express or implied terms of that agreement – the 'breach of contract' scenario. In the second, the supplier has not collected or acquired the material legally, but nonetheless goes on to supply the material to the recipient – the 'illegal acquisition' scenario. The implications of each of these scenarios will be discussed briefly in turn.

The 'breach of contract' scenario

A typical example of this scenario would be the case where an academic collector obtains a 'scientific' collecting permit (and various export and phytosanitary permits) to collect and take home certain samples, but then supplies these to a company for commercial purposes. Let us imagine that the company derives a highly successful commercial product from its work on the samples. The collector would, arguably, be in breach of the terms of the permit even if the permit contained no explicit prohibition of commercialisation. The courts might decide that it is an implied term of such a contract that a 'scientific' permit granted to a 'scientific' institute, ostensibly for non-commercial purposes, prohibits the use of the materials for commercial purposes. Collectors frequently complete detailed forms explaining the purposes of collection and the use to which they intend to put the material. If there was no mention in the application of commercialisation, the permitting authorities might argue that they granted the permit in reliance upon the applicant's assurance that the materials would be used only as described in the application. In a clearer case, the permit may explicitly prohibit the commercialisation of the samples or their derivatives, or their supply to third parties for commercialisation or any other purposes.

In this scenario, the permitting authorities could sue the collector for breach of contract. This might not avail them much, if the collector was penniless and had already parted with the materials to the company. The authorities could not sue the recipient company for damages or a 'share in the benefits' since they were not parties to the same agreement. The authorities might conceivably have some ground for action against the recipient company if they could prove that the company was well aware that it was receiving the materials on terms that breached the agreement between the authorities and the collector. In practical terms, this would be extremely difficult to show. In order to avoid the risk of any such action, however, companies are increasingly likely to require their suppliers not only to guarantee that they have good title to the materials, which they are entitled to pass to the company, but also to indemnify the company against any possible actions by aggrieved source country authorities or others.

Faced with these circumstances, a company might claim damages against the collector for supplying it with materials in circumstances in which it was not entitled to do so, but the ability of the collecting institution to pay damages would depend on its standing. For this reason, companies may choose to receive materials only from well-established, reputable organisations that will examine their contractual obligations with care, and, in the worst case, would have the assets or insurance to make good any orders for damages.

However, an institution is likely to be reluctant to give any form of indemnity to a company and may well refuse to do so. The company may therefore need to verify for itself that the supply of materials was permissible.

If a challenge arose, it is questionable whether or not the company could continue with the commercialisation. This would depend on the law of the country in which the company is established. That law might well permit the source country or entity to prevent this commercialisation on grounds akin to intellectual property rights enforcement, namely that no licence was granted to the company to use the materials for that purpose. The extent to which such preventative action would be possible is likely to depend upon the nature of the laws introduced in response to the CBD in 'user' countries, since the chances would be much higher if legal provisions in user countries complemented and reinforced those introduced laws by 'provider' countries (see below).

The answer may also depend on the law of the source country. In all contractual (and indeed other) scenarios, the rules on conflict of laws in the countries involved would need to be considered. To give an illustration, consider a material transfer agreement between a Peruvian institution and a British company which leads to a dispute. If the material transfer agreement between the parties contained a provision such that it was governed by the laws of Peru, an English court would apply Peruvian law to any contractual dispute, subject to public policy considerations. The third transitory Provision of Andean Pact Decision 391 states that 'The Member States may take any legal action that they deem appropriate to reclaim genetic resources, derivatives and related intangible components for which they are countries of origin and to obtain any indemnities and compensation that may be due.' It is possible that an English court could grant a remedy under that law even if the same remedy was not available under English law. However, the court would not be likely to do so if the remedy were punitory or otherwise contrary to English law.

A further practical issue that a company will need to consider is that it may have operations in many countries, including the source country. In these circumstances, it could be susceptible to the laws of, and legal action in, the source country. A company will thus need to exercise caution in its receipt of materials, in order to ensure that no possible claims arise in the countries where it operates.

The 'illegal acquisition' scenario

In this scenario, the collector may not have obtained any consent to take the materials, or may not have obtained the consent of all those from whom it was required by the relevant laws. However, the collector provides some of the material to a company, and a commercial product results. As mentioned above, access legislation frequently provides for sanctions against the collector in these circumstances. In the absence of access legislation, the source country may be able to prosecute the collector for theft, obtaining by deception, fraud, or any other associated crime. Furthermore, this scenario differs from the first in that the possibility of civil and possibly even criminal actions against the recipient company itself is higher, although remedies for dealing in stolen property, whether knowingly or unknowingly, vary from country to country, and questions of jurisdiction arise where the goods concerned are involved in international trade. The risk is, however, that the company may not obtain clear title to the materials, and may face claims for their restitution by the source country authorities. Furthermore, the source country may be able to obtain an injunction preventing the company from commercialising the resulting products without its permission, which would provide the setting for benefit-sharing negotiations.

These considerations about the actions of private individuals and organisations raise a broader question concerning the obligations of the countries where the commercialisation takes place. So far, *de facto*, the burden of regulating access has fallen on the countries providing genetic resources. Since all Contracting Parties are equally bound by the provisions of the Convention, it can be argued that governments are obliged to take legal, administrative or policy measures to ensure that companies and other users of genetic resources have acquired them in accordance with the Convention, and that the benefits arising from access to genetic resources are shared fairly and equitably. Several possible approaches to such measures have been mooted, including preventing genetic resources from entering countries without proof of prior informed consent; requirements to disclose the country of origin and demonstrate prior informed consent in patent applications; and national legislation which would require companies to obtain genetic resources in accordance with the Convention, and to facilitate the enforcement of claims against collectors or commercialising entities (ERM, 1996/7; Tobin, 1997; Glowka et al, 1998).

Thus, in some circumstances, a source country could bring an action based either in criminal or civil law against not only the collector, but the recipient of genetic resources that commercialises them. Where the collector did not literally 'steal' the materials concerned, and where the company is the passive, innocent recipient of derivatives of genetic resources in a country other than the country of origin of the resources, the chances of success and successful enforcement of such claims under present laws are questionable. However, the acquisition of samples by a company in either of the two scenarios described above throws sufficient doubt over its title to the genetic resources, their derivatives and the profits arising from their sale as to provide good legal arguments for avoiding the situation. Furthermore, there are good business arguments for doing so, namely to maintain the company's reputation with its customers and other stakeholders, and to protect the company's ability to acquire raw materials from countries around the world in the future.

The best way to avoid the risk of legal actions, loss of reputation and the risk of losing channels for the supply of raw materials in the future is to ensure that genetic resources are acquired under material acquisition agreements containing terms that clarify the rights and responsibilities of the supplier from whom the company acquires the genetic resources, and the recipient company.

Collectors and companies might feel that the collecting permits issued by many countries resolve these questions, since they appear to be proof of government agreement for access to genetic resources. There are two main points to be made about collecting permits: first, they are legal agreements between the permitting authority and the individual or organisation granted access, and may set detailed terms for access. Second, they may, however, not grant sufficient authority for applicants to use the materials obtained for commercial purposes. Each of these points will be discussed in turn.

Many permits are granted in reliance upon quite detailed information provided in application forms and project proposals submitted to the permitting authority. There are strong legal arguments to suggest that the granting of the permits thus concludes contracts, whose implied terms might incorporate restrictions on the use of the materials to those described in the applications. Permitting authorities might also be able to sue for misrepresentation if an applicant subsequently used samples in a manner other than described in the application which resulted in the permit being granted.

However, collecting permits suffer from two major shortcomings as far as avoiding the legal doubts described in this section are concerned. First, they are often quite vague on the conditions under which materials acquired may be used. Frequently, they simply list the material, the collector's name, and the location where the resources were collected. A stamp shows that permission has been granted. Sometimes the permits are headed 'scientific collecting permit', or 'research permit', but often they are silent on whether the permit allows the recipient to conduct research for commercial purposes, whether the materials can be used for other specific purposes (for example, public display, DNA analysis or biochemical screening, even for ostensibly non-commercial purposes), or whether the material can be passed to third parties for these or other purposes. Second, collecting permits may be just that: permits to enter a protected area to collect. Separate permits from other departments of government may be needed to export or otherwise use the samples. For these reasons, organisations such as university departments and botanic gardens that collect genetic resources, and companies that obtain samples from them, may prefer a material acquisition agreement, rather than simple evidence of a collecting permit.

The basic terms that a company (or other organisation acquiring genetic resources) might therefore wish to see reflected in a material acquisition agreement are:

- *The material was legally acquired:* confirmation that the material was collected in accordance with the law of the host country.
- *The government has authorised access:* confirmation that the supplier of that material is authorised under the laws of the host country to grant access to and to provide the material to the recipient (ie confirmation of prior informed consent); and
- *Terms for use of the material:* a clear statement of the terms and conditions attaching to the use of the material or, if there are none, a clear statement that this is the case.

There are two main barriers to obtaining such agreements: identifying the competent national authority, and identifying local level stakeholders whose consent is also needed.

The fundamental difficulty faced by companies and the intermediary organisations that collect and supply genetic resources is that it is often practically impossible to determine precisely who is the 'gatekeeper' authorised to grant access, and exactly what are the requirements one must follow. This makes it difficult for suppliers to give absolute guarantees that materials were legally acquired or that the correct government body has authorised access for the purposes for which the recipient company wishes to use the resources.

In many countries, different government bodies are authorised to deal with different aspects of regulating access. For example, in China the Ministry of Forests is responsible for access to plants in forests and protected areas, and the Ministry of Agriculture for access to plants elsewhere. Access to insects and microorganisms is regulated by the Chinese Academy of Sciences (personal communication, Prof Song Da-Kang, February 1999). Further permissions may also be needed from Ministries (such as the Ministry of Public Health, in the case of medicinal plants) where work involves organisations under their jurisdiction. Until national access legislation has been introduced, it is difficult to tell which single government body is authorised to grant access, whether a number of different bodies are empowered to grant access and any one will do; whether different bodies grant access in different circumstances for different uses; or whether permission may be needed concurrently from more than one body. The challenge of finding clear answers to these questions is compounded by the fact that many representatives of the governments concerned do not themselves know the answer to the questions and also by the 'diplomatic' difficulty of raising the need for central government approval with other institutions (such as research institutes, genebanks, botanic gardens, herbaria and universities) which have, *de facto*, granted access to genetic resources for many years.

The challenge of identifying the local level stakeholders from whom consent is needed is described in Whose consent is required?, on page 27.

Permitted use of genetic resources

One useful purpose that can be served by MTAs is to clarify the use which the recipient is entitled to make of the materials received. If the recipient is a company, it is likely that the materials will be used for research and development which may (but more likely will not) lead to successful commercialisation. In that case, the agreement should state explicitly that the recipient is entitled to commercialise the material transferred, or its derivatives. While a court could, in the event of a dispute, decide upon what activities constitute 'commercialisation', it is preferable to avoid any doubts by including a definition of 'commercialisation' or 'commercial use' in the agreement. Many such definitions bear some similarity to the following one:

> Commercialisation *shall include, but not be limited to, the following activities: sale, filing a patent application, obtaining or transferring intellectual property rights or other tangible or intangible rights by sale or licence or in any other manner, commencement of product development, conducting market research, and seeking pre-market approval.*

Definitions such as this serve not only to prescribe the activities which the recipient may conduct on the material acquired, but also to set various 'milestones' in the use of material by the recipient. Thus other clauses in the agreement can refer to the events listed in the definition of commercialisation, which can be used as the 'trigger' for notification of the supplier, for the transfer of benefits such as milestone payments, or for further negotiations, either with the supplier, or with source country governments. (For example, see the mechanism used by the National Cancer Institute (NCI) to ensure that a licensee company negotiates mutually agreed terms with the source country government, ten Kate and Wells, 1998.)

Where the main purpose of the transfer of material does not foresee commercialisation, the provider may wish explicitly to prohibit commercialisation of the material by the recipient, using a definition such as the one above. Alternatively, the parties may agree a list of other activities which the recipient may or may not undertake on the materials. For example, a government granting a permit to a botanic garden could stipulate that the botanic garden may only use the materials for the creation of herbarium specimens and not for public display; or for public display but not for screening or DNA extraction, etc.

Restrictions on supply

Recipients of material need to ensure that they are clear about their rights to supply the material to third parties, or otherwise dispose of it. If the recipient is an intermediary institution such as a small company, genebank or a university department, it should establish whether it is entitled to supply the materials, or products derived from them, to third parties such as companies or other genebanks. Even a company that is likely to manufacture a final product from the materials may license out access to its libraries of extracts and compounds, or may merge with or be acquired by another company, so terms on the supply of materials to third parties may also be relevant.

Benefit-sharing

A range of monetary and non-monetary benefits can be shared by the provider and recipient of the material transferred at several times that can be mutually agreed when the agreement is signed; for example, at the time the materials are first transferred; at various stages during the research and development process (see 'triggers' under the definition of commercialisation, above); and once a product is successfully commercialised.

As well as stipulating the stage at which benefits are to be shared, the nature of the benefits and the basis for calculating them can also be set out in an MTA.

Aspects of benefit-sharing that are listed in national access legislation for incorporation into material transfer agreements, and that are common in MTAs for the commercial use of biodiversity, include (see UNEP/CBD/COP/3/Inf.53 and UNEP/COP/3/20):

Monetary benefits

- up-front fees, either for access to genetic resources, or to cover the costs of any preparation of samples, research conducted on them, and handling and shipping costs;
- milestone payments when various stages in discovery and development are reached (either independent payments, or set-off against any royalties that may be incurred in the future); and
- royalties. It is important to clarify the basis of royalty payments, for example whether they are calculated on gross or net sales.

Non-monetary benefits

- participation of source-country scientists (who may be third parties) in research;
- transfer of equipment, software and know-how;
- exchange of staff and training;
- in-kind support for conservation;
- acknowledgement of provider in research publications, patents and other forms of IPR;
- sharing of research results, including notification of discoveries and ensuring that copies of publications concerning research on the genetic resources provided are sent to the source country;
- voucher specimens to be left in national institutions; and
- terms for the licensing of technologies developed from research on the genetic resources transferred.

Other terms

In addition to terms on legal acquisition, authorised uses of materials, transfer to third parties and benefit-sharing, it is common to include a number of other clauses in material transfer agreements, which may deal with the following:

- definitions;
- duty to minimise the environmental impact of collecting activities;
- duty to maintain samples in good condition;
- ownership of materials;
- representations and warranties concerning the quality, identity of the materials transferred and any liabilities arising from their use;
- indemnity against potential liability, such as that arising from civil or criminal actions connected to access and benefit-sharing brought by source countries and other providers or recipients of genetic resources;
- length of agreement;
- notice;
- the fact that the obligations in certain clauses (e.g. benefit-sharing) survive the termination of the agreement;
- independent enforceability of individual clauses in the agreement;
- events limiting liability of either party (such as Act of God, war, fire, flood, explosion, civil commotion, industrial disputes, impossibility of obtaining gas or electricity or materials), and requirement to notify the other party in the case of any such event;
- arbitration;
- assignment or transfer of rights; and
- choice of law.

2.2.3 Prior informed consent

Article 15 (5) of the CBD provides that access to genetic resources shall be subject to the prior informed consent of the Contracting Party providing such resources, unless otherwise determined by that Party. As required by Article 15(2), Parties must endeavour to facilitate access, thereby ensuring the continued exchange of genetic resources. National measures must therefore strike a balance controlling access so as to arrive at mutually agreed terms for access and to ensure benefit-sharing, and the need to ensure that access procedures and requirements do not block access.

Prior informed consent is not defined in the Convention, but several commentators have identified the key elements as being:

- *'prior'* – before access takes place;
- *'informed'* – based on truthful information about the use that will be made of the genetic resources that is adequate for the authority to understand the implications; and
- *'consent'* – the explicit consent of the government (and possibly other stakeholders, according to national law) of the country providing genetic resources.

(See, for example, Glowka, 1994; Posey and Dutfield, 1996.)

In those cases where countries have introduced explicit access legislation after the CBD, they require consent from the state and also from certain other individuals, communities and organisations.

Whose consent is required?

Prior informed consent from the State

Although there is some debate as to the meaning of the phrase 'unless otherwise determined' in Article 15(5), the CBD is fairly unambiguous that informed consent is to be sought from national governments prior to access. In practice, identifying the appropriate organ or organs of government from whom consent must be sought is often extremely difficult. Most countries have not yet taken legal, policy or administrative steps to clarify access to genetic resources in the light of the CBD. In these circumstances, access to genetic resources is generally governed by a patchwork of different laws concerning access to protected areas, collection of specimens in various other geographical locations, scientific research, export, the Convention on International Trade in Endangered Species of Wild Fauna and Flora (CITES), phytosanitary provisions, biosafety, the laws of private property, and the rights of local and indigenous communities. These laws often overlap, but also often reveal important 'gaps' – such as no regulation of access to microorganisms or to marine resources.

Access is often administered by an equally varied group of organisations, including several different ministries (typically, the ministries of environment, forests, agriculture, health, science, and commerce), local or regional government, para-statal organisations, institutions and individuals.

Governments can play a variety of roles in access and benefit-sharing arrangements established pursuant to access legislation, and these roles are reflected in the function designated for the state in terms of oversight, negotiation, approval and monitoring of access agreements. At one extreme, a government might be party to each commercial agreement. At the other extreme, a government might establish laws that guide the development of access and benefit-sharing arrangements but then remain distant from all negotiations and transactions, leaving private institutions to enter into their own agreements, consistent with the law. A range of roles for government bodies is spelled out in access regulations, and these often include assessment of the merit of each proposed access and benefit-sharing arrangement, and subsequent approval (or rejection) of the terms of access and benefit-sharing contracts. Several countries have established or are considering cross-sectoral, representative bodies to perform such tasks. These may be inter-ministerial or inter-agency committees, which include the representation and participation of indigenous and local communities, the private sector, the research community, non-governmental organisations, and other stakeholders. In several countries, such a committee is supported by one or more technical advisory committees.

Local-level prior informed consent

Several access measures establish the requirement for the approval of local and indigenous communities prior to access either to genetic or biological resources within their territory, or to their knowledge, innovations and practices. For example, the Philippines law requires the prior informed consent of indigenous cultural communities, in accordance with customary laws, for the prospecting of biological and genetic resources within their ancestral lands and domains (PEO, Section 2). It does not, however, refer to access to their knowledge, innovations and practices.

While there is no clear provision requiring the prior informed consent of local and indigenous communities, Article 7 of Decision 391 states that the five member states of the Andean Commission 'recognise and value the rights and decision-making capacity of indigenous, Afro-American and local communities with regard to their traditional practices, knowledge and innovations connected with genetic resources and their derivatives'. The Decision states that an applicant wishing to access genetic resources or derivatives with an 'intangible component' (Box 2.4, page 19) must identify the supplier of the intangible component (Article 26(b)) and negotiate an annexe to the access contract providing for the sharing of benefits with them (Article 35). Furthermore, the Eighth Complementary provision refers to the future development of a special system or harmonisation standard 'to provide greater protection of the knowledge, innovations and traditional practices of indigenous, Afro-American and local communities'. The language itself is a little unclear, but it is plain that the intent of the legislation is to ensure that the approval of local and indigenous communities should be sought prior to access to their knowledge. Their prior informed consent may also be needed (as landholders) for access to biological resources on the land, whereas access to genetic resources in the Andean Pact is a question for the member states alone.

According to the law in each country, the consent of other individuals and organisations may also be required. For example, Section 2 of the Philippines Implementing Regulations includes in its definition of prior informed consent both local communities and private land owners:

> *Prior informed consent refers to the consent obtained by the applicant from the Local Community, Indigenous Cultural Communities or Indigenous Peoples (IP), Protected Area Management Board (PAMB) or Private Land Owner concerned, after disclosing fully the intent and scope of the bioprospecting activity, in a language and process understandable to the community, and before any bioprospecting activity is undertaken.*

As well as its provisions that refer to annexe agreements with any indigenous, African American and local communities that provide knowledge, or the 'intangible component' associated with genetic resources, Decision 391 also refers to accessory contracts between the applicant for access and the owner, holder or administrator of the biological resource containing the genetic resource, or of the property on which the biological resource containing the genetic resource is found (Article 41). In this way, access laws effectively grant rights to prior informed consent of individuals and organisations with property rights over biological resources and the land where they are found, land-owners and other private citizens, thus granting a level of prior informed consent to private land-owners as well.

The challenge for the collector is to identify, at the local level, exactly whose consent is required. While it may be possible to identify local landowners, it is often quite a challenge for scientists from foreign countries who are neither anthropological experts nor conversant in local languages or customs to identify from which local and indigenous communities consent is needed,

Box 2.5 *Examples of the Challenges of Obtaining Prior Informed Consent and Negotiating Benefit-Sharing at the Local Level: the Cases of the Peru ICBG and the Kani Agreement with TBGRI*

ICBG, Peru

The Peru International Cooperative Biodiversity Group (ICBG) involves Washington University, several organisations of the Aguaruna and Huambisa communities, the Universidad San Marcos, the Universidad Peruana Cayetano-Heredia in Peru, and Monsanto-Searle Co. The partnership involves three separate agreements:

1 Biological Collecting Agreement between the Aguaruna and Huambisa Peoples (as represented by four organisations: three Aguaruna federations – OCCAAM, FECONARIN and FAD – and the national organisation CONAP to which the three are affiliated) and Washington University. This outlines the basic terms of collaboration and sample collection, as well as benefit-sharing with the Aguarunas and Huambisas, and obliges the signatory community organisations to establish a mechanism to ensure equitable benefit-sharing within three years of its signature;

2 licence option agreement between Washington University and Monsanto-Searle Co that covers financial benefits, including royalties; and

3 know-how licence between the Aguaruna and Monsanto-Searle Co that outlines Searle's use of the traditional knowledge and specific benefits associated with that use.

The Aguaruna and Huambisa communities are part of the Jivaro peoples of the lowland Andean rainforest, of which there are five along the border between Peru and Ecuador, also including the Shuar and Achwar which are in Ecuador. The Aguarunas and Huambisas reside primarily in over 140 communities along a series of river basins. These communities share a common language and cultural heritage, including medicinal plant knowledge, and have somewhat fluid alliances with various Aguaruna political organisations. In detailing their collection and benefit-sharing agreements, the Peruvian ICBG struggled to balance competing claims of representation, the need to compensate those most actively involved and the ideal of providing benefits to all of the relevant communities and individuals. The balance that has been struck is to dedicate short- and medium-term benefits from research activities and advance payments to the organisations and communities actively involved in the project, and to share long-term contingent benefits (such as any royalties that arise from sales of any successful products) with all Aguaruna communities.

What began as a negotiation primarily between cooperating individuals, a couple of organisation leaders and permit-granting authorities in the government, turned into a more protracted disclosure and consensus-building process, including many community individuals, village leaders, and others. OCCAAM, FECONARIN and FAD are three of the smaller Aguaruna federations, and the lack of national law to govern the negotiating process offered no means for Aguaruna federations opposed to the negotiations to impede its signature. This led to the establishment in Peru of a working group on *sui generis* rights.

Kani, India

The Kanis, with a population of 16,181, are a traditionally nomadic tribal community who now lead a primarily settled life in tribal hamlets, each consisting of 10 to 20 families, in and around the forests of the Western Ghats (south-western India). A lead provided by the Kanis relating to the anti-fatigue properties of a wild plant, *Trichopus zeylanicus*, led to the development of the drug 'Jeevni' by the Tropical Botanic Garden and Research Institute (TBGRI), which transferred the manufacturing licence to the Aryavaidya Pharmacy, Coimbatore Ltd for a licence fee of Rs 10 lakhs (one million rupees, approximately US$25,000). The TBGRI agreed to share 50 per cent of the licence fee and royalty with the tribal community.

The traditional system of governance among the Kanis has been eroded, and their day-to-day activities and system of governance are now linked to that of non-tribal local communities, rather than governed by the tribal chief. This system of governance, referred to as the *Panchayati Raj* system, is based on the principle of devolution of administrative powers to the local village level and has been institutionalised under the Constitution of India. Each *Panchayat* area consists of a number of wards with assemblies of all the adult members, called the *Gram Sabha* (village council). There are 1,000 to 2,000 members in each Gram Sabha, inclusive of both non-tribal and tribal members, but predominantly non-tribal. Members of the Panchayat's decision-making body are elected by the members of all the Gram Sabhas constituting the Panchayat.

Kanis in different areas of Thiruvanathapuram district have differing opinions on the arrangement with TBGRI, which has interacted primarily with Kanis from the Kuttichal Gram Panchayat area, and hired as consultants the two Kanis from this area who imparted the knowledge of *Aarogyappacha*. This section of Kanis has been supportive of and appreciative of TBGRI's role. However, Kanis in other Panchayat areas have expressed offence at the fact that TBGRI has not been working with the Kanis from their areas. Some feel the benefit-sharing arrangement is a superficial exercise since the Kanis have neither been consulted nor involved in the exercise. In September 1995, a group of nine medicine men (called *Plathis*) of the Kani tribe wrote a letter to the Chief Minister of Kerala, objecting to the sale of their knowledge to 'private companies'. TBGRI acknowledges that it has not reached out or communicated to all the members of the Kani tribe, but is establishing a Trust to administer the benefit-sharing arrangement as constituted, which may involve more Kanis.

Sources: (Peru) personal communication, Brendan Tobin, 10 March 1999; Rosenthal, 1998. (India) personal communication RV Anuradha, 15 March 1999; Anuradha, 1998, and case study submitted to the CBD Secretariat by the Government of India, 1998.

and who can negotiate on their behalf (eg, Box 2.5). The necessity to obtain consent from local and indigenous communities raises a number of important questions, such as: how many distinct communities are involved? Does the legal system of the country concerned grant rights to individuals, or collective rights to communities? Does the community have legal standing to enter into agreements? Who is authorised to negotiate on behalf of the community? What happens if a certain proportion of members of a community agree to negotiated terms, but the rest do not? What happens if certain communities break away from umbrella groups that represented them and with whom access agreements were negotiated? Collectors are likely to need help from local experts, for example, from the government, that will require evidence of agreement with these communities prior to approving an access application, or from local NGOs.

The PIC procedure

Procedures which an applicant for access must follow vary from country to country, but in access regulations such as those in the Philippines and the Andean countries, they generally involve:

- submission by the applicant to the competent national authority of a proposal (see below);
- (in some procedures) public notification through publication of elements of the proposal in national and local newspapers. The Philippines Implementing Regulations, for example, call for public notification, through various media, of any collection activities, as well as local community and relevant sector consultation and notification;
- negotiation of contracts between the applicant and the competent national authority and also, if appropriate, between the applicant and other parties such as private landowners, indigenous groups providing traditional knowledge or *ex situ* collections providing access to specimens (see below);
- approval by government of the contracts and the proposal, and procedures for appeal if access is refused; and
- implementation, for example periodic reporting by the applicant on the use it is making of the resources accessed.

Box 2.6 *Flow Diagram Summarising the Procedure to Obtain a Commercial Research Agreement in the Philippines*

Step 1 The applicant submits the following requirements:
Letter of Intent and Three Copies of the Research Proposal to the IACBGR through the Technical Secretariat; Filing Fee of P25.

Step 2 The Technical Secretariat conducts an initial screening of the application to determine whether the research is within the coverage of EO No 247.

Step 3 If the proposed research is within the coverage of the EO:
(a) the Technical Secretariat provides the applicant with a checklist of additional requirements to be submitted which include, among others, the following:
Duly accomplished application form;
Company/Institution/Organisation/Agency Profile;
Environmental Impact Assessment (EIA), as determined by the Technical Secretariat; and
Processing fees in the following amounts:
Philippine national – P1,000 per application
Foreign national – P2,000 per application;
(b) the applicant submits a copy of the summary of the research proposal, duly certified by the Technical Secretariat, to the recognised head of the indigenous peoples, municipal or city mayor of the LGU, PAMB, or private land-owner concerned for the required PIC Certificate (2.2.3, p28).

Step 4 If the research proposal is not within the coverage of EO No 247, the Technical Secretariat issues a Certificate of Exemption and refers the proposal to the government agency that has jurisdiction over the project.

Step 5 The applicant submits the PIC Certificate signed by the recognised head of the IP, municipal or city mayor of the LGU, PAMB, or private land-owner concerned, together with proofs of public notification and sectoral consultation.

Step 6 The Technical Secretariat conducts an initial review and evaluation of the application and documents.

Step 7 The Technical Secretariat submits the result of its evaluation, including the draft Commercial Research Agreement, to the IACBGR within 30 days from receipt of all requirements.

Step 8 The IACBGR conducts a final evaluation of the application.

Step 9 The IACBGR submits its recommendation to the Agency concerned.

Step 10 The Secretary of the Agency concerned considers the approval or disapproval of the Research Agreement.

Step 11 Upon approval of the application, the applicant pays the bioprospecting fee as determined by the IACBGR.

Step 12 The Agency concerned transmits the signed Research Agreement to the Technical Secretariat who shall furnish a copy to the applicant, IP, local community, PAMB, or private land-owner concerned.

Key: EO – Executive Order 247; IACBGR – Inter-Agency Committee on Biological and Genetic Resources; IP – indigenous peoples; LGU – local government unit; PAMB – Protected Area Management Board; PIC – prior informed consent.

Source: La Viña et al, 1997.

Box 2.7 *Different Contracts Envisaged in the Andean Pact Common Regime*

Access contracts

Parties: the Competent National Authority (CNA) and the applicant for access.
Covers: access to genetic resources.
Note: must 'take into consideration the rights and interests of suppliers of genetic resources and their derivatives, and of biological resources and their intangible components', in accordance with the other contracts mentioned below (Article 34).

Access contracts with ex situ centres

Parties: the CNA and *ex situ* conservation centres or other bodies.
Covers: activities involving access to genetic resources, their derivatives or, as applicable, their related intangible component.
Note: Collections obtained prior to the Decision and prior to the entry into force of the CBD are under the scope of the Decision. The CNA could enter into access contracts with third parties concerning genetic resources of which the member state is a country of origin, and which are maintained in *ex situ* centres. It would appear that these contracts could involve either *ex situ* collections and third parties in the Andean region, or those outside it. Such contracts must take into consideration the rights and interests of suppliers of biological resources, just as for access contracts.

Annexe to the access contract

Parties: the supplier of the intangible component and the applicant for access.
Covers: 'intangible component' (information concerning genetic resources).
Note: where the CNA is not a signatory, the annexe must include a 'suspensive' term making it subject to the access agreement. Failure to comply with such an annexe agreement is grounds for termination of the access contract.

Accessory contracts

Parties: the CNA and:

- owner, holder or administrator of the property on which the biological resource containing the genetic resource is found;
- *ex situ* conservation centre;
- owner, holder or administrator of the biological resource containing the genetic resource; or
- the national support institution, in connection with activities which it is to carry out and which are not included in the access contract.

Covers: biological resources and activities connected with access to genetic resources.

Framework contracts

Parties: the CNA and universities, research centres or recognised researchers.
Covers: various field or collaborative projects.
Note: universities, research centres or other researchers could enter into 'deposit contracts' for the maintenance of resources *ex situ* or 'administrative contracts' with the CNA. Could be used either to deposit biological or genetic resources and their derivatives for purely custodial purposes, with the CNA maintaining jurisdiction and control over them, or for administrative purposes rather than for access, such as managing genetic resources, their derivatives or synthesised products.

Contracts with organisations involved with access

Note: the CNA could use such contracts to clarify the rights and obligations of such institutions, even with respect to collections obtained prior to the Decision and prior to the entry into force of the CBD.

Sources: Comisión del Acuerdo de Cartagena (1996) and personal communication with Manuel Ruiz, February 1999.

Typical of the requirements for the information to be contained in an access application so that any consent granted by the state can be 'informed' are those set out in the Philippines Executive Order and Implementing Regulations. This asks applicants to provide a letter of intent and an application form setting out the kind, number and quantity of specimens sought, the purpose and places of collection, and the foreign and local-counterpart researchers (Annexe B). Applicants must also attach a research proposal setting out the objectives and date of the project, the nature of the bioresources involved, methodology, the manner of collecting, the anticipated intermediate and final destination of the bioresources, how they are to be used, a description of funding and budget, foreseen impact on biological diversity, details of immediate and long-term anticipated compensation, and a list of in-country entities likely to receive compensation (Annexe A). Other accompanying documents must include a letter of acceptance from Filipino counterparts, a letter of endorsement from the head of the institution with which the applicant is affiliated, an institutional profile of the applicant, and any others that the concerned government agency may require (Article 6.1).

The kind of terms that may be mutually agreed in access agreements are described in Section 2.2 above. This section will summarise briefly the different kinds of access agreements envisaged in various models of access legislation.

As Box 2.6 on Commercial Research Agreements suggests, The Philippines Executive Order distinguishes between Academic and Commercial Research Agreements. The former are to be used by duly recognised Philippine universities and academic institutions, domestic governmental entities and intergovernmental entities where the genetic resources are intended for academic and scientific purposes. In contrast, Commercial Research Agreements are to be used by private persons, corporations and foreign international entities when the genetic resources are intended directly or indirectly for commercial uses. Box 2.7 shows the broader range of contracts envisaged in Andean Community Decision 391.

2.3 Conclusions for individuals seeking access to genetic resources and benefit-sharing

Many countries regulate access to genetic resources, whether in explicit laws on access and benefit-sharing introduced nationally or regionally in response to the CBD, or in a range of other legal measures that may have existed for many years prior to the entry into force of the CBD (Section 1.5). The new wave of laws introduced since the CBD in the Philippines and the Andean Pact, and laws and drafts in several other countries, also regulate aspects of access and benefit-sharing for derivatives of genetic resources and traditional knowledge concerning them. The scope of resources and exact activities regulated by these laws (especially with respect to derivatives and the status of *ex situ* collections) is often unclear. The processes set out in access laws are elaborate, and many domestic and foreign scientists and companies have reported finding them cumbersome, time consuming and costly to follow (Chapter 10).

In the absence of such laws, however, those acquiring genetic resources face another problem, namely the challenge of identifying which government agency (or agencies) is the competent national authority with the power to grant prior informed consent for access to genetic resources. Also, it is hard, particularly for foreign organisations, to establish from which individuals, organisations and communities (such as land-owners, protected area management boards and local and indigenous communities) prior informed consent must be obtained. Both factors combine to make it difficult for collectors and subsequent recipients to be confident that they have acquired materials legally. It is thus difficult to arrive at a position of legal certainty with respect to title to material; a situation which is in the interest neither of a country wishing to protect its sovereign rights and those of its citizens, nor of a company which wants to avoid financial risk when investing large sums in the commercialisation of genetic resources.

In addition to legal requirements imposed by the CBD and national laws related to access to genetic resources, there are compelling business reasons for those who may well not be bound by specific legal requirements to ensure they are acquiring materials legally and are authorised, should they wish to do so, to commercialise them. These include avoiding the risk of law suits (whether ill- or well-founded), maintaining the company's reputation with its customers and other stakeholders, and protecting the company's ability to acquire raw materials from countries around the world in the future. These arguments apply to the many companies and research institutions based in countries that have not adopted laws on access to genetic resources and benefit-sharing, or are not even parties to the Convention. The most practical way for organisations in this position to ensure they are acquiring materials legally and are authorised to commercialise them is to use material acquisition agreements (Section 2.2). Another useful tool to build trust and clarify an institution's approach is for the institution to develop its own, voluntary policy on access and benefit-sharing (Chapter 10).

As we saw in Section 1.5, few countries are entirely devoid of law that relates to access to genetic resources, but the patchwork of legal provisions on issues such as conservation of wildlife, trade in endangered species and phytosanitary regulations often does not clarify institutions' rights and responsibilities with respect to gaining access to genetic resources or providing access to genetic resources held within their own collections. Many countries have used collecting permits to regulate access for several decades. Organisations providing samples obtained under collecting permits will need to ensure that they do so on terms consistent with those set out in the permits and other legal agreements (such as export permits), but may also need an additional agreement to clarify their rights and responsibilities with respect to the material.

The approach that individuals and institutions take in such circumstances will depend on their motivation and strategy.

Legal compliance

If the motivation of a particular individual or organisation is strict legal compliance, they might establish whether each country with which they work is a party to the CBD, the date on which the CBD entered into force in that country, and what exactly are the legal requirements in the country concerning access and benefit-sharing. The institution would then enter into quite different partnerships in different settings, reacting to the legal requirements made of it in each setting. Taken to its logical extreme, an institution adopting this approach would feel free to help itself without prior informed consent to genetic resources in country A (which is not a party to the Convention), and commercialise the results without sharing any benefits, while in country B, it might satisfy rigorous prior informed consent requirements and enter into an elaborate benefit-sharing arrangement.

Best practice

If the motivation of a particular individual or organisation is to encourage others to grant access to genetic resources, the organisation in question will need to establish a policy acceptable to its different stakeholders, such as its own board of directors and staff, the government of the country where it is based, its partner institutions both at home and abroad, and governments of countries from where it hopes to collect genetic resources in the future. It may then choose to apply this policy in all countries, regardless of the nature of their national access legislation. An analogy is the practice of some multinational corporations which, in order to be 'good corporate citizens' voluntarily apply the highest common denominator of the standards in which they find themselves operating around the world, whether or not legislation requires this. If institutions were to take the 'good corporate citizen' approach, they might not wish to 'take advantage' of countries that do not yet have access legislation, but might extend to them the same standards as they would where obliged to obtain prior informed consent and to share benefits fairly and equitably. Similarly, they would be unlikely to make a distinction between parties and non-parties to the CBD.

A strategy of basic legal compliance might, at first sight, appear to offer the advantages of minimising costs by involving the organisation in taking only such steps as are required in law. However, it should be clear from this chapter that, for any organisation working internationally, to know precisely what measures are required for 'strict legal compliance' is not easy. In addition to fostering the trust and goodwill needed to forge access and benefit-sharing agreements, the 'best practice' approach may also save costs, particularly for trans-national organisations. It could allow organisations to harmonise practice internationally, rather than having to concern itself with the implications of distinctions between the plethora of different access rules around the world; to anticipate rather than react to changes in regulations; and to invest in practices and procedures that will secure medium- and long-term goals as well as their short-term ones, thus avoiding the need constantly to retrain staff and rewrite the organisation's guidelines.

As the rest of this book will show, while it is essential, as a bare minimum, to follow the laws of the country where one is working, successful access and benefit-sharing arrangements are more often the result of best practice than of basic legal compliance.

Note

1 Decision 391 of the Andean Commission Common Regime goes a little way towards clarifying the distinction. Article 14 states (within the limited scope of free movement of biological resources within the Andean area) that the decision shall not impede the use and free movement of biological resources 'provided that access is not made to the genetic resources contained in biological resources'. In addition, the Fourth Complementary Measure provides that certain exports of biological resources shall bear the inscription 'not authorised for use as a genetic resource'.

Sarah A Laird and Kerry ten Kate

Chapter 3
Natural Products and the Pharmaceutical Industry

3.1 Introduction

The public imagination has been fired by the concept of the 'medicinal riches of the rainforest', and the argument these riches make for forest and biodiversity conservation. However, more than seven years after the Earth Summit in Rio de Janeiro where the CBD was opened for signature, our understanding of the size, structure, and nature of pharmaceutical research and development, and the role of natural products in this enormously complex industry, remains poor. The pharmaceutical industry is the most research-intensive industry in the world. Global expenditures on research and development totalled around US$21.1 billion in 1998 alone (PhRMA, 1998). The costs of developing a single drug are estimated at between US$231 and US$500 million (Di Masi et al, 1991; PhRMA, 1998). The odds of a single compound becoming a drug once it enters the discovery process are generally estimated at one in 5,000–10,000. More than half of all prescriptions filled in the USA in 1993 contained at least one major active compound 'now or once derived or patterned after compounds derived from biological diversity' (Grifo et al, 1997). 42 per cent of sales of the 25 top-selling drugs world-wide are either biologicals, natural products, or entities derived from natural products (Newman and Laird, Appendix A).

But what do these figures tell us about the demand for access to new natural products today, and the types of benefit-sharing arrangements that are current practice? What is the wider market, scientific, and regulatory context in which access and benefit-sharing takes place?

This chapter examines some of the underlying issues associated with pharmaceutical natural product research and development. It begins with an overview of international markets, the structure of the industry, the major actors involved, and the costs of research and development and satisfying product approval regulations, as well as factors involved in investment decision-making. This is followed by an outline of scientific and technological trends that affect demand for natural products, and a review of current practices for the acquisition of natural product samples. Finally, the chapter concludes with an examination of current benefit-sharing practices in the pharmaceutical industry, including the development of 'equitable partnerships' of the kind envisioned in the CBD. In these partnerships, involving significant non-monetary, as well as monetary, benefits we can see new trends emerging in the partnerships associated with natural products.

3.2 The pharmaceutical industry

3.2.1 Global markets

Annual global sales of medical drugs are currently some US$300 billion a year, a figure that is likely to grow by about 6 per cent each year until 2001 (*The Economist*, 1998; *C&EN*, February 1998). The USA is the largest market for pharmaceuticals. Sales there have more than quadrupuled since 1985, when they were US$31.6 billion. Projected sales in the USA for 1998 are US$124.6 billion (PhRMA, 1998). Other major markets within the OECD include Japan, where the 1996 sales of

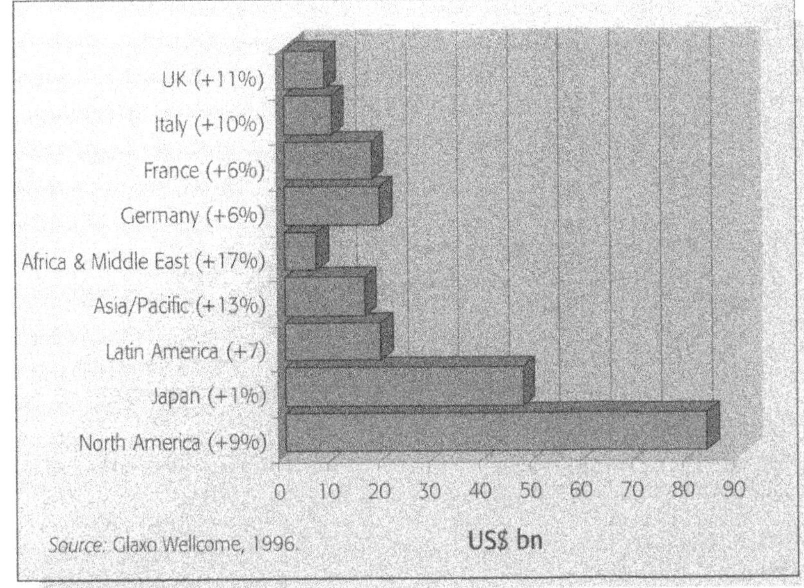

Figure 3.1 *World Pharmaceutical Market, 1996*

pharmaceuticals came to US$47 billion – twice those for the whole of Latin America. The UK market was US$8 billion, and the German market, the biggest in Europe, US$19 billion (Glaxo Wellcome, 1996). By comparison, the smallest regional market is in Africa and the Middle East, where annual sales of pharmaceuticals in 1997 were just over US$8 billion. Within the Asian Pacific region, 1997 sales were estimated at US$18 billion. Within this region, China is an important market, with 1996 sales of pharmaceuticals at US$10 billion, a figure that is expected to grow to US$25 billion by the year 2000 (Kalorama, 1996, 1997; *Chemical Week*, 1995; *MedAd News*, September 1998) (Figure 3.1).

The pharmaceutical industry is increasingly global in scope. Previously, companies might launch a number of products in one or two of the three major markets (Europe, Japan, and the USA). Today, in order to derive a satisfactory return on research and development, pharmaceutical companies generally launch products in all three markets (Shearson Lehman Brothers, 1991).

In developing countries, such as Brazil, Chile, Morocco, Pakistan, the Philippines, and sub-Saharan Africa, multinational companies hold the majority of the local market, since national policies have welcomed the local manufacture of their drugs as a form of import-substitution. In other countries, such as China, Egypt and India, national policy has promoted locally-financed pharmaceutical industries, almost exclusively involved in manu- facturing generic drugs. Companies in these countries do little or no primary research on NCEs (new chemical entities), although India is an exception. In this group of countries, multinational pharmaceutical companies hold a smaller market share. For example, in India the market share of multinational companies is just 20 per cent, and in China foreign joint ventures account for 13–25 per cent of the market (Dumoulin, 1998; Kalorama, 1996; *Chemical Week*, 1995).

3.2.2 Structure of the industry

This section will introduce some of the major actors involved in the discovery and development of pharmaceuticals. Given the vast sums required to bring a pharmaceutical to market, the bulk of pharmaceutical development is conducted by the private sector (Table 3.1). Pharmaceutical companies range in size from start-up enterprises whose staff can be numbered on the fingers of one hand, to the pharmaceutical divisions of life science multinationals which may employ over 100,000 people. Despite the predominant role of the private sector in pharmaceutical development, government agencies continue to play an important role in many countries, whether by conducting basic research,

Table 3.1 *The World's Top Ten Pharmaceutical Companies, 1998*

	1998 Forecast sales ($US bn)	
Company	Pharmaceuticals	Group
Merck & Co** (USA)	15,971	27,439
Glaxo Wellcome** (UK)	14,010	14,010
Novartis** (Switzerland)	13,433	23,685
Bristol-Myers Squibb** (USA)	12,220	18,341
Pfizer** (USA)	10,689	13,682
Roche** (Switzerland)	10,553	18,784
American Home Products* (USA)	10,549	14,657
Johnson & Johnson* (USA)	8,925	25,365
SmithKline Beecham** (UK)	8,688	14,100
Hoechst** (Germany)	8,173	30,072

** = companies with natural products discovery programmes
* = companies with wholly-owned subsidiaries conducting natural product discovery

Sources: interviews; *Financial Times*, 1998.

providing funding for product research, or introducing and administering the product approval regulations which a pharmaceutical must pass if it is to be introduced to the market. In addition, a panoply of non-profit institutions are involved in the picture, mainly collecting and supplying the samples from which natural product genetic resources are derived, and sometimes conducting extraction, screening and more value-added research.

Box 3.1 *Top Ten Pharmaceutical Companies, USA, Europe and Japan, 1997 (Prescription Sales Only)*

USA	Europe	Japan
Merck & Co	Glaxo Wellcome	Sankyo
Bristol-Myers Squibb	Novartis	Takeda
Pfizer	SmithKline Beecham	Yamanouchi
Eli Lilly	Roche	Dalichi
Johnson & Johnson	Astra	Eisai
American Home Products	Rhône-Poulenc Rorer	Shionogi
Schering Plough	Zeneca	Fujisawa
Pharmacia & Upjohn	Hoechst Marion Roussel	Chugai
Abbott	Bayer	Banyu
Warner Lambert	Schering AG	Ono Pharmaceuticals

Sources: Datamonitor, 1998; *Argus Research*, 1998; *Scrip's 1997 Year Book*.

Multinational companies

Pharmaceutical companies are traditionally large, vertically-integrated concerns that conduct the full range of activities from creating libraries of compounds to marketing the drugs that emerge from their pipelines. There are few companies that produce pharmaceuticals alone. Most also manufacture and sell a combination of nutritional products; medical or laboratory products, devices, or diagnostics; consumer health products; agricultural products; cosmetics or beauty care products; flavours or fragrances; vitamins and fine chemicals; or animal health products (*C&EN*, February 1998).

Many of the large multinational companies, such as Hoechst, Warner Lambert, American Home Products, Novartis, and Johnson & Johnson, do not earn the bulk of their income from pharmaceuticals. These companies might conduct research through wholly-owned subsidiaries (eg Wyeth Ayerst and Lederle, which are owned by American Home Products), or pharmaceuticals might make up only one division of 'life science' companies, which combine businesses in pharmaceuticals, agricultural chemicals and products, and food and nutrition. They include Monsanto, with some 32 per cent of its annual sales in pharmaceuticals; Rhône-Poulenc, with around 37 per cent of pharmaceutical sales, and Novartis, with around 55 per cent of sales in pharmaceuticals, generics, and consumer health (*C&EN*, February 1998; *MedAd News*, September 1998).

In contrast to life sciences giants, or larger companies of which pharmaceuticals are only one part, there also exist a number of what are known as 'pharmaceutical houses'. These companies concentrate on pharmaceuticals as their primary line of business. For example, in 1997 Glaxo Wellcome PLC earned 100 per cent of its income from pharmaceutical sales, Eli Lilly derived 87 per cent of its income from pharmaceutical sales, Sankyo Co Ltd 75 per cent, and Pfizer 74 per cent (*MedAd News*, September 1998).

In 1996, the top ten multinational pharmaceutical companies (Table 3.1) accounted for approximately 36 per cent of world pharmaceutical sales, and had an average profit margin of 30 per cent (*The Economist*, 1998; Dumoulin, 1998). In the USA, the ten largest companies accounted for around 53 per cent of the retail prescription market in 1997 (Bossong-Martines, 1998). Although the top 20 companies have annual turnovers of many billion dollars each, they each hold only about 2–5 per cent of the market (*C&EN*, February 1998).

Over the last few years (and most strikingly in 1994 and 1995), there has been a series of mergers and acquisitions by the largest pharmaceutical companies, leading to the creation of some life science giants with turnovers in the range of US$20–30 billion. For example, in 1996 the giant Swiss pharmaceutical companies Sandoz and Ciba Geigy merged to form Novartis (Box 3.2). 1998 saw considerable speculation over a reported £100 billion merger between SmithKline Beecham and Glaxo Wellcome, another between American Home Products and SmithKline Beecham, and yet another between American Home Products and Monsanto – none of which transpired (*Financial Times*, 1998; *The Economist*, 1998; 1998b; *C&EN*, 1998).

In March 1999, the European Commission conditionally cleared a US$35 billion merger of Zeneca Group PLC and Astra AB, which will create the world's third largest drug company, and will be the biggest ever deal between two European companies (Dow-Jones *Newswire*, 1999; *The Economist* 1998b). On track for completion by May 1999 is a US$10 billion merger between French pharmaceutical companies Sanofi SA and Synthelabo (Dow-Jones *Newswire*, 1999). Hoechst AG of Germany and Rhône-Poulenc SA of France agreed in late 1998 to a two-step corporate merger that would create the world's largest life science group (Moore, 1999).

The current 'urge to merge' in the pharmaceutical industry, particularly in Europe, is the result of a range of factors. These include a search for economies of scale; better access to US markets; greater funds

Box 3.2 *Mergers and Acquisitions that have Reshaped the Drug Industry over the Past Decade*

Year	Merger or acquisition	New company name
1998–9	Astra AB and Zeneca Group	AstraZeneca
	Sanofi and Synthelabo	
	Hoechst and Rhône-Poulenc	Aventis
	American Home Products and SmithKline Beecham (collapsed)	
	SmithKline Beecham and Glaxo Wellcome (collapsed)	
	American Home Products and Monsanto (collapsed)	
1997	Roche and Boehringer Mannheim	Roche
1996	Ciba Geigy and Sandoz	Novartis
1995	Glaxo and Burroughs Wellcome	Glaxo Wellcome
	Hoechst-Roussel and Marion Merrell Dow	Hoechst Marion Roussel
	Upjohn and Pharmacia	Pharmacia & Upjohn
	Rhône-Poulenc Rorer and Fisons	Rhône-Poulenc Rorer
1994	American Home Products and American Cyanamid	American Home Products
	Hoffmann-La Roche and Syntex	Hoffmann-La Roche
	Pharmacia and Erbamont	Pharmacia
	Sanofi and Sterling (prescription drugs)	Sanofi
	SmithKline Beecham and Sterling (OTC)	SmithKline Beecham
1991	SmithKline and Beecham	SmithKline Beecham
1990	Pharmacia and Kabi	Pharmacia
	Rhône-Poulenc and Rorer	Rhône-Poulenc Rorer
1989	American Home Products and AH Robins	American Home Products
	Bristol-Myers and Squibb	Bristol-Myers Squibb
	Merrell Dow and Marion	Marion Merrell Dow

Sources: *C&EN*, 23 Feb 1998; *The Economist*, 1998b; *Dow-Jones Newswire*, 1999; *Moore*, 1999.

available for expensive research and development programmes; and sometimes the desire to acquire a new research 'pipeline' to compensate for any shortcomings in existing, internal research (*The Economist*, 1998b).

Small and medium-sized companies

Scientific and technological developments (discussed in Section 3.5) have changed the structure of the pharmaceutical industry. Barriers to entry have been lowered and economies of scale are no longer as important as they once were. Small companies are springing up that specialise in individual stages of the research and development process. The first 'breaching' of barriers into the pharmaceutical industry occurred in the 1980s with the rise of small pharmaceutical biotech (or biopharmaceutical) firms such as Amgen, Genentech, Chiron and Genzyme, based on therapeutic proteins and a new technology called 'recombinant DNA' (Table 3.2). This represented the first experience of large pharmaceutical companies buying in drug candidates rather than developing them in-house (*The Economist*, 1998). Today, there are some 1,700 US biotechnology companies, and 1,000 such companies in Europe. Some 60 per cent of European biotechnology companies (Ernst and Young, 1998), and 68 per cent of US biotechnology companies (Ernst and Young, 1995) work in fields of human and veterinary products, diagnostics (*in vivo* and *in vitro* tests to detect disease and measure bodily functions), vaccines and other products to administer and deliver drugs.

Small companies have advantages of flexibility and innovativeness, but suffer from certain financial disadvantages. As David Corley, a natural products chemist at Monsanto, said, 'For discovery, small and flexible is the way to go. For penetrating markets, and taking products to market, being big is everything.' Companies have developed different approaches to carving a niche in the pharmaceutical market. Some of the smaller start-up companies – such as Amgen and Genzyme – are set on becoming fully integrated pharmaceutical companies, capable of conducting drug discovery, development and marketing. Others, by contrast, are reluctant to take the risks or bear the costs of clinical trials, and choose instead to develop their products jointly with the large companies. These companies tend to specialise in moving lots of molecules along short segments of the pipeline for other people, rather than taking a few of them all the way through on their own (*The Economist*, 1998).

An example of such a partnership is that between Phytopharm PLC and Pfizer. In August 1998, Phytopharm PLC entered into a licensing agreement with Pfizer for exclusive global rights to an experimental drug against obesity derived from extracts of a plant native to South Africa. Phytopharm receives US$32 million in research funding, licence fees and other payments (*The Wall Street Journal*, 1998). The identity of the species from which the compound is derived is not publicly available at this time, nor are the terms of Phytopharm's relationship with parties in South Africa.

There is a growing opportunity for such partnerships, since large, traditional drug firms increasingly out-source research and development through alliances, collaborations, and joint ventures with smaller

Table 3.2 *Top 20 Biopharmaceutical Companies, 1997*

Company	Revenues 1997* (US$ m)	Earnings** (US$ m)
Agouron Pharmaceuticals	296.0	-7.3
Amgen	2,401.0	736.8
BioChem Pharmaceuticals	193.7	55.7
Biogen	434.0	89.2
Bio-Technology General	65.3	14.5
Centocor	200.8	11.1
Chiron	1,162.1	71.2
Genentech	1,016.7	129.0
Gilead Sciences	40.0	-28.0
IDEC Pharmaceuticals	44.6	-15.5
Immunex	185.3	-15.8
Incyte Pharmaceuticals	88.4	10.4
Isis Pharmaceuticals	36.5	-31.1
Ligand Pharmaceuticals	51.7	-35.2
Liposome Co	65.1	-26.4
Medimmune	81.0	-36.9
Millennium Pharmaceuticals	89.9	2.6
Regeneron Pharmaceuticals	33.1	-11.6
Scios	47.4	-38.7
Sequus Pharmaceuticals	40.0	-23.6
Total	6,572.6	849.8

* Revenue results include collaborative or contract research funding, royalties, and interest income as well as product sales
** After-tax earnings from continuing operations, excluding significant extraordinary and non-recurring items.

Source: *C&EN*, 23 February 1998.

drug discovery companies, academia, and research institutions. To fill this niche, a wealth of small pharmaceutical 'service companies' are springing up to provide genomic information and combinatorial chemistry libraries, and to run pre-clinical or clinical trials as 'subcontractors' to major pharmaceutical companies.

These relationships allow large companies to access new technologies, breakthroughs and product leads in a cost-effective manner. In the USA, for example, more than half of the substances currently in clinical trials originated outside the laboratories of the biggest pharmaceutical companies. In 1997 alone, around 374 agreements were set up between large pharmaceutical firms and small drug discovery companies, with some pharmaceutical firms having as many as two or three dozen collaborations at one time (*C&EN*, February 1998).

In addition to bringing in new technologies and ideas, funding external research allows companies to undertake work that their own staff and facilities cannot handle. As Dwight Baker, Senior Project Manager in Drug Discovery Services at MDS Panlabs, a contract research and development company based in Washington state,[1] said, 'MDS Panlabs can do research and development projects in the short term that companies would have to build infrastructure to do. Their own research groups are just too busy doing other things, and if they have an exciting lead and don't have the capacity to deal with it, they send it to people like us – they call this "overflow research".'

In one of the largest agreements last year, Lilly established a metabolic disease-related discovery and development agreement with Ligand Pharmaceuticals worth US$190 million in equity, fees, research and development funding, and potential milestone payments over eight years; it is also working with a new subsidiary of Millennium Pharmaceuticals to use genomics to find therapeutic protein-based drugs. Wyeth-Ayerst, the research arm of American Home Products, set up a five-year US$100 million agreement with ArQule, a combinatorial chemistry firm. In another example, Bristol-Myers Squibb, Affymetrix (one-third owned by Glaxo Wellcome), Millennium Pharmaceuticals, and the Whitehead Institute/MIT Center for Genome Research formed a consortium in April 1997 to focus on functional genomics (*C&EN*, February 1998).

In 1993, there were 69 alliances formed between biotechnology and drug companies, through collaborations valued at US$1.3 billion. By 1997, this number had grown to 227, with a value of US$4.3 billion (Purcell, 1998). The pharmaceutical industry is likely to increase the amount of its research budget invested in collaborations with third parties from the current 16 per cent of today, to around 30 per cent within the next five years. (Drews, 1998). Specifically, the proportion of drug discovery 'outsourced' is predicted to rise from 10 per cent today to 30 per cent within five years. (Larvol and Wilkerson, 1998). While this provides numerous opportunities for small companies, the large number of competitors for each component of the drug discovery process is increasingly making it a buyer's market (Drews, 1998). (Box 3.3)

Government agencies and non-profit research organisations

Government bodies around the world remain involved in pharmaceutical research. For example, the NCI is one of the seventeen US National Institutes of Health (NIH) under the auspices of the Department of Health and Human Services (DHHS) of the US Government. Over the past 40 years, the NCI has facilitated the preclinical anti-tumour screening of more than 400,000 compounds and materials submitted by a wide range of grantees, contractors, pharmaceutical and chemical companies, and other private and public scientific institutions world-wide (ten Kate and Wells, 1998). NIH has

Box 3.3 *Biopharmaceuticals*

Biotechnology pharmaceuticals ('biopharmaceuticals' or 'biologicals') refer to human therapeutic products produced using biotechnology techniques. Major categories of biopharmaceutical products include proteins and vaccines made using recombinant DNA (rDNA) techniques, and monoclonal antibodies, produced by cell fusion.

In 1998, there were 75 biopharmaceuticals on the market, and 35 more slated for FDA approval. Over 350 biopharmaceuticals produced by 140 pharmaceutical and biotechnology companies are currently under development (see too Table 3.2, p38). Combined annual sales of therapeutic proteins in 1998 were some US$13 billion. Biopharmaceuticals currently contribute 11 per cent of the sales of 'blockbuster' drugs (8.2 per cent of the top 25 sellers). However, the proportion and value of biopharmaceuticals in the market is likely to increase rapidly. US biopharmaceutical sales are expected to grow at an average annual rate of 12 per cent, from US$6.5 billion in 1995 to US$20.4 billion in 2005.

The therapeutic area of biopharmaceuticals that is currently the most significant is cancer, as can be seen from two of the most important cancer products currently on the market. The first, **alpha-interferon,** is used to treat hairy cell leukemia, AIDS-related Kaposi's sarcoma, hepatitis A and hepatitis C. The second, **granulocyte colony stimulating factor (G-CSF)** is used to treat neutropenia in cancer patients undergoing radiation and chemotherapy treatments. Sales in the anti-cancer category are expected to grow at an average annual rate of 13 per cent, from US$2.2 billion in 1995 to US$7.4 billion in 2005. A third blockbuster biopharmaceutical, **erythropoietin (EPO)**, is used to treat anaemia in kidney dialysis patients. Sales in the blood and cardiovascular area of biopharmaceutical therapy are expected to grow at an average annual rate of 11 per cent, from US$2.1 billion in 1995 to US$6.2 billion in 2005. The combined sales of alpha-interferon, G-CSF and EPO represent close to half of total biopharmaceutical sales.

There are over 2,000 biotechnology companies in the world, including 1,300 in the USA and 700 in Europe. At least half are involved in biopharmaceutical discovery and development. Roughly three-quarters have 50 employees or less, although there are some biopharmaceutical companies with multinational scope. For example, Chiron created the first genetically engineered human vaccine for Hepatitis B, and has cloned the Hepatitis C genome and sequenced the entire HIV genome. The company has over 7,000 employees, facilities on four continents, and markets products in 97 countries. In 1997, its revenues were US$1.2 billion. Biopharmaceutical products – such as insulin, interferon, human growth hormone, and hepatitis B vaccine – also contribute significantly to revenues at large pharmaceutical companies like SmithKline Beecham, Merck, Eli Lilly, Schering-Plough, and Roche.

Many analysts believe that molecular biology and genetic engineering will become an important, and possibly, the predominant, approach to drug discovery and development in the 21st century. Based on their analysis of the number of pharmaceutical products in the discovery phase in biotechnology companies, the analysts Lehman Brothers predict in their *PharmPipelines* report that biopharmaceutical products will supply at least half of the technology for new drugs in the next decade. Biotech products are forecast to increase their proportion of blockbuster sales from 11 per cent to 17 per cent of the pharmaceutical market by the end of the decade.

Biopharmaceuticals require access to genetic resources, but in different ways from traditional natural products discovery programmes. The resources in question are usually human genes, or genes from domestic animals. These genes are readily available in collections, laboratories, and hospitals world-wide, and there is usually no need to collect exotic genetic resources from around the world.

Sources: The Economist, 1998; Champsi, 1998; *Scrip's 1997 Year Book*; *Pharma Business*, 1995; Ernst & Young, 1997; Chiron, 1998; PhRMA, 1998; *C&EN*, February 1998; Smith, 1996; Shearson Lehman Brothers, 1991.

also contributed substantial funds to the private sector for research (Purcell, 1998).

Public sector research continues to produce results that become the basis for pharmaceutical development in the private sector. For example, government-sponsored research will ensure that the primary sequences of most therapeutically relevant organisms will be determined and in the public domain within the next decade (Crooke, 1998). 15 per cent of biotechnology companies interviewed in EuropaBio's survey of biopharmaceutical companies said that the scale and quality of public research and development were significant factors affecting their investment decisions (EuropaBio, 1997).

Another important role of government is to finance pharmaceutical research and to offer grants and loans to support start-up pharmaceutical companies. For example, the UK Department of Trade and Industry offers matching grants of up to US$0.5 million to start-up biopharmaceutical companies. In Germany, if a biotechnology company can raise US$1 million in venture funding, it is likely to obtain matching grants of US$1 million from state governments, another US$1 million from the Federal government, and a further US$1 million of very soft loans from government (Haycock, 1998).

Academia and research institutions

Academic and research institutions have made important contributions to pharmaceutical discovery and development, and continue to do so. For example, recent insights into molecular targets, which form the basis of much pharmaceutical research today, result from research undertaken over many years in molecular biology and pharmacology departments in universities (Crook, 1998).

To capitalise on this expertise and innovation, large pharmaceutical companies regularly enter into partnerships with academic institutions. These collaborations can be worth many tens of millions of dollars. For example, in addition to the commercial partnerships like those with Chiron, Isis Pharmaceuticals, and Oncogene

Sciences, Novartis has research relationships with Johns Hopkins University, New York University, and Scripps Research Institute (Popovich, 1997). However, for the most part today, the biotechnology industry has replaced academia as the primary collaborator for pharmaceutical companies (Steinmetz, 1998).

Academic institutions also often form the launching-pad for small, start-up pharmaceutical companies. For example, a number of drug discovery companies have been founded in recent years by Oxford University scientists, including Oxford GlycoSciences, Oxford Biomedica, Synaptica, and Oxford Asymmetry International, which has refined combinatorial chemistry techniques to produce large numbers of single compounds, rather than compounds as mixtures – the usual product (Kozlowski, 1998).

In Costa Rica, the National Institute of Biodiversity (INBio), in collaboration with local universities, museums, and government agencies, has developed a role as collector, manager, and supplier of genetic resources, their derivatives, and information associated with these resources, to industry. Strathclyde Institute for Drug Research (SIDR) in the UK is also capitalising on the growing trend towards outsourcing in the area of natural products. SIDR has established itself as an intermediary organisation conducting interdisciplinary research aimed at drug discovery using natural products as the source of chemical diversity. Based at the University of Strathclyde, SIDR acts as an intermediary between the private sector and institutions located in countries with high biodiversity (Waterman, 1997).

3.3 The role and value of natural product-derived drugs

There are some 3,000 metabolic drugs on the market (*The Economist*, 1998), but these can be simply classified into two main categories. One group consists of compounds of small molecular weight which are either 'synthetic compounds', being manmade in origin, or 'natural products', derived from compounds isolated from plants, animals or microorganisms. These products, known in the pharmaceutical trade as 'small molecule drugs' have a molecular weight generally less than 500 daltons.

The second category of products is 'biopharmaceuticals', a term comprising protein drugs, generally known as 'therapeutic proteins' and vaccines, both produced by recombinant DNA technology, and monoclonal antibodies, produced by cell fusion. Biopharmaceuticals generally have molecular weights of thousands or even tens of thousands of daltons, and are thus considerably bigger than 'small molecule' drugs, whose molecular weight tends to lie between 300 and 500 daltons. In addition to these two classes of drug, a third category is likely to join them in the near future, when gene therapy leaves the laboratory and becomes an accepted method of clinical treatment.

Natural products contribute in a range of ways to the discovery of a drug (Newman and Laird, Appendix A). These include:

- *biologicals or biopharmaceuticals* – an entity that is a protein or polypeptide either isolated directly from the natural source or more usually made by recombinant DNA techniques followed by production using fermentation;
- *natural products* – an entity that, though occasionally manufactured by semi-synthesis or even total synthesis, is chemically identical to the pure natural product;
- *derived from a natural product* – an entity that starts with a natural product that is then chemically modified to produce the drug; and
- *structural class from a natural product* – this is material where a parent structure came from nature and then materials were synthesised *de novo* but following the natural template.

Natural products have long formed an integral part of drug discovery. Although interest in them as a source of leads for product development has proved cyclical, cumulatively natural products have yielded more than half of the drugs in many therapeutic categories in use today. Whether and how this trend will continue is the subject of Section 3.5. However, the importance of natural products as a source of drugs to date is borne out by the studies mentioned below.

Grifo et al (1997) analysed the top 150 proprietary drugs from the National Prescription Audit for the period January–September 1993, published by IMS America (Table 3.3). The audit is a compilation of virtually all prescriptions filled in the USA during this time.

Table 3.3 *Origins of Top 150 Prescription Drugs in the USA*

Origin	Total no of compounds	Natural Product	Semi-synthetic	Synthetic	%
Animal	27	6	21	–	23
Plant	34	9	25	–	18
Fungus	17	4	13	–	11
Bacteria	6	5	1	–	4
Marine	2	2	0	–	1
Synthetic	64	–	–	64	43
Total	150	26	60	64	100

Source: Grifo et al, 1997.

The data were based on the number of times a prescription was filled. They found that 57 per cent of the prescriptions filled during this period contained at least one major active compound 'now or once derived or patterned after compounds derived from biological diversity'.

Cragg et al (1997) analysed data on new drugs approved by either the US FDA or comparable entities in other countries, and reported mainly in the *Annual Reports of Medicinal Chemistry* for the years 1983–94, as well as other data covering potential anticancer compounds reported to be in the pre-New Drug Application (NDA) phase up to the end of 1995 (Box 3.4). Their analysis focused on the areas of cancer and infectious diseases, and their results demonstrate that – despite purported trends away from natural products as a source of potential, new chemotherapeutic agents in favour of other approaches to drug discovery – natural products have played, and continue to play, a vital role in the drug discovery process related to all disease types. In particular, natural products play an important role in the areas of cancer and infectious diseases. Of the 87 approved cancer drugs, 62 per cent are of natural origin or are modelled on natural product parents. Furthermore, of the 299 pre-NDA anticancer drug candidates (ie in preclinical or clinical development for the period 1985–95) 61 per cent (excluding the biologics) have a natural origin.

Results vary by disease indication, with analgesic, antidepressant, antihistamine, anxiolytic, cardiotonic, and hypnotic drugs, together with antifungal agents, being exclusively synthetic in origin, as are 67.5 per cent of the anti-inflammatory drugs. However, of new approved drugs reported between 1983 and 1994, drugs of natural origin predominate in the area of antibacterials (78 per cent), while 63 per cent of the 93 new approved anti-infectives are of natural origin (Cragg et al, 1997).

In the first study of its kind, Farnsworth et al (1985) reported that at least 119 compounds derived from 90 plant species can be considered as important drugs currently in use in one or more countries, with 77 per cent derived from plants used in traditional medicine. In the USA, 25 per cent of all prescriptions dispensed from community pharmacies from 1959 to 1980 contained plant extracts or active principles prepared from higher plants. Duke (1993) clarifies the latter statement by saying that '25 per cent of modern prescription drugs contain at least one compound now or once derived or

Box 3.4 *Examples of Plants Yielding Pharmaceutical Compounds*

Catharanthus roseus

This species yielded the alkaloids vincristine and vinblastine. Traditionally used by various cultures for the treatment of diabetes, these compounds were first discovered as part of an investigation of the plant as a possible oral hypoglycemic. Originally native to Madagascar, this species is now widespread and common. The vinca alkaloids are used in the treatment of childhood leukemia and Hodgkin's disease.

Podophyllum spp

The two active ingredients – etoposide and teniposide – are semisynthetic derivatives of the natural product epipodophyllotoxin, which was isolated as the active anti-tumour agent from the roots. These species were long used by indigenous peoples in America and Asia, including for the treatment of skin cancers and warts. Etoposide generates US$100–200 million in revenues for the manufacturer, Bristol-Myers Oncology. The combination of cumulative demand and loss of habitat led to the placement of *Podophyllum hexandrum* on Appendix II of the CITES list in January 1990.

Taxus brevifolia

Collected in Washington State, USA as part of a random collection programme by the US Department of Agriculture (USDA) for the NCI, paclitaxel was isolated from the bark of this species. Indigenous peoples in North America used this species for the treatment of some noncancerous conditions. The needles of *Taxus baccata*, in which paclitaxel is found, are used in traditional Ayurvedic medicines, with one reported use being for cancer. Paclitaxel is now marketed under the brand name Taxol by Bristol-Myers Squibb.

Camptotheca acuminata

This species is an ornamental tree in China, and has yielded the clinically active agents topotecan, irinotecan, and 9-aminocamptothecin, which are semisynthetically derived from camptothecin, isolated from this tree. Camptothecin is now marketed under the generic name 'Topotecan' and the brand name 'Navelbine' by SKB.

Chondrodendron tomentosum

This species, used by indigenous peoples in South America in an arrow poison known as *curare*, yields the compound d-tubocurarine, which is used as the model for a series of similar synthetic neuromuscular-blocking agents, such as succinylcholine, now commonly employed as an anesthetic in surgical operations. Recently, synthetics such as vencuronium and atracurium have completely replaced the natural product for clinical use.

Rauwolfia serpentina

Known as the Indian snakeroot, with a long history of traditional medical use, including for the treatment of mental disorder, snake bites, and as a tranquiliser, this species yields the antihypertensive compound reserpine. The alkaloid reserpine revolutionised Western medical treatment of hypertension in the 1950s, and caused massive over-harvesting of wild populations in India, leading to its inclusion on CITES Appendix II in 1990. Estimated sales of reserpine in 1989 were US$42 million in the USA alone. The Indian Ministry for Environment and Forests estimated sales of antihypertensives derived from the Indian snakeroot at more than US$260 million in 1994.

Chincona spp

The bark of this species yields quinine, a treatment for malaria. In the 1940s the principal alkaloids were isolated and synthesised for the pharmaceutical market, but its use in international markets had been established hundreds of years earlier.

Source: Cragg et al, 1997; Sheldon et al, 1997.

patterned after compounds derived from higher plants', since many of these drugs are produced for sale by synthetic or semi-synthetic means, although they were originally discovered from natural sources.

Estimates for the value of natural product pharmaceuticals vary considerably. 1995 world-wide sales of the following plant-derived pharmaceuticals were significant: opiates, US$1.5 billion; taxanes, US$400 million; digoxins and related compounds, US$200 million; Ergot alkaloids, US$150 million; and Catharanthus derivatives, US$100 million (CMR, 1997). In 1976, Farnsworth and Morris (1976) estimated that the 25 per cent of plant-derived global pharmaceutical sales would come to a US$5 billion market. Farnsworth et al (1985) revised this up to US$8 billion. In 1986, Elliott and Brincombe (1986) put the value of plant-based pharmaceuticals at US$10–20 billion per year.

Top-selling drugs of natural origin include Bristol-Myers Squibb's cholesterol-reducing drug Pravachol (US$1.437 billion in 1997 annual sales), and Taxol, an anti-cancer agent derived from the Pacific yew tree (US$941 million), the 30th best-selling drug world-wide in 1997. Merck's top-selling drugs with natural origins include Zocor (US$3.56 billion) and Mevacor (US$1.1 billion). Novartis' Sandimmune and Neoral had 1997 sales of US$1.3 billion (Newman and Laird, Appendix A; *MedAd News*, May 1998, September 1998; Cragg et al 1997a).

Natural products continue to be a major player in the sales of ethical pharmaceutical agents approved for use through 1997. As Newman and Laird (Appendix A) demonstrate, 11 of the top 25 best sellers in 1997, representing 42 per cent of sales, are either biologicals (8.2 per cent of top 25 sales), natural products or entities derived from natural products (28.1 per cent), or are synthetic versions based on a natural template (6.1 per cent). Their total 1997 value was US$17.5 billion. This study also found that a significant portion – between approximately 10 per cent and 50 per cent – of the 10 top selling drugs of each of 14 top pharmaceutical companies are either natural products, or entities derived from natural products (Appendix A).

It is also the case that all ten top pharmaceutical companies (Box 3.1, p36) are engaged in natural product discovery, either through in-house programmes, or through research undertaken by their subsidiaries. Companies with in-house natural product discovery programmes include Merck & Co., Glaxo Wellcome, Novartis, Bristol-Myers Squibb, Pfizer, Roche, Rhône-Poulenc Rorer, and SmithKline Beecham (SKB) (although SKB recently closed its natural products research programme in the USA, it continues a fermentation programme in Spain.) Companies conducting natural product research programmes through wholly- owned subsidiaries include American Home Products (through Wyeth Ayerst and Lederle) and Johnson & Johnson (through Jansen and Ortho).

In order to arrive at a rough estimate of the value of natural product-derived pharmaceuticals, we might employ the Grifo et al (1997) figure of 57 per cent of all prescriptions filled in the USA (the world's biggest market); or the Farnsworth et al (1985) figure of 25 per cent of all US prescriptions being based on plants; or the 42 per cent of top 25 sales reflected in the Newman and Laird (Appendix A) study; or one might use the percentages arrived at by therapeutic category in Cragg et al (1997). Percentages employed will vary by therapeutic category, and whether one uses prescriptions filled or total number of drugs on the market. 25 per cent of total sales would come to US$75 billion, while 40 per cent would mean that natural product-derived

Box 3.5: *Agents from Natural Sources Approved for Marketing in the 1990s, in the USA and Elsewhere*

Generic	*Brand name*	*Manufacturer*
In the USA and elsewhere		
Cladribine	Leustatin	Johnson & Johnson (Ortho Biotech)
Docetaxel	Taxotere	Rhône-Poulenc Rorer
Fludarabine	Fludara	Berlex
Idarubicin	Idamycin	Pharmacia & Upjohn
Irinotecan	Camptosar	Yakult Haisha
Paclitaxel	Taxol	Bristol-Myers Squibb
Pegaspargase	Oncospar	Rhône Poulenc
Pentostatin	Nipent	Parke-Davis
Topotecan	Hycamtin	SmithKline Beecham
Vinorelbine	Navelbine	Lilly
Only outside the USA		
Bisantrene		Wyeth-Ayerst
Cytarabine ocfosfate		Yamasa
Formestane		Ciba-Geigy
Interferon, gamma-la		Siu Valy
Miltefosine		Acta Medica
Porfimer sodium		Quadra Logic
Sorbuzoxane		Zeuyaku Kogyo
Zinostatin		Yamamouchi

These agents are either pure natural products, semi-synthetic modifications, or the pharmacophore is from a natural product.

Sources: Cragg et al, 1997b.

pharmaceuticals contributed US$120 billion to global sales of pharmaceuticals in 1997.

A range of natural products has contributed to the discovery and development of drugs, including plants, microorganisms, fungi, marine organisms, insects, animal genetic resources, and human genetic resources. The following is a brief discussion of these natural products, their role to date in drug discovery, and their potential role in future research efforts.

3.3.1 Plants

Despite the historical and current prevalence of plants in the pharmacopoeia (Box 3.5), only between 5 and 15 per cent of the approximately 250,000–500,000 species of higher plants have been investigated for the presence of bioactive compounds (Farnsworth et al, 1985; Balandrin et al, 1993; Tempesta and King, 1994; Joffe and Thomas, 1989; Balick et al, 1996). Plants are a complementary source to microorganisms, since they are unlikely to produce the same kinds of secondary metabolites, so in order to maximise the scope of chemical diversity evaluated in a screening programme, many companies regard both plants and microorganisms as important (Borris, 1996).

The primary metabolites of photosynthesis in plants are converted through enzyme pathways into all of the major classes of natural product compounds, including alkaloids, glycosides, terpenoids, steroids, and flavonoids. Within the basic framework of a nitrogen atom in a heterocyclic ring, alkaloids exhibit an enormous diversity of structure and substantial pharmacologic activities. Alkaloids include atropine from *Atropa belladona*, an anticholinergic. Glycosides are molecules formed by a linkage between a sugar molecule (the glycone) and a non-sugar molecule (the aglycone); they include digitoxin, a heart drug from *Digitalis*, and salicin, found in the bark of willows (*Salix* spp) and the basis for aspirin (Gruber and DerMarderosian, 1996).

Plant samples are rigorously documented with voucher specimens, photographs, and written notes, which allow for botanical identification. They are processed, perhaps by drying and milling, and then extracted to remove the compounds of interest from the biomass. These extracts are screened in a primary panel of assays. Re-collection of promising species is more complex than for microorganisms, since plants must be collected at the same stage in their life cycle, in the same season, and preferably in the same location to be confident of acquiring comparable material (Borris, 1996; Soejarto, 1994). Scale-up considerations for testing and manufacture are also significant for plants, particularly if the lead is based on a slow-growing species, or a tissue that is difficult to replace like bark or roots, rather than leaves (J Rosenthal, personal communication, 1999). However, as one representative from a large pharmaceutical company said, 'the advent of plant cell culture is one of the solutions to the problem – if a cell line is developed from a plant it can be scaled up with plant cell culture'.

3.3.2 Microbial sources

> *'I believe plants are primarily the scaffolding that supports the bulk of biodiversity. The most promising organisms for drug discovery are probably microorganisms.' (Georg Albers-Schönberg, Merck Sharp and Dohme Research Laboratories, 1996)*

Microorganisms are a prolific source of structurally diverse bioactive metabolites, and have yielded some of the most important products of the drug industry, including penicillins, aminoglycosides, tetracyclines, cephalosporins, and other classes of antibiotics. Microorganisms are also useful for activities other than antibiotic action, including antitumour agents (eg mitomycin, bleomycin, daunorubicin); immosuppressive agents (eg cyclosporin, FK-506, rapamycin); hypocholesterolemic agents, enzyme inhibitors, and antimigraine agents (Demain, 1998). Microbial agents continue to play a major role in drug discovery and development within the pharmaceutical industry, and at most large companies are a mainstay of pharmaceutical natural products research (Borris, 1996). As of 1992, however, less than 5 per cent of microorganisms were even described, let alone screened for their commercial potential (Hawksworth, 1992). As Terrance J Bruggeman, former CEO of Diversa, said in 1998, on their three-year biodiversity prospecting agreement with the Institute of Biotechnology at the National Autonomous University of Mexico (UNAM):

> *Less than 1 per cent of all the microorganisms in our world have been identified. Yet from that small percentage, scientists have developed a large number of important drugs and industrial products that have changed the world we live in. By searching the 99 per cent of the microbial world that is still unknown, we're hoping to identify compounds that will be useful to pharmaceutical, agriculture, and industrial companies.*

Pharmaceutical companies' research with microorganisms has generally focused on the *Actinomycetes*, but also includes research on other taxa, and is expanding into the study of organisms from diverse environments, such as shallow and deep marine ecosystems and deep terrestrial subsurface layers. Genetic engineering also offers the possibility of conducting rational drug design of novel structures by exploring microbial metabolites.

Work on bacterial polyketides has already demonstrated the scope in this area (Cragg et al, 1997).

According to Novo Nordisk, microorganisms have a number of positive features that distinguish them from other natural products for their discovery programme:

- they provide a broad diversity of compounds;
- they are easy to preserve and maintain in *ex situ* collections (although they can change genetically over time, which can influence the expression of their metabolites);
- they can be cultivated in laboratories to provide suitable amounts for screening; and
- they require only a negligible part of *in situ* populations to be samples; because it is only necessary to collect a very small portion of the whole organism to have a viable sample, the danger of unsustainable collection is reduced.

Microbial samples are collected from a wide range of environments, from traditional soil samples to leaf litter, tree branches, animal dung, and beetle carcasses. Each type of sample has its own characteristic spectrum of organisms. The number of organisms in each sample varies, but more than 100 species from a single sample is not unusual (Borris, 1996). Microorganism samples are notoriously difficult to track and monitor, since they are easily collected, with minimal infrastructure or local collaboration required, can be easily packed and shipped or carried out and their taxonomy and geographical distribution are poorly known. Together, these factors mean that it is often difficult to tell from where a microbial sample originated. Some companies still ask employees to take plastic bags on their holidays in other countries to fill with soil samples.

3.3.3 Marine organisms

Marine organisms represent a valuable resource for potential chemotherapeutic agents. The pace of investigation of marine invertebrates has increased over the past few decades, but remains the smallest component of natural products screening to date. The systematic investigation of marine environments for sources of novel biologically active agents began in earnest only in the mid-1970s. From 1977 to 1987, about 2,500 new metabolites were reported from a variety of marine organisms. Marine organisms have been shown to be a rich source of bioactive compounds, many from novel chemical classes not found in terrestrial sources. As of today, no compound isolated from a marine source has advanced to commercial use, although several are in various phases of clinical trials. Examples of marine chemicals under investigation include (Cragg et al, 1997b; Faulkner, 1992; Cragg et al, in press; Mestel, 1999):

- Bryostatin, an anti-cancer agent derived from the US West Coast bryozoan *Bugula neritina;*
- Dolastatin 10, an anti-cancer agent derived from an Indian Ocean mollusk, *Dolabella auricularia;*
- Ecteinascidin 743, an anti-cancer agent derived from a sea whip, *Ecteinascidia turbinata;*
- Manoalide, a compound from *Luffariella variabilis,* a sponge collected in the Pacific Island of Palau in 1979;
- Discodermolide, an immunosuppressant agent from the Bahamian sponge, *Discodermia dissoluta;* and
- Didemnin B, an anti-cancer agent isolated from a Caribbean tunicate of the genus *Trididemmum.*

Collection of marine specimens is usually more complex and costly than expeditions to collect terrestrial plant materials. Materials are collected by scuba diving, deep water trawling or dredging, and the use of manned submersibles or remotely operated vehicles. The latter techniques are expensive, however, and are not employed for routine collections. Dredging and trawling are not ideal techniques, since they lead to non-selective sampling and cause environmental damage (Cragg et al, 1997b; Faulkner, 1992). Marine organism samples are frozen, and are usually around 1 kg wet weight.

3.3.4 Insects

A major drug has yet to be developed from work on insects, but very little research has taken place in this area to date. Some of the pharmaceutical companies interviewed expressed a growing interest in invertebrates, and it appears that the massive diversity in insect species and the novel chemical compounds they use for defence and other purposes will receive more attention in the near future. Many of the chemical challenges faced by invertebrates are also important to humans, including protection from microbial invasion. Researchers believe insects could prove of great medical value (Rouhi, 1997).

3.3.5 Animal genetic resources

Animal genetic resources have a limited but significant history in natural product drug discovery. Grifo et al (1997) found that 23 per cent of all compounds contained in prescription drugs dispensed in the USA are derived from animals (Table 3.3, p40). Of the 25 drugs with the greatest sales in the UK in 1993, three products, accounting for 13 per cent of the sales of the top 25 drugs surveyed, were derived from animals (ten Kate, 1995). These were Premarin, Capoten and Vasotec. American Home Products's Premarin (a treatment for osteoporosis, with 1993 sales of US$481 million) is a natural product using conjugated oestrogens derived from the urine of pregnant horses. Bristol-Myers Squibb's cardiovascular drug Capoten (with 1993 sales of US$1.27 billion) is synthetic, but its mode of action was inspired by work with snake venom, and Merck's cardiovascular drug Vasotec (with 1993 sales of US$1.99 billion) has a similar mode of action (Glaxo figures, cited in *The Sunday Times*, 8 May 1994).

Polypeptide toxins in purified form from venomous animals – snakes, spiders, insects, scorpions, snails, etc – frequently produce highly selective actions on a specific component of a biological system. Dendrotoxins from the green mamba snake, for example, bind to only certain types of potassium channels in neuronal membranes. These selective toxins are extremely useful tools in experiments, and may also provide starter compounds for highly selective drugs (Bowman and Harvey, 1995). Abbott Laboratories in the USA is working with the compound epibatidine, which is secreted in the skin of the poisonous frog *Epipedobates tricolor*.

Indigenous peoples in the neotropics have long used these poisonous compounds to dress their spears for hunting (Bravo and Gallardo, 1998). An analogue of the toxin epibatidine has been shown to be as effective as morphine in dampening pain, but it binds to a different molecular receptor from morphine, suggesting that it might be non-addictive – 'another example of the importance of natural products in discovering novel *prototypes* of medicinal agents' (Capson, 1998).

Animals are also the source of hormones and other metabolites used in biotechnology and may be used in the future for xenotransplantation (eg breeding genetically engineered pigs to produce hearts for transplantation into humans). Animal genetic resources tend to form the basis of specialised research and development programmes, rather than acting as a component of broader screening initiatives.

3.3.6 Human genetic resources

The use of human genetic resources is increasing rapidly in the pharmaceutical industry. The growing market for biopharmaceuticals and genetic products signifies an increasing interest in access to human genetic resources. In 1993, world net sales of human proteins produced by rDNA techniques amounted to US$7.7 billion (Straus, 1995).

The Human Genome Project – an international, loosely-organised and largely government-sponsored initiative – is intended to unravel the human 'genome' (which comprises between 80,000 and 150,000 genes, depending upon the estimate). Incyte Pharmaceuticals in California and Human Genome Sciences in Maryland are running their own private genome projects using a technique known as expressed-sequence tagging of genes. The information gathered by these companies is extremely valuable. Access to Incyte's database, for example, costs around US$15 million a year. 20 companies, including those controlling nine of the world's top ten pharmaceutical research and development budgets, have made use of Incyte's database to date (*The Economist*, 1998).

Some of the companies interviewed have entered into agreements with institutions in other countries to provide them with access to a variety of human genetic material. Although a decision at the second meeting of the Conference of the Parties (Decision II/11 para 2) interpreted the Convention in such a way as to exclude human genetic resources from its provisions on access and benefit-sharing, this facet of access to genetic resources raises profound ethical questions and remains a largely unregulated area.

3.4 Pharmaceutical industry investment in research and development

Decisions relating to investment in research and development on the part of the private sector are both scientific and economic. Science defines the opportunities and constraints, but economics determines which opportunities will be investigated through research programmes, based on an assessment of which investments are likely to lead to financial gain in the future (OTA, 1993). This section will explore some of the economic factors involved in pharmaceutical research and development, and decisions that companies must take relating to the development of new products.

3.4.1 An overview of pharmaceutical industry research and development expenditures

Pharmaceutical companies invest a higher proportion of sales in research and development than most other industries, including electronics (6.4 per cent), aerospace and defence (3.9 per cent), office equipment (including computers) (7.2 per cent), automotive (4.3 per cent), and telecommunications (3.9 per cent) (PhRMA, 1998). In the UK, pharmaceutical research and development made up 20 per cent of all industry research and development in 1997 (ABPI, 1998). However, the US Congressional Office of Technology Assessment found in 1993 that, over time, economic returns to the pharmaceutical industry as a whole exceeded

Table 3.4 *Increase in Healthcare Research and Development, 1996–97*

Company	1997 Healthcare R&D expenditures (US$ bn)	% increase over 1996
Boehringer Ingelheim	0.83	22.5
Bristol-Myers Squibb	1.9	11.8
Hoechst Group	1.3	(26.6)
Eli Lilly	1.4	16.2
Glaxo Wellcome	1.9	(1.1)
Merck & Co	1.7	13.2
Novartis	2.0	19.5
Pfizer	1.9	14.5
Pharmacia and Upjohn	1.2	(3.9)
Rhône-Poulenc	1.3	30.8
Sankyo	0.44	(0.2)
SmithKline Beecham	1.4	10.1

Source: MedAd News, September 1998.

returns to corporations in other industries by 2–3 percentage points per year from 1976 to 1987, after adjusting for differences in risk among industries. Rapid increases in revenues from new drugs in the 1980s encouraged pharmaceutical companies to invest more in research and development, resulting in growth in research and development expenditures during this period of more than 10 per cent per year (OTA, 1993).

Pharmaceutical industry research and development is currently focused on developing more effective drugs for a wider range of diseases; making research and development less expensive, and speeding it up so that the industry can benefit from patent protection for longer. This has resulted in cost-cutting, by eliminating operational redundancies, by combining existing lines to increase the flow of products into the market place, and by increasing support for research and development activity. There has been less emphasis on a few big hits and more on the development of a broader range of drugs. Increased investments in research and development are seen as critical to keep the pipeline flowing and maintain a competitive position (*The Economist*, 1998) (Table 3.4).

Research-based pharmaceutical companies invested US$21.1 billion in research and development in 1998, a 10.8 per cent increase over 1997. Research companies have more than doubled their research and development expenditures since 1990, when spending was US$8.1 billion (PhRMA, 1998; OTA, 1993). Over the last two decades, the percentage of sales allocated to research and development has increased from 11.9 per cent in 1980 to an estimated 20 per cent in 1998 (PhRMA, 1998; ABPI, 1998). In 1997, UK companies spent 20 per cent of all sales on research and development, and in Japan, the figure for member companies of the Japanese Pharmaceutical Manufacturers Association, was 12.5 per cent – both significant increases over the decade (ABPI, 1998; JPMA 1998). Most pharmaceutical companies believe greater investment in research and development will increase productivity, decrease the costs and time for drug development, lead to innovative products, and promote continued growth in sales (*C&EN*, February 1998).

Approximately 36 per cent of pharmaceutical research and development conducted world-wide is performed in the USA. Of the 152 major drugs developed between 1975 and 1994, 45 per cent originated in the USA, 14 per cent in the UK, and 9 per cent in Switzerland. In the biotechnology sector, of the 150 healthcare patents issued by the US Patent and Trademark Office in 1995, 122 went to US applicants (PhRMA, 1998). In 1993, the US-based industry had introduced roughly one out of every four new chemical entities (NCEs) to the world market since 1961 (OTA, 1993).

Much pharmaceutical research and development is devoted to innovation. Nearly 84 per cent of research and development in the USA is oriented towards NCEs, with just 16 per cent devoted to improvements or modifications of existing products (PhRMA, 1998). For a number of reasons, companies are anxious to develop novel products, not just 'me too' products that follow the inspiration of a product released earlier by another company. First, 'me-too' drugs may have little benefit over the original product, so it might be difficult to

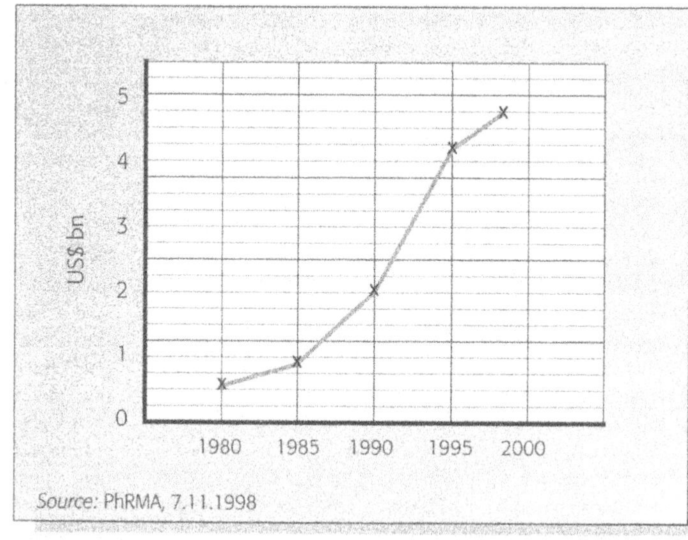

Source: PhRMA, 7.11.1998

Figure 3.2 *Pharmaceutical Industry R&D Expenditures, 1980–98*

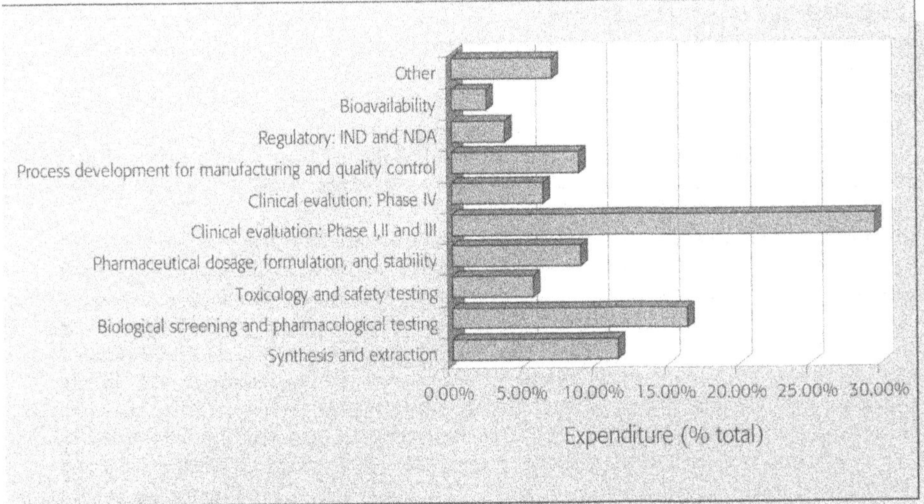

Figure 3.3 *Allocation of Domestic US Research and Development by Function, 1996*

Source: PhRMA, 1998.

carve out a market. Second, they might be difficult to patent, since the original patents are often very widely drawn; and third, it is increasingly difficult to satisfy regulators that a 'me-too' drug is a significant improvement on the original (P Hylands, personal communication, January 1999).

3.4.2 Investment and time to discover and develop a new drug

Cost

It is difficult to evaluate the costs and time required to reach specific milestones in the research and development process and rates of success or abandonment along the way, since much of the data is proprietary (OTA, 1993). It is clear, however, that the costs of research and development on NCEs have soared in the last 30 years (PhRMA, 1998). In 1991, Di Masi et al (1991) reported the average cost of developing a NCE as US$231 million. The Pharmaceutical Research and Manufacturers of America (PhRMA) estimate that it costs more than US$500 million to develop a NCE, including the costs of failures as well as interest costs over the entire period of the investment (PhRMA, 1998). The Association of the British Pharmaceutical Industry (1998) estimates that it takes US$570 million to develop a new medicine. About 37 per cent of research and development budgets in the USA are allocated to discovery-related research, and the remaining 63 per cent to the development stage (Figure 3.3). It is standard within the EU, Japan, and the USA for one-third of the research and development budget to be devoted to clinical trials, which absorb a larger portion of research and development expenditures than any other function (Shearson Lehman Brothers, 1991).

Time

The time it takes to develop a drug has lengthened during the past few decades, partly due to increased research and regulatory complexity, including increased understanding of biological processes at a molecular level, and a greater desire to characterise new drugs fully (Christofferson and Marr, 1995). Today, it takes on average 15 years to bring a product to market (PhRMA, 1998), compared with 1993 estimates of only 12 years (OTA, 1993). The Japanese Pharmaceutical Manufacturers Association (1998) estimates drug discovery and development times at between 10 and 18 years, with clinical trials absorbing from three to seven years.

Increases in drug development times over the last 40 years (Table 3.5) are due in large part to the lengthening of the clinical evaluation phase of drug development, which must meet increasingly rigorous requirements. Clinical testing now involves greater numbers of clinical trials per drug, a larger number of patients per trial, and more medical procedures per trial. The average number of clinical trials per drug, for example, has doubled since 1980, and the number of patients in clinical trials per NDA has tripled (PhRMA, 1998; *The Economist*, 1998; OTA, 1993).

In 1997, in the USA, the Food and Drug Modernisation Act was passed in order to extend the Prescription

Table 3.5 *Total Drug Development Time: from Synthesis to Approval, USA*

	1960s	1970s	1980s	1990–96
Approval phase	2.4	2.1	2.8	2.2
Clinical phase	2.5	4.4	5.5	6.7
Preclinical phase	3.2	5.1	5.9	6.0
Total (years)	8.1	11.6	14.2	14.9

Source: Tufts Center for the Study of Drug Development, 1998.

Drug User Fee Act of 1992. Under this latter Act, the pharmaceutical industry agreed to pay US$327 million between 1993 and 1997 to allow the FDA to hire 600 additional reviewers, and to improve the drug approval process. As a result, the FDA was able to cut drug approval times (not including clinical testing) almost in half (Tufts, 1998). Drug review times in most major markets – including in the USA, the UK, France, Germany and Japan – are moving towards an average of two years (Scrip's 1997). In Japan, the time required for the standard approval procedure, now estimated at 1.5 years, will be shortened to one year (JPMA, 1998).

Return on investment

The Pharmaceutical Research and Manufacturers of America (PhRMA) (1998) claims, based on a Duke University study, that only 3 out of every 10 NCEs introduced from 1980 to 1984 made a profit, resulting in returns greater than their average after-tax research and development costs. 20 per cent of the products with the highest revenues generated 70 per cent of returns during the period studied (PhRMA, 1998; Grabowski and Vernon, 1994). According to Andersen Consulting, between 1990 and 1994 the top 10 drug companies launched an average of only 0.45 truly new drugs (ie novel molecules) a year each. To maintain their current annual revenue-growth rate of 10 per cent without resorting to yet more mergers (whose principal benefits are often one-off cost cuts), Andersen estimates that these companies need to increase their productivity tenfold, launching five new compounds a year, each with an annual sales potential of US$350 million. Half of newly introduced drugs are reported to make less than US $100 million a year (*The Economist*, 1998). However, the OTA (1993) found that each new drug introduced to the US market between 1981 and 1983 returned, net of taxes, at least US$36 million more to its investors than was needed to pay off the research and development investment.

In 1987, some ten products achieved sales in excess of US$500 million, and these products were all found in one of four therapeutic categories: anti-ulcer, cardiovascular, anti-arthritic, or antibiotic. In 1990, the number of 'blockbuster' drugs more than doubled, and included drugs in several therapeutic categories new to the 'blockbuster' list, including CNS, respiratory, anti-viral, and immunosuppressant (Lehman Brothers, 1991). In 1997, 71 drugs earned more than US$500 million, and 27 more than US$1 billion. Therapeutic categories addressed by the top ten selling drugs include anti-ulcer, cholesterol-lowering, hypertension, antidepressant, hematologic, antibacterial, and antihistamine (*MedAd News*, May 1998; Appendix A).

Probability of success

Of 5,000–10,000 molecules screened, only one becomes an approved drug (PhRMA, 1998; McChesney, 1996). According to the Tufts Center for the Study of Drug Development, only 23.5 per cent of the drugs that entered clinical trials during 1980–84 are expected to be approved for marketing, and only 18.3 per cent will actually be marketed (PhRMA, 1998; Box 3.6 and Figure 3.4).

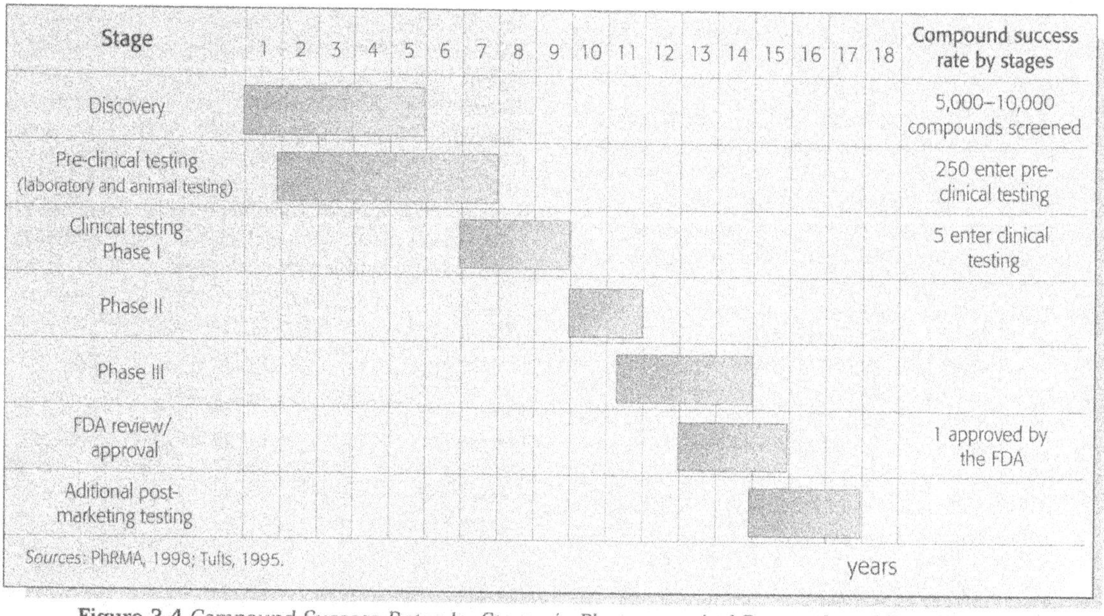

Stage	1	2	3	4	5	6	7	8	9	10	11	12	13	14	15	16	17	18	Compound success rate by stages
Discovery																			5,000–10,000 compounds screened
Pre-clinical testing (laboratory and animal testing)																			250 enter pre-clinical testing
Clinical testing Phase I																			5 enter clinical testing
Phase II																			
Phase III																			
FDA review/ approval																			1 approved by the FDA
Aditional post-marketing testing																			

Sources: PhRMA, 1998; Tufts, 1995.

years

Figure 3.4 *Compound Success Rates by Stages in Pharmaceutical Research and Development*

Box 3.6 *Probabilities of Success of Drug Candidates at Various Stages in the Research and Development Process*

Of the drugs which entered Phase 1 clinical trials in 1991:

- 75% entered Phase 2;
- 36% entered Phase 3;
- 23% received FDA approval.

The probability of a drug launch (%) in 1991, therefore, was:

- Pre-clinical R&D: 5–10%;
- Phase 1: 15–20%;
- Phase 2A: 30%;
- Phase 2B: 40%;
- Phase 3: 70%;
- Regulatory Review: 90%

Source: Shearson Lehman Brothers, 1991.

3.5 Pharmaceutical drug discovery and development

Pharmaceutical research and development is the process of discovering, developing, and bringing to market new ethical drug products. Drug discovery is defined in a number of ways, but for the purposes of this chapter will include the process by which a lead is found, including the acquisition of materials for screening; identification of a disease and therapeutic target of interest; methodology and assay development; advanced screening; and identification of active agents and chemical structure. The basic research that acts as the foundation of drug discovery is often funded by governments, and is conducted in government and university, as well as company, laboratories (Christofferson and Mar, 1995; OTA, 1993; PhRMA, 1998; Grifo et al, 1997).

Drug development includes chemical improvements to a drug molecule, animal pharmacology studies, pharmacokinetic and safety studies in animals, followed by Phases I, II, and III clinical studies in humans. Drug discovery and development activities were once carried out sequentially, but companies now take a parallel or overlapping approach that some estimate will contract research and development times by half. Drug development is funded almost solely through commercial research and development budgets, although it may be conducted in contract research organisations and universities as well as in company laboratories (Christofferson and Mar, 1995; OTA, 1993; PhRMA, 1998; *C&EN*, 1998; Grifo et al, 1997).

3.5.1 Discovery

Drug discovery through screening

Screening – a highly sophisticated approach to testing drug candidates – contributes significantly to the development of new medicines today. The development of a drug by screening consists of two parts: developing a 'screen' to detect biological activity of interest, and finding the chemical compounds (both synthetic and of natural origin) to test in the screens.

Developing screens

Developing a screen involves selecting a target such as an enzyme or receptor cell in the human body upon which it is hoped that a new medicine will act. As the section on rational drug design and genomics (page 51) explains, developments over the last decade in molecular biology, pharmacology, genomics and bioinformatics have enabled scientists to gain a better understanding of the biological and chemical pathways and reactions in living organisms, which has led to the identification of many new targets for drugs. As well as finding a suitable target, an important part of the challenge of designing a screen is to find some way to detect whether the compound being tested for its potential effect as a drug does or does not produce the desired result on the target. Thus a screen generally involves some indicator chemical which changes colour or reveals in some other simple manner whether the potential drug molecule has interacted biologically or chemically with the target, for example by killing a cell or rendering an enzyme inactive.

There are two main kinds of screen. '**Mechanism-based**' screens (Figure 3.5) – the most common approach in the pharmaceutical industry – use individual biochemicals, isolated from compounds, enzymes or receptors, that will reveal specific biological activity (the 'mechanism' concerned) when combined with the chemical to be screened. With this approach, for each possible mechanism of action, a screen with the relevant target must be created. A collection, or 'library', of chemicals to be tested will be built up and maintained. Each of these chemicals will be tested many times against an ever-changing array of mechanism-based screens. Companies using mechanism-based screens often change them every three months. By contrast, '**whole-organism**' screens operate *in vivo*, and expose an entire human cell to the chemical being screened, enabling the potential drug to operate through a range of different mechanisms during the one test. In this approach, the screen remains the same.

At the same time that a greater understanding of targets and receptors was emerging in the 1980s, the

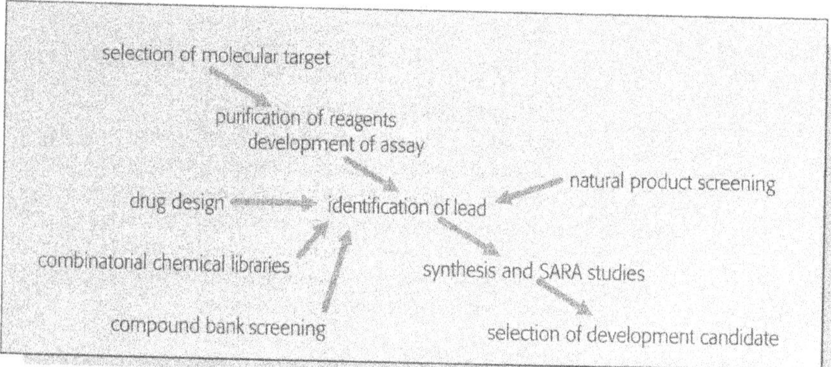

Figure 3.5 *Mechanism-Based Screening*

Source: Carte 1997.

miniaturisation and automation of the technology used for screening meant that many more compounds could be screened more rapidly and more cost effectively. The advent of 'high throughput', and, more recently, 'ultra high throughput' screening has dramatically increased the annual screening potential of companies. Plastic trays known as micro-titre plates, containing 96 depressions or 'wells' arranged in 12 rows of eight, are filled with small quantities of the molecular targets – often the result of genomic research. A robotic apparatus with 96 pipettes in the same array as the plate then injects eight sets of 12 slightly different compounds to be screened. Small quantities of compounds such as radioactive labels or fluorescent dyes are added to identify active compounds (*The Economist*, 1998).

Box 3.7 *Combinatorial Chemistry*

Combinatorial chemistry enables researchers to generate rapidly a huge number of chemical compounds for screening, to speed the process of drug discovery, and to ease the identification and production by chemical synthesis of the active compound. It is based on the idea that all but the smallest organic molecules can be thought of as made up of modules, which can be put together in different ways. By going through all the possible combinations, a huge number of molecules can be created from a small number of starting modules. Just like high throughput screening, combinatorial chemistry is carried out on micro-titre plates with 96 wells (arranged in 12 rows of 8), each row being filled with polysterene beads impregnated with one of eight slightly different, but chemically related, substances. 8 sets of 12 slightly different compounds of another chemical class that is known to react with the first group are injected on to the plate, resulting in 96 'dimers', or molecules consisting of two parts. These are then split into several plates and the process is repeated with a third set of molecules, which are added to the dimers forming an even larger number of 'trimers'. The large numbers of different compounds that result from iterations of this process can be screened against molecular targets to gauge their biological activity.

Source: *The Economist*, 1998.

High throughput screening is so rapid that, according to a researcher at Pfizer, the company's entire collection of 1.1 million compounds can be screened against a single target protein in six months at a 'normal pace of work', and within three weeks if all the company's resources were thrown into the exercise. The result of this new technology is an increasing capacity to develop new screening targets and to produce chemical diversity. This has considerably reduced the cycle times in drug discovery.

Compounds to be screened

The chemicals that are tested in screens may be man-made, 'synthetic' chemicals or 'natural products'.

a) Synthetic compounds and combinatorial approaches

Chemists in pharmaceutical and biotechnology companies create and screen large libraries of synthetic compounds from basic inorganic and petroleum-based chemicals, or purchase them from other companies or researchers. Companies running high throughput screens may need hundreds of thousands of compounds to screen each year, in order to keep this expensive screening process occupied and productive.

Libraries of synthetic compounds have been built up by companies over many years (see Section 3.6.2). However, the development of combinatorial chemistry in the 1980s enabled chemists to rapidly combine chemicals, giving rise to an extremely large number of compounds with a similar basic structure, or 'backbone', but whose precise structure reveals many different permutations and combinations. Theoretically, chemists could manufacture any one of at least 10^{60} compounds of the right sort to be small-molecule drugs (although, using all the matter in the universe, there would only be one thousandth of a gram of each) (*The Economist*, 1998). This technology allows companies to synthesise vast numbers of molecules for screening, and offers the advantage that the structure of the resulting compounds can be predicted in advance. The structure of synthetic compounds, including those produced by combinatorial chemistry, is therefore known at the time of screening – an advantage over natural products. By contrast, elucidating the unknown structures of compounds from natural extracts is one of the more expensive aspects of working with natural products. Combinatorial chemistry is currently supplying many of the compounds for the increased screening capacity of micro-titre plate screens (Box 3.7).

b) Natural products

Natural products are extracts from plants, fungi, microbes, insects, marine organisms, or animals. Of the companies interviewed, between 1 and 4 per cent of research and development budgets were allocated to natural products discovery. Companies establish criteria for the selection of samples depending upon their screening strategy. Companies' preferences for a random, ethnobotanical, biorational, or taxonomic basis for the selection of natural product samples for screening are discussed in Section 3.6.5. Once the samples have been obtained, extracts are prepared, often both in water and, separately, in organic solvents, since these mixtures may reveal different biological activity. Each crude extract is likely to contain a large number of naturally occurring compounds. In a crude plant extract, probably in excess of 100 compounds will be present, each of which could be responsible for signs of biological activity in screens. Fermentation broths of microorganisms will also contain many chemicals, the structure and quantity of which vary with the fermentation medium and the time of harvest. Extracts are screened against the targets selected, as described above, and those that reveal promising activity, or a 'hit', progress to further, advanced screening. The various components of the extract are separated, or 'fractionated', and the different fractions may be screened again in order to establish which components account for the biological activity revealed in the early screens (Box 3.8). The chemical structure of the compound responsible for interesting biological activity is determined, using techniques such as gas chromatography and nuclear magnetic resonance, and compared against the structures of known compounds, to prevent companies from investing in expensive research which would simply 'reinvent the wheel' rather than resulting in a novel product.

Box 3.8 *Elements of Natural Product Random Screening-Based Discovery*

ACQUISITION

raw material: field collections, culture collections, screening libraries, etc

EXTRACTION

sequential extraction

PRIMARY SCREENS

extracts are screened in a primary screening panel of assays

ISOLATION AND CHARACTERISATION

active extracts are subjected to bioassay-guided fractionation to isolate and characterise the pure, active constituents

SECONDARY SCREENS

agents that show significant activity in the primary screens are selected for secondary testing in several *in vivo* systems, in secondary screens

STRUCTURAL ELUCIDATION

PRE-CLINICAL DEVELOPMENT

compounds exhibiting significant *in vivo* activity are advanced to pre-clinical and clinical development

Rational drug design and genomics

Before 1980, an understanding of the mechanism of action of a new drug was often not developed until the compound had demonstrated its value in the clinic (Crooke, 1998). Companies could create many new chemicals, but had little knowledge or understanding of how these might interact with the body, and thus of how to use them (Crooke, 1998). Of the 3,000 or so human metabolic drugs on the market, 15 per cent act on a target molecule that is still unknown, and the rest interact with just 417 different target molecules (*The Economist*, 1998). At the heart of **rational drug design** is a much greater knowledge and understanding of the molecular targets on which drug molecules operate. The approach is based on recent advances in molecular biology and molecular pharmacology. Numerous molecular targets that may be appropriate for drug discovery have now been identified, and work to map the human genome is likely to provide 5,000–10,000 targets for diseases (Drews, 1998).

'Drug discovery needs to ask two main questions: Do I have the right targets? and Do I have the right compound to hit that target? ... In the last 15 years, research has shifted towards the first question. Companies sit on large piles of compounds. Most big companies have in-house libraries of up to a million compounds, so they are less interested in the second question.' (Reimar Bruening, Director of Natural Products, Millennium Pharmaceuticals)

Rational drug discovery centres on finding drugs which will interact with molecular targets that research suggests are relevant to particular diseases. The mechanism of action is thus evaluated very early in the discovery process (Crooke, 1998). Drug candidates tend to fail because they are aimed at the wrong target, do not reach the target, or have unexpected side effects. With human genetic or molecular targets, researchers are able to produce more precise, powerful, and safer drugs (*C&EN*, 1998; PhRMA, 1998).

As Steven Paul of Eli Lilly said, 'We lose drugs at various stages ... , so what we've [tried to do] is figure out what we need to do to reduce these enormous attrition rates. Getting something through Phase II trials, which is where we look for efficacy, [requires] having a greater degree of assurance that what the drug is doing from a mechanism point of view is likely to be very relevant to the disease you are looking at ... . The great advances in and utilization of genomics are going to be helpful in actually identifying and validating targets in a much more definitive way.' Frank Douglas of Hoechst Marion Roussel (HMR) echoes this: 'When you create an environment where scientists produce quality data in a timely fashion, [that] enables decisions to be made much more rapidly. The sooner we can say no to something, the sooner we can apply those resources to something else that might be more promising' (*C&EN*, 1998).

The pace at which new targets are being discovered now exceeds industry's ability to exploit them, so the main challenge of rational drug design is how to decide which of the vast number of potential targets in the full range of human cells is most likely to be relevant to disease pathways and thus worthy of investment. This process is sometimes known as 'target validation' (Crooke, 1998; Poste, 1998).

When the target has been established, the next task of rational drug design is to design a drug molecule that will fit it. As part of the growing field of **bioinformatics**, computer programmes can generate vast numbers of molecular designs which can be run through programmes created to screen and select 'virtual' compounds that are likely to fit well with the target molecule. Computer programmes such as these often include virtual, three-dimensional representations of the position of each atom in the target molecule. Chemists can view computer representations of both the target molecule and the candidate drug molecule on the screen, and alter the 'virtual' drug compound atom by atom to achieve a better fit (*The Economist*, 1998). Computer-aided rational drug design has drastically reduced the time required from finding a 'lead' to identifying and synthesising its chemical structure.

Once the model of the ideal compound has been designed, chemists search databases linked to libraries of real compounds, looking for the closest fit. The compounds screened for fit against the target could be natural products, but are more likely to be small, synthetic molecules. These synthetic compounds may have been constructed one by one by chemists, but are now likely to have been produced by combinatorial methods. In the past, a busy chemist might have synthesised 50–100 new compounds a year. Now, using standard combinatorial chemistry (Box 3.7, p50), it is possible to produce anything from a couple of thousand to 50,000 compounds a year (*The Economist*, 1998). The majority of multinational companies use combinatorial chemistry to generate synthetic compounds, and several smaller biotechnology companies now specialise in generating combinatorial libraries, which they license to pharmaceutical clients. Pharmacopeia, for example, has 3.3 million molecules on its books, and may hire a library of 100,000 molecules to pharmaceutical companies for US$1 million (*The Economist*, 1998).

As with the virtual compound on the computer screen, any compound with a suitable structure found from a compound library may need to be altered atom by atom until it has the optimum structure. A pharmaceutical produced by just such a method is now on the market. Viracept, an HIV protease inhibitor, was produced by the company Agouron in just six years – half the average time for a product to come to market (*The Economist*, 1998).

Genomics (Box 3.9) is another of a suite of important new technologies that has emerged to support rational drug design. In less than five years, genomics has transformed drug discovery and has been incorporated into the programmes of almost every biotechnology and pharmaceutical company (Haseltine, 1998). A process known as 'database mining' uses genomics to identify and develop a molecular target, which then becomes the subject of a programme of rational small molecule design, as described above. Database mining now accounts for the majority of small molecule drug discovery targets used by companies that access such databases (Haseltine, 1998). No product using genomics for the rational design of small molecules is yet on the market. The bottleneck lies not in finding compounds that appear to act on the targets identified, since high-throughput and ultra-high-throughput screens turn up many potential drug candidates. Rather, the limitation lies in medicinal chemistry, in creating compounds with suitable absorption, distribution, metabolism and elimination from the human body (Haseltine, 1998).

These limitations do not, however, affect another major application of genomics, namely its use in the design and genetic engineering of recombinant therapeutic proteins. The use of genomics in the development of therapeutic proteins is rather different from the design of small molecules. It does not require prior identification of a molecular target, since the proteins are identified instead by comparison with proteins already known to have the desired activity. By comparing the DNA sequences of new genes with genes of known function from other organisms, geneticists can predict what function the new proteins are likely to perform. This approach has been proved by a number of companies such as Amgen, Genentech, Chiron and Genzyme, whose rise in the 1980s was based largely on their success in using recombinant DNA technology to

Box 3.9 *Genomics*

The term 'genomics' generally refers to technology for the automated, high-throughput sequencing of DNA for the study and interpretation of the 120,000 genes currently estimated to form the human genome (*The Economist*, 1998). The launch of the Human Genome Project in 1990 marked the beginning of what is likely to prove a radical change in methods of drug discovery, as world-wide efforts were put underway to sequence and map human genes. Drews (1998) predicts that genomics will provide pharmacology with some 10,000 disease targets, compared with less than 500 in current use. Also, while no large company has yet entered genome-based protein drugs into clinical trials (Haseltine, 1998), the success of existing biopharmaceuticals (Box 3.3, p39) suggests that genomic research will lead to the discovery of some 10,000 soluble proteins, some of which will inevitably qualify as drugs in their own right.

Genomic databases offer several different ways of identifying new target molecules. Many genes fall into families, some of which are particularly rich sources of potential target molecules. Databases also offer information on where and when particular genes are active, enabling designers to identify protein targets specific to the tissues they wish to affect. 'Functional genomics' provides information about the role of the target in the body's metabolism, and allows scientists to determine where the target fits into the various pathways in which it participates, and its position in the various hierarchies within a cell. Genomics can also help scientists to understand the regulatory processes that control the target. Scientists can then guess how proteins in particular tissues interact to form metabolic pathways and can compare healthy tissue with diseased tissue. Advances in large-scale population genetics and individual genotyping also enable researchers to establish correlations between specific genes and an individual's predisposition to major diseases (Post, 1998; Crooke, 1998; *The Economist*, 1998).

For genomics to deliver on its promise, it will need to be integrated into the full panoply of drug discovery techniques, including signalling pathway analysis, target validation, combinatorial chemistry, whole-cell functional assays, fluorescence techniques, robotics, miniaturisation and drug screening (Gower, 1998). It may lead to changes in the entire drug discovery and development process, to find more specific ways of assessing the effects of altering chemical pathways and the more careful selection of patients with disease amenable to treatment based on a particular molecular target.

If predictions for genomics are well-founded, it is likely to lead to a whole new generation of 'preventive' and 'targeted' care, in which genetic profiling will identify the disease predisposition risks faced by individuals, and later enable therapy to be tailored to individual patients' needs.

Sources: Poste, 1998; Drews, 1998; Haseltine, 1998; Crooke, 1998; Gower, 1998; *The Economist*, 1998.

create proteins similar to those whose function was already known. They achieved this by manipulating existing proteins, without recourse to huge screening libraries (*The Economist*, 1998).

Proteins are much larger and more complicated molecules than 'small molecule' drugs, and are thus difficult to synthesise. Instead, the human gene carrying the instructions for making the desired protein is generally identified and 'recombined' into the DNA of non-human cells that can be grown rapidly, such as those from the bacterium *E coli*. In these cells, the human gene is translated into protein in just the same way as it would be in the native human cell. The protein can then be extracted and purified (*The Economist*, 1998).

One way to obtain useful proteins is to obtain them directly from human subjects. Another, to express them in genetically manipulated cells grown in the laboratory, is described above. A third way, however, offers the means of obtaining hundreds, and even thousands, of times more protein than through this method. Several transgenic species have been developed to secrete human protein in their milk. For example, Genzyme Transgenics Corporation is harvesting Antithrombin III, a protein that controls blood clotting, from goats' milk. The protein appears, from clinical trials currently underway, to behave much like the ordinary human protein. Alpha-1-antitrypsin, a protein that may help cystic fibrosis patients, is currently being produced by transgenic sheep at PPL Therapeutics and the Roslin Institute in Scotland (where the famous Dolly was bred), and is in clinical trials on healthy volunteers. PPL Therapeutics has also produced transgenic cows whose milk expresses human alpha-lactalbumin, a food supplement protein for nutraceutical use. Transgenic pigs at the Virginia Polytechnic and State University are producing human blood products such as Factor VIII and fibrinogen in their blood (*The Economist*, 1997).

3.5.2 Development

Drug development includes chemical improvements to the drug molecule, pre-clinical animal studies (toxicity, dose, bioavailability, and formulation), clinical trials (Phases I, II and III), and the process by which approval is sought from the regulatory agency.

Chemical and pre-clinical studies

A compound – whether of natural or synthetic origin – which appears to be sufficiently novel, and has demonstrated interesting activity in a screen, becomes a 'lead'. It is then put through a sophisticated chemistry programme, where attempts are made to improve its activity by modifying its chemical structure. The compound itself, or that part of the compound responsible for biochemical interaction with the target, may be altered. Different related compounds known as 'analogues' – each a unique structural modification of the original – are often synthesised by chemists in the laboratory. The aim is to produce the most potent, least toxic, and smallest molecule that exhibits high selectivity towards the target organism and has a novel chemical structure and, preferably, a new mechanism of action.

A range of 'pre-clinical' biological and pharmacological studies may then be conducted on various analogues of the lead molecule in order to assess which version of the molecule produces the highest level of pharmacological activity and demonstrates the most therapeutic promise, with the smallest number of potentially harmful biological properties. These tests may involve the use of animals, isolated cell cultures and tissues, enzymes and cloned receptor sites as well as computer models. Individual companies may take all these stages in a different, or overlapping, order.

At this point, several drug candidates may have emerged. Since the research from here onwards becomes much more expensive, companies need to make well-informed decisions about which drug candidates to take forward through development. Further improvements of the structure of the lead molecule, and more pre-clinical studies might be conducted. At this stage, the tests are likely to involve animals. They test the putative drug for its toxicity (how poisonous it is and what sort of side-effects might be expected); its bioavailability (how effectively it is taken up into the body and delivered to the tissue where it is needed); and for pharmacokinetics (how it is metabolised, and therefore how long it stays in the body). The company developing the pharmaceutical then files a NDA to the relevant regulatory authorities (see below) and the drug candidate embarks upon the most expensive phase of development: clinical trials.

Clinical trials

Pre-approval clinical trials are divided into three phases in many countries (Box 3.10). Phase I involves tests on a limited number of healthy volunteers, to check that the candidate pharmaceutical is safe for human use, since the human body sometimes reacts differently from the laboratory animals used in pre-clinical trials. If the volunteers suffer no ill-effects, a series of Phase II trials is conducted in small populations of people suffering from the disease the drug is intended to treat, in order to show if it is safe for sick patients to take, and to indicate whether it will be effective as a treatment. If the drug candidate passes this stage, it is put into Phase III trials, in which its efficacy to treat much larger numbers of patients is tested. Ten years ago, a drug candidate undertook some 40 clinical trials prior to product approval in the USA. Today, the number has increased to 60. Furthermore, the number of procedures carried out in each trial in the USA has risen by 50 per cent in a decade, and the average number of patients involved in each trial has doubled (*The Economist*, 1998).

Once the company has collected all the data on how the drug candidate performed in the three phases of clinical trials, a NDA is filed with the relevant regulatory authorities (see below for the US, Europe, and Japan). NDAs can run to 100,000 pages or more, and the average review time for all NDAs approved in the USA in 1997 was 16.2 months (PhRMA, 1998).

If and when the regulatory agency grants product approval for the pharmaceutical, the dosage formulation and testing for stability is the final stage of product development. Dosage formulation is the process of turning an active compound into a form and strength suitable for human use. A pharmaceutical product can take any one of a number of dosage forms (for example, liquid, tablets, capsules, ointments, sprays, patches) and dosage strengths (eg 20, 100, 250, 500 mg). The final formulation includes substances other than the active ingredient, called excipients. Excipients are added to improve the taste of an oral product, to allow the ingredient to be compounded into stable tablets, and to delay the drug's absorption into the body.

Clinical trials are the part of the pipeline most frequently out-sourced by both big and small pharmaceutical companies. This is largely due to the increased complexity and difficulty of conducting trials. Saving time on clinical trials is extremely important to drug companies because it directly influences the profitability of a drug. Companies do not put a drug through an Investigational New Drug (IND) Application without a patent. So the longer the clinical trials, the shorter the product's commercial life under a 20-year patent: every day saved is an extra day of patent-protected sales. By speeding up trials, one contract research organisation (CRO) estimated that it saved almost US$580 million in patent-protected sales for an anti-Alzheimer's drug Aricept, for Eisai, a Japanese firm, and Pfizer, its American partner. The drug was approved in early 1997, 5.5 years after trials began (*The Economist*, 1998).

Because of the enormous importance of regulation and its implications for earnings in the pharmaceutical industry, drug regulators from the USA, Europe, and Japan are participating in the International Conference on Harmonisation of Technical Requirements for Registration of Pharmaceuticals for Human Use (ICH). The ICH is intended to achieve greater harmonisation in the interpretation and application of technical guidelines and requirements for product registration in order to reduce or obviate the need to duplicate testing (ICH, 1998). Parties to the ICH include national regulatory bodies and trade associations (ICH, 1998; JPMA, 1998; EMEA, 1998) as follows:

- *The European Agency for the Evaluation of Medicinal Products (EMEA)*, established by the European Commission (EC), is based in London. It is the European agency to which companies submit applications under the new 'centralised procedure'.

Box 3.10 *Clinical Trials Required for Drug Approval*

Clinical evaluation Phase I – safety studies and pharmacological profiling

This phase determines the drug's pharmacological actions, its safe dosage range, how it is absorbed, distributed, metabolised and excreted, and the duration of its action. These tests involve a small number of healthy subjects (not patients). Phase I clinical testing can usually be conducted in less than one year.

Clinical evaluation Phase II – pilot efficacy studies

This phase consists of controlled studies in approximately 200 or 300 volunteer patients (people suffering from the illness) to assess the drug's effectiveness. Simultaneous animal and human studies continue to determine the drug's safety. Phase II clinical trials last about two years.

Clinical evaluation Phase III – extensive clinical trials

Here the testing moves to larger numbers of volunteer patients, usually 1,000 to 3,000, in clinics and hospitals. The drug is administered by practising physicians to those suffering from the condition the drug is intended to treat. These studies must confirm earlier efficacy studies and identify low-incidence adverse reactions. Phase III clinical trials last about three years.

Clinical evaluation Phase IV

Those studies conducted after approval, during general use of the drug by medical practitioners. Also referred to as post-marketing studies. A growing number of drugs go into post-approval Phase IV trials in order to extend the range of conditions for which they can be used, and reformulated in ways that make them more effective, and therefore more profitable.

Sources: PhRMA, 1998; *The Economist*, 1998; OTA, 1993.

Technical and scientific support for EMEA activities is provided by the Committee for Proprietary Medicinal Products (CPMP) of the EMEA.

- *The European Federation of Pharmaceutical Industries' Associations (EFPIA)*, situated in Brussels, has a membership comprised of manufacturers of pharmaceutical medicines. Much of its work is concerned with the EC and the new EMEA.
- *The Ministry of Health and Welfare (MHW), Japan*, within whose Pharmaceutical Affairs Bureau is the Pharmaceuticals and Cosmetics Division, which reviews and licenses all medicinal products and cosmetics. The Drug Organisation provides technical consultation services on clinical trials. The Pharmaceutical and Medical Devices Evaluation Center was established in 1997 within the National Institutes of Health Sciences to expedite the approval system.
- *Japan Pharmaceutical Manufacturers Association (JPMA)*, which represents 90 member companies.
- *US Food and Drug Administration (FDA)*, which regulates drugs, biologicals, medical devices, cosmetics, and radiological products. It is the largest of the world's regulatory agencies. The FDA consists of administrative, scientific, and regulatory staff organised under the Office of the Commissioner and has several centres with responsibility for the various regulated products.
- *Pharmaceutical Research and Manufacturers of America (PhRMA)*, which represents the research-based pharmaceutical industry in the USA. It has 67 member companies, and 24 research affiliates. The PhRMA was previously known as the US Pharmaceutical Manufacturers Association (PMA).

For the last eight years, the ICH has worked to eliminate duplicative requirements for drug development and approval in the three regions. In the year 2000, the group will be presented with a Common Technical Document which is intended to form the basis for drug approval in the three regions. In the meantime, in 1998, the USA and the EU negotiated a pharmaceutical Mutual Recognition Agreement (MRA) to eliminate regulatory barriers, including recognition of each other's inspections of manufacturing facilities for drugs (PhRMA, 1998; ABPI, 1998). Some companies, like Hoechst Group, are developing uniform procedures for collecting and preparing clinical data to meet simultaneous regulatory filings in key countries, in order to save on costs associated with this most expensive phase of drug development (*C&EN*, 1998).

3.5.3 Trends in pharmaceutical research and development and the role of natural products

Scientific developments in the fields of biochemistry, molecular biology, cell biology, immunology, and information technology are transforming the process of drug discovery and development. Advances in molecular biology and genomics produce a previously inaccessible range of disease targets for the development of new drugs. New scientific technologies – such as combinatorial chemistry, high-throughput screening, and laboratories-on-a-chip – provide unprecedented numbers of compounds to test in high-throughput screens, and better ways to turn the new knowledge into conventional molecules and those produced by biotechnology for testing. Molecular medicine is also likely to shift the emphasis in healthcare progressively away from costly, inefficient intervention after disease is evident to proactive prediction and the prevention of disease risk (Poste, 1998).

Biological discoveries which would once have taken years to develop now take days or months. Eli Lilly, for example, aims to use the new technologies to shorten the research and development process from an average of 4,800 days to 3,000 days. The new methods produce not only more drug candidates, but a corresponding increase in the amount of data to collect, handle, and interpret (*C&EN*, 1998; PhRMA, 1998; *The Economist*, 1998). 'Information management is one of the key issues that the industry is facing ... . We are learning how to convert data to knowledge because, unless you can do that, you won't be able to take advantage of this vast amount of information,' says Dr Koster of BMS (*C&EN*, 1998).

These developments have been accompanied by a shift in research away from traditional functional areas such as 'chemistry' or 'biology' and towards multifunctional teams that focus on interdisciplinary areas such as 'lead optimization' and 'lead generation'. This can cut discovery time in half and increase the probability of success, according to Steven Paul, Vice President for Lilly Research Laboratories (*C&EN*, 1998). At the same time, companies are beginning to coalesce around a new, three-part 'holy grail' for drug discovery and development. This involves increasing the number of novel molecules from 0.5 per year to 2–4; achieving blockbuster status for 30 per cent of all drugs, with peak sales of greater than US$1 billion; and compressing the time taken for the discovery and development cycle from 12 to 7 years or less (Larvol and Wilkerson, 1998) (Box 3.11).

Where do these trends leave natural product drug development? The result of these scientific and technological advances is a growing perception within industry that natural product drug discovery is too slow, too costly, and too problematic to compete with the new methods. LH Huang, Principal Research Investigator at Pfizer, for example, feels that some companies are scaling down their natural product discovery programmes because synthetic compounds are 'cleaner', and produce less interference in high-throughput screens. Natural products also require more staff time and resources for the multi-disciplinary teams needed to isolate, process, and elucidate the structure of the compounds, so natural products research is slower and more costly than purely synthetic approaches. In the current research environment in which management wants faster and more cost-effective results, natural products researchers find themselves under tremendous pressure to compete with other approaches to drug development.

Predictions as to the fate of natural products change constantly. As recently as 1997, commentators suggested that, after an upswing over the past decade, natural products research was once more on the decline, in the face of more attractive approaches to drug discovery such as combinatorial chemistry and computer-based molecular modelling designs (Cragg et al, 1997; Borris, 1996). One natural products researcher at a large pharmaceutical company reported, 'We gave natural products a long hard run, but it didn't prove extremely productive. Technologies have changed, they can produce many more compounds synthetically than they could before. The time frame for turning things into drugs is moving so fast, natural products just do not fit in anymore and are no longer of interest in the large companies'.

However, by 1998, commentators were a little more optimistic in their prognosis for natural products, and less sanguine about the chances that combinatorial and synthetic chemistry could in fact replace the diversity, complexity, and novelty of natural products. As the initial enthusiasm about the new techniques are tempered with experience, the limitations of synthetic and combinatorial chemistry emerge. According to Dwight Baker, Senior Project Manager in Drug Discovery Services at MDS Panlabs, ' In the past 4–5 years there has been great interest in synthetic and combinatorial chemistry. As a new technology, combinatorial chemistry got a lot of people excited. But the enthusiasm is now on the wane because there have not been a great number of success stories. Now companies are going back to natural products because they know natural products work.' Another reason for industry's resumed interest in natural products is the improvement in technologies for identification, purification, and fractionation of compounds that can speed natural product drug discovery and help it to be competitive.

Box 3.11 *Key Factors Influencing Management Decision-Making in the Pharmaceutical Industry*

- The need for significant breakthrough therapies:
  - no more 'me too' drugs;
  - therapies against unmet medical needs; and
  - need for 'pioneer' medicines.
- Pressure to reduce costs:
  - control rising costs of medical care;
  - meet business realities; and
  - exponential increase in cost of new technologies.
- Pressure to reduce cycle time:
  - extend patent lifetime;
  - beat competing therapies to the market; and
  - faster, better, more.
- Phenomenal pace of technological advances:
  - biology (genomics, bioinformatics);
  - combinatorial chemistry;
  - automation and computing power; and
  - consolidation and outsourcing.
- Biotech acquisitions and mergers:
  - large pharmaceutical mergers; and
  - global markets – only the large can survive.

Source: Carte, 1997.

Some argue that combinatorial chemistry has simply 'increased the size of the haystack where you have to find a needle' (Crooke, 1998), and there are questions as to whether combinatorial chemistry can deliver the diversity of molecules needed to cater for the full range of molecular targets available today. Some believe that it cannot conveniently produce enough types of molecule to cater for the 32 molecular families which form the targets of half of the drugs currently on the market (*The Economist*, 1998). Also, tools such as combinatorial chemistry require structures on which to base the many combinations that they produce, which, according to David Corley, natural products chemist at Monsanto, is where natural products fit in. 'Combinatorial chemistry is a great technology, but we always need new leads to base it upon. We cannot fathom the types of molecules that have evolved over millions of years in nature. Synthetic chemists can then apply their skills to make compounds better for human use.' Combinatorial chemistry (Box 3.7, p50) and natural products are in fact complementary approaches to drug discovery (Box 3.12). Combinatorial chemistry can provide large quantities of related compounds. Natural products can supply an array of diverse compounds.

While there is thus continuing interest in natural products, there is obvious concern about their ability to be competitive, compared with other methods of drug development. As Reimar Bruening, Director of Natural Products at Millennium Pharmaceuticals put it, 'natural products represent diversity and complexity of structure – they should be part of a company's research programme. The problem is that historically it is a slow and costly process ... natural products chemists have to make the case to managers who ask: what can you show me for the money?' If natural products are expensive, or difficult to acquire, some researchers feel that their companies might do without them. Bruening adds, 'If you can produce a synthetic compound for 10 cents each, and can make hundreds a day, and then people ask for hundreds of dollars for natural product samples, it is clear that companies will go elsewhere.'

Box 3.12 *Comparison of Inputs for High Throughput Screening*

	Advantages	Disadvantages
Combinatorial and synthetic libraries	Large numbers Low cost/sample 'Drug-like' molecules Fast structure identification Ease of synthesis Reliable scale-up production	Limited chemistry available Restricted structural diversity Lack of structural novelty
Natural products	Large structural diversity Novel structures Biologically active molecules Low numbers	Slow process Slow structure identification High cost/sample Nuisance compounds Re-supply and production issues Difficulty of synthesis Potential controversy with sample supply issues

Sources: Carte, 1997; interviews.

Probably the most important factor influencing industry investment in drug discovery, however, is recent success. Following the success of Taxol and other plant-derived drug candidates over the past decade, most large companies have put 'at least pocket change' into plants, and more than half of the world's 250 drug companies have reintroduced plant-research programmes (Duncan, 1998). Many natural products researchers interviewed felt that if one or two significant candidates were commercially developed from natural products, interest in this area would increase dramatically.

3.3.4 Conclusions

Biotechnology will speed bioassays and the production of new drugs, explain more accurately how drugs act in the human system, and may reduce the vast costs of pharmaceutical research and development. Most natural products researchers feel that natural products and combinatorial chemistry are complementary approaches to drug discovery, with combinatorial chemistry providing large quantities of similar compounds, and natural products providing diversity. In order for natural products to compete with newer methods of drug discovery, advancement is required in logistical, legal, political, and technological areas (Carte, 1997). Natural products are likely to remain an important, although minor, component of drug discovery, especially if some new and important natural product pharmaceuticals emerge soon on to the market from companies' pipelines. However, the techniques of molecular biology and genetic engineering are likely to become the dominating factor in drug discovery, design and development. Genomics and informatics are set to become the dominant growth industries of the early 21st century, and small molecule drugs directed toward targets discovered by genomics will eventually account for the great majority of drugs (Haseltine, 1998).

It is reasonable to infer from these predictions that natural product drug discovery will continue, on a modest scale, and that there will continue to be a demand for access to genetic resources. However, as drug discovery evolves, so too will the nature of demand. Assays are now very specific, and researchers are looking for action against very specific targets. Demand by industry is likely to be for quantitatively smaller numbers of samples, but these will be much more carefully chosen. As Murray Tait, Director of Natural Products Drug Discovery at AMRAD, put it, 'demand for access to more genetic resources could be significantly less in numerical terms (how big does a library of extracts need to be?) but more highly-focused on specific collections (taxonomic gap-filling and/or specimens of specific interest to particular drug development programmes)'.

3.6 The acquisition of natural product samples

The acquisition of genetic resources involves a range of parties and approaches. This section will examine the 'who' (types of genetic resource collectors), 'what' (the nature of samples acquired, collecting strategies employed, and the acquisition and use of traditional knowledge), and 'where' (the value of geographic diversity to industry research programmes) of the acquisition of natural products by industry.

3.6.1 Who collects genetic resources?

Companies

The staff at a few multinational pharmaceutical companies conduct their own field collections. At Pfizer, for example, staff collect 10 per cent of its natural product samples (60 per cent are collected by botanic gardens, and 30 per cent acquired from culture collections). Sankyo Co Ltd staff also collect a portion of the samples on which the company conducts research. Small and medium-sized companies fall into two groups. A small proportion undertake specialised collecting expeditions. Japan's Marine Biotechnology Institute Co Ltd collects 100 per cent of its samples, 50 per cent of these from other countries. Shaman Pharmaceuticals' staff collect 60 per cent of the company's material (95 per cent of it from overseas). But the majority of small companies do not conduct field collections. As John Elwood, Associate Director of Business Development at Tularik, said, 'For a smaller company without an explicit focus on natural products or traditional knowledge, it is not worth investing in elaborate partnerships in high biodiversity countries. It makes sense for companies like Shaman Pharmaceuticals, but Tularik would prefer to use middlemen, or groups like the NCI's library.'

Intermediaries

The majority of companies do not conduct field collections, and rely instead on existing in-house collections of material, or buying-in compound or culture collections. Most companies 'outsource', or contract to others, the acquisition of samples for their screening programmes. They obtain samples through brokers, agents who collect on their behalf, or through specific deals with supplier organisations. The majority of companies interviewed for this study out-source the collection of natural product material, acquiring it through intermediaries who are responsible for collecting activities. Monsanto, Pfizer, AMRAD, Xenova, Glaxo Wellcome, and Merck, for example, all cite third parties as their main source of materials. The bulk of collecting activities are conducted by non-profit organisations. These intermediaries collect, taxonomically identify, ship, and re-supply materials as needed, and include:

- botanic gardens and universities that collect in source countries other than where they are based;
- partners within source countries, including research institutes, universities, botanic gardens and commercial collectors; and
- commercial brokers and importers of material based outside the provider country.

Some intermediaries collect initially for their own academic research programmes, but may subsequently allow pharmaceutical researchers to access specimens from their collections.

A large proportion of collections are still made by institutions based outside source countries, typically in the country where the company itself is located. However, a few companies have formed strategic alliances with source country institutions for the sourcing of materials, the best-known of these being Merck's arrangement with INBio in Costa Rica (Sittenfeld, 1996). Shaman Pharmaceuticals has developed a number of partnerships with groups in high biodiversity countries, and as part of an ICBG International Cooperative Biodiversity Group), Monsanto, American Home Products, and Bristol-Myers Squibb are also actively involved in collaborations with source country partners (Rosenthal, 1998; Baker et al, 1997).

In response to the CBD, companies are collecting in fewer countries than formerly, and rely to a greater extent on other organisations to collect for them. For example, Phytera Vice President John McBride was quoted in 1996 as saying that the company had been trying to negotiate access to plants with the governments of 'five or six' developing countries, but because of difficulties in reaching agreement, Phytera became 'more dependent than we would like' on acquiring plants from botanic garden intermediaries (*New Scientist*, June 1996).

The pharmaceutical companies interviewed in this survey indicated that the most important criterion for selecting partners from whom to obtain samples and with whom to collaborate on the research is the calibre of the collaborators and quality of the institution. The second most important criterion is the biological diversity to which they will have access; and the third, the quality of the samples supplied. After these factors, companies listed three other criteria of roughly equal importance, namely the ability for the collaborator to identify a source for re-supply; the cost of samples; and the ease of obtaining the correct permits for collection. Companies were not particularly concerned about the availability of intellectual property rights in countries where they conduct collaborative research, and 'IPR' was not a significant factor in the companies' selection of collaborators.

3.6.2 Sourcing samples from libraries and *ex situ* collections

As they collect or acquire samples from third parties over decades, many pharmaceutical companies build and maintain libraries of compounds, extracts, or dried plant material that they can use in their screening programmes. The bulk of material in culture collections and screening libraries today was collected prior to entry into force of the CBD. Virtually every company interviewed possesses or intends to build a library, except for the smallest of the start-up companies.

The extent to which companies draw upon their existing corporate collections for screening, and the extent to which they seek to renew or supplement the content of their libraries, depends upon their individual strategies for research and development, and, in particular, on their approach to screening.

Companies' libraries are a valuable resource. Not only are they used for in-house research, but companies may license access to their libraries to customers, or exchange and share them with commercial partners. An exclusive library of around 100,000 molecules can be hired for US$1 million today, which means that high-throughput screening of large numbers of compounds is more widely available (*The Economist*, 1998). Once only affordable to the largest pharmaceutical companies, combinatorial chemistry has made it possible for small, highly-specialised drug firms to have their own libraries of synthetic molecules. Axxys Pharmaceuticals in San Francisco, for example, has a library of over 200,000 compounds. Pharmacopeia in Princeton, New Jersey, has 3.3 million molecules. Pfizer has a compound library of 1.1 million compounds (*The Economist*, 1998; Table 3.6).

Table 3.6 *The Average Compound Library Size*

1994: 100,000
1996: 200,000
2000: more than 500,000

Source: Andersen Consulting in *The Economist*, 1998.

In comparison to libraries resulting from combinatorial chemistry, libraries of naturally-derived substances – often the result of painstaking collections of small numbers of specimens over many years – are often modest. Novo Nordisk's library of microbial extracts numbers 57,000, and their screening library 30–35,000 (30 per cent of which were collected since entry into force of the CBD). Millennium Pharmaceuticals' fungal isolates collection numbers 35,000, but is growing quickly. The US NCI has a library of 150,000 extracts and 400,000 compounds. Tularik has a collection of 116,000 natural product extracts.

It should be made clear that numbers of samples in a library do not equal the number of species represented. For example, according to Murray Tait, Director of Natural Products Drug Discovery, in AMRAD's collection, 'there are typically 3–5 different plant parts per species of plant in the collection, and microbes are typically fermented on ten different media – giving rise to ten different chemical extracts per microbial isolate. Marine organisms are typically present at 1 extract per species'.

3.6.3 The scale of acquisition programmes

The number of samples collected by companies per year falls within a wide band, depending on the company's research and development strategy. Smaller companies with more directed research approaches might collect and research anywhere from 10 to 100 samples per year. Larger companies collect on average 2,000–10,000 samples per year (with as many as 80,000 microorganisms per year in some companies). Between 1960 and 1982 the NCI, for example, collected and screened for anti-tumour activity more than 180,000 microbial, 16,000 marine, and more than 114,000 plant extracts.

Screening capacity varies greatly by company. Companies with large-scale, high throughput screens explore millions of samples a year. In the middle range, companies might screen 100,000 samples a year, with small specialist companies screening around 1,000 samples a year. The number of screens against which these samples are tested within a given year typically ranges between 10 and 50.

3.6.4 Types of material acquired

To maximise the scope of chemical diversity in a screening programme, companies look to a range of natural products – plants, microorganisms, marine organisms, insects, and animals – although the priority categories for most companies are plants and microorganisms. Half the companies and research institutions interviewed as part of this study consider plants their priority, usually with microbes a close second (Table 3.7). These include Monsanto, the NCI, and Glaxo Wellcome (90 per cent of their samples are from plants). For one-third of companies, the order is reversed, with microorganisms the priority and plants the second choice. This groups includes MDS Panlabs (90 per cent of their collection is in microorganisms); Novo Nordisk; and Xenova. In most cases, however, companies have significant programmes for both types of organisms.

Pfizer's main area of interest in natural products is fungi. A number of companies expressed a growing interest in marine organisms, and four in invertebrates. All of the companies interviewed included plants in their research and development, with the exception of highly specialised firms such as Magainin, which focuses on shark and frog resources; Millennium and Biodiversity which emphasise fungi; and the Marine Biotechnology Institute of Japan, which studies exclusively marine organisms.

Few of the companies interviewed specialise in animal genetic resources which may, however, play a larger role in biopharmaceuticals in the future. One company, Magainin Pharmaceuticals, has developed two antibiotics known as 'Magainins' from the skin of frogs, and has subsequently become well-known for its compound squalamine, isolated from sharks (Coghlan, 1997), which has entered Phase I clinical trials for its anti-cancer properties (http://www.squalamine.com/magaininnews2.htm). Research into insects is still comparatively rare in the pharmaceutical industry, although certain intermediary institutions are collecting insects. INBio in Costa Rica is compiling large collections of insects, some of which they supply to companies and government research institutions, including the US NCI, which recently initiated a programme with INBio for the supply of insect samples.

Table 3.7 *Ranking of Importance of Types of Genetic Resources: Results from Interviews with Pharmaceutical Companies*

	Ranking of priority (number of companies)				
Organism	1	2	3	4	5
Plants	12	5	–	1	1
Microorganisms	8	6	–	–	–
Marine	2	6	2	–	–
Fungi	3	–	1	–	–
Insects	–	–	4	–	–

These materials are supplied to the companies interviewed in a wide range of forms, reflecting the diversity in industry research programmes. Many companies prefer **raw samples** in the form of dried plant and soil samples, since any processing on the part of collaborators can affect the way in which a sample responds to a screen. Most companies acquired material in the form of **extracts (organic or acqueous)** and some materials accompanied by **ethnobotanical** information (discussed below).

It is comparatively rare for companies to acquire samples supplied with the results of screening, **identified bioactive compounds** (ie compounds which have proven activity, and whose chemical structure is already known), or indeed with data emerging from trials on animals or clinical trials. This kind of value-added product is more likely to be acquired by companies in the context of collaborative research with a partner institution rather than 'off-the-shelf' purchase of samples. To date, however, most pharmaceutical companies are not entering into collaborations involving high levels of discovery with source countries (see the discussion of NCI's approach in the next section for an exception). Companies tend to emphasise high quality sample collection, processing into a range of extracts, and solid re-collection strategies. In response to improved capacity and infrastructure in source countries, however, there appears to be a trend towards conducting higher levels of research in source countries, although at this stage these efforts are primarily supported by government-sponsored research programmes (eg the NCI; the ICBGs).

3.6.5 Criteria for sample collection

According to Merck's Bob Borris, important questions for a company when deciding from where and on what basis to sample genetic resources include 'Where can one practically collect a large number of diverse species?'; 'What criteria will be used to select species for collection?'; 'What selection criteria best fit the overall goals of the screening programme?'; and 'Is there enough local infrastructure to support collections?' (Borris, 1996).

Depending upon the objectives of a particular screening programme, a number of different approaches may be adopted for the sampling of natural products. These include:

- *random/ 'blind'* – collections are conducted on a random basis within a given geographical area in order to obtain a representative, but random, sample of local diversity;
- *ecology-driven/biorational* – collections based on an understanding/observation of ecological relationships between species which might lead to the production of secondary compounds;
- *chemotaxonomic* – collections based on knowledge of taxa with certain classes of compounds of value in natural products screening; and
- *ethnobotanical* – collections based on indigenous peoples' or local communities' uses of species.

Through their contracted collectors, most companies employ a combination of these approaches. Ethnobotanical collection strategies appear to be most valuable for companies with research goals focused on a narrow range of disease targets. Chemical taxonomy is useful if a company knows at the outset what kinds of chemicals are of interest. Broad-based screening approaches tend to employ either random/empirical selection, or a more focused biodiversity-driven selection based on taxonomy (Borris, 1996).

Of these collecting strategies, the taxonomic approach is most popular among the companies interviewed (six cited this as their preferred strategy). Monsanto, for example, has contracted with the Missouri Botanical Garden for a number of years to collect taxonomically diverse samples. The ecological/biorational approach was the next favoured (four companies cited this as their preferred strategy). A number of companies employ 'random' methods, and a few companies use ethnobotanical collection strategies. Some companies do not conduct field collections at all, however, and rely instead on existing in-house collections of material, or buying in compound or culture collections.

3.6.6 The ethnobotanical approach to drug discovery

The ethnobotanical approach to drug discovery – the use of people's knowledge and experiences of the medicinal properties of plants and other genetic resources to guide drug discovery – has yielded most of the plant-based pharmaceuticals in use today. Of the approximately 120 pharmaceutical products derived from plants in 1985, 75 per cent were discovered through the study of their traditional medical use (Farnsworth, 1985). Grifo et al (1997) demonstrate that the base compound in most of the top 150 plant-derived prescription drugs and their commercial use correlate with traditional medical use. Species with related commercial and traditional uses include: *Ephedra sinica* (yielding ephedrine); *Melilotus officinalis* (coumarin); *Ammi visnaga* (khellin); *Digitalis lanata* (digoxin); *Papaver somniferum* (codeine); *Camellia sinensis* (theopylline); *Atropa belladonna* (atropine); *Paullinia cupana* or *Coffea arabica* (theobromine); and *Spirea/Salix alba* (salicylic acid).

Numerous researchers continue to demonstrate the value of ethnobotanical information as a lead for drug discovery (eg Balick et al, 1996; Iwu, 1994; King, 1996; Lewis, 1994; Cox, 1994). Cox (1994) describes three primary ways in which ethnopharmacological information can be used in the drug discovery process:

1. as a general indicator of non-specific bioactivity suitable for a panel of broad screens;
2. as an indicator of specific bioactivity suitable for particular high-resolution bioassays; and
3. as an indicator of pharmacological activity for which mechanism-based bioassays have yet to be developed.

Probably the best known example of a company employing traditional knowledge as a lead for drug discovery is Shaman Pharmaceuticals. Shaman's drug discovery efforts integrated traditional use of species with pharmaceutical scientific approaches. A combined ethnobotanical/chemotaxonomic approach to collecting tropical medicinal plant species allowed the company to produce a highly focused selection of plant candidates for screening and development (Tempesta and King, 1994; CRS, 1993; Duncan, 1998). A compound from one species – Sangre de Drago (*Croton lechleri*) showed particular promise in recent years, and was in phase III clinical trials in late 1998. Sangre de Drago yields a red latex that is one of the most common traditional medicines in all of Latin America (Ubillas et al, 1994).

In February 1999, Shaman Pharmaceuticals ceased its pharmaceutical operations, laying off 65 per cent of its work force, and moving its assets into operations of the privately-held Shaman Botanicals (Pollack, 1999). Shaman had run up against difficulties in getting the Sangre de Drago product through clinical trials : 'Due to our recent conversations with the FDA, we felt we would face a significant delay in bringing our first pharmaceutical to market,' reported company President Lisa Conte (Shaman, 1999). Some interpreted Shaman's closure as a failure of the ethnobotanical approach to drug discovery (*The Economist*, 1999); however, others identified Shaman's problems as indicative of issues typical to small biotechnology companies: 'Shaman is just one of many biotechnology companies that have virtually collapsed after their first product hit setbacks. It is also one of many that are running out of money and unable to raise new capital because of depressed stock prices' (Pollack, 1999).

In total, around half of the companies interviewed make use of traditional knowledge, with less than half of these conducting or commissioning ethnobotanical field collections. For example, Monsanto has obtained materials collected on the basis of the knowledge of the Aguaruna in Peru as part of an ICBG collaboration with Washington University and institutes in Peru. David Corley, Natural Products Chemist at Monsanto, explains how the company uses ethnobotanical information: 'We use information on traditional uses to guide us in testing, and to determine what models might be most appropriate. For example, every sample will go through high throughput, randomized screens, but if we think there is good ethnobotanical data, we may go to a higher level of testing with a sample. Separately, literature and databases with information on traditional uses might be consulted after some activity has been demonstrated'.

The Monsanto example illustrates the diverse ways in which ethnobotanical information is used in pharmaceutical research and development. Some companies – such as Pfizer, Monsanto, or previously Shaman Pharmaceuticals – use traditional knowledge (gathered through field collections, literature, or from database searches) to identify leads, and possibly to eliminate materials from screens. Others – representing the majority that make use of ethnobotanical data – consult databases and literature following identification of activity. Most companies interviewed – such as MDS Panlabs and Novo Nordisk – consider their use of ethnobotanical data to be 'minor' and 'occasional'.

Traditional knowledge is not considered a good lead for high throughput screening, which would quickly exhaust plants with ethnobotanical uses. Additionally, as Robert Borris of Merck Sharp and Dohme Research Laboratories said (1996), 'If one deals with a narrow range of target diseases, it may be possible to develop a selection programme using ethnobotanical or ethnomedical information as a primary criterion. As the range of targets increases, this approach becomes more difficult to apply.' Companies focused on microorganisms or fungi had little or no interest in ethnobiological data, since it is not well-documented. Similarly, in marine organism collection, ethnobiological data is rarely available to enrich research.

80 per cent of all companies that use ethnobotanical knowledge (only half of those interviewed) rely solely on literature and databases as their primary source for this information. This fact has significant implications for benefit-sharing, and suggests that academic publication and transmission of knowledge into databases – rather than field collections on behalf of companies – are the most common route by which traditional knowledge travels from a community to the commercial laboratory. Companies therefore have access to knowledge in ways that do not trigger benefit-sharing.

3.6.7 Representation of geographic diversity in collections

Most companies receive materials from a large number and wide range of countries. The diversity ensured by broad geographic and ecological sampling is considered of high value to screening programmes. Of the companies interviewed, only one company does not work to represent a range of diversity in its natural product acquisition programme. As David Corley, natural products chemist at Monsanto, said, 'We try to have as much taxonomic and genetic diversity as we can in our screening programmes because we are looking for serendipity to happen. As a result, areas with lots of biodiversity are extremely important sites for collection.'

John Elwood, Associate Director of Business Development at Tularik, similarly sees a value in diversity: 'We think in terms of chemical diversity – that is the primary goal. Having said that, biological diversity promotes chemical diversity, and specific niches where organisms grow and compete would foster greater generation of secondary metabolites, so we can look for specific niches as a way of generating chemical diversity.' At Novo Nordisk Natural Products Discovery, 'the collection is constantly being expanded with new organisms to ensure an ever increasing taxonomic, ecological, and geographical diversity of the collection'.

As Murray Tait, Director of Natural Products Drug Discovery at Australian-based AMRAD put it, 'We aim to have as much taxonomic diversity as possible represented within the collection. Thus, the goal is randomness, but the means by which this is achieved often involves targeting specific taxonomic groups ... While much remains to be done in sampling the extensive plant biodiversity found in Australia (approximately 23,000 species of higher plants) ... arrangements in South-East Asia are important to ADT in providing samples from other countries in a region which also has high biodiversity, and in which the botanical diversity is distinct from that found in Australia.'

The matter is quite different for microorganisms, however. He continues: 'Taxonomic diversity does not necessarily lead to functional diversity (in terms of secondary metabolite production) and novel organisms *per se* are of no value unless they produce something useful. Therefore, the need to fully explore the biosynthetic potential of all distinct microbial isolates (eg through the use of a diverse range of fermentation media and conditions) is judged to be more important than simply collecting a range of novel organisms.'

Dwight Baker, Senior Project Manager in Drug Discovery Services at MDS Panlabs, also does not see geographic diversity as an important element for microroganisms: 'Panlabs could collect in a few states of the USA the broad diversity of most microorganisms found in the world. But clients perceive that geographic diversity equals metabolite diversity, which it does not. For microorganisms, a collector does not really need to go to regions with high biodiversity. There are very few microorganisms that one can point to and say "that is endemic". They are very different from plants in that way. For example, in the Pacific North West [USA], they are now talking about the bacterial load in the air from dust storms in China.' (This issue is discussed further in 8.2.4, p245.)

One natural products researcher at a large pharmaceutical company feels that the case has not been made for plants either: 'Diversity is very important, but I don't know about geographic diversity. We haven't documented that geographic diversity is important – no more so than taxonomic diversity – so one might be able to get what is needed in the USA if one collected broadly. In general the geographic diversity hypothesis remains untested.'

However, most large-scale natural products programmes collect materials from 20–30 countries (although this is changing with the CBD, see Chapter 10). Over time, most significant natural product programmes have accessed material from more than a hundred countries, and this is reflected in their libraries/collections. A common response from companies interviewed to the question 'From which countries do you source materials?' was 'all over the world'. Many felt the countries were so numerous they were not worth listing. In fact, those countries which company representatives did bother to list as part of this survey came to more than 45, and included Antarctica.

3.7 Benefit-sharing

The pharmaceutical industry is increasingly embracing an integrated approach to benefit-sharing that includes a 'package' of monetary and non-monetary benefits, and attempts to distribute them over the short-, intermediate-, and long-term (King and Carlson, 1995; Iwu and Laird, 1998). Emerging best practice from a growing number of partnerships reveals a typical 'benefit-sharing profile' for partnerships of a similar kind (Figure 3.6, p65; Table 3.8, p64).

Inevitably, details of commercial partnerships are often confidential. The models about which the most information is publicly available are perhaps the partnerships established under the US government-sponsored ICBGs, and the arrangements established by other organisations such as the US NCI which have been described in a series of 16 benefit-sharing case studies prepared for the Secretariat of the CBD (1998).

In addition to government-sponsored initiatives such as the ICBGs and the approach to benefit-sharing taken by the NCI, the private sector has established a number of purely commercial benefit-sharing arrangements. These include the recently announced agreement between Diversa and the National Autonomous University of Mexico (Business Wire, 1998); Shaman Pharmaceuticals' distribution of benefits across time and among the communities who have contributed knowledge to its programme; the establishment of a trust fund in Nigeria through the Healing Forest Conservancy with funding from Shaman Pharmaceuticals; and the 'original' benefit-sharing package negotiated between INBio in Costa Rica and Merck. Most companies now employ an approach that includes 'financial payment, training of scientists and technology transfer', as Novo Nordisk (1997) describes its partnerships with universities and research institutes on five continents.

Xenova Discovery Ltd, in its *Policy for the Acquisition of Natural Product Source Materials,* also provides for a range of benefits. In addition to 'payments for the supply of materials' and 'additional short term and/or long term compensation', examples of compensation provided to source countries include salaries for researchers in the source country; funding of laboratory modifications/maintenance; funding of computer systems; provision of funding of laboratory equipment and consumables; provision of field work, communication, and shipping costs; scientific training; and access to scientific literature. For all the perceived faults of specific agreements, a trend is apparent. As David Corley of Monsanto said, 'We are trying to get away from the old model of only paying for samples.' Today, more companies believe that greater returns to source countries are appropriate than was the practice just five years ago.

The benefits that arise from the development of a natural product pharmaceutical take many forms. Monetary benefits might include fees per sample, grants to cover agreed research programmes, profit-sharing, stakes in equity, joint ventures, royalties and the prospect of local employment opportunities. Access legislation and the negotiating position of individual provider institutions increasingly prioritise non-monetary benefits such as the sharing of research results, participation in research, technology transfer, training and capacity-building. Some partnerships offer help in kind, such as medical assistance and investment in local infrastructure. Others support conservation projects in the field.

3.7.1 Monetary benefits

A range of monetary benefits have become standard practice in natural products drug discovery. These include fees, milestone payments, advance payments in the form of research grants, and royalties. Virtually all programmes involve fees for samples, and the vast majority some kind of royalty. Perhaps half those interviewed also use milestone payments, and perhaps a third advance payment in the form of research grants.

Fees for samples

Interviews with companies revealed a range of prices for samples. Plant samples usually take the form of dried plant material of between 0.1 and 1kg. Prices range from US$25 to US$200/kg dry weight, with the average falling between US$50 and US$100. The price for extracts (with 100mg plant solvent extracts the minimum required for screening) falls between US$100 and US$200 per 25g sample. The cost of a sample might include field collection, documentation and packaging, data processing, literature searches, shipping, and staff salaries. The costs that a provider institution faces in preparing a sample vary according to the labour and technology needed to prepare them. In Fiji, for example, collection and processing fees in one case come to about US$20 per sample, while the cost of machinery used in the grinding of material and extraction comes to about US$5,000 – or an additional US$10 per sample for 500 samples (Aalbersberg et al, 1998).

The cost of microbial cultures typically ranges between US$20 and US$140 for each sample of a living organism. Negotiations for prices are based on the prices of microorganisms in a culture collection. The prices may rise about 20 per cent during an agreement lasting two years. Fungal samples cost on average between US$60 and US$100. For all organisms, samples cost more if they are difficult to collect, or if recipient companies are granted exclusive use.

Table 3.8 *Comparison of Different ICBG Partnerships, Showing the 'Packages' of Different Benefits Involved*

	Suriname	Chile Mexico Argentina	Costa Rica	Peru	Cameroon Nigeria
Monetary benefits					
Per sample fee			■		■
Advance payment	■			■	
Milestones				■	
Royalties (1–5%)	■	■	■	■	■
Non-monetary benefits					
Salaries	■	■	■	■	■
Commitment to source locally	■	■	■	■	
In-kind					
Medicines	■			■	
Infrastructure	■				■
Informational					
Biodiversity inventory data	■	■	■	■	■
Screening data	■	■	■	■	■
Scientific literature	■	■	■	■	■
Local distribution of information	■	■	■	■	■
Technology transfer					
Equipment	■	■	■	■	■
Know-how	■	■	■	■	■
Training					
Scientific	■	■	■	■	■
Information management	■	■	■	■	■
Capacity-building					
Knowledge base	■	■	■	■	■
Collections	■	■	■	■	■
Medical research skills	■	■	■	■	■
Study of traditional medical systems	■				■

Sources: J. Rosenthal, personal communication, 1999.

years: 1 2 3 4 5 6 7 8 9 10 11 12 13 14 15 16 17 18

LEAD DISCOVERY
6 months–1 year ~US$5–10 m

Collection → Extraction → Primary screening **(5 million compounds)** → *HITS* (1,000 hits)
Structural elucidation → Dereplication → Secondary screening → *NOVEL LEAD* (10 leads)
Patent filing

LEAD OPTIMISATION 1–2 years ~US$10–20 m

Analogue synthesis Improve yield Improve potency and bioavailability
Reduce steps in synthesis Patent registration

DRUG CANDIDATE (5 candidates)

DEVELOPMENT
5–15 years ~US$100s m

Biological studies (ADME)
↓
Scaling-up of production
↓
Pre-registration
↓
Phase I clinical trials
Tests on 10s of people; c 2 years; US$10 m
(5 candidates)
↓
Phase II trials
Tests on 100s of people; C 2–3 years; US$10s m
(5 candidates)
↓
Phase III trials
Tests on 1000s of people; 5 years min; US$100s m+
(2 candidates)
↓
Filing of NDA or equivalent
↓
Request approval from FDA
↓
PRODUCT **(1 product)**
↓
Registration and marketing
↓

Benefits

Non-monetary benefits

LEAD DISCOVERY

Share taxonomic and ecosystem information; collection technology; employment of local people; complete set of voucher specimens left with local institution.

In-country extraction by collaborating partner.

Share results of screen and/or technology transfer of screening techniques.

Joint research in chemistry; donation of equipment.

Share results of research on previous uses and screens; collaborate to promote screening of samples supplied for local diseases.

Possible joint authorship on patent. Share information on discovery with collaborating scientists and with local and indigenous peoples, as appropriate.

LEAD OPTIMISATION

Exchange of staff and joint research on lead optimisation.

DEVELOPMENT

Share results of biological studies.

Any re-collection of larger quantities of genetic resource should employ local people, and should, where possible, be in the context of joint research and/or joint ventures with local institutions on product development.

Grant licence to manufacture product to provider country company.

Continuing supplies of bulk, cultivated raw materials, or value-added, processed materials from provider country, where possible.

Monetary benefits

LEAD DISCOVERY

May include, singly or in combination:

- collection fee
- research grant (upfront or in instalments)
- consulting fee
- annual payment for supply

Milestone payment take place on:

- filing of patent

DEVELOPMENT

Milestone payments take place on:

- start of Phase I
- start of Phase III
- filing of NDA or equivalent

Twice-yearly or other periodic royalty payments based on net sales, until patent expiry, or fixed period or maximum sum

Figure 3.6 *Benefit-Sharing Profile for the Development of a New Drug by a Major Pharmaceutical Company*

Source: K ten Kate, Royal Botanic Gardens, Kew.

Advance payments

Advance payments are generally supplied to support an agreed and well-defined work plan, and to cover operational costs. The nature and scale of payment depends upon the collaboration developed between suppliers and companies. Companies rarely offer advance payments outside the context of a partnership or agreed work plan, although in some cases a lump sum advance payment is provided rather than payment on a fee per sample basis.

While advance payments may involve a trade-off in lower royalty rates, they ensure some up-front benefits for source country institutions instead of waiting for royalties that will only transpire if a commercial product is developed. Advance payments provide financial revenues that might be applied to the establishment of trust funds, for example. The five ICBG projects revealed that different companies have different approaches to requests for advance payments, some favouring per sample payments; others finding a lump advance sum a good public relations and bargaining tool; and yet others reluctant to provide significant monetary benefits until the partnership is showing results (Rosenthal, 1998) (Table 3.8).

Milestone payments

Several companies offer milestone payments during the research and development phase. For example, in the case of Novo Nordisk, these might be made:

- on identification of an active compound from the source material;
- in the event that Novo Nordisk approves a drug candidate for development on the basis of the structural information obtained from the active compound;
- upon initiation of Phase II clinical trials;
- at NDA approval.

The magnitude of milestone payments varies from case to case, and depends upon a similar range of factors to those discussed below for royalties. Some milestone payments are made in addition to (lower) royalty commitments. Others are treated as 'advances' against royalties which are only paid once the amounts received as milestone payments have been deducted (Table 3.9).

Royalties

Royalties are now generally incorporated into most company-collector agreements, whereas five years ago they were offered less frequently. They are also required in most national access legislation introduced in response to the CBD.

Some royalty payments are fixed in contracts signed at the time of collection, but at this stage in the research process it is difficult to predict the nature of any product that may result, and how a given sample will contribute to drug discovery and development. Consequently, it is common for contracts to express royalties as a range, and to provide for negotiation of the specific rate at a later stage in the research and development process, when the nature of the product and contribution of the sample are clearer. Sometimes a contract stipulates the factors that will be taken into consideration in deciding where, within the range, the final royalty payment will lie. On other occasions, the company and the provider are simply left to negotiate based on whatever arguments each can marshall.

Table 3.9 *Example of Milestone Payments (US$)*

First patent filing	5,000
Initiation of Phase I clinical study	10,000
Initiation of Phase III clinical study	25,000
Filing of NDA or equivalent	50,000

Source: interviews.

The exact royalties in contracts are generally highly confidential, since companies and providers do not wish to tie their hands for future negotiations by publicising rates that could be used against them. However, a given company is likely to negotiate royalties with several providers, and individual providers may well supply genetic resources to a number of different companies. In this way, despite the lack of publicity of royalty rates, individual institutions build up an appreciation of the current 'going rates', or ranges of royalties, available in the market-place.

A number of factors commonly influence the magnitude of royalty payments, and these vary from company to company. For example, Rosenthal (1998), discussing factors considered in establishing royalty rates for the ICBG projects, states that some issues to consider in royalty structures include: a) relative contribution of partners to invention and development; b) information provided with samples; and c) novelty or rarity of sample organisms. While these are common factors considered in establishing royalties, some companies consider other factors to be important, such as the degree of derivation of the final product from the genetic resources supplied; the current market rate for royalties; and the likely market share of the final product.

Companies rarely offer royalties that fall outside their estimation of the current market rate, which, according to our interviews, exerts a strong influence on their rates. Another factor influencing royalty rates is the likely market share of a given product. Even if it seems counterintuitive to those basing their calculations on a fixed proportion of the final sales, a multinational company whose product is likely to have blockbuster sales and occupy a large share of the available market may offer lower royalties than a small company producing a more speculative product, based on the argument that the income in the former case will, in any event, be extremely high.

The relationship of the final product to the lead supplied is another important criterion (Rosenthal, 1998), and, indeed, several of our interviewees stated that the degree of derivation of the final product from the genetic resource supplied is one of the key considerations in fixing a royalty rate. Thus it would be common for a company to offer higher royalties for a compound directly derived from the genetic resource and partly altered by semi-synthesis, than for a wholly synthesised analogue which was rationally designed following mode of action studies 'inspired' by research on a natural product. A range for royalties based on the relationship between the final product and the sample is suggested by Biotics (1997) as follows:

- Raw plant material or basic extract is the source of the natural product which is marketed: 3–5 per cent.
- The natural product is converted to a chemical derivative which is marketed: 2–3 per cent.
- The marketed natural product or derivative is wholly manufactured by synthesis: 0.5–1 per cent.

However, this is just one parameter influencing the magnitude of royalties, and many companies do not propose bands of royalties so closely tied to the degree of derivation.

Another important factor influencing royalty rates is the relative role of the various partners involved in the inventive step leading to patent protection of the product. The value of the biological and intellectual information provided by a source country is weighed against the intellectual and financial investment a company has made to develop a useful product (Laird, 1993).

Samples collected using ethnobotanical data often receive a higher proportion of royalties than those collected randomly. Table 3.10 shows how the Suriname ICBG established a benefit-sharing plan for monetary benefits from advance payments and royalties provided by Bristol-Myers Squibb Pharmaceutical Research Institute to 'The Forest Peoples Fund' and five local non-governmental and governmental organisations, including Bedrijf Geneesmiddelen Voorziening Suriname (BGVS) – a pharmaceutical company owned by the Surinamese government. Reflecting the importance of community stewardship of those resources, the Forest Peoples Fund is the largest single recipient of any financial return, and its share is even greater in cases for which there is a documented relevant traditional use for the species that produces a commercial discovery (Rosenthal, 1998). The Forest Peoples Fund was established in 1994 with a US$50,000 contribution from Bristol-Myers Squibb, followed by another US$10,000 donation in 1996. Any tribal person in Suriname, community or foundation can submit a proposal for project funding (Guérin-McManus et al, 1998; Box 3.14).

Table 3.10 *Monetary Benefit-Sharing, Suriname ICBG*

Institution sharing royalties	Share of royalties for ethnomedical collection, %	Share of royalties for random collection, %
Forest Peoples Fund	50	30
Bedrijf Geneesmiddelen Voorziening Suriname	10	10
University of Suriname Herbarium	10	10
Stichtung Natuurbehoud Suriname	5	10
Conservation-International-Suriname	10	10
Suriname Forest Service	5	10
Future collaborating institutions	10	20
Total	100	100

Source: Guérin-McManus et al, 1998.

Another ICBG partnership has established a system of 'know-how' licences which will enable Monsanto to compensate the Aguaruna peoples of Peru, should their knowledge concerning plants result in a product. The financial benefits involved, including royalties, would derive from the Licence Option Agreement between Washington University and Searle Co. In addition, a know-how licence was negotiated between the Aguaruna and Searle Co that outlines Searle's use of the traditional knowledge and specific benefits associated with that use (Rosenthal, 1998) (Figure 3.7, p68).

Both the Suriname royalty-sharing agreement, and the Peruvian ICBG agreement structure demonstrate that there are often several parties involved in benefit-sharing negotiations, all of whom may receive a share of royalties. The examples may also serve to clear up a common misconception about royalties. Under the Suriname ICBG agreement, for example, the Forest Peoples Fund will receive 50 per cent of the royalties (in the case of commercialisation of a lead selected using ethnomedical information). This is often misunderstood as '50 per cent of sales'. In fact, the Fund will receive half of any royalties eventually paid by Bristol-Myers Squibb. The royalty ranges for the ICBG projects have not been disclosed, but if we assume that Bristol-Myers Squibb agreed to share 2 per cent of net sales, then the Forest Peoples Fund would receive 1 per cent royalties on sales. If sales of a product were to reach US$100

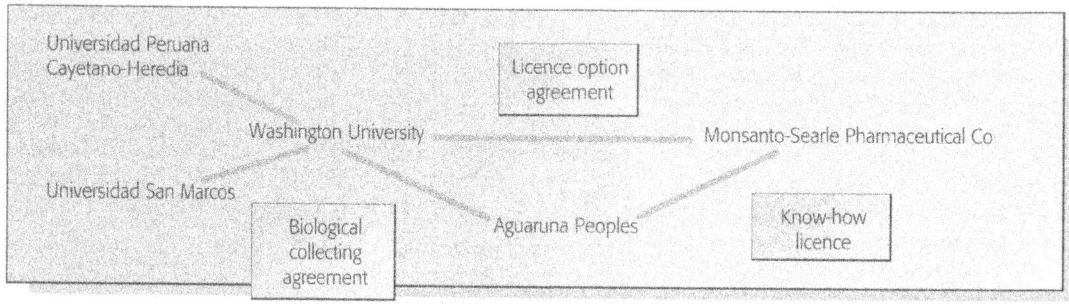

Figure 3.7 *Peru ICBG Agreement Structure*

million, the Fund would receive US$1 million, not US$50 million.

The category of the genetic resources also influences the payment of royalties. Companies frequently offer lower royalties, and sometimes even no royalties, for soil samples and microorganisms. A number of companies interviewed felt that fungal and microorganism samples should not involve royalties. As Dwight Baker, Senior Project Manager in Drug Discovery Services at MDS Panlabs, said, 'The value of any one microorganism sample is low when you consider the amount of research time to get to an end product. Plants are different, because they might be the source of the compound. But with microorganisms, the company does fermentation and gets the microorganism to produce something.' Reimar Bruening, Director of Natural Products at Millennium Pharmaceuticals, agrees: 'No company in its sane mind would sign a royalty agreement. Plants are slightly different. But with fungi, we buy them by the hundreds and thousands, pay a fee, and that is the end of the story. It would be totally impossible to track one particular compound to one particular sample from God knows where. Within a soil sample there are more organisms than stars in our galaxy, most of which are found in other soil samples.'

However, many other companies do offer royalties for microorganisms, and believe that under the CBD companies are required to do so. As David Corley, Natural Products Chemist at Monsanto, said, 'Monsanto has no trouble in documenting where our microbes come from and we feel that they fall under the same principles as plants. Yes, many microbes can be found most anywhere and it's difficult to say that a country "owns" a microbe. But so what? You had to collect it somewhere and the same sovereignty rules apply.' Access to unusual fungi have elicited commitments from companies to negotiate royalties in recent ICBG partnerships as well (Rosenthal, personal communication, 1999).

Box 3.13 gives an indication of current market rates for different kinds of materials based on interviews with companies and other sources. It should be clear from the foregoing discussion that these rates are indicative only, and that there is significant variation among companies and from case to case. Several companies interviewed felt that far too much attention is paid to the issue of royalties, and that expectations on the part of providers of genetic resources are often well above what is reasonable within the given market. In some cases, interviewees felt that excessive demands for royalties were a deciding factor that tipped companies into moving out of natural product research altogether.

Box 3.13 *The Market Rate: Average Range of Royalties on Net Sales*

'Raw' materials or early research: 0.5–2%

Raw material (eg dried plants, soil samples) and basic extracts (organic or aqueous): 0.5–2%
Extracts (organic or aqueous): 0.5–2%

Value added data: 1–4%

Ethnobotanical information: 1–4%
Material supplied with some results from screening: 2–3%
Identified bioactive compound (with known structure and test tube activity): 1–4%

Clinical data: 2–15%

Animal model data supplied with identified bioactive compound: 2–6%
Clinical data supplied with identified bioactive compound: 5–15%

Source: interviews.

3.7.2 Non-monetary benefits

The term 'non-monetary' benefit is used to describe a range of benefits, not all easily grouped together. Some – such as 'participation in research' – describe a relationship within which a variety of benefits may be shared. Other 'non-monetary' benefits describe specific products of a research relationship, such as training, technology transfer, and information. Still others describe wider societal benefits, such as research on tropical diseases and conservation projects. The

Box 3.14 *'Sharing' Financial Benefits: The Growing Interest in Trust Funds*

Royalties will not accrue to beneficiaries for a number of years after original collections. At a minimum, the time taken for drug discovery and development – estimated now at 15 years – is required for a natural product sample to become a commercial product that generates royalties. Many compounds sit 'on the shelf' or in libraries for many more than 15 years. It is important, therefore, that provisions be made at a local level for distribution of financial benefits over time and according to agreed-upon objectives. To this end, a number of companies and research institutions are investigating the potential of trust funds (Laird, in press). A number of the ICBG programmes – including those in Nigeria, Suriname and Panama – have employed trust funds as a tool for distributing financial benefits in the future (Iwu and Laird, 1998; Guérin-McManus et al, 1998). The Healing Forest Conservancy has developed a model trust fund Constitution, and has assisted the Bioresources Development and Conservation Program (BDCP) to launch the Fund for Integrated Rural Development and Traditional Medicine in Nigeria. In this way, the Healing Forest Conservancy was working to prepare a structure for the distribution of financial benefits resulting from any product developed by Shaman Pharmaceuticals (Moran, 1998).

discussion here is intended to provide an idea of the scope and breadth of benefit-sharing, and is not exhaustive.

Most companies interviewed made use of a package of non-monetary benefits to complement financial returns in the short term. For example, in its agreement with the State of Sarawak Government, according to Murray Tait, Director of Natural Products Drug Discovery, AMRAD, it has included 'provisions for training of Sarawak scientists at ADT's facilities and technology transfer to Sarawak. Assistance with the development of drug discovery facilities in Sarawak is also anticipated in the agreement. All sample supply collaborators also receive information on the key results of research with their samples.'

In addition to monetary benefits such as royalties, Diversa's agreement with the National Autonomous University of Mexico (UNAM) includes the provision of training and equipment for UNAM scientists to enable them to collect and isolate environmental samples taken from various habitats in Mexico, and the establishment of a centre for microbial diversity studies at UNAM (the first of its kind in Mexico). Xavier Soberon, Director of the Institute of Biotechnology at UNAM said, '[Diversa] were ... interested in doing terrific science and we decided that the proper way to work an agreement would be to have technology transferred to the National University and royalties directed toward the preservation of protected areas' (*Business Wire*, 1998).

Participation in research

A growing number of companies are establishing joint ventures with source country institutions and commercial partners (Baker et al, 1997). Within a few high biodiversity countries with the requisite scientific capacity and infrastructure, joint ventures allow most of the discovery phase to be conducted in source countries. This type of partnership is likely to achieve far greater benefits for provider countries and institutions, as it can provide the foundation for a domestic industry based on natural products. This, in turn, may create incentives for conservation. By contrast, it is unlikely that the supply of raw material and extract samples will ever generate revenues of a magnitude likely to promote conservation. As Edgar Asebey, President of Andes Pharmaceuticals, said, 'The extractive approach of merely providing biological resources (eg raw material, crude extracts, etc) will not generate significant revenues for provider countries. Only when value is added do you start getting into interesting profit-margins that would justify the activity.'

An example of such a partnership, in which the company and the provider country government share in both the risks and higher rewards associated with value-added research, is the joint venture between Medichem Research and the State Government of Sarawak, described in Box 3.15, p70.

The NCI is establishing partnerships that shift a greater proportion of research to source countries. In those countries with both high biodiversity and scientific capacity – such as Brazil, China, Costa Rica, Malaysia, Mexico, Panama, the Republic of Korea, and South Africa (Boxes 3.15 and 3.16) – NCI aims to use a new approach that involves local collaborators collecting, extracting, and screening material with NCI assistance. NCI supplies the cell lines, and the in-country institution does the isolation work. If something interesting emerges, NCI uses its resources to help with development. However, because discovery took place within the source country, the in-country institution will hold the patent on compounds of interest. Companies will then negotiate directly with the in-country institutions, rather than with NCI (Cragg, pers. comm., 1997).

Information exchange

Most companies are involved in a form of information exchange, which may include the provision of research results arising from collaborations with in-country partners. Companies are cautious about providing any information which could breach confidentiality and thereby

Box 3.15 *The Joint Venture Between Medichem Research and the State Government of Sarawak*

In 1987, initial collections of *Calophyllum lanigerum* in Sarawak were carried out by botanists from the Sarawak State Forestry Department and the University of Illinois at Chicago (UIC). The NCI subsequently isolated the anti-HIV compounds (+)-Calanolide A and (-)-Calanolide B from the samples, and in 1994 signed a 'Letter of Collection' with the Sarawak State Government. The NCI worked on the Calanolide compounds with Medichem Research Inc, a pharmaceutical company, based in Illinois, USA, and in 1995 granted the company rights to all further development under a licence. The licence specified that Medichem Research was obliged to negotiate an agreement with the Sarawak Government, thus fulfilling the NCI's obligations under the 1994 Letter of Collection agreement. Medichem Research entered into a joint venture with the Sarawak Government in 1996, under which it provides technical expertise, research facilities and training opportunities for Sarawak scientists. The joint venture, called Sarawak Medichem Pharmaceuticals Incorporated (SMP), has a number of goals:

1 To bring (+)-Calanolide A (the primary candidate) to clinical development with the objective of obtaining FDA approval as an anti-HIV drug in the USA.
2 To develop (-)-Calanolide B (Costatolide) as a candidate for clinical development.
3 To facilitate the investigation of other drug candidates from the Malaysian rainforest (as part of Sarawak's approach to sustainable development).
4 To operate as a platform for the training of Malaysian scientists.

SMP has the right to file patents on all subsequent innovations arising out of this work. As with rights under the exclusive licence, Medichem Research and the Sarawak Government will own such patents jointly.

The State Government of Sarawak is sharing in the risks, as well as the rewards, of the joint venture. It is providing funding up to an agreed point in the clinical development of (+)-Calanolide A (ie completion of Phase I). After this stage in the research, additional funds will be raised by the two parties, through a variety of mechanisms, to permit the drug's passage through the next, and more expensive, phases of clinical trials. Currently, royalties that arise once the drug is marketed will be split 50:50, based on the contribution of chemical knowledge and expertise by Medichem Research and the contribution of investment by Sarawak. The sharing of benefits may change depending on how the current investment patterns develop over time.

A Malaysian PhD Chemist is treasurer of the joint venture and is based in the SMP offices in Illinois, where he is observing clinical trials and conducting preclinical studies and toxicological work on two back-up compounds. Two other Sarawak physicians are also to be assigned to participate in the clinical work.

Source: ten Kate and Wells, 1998.

endanger the prospect of obtaining patent protection, but provided these concerns are addressed, they are often prepared to make available 'user-friendly' (ie interpreted, rather than raw data) research results, and information in a form that could be used by provider country partners in their domestic research.

Among the kinds of information shared by companies are:

- scientific data on biodiversity, including botanical and ecological information on economic species important for management, contributions to country flora, and national inventories;
- research results (eg results of screens, results from standardisation studies) of wider use in the standardisation of traditional medicines and the building of domestic industries based on medicinal and other economic species;
- provision of scientific and technical literature available within national research institutions and universities; and
- draft handbooks and informational materials for applied and wider local use, developed from research results.

An example of the latter is a manual developed by Shaman Pharmaceuticals for the sourcing of Sangre de Drago in Latin America. Resulting from ecological and management studies funded by Shaman as part of its efforts to identify sources of raw materials for its leading drug candidate, this manual, *El Manejo Sostenible de Sangre de Drago o Sangre de Grado Material Educativo*, is intended as a practical guide for the propagation, cultivation, and sustainable management of Sangre de Drago (Baltazaretal, 1998). Shaman was also active in returning scientific data from their *in vivo* biological studies to countries and cultural groups with whom they collaborated (Richter and Carlson, 1998).

Technology transfer

Technology transfer can take many forms in biodiversity prospecting agreements. It can include the transfer of equipment for field and laboratory, and the provision of offices for use in the research process. Field equipment may include lights, nets, mobile refrigerators, cameras, and global positioning systems (GPS). Laboratory equipment might include milling machines, freeze-dryers, chemicals, microscopes, balances, containers, automated screens, and high-performance liquid chromatography (HPLCs). Inventory and collection equipment might include plant presses, mounting boxes, cabinets, and solvents. Computer software, such as compound and collection databases, and collection management tools might also be transferred.

Cell lines, proprietary screening technology, and know-how needed to set up and operate these

Box 3.16 *Institutions Collaborating with NCI on Natural Product Drug Discovery and Development*

- South American Office for Anticancer Drug Development, Porto Alegre, Brazil
- Universidad Paulista, São Paulo, Brazil
- Instituto Nacional de Biodiversidad (INBio), Costa Rica
- Instituto de Quimica, Universidad Nacional Autonoma de Mexico
- Facultad de Farmacia, Universidad de Panama
- University of Dhaka, Bangladesh
- HEJ Research Institute of Chemistry, University of Karachi, Pakistan
- Kunming Institute of Botany, Yunan, People's Republic of China
- Korea Research Institute of Chemical Technology, Taejeon, Republic of Korea
- National Instititute of Water and Atmospheric Research, New Zealand
- Division of Food Science and Technology, Council for Scientific and Industrial Research, South Africa
- Zimbabwe National Traditional Healers Association (ZINATHA), University of Zimbabwe, Harare
- University of Iceland
- Hong Kong University of Science and Technology

In addition, NCI has Letter of Collection agreements with organisations in Bangladesh, Ecuador (Awa Peoples Federation), Gabon, Ghana, Laos, Madagascar, Malaysia (Sarawak), Philippines, Tanzania, South Africa, and Vietnam.

Source: Cragg/NCI 1998.

technologies are also transferred. Recipients can improve their capacities in the development of high quality extract preparation and formatting; running and developing new screens; dereplication of hit extracts, isolation and identification of lead compounds; rational design and medicinal chemistry; and drug development. In the ICBG partnerships, for example, companies have donated used equipment to source country institutions, particularly machines that aid extraction, characterisation, and data management associated with the project (Rosenthal, 1998; Table 3.8, p64).

Many of the companies interviewed in this survey provide some form of technology transfer to source country collaborators. For example, Novo Nordisk provides 'support for the establishment and maintenance of a microbial culture collection, including introduction to isolation and preservation techniques, ... safety procedures for handling unknown microorganisms (mycotoxins, etc), sterile techniques in general, and fermentation technology'. Shaman Pharmaceuticals provided the Phytotherapy Research Laboratory, University of Nigeria, Nsukka with US$40,000 over the course of a few years to support the pharmacognosy laboratory, including chemical and pharmacological reagents and equipment; assistance with establishing biological assays for a variety of microbial pathogens; and a botanical herbarium and collection supplies (Carlson et al, 1997).

A licence to manufacture and sell a product in the source country, particularly a product developed from an endemic species, is another form of technology transfer which could generate significant monetary benefits for a source country. Many companies have voiced concern over a clause in the Philippines Executive Order on access to genetic resources (Chapter 10), which is broadly interpreted as supporting concessional compulsory licensing. Some companies, however, felt that if licensing were restricted to the source country and not – as may be inferred from the Philippine Executive Order – applied more widely, this could be an acceptable and reasonable form of benefit-sharing.

Training and building of source-country institutional capacity

Training in the context of research collaborations can be a significant benefit for the staff of source-country institutions. Almost all of the companies interviewed engaged in some form of training, if only in techniques of immediate relevance to the agreed collection and study of natural products. Dwight Baker of MDS Panlabs, for example, might provide to collaborators in Pakistan and Australia 'training in relation to isolation protocols, medium preparation, and general laboratory training ... it depends upon the country, but the priority should not be money in the future but building capacity now'.

Access and benefit-sharing arrangements have supported training and capacity building in a broad range of disciplines:

- *scientific:* collecting techniques and preparation of specimens; sustainability of sourcing; *ex situ* collection management; extraction; fermentation; screening technology; systematics; biochemistry; molecular biology and microbiology; ecology; ethnobotany; micro-propagation; plant breeding;
- *resource management:* applied ethnobotany training courses; resource management related to disciplines involved in collection, inventories of useful species, and ecological studies; and
- *information management:* biodiversity inventories; developing GIS systems; inventories of extraction procedures and samples; tracking and monitoring transfers and use of ethnobotanical information; herbarium management.

The most common form of training is in scientific disciplines involved in the development of a pharmaceutical. For example, three Nigerian scientists visited Shaman Pharmaceuticals' laboratories as part of the company's agreement with institutions in that country. One

Nigerian natural products chemist worked in the Shaman laboratories from January 1994 to March 1995 to learn modern analytical techniques (Carlson et al, 1997). However, as well as these scientific skills, benefit-sharing arrangements can be structured to build the legal and business skills needed to support domestic natural products industries. For example, one element of the Panama ICBG is the development of contractual models for other biodiversity prospecting programmes. Staff on this project believe that assisting the Government of Panama in the design of equitable biodiversity prospecting partnerships will help build a basis for effective national access and benefit-sharing measures (T Capson, personal communication, 1999). Experience of commercial partnerships can help to build local know-how on the terms and nature of business partnerships, and the content and negotiation of agreements.

Employment and sourcing of raw materials

Biodiversity prospecting collaborations can result in the employment of staff for field collections, processing of samples, laboratory research, and the production of raw materials for the manufacture of products. A number of large companies employ collaborating researchers in laboratories in source countries, and field collections often require several staff. For example, the Suriname ICBG project employed and trained ten Surinamese botanists in collecting, vouchering, and drying of the plant samples (Guérin-McManus, 1998) .

Increasing amounts of raw material are required as a compound moves through discovery and development. Re-collection of material is required for preclinical studies, and much larger quantities are needed for clinical trials. For example, preclinical and early clinical studies on Taxol required 2,000–15,000 pounds (909–6,818 kg) of Pacific yew bark (yielding a total of approximately 1.3 kg of the drug). As clinical trials advanced, however, annual requirements leapt to 60,000 pounds (27,273 kg) of bark per year, creating a crisis in supply (Cragg et al, 1993), as will be seen in the Taxol Case Study (p73). A number of companies interviewed have returned to source countries for additional raw material for research and development purposes. For example, in the last two years, LH Huang, Principal Research Investigator at Pfizer, reported that the company has returned twice to a source to acquire more material for development research.

Roughly half of new products can be wholly synthesised, removing any need for the sourcing of raw materials. However, of the 76 compounds from higher plants present in US prescription drugs, only 7 per cent are commercially produced entirely synthetically (Farnsworth, 1988). Some drugs, such as the alkaloids from *Catharanthus roseus* are still produced by isolation from raw materials, in this case grown in plantations in Texas, Spain, and other areas. Others are semi-synthetically produced from natural precursors, such as etoposide and teniposide, isolated from *Podophyllum emodii* and harvested in northern India and Pakistan (Sheldon et al, 1996).

The resupply of raw materials for research purposes and subsequent industrial-scale collection or cultivation of plants for the manufacture of commercial products can yield significant benefits for local communities and source country companies, primarily by providing an alternative source of income, and building domestic capacity and industries based on local biodiversity. For example, as far as income is concerned, in the early 1990s, pharmaceutical organisations in India quoted to the NCI prices ranging from US$20,000 to US$38,000 per kilo of Camptothecin (at 92 per cent purity). When extraction technology improved soon after, prices rose to US$85,000 per kilo (for 98 per cent purity) (ten Kate and Wells, 1998). However, this form of benefit-sharing must be carefully assessed in light of threats that this demand poses to wild species in the form of over-harvesting, and the effects of dependence on commercial production on local communities (Box 3.17).

The resupply of materials for clinical trials requires surveys of distribution and abundance, determination of drug content in various plant parts, and studies of the fluctuation of content with the time and season. However, large-scale cultivation, microbial fermentation or aquaculture of the source organism is likely. Such activities do not necessarily result in benefit-sharing strategies that contribute to local conservation and sustainable development efforts over the long-term (Laird, 1995; Cragg, 1997).

In the case of *Ancistrocladus korupensis* in Cameroon, the NCI invested heavily in cultivation options for raw materials as a source of the active compound michellamine B. Following the supply problems associated with *Taxus brevifolia*, NCI sought to meet potential demand for large quantities of raw material in a relatively short time frame. This process did result in benefits in the form of some employment, training, funding of scientific research, and limited infrastructure and equipment. However, the more significant benefits anticipated through the widespread cultivation of *A korupensis* in agroforestry, plantation, and other systems have not materialised, due to a stall in research resulting from toxicity of michellamine B (Laird and Lisinge, 1998; Laird et al, in press; Cragg et al, in press).

Similarly warned by the case of Taxol, Shaman Pharmaceuticals invested for a number of years in long-term sustainable harvesting and management of the SP303 source species, Sangre de Drago (*Croton lechleri*) in Latin America. This has involved ecological and distribution studies at sites from Mexico to Paraguay, as well as examining the yield and sustainability of traditional methods of harvesting (slashing bark to release latex from the standing tree) (King et al, 1997; Ubillas et al, 1994; Baltazar et al, 1998). Shaman's investment in developing sustainable harvest and management systems for Sangre de Drago came to US$1 million (Meza et al, 1998; Meza, 1999; King et al, 1998).

Sourcing issues for marine organisms can pose significant problems, since large-scale production can be difficult to achieve. Until recently, aquaculture of marine organisms for the production of potential drugs had not been investigated. There are now new efforts off the coast of California to address sourcing issues with the colonizing bryzoan, *Bugula neritina*, and to investigate the sourcing of several sponges, including a *Lissondendoryx* species off the coast of South Island, New Zealand (Cragg et al, 1997b).

Microorganisms do not raise supply issues of anything like the same magnitude. Microorganisms can generally be produced in bulk cultures, and fermentation allows the manufacture of microbial-derived intermediate compounds. For example, Azythromax can be synthesised from the precursor erythromycin, or companies can produce it in bulk from their own proprietary microorganisms.

Box 3.17 *Gathering Jaborandi Leaves: A Case of Benefit-Sharing?*

Pilocarpine is derived from the shrub *Pilocarpus jaborandi* found in north-east Brazil. This case illuminates some of the difficulties associated with developing sustainable supplies of raw materials, and in linking this to sustainable and equitable development. Pilocarpine is used in ophthalmology to reduce intra-ocular pressure, and is the treatment of choice for several forms of glaucoma. It was also granted approval by the FDA, under the trade name Salagen, for use in the treatment of radiation-induced xerostomia (dry mouth). It is the first pharmacologic treatment for dry mouth, and was developed as an orphan drug. Salagen is marketed by the company MGI Pharma, headquartered in Minnesota (MedSc Bull, 1994). This recent pharmaceutical application for the treatment of dry mouth parallels traditional uses (recorded by Europeans more than a hundred years ago) of this plant by the Tupi in Brazil. The Tupi name 'Jaborandi' literally means 'what causes salivation or produces saliva' or 'what causes slobbering' (Holmstedt et al, 1979).

The manufacture of pilocarpine from the leaves continues to be cheaper than synthesis, and for more than 20 years jaborandi leaves have been collected from the wild by an estimated 25,000 people in the north-east of Brazil (Pinheiro, 1997). These collectors are primarily local communities and indigenous people. Although leaves are generally considered a more sustainable plant part to harvest than roots or bark, populations of jaborandi appear to be under threat from excessive harvesting of wild populations (Pinheiro, 1997). The German company E Merck and Co, through its subsidiary Vegetex-Extratos Vegetais do Brasil Ltda, developed a system for 'sustainable' collection of leaves in the wild. At the same time, the company spent more than a decade trying to produce jaborandi with the required alkaloid content, and has large plantations established in the state of Marnahao.

On the one hand, conditions of collection of wild jaborandi leaves appear harsh and culturally-inappropriate, if not destructive. Shelton Davis (1993) reports that the people he visited in the Guajajara reserve of Arariboia, one of the major centres of jaborandi leaf collection, 'had become totally dependent on commercial plant extraction to the detriment of other aspects of their local economy and the general social welfare and psychological well-being of their community'. Benefits promised to the community by Vegetex agents – steady income, roads, schools, clinics – never materialised (Davis, 1993). On the other hand, should a large portion of the 25,000 people involved in wildcrafting jaborandi be put out of work by plantations, they might lose an important, and by now relied upon, source of income (Pinheiro, 1997).

Taxol: raw material sourcing and benefit-sharing in the North – a case study

Taxol is the generic trade name for the compound paclitaxel, developed from the Pacific Yew tree, *Taxus brevifolia*. In 1997, taxol was the 30th top-selling drug in the world, with sales of US$941 million, up 15.6 per cent over 1996 sales (*MedAd News*, May 1998). These are unusually high sales figures for an anti-cancer drug. Taxol's development was a long and circuitous one, facilitated and funded by the US NCI, which spent close to US$40 million on its discovery and development. Because taxol is a highly complex compound, a synthetic option for production has not been economically feasible, and developing a sustainable supply has proven expensive and complicated.[2]

The original collections of *Taxus brevifolia* were made in 1962 in the Pacific Northwest of the USA. Collections were undertaken by researchers at the USDA for the NCI as part of an inter-agency agreement. At that time the USDA collected samples for the NCI primarily on government land. The sample was extracted and tested at the NCI, and showed some activity against an *in vivo* mouse leukemia model, but nothing more remarkable than other natural products in the programme at the time. Taxol was isolated in 1969 by the research group of Wall and Wani at Research Triangle Institute, but it sat on the shelf until 1975. At that time NCI tested it against new *in vivo* models for mouse melanoma and mammary tumours. It showed promising activity, and NCI started formulation, toxicology and other preclinical studies.

In 1979, Dr Susan Horwitz and researchers at the Albert Einstein School of Medicine discovered the unique mechanism of action of taxol in cancer cells. Clinical development accelerated, and in 1983 taxol went into clinical trials. A compound used in taxol's formulation – Cremophore EL – caused massive allergic reactions in patients, but taxol was considered the culprit. In 1984–5 taxol was all but dropped. A clinician at the Albert Einstein School of Medicine persisted in clinical studies, however, and developed an infusion of taxol that allowed patients to tolerate it. By 1987–8, clinical studies at Johns Hopkins University were demonstrating its value against ovarian cancer.

Up to 1985, between 5,000 and 15,000 pounds (2,273 and 6,818 kg) dry weight of yew bark had been collected in Oregon to supply preclinical and early clinical studies (yielding approximately 1.3 kg of the drug in total). But in 1988, with the results of the Johns Hopkins studies, demand rocketed for supply of the raw material. In 1987–8, 60,000 pounds (27,273 kg) were collected to supply Phase I clinical trials.

Initiation of a second 60,000-pound (27,273 kg) collection in 1989 raised concerns about the environmental impact of the harvest. *Taxus brevifolia* is an understorey tree with a native range from northern California into British Columbia. Previously treated as throw-away scrub that was burned following clear-cut logging, it was not clear to what extent the population could withstand massive, directed bark harvest. Collection of bark from *Taxus brevifolia* is extremely difficult, except when the sap is running in spring and summer, and the bark can readily be peeled away. A survey by the Forest Service and the Bureau of Land Management (BLM) in 1990, however, determined that the tree was abundant, with estimates of more than 130 million trees on 1.8 million acres of National Forest lands in Washington and Oregon.

In early 1991, Bristol-Myers Squibb (BMS) was awarded a CRADA (Cooperative Research and Development Agreement) for taxol's development. BMS was now in charge of bark collection, extraction, and isolation. Under contract to BMS, Hauser Northwest, a subsidiary of Hauser Chemical Research, collected 1.6 million pounds (727,273 kg) of bark in both 1991 and 1992, supervised by the Forest Service. Part of the NCI agreement with BMS included their funding of an environmental impact assessment conducted in conjunction with the Forest Service, Bureau of Land Management and USDA.

It was clear, however, that alternative sources of taxol must be developed. NCI funded surveys of *Taxus* species around the world, including in Canada, Mexico, Russia, the Ukraine, Georgia, and the Philippines. The content of paclitaxel and key baccatin precursors in various *Taxus* species and cultivars was determined. Needles, more sustainable to harvest than bark, showed promise in several species as a source of baccatin precursors to paclitaxel . French researchers led by Greene and Poitier pioneered a way to convert baccatins to taxol from the European yew species *Taxus baccata*. Eventually, BMS developed an agreement with the Italian firm Indena to supply baccatins from *Taxus baccata* sources, for conversion to paclitaxel using an improved method developed by Holton and co-workers in Florida.

Indena's initial forays into sourcing *Taxas baccata* needles brought them to India, where they purchased needles from local groups as part of a 'renewable resource' strategy. In fact, suppliers felled trees, selling the wood to one party, and the needles to Indena. Indena then transferred their sourcing to Europe, and currently acquires its raw material from local sources.

Taxol is marketed by BMS, its patent due to expire in 2004. The French company Rhône-Poulenc is marketing Taxotere, a related compound used for the same indication, and also developed from baccatin precursors.

The discovery and development of taxol largely pre-dated the CBD and considerations of benefit-sharing. Benefits generated by the sourcing of the species appear limited. While BMS invested hundreds of millions of dollars in developing a sustainable supply, an argument cannot be convincingly made that these benefits contributed directly to local economic or conservation priorities. Original collectors of yew bark in the Pacific Northwest of the US were mainly local people, for whom bark harvest provided a good, if brief, source of income. But a company seeks the most cost-efficient and effective way to source raw materials. This case demonstrates the inherent fickleness of sourcing relationships – if a cheaper, more reliable, or more effective alternative source emerges, companies naturally take advantage of it.

But perhaps most strikingly, this case demonstrates the marked difference between most northern countries' approach to access and benefit-sharing, and that outlined in the CBD. Taxol is primarily the product of US government and research institutions, and *Taxus brevifolia* is an endemic species. While BMS is a US company, Rhône-Poulenc is also marketing a product based on the lead provided by *T brevifolia* – albeit based in part on French researchers' discoveries, and sourced from the European species, *T baccata*. Sourcing of raw materials for the manufacture of taxol shifted first to Europe (Indena) and India – away from US supply companies, and US raw material sources – and then entirely to Europe.

The American public did respond strongly to the case of Taxol (for example, articles in *Newsweek*, 1991; *Time*, 1993; *US News and World Report*, 1993). But the issues they emphasised were: 1) the security of the species (prompting Congressional hearings and a great deal of press coverage); and 2) the need to access large quantities of raw material – regardless of species conservation – to help develop a treatment for cancer and save lives. At times this meant confrontations between environmentalists and cancer patients. Although the USA clearly benefits in the form of a powerful new anti-cancer drug, and BMS, an American company, is benefiting handsomely from taxol sales, 'benefit-sharing' as provided for under the CBD – for the region, for the USA, as an incentive for wider biodiversity conservation – has not featured prominently, if at all, in the public discourse.

It is true that the National Park Service and the Forest Service in the USA are showing increased interest in biodiversity prospecting agreements as a way to generate revenue, largely in response to the Diversa-Yellowstone agreement, but also in part to the Taxol case. However, this interest emphasises relationships in which access is linked to benefit-sharing. In the Taxol case, the bond between access and benefit-sharing was severed, since collecting activities pre-dated both the CBD and commercialisation by decades, and it would appear that wider issues relating to 'equitable benefit-sharing' did not catch the public imagination or find widespread recognition even among environmental groups.

3.7.3 Public benefits

In addition to the direct benefits that accrue to source country partners such as those described above, access agreements can result in wider societal, or 'public', benefits for source countries. These include the opportunity for national programmes to study tropical diseases, build collections and inventories of national biodiversity, and broader conservation benefits.

Research on tropical diseases

Pharmaceutical companies invest in disease categories that afflict the affluent. As *The Economist* (1998) recently said, 'everyone with a half-decent research laboratory seems to be looking for treatments for osteoporosis, Alzheimer's disease and Parkinson's disease'. According to Mark Mozenson, a partner with Deloitte & Touche Consulting Group, 'There must be a strong population base at an appropriate income level to afford brand-name prescription drugs'. Until this has occurred (as is likely to happen in China in the next 5 to 20 years), 'the country will continue to be a generic marketplace' (Popovich, 1997).

Since 1975, among the 1,219 newly-developed drugs that entered the world market, only 11 focused on tropical diseases such as malaria, river blindness or worm infections (Dumoulin, 1998). WHO estimates suggest that the proportion of total global research and development devoted to disease conditions that primarily afflict the developing world is a mere 4 per cent (Chetley, 1990). Some of the most effective medicines used to treat tropical diseases were developed by the public sector (eg the US Army), rather than by pharmaceutical companies, or were initially developed for other purposes. Ivermectin (developed from a microorganism found on a Japanese golf course) is used to treat river blindness (onchoceriasis), a parasitic disease that afflicts millions of people in Africa, the Middle East and parts of Central and South America. Ivermectin was commercially developed, however, for use in the more profitable veterinary medicine business to treat lifestock (Chetley, 1990).

It is unlikely that the commercial sector will invest in tropical disease research in the near future, so partnerships that enhance domestic capacity to research tropical diseases can be a major benefit. A central objective of the African ICBG, for example, is to include tropical diseases such as malaria, leishmaniasis and trypanosomiasis as target therapeutic categories. In this way, the ICBG intends to apply the scientific expertise of the pharmaceutical company partners to under-researched diseases afflicting the poor in developing countries. The benefit-sharing package is designed to increase domestic capacity for research into tropical diseases. The researchers involved feel that most such partnerships with industry should be structured to accomplish this objective (*C&EN*, 1997; Iwu, 1994; Iwu and Laird, 1998).The Panama ICBG has also identified research on tropical diseases as a priority, and the Suriname ICBG is assisting the Malaria Task Force of the Suriname Ministry of Health (McManus et al, 1998; Capson, 1999).

Conservation

Benefits involving the conservation of biodiversity are diverse, and often indirectly related to biodiversity prospecting partnerships. They include:

- building scientific capacity in areas relating to biodiversity management, including national inventories, and ethnobotanical, taxonomic and ecological studies associated with economic species;
- building information-management capacity for biodiversity, including databases, software, herbaria and other *ex situ* facilities;
- infrastructure support for national parks, for example through funds for national parks, as in the case of the INBio-Merck agreement;
- development of domestic industries which are dependent upon biodiversity, thereby creating local economic incentives for conservation;
- creation of policies protecting biodiversity, encouraged by studies that have calculated option values of potential commercial products and partnerships; and
- generation of wealth for domestic economies, thereby alleviating some of the economic pressure on biologically diverse ecosystems.

Despite the fact that 'bioprospecting' is much heralded as a means to create economic incentives to conserve biological diversity, comparatively few access laws and fewer benefit-sharing partnerships explicitly contain clauses requiring the dedication of a share of monetary or other benefits to conservation. Exceptions include those in Costa Rica and Western Australia, as well as the ICBGS (Box 3.18).

In Costa Rica, the work of INBio is based on a cooperative research agreement with the Ministry of Environment and Energy (MINAE), which specifically sets the terms and conditions for INBio's biodiversity inventory and bioprospecting activities. According to this agreement, INBio will donate 10 per cent of all bioprospecting budgets and 50 per cent of all income from royalties to the MINAE, for conservation purposes. To date, INBio's bioprospecting agreements have contributed over US$390,000 to MINAE, US$710,000 to conservation areas, US$710,000 to public universities, and US$740,000 to other groups at INBio, particularly the Inventory Programme (8.4.1, INBio/Diversa case study). In Western Australia, the agreements between the Department of Conservation and Land Management (CALM), the NCI and AMRAD resulted in the investment of US$380,000 in conservation projects in Western Australia.

3.8 Conclusions

For the last 50 years, interest in natural products within the pharmaceutical industry has been cyclical. This has been largely the result of economic and scientific factors, and reflects the constant flux in natural products research and development. Ten years ago, natural products research and development was on the up-swing, but over the last five years, its relative importance as a method of drug discovery and development has been thrown into question by the emergence of combinatorial chemistry, genomics and biopharmaceuticals. As a result, it is hard to predict the prospects for natural products discovery among the different approaches that compete for research dollars.

However, despite all mankind's ingenuity, technology cannot yet rival the diversity and novelty of the compounds occurring in nature. Natural products are unlikely to become obsolete for the foreseeable future, and are set to contribute a significant proportion of the new drugs that will be added to the pharmacopoeia in the coming decades. Companies continue to see value in natural products and will continue to seek new samples

Box 3.18 *The Western Australian Smokebush: Benefit-Sharing Arrangements between NCI, CALM and AMRAD*

The Department of Conservation and Land Management (CALM) in Western Australia (WA) is responsible for granting permits to collect and conduct research on WA flora. In 1981, a US Department of Agriculture botanist collected around 1,200 plant specimens in WA. These were processed by the WA Herbarium and sent to the US NIH for screening. In the late 1980s, a species of smokebush (genus Conospermum) from WA showed promising activity in the NCI's anti-cancer screens. The NCI obtained a patent on conocurvone, the active compound of the smokebush. At the same time, research by scientists in WA also revealed the potential anti-HIV activity of the smokebush.

CALM wished to ensure that WA would receive the maximum benefit from the use of its own biological resources. Not content with simply receiving royalties on production of any drug from the smokebush plant, CALM was determined that the development and production of any potential drug should be based and coordinated in WA. CALM entered into negotiations with AMRAD, an Australian pharmaceutical company. Drawing up the agreement required an amendment to the Conservation and Land Management Act, to allow CALM to grant exclusive rights to one company for the commercial development of a product derived from WA flora. CALM agreed to grant AMRAD access to the smokebush and permission to develop it commercially. In return, AMRAD agreed to provide US$730,000 to CALM, a share in royalties, and the right of first refusal for CALM to conduct any research on the active compound. In addition, AMRAD provided US$320,000 for further research by a consortium of 26 WA scientists, in collaboration with the NCI, on some eight smokebush patents lodged by CALM. The research would explore the chemical structure of conocurvone and the synthesis and development of analogues. The CALM-AMRAD agreement supported conservation and benefit-sharing.

Conservation: under the agreement CALM was responsible for ensuring the sustainable collection of the raw material. CALM used the funds received from the AMRAD agreement to support the conservation infrastructure in WA as follows:

- US$380,000 from the AMRAD agreement was put directly into conservation projects in WA:
  - US$190,000 for the conservation of rare and endangered WA flora and fauna;
  - US$190,000 for other conservation activities, including information technology such as geographical information systems, data capture and management to study population dynamics, etc.

Benefit-sharing: *Benefit-sharing* took two forms: *research funding* covering joint research and technology acquisition, and a share in royalties:

- *Joint research*: The consortium of WA scientists (ecologists, geographers, botanists, chemists, pharmacologists and immunologists) received Aus$150,000 to cover research conducted prior to the agreement that had led to several WA patents on conocurvone. Over the year following the agreement, the consortium received an additional Aus$500,000 for its further research.
- *Technology acquisition*: government and university laboratories of consortium members were equipped with technology such as HPLC machines.
- *WABEL*: The remaining funds were used by CALM to establish the WA Biotic Extract Library (WABEL), a library of biotic extracts for drug discovery.

Source: ten Kate, in OECD (1997).

from field collections, but demand for natural products is always balanced against alternative sources of material such as synthetic and combinatorial chemistry, and the vast *ex situ* libraries and collections already available to screening programmes.

Attitudes to natural product discovery and development are changing. The number, nature, quality, variety and source of samples that companies seek, and the kinds of collaboration which they are looking for, differ from those of just a few years ago. If source countries wish to benefit from partnerships involving the discovery and development of natural product pharmaceuticals, they will need to follow these economic and scientific developments closely. Only through a good understanding of industry and the market, on the part of source country institutions, and of the aspirations of the latter, on the part of companies, will it be possible for all concerned to arrive at mutually beneficial partnerships. Familiarity with the commercial use of natural products will also be essential for regulators to strike the right balance between facilitating access to genetic resources, and controlling it in such a way as to encourage – rather than discourage – such partnerships.

Despite a feeling on the part of both source countries and the private sector that the other has unreasonable expectations, there has been good progress in the development of more equitable relationships involving the acquisition of natural products. A number of innovative partnerships and approaches have emerged over the last five years that offer insight into how relationships can be designed to balance commercial and source country needs.

Notes

1 The drug discovery unit of MDS Panlabs was acquired by Thetagen in February 1999. The natural product discovery services provided to the pharmaceutical industry by MDS Panlabs, as described in this document, are now provided from the discovery division of Thetagen.

2 *Sources for the history of taxol' s development:* Cragg, personal communication, 1999; Cragg et al, in press; Cragg et al, 1993; Sheldon et al, 1997.

Sarah A Laird

Chapter 4
The Botanical Medicine Industry

4.1 Introduction

The botanical medicine industry is experiencing rapid growth world-wide. Annual growth rates are between 10 and 20 per cent in most countries, with the highest rates in relatively immature markets like that of the USA. One feature of this growth in both Europe and the USA is the sale of botanical medicines in a wider range of outlets, including those of the mass market like drugstores and supermarkets. At the same time, the industry is experiencing consolidation of distribution channels, and vertical integration, as larger companies acquire smaller companies along the production and marketing chain.

Botanical medicines, as distinct from pharmaceuticals, are produced directly from whole plant material. As a result, they contain a large number of constituents and active ingredients working in conjunction with each other, rather than a single, isolated active compound. Because the drug approval process and patenting systems do not provide incentives for companies to conduct (expensive and time-consuming) research on the synergistic and collective function of active ingredients in whole plants, or plant formulas, botanical medicines are often scientifically poorly understood (Steinhoff, 1998). However, most botanical medicines have long histories of traditional use, which confirm safety and efficacy, and as documented are used in many regulatory systems to guide approval of commercial products.

The global botanical medicine industry is marked by diversity. National industries reflect the scientific, economic, and cultural roots from which they grow. In the 1970s in the USA, for example, the industry was made up of 'a handful of dedicated individuals [who] began selling herbs to each other and a limited consumer following' (Brevoort, 1998). This trade was based only in part on species and knowledge of people indigenous to the USA. In contrast, European and Asian markets reflect long historical traditions of medicinal plant use. Botanical medicine markets in China and India today are current expressions of formal medical systems developed over thousands of years.

Diversity within the industry is also apparent in the structure and nature of participating companies. As reflected in the interviews conducted as part of this study, company size and function vary widely, with some companies employing only a handful of staff, and others a few thousand. Companies might cultivate raw plant material; process material into bulk ingredients, including standardised extracts; manufacture and market finished products, or broker the exchange of raw materials or products. Company philosophies and marketing strategies vary widely too. Some companies emphasise a standardised, scientifically proven effective and safe product; others are primarily in the packaging and marketing business, placing little emphasis on proven product efficacy (and sometimes quality); still others incorporate environmental and social concerns into their business practices. Although a trend exists towards uniformity in the global botanical medicines market, as a result of increased emphasis on quality control, safety, and efficacy, diversity in this complex, heterogeneous industry is likely to remain marked.

One result of the growing emphasis on quality control, safety and efficacy is closer relationships between processing and manufacturing companies and the sources of their raw material. Increasingly, companies seek high quality, reliable supplies of cultivated material, although wildcrafted (wild harvested) material continues to play a significant role in the industry. The impact on supplies of raw material of markets growing at 10–20 per cent per year is significant: species are over-harvested in the wild (eg pygeum, devil's claw, and golden seal)[1]; cultivated sources developed for regional or subsistence use cannot supply international demand (eg kava); and immature, low-quality, improperly identified, or adulterated raw materials make their way into the trade (eg pao d'arco).

A large proportion of products on national markets tend to be native to or naturalised in the region in which they are sold. Demand for access to species 'new' to the market is limited, but likely to increase in coming years due to a variety of marketing, scientific, and regulatory factors, which will be discussed below. To date, benefit-sharing for high biodiversity countries is generally associated with the sourcing of raw materials, but it is limited. Raw plant material is traded in bulk as a commodity and investments in supplies are small compared with company marketing and other budgets. However, the potential for capacity-building and technology transfer resulting from commercial raw material sourcing and research and development partnerships is significant, if poorly developed to date.

4.2 The botanical medicine industry

Botanical medicines are part of larger markets, referred to in the USA, for example, as the 'dietary supplement' market. Dietary supplements encompass vitamins, minerals, herbs/botanicals, and other natural medicines. Often joined in analysis with dietary supplements are nutraceuticals, or foods with health-enhancing properties. This chapter will focus primarily on herbal or botanical medicines, sold in a variety of forms and reflecting differing levels of processing. These include capsules, tablets, herbal teas, extracts, tinctures, and dried and fresh herbs sold in bulk. Some of these products may be standardised active ingredients in a herbal base, known as phytomedicines, while others are sold without specification as to the amount of 'active' chemical compound or chemical group. This chapter will touch briefly on nutraceuticals, which are a rapidly growing market, with demand for access to new natural products.

4.2.1 Global markets

Market data on the international botanical medicine market are difficult to analyse due to the complexity of the industry, overall lack of vertical integration, and the large number of small and medium-sized companies that often do not share sales data. Additionally, there exist a plethora of terms used to describe and group various sectors of the industry, which make comparisons across regions and studies difficult (eg nutraceutical, dietary supplement, vitamin, phytomedicine, raw botanical material, bulk ingredients, health food, and natural food).

The world trade in raw materials for botanical medicines, vitamins, and minerals in 1997 was estimated at US$8 billion; global consumer sales were an estimated US$40 billion (*NBJ*, June 1998). The largest global markets for botanical medicines are found in Germany, China, Japan, the USA, France, Italy, the UK, and Spain. Germany is the world's largest consumer of commercial botanical medicine products (this does not include subsistence or non-marketed use of medicinal plants), with 1996 sales of US$3.6 billion representing 26 per cent of the market. Together, Asian and European markets combine to form 86 per cent of the world market (Gruenwald, 1997; Zhan, 1997; Yuquan, 1998).

In 1996, combined markets for Germany (US$3.6 billion) and France (US$1.8 billion) accounted for 75 per cent of the European Union market (Gruenwald and Buettel, 1996). In 1994, Italy was the third largest market in Europe (11 per cent of the global market), followed by the UK (5.5 per cent), Spain (4.5 per cent), The Netherlands (2 per cent), and Belgium (1 per cent) (IMS, 1994; Walluf-Blume, personal communication to Lange, 1998). By 1996, the UK had grown to 13 per cent, Italy was at 9 per cent, and Spain at 8 per cent of world markets (Kalorama, 1997). Although consumption of 90 per cent of branded botanical products in Europe is accounted for by the EU, demand is increasing in Eastern Europe as well (Gruenwald, 1997).

According to the American Herbal Products Association, 1982 retail sales in the USA botanical medicine industry were around a half billion dollars, a figure that grew to US$1 billion by 1991. By 1995, consumer sales of botanical medicines were up to US$2.5 billion (*NBJ*, October 1996). The following year the market had reached US$3.1 billion; by 1997 it had grown to US$3.6 billion (*NBJ*, June 1998); and 1998 retail sales of botanical medicines were an estimated US$3.87 billion (Brevoort, 1998).

In broader US markets, of which botanical medicines form a part, the following figures emerge. The 1994 'natural foods industry' (which includes health food store groceries, natural cosmetics, and all dietary supplements) was estimated at US$7.5 billion in retail sales

(Brevoort, 1996). In 1995, Americans spent US$8.9 billion on 'dietary supplements', of which botanical medicine sales were US$2.5 billion; 'natural foods and beverages' accounted for US$6.2 billion in sales; and 'natural personal care products' were US$2 billion. Total 'nutritional products sales' encompassing these and other categories were US$17.2 billion (*NBJ*, October 1996).

Botanical medicine markets in Japan in 1996 were estimated at US$2.4 billion (Gruenwald, 1997). Combined sales in 1996 for botanical, vitamin, diet, and energy concoctions in Japan reached US$5.2 billion (Zhan, 1997). Japan has the highest per capita consumption of botanical medicines in the world. Botanical medicine sales have grown rapidly in recent years in part because physicians increasingly incorporate Traditional Chinese Medicine (TCM) as a complement to western medicine. In 1983, 28 per cent of physicians used TCM, but by 1989 this figure had risen to 69 per cent (Steinhoff, 1998).

The total 1996 output in China of the finished TCM sector was US$3.7 billion. Thirteen of the top 50 TCM-producing companies are listed publicly on the Chinese domestic stock exchange; 14 TCM companies are state-owned, and the government intends to form a multi-national TCM company with sales of more than US$1 billion by the year 2010 (Yuquan, 1998). These figures represent production from large commercial companies; estimates including all domestic consumption of botanical medicines would be far higher. One estimate for 1995 sales of botanical medicine products within China is US$5 billion (Iwu, 1996).

Around the world, demand for botanical medicines has increased dramatically during the last ten years. In the USA, annual growth rates over the last few years have averaged between 15 and 18 per cent (Richman and Witkowski, 1998; *NBJ*, 1997; Brevoort, 1996). The European market is growing more slowly, due to its relative maturity. Average annual growth rates in Europe between 1985 and 1995 were 10 per cent, but are expected to slow to 5–10 per cent over the next few years. Japan's is a mature market, but growth has averaged 15 per cent in recent years, reflecting the expanded use of TCM (Gruenwald and Buettel, 1996; Gruenwald, 1997; Steinhoff, 1998) (Table 4.1).

Table 4.1 *Historic and Projected Growth for Botanical Medicine Markets*

Country	1985–90	1990–95	1995–9
Germany	8	6	7
France	6	8	9
UK	15	14	16
Italy	15	11	13
Denmark	8	10	14
Spain	15	9	11
Netherlands	8	13	16
Belgium	15	8	10
Portugal	15	9	11
Greece	5	12	15
Ireland	12	10	9
Luxembourg	5	6	8
India	20	25	15
Pakistan	10	8	12
Japan	15	15	10
Taiwan	12	15	15
Korea	15	13	15
Singapore	12	9	10
Malaysia	10	8	10
Indonesia	10	8	10
Philippines	0	0	0
Thailand	8	5	10

Sources: McAlpine, Thorpe, and Warrier Ltd, November 1996.

4.2.2 Botanical medicine products

Botanical medicines represent a range of product types. These include products sold as the **raw herb – dried or fresh** – and others that are processed to varying degrees, including: **tinctures** (an infusion of herbs in alcohol); **extracts** (greater concentration of the original material produced through separation of the active material from the plant with the aid of a solvent); and **standardised extracts** (plant material 'standardised' to one or more chemical 'markers') (Brevoort, 1998). Products made with the whole herb or a 'total extract' utilise the entire range of compounds available in a plant (or plant part), or combination of plants. Total extracts can be expressed in conventional strengths (quantity of plant material per quantity of extract) or can be standardised against chemical markers or biological activities. Products based on whole plants or total extracts represent the vast majority of botanical medicine products sold today (Leung, May 1997).

Increasing interest is currently paid to **phytomedicines** – products which are standardised to only a few groups of active marker compounds, and sometimes eliminate other compounds found in the original botanical material (Brevoort, 1998). These products are based primarily on modern scientific findings, such as activity found in a chemical or chemical group present in the herb, or known chemicals in the herb that possess specific pharmacological effects. Phytomedicines fit easily into Western concepts of medicine because their effects have been 'proven' through modern research, their active components can be chemically identified and controlled, and they can be easily manufactured. As extraction technology advances, separation of active botanical chemicals, to be used in their

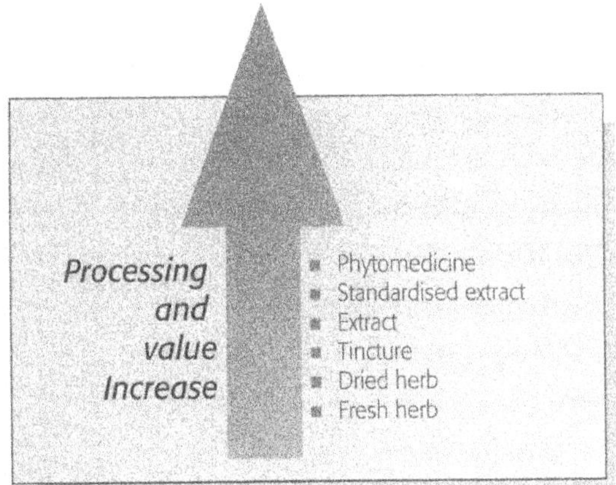

Figure 4.1 *Botanical Medicine Products, in Order of Degree of Processing and Value Increase*

Source: Simone and Associates, October, 1997, as in Brevoort, 1998.

pure forms – one step further along the processing chain – is likely to become economically feasible and more common (Leung, May 1997). (Figure 4.1)

Delivery formats for botanical medicine products vary widely. In the USA, for example, capsules and tablets are the most common mode of delivery (68.4 per cent), followed by teas (11.3 per cent), extracts (6.9 per cent), tinctures (6.3 per cent), and bulk herbs (5.4 per cent) (Richman and Witkowski, 1996). Capsules and tablets include: compressed tablets (uncoated), coated tablets, time-release tablets and capsules, two-piece hard gelatin capsules, soft gel capsules, and vegetable-based capsules. Other product forms include bulk dried whole herbs, cut herbs, powdered herbs, liquid extracts, tinctures, oils, syrups, sprays, inhalants, and gums. Herbal teas come in loose bulk, tea bags and extracts (Kalorama, 1997). The development of new delivery systems is ongoing, with recent introductions including a St John's wort transdermal patch, yam cream for oestrogen therapy, and sustained-release formulae (Brevoort, 1998).

In the USA there currently exists a trend towards vitamin/mineral/herb combinations, as well as herb/phytonutrient combinations. Multi-herb botanical medicine products accounted for 27 per cent of sales in the USA in 1997 (*NBJ*, September 1998). In Europe, the trend is in the other direction, with mono-preparations (single herb products) assuming a greater share of the market. In most European countries, with the exception of a few including The Netherlands and the UK, multi-herb combinations are no longer permitted (Gruenwald and Goldberg, 1997). Of the 130 botanical medicine products launched in Germany in 1997, 126 were mono-preparations, and only four combination preparations (BAH, 1998). Traditional plant use in Europe usually involved minimal processing, and simple delivery as infusions or tinctures. However, TCM and Ayurvedic botanical medicines generally involve complicated harvesting, processing, and formulation techniques, and combination formulae are the rule rather than the exception (Leung, February 1997).

Top-selling species

Top-selling species used in commercial botanical medicine products vary by country and region. The bulk of the Japanese and Chinese markets, for example, are based on TCM, with its own suite of species. European and North American markets tend to follow similar species, but regional variation in product preference continues, in part due to cultural variables, including concepts of health and disease, but also due to knowledge and availability of raw materials. Many products popular in the USA – such as golden seal and previously

Box 4.1 *Claimed Activity of Selected Top-Selling Botanical Medicines*

Product	Plant part	Activity
Ginseng	root	increases energy and sex drive
Siberian ginseng	root	defuses nervous tension and fights fatigue
Kava	root	combats anxiety and stress
Green tea	leaves	a powerful anti-oxidant and cholesterol-reducer
Milk thistle	fruit	protects from toxins – eg alcohol, pesticides
St John's wort	herb	anti-depressant
Psyllium	seeds	anti-constipation; helps with weight loss
Hawthorn	fruit	lowers blood pressure and fights arthritis
Saw palmetto	seeds	treats prostate problems
Valerian	root	relieves insomnia, anxiety, menstrual cramps, headaches
Liquorice	root	treats ulcers and stomach disorders
Wild yam	roots	alleviates PMS and menopausal symptoms
Aloe	leaves	treats wounds and skin problems
Camomile	flowers	alleviates moods and skin problems; calming
Feverfew	leaves	relieves migraine headaches
Bilberry	leaves	improves eyesight
Cranberry	fruit	keeps the urinary tract healthy
Garlic	bulb	boosts the immune system; lowers cholesterol
Calendula	flowers	soothes skin; fights bacterial, viral and fungal infections
Echinacea	roots, flowers	boosts immune system; prevents colds
Ginger	rhizomes	treats nausea; inflamed joints
Elderberry	flowers, fruit	remedy for head colds
Ginkgo	leaves	improves energy, mood, and brain function

ma huang – have only minor markets in Europe (Gruenwald and Goldberg, 1997). Within Europe pronounced variations in product preference also exist. For example, in the UK cascara buckthorn is the preferred laxative species, but elsewhere in Europe the preference is for senna (Dennis, 1997). As Mark Van Diest, Research and Development Manager at Qualiphar in Belgium, said, 'There is a time difference in products coming to market between the USA and Europe – sometimes a 5–7 year lag. Products might move back and forth between them, but some products always remain separate.' (Box 4.1, p81).

Due to scientific advances, however, European and North American markets are increasingly in synch. For example, St John's wort, ginkgo, ginseng, garlic, and echinacea species feature as major products in both markets. St John's wort has been the number one anti-depressant medication in Germany since 1995 (Box 4.2), when it out-stripped the pharmaceutical Prozac (Gruenwald and Buettel, 1997). In 1997, St John's wort hit the US market and sales grew in the next year more than 10,000 per cent (*Herbalgram* 43, 1998; *NBJ*, 1998). US markets tend to follow the more research-intensive European market. Most of the current big sellers in the USA were previously sold in Europe, based on results from research and clinical trials (Brevoort, 1998).

The leading product in Europe is Tebonin, manufactured by Schwabe, with 1996 sales of US$200 million (Box 4.3) (Gruenwald and Buettel, 1996). A total of 2.4 million prescriptions for Tebonin were filled in 1996, making it the 33rd most-prescribed pharmaceutical product in Germany (Schwabe, 1997). A ginseng product is the second top-seller in Europe, with US$50 million in annual sales, followed by garlic, evening primrose, and echinacea products (Gruenwald and Buettel, 1996).

In the USA, approximately 800 plant species are found in thousands of commercial botanical medicine products (Foster, 1995). The best-selling species in 1996 and 1997 was echinacea, followed by garlic, ginkgo, and golden seal. The top ten sellers in 1997 accounted for 57 per cent of sales. While echinacea remained the top seller, St John's wort rose from number 17 to number 2 top seller by 1998 (Table 4.2). Recent growth of the mainstream market was reflected in dramatically increased sales of the following species in 1997–8: echinacea and golden seal: 95.5 per cent; garlic: 23.7 per cent; ginkgo: 142.5 per cent; ginseng: 25.8 per cent; grape seed: 49.8 per cent; St John's wort: 10,001.3 per cent; other herbs: 85.7 per cent (*Herbalgram* 43, 1998). Green tea, black cohosh, elderberry, kava, soy, and saw palmetto were also among the fastest growing products in the mass market in the year preceding July 1998 (IRI, 1998).

Box 4.2 *Top-Selling Species in Germany, 1995*

Species	Category	Retail sales (DM m)
Ginkgo	circulatory	461
Horsechestnut	circulatory	114
Hawthorn	cardiac	65
Yeast	anti-diarrhoea, acne	59
St John's wort	anti-depressant	59
Myrtle	cough	35
Stinging nettle	urologic	35
Echinacea	immunostimulant	34
Saw palmetto	urologic	30
Milk thistle	urologic	29
Ivy	cough remedy	26
Mistletoe	cancer	22
Soybean	dermatological	15
Camomile	dermatological	15
Comfrey	dermatological	14
Kava kava	tranquilliser	13

Sources: Schwabe/Pfaffrath, 1995; Gruenwald and Buettel, 1996.

Geographic distribution of botanical medicines in trade

A large proportion of top-selling botanical medicines are naturalised or native to the regions in which they are sold. Most of the top-selling botanical medicines in the USA in 1996 are native to Europe or North America (Box 4.4). Of the more than 2,000 medicinal and aromatic species in trade in Europe, two-thirds (1,200–1,300) are native to Europe (Lange, 1998). While markets tend to reflect locally-available and known species, it is also the case that many species are found around the world, and have crossed many borders. For example, most sub-tropical and tropical country traditional pharmacopoeias share a number of weedy medicinal species in common, like senna, guava, and mango. Europe and North America share many species which made their way to North America with European

Box 4.3 *Top Retail Products in Europe*

Product	Species	Company	Annual sales (US$ m)
Tebonin	Ginkgo	Schwabe (Germany)	200
Ginsana	Ginseng	Pharmaton/Boehringer Ingelheim (Germany)	50
Kwai	Garlic	Lichtwer Pharma (Germany)	40
Efamol/Epogam	Evening primrose	Scotia (UK)	30
Echinacin	Echinacea	Madaus (Germany)	30

Source: Gruenwald and Buettel, 1996.

Table 4.2 *Top-Selling Botanical Medicines in the USA*

1997 Rank	Product	% 1996 sales	% 1997 sales
1	Echinacea (1)*	9.6	11.93
2	Garlic (4)	7.2	8.52
3	Ginkgo (3)	5.1	6.8
4	Golden seal (7)	4.7	5.95
5	Saw palmetto	3.1	4.87
6	(tie) Ginseng (6)	6.4	4.76
6	(tie) Aloe (8)	2.4	4.76
8	Cat's claw	2.1	3.49
9	Astragalus	1.3	3.07
10	Cayenne	2.5	2.83
11	Siberian ginseng (9)	3.5	2.7
12	Bilberry	1.6	2.61
13	Cranberry	1.7	2.47
14	Dong quai	1.8	2.13
15	Grape seed extract	2.0	2.07
16	Cascara sagrada	2.8	1.92
17	St John's wort (2)	n/a	1.87
18	Valerian (10)	2.2	1.73
19	Ginger	1.7	1.69
20	Feverfew	1.6	1.59

*(no) = 1998 ranking in Richman and Witkowski, 1998; saw palmetto ranked (5) in their 1998 listing

Sources: *Herbalgram* 41, January 1998; Richman and Witkowski, 1998.

settlers (eg St John's wort, valerian, and burdock, not to mention 'White Man's foot' (*Plantago* spp), thus named by native Americans who observed this plant follow in the foot of the Europeans). Of the 214 Eurasian medicinal and aromatic plant species in trade in Germany which are also found in North America, 146 were introduced and are naturalised (Lange, 1997).

Lange (1997) found that 20 per cent of medicinal and aromatic species in trade in Germany had been introduced or naturalised outside their native range, and only 40 per cent of species in trade had a native range restricted to one continent or a part of it. 16 species occur only in Europe; 63 in Africa; 90 in tropical Asia; 8 in Australia and New Zealand; 124 in North America; and 106 in South America. 248 species originate in the temperate regions of Asia (Box 4.5, p84).

Historically conservative in its incorporation of foreign species 'new' to the market, the botanical medicine industry is increasingly cosmopolitan in its approach. The large number of temperate Asian species in trade (248) in Germany reflects the widespread use of TCM in Europe. In the USA, Chinese herbs that 20 years ago were rarely seen outside Asian communities generated US$30 million in 1996 retail sales, and incorporated around 400 species and formulations (Brevoort, 1996). Examples of species sold widely in Europe and North America, but originating in other regions, include cat's claw and pao d'arco from South America, pygeum and yohimbe from Africa, and more recently kava from the South Pacific. Found in shops from Buenos Aires to Madrid, Tokyo, and New York, these and other species have made rapid progress in markets in recent years.

4.3 The structure of the botanical medicine industry

The structure of the botanical medicine industry is complex and in flux, moving towards greater globalisation and consolidation of companies. Through vertical integration, companies are diversifying their activities, and increasingly operate at a number of different levels of the trade. Despite these trends, the botanical medicine industry remains fragmented, and it is common for plant material to pass through many companies before arriving as finished products at retail outlets. The broad categories of companies through which botanical material moves (Box 4.6, p84) include:

- *supply companies:* cultivators or wildcrafters of raw plant material; wholesalers of raw plant material, including exporters, traders, brokers, and agents; and bulk ingredient and processing companies;

Box 4.4 *Top-Selling Botanical Medicines in the USA, 1996, and Geographic Provenance*

Latin name	Common name	Geographic origin
Echinacea spp	coneflower	North America
Allium sativum	garlic	Central Asia/Europe
Hydrastis canadensis	golden seal	North America
Panax spp	ginseng	North America/East Asia
Ginkgo biloba	ginkgo	China
Serenoa repens	saw palmetto	North America
Aloe vera	aloe	South Africa, Madagascar, Arabia
Ephedra spp	ma huang	Warm America, Mediterranean, China
Eleutherococcus senticosus	Siberian ginseng	North East Asia
Vaccinium spp	cranberry	North America
Rhamnus purshiana	cascara sagrada	North America
Plantago spp	psyllium	Europe/Asia
Valeriana officinalis	valerian	Europe/Asia

Sources: Based on data from: Brevoort, 1996; Richman and Witkowski, 1995; Mabberley, 1987.

Box 4.5 *Geographic Range of 1,464 Medicinal and Aromatic Plants in Trade in Germany, 1997*

Geographic range: regions	Number of medicinal and aromatic plant species found only in the region	Introduced species	Total number of medicinal and aromatic plant species found in the region
Europe	16	71	605
Africa	63	16	343
Asia – temperate	248	13	849
Asia – tropical	90	10	318
Australasia	8	18	55
Pacific	1	1	13
North America	124	186	454
South America	106	25	207

Sources: Lange, 1997; geographical regions according to Hollis and Brummitt (1992).

- *manufacturing and marketing companies:* some high-level processing companies, and manufacturers (including labelling and packaging) of finished products; and
- *consumer sales:* brokers and distributors of finished products to retail outlets; retail outlets, including those of mass and speciality markets; and direct sales in the form of mail order, multi-level marketing, and sales through healthcare providers.

Supply companies

Raw plant material is sourced from around the world. Most large companies receive material, through direct and indirect routes, from dozens of countries. China is the largest exporter of plant material, followed by India (4.8 below). In 1997 global sales of raw materials for botanical medicines, vitamins and minerals were around US$8 billion, with 27 per cent of this (US$2.2 billion) in the USA (*NBJ*, June 1998).

European-based companies, and German in particular (Box 4.7), dominate the global botanical supply industry. The biggest botanical raw materials group is Martin Bauer Group, a German-based corporation with annual sales of over US$250 million. Martin Bauer Group owns Finzelberg, Plant Extract, PhytoLab, and Phtyocon, and acquired Muggenberg, another large German supply company, in 1998. Other leaders include the pharmaceutical giants Madaus of Germany (US$400 million annual sales from botanical medicines alone), Schweizerhall of Switzerland (US$600 million), and Indena of Italy (US$200 million) (*NBJ*, June 1998).

Within Europe, internal and external trade in European-grown botanical materials is dominated by a few

Box 4.6 *Structure of the Botanical Medicines Industry*

Cultivation or wildcrafting of plants

Plants are cultivated or wildcrafted. Plant material is cleaned (extraneous matter removed) and dried. The majority of plant material in trade is in dried form. Drying methods must bring moisture content down to <14 per cent, while retaining the chemical composition of the plant. A minority of material is traded fresh, or preserved in alcohol.

Exporters/importers/wholesalers/brokers/traders

Plant material is purchased either directly from wildcrafters or cultivators, or after it has passed through a number of traders (eg local dealers, village cooperatives, district traders). Brokers and agents act on behalf of purchasing companies. Wholesalers, importers and exporters may specialise in a few raw materials, or in a few thousand, which they sell as commodities to a number of different companies. Wholesalers/traders may also process plant material. Some companies employ testing, or use voucher specimens at this stage, to ensure correct species identification and quality.

Bulk ingredient suppliers and processing companies

Plant material is tested for contamination (eg pesticides). It is formed into bulk ingredients, either coarsely cut, rasped, or ground into powdered form (for use in crude herbal products and in the preparation of extracts). Due to consolidation in the industry, the production of bulk ingredients is often undertaken by wholesalers/traders. Further processing in the form of extraction, particularly standardised extracts, is undertaken by processing companies, many of which also produce branded lines which they sell directly to distributors or retail outlets.

Manufacturers of finished products

Bulk and processed ingredients are supplied to companies that manufacture (eg might add excipients to extracts to make tablets and capsule products, based on in-house formulae), label, and package products for retail sales. Some sell lines directly to health professionals, others sell directly to consumers through multi-level marketing and mail order. Some companies use brokers or distributors to supply their products to retail outlets, others market directly to mass and speciality outlets.

Distributors

Some manufacturers (usually smaller companies) use distributors to sell finished products to retail outlets.

Retail/consumer sales

The bulk of finished products are sold through retail outlets, either mass market (eg chain pharmacies, supermarkets, grocery stores) or speciality (eg healthfood stores, pharmacies), although direct sales hold a significant portion of the market.

Sources: interviews; Lange, 1998; Brevoort, 1998; Smith, 1998.

wholesalers (eg Germany, 20; Bulgaria, 10; Albania, 4). Plant material is usually purchased from collectors and cultivators, and passes through several traders before arriving in the hands of wholesalers, who sell it to bulk ingredient and retail manufacturers. In former Eastern Bloc countries, the internal and external trading system has changed dramatically in the past few years, as countries move from state-controlled centralised systems to free market trade with increasing numbers of private companies (Lange, 1998).

In the USA, 150 companies are involved in the supply of botanical raw materials. The North American botanical raw material market is growing rapidly, with a current value of US$500–600 million per year. Some ten players account for 80 per cent of revenues in this area. Major US botanical suppliers include Triarco, Hauser, Botanical International, and Bio-Botanica (Table 4.3). The trend in the US botanical supply industry is away from large raw material suppliers that simply warehouse ingredients and fill orders to meet limited demand, towards consolidation, partnerships and diversification. Larger manufacturing companies increasingly deal directly with suppliers, cutting out the role of brokers, and some have acquired supply companies as part of the wider trend towards vertical integration (*NBJ*, June 1998).

Box 4.7 *Structure of the Botanicals Trade in Germany*

Drug brokers

Seven brokers or agents are involved in the trade in Germany. Most are active on a global scale, although some specialise in specific countries. Brokers represent foreign import-export companies, traders, farmers, and manufacturers. They deal mostly for wholesalers, and to a lesser extent for pharmaceutical companies or herbal tea companies. Most brokers also trade in spices.

Wholesalers (traders in bulk material)

In Germany, the mainstream bulk trade in botanicals is dominated by about 20 wholesalers (with some transactions undertaken via seven large brokers), with further consolidation of the trade in the past few years. 95 per cent of plants sold by German wholesalers are sold as dried plants and plant parts, with the remaining 5 per cent comprised of plants preserved in alcohol, mainly for use in homeopathy. Traders deal with a range of products, and with a range of customers including the food industry, pharmaceutical companies, cosmetics, liqueur, extract producing companies, and colouring agent companies. Overall volumes imported by individual traders range from 1,000 to 30,000 tonnes annually. On average, each company trades in 400–500 botanical species.

Processing

Wholesalers are often responsible for processing the plant material before sale, including cleaning, cutting and grinding it into a powder. Some wholesalers are also involved in producing extracts, herbal teas, or herbal mixtures.

Manufacturing

Processed material is supplied to manufacturers of pharmaceuticals, plant extracts, cosmetics, liqueurs, dyes, etc, as well as to second-level retail suppliers, and to other wholesalers, and tea-packing companies. Bulk extract producers and pharmaceutical companies often manufacture intermediary products which are then sold to cosmetics, pharmaceuticals, or food companies that manufacture finished products.

Sources: Lange 1997; Lange and Schippmann, 1997; D Lange, personal communication, 1999.

Table 4.3 *Leading Suppliers of Botanical Raw Materials in the USA*

Company	HQ	US sales (US$ m)	Global sales (US$ m)
Indena	Italy	>50	>200
Botanicals International	USA	>50	>50
Henkel	Germany	30–50	>50
Hauser	USA	30–50	>50
Optipure (Chemco)	USA	30–50	<50
Flachsmann	Germany	30–50	>50
Martin Bauer (Finzelberg, etc)	Germany	20–30	>250
Muggenberg Extrakt	Germany	20–30	>20
Schweizerhall	Switzerland	15–20	>200
Folexco	USA	15–20	<20
East Earth Herbs	USA	15–20	<20
Euromed (owned by Madaus)	Spain	15–20	>50
Mafco Worldwide Corp	USA	15–20	<20
Triarco Industries	USA	15–20	<20
Sabinsa	USA	15–20	<20
MW International	USA	10–15	<20
AYSL	USA	10–15	<20
Quality Botanical Ingredients	USA	10–15	<20
SKW Trostburg	Germany	10–15	>100
Pure World (Madis)	USA	10–15	<20
Technical Services Intl (Inabata)	USA	10–15	<20
Arkopharma	France	10–15	>100
Starwest Botanicals	USA	10–15	<20
Trout Lake Farm	USA	10–15	<20

Note: 'This list is the result of independent research, interviews with company executives, industry presentations and reputable published business sources. Although *NBJ* has made every effort to be accurate, revenue figures are not the result of internal or external audits and therefore are not guaranteed to be an accurate representation. Errors and omissions are unintentional.'

Sources: NBJ (San Diego, California), June 1998.

In response to these trends, supply companies are developing new ways to diversify their services and add value to products, primarily through the development of (sometimes trademarked) lines of extracts. As Tim Avila of Systems BioScience (a consulting firm that helps to locate quality raw material suppliers) said: 'It's a value migration. As the industry grows up, there will be different ways to produce value ... . In the old days, many of the distributors or ingredient suppliers were just resellers. These distributors are changing their business model, either getting into semi-finished goods or doing private label' (*NBJ*, June 1998). Kay Wright, Director of Botanical Purchasing at Celestial Seasonings, finds that as a result of these variables, 'There are more extractors in the game now. It used to be mainly Europe and the USA. Now China is big ... there are a lot of players coming in and out. It is a volatile, exciting time.'

Table 4.4 *Top 20 US Supplement-Manufacturing and Marketing Companies, 1997 (Vitamins, Minerals and Botanical Medicines)*

Company	HQ	Supplement revenues (US$ m)
Leiner Health Products (Your Life)	Carson CA	425
Pharmavite (Nature Made)	San Fernando CA	340
American Home Products/Lederle (Centrum)	Madison NJ	325
Rexall Sundown	Boca Raton FL	291
NBTY (Nature's Bounty)	Bohemia NY	281
General Nutrition Corp	Pittsburgh PA	260
Weider Nutrition Group (Schiff)	Salt Lake City UT	219
Twin Labs (Nature's Herbs)	Hauppauge NY	213
Abbott Labs (Ross Products)	Columbus OH	170
Perrigo (Private Label)	Aliegan MI	152
Solgar Vitamin & Herb Co	Hackensack NJ	120
Bayer Corp (One-a-Day)	Pittsburgh PA	110
International Vitamin Corp	Freehold NJ	109
Experimental & Applied Sciences (EAS)	Golden CO	108
Nature's Way Products (MMS)	Springville UT	100
Nutraceutical (KAL, Solaray)	Park city UT	98
Pharmaton Natural Health Products	Stamford CT	90
PowerFoods	Berkeley CA	90
MET-Rx USA	Irvine CA	88
Country Life Vitamins	Hauppauge NY	70

Revenue distribution of supplement-manufacturing companies

Size category	No of companies	revenues	% total market
Greater than US$100 m	16	3320	55
US$20–100 m	38	1540	25
Less than US$20 m	996	1190	20

Source: NBJ, September 1998.

4.3.2 Manufacturing and marketing companies

Manufacturing and marketing companies produce private-label products, which they market for retail sale. Companies may also produce products for a commercial customer which labels, packages, and markets the product. In Germany, most botanical medicine manufacturing companies are privately owned, with sales averaging between US$20 and 100 million (Yuan and Gruenwald, 1997). In the USA, there are more than 1,000 dietary supplement companies, of which only 15 had annual sales of over US$100 million in 1997, and only 54 had annual sales of over US$20 million (Table 4.4). The vast majority of US companies are privately held, small to medium in size, and oriented towards speciality outlets, such as health food stores. However, increasing numbers of large pharmaceutical and over-the-counter (OTC) companies are entering the botanical medicines market, selling primarily to mass market outlets like chain drugstores and supermarkets (Kalorama, January 1997).

4.3.3 Consumer sales

Distributors

Distributors supply retailers with manufactured products. In the USA, most distributors buy products from small manufacturing companies and sell them to independent natural products retailers. Included in this category are Tree of Life; Stow Mills, and Nature's Best (Kalorama, 1997). In Germany, distributors deliver plant-based medicines to pharmacies. 17 distributors are members of the PHAGRO (Bundesverband des Pharmazeutischen Grosshandels), but this number is declining due to on-going consolidation (in 1980 there were almost 50 wholesalers). These 17 distributors have 102 branches in Germany, and a total turnover of DM 28.8 billion in 1996. Phoenix Pharma Handel is the leading distributor, followed by Gehe Pharma Handel

Table 4.5 *Retail Sales of Medicine in Germany (DM bn)*

	1991	1992	1993	1994	1995	1996	1997
Pharmaceutical market total	42.9	46.9	42.9	44.6	46.2	49.2	50.3
Pharmaceutical market in East Germany	6.5	8.1	7.9	8.2	8.9	9.4	9.5
Pharmaceutical market in West Germany	36.4	38.8	35.0	36.4	37.5	39.8	40.8
Rx-bound preparations	28.4	31.2	27.8	29.1	30.4	32.5	34.1
Non Rx-bound preparations in pharmacies*	13.2	14.5	13.8	15.5	15.2	15.5	14.9
Medicinal preparations outside pharmacies	1.2	1.2	1.3		0.8	1.2	1.3

* This category includes the bulk of botanical medicine sales.

Sources: IMS and BAH (1992–8).

(which owns the UK-based company AAH, and OCP, which covers 90 per cent of the French market), Anzag, Andreae-Noris Zahn AG and Sanacorp (PHAGRO, 1997; ABDA, personal communication to Lange, 1998).

Retail

Consumer sales in the US$3.87 billion US botanical medicine industry break down into a number of channels. The mass market is the fastest growing segment, and represents 17 per cent of retail sales. Speciality and natural foods outlets represent 34 per cent of sales, and mail order and multi-level marketing 35 per cent. Health practitioners account for 7 per cent of sales, and teas, through all channels, 7 per cent (Brevoort, 1998). Dramatic growth in the US market is attributed to the expanded sale of botanical medicines in mass retail outlets, rather than solely in speciality stores, and the expansion of retail natural food stores (*NBJ*, June 1998).

Mass retailers that have recently moved into botanical medicines sales include Walmart, Walgreens, and K Mart. In 1997, Walmart sold around US$500 million of natural products (Greenwald, 1998). The entry into the marketplace of large pharmaceutical companies will further expand the mass market, which is likely to dwarf sales in health food and speciality outlets in the near future. Mass market growth between May 1997 and May 1998 was 101 per cent (*Herbalgram* 43, 1998). However, health food store outlets continue to offer the widest range and variety of botanical medicine products, and provide a speciality niche not easily filled by the mass retailers (Brevoort, 1998).

Botanical medicines are attractive retail products, not only because of their popularity with the public, but also due to retail margins ranging in the USA, for example, from 40 to 45 per cent in speciality outlets, and from 30 to 40 per cent in the mass market (Kalorama, 1997). The USA National Association of Chain Drug Stores, for example, reports that while gross profit margins on pharmaceutical drugs are shrinking to around 17 per cent, for alternative therapies including botanical medicines they typically average 30 per cent (*Daily Camera*, 1997). Over 90 per cent of independent and chain pharmacists responding to a Drug Topics survey felt that 'alternative medicine has merit and is here to stay' (Portyansky, 1998). In 1996, dietary supplements were the second fastest growing category in drugstores (Kalorama, 1997). As Chris Kilham, an advisor to the botanical medicine industry, said, 'Pharmacies are getting hit on both sides as insurance companies, managed care organisations and other entities have reduced profitability for prescription drugs, and mass market retailers have reduced prices for OTC medicines ... . For pharmacies, herbal products are truly a godsend.'

In Germany, health food shops sell more than 5,500 different botanical medicine products. In 1997, the Association of Healthfood Shops (neuform Vereinigung Deutscher Reformhaeuser eG) reported an annual turnover of DM 1.3 billion. Packaged food accounted for 49.47 per cent of sales; fresh food 7.68 per cent; dietetic food 9.88 per cent; cosmetics 14.44 per cent; other products 2.14 per cent; and natural medicinal remedies 16.39 per cent. Mass market outlets in Germany have experienced annual average growth in botanical medicine sales of between 7 and 12 per cent (BAH, 1992–8) (Tables 4.5, 4.6 and 4.7).

Table 4.6 *The Non-Prescription Medicines Market – German Pharmacy Retail Sales (DM bn)*

Non-prescription pharmaceuticals	(1993)*	(1994)*	1995	1996	1997
Total	(11.5)	(11.9)	15.2	15.5	14.9
Non-botanical drugs	(8.4)	(8.4)	10.7	10.8	10.5
Botanical drugs	(4.5)	(3.5)	4.5	4.7	4.4
Prescribed botanical drugs	(1.7)	(1.9)	2.4	2.6	2.2
Self-medicated botanical drugs	(1.4)	(1.6)	2.1	2.1	2.2

* 1993 and 1994 figures are available for West Germany only, and so are not comparable to those of the previous and following years.

Source: BAH (1994–1998). Based on IMS-figures.

Table 4.7 *Market Shares of Selected Outlets for Medicinal Preparations (DM m)*

Outlets	1991[1]	1992[1]	1993[1]	1994[1]	1995[1]	1996	1997
Pharmacies	477	450	456	427	401	630	922[2]
Mass market drugstores	440	431	469	467	501	876	942
Drugstores	84	80	75	73	na	na	na
Mass market retailers	203	216	231	205	197	298	305
Warehouses	40	40	41	38	na	na	na
Food shops, others	66	63	64	63	na	na	na
Health food shops	325	320	327	339	196[2]	259	na
Others	45	46	45	44	na	na	na

1 1991–5 figures are available for West Germany only, and so are not comparable to those of 1996 and 1997; 1996 and 1997 figures are for the whole of Germany.
2 This figure is not comparable to that of the previous years.

Source: BAH, 1992–8.

Consumers

US consumption of botanical medicines has increased dramatically over the last six years. Brevoort (1998) cites a number of surveys that show consumer usage increasing from 3 per cent in 1991 (Eisenberg); to 15 per cent in 1995 (Gallup); 19 per cent in 1996 (Gallup); 32 per cent in 1997 (Gallup); and 37 per cent in 1998 (Harris). The Gallup study of US *Attitudes Toward and Use of Herbal Supplements* showed that 39 per cent of all botanical medicine consumers are 'daily dosers'; 31 per cent take botanical medicines weekly; and 29 per cent consume them 'often'. A survey of alternative medicine use in the USA published in the *Journal of the American Medical Association* in November 1998, found that between 1990 and 1997, botanical medicine use increased by 380 per cent. One in five individuals taking prescription medications were also taking botanical medicines, high dose vitamin supplements, or both (Eisenberg et al, 1998).

An opinion poll carried out in Germany by the Institut für Demoskopie Allensbach found that in 1997, 65 per cent of the population used plant-based remedies, 13 per cent higher than in 1970. 74 per cent of consumers were women, and 72 per cent had higher education. The highest growth rates in usage were found among people between the ages of 16 and 29. There was no significant difference found in usage between formerly West and East Germany (Institut fur Demoskopie Allensbasch, 1997).

4.4 Market trends

Market growth averaging 10–20 per cent per year, and trends towards increased safety, efficacy, and quality of products have produced changes in the way botanical medicine companies do business. European companies – part of a significantly more mature market than that in North America – conduct business under relatively long-term timeframes, and operate in a system with well-established research, testing, and quality control procedures in place. In North America, companies are still largely adaptive, rather than strategic, in response to a market in dramatic flux. As one company representative put it: 'In Europe they take their time. In the USA, everyone is in a hurry.'

Market trends in the botanical medicine industry vary by region. Some of the trends discussed below – like the entry of large pharmaceutical companies, increased media attention, and rising consumer demand – are global in scope. Others – like transitions in regulatory environments – are national:

1 **Consumers seek an alternative or complement to pharmaceutical drugs and modern healthcare, which are perceived as limited in scope.**

Consumers increasingly perceive Western healthcare systems as remote, cold, and expensive, and pharmaceutical drugs as having too many side effects. Additionally, the increasing life expectancy of the population, including the large number of ageing baby boomers, has led to a rise in chronic conditions for which conventional treatments are unsatisfactory. Illnesses poorly understood and treated by modern medicine – such as viruses, chronic skin disorders, mental disorders, and compromised immune systems – are increasingly treated with alternative therapies, including botanical medicines.

Almost 30 per cent of the West German self-medication market in 1997, for example, was allocated to plant-based drugs (BAH, 1998). The opinion poll carried out by the IDA (1997) found a trend towards increased self-medication: 56 per cent of consumers self-medicate with plant-based medicine, while 22 per cent use plant-based medicine under a prescription (Table 4.8). Consumers also increasingly emphasise prevention, and see botanical medicines as an important aid in this regard.

In Germany, interest in medicine as a tool for prevention has grown from 24 per cent of people interviewed in 1989, to 31 per cent in 1997 (IDA, Germany 1997). In a Gallup study (Table 4.9), 23 per

Table 4.8 *Retail Sales of Self-Medication Botanical Medicines in Germany (DM m)*

Categories	1996	1997	% change
Respiratory	357	509	+43
Digestives	284	296	+4
Tonics and geriatrics	438	272	-38
Cardiac and circulatory	278	270	-3
Hypnotics/sedatives	205	226	+10
Others	611	632	+3
Total	2,173	2,206	+1.5

Source: BAH: 1996–8.

Table 4.9 *US Consumer Herbal Usage Surveys (as in Brevoort, 1998)*

	Gallup %	Harris/Celestial Seasonings %
Treat health problems	22	51
Prevent health problems	23	19
Enhance overall health	45	30

Sources: Gallup Study, 1997 (704 households, published October 1997); Harris/Celestial Seasonings (500 households, published February, 1998).

cent of Americans are reported to take botanical medicines for prevention (Brevoort, 1998).

2 **Acceptance of botanical medicines by national and commercial insurance companies.**
The national health services in Germany and Japan reimburse patients for botanical medicines, which form an integral part of the healthcare system (although a decrease in reimbursements in Germany might dampen market growth there). In the USA, the expense and red tape of managed care have driven many consumers to purchase botanical medicines out of their own pockets. A Harvard Medical School study found that 75 per cent of the approximately US$13.7 billion paid for alternative care in 1990 was spent out-of-pocket and not reimbursed by insurance companies (*NEJM*, 1993).

At the same time, however, an increasing number of US insurance companies and managed care organisations are accepting alternative therapies and botanical medicines. Since 1993, more than a dozen national insurance companies have begun to cover services for acupuncture, naturopathy, and chiropractic care, which in some cases includes reimbursement for doctor-prescribed botanical medicines. Increasing insurance company acceptance of alternative therapies parallels a rise in acceptance within the medical community, which could further stimulate growth in the botanical medicines market (Brevoort, 1998).

3 **The enthusiasm for, and financial hype associated with, the biotechnology industry has made natural medicines respectable again.**
Some claim that the innovation and expansion of the pharmaceutical biotechnology sector (Chapter 3), which builds upon natural materials, has produced a scientific and financial environment open to the potential medical benefits of other natural products, including botanicals.

4 **Expanded research has improved the legitimacy of botanical medicines.**
Research demonstrating the effectiveness of botanical medicines like St John's wort and ginkgo has validated their safety and efficacy, and improved quality control. This has led to rapidly expanding markets for these products. Many of the top-selling botanical medicine products today, for example, grew out of many years of research conducted on safety, efficacy, and quality control.

5 **The rise in green consumerism has increased demand for ' natural' medicine.**
Consumers are interested in 'natural' products, as part of an increased focus on health, fitness, and well-being. Botanical medicines are perceived as more 'natural' and better for the patient and the environment than pharmaceutical and synthetic OTC drugs.

6 **Increased advertising budgets and media attention have led to rapid growth in consumer demand for botanical medicines.**
Advertising budgets and media coverage of botanical medicines in Europe and the USA have increased dramatically over the last few years. Advertising expenditures by US marketers of botanical medicines reached an estimated US$36 million in 1996, almost double the 1994 figure of US$19 million (Kalorama, 1997). The entry of large pharmaceutical and OTC companies has meant a further expansion of advertising budgets over the past few years. In 1998, mass market advertising for botanical medicines was estimated at US$204 million, with large companies like American Home Products (US$12 million), Bayer (US$35 million), and Warner Lambert (US$15 million) spending heavily to launch new product lines (Brevoort, 1998).

The mass media has also increased its coverage of botanical medicines. In the USA – a market still driven largely by media attention – media coverage has immediate and concrete effects. For example, Nature's Herbs (a division of Twin Laboratories Inc) introduced St John's wort into their product line in 1992. But sales for St John's wort only shot up following a *New York Times* article in 1997, a *Newsweek* special story, and a television programme on ABC's 20/20. The *Wall Street Journal*'s (20 January 1997) front-page profile of echinacea, and another on kava, as well as a 20/20 television programme on kava, yielded similar results for other botanical medicine products (Brevoort, 1998).

Media attention on pharmaceutical drugs may also have a knock-on effect for botanical medicines. As Cindy O'Connor, former Director of Marketing at QBI, said, 'Whenever a product comes out on the market that is popular, like Viagra, it creates demand for an herbal equivalent. In the case of Viagra, for example, yohimbe (from Africa), and damiana (from Mexico) have begun to receive attention.'

7 **Changes in the regulatory environment in the USA have made the manufacture and marketing of botanical medicines more attractive.**
Passage of the US Dietary Supplement Health and Education Act (DSHEA) of 1994 dramatically improved the marketing environment for botanical medicines. By allowing for a broader range of claims for efficacy, it increased opportunities for marketing and advertising products to consumers.

As Grace Lyn Rich, Director of Marketing at Nature's Herbs, said, 'Since DSHEA, a lot more people have entered this market. DSHEA opened doors for some health claims, and the amount and types of information companies could supply to consumers. Herbs are the fastest growing section of the dietary supplement industry. Vitamin companies that have never been involved in the manufacture of herbs are now adding them to their lines. New companies are starting up. You also see line extensions from traditional players, such as Nature's Herbs. Pharmaceutical companies are also interested now.'

8 **The entry of large pharmaceutical and OTC companies has helped spur expansion of the botanical medicine industry into the mass market.**
Pharmaceutical and OTC drug companies are moving into the botanical medicine industry, bringing with them research and marketing budgets many times those of the traditional botanical medicine firms. Within the US botanical medicine industry, feelings are mixed on this development, but a 'shake-out' of the industry appears certain (Greenwald, 1998; Brevoort, 1998). Large pharmaceutical and OTC companies have greater access to mass markets, and place greater emphasis on safety, efficacy, and quality-control, including clinical testing (*NBJ*, June 1998). Larger research budgets mean that companies might research more new products, and demand may increase for access to new species and information on their traditional use.

Large companies are also entering the botanical medicine industry as bulk ingredient suppliers. 'The major trend in the industry we've seen is multinational companies getting more involved in natural ingredients,' said Tim Avila of Systems BioScience (CA). Most of these multinational companies have backgrounds in fine chemicals, agricultural chemicals or pharmaceuticals, and 'now they are taking their chemical expertise and applying it to other areas' (*NBJ*, June 1998).

Large pharmaceutical and OTC companies are accustomed to following required Good Manufacturing Practices (GMP), and their entry into the market is also likely to result in higher standards set for suppliers and manufacturers (*NBJ*, June 1998). Most European companies already follow protocols for quality control and GMP, but many US raw herb and bulk ingredient suppliers do not. However, a consortium of US trade associations has proposed GMPs for the botanical medicine industry to the FDA (Brody, 1999). In addition to manufacturing practices, GMPs for commercial botanical medicine products cover collection, preparation, and production of the ingredients, including raw herbs and extracts (Leung, May 1997).

Examples of large pharmaceutical and OTC companies entering the botanical medicine industry include Bayer AG, which has launched a 'One-A-Day Specialised Supplements' line of eight products consisting of vitamins, minerals, and botanical ingredients. American Home Products launched a line of botanical products under the Centrum label in late 1998, and has acquired Solgar products for sale in health food stores. SmithKline Beecham introduced a 100-year-old line of German herbal products called ABTEI, which include saw palmetto, St John's wort, ginkgo, and ginseng. Warner-Lambert recently introduced Quanterra single-ingredient botanical products, and a number of other companies have botanical medicine products in development (*NBJ*, June 1998; Brevoort, 1998).

9 **Consolidation and globalisation are on the rise.**
The botanical medicine industry is increasingly global in scope, and is undergoing consolidation at all levels, including retail, wholesale, and supply of bulk ingredients and raw materials (Box 4.8). Companies are acquiring partners, setting up sourcing partnerships, and distributing ingredients and products through overseas distributors.

All of the major European plant extract producers, including Indena, Finzelberg, Muggenburg, and Euromed, have branches in the USA (Gruenwald and Goldberg, 1997). Schwabe, Madaus, Lichtwer

Box 4.8 *Mergers and Acquisitions in the US Botanicals Industry*

1989	TwinLabs and Nature's Herbs
1995	Celestial Seasonings and Botalia
1997	Chattem and Sunsource International
1997	Jones Pharma Inc and Crystal Star Herbs
1997	SmithKline and Abtei
1998	Celestial Seasonings and Mountain Chai
1998	Natrol and Pure Gar
1998	Natrol and Laci Le Beau
1998	Quality Botanical Ingredients and Botanical Products International

Source: Bingham, 1998 as in Brevoort, 1998.

Pharma, and Pharmaton have developed US subsidiaries to take advantage of the growing US market (Yuan and Gruenwald, 1997). US-based ExtractsPlus is the exclusive North American distributor for pharmaceutical and nutraceutical botanical extracts manufactured by Emil Flaschmann AG of Switzerland; it also imports and distributes extracts from Japanese companies in the pharmaceutical and functional foods industries. In another example of global ties, TSI, a subsidiary of Japanese based Inabat & Co Ltd, has established research partnerships with the Chinese Academy of Sciences, and distribution arrangements with Pierre Fabre Sante of France (*NBJ*, June 1998).

10 **Increased emphasis on safety, efficacy, and quality have changed the types of product in demand, and requirements placed on suppliers of raw material and bulk ingredients.**

Increased demand for safety, efficacy, and quality control in botanical medicine products has resulted in greater corporate research and development expenditures, a shift towards standardised products with identified chemical markers, and requirements for high quality raw materials. Entry of large pharmaceutical and OTC companies into the botanical medicine industry has further fuelled this trend.

Today, ingredients and botanical medicine products are often processed or manufactured by companies without knowledgeable pharmacognocists, herbalists, or technicians on staff to identify and evaluate plants and their extracts. As a result, there is widespread adulteration or mislabelling of commercial botanical extracts, with a knock-on effect for the finished products offered to consumers (Leung, May 1997).

As Frank D'Amelio, owner of Bio-Botanica, said, 'I see samples all the time, and some of it is just terrifying ... . They'll assay at virtually nothing. The raw material has next to zero or very poor active constituent.' Raw plant material has also been shown to contain heavy metals, and microbiological, pesticide and herbicide contamination. As one executive noted on the latter point: 'Not everything is wild-crafted anymore. More and more botanicals are grown as field crops to insure a constant supply, so they are sprayed with pesticides.' When plant constituents are concentrated into extracts, pesticides and other contaminants may be concentrated as well (*NBJ*, June 1998; Hauser, 1997).

As a result of these concerns, quality control has become a central issue for many botanical suppliers. While in the USA testing for contaminants – such as pesticides, aflatoxins, and solvents – is not common (neither the Environmental Protection Agency (EPA) nor the FDA has established standards for testing), in Europe more stringent standards generally pick up on contaminants. Increasingly these standards are adopted in the USA, as well. Many in industry feel that quality raw material suppliers are critical to the industry's long-term survival, since consumer alarm could damage the industry's reputation (*NBJ*, June 1998). As Staci Eisner, Technical Director of ExtractsPlus, said, 'We must improve the quality and consistency of our products, in order to ensure their legitimate place in the healthcare system.'

Increased interest in quality products in the USA has provided European bulk ingredient and raw material suppliers a marketing advantage, since European companies have long histories of conducting research and processing botanical extracts to high standards. Many were formed decades before the US industry took shape. European companies also operate under tighter European regulations governing botanical extracts, have established quality control and GMP procedures, and decades of clinical data to support their products (*NBJ*, June 1998; Morein, 1998).

Many US companies have created in-house testing facilities and others rely on outside laboratories to conduct increasingly sophisticated quality tests on raw materials. But this means that barriers to entry for reputable suppliers have also risen, with start-up capital for testing, milling, extraction, concentration, and compounding facilities reaching US$200 million. At the same time, a burgeoning market has created openings for a large number of operators, many with little interest in quality control. The number of botanical suppliers has increased by 20–25 per cent in the last three years (*NBJ*, June 1998). As a result of these factors, and despite trends towards greater quality control, incentives still exist to cut corners.

Resolving quality control, safety, and efficacy issues are vital to the long-term success of the botanical medicine industry. In the USA, where the industry is heavily dependent upon media to promote its products, the press has begun to focus a more critical eye on what is now a booming industry. In November 1998, for example, *Time Magazine* ran a cover story, 'The Herbal Medicine Boom: It's great business, but is it good for what ails us?'. Greenwald (1998) reports in *Time* that while 'the top US and European manufacturers pay close attention to the safety, efficacy, and consistency of their products, parts of the industry resemble a Wild West boom-town, where some 800 lightly-regulated US companies compete ferociously with fly-by-night hucksters'.

A March 1999 *New York Daily News* headline caught the reader's eye with 'Herbal Sex Drink Scare', and in another example, in February 1999, *The New York Times* ran an article titled 'Americans Gamble on Herbs as Medicine: with few regulations, no guarantee of quality'. In this article, Brody (1999) comments: 'consumers have no assurance that an herbal product contains what the label says it does or that it is free from harmful contaminants. Independent analyses of some products, particularly those containing costly or scarce herbs, revealed that some have little or none of the purported active ingredient listed on the label'. The botanical medicine industry in the USA will need to 'mature' quickly in response to this type of press coverage, in order to ensure its continued growth and further consolidate its new role in healthcare.

4.5 Scientific and technological trends

Botanical medicines contain multiple compounds, so do not lend themselves to patent protection. As a result, commercial research on botanical medicines in the USA has been limited to date. Companies have little incentive to conduct expensive and time-consuming research if they cannot gain exclusive rights to sell a product for a given period of time. Plant research in the USA, therefore, tends to focus on single compound drugs which lend themselves to patents, and is based in the pharmaceutical, not the botanical medicines sector (Greenwald, 1998; Foster, 1995). In contrast, many European and Asian countries have linked government, academic, and industry botanical medicine research traditions that create strong foundations from which commercial product development grows. This system encourages a kind of conservatism in new product development, but has allowed for an affordable and timely approval process for botanical medicines, while ensuring safety and efficacy.

Across the board, expenditure on botanical medicine research and development is likely to increase in coming years. As we have seen, large pharmaceutical and OTC companies with vast research and development budgets, dwarfing those of traditional botanical medicine companies, will require some kind of clinical testing, reputable suppliers and manufacturers, GMPs, and proper testing to assure quality and potency. In their quest for innovative, efficacious products, bulk ingredient suppliers are also undertaking increased research and development (*NBJ*, June 1998).

4.5.1 Developing a new product

It takes from 10 to 15 years to develop a pharmaceutical drug from concept to appropriate dosage, but it is possible to convert crude plant medicines into standardised phytomedicines in less than two years. Pharmaceutical drug development begins with the identification of active lead molecules, followed by detailed bioassays and formulation, and clinical studies which establish safety, efficacy, and the pharmacokinetics profile of the new drug. In contrast, botanical medicine research and development usually begins with clinical evaluation of the treatment modalities and therapy as administered by traditional healers or used by local communities. This is followed by acute and chronic toxicity studies and, if the substance proves safe, detailed pharmacological and biochemical studies (Iwu, 1996).

Traditional knowledge

Traditional knowledge is widely used in the botanical medicine industry as the basis for determining safety and efficacy, to develop agronomic practices for cultivation of materials, and to guide the development of new products. Every company interviewed made use of documented traditional knowledge associated with species in some form. All consult scientific journals, databases, the internet, university research results (including unpublished PhD dissertations), professional literature, or other secondary sources. A number of companies also have direct research collaborations with universities or research institutes for the study and testing of traditionally-used products.

A few companies conduct ethnobotanical field research and use the knowledge gathered to develop commercial products. For example, Axxon Biopharm uses a product discovery and development tool called 'Clinical Outcome Based Ethnomedical Leads'. This method involves an ethnobotanical survey to identify traditional uses of plants used for the treatment of diseases with clearly identified symptoms or symptom-complexes. This is followed by ethnomedical analysis, evaluation of plants through biological and chemical de-replication, and selecting only those with widespread and common use across regions. Clinical outcome evaluation is conducted, followed by standardisation and manufacture of selected products. Axxon Biopharm collaborates with the non-profit organisation Bioresources Development and Conservation Program for the collection of samples in Nigeria, Cameroon, Guinea, Ghana, and South Africa.

Shaman Botanicals has identified four potential botanical medicine products to treat gastro-intestinal problems, anxiety, sexual frustration, and an energy booster, based on traditional knowledge collected as part of collaborations with communities and research institutions in high biodiversity countries. The latex of *Croton lechleri* (Sangre de Drago), previously under development as a pharmaceutical (Chapter 3), is now being developed as a botanical medicine product. This species is widely used by local communities throughout Latin America. Shaman Botanical's research strategy involves targeting five key market categories with significant commercial potential, and identifying candidate plants with a documented ethnomedical history found in the company library and database compiled over nine years of Shaman Pharmaceuticals field work in 30 countries.

Traditional knowledge is used by many companies as an important complement to scientific testing, and many consult it to confirm research results produced in the laboratory, including safety and efficacy. As Dr Mark van Diest, Research and Development Manager of Qualiphar, said, 'Toxicity is a very important issue for companies. Even if there are good claims for a species, a company cannot be sure about toxicity. If a species has been used for years in traditional medicine, then companies can feel more comfortable.'

Traditional knowledge is also widely used for marketing purposes. Most botanical medicine products include in their packaging some reference to the traditional use. For example, Yogi Tea Company's Green Tea with Cat's Claw carries the following information: 'Grown in the highlands of the Peruvian Amazon, Cat's Claw is an ancient herb traditionally used by generations of native Ashanica Indians to support the self defense and digestive systems'. Nature's Way Kava Root product informs the consumer that 'Kava (*Piper methysticum*) is native to the islands of the South Pacific. It was originally used by Polynesian chieftains as a ceremonial beverage. Cultivated throughout the Pacific Islands, the large, knotty root stock is harvested at 3–4 years of age for optimal potency'. Source Natural's Kava Kava Root Extract label similarly states: 'For centuries, South Pacific Islanders have relied upon the root of the Kava plant for physical as well as mental relaxation'. Traditional Medicinals Ginger Aid Herb Tea states that ginger's 'rich and colorful history can be traced back to the earliest written records of India and China when it was revered as a gift from the gods'. It is rare to find a botanical medicine marketed without reference to traditional use.

Demand for access to 'new' species

Botanical medicine companies rarely conduct field or laboratory research to develop new products. 'Research' generally involves safety, stability, formulation, and quality control studies, and in both the USA and Europe, companies widely reported 'staying away from anything controversial'. In Germany, most companies only develop products from species with monographs produced by the Commission E (see Appendix C), arguing that it is not economically feasible to develop commercial products from previously unstudied natural products. Over the last few years, none of the new preparations launched on the German market was based on plant species 'new' to the market (BfArM, letter to D Lange, December 1998). One large German company interviewed estimated the cost of developing a commercial product from a species new to the market at DM 20 million, which it considered too expensive. As a result, they prefer to focus on new applications of known species.

In the USA, most of the 'new' product launches are also new formulations, or blends of known species. As Grace Lyn Rich, Director of Marketing at Nature's Herbs, said, 'We introduce 12–30 new products a year, and have done so for the last 5–8 years. Many of these products are a new combination of traditional products – eg new blends that provide synergistic benefits, new herb-vitamin combinations, or line extensions like a cardio ginkgo, or an energy ginkgo.'

However, products based on species new to the market can be an important way for companies to gain market differentiation. Nature's Way, for example, invested significant time and resources in developing a new ginseng product (see Case Study 4.11, *Panax vietnamensis*). In searching for 'new' species, companies seek out those with well-documented traditional uses, supported by scientific studies, which are not marketed widely, if at all. As Dr Qun Yi Zheng, Executive Vice President of Pure World (formerly Madis Botanicals), said, 'People are always looking for new products. Companies jump on the bandwagon once a new product is introduced to the market. Pure World is constantly working on new products from an R&D point of view ... Since 1996 we have invested more than two million dollars to build our R&D capacity.'

New commercial botanical medicine leads come from a variety of sources including trade shows, consultation of literature and databases, and professional trade organisations, or from promotion of a product by a trader. Interested companies then investigate documentation of traditional use and scientific literature, while marketing departments determine whether the product would appeal to consumers. Suppliers are contacted to ensure that sufficient quantities of the material are available. As Rod Lenoble, Manager of Scientific Affairs at Hauser Inc, said, 'Safety and efficacy, as determined by scientific studies, and availability of starting materials, are important issues in deciding which new products should be developed.' W Letchamo, formerly Director of Phytopharmacy at Trout Lake Farm, echoes this: 'Sourcing is critical for the introduction of new species to the market – Devil's claw (*Hapagophytum procumbens*), for example, could be extinct in the wild in the next five years – we have to figure out a way to cultivate it; currently tons are being taken out of the wild.'

Only after a species has passed through these stages will it be considered for commercial development. Even then, companies might wait to see how an 'esoteric' or 'obscure' new product plays out on the market before they launch a product line. Many products which appear 'new' to the market, like St John's wort, cat's claw, and kava, have been on the market for years, but have recently leapt into the public spotlight. As Kay Wright, Director of Botanical Purchasing at Celestial Seasonings, said, 'Celestial Seasonings has been aware of these products for a long time. Demand for them now is driven by consumers hearing through the media that these are important products. The industry then responds to consumer demand by marketing these products.'

But the window of opportunity for companies to market exclusively 'new' species is open only briefly in the botanical medicine industry. As Grace Lyn Rich of Nature's Herbs said, 'New products set the company above competitors, so we are always open to them – a company may be the first western or US company to market a Chinese or Ayurvedic medicine. But herbs are commodity products, so once we have marketed them, anyone else can too, and often does ... . But new *processes* for delivering herbs can be an important marketing tool, as well. While we can't patent a natural substance, we can claim exclusive rights to a process, such as time-release delivery – so there is an incentive to invest in that side of new product development.'

Industry demand for access to 'new' species from around the world is likely to grow in coming years, however. As government-funded and academic scientific research, largely based in the North, increasingly address species from other regions, the scientific base that companies require to develop commercial products is expanding. As Dr Mark van Diest, Research and Development Manager of Qualiphar, said, 'There is a much longer tradition of studying the plants used in traditional medicine in Europe, than those used in, say, Africa and Asia. Now the universities in Europe are much more interested in African or Asian plants, since the European species have been widely studied. But there is a lag between this research, publication, and subsequent commercial interest.' Many of the best-selling botanical medicines and the media interest which has driven the growth in sales came out of academic and industry-sponsored research (primarily in Europe). As Rod Lenoble of Hauser Inc said, 'St John's Wort, ginkgo, and kava, for example, have been around for a while, but the recent boom in demand has come about in response to scientific studies demonstrating efficacy, which were picked up by the popular press.'

4.5.2 Identification, extraction and standardisation

Identification

Botanical medicine products undergo varying levels of processing, and scientific and technological input. The first step in the production of a commercial product is correct identification of the plant material. Up to 10 per cent of all botanical medicine products on the market contain mis-identified herbs, and at least another 20 per cent contain adulterated herbs and herbal extracts (Leung, February 1997). There exist few fast chemical methods for identifying botanical medicines. Most plant species contain multiple active components, and frequently these components are also found in other herbs that may or may not be taxonomically related. As a result, chemical analysis can only be used as one of various means of identifying an herb. An experienced ginseng dealer can accurately assess the quality (level of gingenosides to a chemist) of a ginseng specimen by colour, appearance, smell and taste, whereas it would take hours for a competent chemist to analyse for gingenosides. Traditional methods – organoleptic, macroscopic, and microscopic evaluation – are still required for herb identification (Leung, February 1997).

Voucher specimens work in many cases, and are employed by companies like Botanical Liaisons (which offers a voucher specimen reference collection service to other companies) to ensure proper species identification. Voucher specimens cannot be used for TCM, however, which employs processed plant parts. As Dr Zhong, Director of UK operations, East West Biotech, said, 'Botanists have difficulty identifying Chinese herb material because they are never whole herbs – but cut and dried pieces. So botanists can do little with them. Chemists would face similar problems for different reasons – there is not any chemical taxonomic system established for plants yet.'

Extraction

Extraction involves separation of the active material from the plant with aid of a solvent (eg water, ethanol, methanol, and hydroalcoholic mixtures). Extracts come in the form of fluid extracts, solid extracts, and powdered extracts, the latter produced when the solvent is evaporated. Although there are many liquid forms of extracts on the market, the preferred extract for use by industry is solid or powdered. Solid and powdered extracts provide greater chemical stability and are cheaper than fluid extracts, since the alcohol can be more expensive than the herb itself (Smith, 1998; Leung, February 1997). Extracts may be 'total' extracts containing the full spectrum of compounds found in a plant or plant part, or might concentrate pharmacologically active compounds:

- *Whole or total extracts* retain a balance of the positive and negative, and synergistic effects of several and diverse compounds, as found in traditional uses of the botanical medicine. Although an 'active' principle could be identified for a particular pharmacological effect, this compound may be only one of several other 'active' compounds that are present but have not yet been identified. Supporting compounds might reduce toxicity associated with the active compound, or act as preservatives, among other effects. Based on long histories of traditional use, these extracts are usually safe and effective although the active principles are not well defined (Leung, February 1997; May, 1997).
- *Extracts containing high concentrations of pharmacologically active chemicals* result when bioactive compounds are extracted from herbs, usually based on scientific data. Although a herb may have a long history of traditional use, the selectively isolated chemical components may not represent the whole herb as it exists in nature, or its traditional properties. For example, ginkgo leaf extracts are standardised to only two groups of marker compounds, and during the process other compounds found in the plant may be lost. Ginkgo standardised extracts are based solely on pharmacological studies of its chemical components, and do not correlate with traditional use in China (Leung, February, 1997; Brevoort, 1998).

Appropriate extraction techniques depend upon the nature of the herbs, and the traditional uses. For example, Western herb extraction is usually effected in alcohol or aqueous alcohol. For herbs containing aromatic components (eg mint, angelica, thyme, sage) extraction is conducted at lower temperatures and for shorter periods of time. Most Chinese herbs are extracted in water and at higher temperatures and for longer times to simulate traditional methods (Leung, 1997). As Staci Eisner, Technical Director of ExtractsPlus, said, 'The development of a modern phytopharmaceutical extract is very complex. First, the chemical composition of the traditional preparation is analysed as extensively as possible. Then, extraction methods are developed which will duplicate as fully as possible the composition of the original preparation. This is the only way to ensure the extract will duplicate the safety and efficacy of the original preparation. In some cases, specific known toxins are removed to further ensure safety; however, the chemical composition must not be extensively altered.'

Extract quality is expressed either as *strengths*, or as *standardised* against a chemical or group of chemical compounds. The strength of an extract reflects the amount (kg/pound) of raw herb used to produce one kg/pound of extract. In practice, commercial extracts are frequently mislabelled in response to perceptions that the higher the strength, the better the quality (although this is not necessarily the case). Because of inconsistency and variability in the concentration of phytochemicals found in different batches of raw material, extracts expressed in strengths do not guarantee consistency in active compounds (Leung, February 1997).

Standardisation

In contrast to botanical extracts expressed in strengths, standardised extracts contain specified and replicable amounts of a presumably 'active' chemical compound or chemical group. Individual plants have different amounts of key constituents, depending upon growing conditions, season in which harvesting took place, and other variables. By setting standardised levels of active ingredients, manufacturing companies can ensure that customers are buying the quality and dose required (Leung, February, 1997).

The rationale for using a standardised extract is rooted in our modern approach to drug development and quality control. It offers predictable pharmacological activities of the chemicals involved, and easy quality control via chemical (HPLC) analysis, as well as convenient stepping stones to developing new modern drugs. This approach works best for species with well-established active principles for specific pharmacological functions, such as: cascara bark (cacarosides and aloins); senna leaf and pod (sennosides); aloe (aloins); feverfew leaf (parthenolide); ginkgo leaf (terpenoids and flavonoids); and saw palmetto berry (lipoids) – all of which pertain to specific conditions for which these extracts are therapeutically used. However, the standardised extract approach is not useful for traditional herbal tonics with multiple chemical components that possess multiple pharmacological effects, like ginseng and astragalus (Leung, February 1997).

In response to increasing calls for improved quality and standardisation of products in the USA, trade associations and companies have taken steps to develop guidelines or services to ensure higher standards. For example, the American Herbal Pharmacopoeia has completed a monograph on St John's wort, and plans to publish six monographs in early 1999. The US Pharmacopeia (USP) has prioritised 21 botanicals for monograph development. In addition, a Methods Validation Programme was launched by the Institute for Nutriceutical Advancement (INA) to bring together manufacturing and supply companies to develop protocols for botanicals in the marketplace. Validated methods for ginkgo, echinacea, and ginseng, are near completion. Key issues addressed by this group include:

- recognition that many botanicals are standardised today on marker, rather than active, chemical components; and
- the need for a standard, replicable methodology to analyse and identify the active compounds, in order to ensure a standardised, replicable final product (Brevoort, 1998).

Some companies are actively marketing their scientific and technical capacity. QBI, for example, provides a 'certificate of analysis and material safety data sheet' on all products, and retains lot samples for three years. Hauser markets NaturEnhance products that carry the 'Hauser certified assay trademark', which guarantees measured concentrations (the assay) of one or more selected marker compounds in the final product, ensuring that the level of marker compounds is representative of the botanical's natural ratio of all compounds. The Certified Assay Programme incorporates steps along the entire product development process: high quality raw material sourcing; proprietary extraction that retains the full range of compounds in their natural ratio; analytical chemistry; GMP production; pharmaceutical grade stability testing; and screening for herbicides, pesticides, microbiological, heavy metal, and foreign matter contamination (Hauser, 1997).

4.6 Regulatory trends

The regulatory environment for botanical medicines varies a great deal by country (Appendix C). In Europe and Japan botanical medicines are comprehensively regulated, with standards set for both manufacturing and ingredients. Because botanical medicines are integrated into the healthcare systems in these countries, regulations reflect norms for safety, efficacy, and quality control. On the other hand, in the USA botanical medicines have been in regulatory flux for many years, and have yet to find a home in the mainstream healthcare system. Once regulated as food additives, since 1994 botanical medicines are considered 'dietary supplements' – another food category. The change in regulatory structure has, allowed the botanical medicines market to expand rapidly, but botanical medicines are more lightly regulated than OTC medicines (which must be proven safe and effective) and packaged food (which must demonstrate purity). In most developing countries, traditional and herbal medicines are used by the majority of the population, and are rarely regulated, although there are moves towards the standardisation of traditional medicines in some countries (Steinhoff, 1998; Balick et al, 1996).

Regulatory frameworks set standards for proof of safety, efficacy, and quality; determine the scope of claims made about products; the information included on labels, and the content of advertisements. As a result, they help determine the nature of the industry, including the demand for 'new' materials. In most of Europe, and Japan, monographs are produced for botanical medicines in trade, and research and testing supporting claims to safety and efficacy is required. Materials 'new' to these markets will therefore take a slower route to the consumer than in the USA, where products are currently considered safe unless proven otherwise. However, the cost of conducting tests in Germany to establish 'reasonable certainty' that a botanical medicine has the desired effect and is safe might be only US$1–2 million, and may take only a few years – a great deal less time and cost than for a pharmaceutical product (Brody, 1999). Short-cuts are possible, as well. Germany and France, for example, have put in place new regulations that allow establishment of 'reasonable certainty' based in part on documented traditional use.

However, an incentive exists in Europe and in Japan to stick with known species. As one researcher in a European botanical medicines and personal care company said, 'We focus most of our research on improving medicaments, rather than trying to find a new one. With the cost of conducting clinical trials on new medicaments coming to at least US$1 million, it is better to try to modify existing ones. In the medicaments area there is really not much room for research on new agents.'

Although the market in the USA is smaller and relatively more immature, the regulatory environment today is more conducive to the introduction of 'new' species from high biodiversity countries, than is the case in Europe or Japan. As Trish Flaster of Botanical Liaisons said, 'Many raw materials are grand-fathered under DSHEA. With new products, all a company has to do is notify the FDA that new plant material will be introduced into a product. If they don't respond, a company can go ahead with the product. A company doesn't need to do much of anything to get new natural products in the botanicals market, as long as the labelling is accurate. It is not costly, which is good, but people can launch poor quality products onto the market, which is bad. It is a trade off.'

4.7 Nutraceuticals: market and regulatory trends

The term 'nutraceuticals' refers to any food ingredient or product consumed for its medical and health benefits, including the prevention and/or treatment of disease. Products include dietary supplements, entire diets (eg macrobiotic), isolated nutrients, and functional or medical foods, including 'designer' biotechnology-enhanced foods, and fortified processed foods such as cereals, soups, and beverages (Kalorama, 1998). Estimates for the 1996 nutraceuticals market include US$16.7 billion (Datamonitor, 1997). Nutraceuticals are an area undergoing rapid growth, with the functional food component alone growing at 12.9 per cent per year since 1992 (Datamonitor, 1997). Because dietary supplements, in particular botanical medicines, are the subject of the remainder of this chapter, and diets are beyond the scope of this book, this section will focus on 'functional' and fortified foods.

Functional (and fortified) foods encompass a wide range of products. In 1996, the largest markets for functional foods and beverages were: soft drinks (35.4 per cent), dairy (32.1 per cent), bakery (16.5 per cent) and cereals (5.4 per cent). Snack foods represented only 4.1 per cent of the total functional food and drink sector in 1996, but experienced the fastest growth (56 per cent) (Datamonitor, 1997). 'Functional' beverages, or nutraceutical drinks, are a rapidly growing segment of the US$20 billion beverage market. Sales were expected to reach US$100 million for 1998, up from US$20 million in 1997 (Greenwald, 1998).

The category of nutraceuticals is expanding and innovating rapidly, and includes anything from 'ginkgo potato chips to ginseng candy bars, Chinese herb cereal, echinacea fruit drink, and kava corn chips' (Brevoort, 1998). Interior Design Nutritionals, a division of NuSkin, launched Splash C with Aloe, a new formulation of a popular drink that provides the benefits of aloe vera 'taken internally'. The company promotes this addition by describing aloe vera as 'rich in phytonutrients and other important compounds including lignin, saponins, anthraquinones, minerals, vitamins, enzymes, amino acids, and monopolysaccharides'. The concept was developed based on 'ancient wisdom' of 'indigenous cultures from India to Africa [that] have been drinking aloe vera juice for centuries'.

'Functional' ingredients include dietary fibres, phytochemicals, lignans, antioxidants, oligosaccharides, and vitamins and minerals. Ingredients in functional foods are not always novel, and may already be associated with a food but specifically developed to be bioavailable – in other words, the body can take full advantage of the active ingredients. Research on new ingredients has been pioneered by Japanese manufacturers. Recently introduced and popular products derived from natural products include: gymnema (derived from the leaves of a plant, it attaches to the taste buds and limits consumption of sweet foods; it is used in diet-related products); monacolin (from an extract of the fungus monascus, considered to be an inhibitor of cholesterol and an effective treatment of several diseases); and cilli (contains massive amounts of vitamin C). The latter two products are marketed by Nichimen, the ingredients arm of Maruzen Pharmaceutical Company, Japan (Datamonitor, 1997).

Food and beverage companies have long 'prospected' in high biodiversity countries in search of alternative ingredients for their products. For example, since the 1960s, scientific and commercial interest in sweet proteins, required in minute quantities, and used as a sugar substitute, has resulted in collecting expeditions around the world. Species yielding intensely sweet plant proteins include *Dioscoreophyllum cumminisii* (West Africa), *Thaumatoccus danielli* (West Africa), *Richardella dulcifica* (West Africa), *Curculigo latefolia* (Malaysia), *Capparis masaikai* (South China), and *Pentadiplandra brazzeana* (Central Africa). Most of these species were used traditionally to sweeten food and beverages (Dansby, 1997). Demand in this industry for access to both species 'new' to the sector and traditional knowledge associated with their use has been significant, and is likely to increase alongside the continuing growth of the nutraceuticals sector.

Market growth in the functional and fortified foods sector has attracted the interest of many major food and drink manufacturers, including Kellogg's, Kraft, Unilever, Nestlé, Nabisco, and Campbell Soup Co. Large pharmaceutical and chemical firms, with in-house research and development capacity easily adapted to these products, are also showing interest. Nutraceuticals offer attractive profit margins, but are less strictly regulated than pharmaceuticals, so the cost of entry is one quarter that for drugs and the time required to introduce a new product a great deal less (Kalorama, 1998; Datamonitor, 1997). Food and pharmaceutical companies are also striking strategic partnerships. Pharmaceutical companies bring funding, experience in product research, and brand-building expertise to the partnership, while food companies provide insight into the marketing of food products featuring diet or disease claims (Kalorama, 1998). The leading functional foods and beverages marketed in the USA in 1997 are shown in Box 4.9, p98.

Government and academic institutions are also undertaking research into nutraceuticals. In the USA, the National Cancer Institute's (NCI) Designer Foods Project is conducting research into the medical benefits

Box 4.9 *Leading Functional Foods and Beverages Marketed in the USA, 1997*

Category	Brand	Manufacturer	Description
Soft drinks	Minute Maid Premium Calcium Rich Orange Juice	Coca-Cola	calcium-fortified juice
	Iced Tea with Ginseng	Ferrolito Vultaggio & Sons	ginseng tea
	Josta	Pepsi Co	carbonated guarana drink
	Citri-Lite	Citri-Lite Co	citrimax-based diet cola
	Aloe Berry Nectar	Forever Living Products	aloe gel with fruit juices
Dairy	Yoplait Fat Free with Bifidus Kefir	General Mills Lifewat Foods	probiotic yogurt
Bakery	Mighty White	Allied Bakeries	white bread fortified with folic acid
	Kellogg's All Bran	Archway Cookies	low-fat, high-fibre biscuits
Pasta and rice	Uncle Ben's with Calcium Rice	Uncle Ben's Inc	calcium-fortified rice
Other	Intelligent Quisine	Campbell Soup Co	fortified frozen meals
	Peter Pan	Procter & Gamble	peanut butter with vitamins and minerals
	Ensure	Abbot Laboratories	protein-based nutritional drink
	PowerBar	PowerFoods Inc	energy bar

Source: Datamonitor, 1997.

of 'functional' plant compounds, such as those derived from umbelliferous (root) vegetables, soybean, and flax seed. The University of Illinois Functional Foods for Health Program is also focusing research efforts on 'functional' phytochemicals (Kalorama, 1998).

4.7.1 US regulatory environment

The nutraceuticals industry in the USA was spawned in part by the 1990 Nutrition Labeling Education Act (NLEA). The NLEA established standardised US food labels, and spurred consumer interest in the nutritional content of food and beverages. However, the growth potential of lessening or removing the negative elements of foods has gradually been exhausted, so companies are now seeking 'functional' health ingredients to differentiate their products (Kalorama, 1998; Datamonitor, 1997).

In order to implement the NLEA, the US Congress directed the FDA to develop regulations governing health claims for foods, requiring that all claims be supported by 'significant scientific agreement'. Although the FDA allowed general categories of claims surrounding dietary fibre, fats, sodium, and calcium, the overall effect was to reduce health claims made for food products (Datamonitor, 1997). The FDA does not officially recognise the term 'functional food', so these products fall under regulations for either 'foods' or 'dietary supplements'. In order to make medical claims for nutraceutical foods and beverages, marketers must submit their products to the lengthy and expensive pharmaceutical drug approval process (Chapter 3). Campbell Soup spent five years and nearly US$30 million to develop Intelligent Quisine, running clinical trials to demonstrate that eating these meals reduces blood pressure and cholesterol (Kalorama, 1998; Datamonitor, 1997).

The absence of patent protection for researched health claims, and the transparency of the claim system, act as deterrents to large investments in research. Research conducted in order to file an application for health claims is made 'transparent', or publicly available, which means that any food having sufficient quantities of the nutrient under study may use the health claim. As a result, few petitions are filed for health claims. One recent exception was Quaker Oats Co, which filed a petition with the FDA in 1995, based on 37 clinical studies conducted between 1980 and 1995 demonstrating the relationship between the consumption of oat products and serum lipid levels. But Quaker Oats commands a large share of oat-product sales in the USA, so will gain a significant return on the investment. Only a handful of companies are in this position. (Datamonitor, 1997)

A division of Johnson & Johnson plans to test market a margarine spread called Benecol, which the company considers a nutritional supplement because it has been shown in studies to lower cholesterol by as much as 9 per cent. Benecol's active ingredient is sitostanol, which is derived from pine trees. Although the product has been sold in Finland since 1995, the FDA requires that rigorous testing be undertaken in the USA if any health benefits are claimed. As a result, the company is marketing Benecol as a nutritional supplement, since such products are exempt from most FDA regulations under DSHEA (Greenwald, 1998).

In March 1999, the advocacy group Center for Science in the Public Interest urged the US government to more effectively regulate functional foods that, while holding promise, could otherwise become 'the snake oil of the next century' (Neergaard, 1999). There are increasing calls for the FDA to reconcile regulatory frameworks for dietary supplements and functional foods, and should this transpire a boom in the functional foods sector in the USA is likely to result. In the meantime, despite recent growth in the US market, it is likely that Europe and Japan will lead this sector for the foreseeable future. European countries and Japan have established research programmes and regulatory environments conducive to investment in nutraceutical product research and development (Datamonitor, 1997).

4.8 Raw material supplies

Global demand for high quality botanical medicine products has focused greater attention on the quality of raw plant material. At the same time, demand for high quality raw material is on the rise; however, massive expansion of markets for botanical medicine products has created openings for poor quality material. Spikes in demand for products such as St John's wort and kava have depleted supplies and driven up prices for raw plant material. As a result, many less reputable suppliers have access to a market which, although increasingly quality conscious, seeks first to fill its demand for raw material.

There is a trend in the botanical medicine industry towards cultivated material, particularly in the case of mass marketed species. Cultivated material usually has the advantage of being reliably available at a consistent quality and price, and genetic improvement can often select for chemical composition or characteristics that might be lost when a plant is removed from its native habitat. Supply of raw materials is such an important issue for some of the larger companies that they will not pursue product development on a species for which they have not secured a reliable supply. In general, companies interviewed prefer easily accessible, commonly-known, and widely-available plant species.

As Cindy O'Connor, former Director of Marketing at QBI, said, 'The smart companies look around and realise that you don't have a business if it is not sustainable ... if you have any kind of vision, companies will make the kind of investment QBI is making in farming.' This includes the development of planting, harvesting, and drying practices that optimise quality, yield, and bioactivity levels. QBI has set up an affiliate, Herb-Tech, which specialises in domestic cultivation. 'Typically, wild harvesters take plant material when it is easiest to identify, not necessarily when it is at its bioactive peak, or after the seed has matured. We wish to conserve wild resources and preserve biological diversity, as well as ensure that plants are harvested with their maximum potency intact' (QBI, 1998).

Traditional wildcrafting systems, however, usually time harvests to take advantage of bioactive peaks and incorporate sustainability levels to supply local demand. Herbalists generally prefer the qualities of wildcrafted material. But for most species sold on to mass markets, cultivation offers the only sustainable sourcing option and guarantee of consistent quality (McCaleb, 1997). As Mark Van Diest, Research and Development Manager at Qualiphar, said, 'Standardisation is very important for herbal medicines. It is much easier to standardise an extract if you start with standardised cultivation methods – the same soils, same plant material, etc. The best companies Qualiphar purchases material from have their own cultivated material. Cultivation also ensures access to raw material over time – if a company launches a product and the extract is no longer available, it is very embarrassing. We need assurance of quality and continuation of delivery of raw material.'

Wildcrafting continues to supply the majority of raw plant material to the industry, however, and some think that it will always have a role, particularly for lesser-traded species and those difficult to cultivate. As Trish Flaster of Botanical Liaisons said, 'There is increased interest on the part of industry in sustainable sourcing, but to do it right, you must look ahead five years. Companies are just discovering sourcing as an issue, because things are getting hard to find. Extraction is becoming more common, but requires larger volumes of raw material – one to three times as much ... Industry's response to shortages has been to stop collecting from the wild, and get material into cultivation, but it is possible to sustainably source a lot of materials from the wild, and some species cannot be cultivated. The quality of wildcrafted material might be better as well.'

Companies employ a range of strategies to source raw materials. Most buy their plant material in bulk from wholesalers, via elaborate and complex trade networks, with little or no information on sourcing strategies, including environmental and social impacts. Other companies purchase farms or wholesale operations to ensure greater control over the quality of raw material supplies. Some companies are working to bring as high a percentage of plant material as possible into organic, small-farm based, or otherwise environmentally- and socially-sound systems.

Consumer demand for organically grown material is increasing, along with pressure on suppliers to comply. W Letchamo, former Director of Phytopharmacy at Trout Lake Farm, one of the largest producers of organic material in the USA, said 'There is a growing demand for organic and properly handled material. Consumers have started to think about the importance of organic products. But organic material is more expensive, because it requires a lot of expensive labour to deal with weed and microbial issues.' It is not clear, however, that the rapidly expanding mass market for botanicals will demand or support more costly organic cultivated raw material. The speciality health food store consumer would be likely to respond best to organic certified botanical medicines, but ironically this group has only recently come to recognise raw material sources as an issue. Additionally, some consumers associate wild harvested material with higher quality, so might reluctantly purchase organic farmed material.

Given the fierce competition in the botanical medicine industry, however, some in industry see quality control and sustainable sourcing as a valuable way not only to ensure quality products, but to differentiate them in the marketplace. According to Marian Burningham, New Products Manager of Nature's Way, 'So many products are from the same material, and now everything is moving towards the same standard of manufacturing, so companies need to find points of differentiation. It could make a difference to the consumer if a label said "has not harmed people, animals, rainforests, etc." Consumers are increasingly aware of these issues. But it is a chicken and egg thing – in the beginning a small and vocal group of consumers heightened awareness within industry, which is now following, and then through industry these issues will reach a mass market.'

Raw materials supplied to the botanical medicine industry are still traded as bulk commodities, with multiple companies handling material before it reaches a manufacturing facility. Usually only a fraction of the cost of a finished product reflects the raw material costs; prices paid by consumers are much more a reflection of costs associated with marketing than of raw materials (*NBJ*, June 1998). The price mark-ups on raw material along the way are also significant. For unprocessed cat's claw in Peru, for example, local collectors are paid between US$0.30 and US$0.65 per kg. Unprocessed bulk cat's claw in the USA sells for around US$11.00 per kg (King et al, in press). Although some companies have supported resource surveys (eg Boehringer Ingelheim for yohimbe in Cameroon), and others have invested in cultivation schemes, for the most part significant investments in sustainable raw material sources are not considered economically warranted, and are not common practice in the botanical medicine industry.

Sustainable supply issues, along with calls for quality control, safety, and efficacy, are increasingly raised by consumers, the media, and within some industry circles, however many companies feel bound by the commodity tradition of sourcing. One US representative from a large company responded to a question about sustainable sourcing with, 'We buy everything in powdered form, already processed. We hope the vendors do it right. If it was a perfect world we would all be playing fair, but it is extremely difficult to monitor sources completely. We are not at the source, so all we can do is try to buy from ethical suppliers.' While this attitude towards sourcing does not reflect most of the companies interviewed, it appears widespread and common not only in the USA, but in Europe as well. Most surprisingly, it is found in companies that market themselves as distinctly 'green'. As a result, many products sold as 'green' are in fact damaging to the environment, and are based on locally-exploitative or culturally-intrusive labour and social practices.

4.8.1 'Boom-bust' cycles

Increased global demand for botanical medicines has meant that supply is tight for most species, which fall into some form of a 'boom-bust' cycle. As described by Landes (1997), the boom-bust cycle works in the following way: a sudden surge in demand for a product leads to a spike in demand for supplies. Good-quality supplies are usually limited, so the market is emptied in days, sometimes hours. Prices for low-quality material skyrocket and companies end up paying high prices for poor quality material. Next, valueless material comes on the market and even that is bought by unsophisticated buyers. Then growers and gatherers are encouraged by extremely high prices and seemingly unlimited demand to cultivate or gather more material. Good quality material is over-supplied just as consumers move on to the next 'hot' botanical medicine. Landes (1997) believes that one can tell where a botanical medicine is in the market boom-bust cycle by looking at price, available quality, and the number of suppliers 'desperately trying to sell it'.

Kava, for example, sold to manufacturers for around US$100 per kg a year ago, but today sells for US$250–300 per kg (see Case Study below). When a hurricane devastated saw palmetto berries in Florida in 1995, the processed oil sold by Indena went from US$250 to US$450 per kg (*NBJ*, 1998). Even sales for common and widely known species like St John's wort can spike when media attention is brought to bear. As Peter Haferman, Vice President of Marketing and Sales at Botanical International, said, 'Who could forecast that St John's wort sales would go from zero to something like 27,000 per cent growth? Everyone's looking for the next hot herb .... Our customer base is beginning to ... recognise the need for forecasting'. Some companies have had to ration St John's wort, and have moved rapidly to get the plant into cultivation (*NBJ*, June 1998). Landes (1997) recommends that marketing departments in US companies communicate closely with purchasing departments to anticipate demand, and that good contact is maintained with reliable suppliers. In Europe, this type of market anticipation and long-range planning is already the norm (*NBJ*, June 1998). Martin Bauer, for example, has plant material for many species in stock for at least 18 months in advance.

4.8.2 Major raw material exporting and importing countries

The 1996 global trade in medicinal and aromatic plant material (commodity group 'pharmaceutical plants'[2]) was more than 440,000 tonnes, valued at US$1.3 billion. China is the leading exporting country, with annual exports between 1992 and 1996 averaging more than 140,000 tonnes (UNCTAD, 1996; Lange, 1998). Hong Kong was the world's leading consumer country, with 80,000 tonnes per year processed in the large local pharmaceutical industry (Table 4.10). The fastest growth in imports is by Japan, which moved from importing 37,698 tonnes in 1992 to 96,140 tonnes in 1996 (UNCTAD, 1996; Lange 1998).

Virtually every company interviewed as part of this study cited sources for raw material as 'global', 'world-wide' or 'too numerous to mention'. For example, Celestial Seasonings sources material from 38 countries, with imported material representing 60 per cent of that used by the company. Martin Bauer sources 70 per cent of its raw plant material (by volume) from the northern hemisphere; 95 per cent of its plant material comes from developed countries (including Eastern Europe), with only 5 per cent from developing countries (eg Sudan and India). In another example, East West Biotics sources 50 per cent of its material from China, purchasing the remaining 50 per cent from wholesalers that source material from around the world.

4.8.3 Wild versus cultivated sources

Although wildcrafting continues to provide the majority of plant material consumed by the botanical medicine industry, the trend is towards cultivated materials that can better guarantee supplies, consistency, species identification, and high levels of post-harvest handling (Box 4.10). Wildcrafting will continue to play an important role in the industry, however, because many herbalists prefer wildcrafted material for their own use; some species are extremely difficult to cultivate; others have small markets so wildcrafting can be sustainable and cultivation uneconomic; wildcrafted material is often cheaper; and species new to the market are unlikely to have been in extensive cultivation previously.

Many species in the botanical medicines trade are common and widespread 'weedy' species found in

Table 4.10 *Top 12 Importing and Exporting Countries of 'Pharmaceutical Plants', 1992–96 (Commodity Group HS 1211 = SITC.3: 29.24)*

	Average annual volume (tonnes) 1992–6	Average annual value (US$) 1992–6
Importing countries		
Hong Kong	80,550	331,700
Japan	57,850	158,300
USA	51,600	118,400
Germany	45,400	107,100
Rep Korea	34,200	53,350
France	19,800	46,350
Pakistan	12,550	12,650
Italy	10,400	39,100
China	9,300	35,950
Singapore	8,500	60,350
UK	7,400	24,450
Spain	7,350	24,400
Exporting countries		
China	140,450	325,550
India	35,650	53,450
Germany	14,900	72,550
Singapore	14,400	62,750
Chile	11,700	26,350
USA	11,650	120,200
Egypt	11,300	13,650
Pakistan	8,500	5,450
Mexico	8,250	9,400
Bulgaria	7,350	12,250
Morocco	7,150	11,970
Albania	7,100	13,750

Sources: UNCTAD COMTRADE database, United Nations Statistics Division, New York; Foreign Trade Statistics of Bulgaria, National Statistic Institute, Sofia; Foreign Trade Statistics of Albania, National Statistic Institute, Tirana, Albania; Lange, 1998.

Box 4.10 *Commercial Collection from the Wild versus Cultivation of Medicinal Plants*

	Wildcrafting	Cultivation
Availability	decreasing	increasing
Fluctuation of supply	unstable	more controlled
Quality control	poor	high
Botanical identification	can be unreliable	definite
Genetic improvement	no	yes
Agronomic manipulation	no	yes
Post-harvest handling	poor	usually good
Adulteration	likely	relatively safe

Source: Palevitch, 1991.

many regions. 20 per cent of those in trade in Germany are naturalised outside their native region, and only 40 per cent are restricted to a single continent (Lange, 19997). As a result, many botanical medicine species can withstand wild collection (in contrast to rare, specialist species found in restricted areas). But 'weediness' is not always sufficient, as the widespread and common St John's wort demonstrates, and many species populations cannot withstand even moderate wildcrafting. The industry-wide trend is to move big selling species into cultivation as quickly as possible, often with the direct involvement of manufacturing and processing companies, and to source those sold in smaller quantities through brokers or wholesalers, from what are often wildcrafted sources.

Of the 1,560 plant species traded in Germany, 50–70 per cent by volume are wildcrafted, representing 70–90 per cent of species. Only about 130–140 species originate from cultivated sources, with just 5–20 per cent of these purchased from German farmers. 'Green' traders purchase a larger proportion of their botanicals from German farmers (up to 70 per cent) (Lange, 1997; Lange 1998). Of the 1,200–1,300 species native to Europe in trade, at least 90 per cent are still wild-collected. Wild harvesting is particularly common in Albania, Turkey, Hungary, and Spain. By volume, 30–50 per cent of medicinal and aromatic plant material in trade in Hungary is wild-collected; 50–70 per cent in Germany; 75–80 per cent in Bulgaria; and almost 100 per cent in Albania and Turkey. The overall volume of wild-collected material in Europe is estimated to be 20,000–30,000 tonnes annually. Cultivated species include peppermint (*Mentha x piperita*), lavender (*Lavendula* spp), opium poppy (*Papaver somniferum*), caraway (*Carum carvi*), and fennel (*Foeniculum vulgare*). Some species, like yellow gentian (*Gentiana lutea*), are both cultivated and wild-harvested. Countries in the EU producing large quantities of cultivated material include France, Hungary, and Spain (Lange, 1998).

Within individual companies, a trend is apparent towards larger proportions of cultivated sources. For example, in 1989 Indena USA acquired 10 per cent of its biomass from cultivated sources, but in response to rapid increases in demand by 1998 that figure had risen to 60 per cent (*NBJ*, June 1998). Of the companies interviewed, few could break down the percentage of wildcrafted versus cultivated material, and figures varied by volume, number of species, or value. By volume, 11 companies estimated that the majority of their material comes from cultivated sources, five cited wild-harvested material as their primary source, and three said it was about 50 per cent for each. For example, Martin Bauer estimates that 70–90 per cent of its raw material by volume comes from cultivated sources, but that this reflects only 20 per cent of the species used. Shaman Botanicals acquires 80 per cent of its raw material from wild sources, but consistent with industry-wide trends is moving to a larger proportion (60 per cent) of cultivated material over the next two to four years. Virtually all companies interviewed expressed the desire to acquire as large a percentage of material as possible from cultivated sources.

4.8.4 Conservation and wise management

The sourcing of raw materials for the botanical medicine industry is not directly linked to the access and benefit-sharing provisions of the CBD (Chapter 2). However, the wildcrafting of botanical medicines can threaten species and biodiversity conservation, and is closely tied to sustainable use and other elements of the Convention. The botanical medicine industry is squeezed by increasing global demand on the one hand, and decreasing wild populations of many species on the other. Wild populations of species like pygeum (*Prunus africana*) and yohimbe (*Pausinystalia yohimbe*) are currently harvested in unsustainable and destructive ways in order to feed international markets (Cunningham et al, 1997; Cunningham and Mbenkum, 1993; Sunderland et al, 1997) (Box 4.11).

Around 200 medicinal plant species have been added to CITES (Convention on International Trade in Endangered Species of Wild Fauna and Flora) appendices. CITES regulates international trade in animals and plants through a system of permits and certificates, protecting species from excessive trade by listing them in Appendices. Appendix I includes species threatened with extinction, and for which trade in wild material is generally prohibited; Appendix II species can be traded with the proper permits. Medicinal species on CITES Appendix II include: *Saracenia* spp, *Aloe ferox, Pterocarpus santalinus, Rauwolfia serpentina, Prunus africana, Panax quinquefolius, Hydrastis canadensis,* and *Podophyllum hexandrum* (Sheldon et al, 1996; Robbins, 1998; Lange, 1998b).

Efforts are underway on a number of fronts to create guidelines for sustainable harvesting and codes of conduct for collectors. For example, the Rocky Mountain Herbalist Coalition was formed in the USA in 1991 to regulate the collection of medicinal plants by creating an official registry of herbalists and collectors, and by establishing guidelines for collection. The United Plant Savers is working to preserve native medicinal plant populations in the USA suffering from loss of habitat and over-exploitation, such as golden seal, American ginseng, kava, and all species of echinacea. The IUCN Species Survival Commission Medicinal Plant Specialist Group also focuses attention on vulnerable species

Box 4.11 *Examples of Species Under Pressure from International Trade:* Prunus africana *and* Tabebuia *spp*

Prunus africana (Pygeum) is a hardwood species native to montane forests in Africa, with a range extending from Cameroon across mainland Africa to Kenya and Madagascar, and as far south as the Cape. Pygeum has been used traditionally for carving, and in medicine. Over the past 20 years, it has been sold on the botanical medicine market in Europe as a treatment for benign prostatic hyperplasia, with total annual sales estimated at US$150 million. Pygeum bark continues to be over-harvested for export markets, and the species was listed on CITES Appendix II as a 'vulnerable species requiring monitoring'. Collections of bark in the wild continue, and sustainable, cultivated sources are few (Cunningham et al, 1997; Cunningham and Mbenkum, 1993; Sheldon et al, 1997).

Another species under pressure from international markets is *Tabebuia* spp (pau d'arco), the inner bark and hardwood of which have been used for centuries by indigenous peoples in South America to treat a range of ailments. Pau d'arco has recently been discovered by international consumers for the treatment of a range of problems, including cancer, leukemia, baldness, allergies, dysentery, malaria, herpes, diabetes, and acne. A portion of the pau d'arco used in the botanical medicines industry is collected prior to felling trees for their timber. However, the bulk of pau d'arco is still harvested from standing trees, unconnected to logging. Increased international demand has placed wild populations under severe pressure. As a result, similar species – which do not possess the desired active ingredients – are substituted for pau d'arco and sold in stores across the USA and Europe. Some of the substitute species themselves are now under threat from over-harvesting (Sheldon et al, 1997; Laird, 1995).

(Sheldon et al, 1997; IUCN SSC Medicinal Plant Specialist Group, 1997; Rocky Mountain Herbalist Coalition, 1995; Cech, 1997).

Within industry, some companies are involved in monitoring the trade in raw materials. East West Biotics, for example, has entered into a partnership with the Royal Botanic Gardens, Kew, and the Institute for Medicinal Plants in Beijing to establish research links, and monitor TCM plants in the UK trade for correct identification, and to ensure CITES species are not marketed. As Marian Burningham, New Products Manager at Nature's Way, said, 'There is a need for documentation of raw materials. Many suppliers say material is sustainably sourced, and they know the right things to say, but a lot are not very conscientious. There should be a penalty attached to selling herbs that are endangered or acquired unlawfully, because if there is a market for something, it will be filled. Eager raw material suppliers and manufacturers will find a way to do it, so the business needs to be regulated at the marketing end. Control over the trade in botanicals is important for both countries and consumers, since it relates directly to quality of raw materials. Some kind of authentication process, labels, and auditing, is needed.'

One of the most commonly-cited ways in which companies apply pressure for sustainability is through the creation of long-term sourcing partnerships. Kay Wright, Director of Botanical Purchasing of Celestial Seasonings, for example, said that 'Celestial Seasonings wants reliable supplies of material year in and year out. Since we are not a trader, and have a production line that must have raw material, we can create incentives for farmers to produce the material according to our specifications, since they can rely on us to come back next year. This includes growers in China, Egypt, Eastern Europe, Turkey, and Thailand. We have been working with some of our suppliers in most of those countries, as well as California and the north west USA, for more than 20 years'.

Rod Lenoble, Manager of Scientific Affairs at Hauser Inc, also feels that long-term relationships with raw botanical suppliers benefit the product quality, the growers, and the environment. He said, 'We like to set up long-term growing contracts, so instead of purchasing raw botanicals on the spot market or seasonally, we make agreements that last for years. This creates an incentive for sustainable farming and harvesting. We also furnish them with incentives to ensure that the material is clean and of high quality and meets our specifications. We try to educate the growers that we work with by offering analytical services to check crops as they grow. We work with growers, collectors, and buyers throughout the US, Central and South America, Europe, and Asia.' Axxon Biopharm is working through partnerships within Nigeria and other African countries to make sourcing relationships equitable and sustainable. As Angela Duncan, Chief Operating Officer, said, 'By being intimately involved in the sourcing of raw materials, the company can ensure the quality of the raw materials, and ... the integrity of the relationship, including respect for local values and individuals, and for the environment.'

4.9

Case study Exploding international demand for a species with widespread traditional use: the case of kava (*Piper methysticum*)

Kava (*Piper methysticum*) is endemic to the South Pacific, and has a long and extensive history of traditional use in Pacific Ocean societies. Over thousands of years, farmers have selected for desired traits, developing 118 cultivars of kava, which are grown in agricultural systems refined over generations. In the 1990s, promotion of kava on the international market led to an explosion in demand for kava products. The result is dramatic price increases, and pressure on supply sources developed to serve regional markets and subsistence use. The case of kava highlights many of the potential benefits and risks involved in the marketing of species 'new' to international consumers, and the ways in which the botanical medicine industry can generate benefits for communities and countries upon whose knowledge and resources commercial markets are based.

The species *Piper methysticum*

Kava (*Piper methysticum*), a member of the pepper family, the Piperaceae, is native to Melanesia, Polynesia, and Micronesia, including the countries of Papua New Guinea, The Solomon Islands, Fiji, and Vanuatu. It is thought that kava originated in northern Vanuatu, which has the greatest diversity of kava zymotypes, morphotypes, and chemotypes. Kava is a hardy, slow-growing perennial shrub that can attain heights of more than three metres (Lebot et al, 1997). Referred to by Lebot et al (1997) as 'one of the classic enigmas of Oceanic ethnobotany', its origins, distribution, and history of migration remain unresolved. While kava does flower, specimens of *P. methysticum* in herbaria around the world do not have seeds, and female parts are generally uncommon (Lebot et al, 1997).[3]

Traditional use

Kava is the preeminent mind- and mood-altering substance of Oceania. It alleviates stress and anxiety, and elevates the mood. Like betel nut in South East Asia, coca leaf in Andean society, and coffee, tea, and alcohol in Western societies, kava occupies a central place in the culture and social customs of the region. It is also used medicinally to treat uro-genital inflammation and cystitis, gonorrhea, asthma and tuberculosis, and to relieve headaches, restore vigour, soothe the stomach, and cure whooping cough. Kava can also be applied topically to treat fungal infections and soothe skin inflammations (Kilham, 1998; 1996; summary of ethnobotanical studies in Lebot et al, 1997). In addition to its secular function, kava is linked to religious inspiration, and its use can involve the invocation of ancestral spirits (Lebot et al, 1997).

Kava consumption has increased in the South Pacific over the last few decades. In Vanuatu, for example, its use as a traditional and socially-acceptable alternative to alcohol has been successfully promoted by the government. Christian church hostility to kava has largely faded, and kava has also acquired a political value as an icon of national and indigenous identity and unity in many South Pacific countries. The number of urban *nakamals* – commercial establishments selling prepared kava – has dramatically increased in recent years (Lebot et al, 1997).

Traditional cultivars

Kava has been in cultivation for more than 3,000 years, with a resulting 118 sterile cultivars found across the South Pacific (Lebot, 1997). Evidence suggests that cultivated kava derives from a wild progenitor, *Piper wichmannii*, a fertile Piper indigenous to New Guinea, the Solomon Islands, and Vanuatu, and that these two taxa of Piper might be wild and cultivated forms of the same species. Farmers in Vanuatu are thought to have been the first to select and develop the species as a vegetatively-reproduced crop, around 3,000 years ago. From Vanuatu, kava cultivars were carried to the other Pacific Islands (Lebot et al, 1997).

There are numerous varieties of kava, with some more highly prized than others; most kava farmers have a preferred variety, linked to its effect. Cultivars are distinguished by the colour of the leaves and stalks, the thickness of stalk joints, stalk length between joints, the presence of spots, and other factors. Varieties of kava are valued for both their effects upon the body, and their ornamental and spiritual worth. Kava plants are exchanged and used at most important ceremonies, and some varieties are cultivated into ritual shapes (Kilham, 1998; 1996). Each cultivar has specific environmental requirements, and a range more restricted than *Piper methysticum* as a whole (Lebot et al, 1997).

Cultivation

Like most traditional Pacific crops, kava is propagated asexually. Kava cultivars are reproduced by vegetative propagation – farmers plant cuttings taken from existing stems. Kava is flexible in its cultivation requirements and thrives in shade, growing well in multi-crop Melanesian gardens, partially shaded by taller crops like bananas and manioc, and trees retained when gardens are cut from the forest. Kava plants are usually harvested after three or four years, but may be left growing for more than 20 years. Most kava crops do not receive the same level of attention as subsistence crops like yam and taro, but some cultivars are intensively cultivated, particularly those important for ritual, display, and exchange (Lebot et al, 1997). For the most part, however, kava requires limited labour, capital investment, and no chemical inputs (Kilham, 1998). Kava is grown traditionally in multi-cropping systems, but now a significant portion is grown as a monocrop, and this portion is increasing in response to rapidly escalating export markets.

Kava has a number of advantages as a cash crop (Lebot et al, 1997):

- the production technology is familiar to farmers;
- there are ample supplies of planting material;
- small-scale production is feasible;
- it is resistant to hurricane damage;
- it can be grown in multi-cropping systems;
- it matures earlier than the major tree crops;
- commercial processing is relatively simple (cleaning and drying);
- the return per work hour is high; and
- the return per hectare is high.

The return per hectare of kava production exceeds that for many cash crops such as coconuts, coffee, and cocoa. Spices such as vanilla, cardamon, garlic, and ginger exceeded the return of kava per hectare in 1986 (see Lebot et al, 1997), but this is unlikely to be the case today, with prices at the farm gate having increased five-fold or more.

Active constituents

Chemical and pharmacological studies of kava have been underway for the past 140 years, but so far the results of these efforts have been inconclusive. The main active ingredients in kava are found in the roots and rootstocks, and are a group of compounds known as the kavalactones. 15 kavalactones have been isolated, and six are found in high concentrations in kava: yangonin, methysticin, dihydromethysticin, kavain, dihydrokavain, and demethoxy-yangonin (Lebot et al, 1997). Scientific research on kava continues in Europe, and is now underway in the USA as well. The American Herbal Products Association, for example, has established a kava committee, and is supporting research on toxicity. The FDA is undertaking research on kavalactones, and companies and academic institutions have initiated clinical trials, such as those underway by Duke University in conjunction with Pure World (formerly Madis) Botanicals.

Market growth

Kava has featured in European, US and Japanese medicine for around a hundred years. For example, the 1914 British Pharmacopoeia included 'kava rhizoma'; the 1950 US Dispensatory listed kava as a treatment for gonorrhea (under the name 'Gonosan') and nervous disorders ('Neurocardin'). Over the last few decades, French companies, in particular, have purchased kava on a consistent basis (Lebot et al, 1997; Kilham, 1998). Kava is the only species from the traditional Pacific Island pharmacopeia that has been marketed by the European pharmaceutical industry (Lebot et al, 1997).

In the last five years, international demand for kava has soared. In 1996, a coalition of 21 botanical medicine producers developed a coordinated plan to market kava aggressively in the USA, with resulting mainstream media coverage, and massive consumer interest in kava products. In the mass market, kava saw 473 per cent growth in the USA over the 52 weeks preceding July 1998, and was ranked the eighth top-selling herb, with US$8 million in sales (Brevoort, 1998; IRI, 1998). In health food stores, in the 12 months preceding June 1998, calmative herbs were the second-fastest growth category, at 47 per cent growth. In addition to kava, this category includes valerian, chamomile, and scullcap (Brevoort, 1998; SPINS, 1998). South East Asian markets for kava are also reported to be expanding (Lebot et al, 1997).

Supply crunch

As with any product that receives sudden media and consumer attention (see above), demand for kava has outstripped supplies. Germany (the main bulk ingredient supplier to world markets) and the USA are absorbing any kava made available to the market. Kava takes from three to four years to reach maturity, so demand initiated in 1996 will take a number of years to fill. Between 30 and 50 kava buyers from the USA and Europe were reportedly in the Fiji Islands in October 1998. A Tasmanian exporter, Ronald Gatti, reported that 50 tons of dried kava root are shipped out of Fiji every week (Adams, 1998). This compares with a reported total 100 tons shipped to Europe for the manufacture of kava products for the entire year of 1996 (Kilham, 1996). However, the main limiting factor in the supply of kava to international markets appears to be the lack of suitable infrastructure to deal with transporting and handling large quantities of the material. Like the cultivation of the species itself, development of appropriate infrastructure takes time (Kilham, personal commun-ication, 1999).

The results are predictable: in more accessible areas, immature plants are harvested; bulk material is adulterated with other species; and availability of kava for domestic markets is impacted. However, because kava is seen as a valuable long-term cash crop and marketable product, local groups, government, and international companies are actively working to get more kava into cultivation.

'Kava's traditional cultural meanings and social functions are now overlaid with new uses in the contemporary Pacific: kava as symbol of Christian atonement; kava as icon of the new state; kava as cultural fetish within developing nationalist discourse; kava as assertion of resistance and indigenous rights; kava as cash crop; kava as ethnic valium or alcohol; kava as fulcrum of ongoing male domination and gender inequalities; kava as camouflage for developing economic equalities and class formation; kava as shared pick-me-up of urban Pacific kava bars.'
(Lebot et al, 1997)

Price increase

In the 1980s, growers received around US$1 per kg of dried kava rootstock (Lebot, 1997). Today, the price is US$5–8 per kg for growers, with the export price for extraction companies and manufacturers being US$22–8 per kg (Box 4.13). Prices paid to farmers vary significantly, but for the most part farmers appear to benefit from higher prices paid by international companies desperately in search of raw material. Kava exports from Fiji earned US$10 million in the first six months of 1998, the same amount as for the entire year of 1997 (Adams, 1998).

In 1997, kava sold for around US$100 per kg to manufacturers; the price today is around US$250–300 (*NBJ*, June 1998). The largest profits made on the kava product market, however, appear to be those of companies selling ground kava root products, without any form of standardisation (and dubious efficacy), to American consumers (Kilham, personal communication, 1998).

Major companies

Most high-quality bulk ingredient supply companies produce kava extracts, including Schwabe (Germany), Finzelberg (Germany), Muggenberg (Germany), Potter's Herbal Supplies (UK), Brenner-Efeka (Germany), Fink (Switzerland), Merrell Dow (Germany), East Earth (USA), Hauser (USA), and QBI (USA) (Kilham, pers. com., 1999; Kilham, 1996). In the USA, the largest producer of kava bulk ingredients is Pure World Botanicals (Box 4.12). Almost all major manufacturing and marketing companies have one or more kava products, including in the USA Nature's Way, Twinlabs (Nature's Herbs), Solgar, and Country Life.

Impact of increased international demand on local communities

Concerns have been raised about the ways in which foreign buyers work with local growers. Although farmers appear to be benefiting from price increases, the types of commercial relationships they arrange with international buyers might not be to their long-term advantage. Adams (1998) quotes Chief Josetika Nawalowalo, chairman of the National Kava Council established in 1998 by growers and traders to protect the kava industry, as saying, 'We've had fly-by-nights from overseas negotiating directly with farmers and it's a dangerous trend. Our people are not educated to deal with multinational companies'. He recommends that the government regulate the kava industry, projected at US$100 million in Fiji in the year 2,000, in order to protect local farmers. Lebot et al (1997) also

Box 4.12 *Pure World (Madis) Botanicals: Bulk Ingredient Supplier of Kava*

According to Qun-Yi Zheng, Executive Vice President of Pure World Botanicals, Pure World became interested in kava in the early stages of the market. After reviewing traditional uses and scientific literature, Pure World decided that a product with kava's claimed attributes would have a significant market in the USA. Pure World then undertook 'research from the field to the finished extract'. This involved Chris Kilham, an early marketer of kava, studying the growth and traditional uses of kava in the South Pacific Islands. In order to obtain quality kava root, Natalie Koether, President of Pure World Botanicals, was also involved in establishing a partnership with a cooperative of Pacific Island growers to source raw materials at an agreed price, for a set period of time. An intermediary wholesaler works directly with communities and acquires permits on behalf of Pure World from the government. In some cases, Pure World supplies equipment like driers to local farmers. The research team at Pure World, under the direction of Dr Zheng, developed an analytical method to determine accurately the level of kavalactones and identify the species. This methodology was published in the *Journal of Chromatography* and is widely used in the industry. The research and development efforts at Pure World resulted in a line of KavaPure extracts, including powdered, fluid, and softgel extracts.

Pure World is currently funding clinical trials for kava at Duke University. The trials will cost the company around US$200,000, but Pure World considers this a worthwhile investment, since they are positioned as a science-based company and want to demonstrate strong scientific support for their products. Demand for kava has greatly increased over the last year, and Pure World is in a leading position to supply to the US market.

argued (and this has been borne out by practice over the last few years) that without government supervision, immature kava would be harvested for the export market, jeopardising quality of the medicinal product. There are numerous reports from Pacific Island countries of European and US companies offering to buy immature kava.

In addition to concerns surrounding the export of raw materials, other concerns relate to the export of local varieties for cultivation in other countries. Fei Tevi of the Pacific Concerns Resource Centre in Suva stated that 'Kava has already been hijacked ... In traditional custom, you do not harvest the kava for money. We want pharmaceutical companies to follow a 10-point plan, respecting indigenous people's culture and their rights to royalties' (Adams, 1998). In some communities, particularly on the islands of Tanna and Pentecost, it is also felt that modern cash and traditional subsistence economies should be kept separate. There are misgivings about the commercial trade in a crop with strong traditional ties (Lebot et al, 1997).

Availability of kava for domestic use has reportedly been impacted by increased demand from export markets. Prices in the Pacific Islands have doubled: whereas one kg used to sell for about US$10, the price is now about US$20 (Adams, 1998). Lebot et al (1997) argued that if exports divert local supplies of rootstock, shortages on local markets could force up domestic prices, which would make alcoholic beverages an attractive substitute.

One positive element of the commercial trade in a product with strong domestic use is that domestic markets can, to some extent, provide a safety net should international demand suddenly decline. Additionally, a kava garden is a source of pride and is always a social, if not an economic, asset. Because the labour input needed to establish a kava garden is minimal, a farmer who does not sell his crop in the international market might suffer less than a farmer of a more labour-intensive crop (although farmers who intensively cultivate kava remain vulnerable) (Lebot et al, 1997).

Conclusions on access and benefit-sharing

It is clear that kava has great potential for Pacific Island countries as an internationally-traded cash crop. Already prices paid to local farmers have increased, and a number of companies have established partnerships with local cooperatives and growers which include the provision of equipment and limited capacity-building, as well as guarantees of price and demand over a set period of time. But further steps are necessary in order to ensure the region gains a greater share of commercial kava revenues, and protects its claim as the 'true' source of the crop.

Because kava was already in trade and its use widely known before the CBD, and because a large number of companies are involved in sourcing bulk material, benefit-sharing resulting from commercial use is not linked to access to genetic resources. However, value-added processing of standardised extracts, and development of domestic industries that work at higher levels in the kava value chain (Box 4.13, p108), would help to ensure that Pacific Island countries capture a greater share of commercial kava benefits. Additionally, access to kava cultivars could be restricted and granted only upon development of 'mutually agreed terms' and adequate benefit-sharing packages.

It is likely that if kava is traded only as a commodity, and not as a distinctive product resulting from long histories of South Pacific use and know-how, it will be grown in plantations outside the region, driving down prices, and removing the region's claim to and control over the species. Already, countries like Guatemala and Australia, as well as Hawaii, are reported to have kava in cultivation.

As a result, at the Kava Symposium, held in Fiji in October 1998, and coordinated by the Kava Forum Secretariat, the potential for developing geographic indications to 'protect' the kava resources was explored. Steps are currently underway to implement a system that would resemble the French 'appellation of origin' system for wine (Lebot, personal communication, 1998; Downes and Laird, 1999). Appellations of origin are used on products with 'long histories of empirical experimentation and experience' (Moran, 1993), and reward the type of community-based historical innovation that has resulted in kava as we know it today. Appellations of origin grow from four key elements:

- distinctive varieties/species;
- yields;
- production methods; and
- processing methods.

All of these apply to kava (Downes and Laird, 1999). A system such as this could result in a 'True Pacific Island Kava' of some form, or a number of 'True Kavas' for each country (Lebot, personal communication, 1998).

Additionally, trademarks might be used to establish quality control, environmental, and social standards for kava production (Downes and Laird, 1999). The potential for adulterated and low-quality materials to flood and discredit the kava market is great, and manufacturing companies, in line with general trends towards quality and sustainability in the industry, are likely to place a premium on guaranteed high-quality sources.

Box 4.13 *The Kava Chain of Value and Production (January 1999)*

GROWERS *grow, dry, basic chopping/processing of the root*

Costs/Inputs
labour
time
land

US$5–8 per kg

WHOLESALERS and EXPORTERS *collect from growers, ship dried root*

Costs/Inputs
government taxes
packaging
shipping

Examples of exporters of kava
Vanuatu Commodities Marketing Board (VCMB)
Carpenters Company

US$22–28 per kg

EXTRACTION COMPANIES *grind root into powder; make extracts*

Costs/Inputs
labour
materials (eg alcohol)
overheads (electricity, heat)

Examples of companies processing kava
Madis Botanicals (USA)
Hauser (USA)
QBI (USA)
East Earth (USA)
Finzelberg (Germany)
Muggenberg (Germany
Schwabe (Germany)

US$250–300 per kg of 30% kavalactone extract

MANUFACTURING COMPANIES *manufacture capsules and tablets*

200 mg 30% kavalactone extract produce 60 mg kavalactone capsule/tablet

Costs/Inputs
labour
equipment and overheads
marketing

Prices obtained for products containing ground root – not based on a standardised extract – vary widely, but are consistently much lower than for standardised extracts.

Examples of US companies manufacturing kava:
Mass market: Rexall-Sundowne; Leiner; Pharmavite
Speciality/health food: Solgar; Twinlabs; Nature's Way; Country Life; Nature's Plus

US$6–12 for 30 capsules/tablets containing 60mg kavalactones

RETAILERS *sell products through the mass market, pharmacies, and natural or health food outlets* average margin on products: 50%

Costs/Inputs
store space
overheads
marketing

Outlets:
natural product/speciality stores
pharmacies
mass market

US$12.95–19.95 retail product: 30 capsules containing 60 mg of kavalactones

Sources: Chris Kilham, personal communication, 1999; QY Zheng, personal communication, 1999; interviews.

4.10 Practices in benefit-sharing

The botanical medicine industry generates a range of monetary and non-monetary benefits, which to date have not commonly been linked to access to resources. In this section, we will discuss some of the monetary benefits associated with the botanical medicine industry (price paid per kg of raw material, and, in some cases, advance payments and royalties), as well as non-monetary benefits like commercial partnerships, capacity-building, technology transfer, and in-kind benefits.

4.10.1 Monetary benefits

Botanical medicine companies pay a price per kg for bulk plant material used in the manufacture of commercial products. Prices paid for materials increase significantly along the value and marketing chain (eg kava, Box 4.13, p108). Other monetary benefits like advance payments and royalties are not common but do occur, for example when species are unique to the trade, or manufacturing techniques or delivery processes have been developed. In the case of *Panax vietnamensis*, Nature's Way offered to provide a royalty for commercial development of the species if export of raw material was restricted for a number of years, in effect providing Nature's Way with a form of exclusivity (case study below). While botanical medicine products are not candidates for the type of exclusivity afforded by patents, other areas of product development have been patented, including plant material (eg unique cultivars), manufacturing and processing techniques, formulations, dosage forms, and unique release characteristics (Brevoort, 1998).

Examples of local communities or source countries receiving royalties from commercial botanical medicine products are few, however. One exception is the Kani tribe in India, who are sharing licence fees and a royalty of 2 per cent on profits with the Tropical Botanic Garden and Research Institute in Kerala. The Botanic Garden facilitated development of a commercial product based upon the Kanis' traditional knowledge (Box 4.14). In the case of kava, Pacific Island countries seek to benefit from the commercialisation of 118 cultivars, and traditional uses, developed over 3,000 years by local communities. The intellectual property tools of geographic indications and trademarks are under study as a means to achieve greater financial benefits (4.9, p107).

As the CBD is more comprehensively implemented in coming years, prospecting for species 'new' to botanical medicine markets is likely to take place within

Box 4.14 *Benefit-Sharing and the Kani: the Case of Arogyapacha*

The Kani are a traditionally nomadic community now living a largely settled life in the forests of the Western Ghats of India. Their use of the local forest species arogyapacha (Trichopus zeylanicus travancoricus) interested researchers from the Tropical Botanical Garden and Research Institute (TBGRI) of the district of Kerala. Researchers identified active compounds in the species, and developed a drug called Jeevani, which rejuvenates and builds strength. TBGRI scientists licensed the formula to the Arya Waidya Pharmacy (AVP) for seven years, with the understanding that 2 per cent royalties on profits, as well as the licence fee, be shared equally between TBGRI and the Kani tribe. The intention was to recognise and reward the Kanis' knowledge and custodianship of the plant. The Kani established a Community Welfare Trust to manage their share of financial benefits, and sales of the drug have been brisk. But the intended benefit-sharing plan has faltered before a range of governmental, community, and institutional obstacles, and its future remains uncertain.

Sources: Anuradha, 1998; Martin, 1998.

national access and benefit-sharing frameworks. These measures might require a package of benefits including advance payments for collaborative research, royalties, and other financial benefits. Financial benefits derived from botanical medicines will always be a great deal less than those derived from pharmaceutical products, however, and because raw material is traded as a commodity, it is unlikely that a single company will control trade in the species for long. As a result, expectations for financial benefits resulting from the botanical medicine industry, while warranted, should be managed to reflect the modest research and development budgets, requirements for large quantities of raw material, and smaller commercial revenues that characterise this industry.

4.10.2 Non-monetary benefits

The most significant benefits shared by the botanical medicine industry to date are primarily non-monetary benefits associated with the sourcing of raw materials. The industry might generate a wider range of non-monetary benefits, however, including:

- training in agronomic techniques, and provision of infrastructure and equipment such as nurseries and drying facilities;
- training and equipment for extraction, standardisation, and other scientific technologies;
- assistance with national programmes working to standardise traditional medical systems;
- improved capacity to launch and manage domestic medicinal plant-based businesses building upon local species and traditional knowledge; and

- in-kind benefits like the provision of healthcare for communities and small-scale income generating projects.

Commercial partnerships

Most commercial botanical medicine companies are located in 'northern' countries. Through partnerships and collaborations with high biodiversity countries – primarily in the 'south' – these companies can contribute significant benefits in the form of capacity-building, technology transfer, and training for business development. These benefits can build a foundation for the development of domestic industries supplying local and regional markets. Commercial partnerships can also assist in building local capacity to study and standardise traditional medical systems. The WHO estimates that 80 per cent of the world's population relies on traditional medicines, and 85 per cent of traditional medicines are based on plants. In many high-biodiversity countries, efforts are underway to study and standardise traditional medicines in order to provide affordable, effective, and culturally-appropriate local healthcare (Balick et al, 1996; Iwu, 1996).

For example, Axxon Biopharm has established a strategic alliance with the NGO Bioresources Development and Conservation Program, and its International Centre for Ethnomedicine and Drug Development (InterCEDD) in Nigeria. The company is 'co-owned from the field to the bottle', and its strategy is based on a partnership approach to product development. Axxon Biopharm markets standardised extracts and botanicals to the international botanical medicines, nutraceuticals and personal care markets. In the domestic Nigerian market, it sells high quality botanical medicines based on the original formulations of healers, but with state-of-the art quality control, addressing critical local healthcare needs. The company's public relations materials describe it as 'a corporate expression of a relationship between indigenous cultures, healing and modern science that is rooted in reciprocity and shared meanings'. Products are based on African species with widespread traditional use, including *Cola nitida, Garcinia* spp, *Aframomum melegueta,* and *Ocimum viride*.

In another example, the German company Sertuerner Arzneimittel has established a partnership in Namibia to develop cultivated sources of devil's claw and other species. Value is added locally through extraction facilities, which are intended to build domestic capacity to process products. Stable jobs are created, and efforts are made to provide farm jobs to wild harvesters of plant material. Another large German company is working with suppliers from an African country on the development of low-cost versions of European commercial products for local markets.

Botanical Liaisons has helped to establish a medicinal plant company for the Seminole tribe of Florida. Interested in diversifying their income, the tribe is producing extracts and capsulated herbs, as well as skin- and hair-care products. 50 per cent of all profits is returned to the cultural partner, while the remaining 50 per cent is divided between the Seminole tribe and their distribution company, Helishwa – River of Grass – Products. In another partnership, Botanical Liaisons and the Ghana firm Bioresources International are developing products based on traditional use, with the intention of building sustainable sourcing, research, and processing capacity within Ghana.

Sourcing raw materials

Activities associated with the sourcing of bulk raw materials from high biodiversity countries offer a number of opportunities for benefit-sharing, and are those most commonly cited by companies when asked about benefit-sharing approaches. Botanical medicine companies, contrary to a widespread misconception, do not synthesize commercial products; rather, they rely heavily on wild and cultivated sources of raw materials, and make decisions regarding product development based in part on availability of supplies. As we have seen, most companies buy raw materials in bulk from intermediaries, and do not interact with the growers or collectors of the material they receive. However, due to integration of the industry, and an increasing emphasis on quality control, and in some cases social and environmental concerns, commercial product producers are moving closer to their sources.

One result of this is companies entering into longer-term agreements with suppliers to purchase materials at set prices and at a consistently high quality over time. Martin Bauer, for example, established a joint-venture in Turkey to source St John's wort. Salus-Haus is promoting its partnership approach to sourcing raw materials as supportive of the objectives of the CBD. The company is working to share benefits in the form of jobs and capacity-building associated with value addition, and supports sustainable development projects in the regions from which it acquires raw materials. According to a company poster entitled 'What do Rio and herbal teas have in common?', in this way the company helps to ensure the 'sustainable and environmentally-friendly use of medicinal plant genetic resources on all continents'.

In another example, Nature's Way is building a partnership with groups in the Andes for the sourcing of a 'Peruvian ginseng', which is likely to take off in the market, but for which there is no existing sustainable source. As Marian Burningham, New Products Manager of Nature's Way, said, 'By working closely with local communities, universities, and governments, and

Box 4.15 *A Supply Agreement between Plantecam Medicam Company and Two Villages in Cameroon: A Small Step Towards Sustainable Management of* Prunus africana *(Pygeum) and Increased Benefit-Sharing with Local Communities*

Collection of *Pygeum* bark in the wild is currently undertaken in an unsustainable manner, and wild populations are threatened. Facilitated by the Mount Cameroon Project and the Ministry of Environment and Forests (MINEF), in July 1997 the French-owned company Plantecam Medicam, the main purchaser of Pygeum bark within Cameroon, signed agreements with two villages on Mount Cameroon in order to improve sourcing of the bark. Although there have been significant problems with implementation of the agreements – *Agreement for the Sustainable Management of the Species and Production of African Prunus at Mapanja [Bokwongo] Village facilitated by Mount Cameroon Project (MCP) and Ministry of Environment and Forest (MINEF)* – they provide an insight into possible benefit-sharing associated with sustainable sourcing of raw materials for the botanical medicines trade. In this case benefits primarily take the form of increased payments for material, and agreement to purchase material from villages over time at an agreed price. In itself this is a positive, but unremarkable arrangement for the botanical medicines industry. However, additional benefits in the form of training, and capacity- and institution-building in the villages have also resulted, due to the involvement of a conservation project (MCP) and governmental agency (MINEF) in the agreement.

Under the agreement, the village Prunus Harvesters' Union harvests bark in place of Plantecam's recruited workers, and sells the bark to Plantecam at the rate Plantecam buys from traders with special permits (209 CFA per kg). These traders formerly bought bark from villagers and other collectors at a much reduced price (100 CFA per kg), so the Union members now receive a higher price for their produce. Article 2(D) of the agreement provides for a maximum monthly tonnage of 10 tons to be supplied by the Mapanja Harvesters' Union to Plantecam. Each villager can harvest a maximum of 30 kg a day, and from this must contribute 2 kg to the Village Development Fund and 1 kg to the Union's Fund. Members each retain 27 kg, for which they are paid 209 CFA per kg. Bark harvesters must also pay 10 CFA per kg for transportation from their village to the company's factory at Mutengene, and 10 per cent of total tonnage as exportation tax. After all deductions have been made, each bark harvester receives a wage of about 35,000 CFA a month, assuming that 27 kg was supplied.

As a result of the 2 kg per every 30 kg bundle payments by each bark harvester to the Village Development Fund, the Fund had a sum of about one million CFA just five months after it was set up. The villagers intended to use this money for the realisation of a long-awaited water project. In addition to the monetary benefits accruing to the village in the form of higher payments per kg, and resulting magnification of these benefits in the form of a Village Development Fund, the following non-monetary benefits have resulted from this agreement:

- **Training:** Bark harvesters have been trained to harvest the bark sustainably, and monitoring is on-going to ensure these methods are adhered to. Monitoring is undertaken by a joint team made up of staff from Plantecam, MINEF, and MCP, as well as villagers. Training has also taken place in financial accounting and management of relationships with companies.
- **Capacity and institution-building:** Considerable institutional and capacity-building has taken place to improve village structures. The Village Development Fund, Prunus Harvesters' Union and its Fund, and the Monitoring Committee were established as part of this process. There is greater awareness of the long-term benefits of sustainable harvests of bark under this new agreement with Plantecam, and capacity to realise the benefits on a local level.
- **Infrastructure and equipment:** In some of the less accessible villages, such as Ekonjo, Plantecam has helped in upgrading the roads to improve accessibility. Assistance has also been provided in the building of community halls. Equipment supplied to communities includes cutlasses and climbing gear for use in collections.

Source: Laird and Lisinge, 1998.

providing a connected group of expert resources and contacts, it is the hope of Nature's Way that the Peruvian sources of this new botanical material will see beyond the immediate sale, and plan for long-term supply and sustainability. Perhaps they will realise that by dealing with an ethical and concerned company, taking steps to conserve and protect their indigenous plants and people can still lead to commercial crops and financial gains.'

Many companies see long-term sourcing relationships as a significant 'benefit' for producers, including those located in high biodiversity regions. For example, Kay Wright, Director of Botanical Purchasing of Celestial Seasonings, said, 'We like to build up partnerships ... We will enter into contracts to buy X amount of plant material for the next X years. This allows our suppliers to get bank loans, to purchase warehouses, drying yards, cleaning and packaging equipment ... . We have done this in China and Thailand. Although we don't give them any funding up front, our long-term commitment allows them to build capacity.'

As part of relationships intended to serve commercial objectives to source high quality raw material, a number of companies have invested in training and the supply of equipment to ensure that cultivation, drying, and other processing adhere to set standards. The result is a form of capacity-building and technology transfer, which could be made a more common practice in industry. As Rod Lenoble, Manager of Scientific Affairs of Hauser Inc, said, 'We provide analytical support to suppliers, allowing them to improve their growing techniques and become more profitable. Hauser also helps to ensure that the farmer or harvester will have business down the road – because they know they will have a viable business in the long term, and will be paid a fair price, they can invest in sustainability.'

Although profits vary significantly on raw materials, in most cases only a fraction of the cost of a finished good reflects the raw material costs. This suggests that a minor investment in sustainable sourcing and capacity-building for growers and collectors would not significantly dent commercial profits. As we have seen, a number of companies are already engaged in this type of benefit-sharing, but it should become standard practice in the industry. Steven King of Shaman Botanicals, for example, thinks that companies must undertake ethical benefit-sharing, and should focus on ecological and social sustainability of their sources. 'Companies need to take more responsibility, and consumers should pay the actual cost of sustainable products,' he said.

Increasing calls from consumer and environmental groups for social and environmental certification of raw material sources might have the effect of increasing benefits for source countries and communities. Certification of high volume, low value botanical products might prove extremely difficult to finance, but growing pressure for organic, sustainable, and 'fair trade' materials suggests that industry priorities might necessarily shift to incorporate these costs in the near future.

In-kind benefits

A few companies are actively involved with local suppliers and communities, and have established benefit-sharing packages that include immediate- and short-term benefits not directly linked to commercial activities. These have included the establishment of healthcare and income-generating schemes, professional training, and research exchanges.

4.11 Case study

The development of a benefit-sharing partnership in Vietnam[4]: *Panax vietnamensis* – a 'new' ginseng[4]

Sarah Laird and Marian Burningham

Overview

For a number of years, Nature's Way worked with contacts in Vietnam to develop a direct research and sourcing relationship for a recently-'discovered' species of ginseng (*Panax vietnamensis*). The company first came upon the use of this species by scanning databases and conducting literature searches, and discovered interesting activity documented in studies conducted in Russia, Poland and Japan. This species had not previously been marketed in the USA and showed great promise. It was subsequently discovered to be endemic to a region of Vietnam where it is used traditionally, and where cultivation for local markets was underway. Nature's Way began a multi-year process of correspondence, meetings, and negotiations with local communities, government-owned pharmaceutical companies, researchers at universities, and the local and central government, in order to design a sourcing and benefit-sharing package.

The main actors

- *Nature's Way* – a dietary supplement manufacturing and marketing company based in the USA. The main categories of products produced are: herbs or botanicals; vitamins, minerals, and amino acids; homeopathic; and a small range of OTC products;
- *Steven Foster* – a botanical author, photographer, and botanical medicine expert;
- *REM Ventures* – a small private concern located in Tukwila, Washington, with nearly a decade of business experience in South East Asia, particularly Vietnam, which conducted intermediary research and provided brokering assistance for the relationship;
- *People's Committee of Kon Tum Province* – represented local Sedang communities in the Central Highlands where the plant is found;
- *Government of Vietnam* – the Ministry of Agriculture in the Central Highlands, etc; and
- *University researchers* – conducting on-going work on *Panax vietnamensis*.

Panax vietnamensis

Panax vietnamensis is an herbaceous plant, perennial by rhizome, commonly 40–60 cm in height. It grows in high densities at altitudes of 1,700–2,000 metres, under the canopy of evergreen forests. It is regenerated easily by rhizomes or by seeds. The saponin composition of the species has been studied extensively, since these compounds are the main active component of *Panax* spp. *Panax vietnamensis* has been shown to have similar pharmacological activity to other ginsengs (Minh Duc and Thoi Nham, 1998).

Panax vietnamensis was 'discovered' by scientists in 1973 during a botanical expedition in the montane forests of Ngoc Lay Region, Dekto District, Komtum Province, in Central Vietnam (Tanaka, 1998). This expedition was carried out by the Fifth Zone Health Service of Vietnam. The *Panax* genus occurs in the northern hemisphere from Central Himalaya to North America and through China, Korea and Japan. It includes *Panax ginseng* and its two cogeners *P notoginseng* (Sanchi ginseng) and *P quinquefolium* (American ginseng), which are widely used in botanical medicines across the globe (Minh Duc and Thoi Nham, 1998). *P vietnamensis* has long been used by the Sedang ethnic group living in the Truong Son Range, and is known as 'Cu Ngai Rom Con'. It is used locally as a secret life-saving medicine, for the treatment of a range of diseases and to enhance physical strength. It was only in 1985, after many 'twists and turns' that the plant was designated as a new species (Minh Duc and Thoi Nham, 1998).

Panax vietnamensis is included in the *Red Book of Plants for Vietnam* which lists 250 rare, threatened, and endangered medicinal botanicals. Because it grows in the shade of the forest canopy, a number of factors have contributed to its endangered status:

- demand for local medicinal use;
- the impacts of slash-and-burn agriculture, and harvest of fuelwood, on its habitat;
- deforestation during the war; and more recently
- 'illegal and improper exploitation' (Minh Duc and Thoi Nham, 1998).

Development of a commercial partnership

Nature's Way routinely scans databases and literature for new pharmacological and clinical studies on herbs. The company has an internal research department and library that facilitate this process. During a routine scan on ginseng, they came upon *Panax vietnamensis*. The documentation and scientific discovery of the species were reported, as well as three separate studies identifying new saponin compounds unique to this species. More interesting still were comments in these studies which indicated that an extract of the plant had been used in clinical and non-clinical trials in Vietnam. Additional research and documentation of the biological activity of these compounds was reported by researchers in Russia, Poland, and Japan. *Panax vietnamensis* has never been marketed in the USA on any scale and was unknown to the Nature's Way staff.

The company subsequently spent nearly 18 months in its attempts to track down researchers and companies in Vietnam – a very laborious process. Nature's Way finally hired an intermediary from Seattle – REM Ventures – to help identify collaborators and sources of material in Vietnam. While REM's Al Davignon was successful in securing samples of a locally-produced *Panax vietnamensis* tablet on his first mission, it was difficult to further identify who actually had control over the plant resource, and to whom the company should speak directly about sourcing partnerships.

Nature's Way eventually took the risk of funding a trip to Vietnam by three individuals (Al Davignon, Steven Foster and Marian Burningham) to locate the source of material and to determine who controlled it, the status of supply, and state of commercial development. It took a great deal of detective work and much trial and error before they finally identified potential contacts in the area where the plant was harvested in the Central Highlands of Vietnam. The group finally arranged to travel to the Central Highlands, not knowing with whom they should meet. With only a letter of introduction to a local Arts minister in hand, they were met in Pleiku by a local driver and interpreter. An initial meeting with the People's Committee of Kon Tum Province led to the finding that this group had already – with provincial government funding – established a technical plan for the cultivation and commercialisation of *Panax vietnamensis*. The Committee had extensive documentation on the ecology of the plant, its historical use, and medical benefits, and lacked only a commercial partner to help them realise their plan.

Benefit-sharing

Over the course of the next months, Nature's Way developed a plan to assist with cultivation of the plant, in order to provide income to poor local communities (350 members of the Se Dang ethnic group), while helping to protect the plant from over-harvesting in the wild. Nature's Way considered a 5–30 year investment in cultivation in the area, but found limited local capacity to manage implementation of such a strategy over time, and to the standards required.

The intended strategy involved a joint venture between Nature's Way and the Government, in which Nature's Way would provide the inputs for growing the crop, and all capital required directly for cultivation and harvesting. Nature's Way would then own 70 per cent of the total produced. The other 30 per cent would be owned by the Government, which would then distribute benefits according to standard practices. In return the Government offered to provide the labour of local communities, land, and seed or root stock material. Nature's Way would then reimburse the Government for the costs of labour, as well as materials such as shade cloth and bedding material. About 25 per cent of the budget would go towards re-establishing the plant in its natural habitat.

Researchers at the university had contributed a great deal over the years to cultivation and medical research on the species, and Nature's Way considered it important to support their efforts, as well. The benefits suggested for this group included:

- equipment and materials needed to implement the project;
- training in relevant techniques;
- support for a graduate student to conduct research towards their degree, while overseeing aspects of the project;
- research exchanges with US universities;
- sponsorship and assistance with the organisation of scientific meetings; and
- guidance in applications for grant monies for ecological, cultural, and environmental research and activities of benefit to the local communities.

Although paying royalties in the herb industry is uncommon, it does happen on occasions when a plant is not successfully grown outside of its natural habitat, and the government can restrict exports – thereby removing the plant from the bulk commodity markets, and providing the manufacturing and marketing company with a unique product. The range is modest – perhaps 1–3 per cent of net sales, sometimes of net profits. This reflects the fact that botanical medicine products are not patented, and any company can make copycat or 'knock-off' versions of a product – the long-term value of access to botanical material therefore has its limits.

In this case, Nature's Way was willing to pay a royalty on *Panax vietnamensis* product sales, and spoke to the Government about the need to restrict the export of seed and planting material. Nature's Way argued that regulation of species use and export would allow the local communities and the country to reap maximum benefits from a 'national treasure'. The company suggested that for plants on the rare, threatened, and endangered list, Government should restrict exports. Government officials were not aware of these issues, and in fact it is easier to get phytosanitary permits to export bulk raw materials from Vietnam than it is to get a permit to ship out herbarium specimens. But Nature's Way staff also argued that rare, threatened, and endangered plants should be regulated at the marketing, as well as the supply, end because if there is a market for a species, someone will find a way to supply it. Consumer nations should attach penalties to selling endangered herb materials, and should require certificates of inspection.

Access and benefit-sharing conclusions

Cultural, commercial, and political factors which made it difficult for the company to commit to a long-term investment, despite a significant desire to do so, include:

- the need for multiple meetings to make even basic decisions;
- the limited value placed on benefits for local communities, and conservation of the species, by some government officials;
- the lack of understanding within government of the value of the species to the country; and
- a sense that the terms of the commercial relationship would keep shifting.

Nature's Way decided that the best way to continue work on this project would be through a non-profit organisation or foundation experienced in this type of project development, including monitoring and administering funds supplied by Nature's Way. The company is currently collaborating on the development of a symposium in Hanoi involving scientists, local communities, government officials, and others with an interest and involvement in the species. This group might then develop a strategy for *Panax vietnamensis* conservation and commercialisation. Nature's Way's role will be smaller than previously envisioned, and might largely be that of expressing commercial interest in the product – ie providing a fair price and reliable market for material over time.

The primary access and benefit-sharing lessons learned from this case include the following:

1 Governments need to be educated about the value of their domestic plant material, and the importance of benefiting local communities for the commercial use of their knowledge. There exists little awareness of the ways in which benefit-sharing packages can be structured to benefit local communities and conservation.
2 Bureaucracy is a major threat to any regulatory system for natural products.
3 Capacity-building at the local community and university level are important elements of commercial partnerships, and will help to ensure long-term benefits for conservation and development.
4 The ways in which botanical medicine companies capture value, and the basis upon which they pay royalties, are very different from the pharmaceutical industry in most respects, and tend to revolve around raw material sourcing issues. The differences between these and other industries must be clear to policy makers prior to the drafting of access and benefit-sharing legislation.
5 Benefits resulting from the botanical medicine industry can prove as, or more, significant than those of the pharmaceutical industry in both the short- and the long-term.

6 Capacity-building and technology transfer should be integrated into raw material sourcing collaborations, in order to allow for greater benefit-sharing with local communities, businesses, and institutions.
7 Benefit-sharing packages must be designed on a case-basis, given the wide range of economic, political, and cultural variables.
8 As with pharmaceutical companies negotiating access and benefit-sharing agreements with governments, botanical medicine companies attempting to set new standards for equitable partnerships usually find that they are acting alone (few, if any, support and advisory services are available), and must invest extensive time and resources in order to reach agreement.

4.12 Conclusion

The global botanical medicine industry is characterised by great diversity. In Japan, China, India, and much of Europe, the industry grows from well-developed systems of traditional medicine, and is largely integrated into mainstream healthcare systems. In most high biodiversity countries, traditional medicine provides healthcare for the majority of the population, although little of this is in the form of commercial products manufactured by the private sector. Exceptions include countries like Indonesia, with its large *jamu* industry, and countries like Brazil, and South Africa, which have commercial botanical medicines industries that incorporate products and species from around the world. Because of this diversity, it is impossible to fully characterise global market and research trends in the industry.

Despite its green image, sustainable and ethical sourcing are only now appearing on the botanical medicine industry's 'radar screen'. Even so, many people involved in the botanical medicine industry are uncomfortable with the practical ramifications of this issue, since raw material is used and viewed as a bulk commodity. Sustainable and ethical sourcing involves a greater investment of funds in developing sources and tracking materials than has been standard industry practice to date. However, the negative impact of current practices on species and wider biodiversity conservation is often significant. Likewise, sustainable development, and the provision of local livelihoods based on native species, is little served by short-term, opportunistic sourcing strategies in which buyers move on when a species is depleted in a particular area.

Historically, access and benefit-sharing have been severed in the botanical medicine industry. Most companies do not have contact with the traditional originators of their commercial products, nor with the growers and collectors of raw materials in source countries. Additionally, product development in this industry is conservative, and leads are drawn from literature, trade shows, or long chains of intermediaries. Companies do not receive large numbers of samples collected in high biodiversity countries. Research and development budgets are small and focus primarily on quality control, safety, efficacy, and formulation, with only limited attention paid to species 'new' to the market.

Traditional knowledge relating to species use, efficacy, safety, and sourcing, is an important element of product development in this industry, and is regularly used in marketing products. Because traditional knowledge is acquired through databases and literature, in this way too the connection between access and benefit-sharing is severed, and few in industry have given thought to this issue. A notable exception is the case of arogyapacha (*Trichopus zeylanicus travancoricus*) in India (Box 4.14, p109). In Europe and the USA, general references to 'intellectual property rights' and 'compensation' of indigenous peoples have surfaced in some industry circles, but are consistently couched in vague terms.

Interest in species new to international markets is likely to increase in coming years due to a number of factors including expanding markets, increased consumer awareness, company strategies for market differentiation, improved scientific techniques to study the medicinal properties of plants, and increasing academic research on geographically diverse medicinal species. As a result of these and other factors, it is possible that there might be an increase in 'prospecting' in high biodiversity regions.

Governments developing access and benefit-sharing measures should be cautious in regulating this industry, however. The financial profile of the botanical medicine industry – whether research and development budgets, or commercial revenues – is a great deal smaller than that of the pharmaceutical, biotech, crop protection, or seed sectors. Additionally, the botanical medicine industry is not dependent upon access to species diversity. Most companies reformulate the same suite of species they have worked on for decades, investing primarily in new processing, delivery, formulation and marketing. As we have seen, there exist a number of scientific, market, and regulatory obstacles to investing in commercial development of species 'new' to the market. Access and benefit-sharing measures should not be too heavy handed, and should emphasise spin-off benefits that can result from commercial partnerships.

Governments should seek as their primary objectives the promotion of partnerships that build capacity within high biodiversity countries to work at a higher level in this industry (and not solely as raw material suppliers). Partnerships should also build capacity in areas of domestic importance (eg standardising traditional medicine; developing affordable plant-based botanical medicines for domestic markets; building industries based on local biodiversity; and conducting research on tropical diseases). In this way, benefit-sharing will more likely lead to the creation of incentives for biodiversity conservation and sustainable development over time.

Notes

1 For reading ease, scientific names of species regularly discussed will not be included in the text; see Appendix B for a guide to common and scientific names of species.

2 The commodity category 'pharmaceutical plants' refers to all plant material which is traded in the form of roots, leaves, flowers, etc, fresh or dried, in whole pieces, cut, or powdered. Plants are used to produce phytopharmaceuticals, extracts, pharmaceuticals, cosmetics, foods and flavourings, domestic cleaning products, and insecticides (Lange, 1998).

3 *Ex situ* collections of kava include those at the Royal Botanic Gardens, Sydney; the Royal Botanic Gardens, Kew; Missouri Botanic Garden; Singapore Botanical Garden; the Arnold Arboretum; the Paris Museum, and the University of Malaysia (Perteru, 1997).

4 This case study is based on the involvement and experiences of Marian Burningham, Nature's Way, and others, in this case.

Kerry ten Kate

Chapter 5

The Development of Major Crops by the Seed Industry

5.1 Introduction

Even today, 1.5 billion people live on less than US$1 a day (IFPRI, 1998). Eight hundred million people are undernourished, and 200 million children under five years of age are underweight (FAO, 1996). A 75 per cent growth in agricultural production will be needed in the span of just one generation, to feed a world population that will have grown from the almost 6 billion of today to 8.3 billion by 2025 (FAO, 1998). The improvement of existing varieties of crops and the development of new crop species will be vital for sustainable development and for food security. Both require access to genetic resources.

Just 30 crops provide 95 per cent of human plant-derived dietary energy. Wheat, rice and maize alone provide more than half of this energy intake (Figure 5.1). With the addition of a further six crops – sorghum, millet, potatoes, sweet potatoes, soybean and sugar – this proportion rises to 75 per cent of the total human energy intake. However, analysis of food energy supplies at the sub-regional level reveals a greater number of significant crops. For example, cassava is reponsible for over half of the plant-derived energy in Central Africa, although it accounts for just 1.2 per cent global energy intake (FAO, 1996).

Of the 300,000 to 500,000 higher plant species thought to exist, about 30,000 are edible and some 7,000 have been cultivated or collected by humans in the 10,000-year history of agriculture. This could suggest, on the one hand, that, after concerted efforts over thousands of years, people are unlikely to find more useful species. On the other hand, one could guess that, aided by new tools such as genomics, researchers will uncover valuable traits from the wealth of germplasm that has not yet been used in agriculture, and use them to develop new varieties, to increase the use of underutilised species, or to improve existing varieties.

There is now wide recognition of the need for a broad range of crops, and for diversification not only of the genetic diversity in crops, but also of the varieties and species grown on farms (eg FAO, 1996). The development of minor crops and related practices in access to genetic resources and benefit-sharing merits a study of its own. However, given the modest scope of the

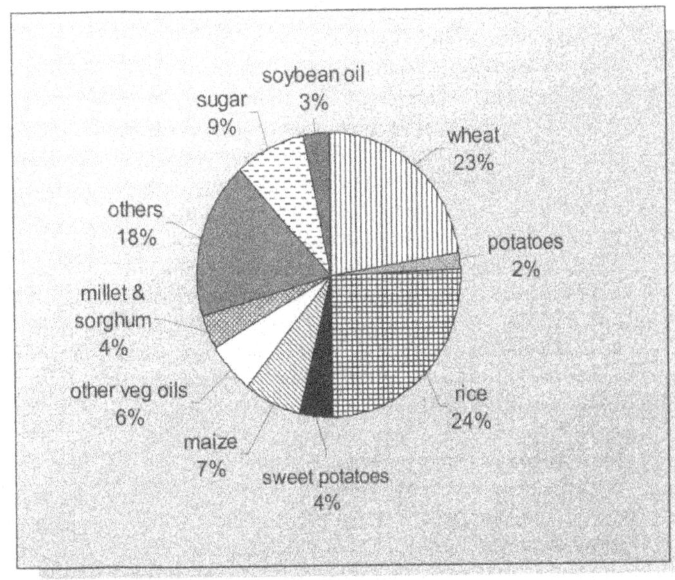

Figure 5.1 *The Most Important Crops for Food Energy Supply*

Source: Global Plan of Action, FAO, 1996.

research leading to this book, it has been impossible to cover the full range of agricultural crops (see Research Methodology and Team, p387). This chapter explores the seed industry producing certain agricultural[1] 'major crops' – the cereals, oil crops, fibre crops, food legumes, forages, vegetables, fruit and roots and tubers which form the basis of the diet of much of the world's population.

This chapter will try to answer the following questions:

- What is the extent and nature of the demand by private seed companies and by plant breeders in the public sector for access to genetic resources?
- What is the role of the different actors involved in the collection and exchange of genetic resources, and in the research and development of new, commercial varieties?
- What kind of benefits arise from the use of genetic resources in plant breeding, and how are these currently shared through different kinds of partnership?
- What conclusions can be drawn about the nature of the policy framework and institutional partnerships that would promote the development of improved crops to meet the world's nutritional needs, ensure a fair and equitable sharing of the benefits arising from access to genetic resources, and support the conservation of biological diversity?

The chapter starts with an introduction to the international legal and policy context for access to plant genetic resources for food and agriculture; then examines the roles of the different actors involved in plant breeding, and the markets for various agricultural products. A brief introduction to the science, technology and economics of plant breeding follows, and an outline of the kind of regulations which control the release of new agricultural crop varieties on to the market. The second section analyses the demand by plant breeders for access to plant genetic resources for food and agriculture (PGRFA), looking at the kind of organisations that seek access; from where they obtain their materials, and the nature of the materials in which they are interested. The third section explores current practice in sharing the benefits arising from the use of PGRFA – both monetary and non-monetary benefits such as information, technology and training. A fourth section provides a case study on the development of two specific crop varieties, and a final section sets out some conclusions about access to PGRFA and benefit-sharing.

5.1.1 Policy context

Since the dawn of agriculture some 10,000 years ago, people have traded seed and foodstuffs. Crops have been widely distributed around the globe, interbred with local varieties and adapted to a host of different conditions. As a result (Table 5.1), production of non-native PGRFA from other regions of the world forms an important part of the agricultural production of every country. The population of each country is dependent for its food on PGRFA once obtained from elsewhere. While this provides an historical perspective, where do we stand today? Are countries now self-sufficient, both for agricultural food production and for the germplasm used in plant breeding, or does our global interdependence on access to PGRFA continue?

Table 5.1 *Percentage of Regional Production of Principal Food Crops Based on Crops Originating in Other Regions of Diversity*

Region	Production % accounted for by non-native crops
West Central Asia	31
Indochina	34
Hindustan	49
Latin America	56
Chino-Japan	62
Africa	88
Euro-Siberia	91
Mediterranean	99
Australia	100
North America	100

Source: Cooper et al, 1994.

As far as food production is concerned, Table 5.2 shows how the global exchange of seed through international trade in commercial seed is growing. And as for plant breeding, the results of our survey of plant breeders from eleven countries (Canada, Chile, Czech Republic, France, Germany, India, Japan, Mexico, South Africa, the UK, and the USA) reveal that breeders still routinely use germplasm obtained from other countries to develop new plant varieties. Thus, both for food production and for crop development, the countries of the world remain dependent on access to each other's PGRFA for food security. However, plant breeders have at their disposal a wealth of national germplasm, and can usually access foreign germplasm from public and private collections held in their own country. This reduces their need to seek germplasm from abroad. It also means that

Table 5.2 *Evolution in the Global Exchange of Seed – International Trade in Commercial Seed*

Year	Exchange (US$ bn)
1970	0.86
1977	1.08
1980	1.2
1985	1.3
1994	2.9
1996	3.3

Source: FIS/ASSINSEL, 1998.

breeders, and others who subsequently obtain material from them, such as seed companies, farmers and consumers, are rarely obliged to share the benefits arising from the use of germplasm held in the collections, since it is usually supplied without any such obligation.

Despite this readily accessible source of international germplasm, free of the burden of seeking prior informed consent or of negotiating benefit-sharing arrangements, breeders are still likely to seek access to germplasm held abroad, for reasons that will be explained later in the chapter. As the FAO *Report on the State of the World's Plant Genetic Resources for Food and Agriculture* explains, even though 'many countries hold a significant amount of plant genetic diversity for food and agriculture in their genebanks and farmers' fields, in the long-term, they are likely to require access to additional diversity from the crop species' centres of diversity' (FAO, 1996). As this chapter will show, access to genetic resources is generally a prelude to lengthy programmes of evaluation and improvement that make germplasm not only accessible, but of practical interest to seed companies. Also, despite the lack of legal obligations to share benefits, networks of breeders have developed over the decades, predominantly in the public sector, and continue to exchange elite germplasm (see, for example, Table 6.1, FAO, 1996). Bilateral aid projects, too, have introduced elite germplasm from developed to developing countries – for example, the transfer of germplasm from the USDA to Brazil. Soybean germplasm from USDA formed the basis for the Brazilian soybean economy, and high lysine maize provided a base for the Brazilian human infant and swine-feed programmes.

Recognising the international interdependence on access to PGRFA, the members of the FAO (at that time, some 150 governments) unanimously adopted the IU, at the 22nd session of the FAO Conference, in November 1983. 113 states have now adhered to the IU.[2] The objective of this voluntary agreement is to 'ensure that plant genetic resources of economic and/or social interests, particularly for agriculture, will be explored, preserved, evaluated and made available for plant breeding and scientific purposes'. Article 5 of the Undertaking refers to the 'universally accepted principle that plant genetic resources are a heritage of mankind and consequently should be available without restriction'.

In the 1980s, private plant breeders in developed countries were concerned that the terms of the IU would oblige them to make their materials freely available without restrictions on their use for further breeding or any other purposes. As a result, the FAO Council took a series of Resolutions between 1989 and 1993 which qualify the 'heritage of mankind' principle and the meaning of 'unrestricted' availability (ten Kate and Lasén Diaz, 1997). FAO Resolution 4/89 introduced an 'agreed interpretation' of the Undertaking which states that Plant Breeders' Rights, as provided for under the International Union for the Protection of New Varieties of Plants (UPOV), are not incompatible with the Undertaking, and that a state may impose only such minimum restrictions on the free exchange of materials covered by the Undertaking as are necessary for it to conform to its national and international obligations. The Resolution also states that 'free access' does not mean free of charge, and that benefits under the Undertaking are part of a reciprocal system for those countries adhering to the Undertaking.

Resolution 4/89 also recognises 'the enormous contribution that farmers of all regions have made to the conservation and development of plant genetic resources, which constitute the basis of plant production throughout the world, and which form the basis for the concept of Farmers' Rights' (para 3), which can be implemented by ensuring 'the conservation, management and use of plant genetic resources, for the benefit of present and future generations of farmers' (para 4), including by the International Fund for Plant Genetic Resources established by the FAO. Resolution 5/89 'endorses the concept of Farmers' Rights', which are 'rights arising from the past, present and future contributions of farmers in conserving, improving and making available plant genetic resources, particularly those in centres of origin/diversity. These rights are vested in the international community, as trustee for present and future generations of farmers, for the purpose of ensuring full benefits to farmers, and supporting the continuation of their contributions, as well as the attainment of the overall purpose of the International Undertaking'.

By 1991, Resolution 3/91 had recognised that the concept of mankind's heritage, as alluded to in the Undertaking, 'is subject to the sovereignty of the states over their plant genetic resources' (para 1). Acknowledging that conditions of access to plant genetic resources need further clarification, Resolution 3/91 endorses the fact that states have sovereign rights over their plant genetic resources; that breeders' lines and farmers' breeding material should only be available at the discretion of their developers during the period of development; and that Farmers' Rights will be implemented by an international fund on plant genetic resources which will support plant genetic conservation and utilisation programmes, particularly, but not exclusively in developing countries.

During the period that Resolution 3/91 was being drafted, many of the same countries were negotiating the text of the CBD, and shortly thereafter, on 22 May 1992, the 'Nairobi Final Act' was adopted at the UNEP Conference for the Adoption of the Agreed Text of the

Convention on Biological Diversity. Resolution 3 of the Nairobi Conference concerns the interrelationship between the CBD and the promotion of sustainable agriculture. It recognises the need to seek solutions to outstanding matters concerning plant genetic resources within the FAO's Global System for the Conservation and Sustainable Use of Plant Genetic Resources for Food and Sustainable Agriculture, and explicitly states 'in particular: *(a)* access to *ex-situ* collections not acquired in accordance with this Convention; and *(b)* the question of farmers' rights'.

In 1993, shortly before the CBD entered into force, the FAO Conference adopted Resolution 7/93 on the follow-up to Resolution 3 of the Nairobi Final Act. Resolution 7/93 requests the Director-General of the FAO to provide a forum for negotiations among governments:

- for the adaptation of the International Undertaking in harmony with the CBD;
- for consideration of the issue of access on mutually agreed terms to plant genetic resources, including *ex-situ* collections not addressed by the Convention, as well as
- for the issue of realisation of Farmers' Rights.

Subsequently, the Commission on Genetic Resources has discussed these issues on eight occasions between November 1994 and April 1999. Negotiations have focused on the nature of a possible multilateral system for access to PGRFA; its scope; conditions for access to the materials within it; benefit-sharing; finance and Farmers' Rights. There has been progress on some elements. In April 1999, the delegates made progress on Farmers' Rights,[3] and also on text stating that, 'In exercise of their sovereign right, Parties agree to establish a multilateral system, which is efficient, effective, and transparent, to facilitate access to plant genetic resources for food and agriculture, and to share, in a fair and equitable way, the benefits arising from the utilisation of these resources'. However, typically for international negotiations, language provisionally agreed thus far on any article or paragraph may still be subject to change, as the IU will ultimately be adopted as a 'package'.

Opinion remains sharply divided on the scope of the multilateral system. Some countries, such as those in the EU, favour a broad system covering – subject to property rights – all taxa of PGRFA. This would not require a list of crops. By contrast, many developing countries would prefer a narrow scope for the multilateral system, with access to all other materials dealt with on a bilateral basis, in accordance with the CBD. The scope of this 'narrow' multilateral system would be defined by a short list covering wheat, maize and rice (which together provide some 60 per cent of food energy supply) and possibly a few more genera. It seems likely that, eventually, a longer list of perhaps 25–30 genera based on the criteria of food security and interdependence will be the compromise position. The list of crops prepared at the Fourth Extraordinary Session of the Commission on Genetic Resources for Food and Agriculture in June 1997 may provide the basis for negotiation, and appears in Appendix D.

Individuals, communities and corporations have property rights over specific genetic resources (ranging from landraces to elite strains subject to IPRs) held *ex situ* in many countries. In addition, some public genebanks are not free to commit collections from other countries which they hold in trust for those countries to the multilateral system. For these reasons, delegations such as the USA and the EU propose a system of 'designation' or 'identification' of accessions which they can commit to the multilateral system. Developing countries fear that such an approach could leave Parties with complete freedom to decide which materials to contribute, and could result in an 'empty' multilateral system and an inequitable position as countries designate different quantities of material.

Farmers, local and indigenous communities and other individuals also own land and hold other rights over genetic resources found *in situ*, so their PIC is likely to be needed before *in situ* genetic resources concerned could be committed to the multilateral system. Private citizens may grant access to *in situ* genetic resources subject to terms that vary from case to case and could easily not be identical to the 'standard' terms for access under the multilateral system.

The inclusion of *in situ* genetic resources is linked to the uses that could be made of genetic resources within the multilateral system and to the opportunities for benefit-sharing within it. Although definitions of terms such as 'commercialisation' have yet to be clarified within the IU, it seems likely that recipients of material from the multilateral system would be entitled to develop commercial food and agricultural products derived from them, but not to take out IPRs on the materials themselves. Breeding new and improved food crops would clearly lie within the use of PGRFA for 'food and agriculture', but some uses of agricultural crops, such as cultivation of medicinal plants, will almost inevitably be excluded. A lively discussion is likely to take place in order to define which other uses – from fodder and litter for animals to industrial and commodity uses of PGRFA such as pulp and paper, fibres and bioremediation – are to be excluded from the multilateral system.

The nature of benefits to be shared and the mechanism for doing so remain undecided, since they are linked to the question of scope of the multilateral system and of commercialisation. Many delegations see reciprocal access to genetic resources as the major benefit to be shared through such a multilateral system, whereas others place more emphasis on the need to transfer technology, to share information and to build capacity in exchange for access to PGRFA. Support appears to be growing for a completely 'multilateral' system of benefit-sharing that would separate benefit-sharing from access transactions. In this model, any benefits arising from the provision of access to specific accessions would not be shared with the particular country that contributed the materials to the system but would go towards more general support for any eligible countries (presumably, developing countries and those with economies in transition). A purely multilateral system for benefit-sharing would obviate the need for costly and bureaucratic 'tracking' systems to monitor the flow of genetic resources within the system, which could well be necessary if the multilateral system were to provide benefits 'bilaterally' to a particular country of origin in the case that a commercial product was produced from PGRFA provided by that country. Developing countries are likely to be accorded priority access to certain benefits, such as technology transfer and capacity building, and the provision of benefits may be linked to the question of financial resources.

A remaining, fundamental question concerns whether there should be a mechanism for sharing the financial benefits arising from use of PGRFA (including commercial use) through a dedicated fund, or rather through the mobilisation of available funds for the implementation of activities such as those described in the Global Plan of Action. The question of a 'Funding strategy' or 'Fund' must be discussed in detail.

All these issues have yet to be resolved. So, too, does the question of the terms of access for private citizens and private organisations within countries that were Parties to the revised IU, and for both governments and private citizens in countries that were not Parties. Finally, another important question will be the Article of the IU that deals with the relationship between IU and other intergovernmental agreements, such as TRIPS and other treaties related to trade and IPRs.

When the IU is finally complete, it is likely to become international law (whether as a Protocol to the CBD, or as some other form of intergovernmental agreement). This will have profound implications for access – whether by companies, public institutions or individual scientists – to PGRFA. To begin with, it may be obligatory for any public collections containing any PGRFA within the scope ultimately agreed for the IU to make them available on the terms for access agreed within the IU. Conversely, any individual or institution seeking access to these PGRFA from such a governmental organisation will be obliged to do so on the terms set out in the IU, which may involve benefit-sharing. Secondly, there may be a mechanism to enable the participation in the multilateral system of private organisations such as companies, genebanks and botanic gardens with collections of PGRFA.

The complexity of the issues at stake and the differences of opinion among the negotiators mean that countries face a considerable challenge in finalising the IU, although the current goal is that the negotiations should be concluded in June/July 2000 and the final text adopted at the 11th session of the FAO Council in November 2000.

Until any revised IU enters into force, the legal framework for access to PGRFA is the CBD and national laws on access to implement it. In practice, as will appear from the experiences shared by plant breeders in the rest of this chapter, it is often necessary to obtain permits when conducting field collections, but the majority of samples of PGRFA are obtained by exchange among institutions. Most countries allow for relatively unrestricted access to their PGRFA from genebanks, research stations and botanic gardens, subject to the phytosanitary controls generally in force.

At least 100 countries now have laws or draft laws on plant variety protection (personal communication, André Heitz, UPOV, 22 February 1999). 81 are members of the International Union for the Protection of New Varieties of Plants (UPOV). Further, under the Code on Trade Related Aspects of Intellectual Property Rights (TRIPS) agreed as part of the Uruguay General Agreement on Tariffs and Trade (GATT) round, the 134 members of the World Trade Organisation (WTO) (as of 10 February 1999) are obliged to provide for the protection of plant varieties, either by patents or by an 'effective sui generis system', or a combination thereof (TRIPS Art 27(3)). Patents and Plant Breeders' Rights affect access to genetic resources in that, during the life of a patent, patent protection prevents scientists from using protected material for plant breeding or commercialisation without permission. The Breeder's Exemption in both the 1978 and 1991 Acts of the UPOV Convention allows breeders to access varieties protected by Plant Breeders' Rights. Negotiations between the breeder of a new variety derived from a protected variety and the holder of the Plant Breeders' Right on the original protected variety are required only when the resulting variety is 'essentially derived', a term whose meaning has not been clarified through judicial interpretation, but can be broadly understood to cover derivatives that include the essential germplasm of the protected variety, or derivatives that are genetically very close to the protected variety (Appendix E, Box E.1).

Although negotiators failed to agree upon a Biosafety Protocol to the CBD in Colombia in February 1999, any intergovernmental treaty that is eventually agreed and regulates the trans-boundary movement of genetically modified organisms, including genetically engineered plant varieties, could also have a profound impact on access to plant genetic resources for food and agriculture.

5.1.2 The roles of actors and markets

The conservation of PGRFA, their use in the breeding of new varieties of major crop and the marketing and sale of the resulting products involve a range of actors, notably the private sector, the public sector, farmers and others such as seed-saver associations and cooperatives.

It was only at the end of the 18th century that the selection and hybridisation techniques of formal plant breeding were developed. Wheat breeding in Europe was well-established in the late 19th century, based in the private sector in countries such as France, and in government institutions in others, such as the Netherlands. In the USA, most plant breeding was carried out by the USDA, initially through the US Land Grant System created in 1862, and the land grant agricultural universities established as part of the system in 1890.

Very little seed-breeding was carried out by the commercial sector until the 1930s, although the American Seed Trade Association was formed by 34 companies in 1883. During the 20th century, commercial seed operations came to replace most government seed production efforts in Europe and North America, and in some countries, most notably France, much commercial seed production was initiated by cooperatives (Tripp, 1997). Until recently, the seed industry has been characterised by small, independent and often family-owned firms. The integration of the seed industry into food and agro-chemical corporations began in the 1970s, and the formation of life science giants has been a phenomenon of the 1990s.

In developing countries, public plant breeding has tended to be accompanied by the establishment of public seed enterprises. For example, in India, the National Seed Corporation was established in 1963, and was followed in various Indian states by the emergence of public-sector seed companies, some of which used the proceeds of their sales to fund plant breeding research. Private seed companies in India have operated only since 1980. The sale of crops such as wheat is still predominantly in the hands of the public sector, but private companies are playing an increasingly important role in seed supply of many other crops such as maize, sorghum and oilseeds.

There are now two distinct seed supply sectors: the 'informal' and the 'formal'. The informal sector involves local seed supply. Farming communities develop crop diversity through local landraces and manage variety selection, seed production, storage and distribution at the community level. The major emphasis of the informal sector is on yield stability, risk avoidance and low external input farming (de Boef and Hardon, 1993). 'Formal' seed supply describes the chain of activities extending from breeding to marketing and distribution, managed by specialists (Tripp, 1997). The system involves international agricultural research centres (IARCs) and other regional and commodity-oriented centres, national research systems and private industry, linked to farmers through extension services and marketing in a linear model of technology development and transfer. The emphasis is on maximising profit through higher yields, higher quality and the use of external inputs, with the main objective of solving the global and national need for more food (de Boef and Hardon, 1993).

The range of activities of public breeders, the private sector and farmers is described briefly in this section, although their detailed role in gaining access to genetic resources is explored in more depth in 5.2.

The private sector

In 1996, Rabobank reported that there were over 1,500 seed companies in the world, 600 of which were based in the USA and 400 in Europe, and that private seed companies competed with government to procure and distribute seed in only 28 per cent of the countries of the world (Rabobank, 1996). Today, there are several thousand companies (depending on whether the term 'seed company' is used to embrace firms involved in trade as well as production). A trend to privatise seed production and distribution, and, in some countries, plant breeding, too, means that the private sector is now producing and distributing seed in more than 50 per cent of the countries of the world (personal communication, Patrick Heffer, ASSINSEL, 21 December 1998).

In 1998, US$30 billion of the US$50 billion total global market for seed consisted of the commercial sale of seed (FIS/ASSINSEL, 1998). 31 per cent of these sales were accounted for by the 23 largest companies (Table 5.3). The 1998 turnover of the top ten seed companies accounted for 23 per cent of the world market for commercial seed, and the three biggest companies alone – Pioneer, Monsanto and Novartis – together accounted for 13 per cent of global sales, with combined turnovers of US$3.9 billion.

Profit margins (net income divided by net sales) vary from crop to crop and company to company, but review of annual reports suggests that the margin of some leading seed companies is currently around 13–14 per cent. (The profit margin before deduction of indirect operating, general and administrative costs is around 24 per cent.) (*Source:* companies' annual reports.)

Table 5.3 *Large International Seed Companies*

Company	Main activity	HQ	Some main subsidiaries and/or main brands	Seed turnover (US$ m)[1]	Some core seed products
Pioneer Hi-Bred International	seeds	USA		1,800	maize, oilseeds, alfalfa, cereals
Monsanto	chemistry	USA	Holdens, Dekalb, Cargill Seed International, Asgrow Agricultural Seeds, PBI, Hybritech, Delta & Pine Land, Agroceres	1,200	maize, oilseeds, cotton, vegetables
Novartis Seeds	seeds	Switzerland	Northrup King, Hilleshög, Sluis & Groot, Rogers	900	maize, oilseeds, cereals, sugar beet, vegetables and ornamentals
Groupe Limagrain	seeds	France	Force Limagrain, Maïs Angevin, Nickerson, Vilmorin, Tézier, Clause, Oxadis, Ferry Morse, Harris Moran	700	maize, oilseeds, cereals, vegetables and ornamentals
Advanta	seeds	Netherlands	Van der Have, Zeneca Seeds, SES, Mommersteg, Pacific Seeds	460	maize, forage crops, sugar beet, cereals, oilseeds
AgriBio Tech	seeds	USA	Clark Seeds, Burlingham & Sons, Olsen-Fennell Seeds, Oseco, Germain's Seeds, W-L Research, Zajac Performance Seeds	460	forage crops and turf
KWS	seeds	Germany	Betaseed, Lochow Petkus, SDME, Great Lakes Hybrids, CPB-Twyford	390	sugar beet, cereals, maize, forage crops, oilseeds, protein crops
Seminis Vegetable Seeds	seeds	USA	Asgrow Vegetable Seeds, Petoseed, Royal Sluis, Bruinsma	380	vegetables, ornamentals
Takii	seeds	Japan		310	vegetables, ornamentals
Sakata	seeds	Japan	Samuel Yates	280	vegetables, ornamentals
Kaneko Seeds	seeds	Japan	Tat Tohumculuk	280	maize, forage crops and turf, vegetables, ornamentals
AgrEvo	chemistry	Germany	Cargill Hybrid Seeds N America, Nunhems, Sunseeds, Leen de Mos, AgrEvo Cotton Seed International	250	maize, oilseeds, cotton, vegetables
Mycogen	biotech	USA	Dinamilho, Hibridos Colorado, FT Biogenetica de Milho	230	maize, oilseeds, alfalfa
Pennington	seeds	USA		180	forage crops and turf
Ball	seeds	USA		180	ornamentals
Pau Euralis	ag trade	France	Rustica Prograin Génétique	175	maize, oilseeds
Barenbrug	seeds	Netherlands	New Zealand Agriseeds, Heritage Seeds, Tourneur, Palaversich	160	forage crops and turf
Sigma	ag trade	France	Semences de France, Ringot, Serasem	160	cereals, protein crops, maize, oilseeds, forage crops and turf, sugar beet
Saatenunion	seeds	Germany		155	cereals, maize, oilseeds, forage crops
DLF	seeds	Denmark		150	forage crops, sugar beet, oilseeds
RAGT	ag trade	France	Semillas Monzon, Joordens Zaaden	140	maize, oilseeds, forage crops, cereals
Svalöf/Weibull	seeds	Sweden	Semundo, New Field Seed, Sursem	140	sugar beet, cereals, maize, oilseeds
CEBECO	ag trade	Netherlands	Procosem, International Seeds, Seed Innovations, Van Engelen, Wiboltt, la Maison des Gazons	140	protein crops, maize, oilseeds, cereals, forage crops and turf, vegetables
Total				9,220	

1 Estimates based on annual reports, press releases, personal communication, FIS. Susceptible to change according to exchange rates

Source: International Seed Trade Federation (FIS), December 1998.

Over the last 20 years, in common with other areas of the life sciences, the commercial seed industry has witnessed a number of major mergers and acquisitions. In 1997, Monsanto acquired Holden's Seeds, said to be the source of 35 per cent of the parental lines used by independent corn breeders, for US$1 billion, and Brazil Agroceres, the largest seed company in the southern hemisphere, for an estimated US$70 million (RAFI, 1998; Bell, 1997). During 1988, Novartis acquired Ciba Seeds and Northrup King Seeds, and Monsanto became the second-largest seed company. Monsanto's family of seed companies, now called the Monsanto Global Seeds Group, owns Agracetus Inc, Agroceres, Asgrow Agricultural Seeds, Cargill Seed International, Dekalb, Delta and Pine Land, Hartz, Holden's Foundation, Hybritech, NatureMark and PBI.

After the life industry giants, RAFI (1998) distinguishes two other tiers of company: large multinational firms, and small and medium-sized enterprises. 'Second tier' companies include multinational firms with interests in agrochemicals and pharmaceuticals, such as Advanta, AgrEvo, Dow AgroSciences, KWS AG, and Groupe Limagrain and several large companies whose primary focus is the commercial seed trade, such as Takii (Japan), Barenburg (The Netherlands) Svalof Weibull (Sweden), Cebeco-Handelsraad (The Netherlands), and Sakata (Japan). This tier of companies has seen its own share of mergers and acquisitions.

The third tier comprises the small and medium-sized, independent seed companies, of which there are several thousand, a small proportion of which are actively engaged in plant breeding.

If recent trends continue, it seems inevitable that there will be more consolidation within the first and second tiers, and there may be some movement from the second to the first tier, as more companies acquire a range of interests in different branches of the 'life sciences'. The effect of these partnerships is to increase the genepool available to seed companies.

Private companies conduct research on a broad range of genetic resources, some of which are within their own collections and some of which they need to obtain from elsewhere. They may sell or license the resulting seeds or other products directly to the consumer, to farmers, cooperatives or other companies.

The companies in each of the three tiers have different scales of operations, international scope, ranges of products, levels of investment in technology and research and development, and profit margins. Consequently, they have varying demands for access to genetic resources and differing views on the most appropriate mechanisms and levels of benefit-sharing. One 'third tier' interviewee in our survey warned of the danger of picturing the seed industry as homogeneous, and worried about the increasing lack of transparency that he perceived in the seed sector: 'Historically, the system was very open. Now it is more closed, which concerns me, especially as the large multinationals' views on the seeds business are very different from ours'.

The market data cited in this section refer to global sales of commercial seed. No reliable data are available on the global market for finished agricultural produce, since to collect or estimate these would be a task verging on the impossible. In a highly simplified description of the supply chain, commercial seed is sold to farmers, who multiply it and sell it to their customers, including distributors and industrial and food-processing companies. These companies either sell produce (such as potatoes) directly to the consumer, or they manufacture products (such as bread) which are then sold to the consumer, often via other retail outlets. This complex chain makes any estimation of the global market for the products arising from global agricultural and horticultural vegetable crops extremely complex. If we assume, for the sake of argument, that once the different products have passed through all these hands, the final value of the produce reaching the consumer is ten times the commercial sales of seed to farmers, this would suggest that the global market for agricultural produce is some US$300 billion. If the multiple is closer to 15, then the value of the global market for agricultural produce is some US$450 billion. All such estimates are highly speculative.

The public sector

Although the commercial sale of seed is growing as a proportion of the total global market for seed, the private sector is the dominant source of seed in less than a half of the countries of the world. In the other countries, government still controls the procurement and distribution of seed. The public sector is responsible for the majority of seed-breeding in many developing and developed countries from India to Japan, despite the growing role of the private sector in developing countries and planned economies. Seed bred and distributed by the public sector accounts for approximately US$10 billion, or 20 per cent of the global market for seed (Rabobank 1994).

While the relative proportions of public- and private-sector research and development in plant breeding, and the rates at which they change vary from country to country, there is a noticeable trend around the world for investment in research and development in plant breeding to increase in the private sector and decrease in the public sector (personal communication Stephen Smith, 14 January 1999; Patrick Heffer, 21 December 1998). For example, the proportion of 'science person years' (SY) devoted to plant breeding research and development in the USA in 1994 by the public sector[4] was 33 per cent, compared with 67 per cent in the private sector. The research expenditure on plant breeding research and development in the USA in 1994 by private companies was US$338 million (61 per cent of the total research and development expenditure), while public sector research and development expenditure that year was US$213 million (39 per cent of the total). Over the five-year period from 1990 to 1994, public-sector research decreased by 2 per cent, while private-sector research increased by 11 per cent (Table 5.4). This is not true for all countries, however, as the increasing role of the state sector in Australia through the coordinated Research Centre programme demonstrates (personal communication, Mike Ambrose, 22 January 1999).

Table 5.4 *Research and Development in the Public and Private Sectors in the USA, 1990–94*

Public/private sector	'Science person years' (SY), 1994	% of whole	Change in SY, 1990–94	Change % compared to 1994 total
Public	749	33	-12.5	-2
Private	1,499	67	+160	+11
Total	2,241	100	–	–

Source: Frey, 1996.

Despite the general decrease in the role of the public sector in plant breeding, its contribution remains important. Within the public sector, there are three main categories of scientists.

The first group includes pathologists, physiologists, biochemists studying biosynthetic pathways, geneticists and biotechnologists identifying markers for desirable traits. The second group is involved with aspects of plant performance in the field and includes agronomists and crop physiologists. They are responsible for making recommendations to plant breeders on changes in practice which may modify their criteria for traits. The third group includes plant pathologists, geneticists studying pathways, and biotechnologists identifying markers. This group of scientists develops the tools and materials for plant breeders. The second group comprises the plant breeders who are actively developing a crop by conducting activities ranging from selection of species to plant breeding, using both traditional seed-breeding techniques and biotechnology. Some plant breeders in the public sector produce finished varieties for planting, while others develop material to a certain point, then turn it over to be finished by plant breeders in international institutions such as the centres of the Consultative Group on International Agricultural Research (CGIAR), or in the private sector.

The kinds of crop produced by the public and the private sectors vary from region to region. For example, in developed countries, maize is bred almost entirely by the private sector, whereas the public sector still plays an important role in maize-breeding in developing countries. In Europe, wheat is largely bred in the private sector, since most farmers do not save wheat seed to replant the following season; but in North America, many farmers save wheat seed and sell it across the fence to other farmers ('brown bag sales'), which decreases its attraction as a product for the private sector. The public sector thus plays an important role in wheat-breeding in North America. Also, oats and barley are often bred by the public sector in North America, whereas they are bred by private companies in Europe.

In countries where the public sector plays an important role in seed-breeding, the work of public sector scientists typically takes some 80 per cent of the time needed to develop a new variety, but just 30–35 per cent of the entire expense, since the breeding work conducted by the public sector involves materials in earlier stages of development. Once the public sector announces that certain material is available and is passed to the private company, the remaining 20 per cent of the time needed to develop a finished commercial variety typically takes 65–70 per cent of the research and development budget for a new variety (personal communication with Bryan Harvey, June 1998). According to our interviewees, breeders in the public sector commonly pass their pre-bred materials to private companies for no more than a nominal fee, such as US$5–20, or for a slightly greater sum that might, for example, enable the public agency to purchase a computer, but certainly not to cover the costs of the research.

Some public-sector breeders, such as those in the US state and territorial agriculture research stations (SAES), have used Plant Variety Protection to protect their finished materials and to qualify these varieties for OECD seed certification (5.2), but, in the past, most have not used patents and claimed royalties (from licensing patented material or allowing breeders to develop varieties 'essentially derived' from those protected by plant variety rights). However, it is now increasingly common for public sector institutions such as universities to license their lines to industry for a fee or royalty.

As well as their role in plant breeding, public institutions are involved in the conservation of PGRFA, both *ex situ* (eg in genebanks, Box 5.4), and in *in situ* conditions.

In some countries, a further role of government is to regulate access to genetic resources (including their import and export, and phytosanitary rules) and the release of new seed varieties on to the market.

Farmers

Since time immemorial, farming communities and individual farmers have adapted germplasm to local conditions and selected the best seed each season, thus gradually improving varieties. Since the advent of a 'formal' seed sector in the 19th and 20th centuries, the emphasis of seed selection and breeding has shifted away from farmers to the public and private sectors. However, in many parts of the world, farmers still play an important role both in maintaining and selecting traditional varieties. Through their role as the 'clients' of public- and private-sector breeders, they also shape the products developed in the 'formal' sector.

As well as their aim for higher productivity and interest in high quality seed, another factor that influences farmers' demand for access to seed is the extent to which they can be self sufficient. In many parts of the world, farmers save a proportion of the seed produced in one year's harvest to plant for the next, and they sometimes sell this seed to other farmers. Sales of farm-saved seed world-wide account for approximately US$10 billion, or 20 per cent of the global market for seed. In the developing countries of Asia, Africa and South America, such 'farm-saved seed' accounts for 80 per cent of farmers' total seed requirements. For example, only about 7 per cent of wheat seed and 13 per cent of rice seed planted in India is from the formal sector.

The practice of saving seed is not limited to developing countries. As Table 5.5 shows, farm-saved seed of small grain cereals (such as wheat, barley, oat, and rye) provides half the seed used in Germany and France, and the proportion in other member states of the EU ranges from 5 to 90 per cent (Rabobank, 1994). In the USA, for example, approximately 50 per cent of cotton and barley, 60 per cent oats, and 70 per cent of wheat planted is farm-saved seed. However, virtually all of the sorghum, maize and vegetables in the USA are planted using commercial seed.

Table 5.5 *Farm-Saved Seed of Small Grain Cereals in the EU*

Country	% of total seed demand
Germany	50
France	50
Italy	70
Netherlands	20–25
Denmark	5
Ireland	20
UK	30
Greece	90
Spain	90
Belgium	35

Source: Rabobank 1996.

A number of factors guide farmers in their choice of whether or not to save seed. For example, harvest conditions may prevent seed germination, and the cost of storage equipment and the effort of collecting and saving seed may exceed the cost of commercial seed. In addition, patents restrict the ability of farmers to sell the seed they have saved and to use it as the basis for breeding new varieties, and companies are increasingly supplying seed under contracts similar to software licence agreements, which prohibit farmers from re-using it. Some farmers have been sued for breach of these contracts. For example, a press release dated 29 September 1998 from Monsanto stated that 'in response to numerous requests from farmers wanting to know details regarding those offenders caught illegally saving and replanting seed containing patented technology, Monsanto Company today announced the specifics of one of its piracy case settlements', and reported on settlements ranging between US$10,000 and US$35,000 paid by farmers who had admitted to illegally saving and replanting Roundup Ready soybeans and other seed patented by Monsanto (Monsanto, 1998).

Companies' interest in producing seed that farmers cannot save from one season to the next (thus maintaining the demand for commercial seed) has led to growing investment in infertile seed and in hybrid varieties, which do not breed true to type and are thus unattractive for seed saving. Hybrids account for nearly 40 per cent of the total global commercial seed business, and are available for many important commercial grains such as maize, sunflower, sorghum, oilseed rape, various vegetables and, to a limited extent, rice (J van Wijk, 1997). Also, self-pollinated seed varieties can now be genetically engineered to produce sterile off-spring. One such approach is the technology developed by Delta and Pine Land Co, which is being acquired by Monsanto, that has been demonstrated in cotton and tobacco and dubbed 'suicide seed' or 'terminator technology'. Chemical treatment of seeds before they are sold activates a gene that produces a protein enabling another gene to 'switch on' a third gene which produces a toxin killing seed at the time when it is maturing.

5.1.3 The development of a new variety

Development and release of a new, modern variety typically takes 8–15 years, and costs in the range of US$1–2.5 million for a traditionally-bred variety, and between US$35 and US$75 million for a transgene used in many varieties, depending on its complexity and the number of regulatory 'events' (Box 5.1).

The research and development cycle of new seed varieties generally involves three overlapping phases:

selection and pre-breeding; breeding; and product approval. These phases differ slightly where the research and development involves GMOs, since the high-technology biotechnological research tools needed for genetic engineering require extra investment, and the costs of regulatory approval are higher.

Plant breeding

Plant breeding involves the identification and crossing of plants each of which exhibit different useful characteristics, to generate populations of genetically recombined individuals. A small proportion of the population will display the desired traits of both parents within individual plants. The breeder's task is then to identify and select those plants with the right combination while discarding the rest. Each plant contains tens of thousands of genes and the breeder endeavours to ensure that genes responsible for useful traits are combined in the improved plant variety and any deleterious genes are selected against. In the quest for new cultivars, breeders cross and back-cross many parents leading to hundreds of different gene combinations represented in thousands of plants (Figure 5.2). Successful plant breeding relies on the number of potential cultivars that can be screened within a breeding programme (Kelly et al, 1998) and how effectively that programme can identify the genetic basis for improved performance, and the ability to choose the most promising parents as quickly as possible.

The seed industry is research-led, since it attaches great importance to the very costly development of new cultivars. Very few of the 1,000 or more 'crossed variants' end up as successful cultivars. Take, for example, the case of a series of 62 sister varieties of wheat known as Veery, released by Centro Internacional de Mejoramiento de Maiz y Trigo (CIMMYT) in the late 1970s, and at one time grown on 3 million hectares, from China to Chile. To develop the Veery lines, CIMMYT used the progenitors (parents) which genealogy traces back to approximately 3,170 different crosses made by breeders around the world, using 51 individual lines from 26 countries. The pedigree for a Veery line printed on A4 paper is more than six metres long (personal communication, Dr S Rajaram, 2 February 1999; IPGRI, 1996). Scale is thus a crucial factor for success in the seed industry and is an important explanation for today's mergers and acquisitions, which have led to a few, giant, multinational seed companies.

Traditional breeding, which remains the mainstay of plant breeding, is sometimes still performed by 'cross pollination', by shaking or brushing

Box 5.1 *Research and Development of a New Variety of Crop*

Phase/Activities	Time	Cost (US$ m)
(For GMOs) Initial development of a transgene Where genetic engineering of this kind is used at all, the cost and time to develop a genetically engineered 'transgene' depends on the individual trait and organism, and thus on the research and biosafety requirements. In Phases I and II, a single transgene might be bred using traditional methods into anything from a few to over 100 varieties.	4 years	30–50
Phase I Generation of genetic material **(Note: Phase I and Phase II often overlap)** In traditional breeding, 100,000s of progeny are generated from 1,000s of crosses, often with the help of genetic markers (which accelerate both Phases I and II), and the use of controlled environment facilities and/or alternate seasons in different hemispheres. Cost and time vary enormously depending on the individual trait and the crop and breeding method concerned.	2–3 years (for either trad bred or GM varieties)	Phase I and Phase II cost 1–2.5 m per variety, including unsuccessful lines
Phase II Evaluation and selection of progeny (In a large company) Over several years, the breeder selects and develops some 5–20 prime cultivars from the initial 100,000 progenies. Cost and time vary enormously depending on the species and the nature of the cross. A wider cross with 'exotic' material increases the time and raises the cost.	Trad breeding: 5–6 years GMOs: 3–6 years	
Phase III Submission of 5–20 prime cultivar to governmental authorities for listing In countries where seed certification is compulsory, tests such as 'distinct, uniform and stable' (DUS) and 'value for cultivation and use' (VCU) are undertaken to determine the potential of new varieties for certification. These include tests for resistance and susceptibility to diseases, purity, use and yield potential. For GMOs, additional environmental and human safety tests.	Trad breeding: 2–3years GMOs: 3–4 years	Trad breeding 1,500+ GMOs: 1–7 per 'event' for approval of a transgene[1]
Total	8–15 years	Trad breeding: 1–2.5 per variety GMOs:[2] 35–75 per transgene

1 Regulatory approval for transgenic varieties costs up to US$1 million (including internal company costs) for non-pesticidal plants, and US$3–7 million for pesticide-producing plants. These costs are for an 'event', a unique insertion of DNA in a plant genome through genetic engineering. Once a transgenic is approved, however, the approval for new varieties incorporating it may be no more than for traditionally-bred varieties.

2 It might cost US$35 million to develop and release a transgene involving one 'event', whereas another involving several 'events' could cost US$75 million. However, the transgene would likely be used in several varieties.

Source: interviews.

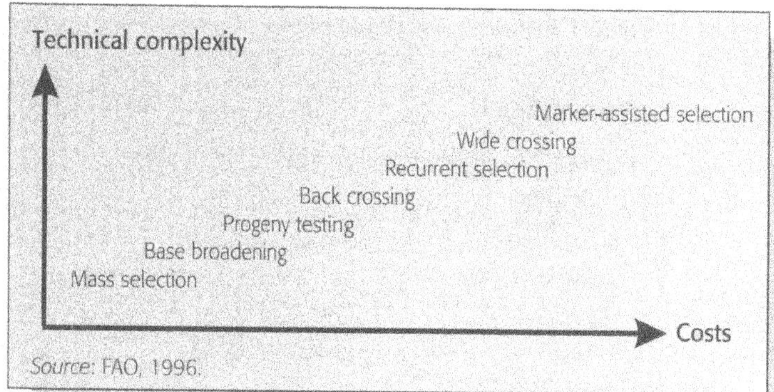

Figure 5.2 *Plant-Breeding Techniques*

pollen from the male plant on to the female plant, but the range of techniques available to the breeder to aid in the crossing procedure has extended significantly over the last 20 years. These techniques include embryo rescue, cell and tissue culture and molecular genetics. Bridging species can be used to enable crosses between cultivated plants and wild species. Marker-assisted selection helps to assess a seedling for a trait that will be expressed only later in the plant's life cycle, thus greatly reducing the time to produce a totally homozygous plant. Technologies such as these influence the numbers of plants tested in the breeding process, the speed with which these numbers can be reduced and the time taken between the initial crossing and the finished variety (Figure 5.3).

Traditional crossing and backcrossing methods are still used to develop the vast majority of new varieties, but the use of biotechnology is growing rapidly. Genetic engineering has already produced a slew of recombinant seed products, including delayed-ripening tomatoes herbicide-resistant cotton and insect-resistant corn, fungus- and virus-resistant crops.

Plant biotechnology can be used for the accurate selection and delivery of desired characteristics, to transfer genes from one species into another, to remove undesirable characteristics such as allergenic and toxic compounds and to produce varieties that require less input of pesticides, fertilisers and even water and energy. Genetic engineering offers the potential to incorporate desirable traits from species that could not be interbred – even those from other taxonomic kingdoms, such as the gene from the flounder fish introduced into wheat to produce a variety able to withstand cold temperatures without freezing. Another common example is the introduction of the Bt gene from the bacterium *Bacillus thuringensis* (Bt) into croplines. The Bt gene generates proteins toxic to common pests, conferring pest resistance on varieties genetically engineered to contain it.

The total commercialised plantings of Monsanto's biotechnology crops in 1998 were more than 55 million acres worldwide, of a total acreage for all biotechnology crops planted in the world during 1998 of more than 73 million acres (personal communication, Diane Herndon, Monsanto, 12 March 1999). In 1998, the market potential for pre-farm gate uses of biotechnology was estimated as in the range of US$50-70 billion. Post-farm gate opportunities were estimated as around US$500 billion (*AgBiotech Bulletin*, 1998). The total world-wide research budget for agricultural biotechnology in 1998 is shown in Table 5.6.

As far as research is concerned, at the most basic level, over half the companies interviewed in this survey are already using basic biotechnologies such as molecular markers and double haploid technology, and a third of those interviewed are already using gene mapping and genome-based breeding techniques. Novartis Seeds, for example, demonstrated how biotechnology is becoming more important. Ten years ago, traditional plant breeding formed 85 per cent of its genetics research, with genetic engineering accounting for the remaining 15 per cent. Today, the proportion is 70:30, and in a few years' time genetic engineering is expected to account for some 40 per cent of the research investments within the company's seed business.

Currently, the process of identifying genes responsible for desirable traits, cloning them and transforming the genome of a variety can take as long as traditional back-crossing, but biotechnology often accelerates the breeding process considerably, and is likely to become more efficient and cost effective as technology develops. As one company put it, 'the attraction of genetic

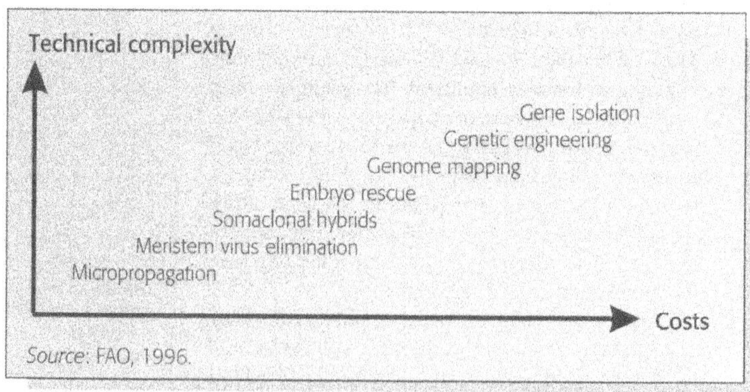

Figure 5.3 *Plant Biotechnology Techniques*

Table 5.6 *Research Budget for Agricultural Biotechnology*

Region	Total public and private budget (US$ bn)	% change, 1996–8
USA	1.26	800
Others	0.69	325
Total	1.95	547

Source: FIS/ASSINSEL, 1998.

techniques lies in the speed of development'. Where single genes are responsible for the trait in question, genetic engineering often reduces the time needed for development and testing, but may not save any time where multiple genes are involved. The proportion of new crop varieties developed through genetic engineering is likely to increase in the future.

Many seed companies are convinced that genetic engineering offers the promise of superior crops, with yields that can meet the demands of growing world population and help the environment by reducing the need for chemical pesticides. However, the technology is still in its infancy. The refinement of techniques and stability of products is likely to be improved in the future. It remains to be seen whether this will allay public scepticism, particularly in Europe, about the health and environmental effects of transgenic crops.

5.2 Regulations relevant to the release of new agricultural crop varieties on to the market

The primary goal of seed regulation is to protect the consumer, who is generally a farmer. The recent advent and increasing importance of GMOs has introduced the further aims of human safety and environmental protection. Before a plant breeder is legally entitled to release a new variety on to the market, it may be necessary for him or her to obtain certification for the seed, to label it, and to obtain various kinds of regulatory approval for the release of GMOs. Lists of seeds authorised for commercialisation exist in many countries. In addition, when plants are imported and exported, there are legal provisions to control the pests and diseases of plants. Finally, while breeders are not required to do so, many choose to take out plant breeders' rights on the varieties they have developed. In Appendix E, various international and national certification and seed-testing schemes, as well as the regulation of GMOs and Plant Breeders' Rights are illustrated through the example of the regimes in the OECD, the EU, the UK, the USA, Japan and Argentina. These are summarised here.

5.2.1 Phytosanitary measures

Phytosanitary measures are designed to prevent the introduction and/or spread of quarantine pests and encompass legislation, regulation or administrative procedures. 'Pests' are broadly considered to be any species, strain or biotype of plant, animal or pathogenic agent injurious to plants and plant products. Two principal international institutions influence policy on international phytosanitary measures:

1 the FAO, through the International Plant Protection Convention (IPPC), which provides the basis under which countries take action and cooperate to prevent the spread of quarantine pests, especially across national boundaries; and
2 the WTO, through the Committee on Sanitary and Phytosanitary (SPS) Measures, whose objective is to ensure that those measures taken to protect human, plant and animal life are consistent with sound scientific evidence.

At the national level, countries have introduced regulations to protect crops and the environment from quarantine pests, and to implement their obligations under the IPPC. The EU has a harmonised plant-health policy.

Typically, if a plant species is imported, the importer consults the national plant quarantine service to determine the phytosanitary regulations for the plant species involved. Import of some plant species is banned altogether. In other cases, the import may be allowed to proceed on issue of an import licence which specifies the conditions under which the material is imported, and may include a requirement for containment, including post-entry quarantine (eg potatoes for planting imported into the EU). For some countries, an import permit, listing specific requirements, is needed before certain material can be imported. For almost all countries, a phytosanitary certificate is required. The certificate is an internationally accepted document under the IPPC, issued by the exporting plant protection organisation. It indicates that the material has been officially inspected or tested and is free from quarantine pests, and that the material complies with the importing country's phytosanitary requirements.

5.2.2 Seed certification and labelling

In many countries, such as Japan, the UK, and Argentina, government controls the certification of seed. National (or international) certification schemes vary. For example, the OECD provides a range of voluntary 'Schemes for Varietal Certification of Seed Moving in International Trade' and the EU prescribes EU-wide minimum quality standards. In the USA, seed certification is voluntary, and is administered by independent authorities. Labelling may be required either for all seed, or for certified seed. Tests for approval of commercialisation relate mostly to criteria such as for distinctness, uniformity, and stability (DUS) – the guidelines set down by UPOV (1.4.3) – and whether or not the new variety has value for cultivation and use (VCU).

5.2.3 Intellectual property rights

Plant breeders' rights

Plant breeders' rights (PBRs) are available in the member states of the International Convention for the Protection of New Varieties of Plants, which constitute the International Union for the Protection of New Varieties of Plants (UPOV). The UPOV member states together agreed upon a system to grant harmonised and exclusive rights of exploitation to the breeders of new plant varieties on an international level. The 1978 Act, which is no longer open for adhesion, required each party to adopt national legislation to give at least 24 genera or species legal protection. The 1991 Act is now in force, and provides protection for all plant genera and species. There is a brief comparison between the two Acts and patent law in Appendix E. A plant variety is subject to protection if it is DUS, and if it satisfies the requirement of 'novelty'. Protection is granted for a minimum of 20 years, and the grant of plant variety rights confers on the holder the exclusive right to sell the reproductive material (such as seeds, cutting or the whole plant). Subject to various limitations, the principle of national treatment prescribes that any Member State must provide to nationals and residents of other Member States the same treatment as it provides to its own nationals and residents.

TRIPS

As a side-agreement to GATT, international legal protection of patents is addressed under the TRIPS Agreement of the WTO. Under the TRIPS Agreement there is a general obligation to comply with the substantive provisions of the Paris Convention of 1967. In addition, all inventions are to be afforded a 20-year patent protection, whether of products or processes, and in almost all fields of technology. This agreement allows for the exclusion from the patentability of, *inter alia*, plants, animals and essential biological processes for the production of plants or animals (other than microbiological processes). However, plant varieties must be eligible for protection either by patents or by an effective *sui generis* system, such as that provided under UPOV. The fundamental principle underlying these provisions is that anything that is already in the public domain cannot be removed and privatised.

Other international agreements addressing intellectual property rights are the 1883 Paris Convention for the Protection of Industrial Property, as amended, and the 1886 Berne Convention for the Protection of Literary and Artistic Works (focusing on copyright), which are both administered by the World Intellectual Property Organisation (WIPO). Another important treaty is the Patent Cooperation Treaty (PCT), under which a successful applicant for a patent receives, for a limited period, patent protection in all the countries listed on the application.

5.2.4 The release and marketing of GMOs

For GMOs, the regulatory requirements vary far more than for the certification and labelling of seed. Regulation of GMOs is still geared towards DUS and VCU testing, but new biotechnology regulation increasingly addresses human and environmental safety and biosafety. There is a move towards harmonisation of biotechnological regulation, through efforts such as standardising interpretations of key terms by the OECD, Memoranda of Understanding on common data elements reviews (eg between the USA and Canada), and the attempt (unsuccessful, to date) to negotiate a Biosafety Protocol.

Consent to release a GMO into the environment is subject to national law. In the UK, for example, this is under the Genetically Modified Organism (Deliberate Release) Regulations 1992 (as amended). Although consent is subject to UK approval, all Member States of the EU are informed of the proposal during the statutory period and have the opportunity to comment on it, in accordance with Directive 90/220/EEC. Consent for marketing, however, is subject to EU approval. Applications are submitted to a single Member State (Competent Authority) which reviews the application under its own legislation. In the UK, an application is reviewed under the Genetically Modified Organisms (Deliberate Release) Regulations by Government Departments and the Advisory Committee on Releases into the Environment (ACRE). If a Member State concludes that consent should be granted, the application is sent to all other Member States (via the EC) who review and vote on it. A final decision as to whether consent should be granted is made under qualified majority vote, according to Article 21 of 90/20/EEC. If the application is successful, a consent is granted by the Member State to which the original application was submitted, to allow marketing in the EU as a whole.

In Japan, the guidelines in respect of production and sale of rDNA organisms are focused on biosafety and risk assessment criteria. Under Argentinian regulations, GMOs must be registered and certified. The cost of a GMO consent ranges from zero (USA) to US$3,260 (UK). The time periods for approval of the release into the environment and/or commercialisation of GMOs vary. Typically, the procedures allow for some two to three months, but the period for approval can take up to two years or longer where more data have been requested. Public registers or notifications of approved GMOs are common in many countries, as are lists of approved GMOs. Labelling of GMOs is required in EU Member States (such as the UK), and in Argentina. The USA, however, merely requires labelling when there is a material health or safety concern. Japan currently has no labelling schemes for GMOs, but requires labelling for transgenic crops (just as for seeds).

5.2.5 The implications for access of seed certification systems, the regulation of GMOs, SPS measures and IPRs

In countries which operate them, seed certification systems prohibit the commercialisation of seeds of varieties that do not appear on the national list. This may prevent the production and sale of seeds of landraces, unless such activities are specifically authorised in the national seed law (as they are in Switzerland, for instance). Seed certification systems affect the nature and range of seed available in the market-place, but do not prohibit or affect the use of all kinds of seed for breeding purposes, although it may, in practice, be harder to find seed that is not certified and thus not commonly used. Access to GMOs for further breeding may be affected by biosafety regulations if these apply to contained and experimental uses as well as to the sale and marketing of GMOs and their international trade.

Phytosanitary measures are not designed to affect access to germplasm for breeding purposes, but the chemical or other treatment of seed as it leaves a country or enters another may affect its viability.

Patents do affect access to genetic resources, since patented material cannot be sold or used for breeding purposes without the permission of the patent-holder. Plant breeders' rights do not affect access for breeding purposes at the time that protected material is initially obtained (since the Breeder's Exemption allows protected material to be used for breeding purposes without the need to ask for prior permission), but, if the variety produced from the protected material is 'essentially derived' from it, the user is required to negotiate compensation with the owner of the protected variety.

5.3 Access to genetic resources

5.3.1 Who needs access to genetic resources?

Many different individuals and organisations are involved in characterising, selecting and improving agricultural materials. Agricultural genetic resources often change hands several times, being altered, improved and bred by public and private organisations around the world before they are commercialised. It is rare for one organisation to conduct all the activities, from collection, conservation and pre-breeding to commercialisation and marketing.

Most collecting activities are conducted by universities, government breeding institutions, genebanks and international organisations. When companies obtain new material, they generally do so via a collector or an intermediary. Several of the companies we interviewed said that their organisations do collect genetic resources from time to time. Breeders may 'collect' material haphazardly when they see something interesting (which is more likely to be a tomato in a local market than a plant in a farmer's field or field border), but this is not part of the formal breeding process. The vast majority of companies stressed that their own collecting activities accounted for a minute proportion of the germplasm they received. Rather than seeking access to landraces which need many generations of breeding before they become commercially viable, commercial breeders take advantage of the free access to germplasm offered by collecting organisations, who provide them with improved material.

Institutions that collect PGRFA may conduct some research on the materials, then pass them on to plant breeders. Some 'intermediary' organisations such as university departments, genebanks, and the international agricultural centres of the CGIAR, supply pre-bred seed to other breeders, who then refine the germplasm and incorporate it into their own products. For example, Dr Jay Scott of the University of Florida described how the university develops a hybrid finished variety and announces the release to various seed companies, one of which might license the variety to develop into a product. Seed companies themselves not only supply finished seed alone, but sell a range of different products. As one interviewee put it: 'We sell lots of different things, from a concept, to a bag of seeds, to a collaboration with a baker'. Others referred to buying and selling the rights to use genes. Some companies sell commercial cultivars directly to farmers, and others supply their products via intermediaries such as dealers, retailers or seed traders. One seed company, for example, has contracted 7,500 dealers in the USA alone to sell its products. Of the plant breeders we interviewed, 61 per cent sold their products direct to farmers, 52 per cent via traders, and 42 per cent passed their materials on to other breeders for improvement.

At each step in the chain of improvement, from initial collection to ultimate commercialisation, different institutions need access to genetic resources of increasing levels of improvement. Thus access to genetic resources is important for the farmer maintaining a landrace in his fields, for the university researcher who collects a specimen from him, for the small, 'foundation' seed company that obtains pre-bred material released by a government breeding institution, and for the multinational that licenses a cloned gene from the foundation company. This section explains how, and by whom, PGRFA are accessed. Section 5.3.2 goes on to explore the kind of genetic resources accessed.

Figure 5.4 illustrates how different actors access and exchange PGRFA and share benefits. It shows three 'stages' in the development of a new variety: collection, pre-breeding and commercialisation (broadly corresponding to Phases I, II and III of the research and development process for a new variety, described in 5.1.3, p126). In reality, the distinction between these three stages is not easy to draw, and, rather than three transactions, it is not uncommon for there to be several. These form a circle from farmer to public breeder to seed company to farmer. One possible scenario is as follows:

- *Collection:* A university researcher collects seed from a local market in country X.
- *Selection:* The researcher grows up the seed, observes it over a few generations, selects the best seed, and passes this to a public genebank that forms part of the national agricultural research system (NARS) in country X.
- *Characterisation and evaluation:* The curator of the collection in the public genebank of country X

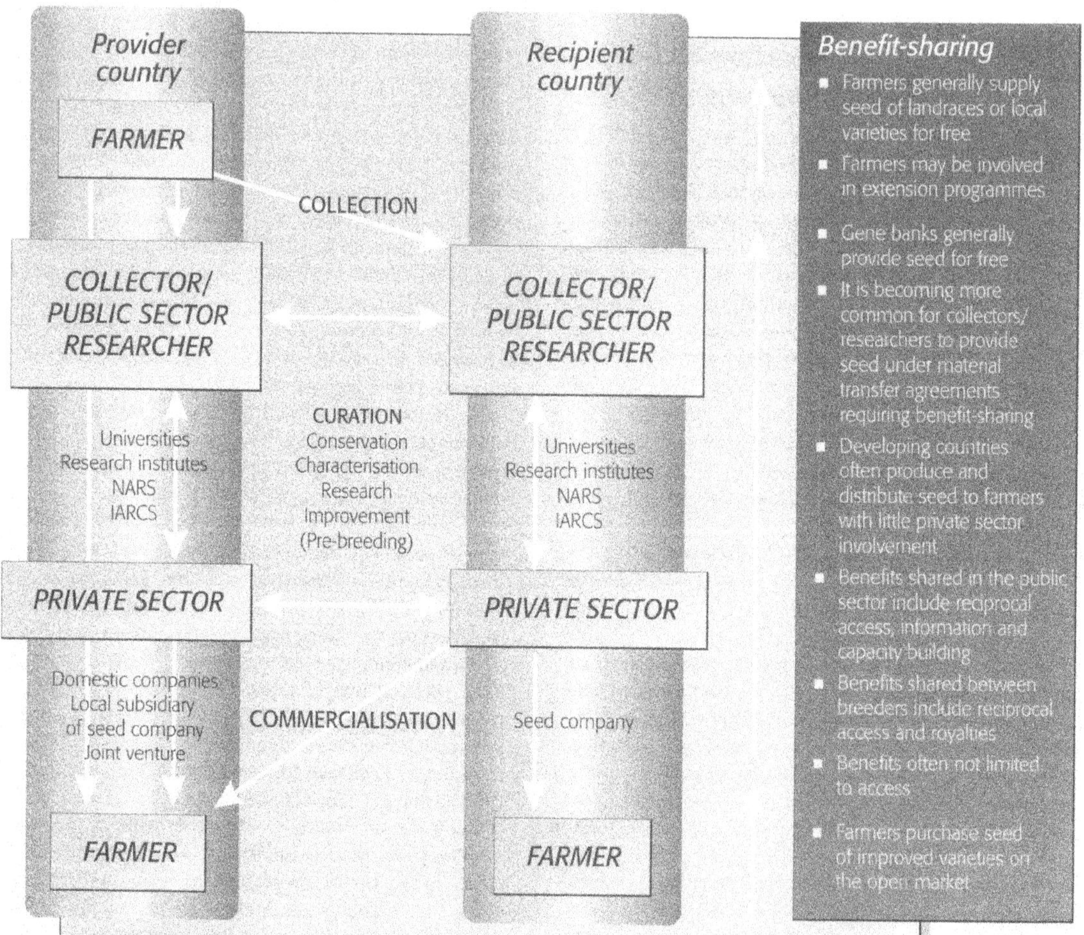

Figure 5.4 *The Access and Exchange of PGRFA, and Benefit-Sharing*

records the characteristics of the material and compares it with other materials in the collection.

- *Exchange:* The genebank in country X exchanges the seed for some other seed with an IARC in country Y.
- *Enhancement:* The IARC in country Y enhances the material and provides the resulting seed to a public body that is part of the NARS in country Z.
- *Pre-breeding:* A public breeder in country Z improves the seed, incorporating genes from this and many other lines into a cultivar which it releases to a foundation company in country Z.
- *Improvement:* The foundation company breeds the cultivar with its own lines and licenses the result to a multinational seed company.
- *Commercialisation:* The multinational seed company crosses the licensed parent and several other materials with its own elite strains, and obtains regulatory approval to release the new commercial cultivar.
- *Marketing:* The original farmer buys seed of the new commercial cultivar from a subsidiary of the multinational seed company in country X.

From this hypothetical example, it is clear that the flow of genetic resources generally takes place through a number of partnerships, with value added to the material at each step of the way. While this example is a possible scenario, it involves a greater number of transactions than the norm. More typically, perhaps, academics would not lodge their material in a genebank, but license it directly to an interested company. Genebanks themselves, conscious of the cost of maintaining material, accept only well-documented material, which they are unlikely to exchange with an IARC, but more likely to provide directly to a company.

Private companies often collaborate with public-sector breeders, from whom they may acquire pre-bred materials which require further improvement prior to marketing. Universities and public breeding organisations use such contracts to support their research. Seed companies also collaborate with the food-processing industry, which may, for example, support research into varieties of tomato suitable for paste, and then acquire such tomatoes on a commercial scale. The recent spate of mergers and acquisitions has resulted in several huge and very diverse companies, whose subsidiaries exchange genetic resources around the world and engage in a range of activities from pure research to commercialisation and marketing.

5.3.2 How do breeders select materials for use in breeding efforts?

How to find useful germplasm

A tremendous amount of genetic diversity exists in both *in situ* and in *ex situ* collections. How, then, do breeders decide which materials to explore for their potentially useful traits? They find out about interesting materials from a number of different sources. According to the results of our survey (Figure 5.5), by far the most common method of learning about useful traits is to consult databases and read relevant literature, for example journals such as *Crop Science*, the *UK Plant Breeding Abstract* and *Plant Breeding News* (California). One company described ordering material on-line from an internet site for next-day delivery. Several interviewees stressed the importance of partnerships with research organisations who offer them the service of keeping them up-to-date with developments, and, in some cases, even pre-select materials for use in the companies' breeding programmes. As one company put it: 'they do the scouting and try it and see if it is useful'.

Companies use a range of different databases to find information on potentially useful traits. Such databases contain information on the characteristics and evaluation of germplasm, which helps them to identify samples for use in research programmes. Even though accessions in genebanks without such information are, theoretically, just as easy for breeders to 'access' as materials accompanied by information on the nature of the material, in practice, accessions without evaluation information are used very rarely, if at all. Breeders would not know which accessions to investigate, nor

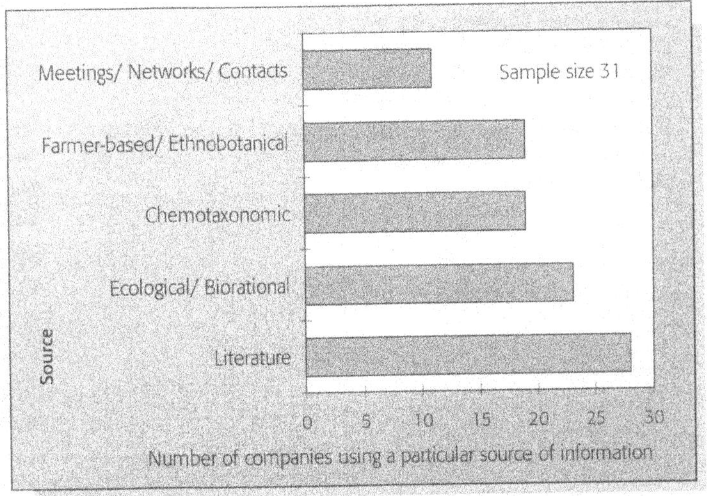

Figure 5.5 *Sources of Information on Useful Traits in Germplasm*

Box 5.2 *The US Germplasm Resources Information Network (GRIN)*

Many breeders from around the world use the Germplasm Resources Information Network (GRIN) database, located within the National Genetic Resources Program of USDA's Agricultural Research Service (ARS).

GRIN is a centralised computer database designed to facilitate the management and operation of the National Plant Germplasm System (NPGS). The NPGS is a joint effort by public (State and Federal) and private organisations 'to preserve the genetic diversity of plants'. Its 26 collection sites (in the US and Puerto Rico) preserve, document and distribute crop germplasm and constantly acquire new crop germplasm. New germplasm enters NPGS through collection, donation by foreign operators or international germplasm collections. Each accession is assigned an identifying Plant Introduction (PI) number and, after a period of quarantine, made available for distribution in small quantities for research purposes only. As of 31 January 1999, the NPGS holdings represented 184 families, 1,488 genera, 10,143 species and 437,903 accessions.

Each of the 26 working collections is responsible for the maintenance, characterisation and evaluation of specific species (eg peas in Washington and desert legumes in Arizona). Descriptor lists, composed of key traits, are used to characterise the collections. Key traits include morphological characteristics, disease and insect resistance, and tolerance to environmental stresses. The database can be searched by crop, country of origin, genus, collection site, PI number, taxon, and can be limited by other criteria such as type of improvement status or type of reproductive uniformity. Descriptor lists within the NGRP gopher cover most of the economically important crops within the USA. Germplasm can be ordered direct from the site.

This can be accessed at http://www.ars-grin.gov/. As well as providing a searchable on-line database, GRIN data for individual crops can be downloaded and searched off-line.

Source: compiled by China Williams from data on http://www.ars-grin.gov/.

see any reason for doing so. For this reason, databases and other sources of information play an important role in plant breeding. An example of a database cited several times by the breeders interviewed was the USDA's GRIN. Among the most useful databases for breeders are those of the national genebanks and IARCs.

How do breeders select the partners from whom they acquire genetic resources?

The breeders interviewed described a number of factors that affect their choice of the countries and institutions with which to collaborate to obtain germplasm (Figure 5.6). The calibre of the scientists and capacity of the institution from where germplasm is obtained were consistently regarded as extremely important factors. This is not only to guarantee the quality of the sample and the reliability of the passport data provided with it, but also for plant health reasons. As one company put it: 'It may be dangerous to work with amateurs, as you run the risk of introducing new diseases and pests that could start epidemics. The CGIAR and other major centres have qualified pathologists.' Several interviewees also stressed the significance of the quality of the sample, the simplicity of the process for obtaining all relevant permits for obtaining and exporting the material, and the freedom of the material concerned from rights such as patents that might restrict a company's ability to use the germplasm for breeding and commercialisation. Biological diversity and the availability of intellectual property rights in the country from which the material was obtained (so that the resulting product could not be 'pirated' in that country) ranked low as considerations when selecting partners, while the cost of obtaining the material was the least important factor.

'If I want something from Turkey or Argentina, I go to GRIN [USDA's Germplasm Resources Information Network] first and see if they have it. So I get my samples from the national collection primarily.' (Dr John Moffat, AgriproBiosciences, USA)

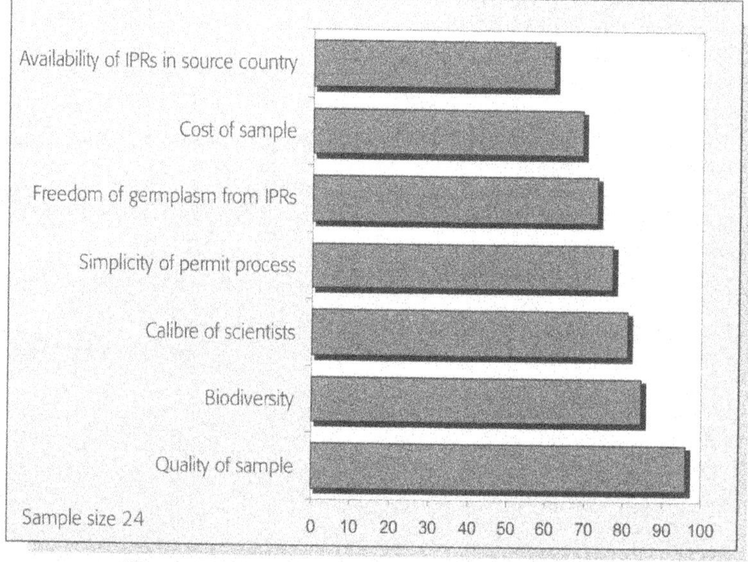

Figure 5.6 *Factors Affecting Breeders' Choice of Suppliers of Genetic Resources*

5.3.3 From where and from whom do breeders obtain their material?

There are two basic sources of material for breeding: materials from *'in situ'* conditions, such as farmers' fields and uncultivated land like forest, savannah, and mountain, and *'ex situ'* collections. Materials from *ex situ* collections are used by plant breeders far more commonly than those collected from *in situ* conditions, so *ex situ* collections are dealt with first here.

Collections of plant genetic resources for food and agriculture

From the results of our survey and earlier work (Figure 5.7 and Table 5.7), it appears that the vast majority of breeding materials used by companies are from their own corporate collections. The next port of call for plant breeders looking for new germplasm is a national genebank, followed by the 'in trust' collections maintained by the International Centres of the CGIAR, and smaller collections held by universities. As one private-sector breeder put it: 'The material we use is mostly from our own collection. Now and again, I'll see something special from a university in an article. . . . Maybe once or twice a year we'll get something from a culture collection. We do get things from seed banks, but hardly ever.'

Collections in companies

Major seed-breeding companies have built up extensive, private libraries of genetic resources by acquiring and breeding plants for many decades. Within these corporate collections are the proprietary, elite strains of seeds which represent the culmination of the companies' breeding efforts to date, and which combine the most desirable traits such as high yield and disease resistance. Companies' collections contain a modest proportion of materials other than seed, such as entire plants or culture collections. As a 1996 survey of ASSINSEL members revealed (Box 5.3, p136), unadapted and 'exotic' materials are represented in companies' collections. However, these materials form a small proportion of the collections, the majority of which consist of 'advanced' germplasm. A company's elite strains are generally protected with intellectual property rights or kept as trade secrets, and the majority of companies do not allow access by others to these materials. However, some companies exchange materials as part of collaborative research projects.

While some private breeders keep large collections of genetic resources, others choose instead to maintain only the lines that are currently under development or growing in their nurseries, or to maintain a representative sample of seed from the last five or so years of breeding efforts. Some companies holding such smaller collections form long-term partnerships with institutions maintaining larger collections. This practice is more common in Europe than the USA.

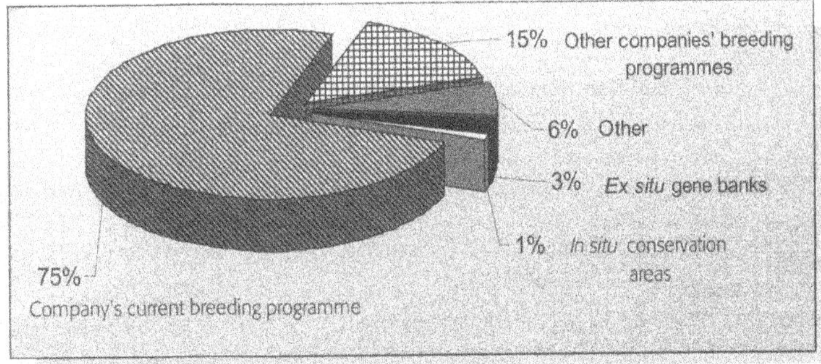

Figure 5.7 *Companies' Source of In-House Germplasm*

Source: Swanson and Luxmore, 1996.

Table 5.7 *Companies' Source of Germplasm Used for All Crops by Crop Group*

Source	All %	Potato %	Cereals %	Oil crop %	Vegetable %
adapted material: **commercial cultivar**	81.5	50	87	78.8	95.7
adapted material: **minor crop**[1]	1.4	8	0.6	1.2	0.3
wild species: ***ex situ* genebank**	2.5	19	1.2	1	1.4
wild species: **maintained *in situ***	1	0	0.7	0.1	0.1
primitive **landrace from genebank**	1.6	1.7	1.7	2.3	1.7
primitive **landrace maintained *in situ***	1.4	0	0.7	7.2	0.3
induced mutation	2.2	3.3	0.7	7.2	0.3
biotechnology	4.5	17.7	3.5	6.8	0.1

1 Minor crop cultivated on a small scale with some improvement over wild ancestors

Source: Swanson and Luxmore, 1996.

Box 5.3 *Companies' Collections of PGRFA*

A 1996 survey among 80 ASSINSEL members, showed that:

- 88 per cent of plant-breeding companies are maintaining genetic resources;
- on average, plant-breeding companies spend 5 per cent of their research budgets for maintaining genetic resources, which represents roughly US$ 50 million per year;
- 84 per cent of breeding companies maintain obsolete varieties, 72 per cent landraces and 53 per cent wild species;
- 80 per cent of plant-breeding companies participate in national programmes and 31 per cent in international programmes on conservation of genetic resources.

Source: personal communication, Patrick Heffer, ASSINSEL, 21 December 1998.

Collections in genebanks, research institutions and universities

A range of plant breeders from the public and private sector makes use of the collections maintained in national genebanks, research institutions and universities. As Box 5.4 shows, the proportion of private-sector users of accessions from the national genebanks in Germany, the USA and Japan ranges from 4 per cent to 33 per cent. Box 5.5 shows the different types of genebanks and collections, and Table 5.8 shows the genebanks and accessions in *ex situ* collections.

The collections held by academic, conservation and research organisations generally contain a higher proportion of primitive and unadapted materials than those within companies. For example, 40,000 of the 130,000 entries for wheat in CIMMYT's collections are landraces. Plant breeders from the public and private sector may use these unimproved materials, but are more likely to acquire lines that the breeders associated with the collections have pre-bred. For example, according to Dr Charles Rick, Director of the Tomato Genetics Resource Center at the University of California at Davis, the Center maintains three types of exotic germplasm: *(1)* wild species related to the cultivated tomato, *(2)* landraces, and *(3)* 'pre-bred' or 'enhanced' lines into which desired traits from the first

Box 5.4 *Summary of Contents and Users of Four National Genebanks*

Genebank	Total number of accessions	Samples provided per year	Foreign/ domestic users, %	Private/public users, %
GERMANY				
Genebank of the Federal Research Centre in Braunschweig	57,000 accessions representing 948 species from 58 families	10,000	44/56	research institutions: 62 public breeders: 16 private breeders*: 22 (*comprising 16% individuals and 6% members of the Association for the Promotion of German Private Plant Breeding (GFP))
INDIA				
National Bureau of Plant Genetic Resources (NBPGR)	175,000 (increasing by 15–20,000 each year)	46,000	35/65	public institutions: 66.6 private institutions: 33.3
JAPAN				
National Institute of Agrobiological Resources (NIAR)	192,860	6,000	4/96	research institutions: 75 universities: 13 private breeders: 4 others: 8
USA				
National Plant Germplasm System (NPGS)	449,000 accessions representing 8,720 species of which 19% originated in the USA	119,664 representing 35% of species in the collection	35/65 foreign users almost all private companies	private companies: 11 private researchers: 'small proportion' State/Federal researchers: 'most'

Sources: German and US country reports for the Preparation of the ITCPGR, Leipzig, 17–23 June 1996 and FAO website; personal communication Dr Shoji Miyazaki, National Institute of Agrobiological Resources, Japan, November 1998; personal communication between China Williams and Dr PL Gautam, NBPGR, India.

Box 5.5 *Types of Genebanks and Collections*

Types of genebank:

- An **institutional genebank** is set up to conserve only the germplasm which is used in the research programmes in its host institute.
- A **national genebank** is set up as a national plant genetic resource centre, maintaining many different germplasm samples of current and potential interest to people working in plant research nationally. Commonly, it contains germplasm which has been collected nationally. Also, it may be closely associated with a research programme or undertake its own research. A national genebank can be a collaborative venture between national institutes, or under the responsibility of one institute which collaborates with other national institutes.
- A **regional genebank** is set up as a collaborative venture between a number of countries in the same geographical region to conserve the germplasm from that region and to support plant research.
- An **IARC collection** is found in all the centres of the CGIAR with a mandate for particular crops. Much of the germplasm is collected world-wide with international collaboration, and is conserved for the benefit of plant genetic resources' activities world-wide.

Types of collection:

- A **base collection** comprises a set of genetically different accessions of a given genepool conserved for the long term, ideally under long-term storage conditions. The international base collection of a genepool can be regarded as the sum total of all genetically different accessions conserved *ex situ* in genebanks around the world. Base collections are not normally used as a routine distribution source.
- An **active collection** is a collection of germplasm for regeneration, multiplication, distribution, characterisation and evaluation. Ideally, germplasm in the active collection should be maintained in sufficient quantity to be available on request. Active collection germplasm is commonly duplicated in a base collection and is often stored under medium to long-term storage.
- A **field collection** or field genebank is a collection of living plants (eg fruit trees, glasshouse crops and perennial field crops). Germplasm which would otherwise be difficult to maintain as seed can be kept in field collections.
- An ***in vitro* collection** is a collection of germplasm kept as plant tissue. In some cases, the tissue is stored at very low temperatures such as under liquid nitrogen (cryopreservation).
- A **core collection** attempts to combine the maximum genetic variation of a species within a manageable number of samples. Core collections are a mechanism to improve the accessibility and increase the use of collections. They are not a substitute for base and active collections.

Source: IPGRI, 1993.

or the second categories have been bred. Although the third group breeds true for the respective traits, they are substandard in other respects, and hence not yet acceptable to growers (personal communication, Dr Charles Rick, 20 January 1999). Other breeders, including those in the private sector, use these materials in further breeding programmes in order to develop lines that are ready for distribution to farmers.

Table 5.8 *Genebanks and Accessions in* Ex Situ *Collections, by Region*

Region	Accessions		Genebanks	
	Number	%	Number	%
Africa	353,523	6	124	10
Latin America and the Caribbean	642,405	12	227	17
North America	762,061	14	101	8
Asia	1,533,979	28	293	22
Europe	1,934,574	35	496	38
Near East	327,963	6	67	5
Total	5,554,505	100	1,308	100
CGIAR Total	593,191		12	

Source: FAO *State of the World* Report, from country reports and WIEWS database, 1996.

Conclusion about sources of materials for breeding

The most common sources of germplasm for companies are the national collections and genebanks, followed by international genebanks, then universities and occasionally botanic gardens and culture collections. These intermediaries acquire genetic resources from around the world, either by mounting collecting expeditions or by being sent samples by similar organisations world-wide.

5.4 What kind of material do breeders want?

5.4.1 The demand for access to landraces and wild relatives for seed breeding

There is a theory that demand on the part of plant breeders for access to agricultural genetic resources will offer countries harbouring high levels of biological diversity an opportunity to gain a range of benefits in return for granting access, and will create an economic incentive for conservation. Since breeders around the world already have free access to the wealth of

'In general, the use of primitive germplasm is now well accepted by most breeders. You will use it when there is nothing else: when you are in trouble.' (Dr Alberto G Cubillos, Agricultural Research Institute Of Chile)

'Typically, exotic germplasm brings along an amazing amount of rubbish.'

agricultural genetic diversity held in *ex situ* collections, an important test for this theory is whether plant breeders do, in fact, desire access to primitive materials such as landraces and wild relatives to which they would need access *de novo*. As we saw above, in Section 5.3, the survey by Swanson and Luxmoore (1996) concluded that material from *ex situ* genebanks contributed 3 per cent of the germplasm used by breeders, and material from *in situ* conservation areas a further 1 per cent. ASSINSEL members believe that less than 5 per cent of the germplasm used in plant-breeding programmes has an indigenous origin (personal communication, Patrick Heffer, 21 December 1998). This section explores the nature and extent of plant breeders' demand for access to primitive materials.

In general, plant breeders are reluctant to turn to unimproved materials, especially 'primitive' materials such as landraces and wild relatives. They already have at their disposal large collections of improved and even 'elite' strains, which are constantly under development and comprise a range of attractive traits, such as high yield and disease resistance. By contrast, the use of 'primitive' materials requires lengthy and expensive breeding programmes to eliminate the negative characteristics that often dog primitive lines. These undesirable characteristics include poor yield and unfortunate morphological traits, such as plants that shatter too easily or do not stand well. They may make the plant unattractive to the small farmer and incompatible with mechanised farming practices. A breeding programme to develop a new commercial variety from elite material can take from five to ten years, but a programme that integrates traits found in primitive germplasm into a commercial cultivar can take from 15 to 20 years. This is uncompetitive in today's cut and thrust market, where companies look to keep the development process down to five to ten years, with between 0.5 and 1 per cent growth in yield per year (depending on the crop). Primitive materials tend to fall between the cracks, as smaller, public institutions claim that only 'big, rich companies' can afford to conduct research on primitive materials, while companies say that only public agencies can afford to conduct basic research which is uncompetitive and far from the market.

According to Patrick Heffer of ASSINSEL, 'If you look at what genetic resources are used by plant breeders, in particular private-sector plant breeders, you see that plant breeders mainly use improved, commercially released varieties, frequently protected by Plant Breeders' Rights. The main reasons are that it is much easier to use improved varieties in breeding programmes than indigenous resources, it is easier to access that kind of material for breeding purposes (without the need for administrative procedures such as Material Transfer Agreements), and information on improved varieties' characteristics is easily available. As a result, less than 5 per cent of the germplasm used in plant-breeding programmes has an indigenous origin' (Heffer, 1998).

A second problem with unadapted materials can be their incompatibility with elite lines, making their merger through traditional breeding programmes difficult. Genetic engineering may overcome such problems in the future (eg Allium, Box 5.7).

Many of the companies we interviewed said they very rarely used primitive material. One, for example, could not name a single one of its varieties to which primitive material had made an important contribution. But despite the popular image of primitive material among breeders as a 'last resort', a large proportion of those interviewed considered its use an important part of seed-breeding. As Dr Ted Givens of NC-Plus Hybrids explained, 'the majority of sorghum hybrids released [contain] exotic germplasm, obtained via the universities'. Responses concerning the extent to which interviewees use primitive germplasm ranged from (one) 'no use' through 'extremely little', 'rare' and 'occasional' to 'fairly heavily' and 'all the time'. Several interviewees described a move, over the last few years, to work on disease resistance by broadening the genetic base – a move that has been supported by new technological advances in seeking and quantifying traits. Similarly, although genes from primitive materials, if used at all, form only a tiny proportion of the overall genome of a modern variety, the contribution they make to the resulting variety may be very important. As Dr Reinhard von Broock of the German company Luchow-Petkus puts it, there has 'never been more than a single gene used from primitive material, and in a plant there are about 200,000 genes. But to introduce resistance in wheat, you might only add one gene responsible for resistance, so that one gene is an important component'.

Dr Beversdorf at Novartis Seeds notes that: 'Germplasm conservation is very expensive. It requires staying power and major investments to collect, maintain and characterise germplasm. We go to CIMMYT, CIAT and similar institutions for reasonably well-characterised exotic materials. We collaborate, but they are the experts. Why should we want to compete with them?'

'Most primitive material does not go directly to the breeders ... To make the material more acceptable for breeders, we do the crossing and back-crossing here and try to standardise material for the breeders.' (Institute For Plant Production, Czech Republic)

Because of the extra time and cost involved in working with primitive materials, few breeders screen them randomly, but are more likely to screen for genes with a very specific end-use. The most common reason to use primitive material is to find resistance or tolerance to disease and pests, then quality traits such as high yield or cold tolerance, characteristics such as increased sugar content and higher soluble solids, and finally to breed sterility into hybrids.

Another important reason that leads plant breeders to turn to primitive materials is the desire to broaden the genetic base of the crop with which they are working. For example, in 1996, the American Seed Trade Association (ASTA) informed a Senate Committee (ASTA, 1996) that: 'Currently, the US$18 billion annual revenue to the US economy from domestic and external consumption of corn and its industrial derivatives is based on using less than 5 per cent of the corn germplasm available in the world. Less than 1 per cent of US commercial corn is of exotic (foreign) origin, and tropical exotic germplasm is only a fraction of that. This situation exists because private sector corn breeders have generally concentrated on genetically narrow based, or elite sources for their breeding efforts, since their use results in getting hybrids to the marketplace faster'. ASTA argued that broadening the germplasm base would provide genes to improve yields and protect against new disease, insect and environmental stresses.

'If borders close and we have to use something exotic to maintain yield growth, we could use the national genebank to keep us going for another 50 years.'

Exotic germplasm could also be a source for changes in grain quality being demanded by export markets, industrial processors and other end users. In order to meet this need, 22 US companies and 19 participants from the public sector in the USA (universities and the USDA) initiated the Latin American Maize Project (LAMP), whose objective is 'to provide the corn industry with materials developed using germplasm enhancement of useful exotic germplasm, with the ultimate aim of improving and broadening the germplasm base of maize hybrids grown by American farmers'.

Box 5.6 gives examples of genetic uniformity in selected crops, and Box 5.7, p140, offers examples of the recent use of primitive material in breeding programmes. Few companies conduct basic research on primitive materials by themselves, and it is typical for such research projects to be conducted in collaboration with public-sector intermediaries who 'pre-bred' primitive materials for use in commercial breeding programmes.

Box 5.6 *The Extent of Genetic Uniformity in Selected Crops*

Crop	Country	Number of varieties
Rice	Sri Lanka	From 2,000 varieties in 1959 to less than 100 today 75% descend from a common stock
Rice	Bangladesh	62% of varieties descend from a common stock
Rice	Indonesia	74% of varieties descend from a common stock
Wheat	USA	50% of crop in 9 varieties
Potato	USA	75% of crop in 4 varieties
Soybeans	USA	50% of crop in 6 varieties

Source: World Conservation Monitoring Centre, 1992.

5.4.2 The demand for access to 'foreign' or 'exotic' material in crop breeding

We asked our interviewees what proportion of their work required access to genetic resources from other countries, and learned that plant breeders still rely heavily on access to genetic resources from other countries. The interviewees came from 11 countries and were working on genetic resources acquired from over 37 countries from 6 continents.

'It would be foolish to use Afghan or Australian material unless you knew the characteristic you wanted, but it is worth it to use French or German material, as it is already adapted.' (Nickerson Seeds, UK)

While exotic germplasm may bring useful traits, breeders much prefer to use germplasm from agroecologically similar conditions, as it will already be adapted to the environment where it must ultimately grow. By contrast, if tropical germplasm is used to enhance a temperate crop, for example, a great deal of work is needed to develop a line adapted to such different conditions. Consequently, the majority of the material obtained from abroad is either fairly local, or from other locations around the world where the agroecological conditions are similar. As one European company explained, 'The wheat we use is about 80 per cent English, and 20 per cent French or German. We rarely go out of this ambit, as the material would be unadapted.'

Box 5.7 *A Few Examples of Recent Use of Primitive Material in Breeding Programmes*

- *Onion*: The wild onion relative *Allium fistolum* possesses commercially attractive characteristics, but cannot be directly crossed with the commercial cultivar *A cepa*. A recent programme by the Centre for Plant Breeding and Reproduction Research, Agricultural Research Department, Wageningen, the Netherlands (CPRO-DLO) first crossed *A fistolum* with *A roylei* (another wild relative), and then crossed the result with *A cepa*. Staining the chromosomes of *A cepa, A fistolum* and *A roylei* using 'Genomic In Situ Hybridisation', made it possible to distinguish chromosomes of each of the three varieties under the microscope, and thus to determine that the offspring of this process contained many useful recombinations of genes from the wild relatives and the commercial cultivar.
- *Potato*: Potato cyst nematodes (*Globodera* spp) can cause very high levels of economic damage in potato fields. CPRO-DLO back-crossed a wild potato species, *solanum spegazzinii*, and the resulting population showed high resistance to nematodes, a trait which was linked to one gene on chromosome 5. Markers have been found for the gene, so it has been possible to show that the resistance was present in nearly all the plants of the backcross population, whereas very few plants susceptible to cyst damage carry the marker. The resistance from the wild species can now be bred into commercial cultivars.
- *Beet*: CPRO-DLO made many crosses between a wild beet and its cultivated relative. Subsequently, it has been able to use the resulting material to isolate a nematode-resistant gene.
- *Wheat*: In one collaboration between CIMMYT, UC Davis and research centres in Iran and Turkey, 13,000 samples of primitive wheat varieties were screened. 150 landraces showed promising results and have been put into breeding programmes.
- *Wheat*: In a separate programme, CIMMYT is working with Kansas State University and Nanjing Agricultural University to graft grass species in the hope of transferring resistance against scab disease to wheat. According to Dr S Rajaram of CIMMYT, 'The research is very costly and futuristic, but if it succeeds, the pay-off could be big. We will know in three years' time whether the resistance can be transferred to wheat.'
- *Rice*: Wild rice seeds collected in 1958 from Thailand by a scientist from the Japanese National Institute of Genetics and his Thai counterpart were stored in Japan, then sent to IRRI in 1980, where *Oryza officinalis* was found to be resistant to insect pests. In 1987, trials of rice hybrids obtained by back-crossing this wild rice with elite lines of cultivated rice demonstrated the highest yield in two seasons. In 1989, rice lines with genes from the Thai wild rice entered international yield trials (Vaughan and Sitch, 1991).
- *Maize*: Since 1987, 22 US companies and 19 US public-sector plant-breeding institutions (universities and the USDA) have been collaborating in the LAMP project to evaluate maize germplasm from Argentina, Bolivia, Brazil, Colombia, Chile, Guatemala, Mexico, Paraguay, Peru, USA, Uruguay, and Venezuela. As part of the programme, 691 Peruvian maize accessions were evaluated for resistance to European corn borer leaf feeding. 11 resistant varieties were identified, with all 11 commonly grown on Peru's northern coast. Analysis revealed that the resistance factor was different from the resistance already found in US Corn-Belt-adapted maize, and thus represented an interesting new trait. Since none of the 11 Peruvian accessions was in the top 5 per cent yielding LAMP accessions, breeding efforts are underway to introduce this alternative resistance factor to the high-yielding Corn-Belt lines (Pollack, 1996).

The diversity available within locally-adapted material limits companies' desire to work with unadapted, foreign material. As one interviewee put it, 'We have found sufficient variability in US germplasm, so there is little motivation to work with exotic germplasm'.

Although access to foreign material remains important for breeders, a wealth of such material from around the world is readily available in national collections. So companies rarely need to deal with the country from where the material originated. Another common method for acquiring and adapting foreign germplasm without negotiating with the country from which it originates is the movement of germplasm between the subsidiaries of multinational companies. Large companies often have subsidiaries in many different countries, where material collected locally is characterised and pre-bred. This modified material can be moved to the company's headquarters or other subsidiaries around the world for adaptation to local conditions.

'The company is so large that samples obtained abroad are generally worked on in that country. The company will then move its own germplasm (ie the modified material) after a couple of cycles.'

Another factor that decreases the interest of companies in obtaining materials from abroad is the increasing difficulty of gaining permission for access to materials. Many interviewees said they felt that the 'greenhouse doors' were gradually being closed. Breeders might know of valuable material, but simply could not access it. The perception of one interviewee is typical: 'Take China. There is a big problem. There is a closed door policy for germplasm leaving the country, and India is starting to go that way, too.' According to Dr Ken Kofoid of KSU Agricultural Research Centre, 'No one is prospecting and this is for political reasons. People can no longer get into Ethiopia, Kenya, the Sudan. Years ago, a lot of the material for drought tolerance was from areas in Sudan where desirable material was known to be. One main requirement today is cold tolerance. This can be found in varieties from the high plateaux of Ethiopia, but they are inaccessible.'

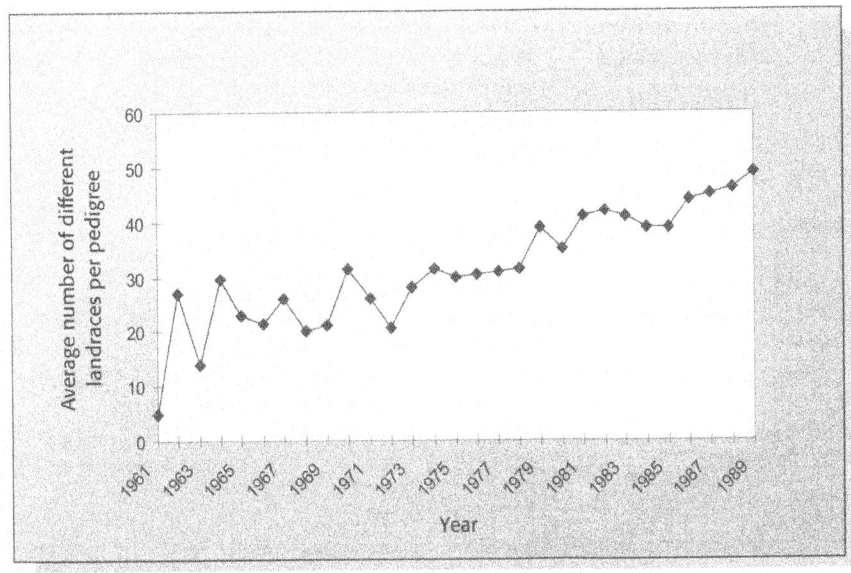

Figure 5.8 *Numbers of Landraces Contained in Pedigrees of Wheats Released in Developing Countries*

Source: CIMMYT Pedigree Management System and Global Wheat Impacts Survey.

make it increasingly easy and cost effective to explore primitive materials. On the other hand, interviewees also cited a number of reasons why demand for primitive material may decrease in the future, including, notably, the reluctance of some countries to grant access to their genetic resources, and the difficulty of unravelling the increasingly complex web of intellectual property rights and material transfer agreements (Box 5.8, p142).

5.4.3 Will the demand for access to primitive and exotic materials grow in the future?

Figure 5.8 shows that the number of landraces in CIMMYT's most important wheat varieties have been increasing over time. The use of parents from around the world leads the pedigrees to become cumulatively more complex, and hence the diversity of the line to increase. Dips in the graph, such as from 1971 to 1973, reveal periods when less diverse parents were used in the crossing programme, for example when breeders are crossing advanced lines, so the number of landraces involved declines. CIMMYT scientists believe that the number of landraces in pedigrees will continue to grow. For example, there are plans to use varieties from China and Kazakhstan in CIMMYT's crossing programme (personal communication, Dr S Rajaram, 3 February 1999; Byerlee and Moya, 1993).

Will the use of primitive and exotic materials in breeding programmes continue in coming years, as Figure 5.8 suggests, despite the inconvenience of using them? Some of the plant breeders interviewed felt that demand for primitive materials would increase in the future, since it was important to broaden the genetic base of breeding, to move from reliance upon chemical pesticides to the identification of genetic traits coding for pest resistance, and since modern technologies

5.4.4 The demand for access to improved material in seed breeding

What kind of 'value-added' products do plant breeders look for?

Since companies rarely obtain primitive materials for use in their breeding programmes, it follows that such material as they use for breeding that is not from within their own collections is material to which 'value has been added' through the provision of information, selection and pre-breeding, and so on. For institutions hoping to provide genetic resources to seed companies for a share of the benefits that result from any eventual commercialisation of the materials, it is important to gain a sense of the relative interest of companies in different kinds of 'value-added' material, which would appear to offer the possibility of greater payment and other benefits.

Figure 5.9 (p143) ranks, in terms of popularity, breeders' demand for 'raw' materials such as wild relatives or landraces, and 'value-added' materials such as strains with proven characteristics, pre-selected or characterised materials, materials provided with associated ethnobotanical information, and elite strains and isolated genes licensed for use in breeding programmes.

Box 5.8 *Will Demand for Primitive Material Grow or Decrease, and Why?*

Reasons why demand for primitive germplasm may grow in the future	Reasons why demand for primitive germplasm may decrease in the future
The need to improve resistance to disease requires access to more genes, and globalisation of research and markets means that pests and disease are transferred faster, increasing demand for access.	Wild/exotic germplasm is not used as much in genetic engineering as in traditional breeding.
Breeders wish to broaden the genetic base of the material they use.	It is becoming harder to access materials from several countries.
The desire to move away from reliance on chemical pesticides to more biological approaches will require access to diverse genetic resources.	Fear of accidental infringement of patents and reluctance to negotiate licences and material transfer agreements will decrease the demand for access to cultivars.
Modern methods make it easier to use primitive materials.	Modern methods mean there is more to be found in your own materials, and less need to turn to primitive materials.
Public funds are drying up so companies will need to access more diverse materials themselves.	Privatisation and commercialisation of research mean that the public institutions that were accessing materials are no longer doing so. It is not competitive for companies to work on unimproved materials.

5.4.5 Conclusions on access

The extent of demand for access to new materials for breeding varies from crop to crop and among companies. The crop breeders interviewed were asked to list which crops they believed would require relatively more access to genetic resources for their continued improvement. They answered that breeders of crops with a 'narrow genetic base' would need greater access to new germplasm for their future development, as would breeders working on 'recently bred crops', 'many horticultural varieties', 'ornamentals', vegetable crops and self-pollinated crops. By contrast, some respondents felt that cereal crops would require comparatively little fresh germplasm for their future improvement.

We asked the companies and institutions interviewed to predict whether they thought the overall demand for access to genetic resources for crop breeding would be greater, less or the same in ten years' time. 53 per cent of the 17 interviewees who answered the question thought that demand would grow in the next ten years; 41 per cent thought it would remain the same; and 6 per cent thought it would decrease.

'We have used plant introductions from Russia and Turkey, and we derived a benefit from it. But to place a value on it is difficult – there is no real value before we get the final material on the producer's field.'

5.5 Benefit-sharing

5.5.1 Introduction

Access and benefit-sharing in the field of crop development are undeniably complex. Any analysis of benefit-sharing must take into consideration the chain of access – frequently lengthy – as germplasm passes through several hands, being improved and adapted at each stage by a number of actors. There is also the question of identifying and placing an economic value on the contributions of each of these actors, and the benefits that arise when a new crop variety is developed and put on the market. The benefits themselves are of many different kinds – economic, environmental and social – and they arise both directly and indirectly. Many benefits arise not from access alone, but as the result of several factors, notably the combination of access to genetic resources and research and development. Benefits are shared not only through arrangements between two parties, but also spread through the constantly changing, informal network of relationships that embraces both public-sector and private-sector breeding organisations around the world, sometimes referred to as a 'multilateral system'. Benefits are often quite disassociated from access to individual accessions, and may not even be tied to access to genetic resources at all, in any but the broadest sense.

Cultural differences add an additional complexity to the picture, as governments, companies, research institutes, communities, farmers and NGOs often hold strong, and quite contradictory, views on access and benefit-sharing. With all these considerations in mind, any attempt to quantify benefits and assess the 'fairness and equity' of benefit-sharing arrangements, faces fundamental challenges at both the 'micro' and the 'macro' levels.

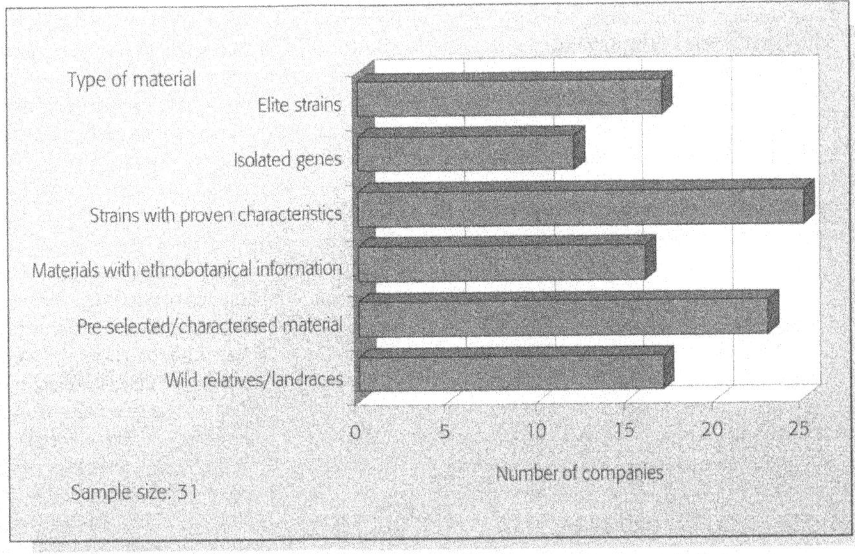

Figure 5.9 *Types of Materials that Companies Acquire from Collaborators*

The 'micro' level challenge lies in quantifying and allocating value to the different genes that are combined in a new crop, contributing to a complex pedigree. If a single gene offers disease resistance – perhaps only one in 100,000 genes in the plant as a whole – is it worth only 1/100,000 of the plant's value? To take an example, one translocation of a gene from a German rye variety collected in 1920 into durum wheat produced a variety with a phenomenal 10 per cent growth in yield. The variety is grown on 60 million hectares in developing countries (personal communication, Bent Skovmand). Would an assessment of benefits suggest that these developing countries owe Germany a share in the revenue from their sales? How could the relative value of genetic resources and the research activities that improve upon them be calculated, and, once the value of the contribution of the genetic resources had been determined, would an appropriate share of benefits for Germany be 10 per cent, or a lesser proportion? Considering the complexity in this relatively 'simple' case, where a single gene can be identified as responsible for a particular trait, how could benefits be calculated for a crop whose new traits were the result of the combination of some 20 or more new genes? Altogether, allocating benefits according to the genetic contributions of the many parents involved and the innovation contributed by several institutions during the development of a commercial product would be quite a feat.

The second, 'macro' level challenge is to quantify the flow of genetic resources around the world, to assess the benefits (both monetary and non-monetary) created by access to genetic resources, and to identify which countries, and within them, which institutions, have benefited and continue to share in these benefits, and to what extent. The prospect of modelling and quantifying even current global flows of genetic resources and the direct and indirect benefits arising from them boggles the mind in much the same way as does the concept of modelling the global climate system, let alone any attempt to unravel the history of access and benefit-sharing. Nevertheless, a number of papers have touched on aspects of the challenge. Taking the value of the contribution of CIMMYT wheat to various national economies as an example, a 1990 study concluded that CIMMYT's annual contribution to the Italian durum wheat crop was not less than US$300 million (INTERAGRES, 1990).

In 1991, Derek Tribe believed that the annual additional contribution of CIMMYT material to Australian wheat was US$75 million (Tribe, 1991, cited in RAFI, 1994); RAFI believed it to be US$122 million (RAFI, 1994), and CIMMYT itself (in 1993) roughly US$40 million (RAFI, 1994). In 1993, CIMMYT reported that at least 34 per cent of the entire US wheat crop for 1984 had been sown to varieties that were either directly from CIMMYT or included substantial CIMMYT germplasm (Byerlee and Moya, 1993). In 1994, then-US Secretary of State Warren Christopher and two of his Cabinet colleagues argued that 'foreign germplasm contributed $10.2 billion annually to two major crops in the US' (Christopher, 1994, cited in RAFI, 1994 and Shand, 1998). According to Hope Shand (1997), Italian scientists estimate that the benefits of exotic germplasm for a single crop, durum wheat, amount to US$300 million per year in Italy alone. From evidence such as this, RAFI concludes that return on 'Northern investment [in the CGIAR] may be as high as tenfold' (RAFI, 1994). However, RAFI also states that 70 per cent of all developing-country wheat lands – some 40 million hectares – are sown to CIMMYT wheat material, and, after conducting similar assessments for several other crops, concludes that 'probably little more than 1 per cent of the wealth created or supported by the CGIAR system accrues directly to the North' (RAFI, 1994).

Both the 'micro' and 'macro' challenges are formidable, and need to be addressed in the future. However, since this study has not gathered evidence on either, their resolution is beyond the mandate of this book. Instead of focusing on the questions of how benefits should be allocated in theory, this section will concentrate on the extent to which, and how, benefits are being shared in practice.

The kinds of benefit that institutions share depend on a range of factors, but predominantly on the use made of the resources and the nature of the institution. Thus public-sector breeders and researchers engaged in non-commercial work are accustomed to gaining access to germplasm without paying for it, and are prepared to share benefits such as the results of their research. Several such institutions have taken steps designed to prevent recipients of genetic resources from taking out patents on them and thus gaining the financial advantage of a temporary monopoly. By contrast, private sector breeders are more accustomed to monetary benefit-sharing, since they often pay for services and germplasm received, but only share non-competitive information, and make widespread use of intellectual property rights. These generalisations are based on a putative distinction between public, 'not-for-profit' and 'for-profit' institutions, a distinction which is becoming increasingly difficult to draw. The fact that hitherto 'public' institutions enter into commercial partnerships and use IPRs is just one of the trends that is causing a shift in benefit-sharing practices in the seed sector.

Having acknowledged these complexities, this section will explore the monetary and non-monetary benefits that can arise from access to agricultural genetic resources and associated knowledge, and how these are currently shared.

5.5.2 Monetary benefits

For many years, the ethos of plant breeding has been one of open access to plant genetic resources, and 'the common heritage of mankind', as reflected in the text of the original IU. Researchers, breeders and even companies, traditionally obtained samples of germplasm from each other without paying, and used them in their breeding programmes. The main 'benefit' shared through the system was generally reciprocal access to other breeders' lines. The increasing use of IPRs (and particularly patents, which do not provide for the 'research exemption' of plant breeders' rights) has distorted the earlier benefit-sharing arrangement of free, reciprocal access, and this trend was one of the factors that led to the agreed interpretation of the IU and, now, negotiations to revise it. Gradually, the principle of national sovereignty over genetic resources and private ownership of innovations, including those encapsulating living matter such as plants, has gained strength. Concepts such as prior informed consent for access, benefit-sharing and intellectual property protection are now enshrined in the CBD and the TRIPs Code.

While these changes have begun to affect the terms of access to plant genetic resources for food and agriculture, it is still common for breeders not to pay anything other than the costs of handling genetic resources, particularly for unadapted materials. Indeed, several of the breeders interviewed – from both the public and the private sectors – had never paid for genetic resources. One company explained how, on most occasions, it did not pay for material, but rather obtains it through a 'line for line' exchange, at no cost to either side. 67 per cent of the interviewees (which included some representatives from organisations such as universities and government research centres and genebanks, as well as companies) said they very rarely or never pay for access to genetic resources (Figure 5.10). The assumption that public institutions are not engaged in commercial partnerships still underlies the attitude of many public-sector breeders. The statement of one interviewee who provides materials 'free ... because we are a public institution' is typical of the seed sector. Indeed, some public collections of PGRFA are required to supply accessions at no charge. As recently as 1990, for example, the US Congress voted into law a provision according to which genetic material assembled by its National Genetic Resources Programme is available 'without charge and without regard to the country from which such request originates'. (Section 1632(a)(4) of Public Law 101-624, 28 November 1990).

However, while it is still relatively common not to pay for access to germplasm, several companies inter-

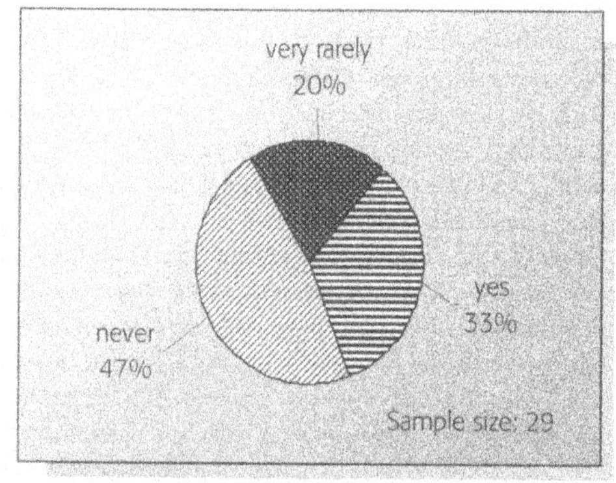

Figure 5.10 *Do You Pay Organisations which Provide You with Access to Genetic Resources?*

viewed pointed to the increasing requirement to pay universities and research centres for access, not to mention companies. As one company put it, 'There is no more free access.'

The use of royalties as a means to compensate those whose germplasm is incorporated into a new variety is now commonplace, and is discussed below. In addition, some private-sector breeders are accustomed to paying not just in the event of successful commercialisation, but 'up front', either on a 'per sample' basis, or by sponsoring a research programme that will guarantee access to the strains that emerge from breeding efforts. The form and method of payment for access to germplasm often seem to depend on the terms on which intermediary institutions choose to provide genetic resources to their customers. The companies interviewed often appeared to be responsive to the manner and magnitude of payment requested by institutions providing germplasm, which suggests that these intermediary institutions may be able to shape access and benefit-sharing in the future, provided the arrangements remain commercially competitive.

We asked the plant breeders interviewed if they would be prepared to pay more for access to genetic resources, if it became more common to do so, or if an additional 'benefit-sharing' element were to be included in the 'price' of germplasm. Several of the companies interviewed explained that they would have no problem paying for genetic resources, nor sharing benefits, but pointed out that the increased cost of producing seed would have to be passed on to the consumer. Several voiced concern about the cost- effectiveness of establishing and maintaining the mechanisms for assessing, tracking, and allocating the benefits arising from the use of PGRFA. As Dr Beversdorf at Novartis put it, 'Will our customers or consumers benefit if we have to maintain an army of lawyers and accountants to collect and pay germplasm fees on the products we develop, produce and sell around the world? Beyond lawyers and accountants, will anyone benefit?'

Fees and research contracts

It is relatively common for companies to make a nominal payment to institutions providing them with genetic resources to cover the handling and transaction costs of access and to distribution of the material, and generally breeders pay a greater sum in the form of a fee for access to germplasm per sample, which appears to vary according to the degree of improvement of the material acquired.

For some higher-value germplasm, particularly the fruits of biotechnology, a common form of payment is a licence fee to use technology in research or breeding, or to use certain strains as parents of a hybrid variety. Some companies license their genetic resources and technology outright, which brings certain risks, as it is subsequently not possible to control nor to tell how the technology has been used or the magnitude of the resulting profits. Another approach is to license the technology based on the frequency of its use, so that, for example, a company receives a certain sum for each test that a licensee conducts using its technology. A third way is for the licensor to receive a royalty on sales of the final product that was made by using or incorporating the technology.

'CIMMYT showed that there were more yield advantages from North to South in wheat than ever occured vice versa. There is solid evidence in support of that. Developed countries receive the material, but it takes 20 years of cost and input from [developed country] breeders before it can be used. So the international centres are basically funded by developed countries.'

Many public breeding institutions are supported by research grants by the private sector. Often, the research grants are made for the institution to conduct an agreed programme of basic research, and the company will have access to the results, which may include a first right of refusal over any germplasm which emerges from the breeding programmes. Sometimes the arrangement is rather looser, such as the practice described by one company: 'We contribute to organisations that provide us with germplasm'. As Novartis said, 'CIMMYT, CIAT and other germplasm centres normally don't charge us for access. We support their research efforts and cooperate when they want assistance'. The same company continued, 'we put several million dollars per year into public institutions to support research, and hope this will double over the next few years.'

Research contracts often involve 'outsourcing' of work, rather than sharing benefits. Nonetheless, this kind of arrangement can provide employment and sponsorship. It is relatively common for companies to pay other institutions to collaborate on field trials such as yield trials that are replicated in several locations. Dr Ted Givens of NC-Plus Hybrids described 'outsourcing' such work to companies in Hawaii and Puerto Rico because the climate there enables them to go through three generations in a year. In another example, field trials were performed under contract in China, Korea, Colombia, Brazil and Japan. Some companies test material in-house for others on a reciprocal basis. Others contract out their genetic engineering work, or secure an independent opinion on products by hiring external organisations to conduct genetic testing. 'They give us a judgement on our methods and cross-check the results.' Others obtain services in genetic mapping. Yet others outlicense more basic research, such as characterisation of germplasm, pre-breeding, screening and training in certain techniques used by the company.

Royalties

Before discussing the frequency with which royalties are used by plant breeders, and providing some information on specific royalty rates that interviewees believed were commonly paid by plant breeders within the seed industry, it is important to distinguish between three different situations in which royalties may be paid, as described in Box 5.9.

Box 5.9 *Different Circumstances in Which Royalties are Paid in the Seed Industry*

Royalties paid by growers	1 ***farm saved seed royalties:*** royalties paid by commercial growers (farmers) to plant breeders when farmers save harvested seed of a protected variety to produce a commercial crop; and
	2 ***certified seed royalties:*** royalties paid to plant breeders by companies that produce and distribute seed for the right to multiply seed of a protected variety for commercial seed production.
Royalties paid by breeders	**royalties or licence fees** paid by one plant breeding organisation to another for the right to use plants protected by patent, or varieties essentially derived from varieties protected by plant variety rights, in a new commercial variety.

Royalties paid by growers

It is now increasingly common for farmers to pay royalties on their farm-saved seed to plant breeders, rather than simply paying an up-front fee per bag of seed ('farm-saved seed royalties' in Box 5.8). In addition, companies that produce and distribute seed ('seed merchants') pay royalties to companies that breed the seed for the licence to multiply seed of a protected variety for commercial seed production ('certified seed royalties'). In their turn, the seed merchants sub-license commercial growers (farmers) to multiply the seed for them under contract.

The basis for calculating the royalties varies. In the UK, for example, farm-saved seed royalties are typically calculated on the basis of the acreage sown to the variety in question, and certified seed royalties on the tonnes of certified seed sold. The royalty rates vary from crop to crop and breeder to breeder. Plant breeders' associations sometimes publish lists of rates, so these tend to be quite well known. For example, the British Society of Plant Breeders (BSPB), acting on behalf of the owners of protected varieties, sends out to farmers lists of current royalty rates for the use of the protected varieties. Thus farmers may be required to pay the holder of the plant variety right on a named variety of spring barley a royalty rate of so many pounds sterling per tonne for first generation seed, and a lesser sum for second generation seed. Similarly, the BSPB periodically notifies potato growers of the current royalty rates per hectare for growing certain varieties of potato seed. (Personal communication, Mr BJ Moore, BSPB, 12 March 1999.)

Royalties paid by breeders

The royalties described above and paid by farmers for the commercial use of protected varieties are different in kind from royalties that plant breeders charge seed companies for the privelege of using their germplasm in the companies' breeding programmes. Since the topic of this chapter is access to plant genetic resources for crop breeding, this latter category of royalty is more relevant here. Royalties paid by companies to use genetic resources for breeding purposes are less well known than those paid in the former category, since they are generally the subject of confidential commercial negotiations between providers and users of germplasm.

The use of royalties for the provision of germplasm by plant breeders is common only in the case of value-added products, such as biotechnologies, but it is growing. One company mentions that it pays royalties if asked to do so, but that it is frequently not asked. Several interviewees explained that the system of Plant Breeders' Rights, with the Breeder's Exemption, meant that there was no call for royalty payments to breeders of protected varieties. In this setting, the practice of paying the providers of germplasm to use materials that were not protected varieties had never arisen.

Those plant breeders interviewed who were accustomed to paying – or charging – royalties for access to germplasm for use in breeding programmes, cited a number of criteria which affect the royalty rates (Box 5.10). The three dominant factors are the degree of improvement of the germplasm supplied, compared with the final product (with a higher royalty rate for value-added material such as a strain with proven characteristics than for unadapted material); the role of the institution providing the germplasm in its improvement (with a higher royalty where the line being licensed contains an inventive step which has been protected by a patent); and the market share of the resulting product (with a higher rate for high value products which occupy a large share of the available market). The 'going' rate for royalties was a less common criterion, and ethnobotanical data provided with the germplasm was rarely cited as influencing the final royalty rate offered to the providers of the germplasm. A range of special factors, which vary according to key features of the crop in question, also affect the ultimate royalty rate. For example, for wheat, royalties vary according to the impact of the trait in question on the hybrid mechanism. Another factor is the number of genes coding for specific, valuable traits, that have been bred into a variety. As one company explained, when there are 15 disease resistant genes, that drives up the price.

Interviewees gave a wide variety of answers when questioned about their experience of royalty rates, so that

Box 5.10 *Factors Important to Interviewees in Determining Royalty Rates*

Factor	Percentage of respondents indicating that factor is important in determining royalty rates (sample size: 19)
Current market rates	68
Degree of improvement of germplasm supplied compared with final product	74
Role of provider institution in value-adding and inventive step	79
Ethnobotanical information provided with germplasm	26
Likely market share of the final product	68

these do not fall into clearly identifiable bands. For example, several reported that companies and public breeding institutions rarely pay royalties for wild relatives, landraces and other 'unimproved' materials, but only for elite materials. Another company, however, said that it would be prepared to pay anywhere between 10 and 30 per cent for 'wild and raw' material, in the context of a research collaboration, and a third company confirmed that a broad range of royalties is paid for wild relatives and landraces.

From companies' responses, royalties typically lie in the range of 0–10 per cent, depending on the research involved and the crop concerned. Consistent with the finding that value-added materials attract higher royalty rates, several companies distinguished between the royalties they paid for 'unimproved' materials and those for value-added materials such as strains with proven characteristics and elite strains. Several felt that royalty ranges between 3 and 5 per cent were 'reasonable' for such value-added materials, particularly, for example, where a finished inbred line is provided and used in a final product. Less developed materials only attracted royalty payments of 0–3 per cent.

5.5.3 Non-monetary benefits

Reciprocal access

Other than payment of research grants and royalties in the event of commercialisation, by far the most common benefit of access to genetic resources remains participation in the informal system of reciprocal exchange of germplasm. Many institutions provide access to their lines as part of an exchange worked out in advance with the recipient, which itself provides access to some of its own germplasm including improved lines developed from the germplasm received. According to Dr Susan McCouch of Cornell University, 'We don't have the cash to pay up front, but we will work on the basis of reciprocity, and give back what we receive. The attitude is to work for the benefit of the whole. If new or important varieties are developed, they are quickly offered to the country of origin'. Another university illustrated the point with the example of Peru, where nematodes have caused problems with potato crops in an extensive area for 30–40 years. A USDA breeding programme has developed 'pre-bred' lines with the required resistance, and these have been given to Peru. According to the university, several samples held in the collections of the USDA can no longer be found in the wild, and the USDA is happy to supply the original material back to the country of origin. This approach is used in the private sector too. Novartis Seeds described how it provided Bt-maize germplasm to CIMMYT and Bt-rice germplasm to IRRI, as the subsistence farmers served by these institutions would not otherwise benefit from the company's efforts.

The benefit of access to germplasm extends, in some circumstances, to the freedom to use it for breeding efforts. As Dr Reinhard von Broock of Luchow-Petkus puts it, 'Our competition is free to do what they wish with our varieties. It is very new, especially in the USA, to get patent protection and prevent others from using varieties as parents in crosses.' Another company confirmed that the breeders' exemption allows other breeders to use the germplasm developed by the company, although following UPOV 1991, the user may need to compensate the provider of a protected variety if the resulting product is esentially derived from it (Box E.1, Appendix E).

However, several interviewees bemoaned the restrictions on access to PGRFA that the increasing use of patents is causing. While it is still comparatively rare for this to happen, now that highly developed varieties may contain a number of gene products all covered by separate patents, seed companies may need to negotiate several licences to use a single variety for commercial development. Horstmeier (1996) gives the example of an insect-resistant maize hybrid genetically engineered by Pioneer Hi-bred, which required access to 38 different patent claims, involoving 16 separate patent holders (in Shand, 1998).

Information, access to research results and training

As one company explained, 'It is an unwritten rule of ethics for breeders that when someone provides genetic resources, breeders will send them information relating to the research done. This happens even if one does not provide royalties, and whether a breeding programme is successful or not. However, not everyone does this promptly.' As another company put it, 'We may share information with those who provide us with materials. They will want to know if their material performs well in a different country.' The IARCs offer information, technologies and products free. According to the International Crops Research Institute for the Semi-Arid Tropics (ICRISAT), information on the characteristics of the materials held in their collections and the results of research programmes are readily available.

Sometimes the company provides results of research that goes beyond that conducted as part of the programme agreed with the provider of genetic resources, and extends to tests conducted as a 'favour'. A relatively common 'in kind' benefit in exchange for access to genetic resources is for companies to screen plants in their nurseries for diseases.

The public sector shares information through databases which are either accessible by the public, for example through the internet, or by correspondence in response to requests for information. An example of such a database is given in Box 5.2 (p134).

Several interviewees from both the public and the private sectors described their involvement in training programmes. One company described how students visit the company for 1–2 days and the staff give lectures as part of a course run by a training institution for developing country

Box 5.11 *Xa21 and the Genetic Resources Recognition Fund Established by the University of California at Davis (UC Davis)*

A specimen of the wild rice species *Oryza longistaminata* from Mali was evaluated by Indian researchers for its resistance to the bacterium *Xanthomonas oryzae* pv. *oryzae* ('Xoo') (rice blight) in 1977. Between 1978 and 1990, the International Rice Research Institute (IRRI) bred the resistance to rice blight from *Oryza longistaminata* into rice variety IR24 and discovered that blight resistance was attributable to a section of DNA, found on a single chromosome, that they termed *Xa21*. In 1990, Professor Pamela Ronald and her colleagues at Cornell University started work to identify the location of *Xa21* on the rice genome, which she continued when she moved to UC Davis in 1992. In 1995, the UC Davis team identified and cloned *Xa21*. UC Davis filed a patent on the *Xa21* gene sequence in 1995, and in June 1996, with assistance from Professor John Barton of Stanford University, UC Davis established the Genetic Resources Recognition Fund (GRRF).

In July 1996 and January 1997, UC Davis entered into agreements licensing the gene to two agricultural biotechnology companies. UC Davis proposed that the financial benefits contributed by companies should take the form of a royalty of a certain percentage of sales of the products marketed by the companies based on *Xa21*. However, from the companies' perspective, *Xa21* would make only a small contribution to the genome and desirable traits of any new crop variety developed, so they were not comfortable with an open-ended royalty commitment. Instead, the university and the companies settled on financial benefits consisting of payment of a single lump sum by each company: US$52,000 in the case of the first company, and US$30,000 in the case of the second company. Given that only a minute proportion of research actually leads to a successful commercial product, the companies and the inventors settled on 'commercialisation' of a successful product, defined as the availability of the product for sale on the market, as the most appropriate trigger for payment of these sums. The benefit-sharing arrangement involves a single payment by each company into the Fund of the agreed sum one year after the commencement of sales by that company of the first new product that makes use of the *Xa21* gene.

Contributions to the GRRF will come from three sources:

1 **Licensee companies:** The option agreements UC Davis signed with the two licensee companies each contain a clause that triggers a licensing agreement in the event that the company 'commercialises' a product containing *Xa21*.
2 **UC Davis:** In a letter from the University's Vice-Chancellor of Research to Professor Ronald, the University has committed to making a payment into the Fund matching that made after the first year of commercialisation by the first licensee company. It is not yet clear if the University will match any further contributions, whether from the second company or from any corporate partners commercialising products containing *Xa21* in the future.
3 **The inventors:** Professor Ronald and her fellow inventors Wen-Yuan Song and Guo-Liang Wang will contribute to the Fund an unspecified amount of the royalties they receive from the companies in the event of commercialisation. This contribution is voluntary, and not required by the terms of any agreement between the companies, the University and the inventors.

If the licensee companies commence sales of a product incorporating *Xa21* and monies are paid into the Fund, the University plans to appoint a committee consisting of UC Davis staff to select students. UC Davis will use the GRRF as a mechanism to share any resulting financial benefits by funding Fellowships at UC Davis for scholars from source countries such as Mali. The Fund will be used to finance Fellowships at UC Davis for students from source countries. In addition, researchers, including institutions in source countries, will be able to access, at cost price, genes and transgenic varieties produced by UC Davis, subject to a material transfer agreement prohibiting commercialisation without the express, written consent of UC Davis.

Source: ten Kate and Collis, 1998.

scientists in that country. Some of the bigger seed companies support scholarships and even professorships, some of which are for developing-country plant breeders. It would appear from the experiences shared by the companies interviewed that most of these opportunities for training are in the countries where the companies are based, rather than in academic institutions in developing countries. Training is rarely conducted specifically in exchange for access to genetic resources, and is more likely to be funded by the philanthropic institutions established by companies rather than taking place in the context of a commercial agreement. The Genetic Resources Recognition Fund established by the University of California, Davis (Box 5.11) is the exception that proves this rule.

Access to technology and joint research

Technology is one of the forms of benefit-sharing most sought after by providers of genetic resources. Several of the plant breeders interviewed provided examples of the transfer of equipment and know-how. Dr Rajaram of CIMMYT describes how CIMMYT uses the 'double haploid' system, which can cut the number of generations of breeding needed from 10 to 4 or 5. It received this technology, perfected for CIMMYT by a Japanese scientist, from the John Innes Centre (UK). One seed company provided a free licence to a European institution conducting research in sugar-beet. Where recipients are not viewed as competitors, companies may give them free access to information under a confidentiality agreement. As one seed company explained, it has a relaxed attitude to developing countries, recently providing training to Pakistani scientists, including training in the use of double-haploid techniques.

The nature and value of the technology transferred depends largely on a close relationship between provider and user, which is often built in the context of joint research. Some companies interviewed described research programmes conducted jointly with some of the institutions from whom they acquire genetic resources. For example, one company has joint research programmes with two leading UK research institutes and various agricultural colleges and receives germplasm from CIMMYT. As the company explained, 'We give them things in kind and research facilities and quality-testing resources.'

The kind of benefit-sharing arrangements that arise from joint research programmes depends, in part, on the policy framework in the countries of the collaborating institutions. If the partnership is between private-sector company and public-sector institute, public law may affect the ability of the public company to 'sell' or license germplasm or technology. Government agencies are often empowered to act by statutory authority. Indeed, specific statutory authority may be necessary in order for an agency to enter into legally binding agreements, to pay for goods and services, to conduct joint research, to own intellectual property and to transfer technology. Statutes that preserve the non-profit-making status of public bodies can restrict their abilities to receive payments for the exploitation of technology, and indeed to share these monetary benefits with collaborators such as providers of genetic resources. Regulations governing licensing arrangements by public bodies often encourage open competition to ensure the cost-effectiveness of activities paid for out of the public purse. The requirement to tender services for competition can prevent public bodies from making commitments to transfer technology to collaborators who provided genetic resources, since other organisations might win the licence to exploit the technology.

Policies such as these shape the benefit-sharing commitments of public bodies not only in the USA, where the Federal Technology Transfer Act of 1986 (15 USC 3710a) and other statutes define the ways in which Federal Government agencies may transfer technology and other benefits (ten Kate and Wells, 1998), but in other parts of the world. One company described how, according to the law in a country where it works, the state owns any technology developed through publicly funded research. Thus, when the company conducts collaborative research with public institutions in that country, the state owns the technology, and the company gains access to genetic resources.

5.5.4 IR64: a case study

IR64 is the world's most widely used rice variety, grown on 10 million hectares. It was developed by staff at the International Rice Research Institute (IRRI) in Los Baños, Philippines, over a 25-year period that started with the organisation's inception in 1960. Since that time, a major activity of IRRI has been the breeding of rice varieties with a combination of desirable agricultural traits, such as pest resistance and high productivity. IRRI, which is one of the IARCs of the CGIAR, conducts some 3,000 crosses each year, using parents from the more than 104,000 samples conserved in its International Rice Genebank Germplasm Centre.

Genetic resources used to breed IR64

Box 5.12 shows the pedigree of IR64. One wild relative (*Oryza nivara*) and 19 traditional varieties were used in the breeding programme that culminated in the development of IR64. The specimen of *O nivara*, the wild species, came from India, and was used to confer resistance to grassy stunt virus. *O nivara* grows in both natural and disturbed habitats in India, Nepal, Sri Lanka, Pakistan, China, Bangladesh, Myanmar, Thailand,

Cambodia, Laos and Vietnam, although the strain from India is the only one resistant to grassy stunt. The 19 traditional varieties came from eight different nations: China, India, Indonesia, Korea, the USA, the Philippines, Thailand and Vietnam. Dr Khush of IRRI, who led the team that developed IR64, believes that none of the traditional varieties used in the pedigree of IR64 was grown by farmers during the 1980s, when IR64 was developed, nor are they grown by farmers today. The varieties most widely grown by farmers before IR64 were IR20, IR36 and IR42.

The other parents used in the breeding programme are either pure line selections (such as GEB24 or CO18) or varieties developed from hybridisation between different parents (such as Peta), by crossing the landraces Cina and Latisail in Indonesia during the 1930s. The pure line selection GEB24 was selected by the late Dr Ramiah, who was to become Director of the Central Rice Research Institute, in Cuttack, India. CO18 was selected at the Paddy Breeding Station in Coimbatore, India. The parents of Cina came from China, and those of Latisail from Indonesia. A Dutch breeder, Dr van den Meulen, selected Peta at Singamerta in Indonesia from this cross, and it was ultimately released in 1941. Two improved varieties, IR8 and IR24 (both IRRI varieties) were also used in the pedigree of IR64.

From many experimental lines selected from the cross of the parents of IR64, IR5657 and IR2061, IR18348-36-3-3 was found to have the most desirable combination of traits that IRRI was looking for. Thus it was released as IR64, and other less promising sister lines were discarded. IRRI has also continued to select and evaluate other lines from different crosses (so called 'cousins'), and some of these have subsequently been released as varieties such as IR66, IR68, IR72 and IR74, whose traits excel those of IR64 in some respects. Varieties such as IR64 may become susceptible to diseases over time, or new diseases may emerge. Soil environments also change, and this can affect productivity. For example, IR8 could produce 9 to 9.5 tons per hectare in the 1960s, but now can produce only 7.5 tons per hectare, while IR72 produced 9.5 to 10 tons in the same trials. For reasons such as these, a continual process of breeding and improvement is important.

IRRI shares the seeds of improved breeding lines such as IR64 with national agricultural programmes around the world. Hearing of superior strains from performance reports, publications and by word of mouth, public-sector breeders from rice-growing countries request seed of improved breeding lines, and IRRI sends them seed that has been matched against their special requirements. IRRI requests recipients to acknowledge that they have used the seed, but the seed is sent with 'no strings attached', in terms of requirements to share with IRRI any benefits arising from the subsequent use of its lines, or any varieties developed from them. The responsibility for testing new varieties of crops and deciding which varieties to release to farmers rests with national organisations – in the case of the Philippines, where IRRI is based, the Philippine Seedboard. IR64 was entered into Philippine national coordinated trials conducted by the Philippine Seedboard as breeding line IR18348-36-3-3, and evaluated at ten experimental stations in the country over a period of two years (four seasons). Finding that it had superior qualities, the Philippine Seedboard approved the line for release on 29 May 1985.

Once a national Seedboard has approved for release a line like IR64, its seeds are produced by organisations such as departments of agriculture, national seed corporations, agricultural universities or extension organisations, sometimes in cooperation with seed growers' associations. These organisations then distribute the seed to farmers.

IR64 is just one of IRRI's high-yielding products. The number of IRRI breeding lines released as varieties by national agricultural programmes around the world now exceeds 300 (Box 5.12). About 1000 improved varieties have been developed by national programmes, 75 per cent of them being progenies of crosses with IRRI-bred varieties of breeding lines. Since 1966, when IR8, the first modern, high-yielding variety to make a major impact on rice productivity was released, the rice-harvested area increased only marginally from 126 to 148 million hectares (17 per cent), while the average rice yield increased from 2.1 to 3.6 t ha^{-1} (Khush, 1995). On average, modern varieties are now grown in more than 70 per cent of the rice areas of the world, providing farmers with an average of 5–7 t of unmilled rice from modern varieties, compared with 1–3 t from traditional varieties. Despite these remarkable developments, rice plant types with higher yield potential and much better grain quality will be demanded of rice breeders in the future. The population of rice consumers is increasing at the rate of 2 per cent annually, but the rate of growth of rice production has slowed to 1.2 per cent (Khush, 1995).

Terms of access and policy framework

Throughout the period during which IR64 was developed, plant genetic resources for food and agriculture were available on the basis of free exchange between scientific and breeding institutions. The accession of *O nivara* was collected from Uttar Pradesh, India in 1963 by a scientist from the Central Rice Research Institute at Cuttack. The accession was sent to IRRI in 1970 as part of a rice collection provided by CRRI on the basis of free exchange. Most of the landraces used involved in the pedigree of IR64 were made freely available by the national collections of the countries from where they

Box 5.12 *The Pedigree of IR64*

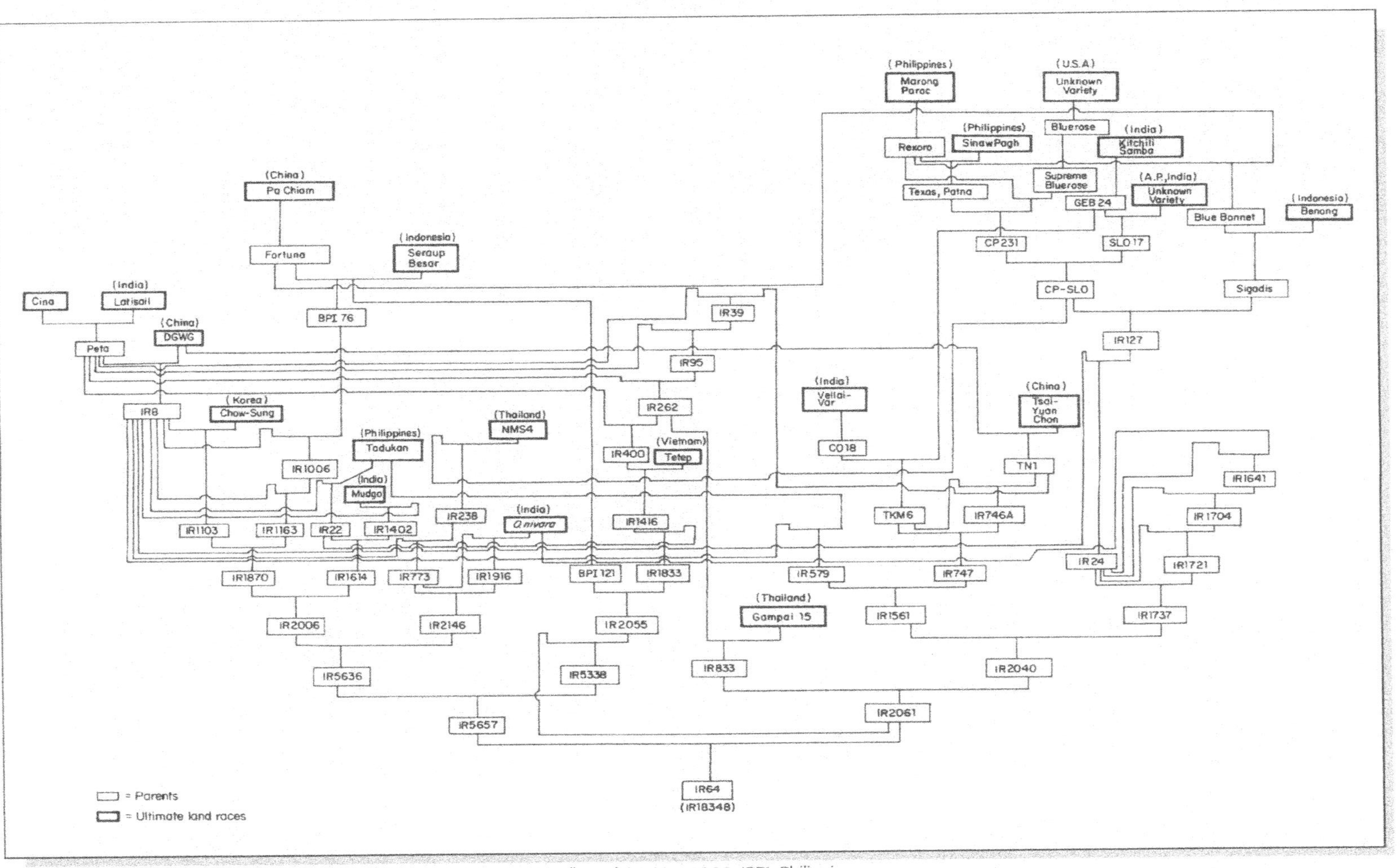

Source: IR64: An Improved Rice Variety Released by the Philippine Seedboard, 29 May 1985, IRRI, Philippines.

originated. Others were collected by IRRI scientists on specific collecting missions. The parent varieties (other than IRRI's own material) used in the breeding of IR64 were donated to IRRI's genebank over the years by national agricultural programmes, under a system of free exchange which allowed different organisations access to each other's genetic material without restrictions and not under any agreement. IRRI was not party to any agreements concerning the use of the parent material for IR64. Much of the material is now stored in IRRI's International Rice Germplasm Centre.

In 1985, when IR64 was released by the Philippine Seedboard, IRRI supplied seed from both its collections and its breeding programme for free, without any restrictions on its use. Any recipient of seed of IR64 was entitled to commercialise it or to use it for breeding purposes, without returning any benefits to IRRI or to the many countries which contributed germplasm to the variety. Subsequently, IRRI's practice has changed considerably for some categories of genetic resources that it supplies.

Towards the end of the 1980s, the CGIAR decided that it needed to respond to the highly politicised international debate about intellectual property rights on plant genetic resources, and published its first policy on plant genetic resources in 1989, describing its role as trustee, rather than beneficial owner of the plant genetic resources held within the IARCs. In 1990, the IARCs began to explore the possibility of bringing the collections held in their genebanks under the governance of an intergovernmental organisation (Hawtin and Reeves, 1997). In October 1994, agreements were signed between the FAO and each of 12 IARCs, including IRRI, by virtue of which each Centre was to hold designated collections of germplasm 'in trust for the international community' (Article 3 of the FAO-CGIAR Agreement, October, 1994, FAO, Rome). Under the terms of the agreement, the Centres will not claim legal ownership of 'designated' germplasm, nor seek any intellectual property rights over the germplasm or related information. They are obliged to make such germplasm and related information available directly to users, or through the FAO, for the purpose of scientific research, plant breeding or genetic resource conservation, without restriction (Article 9), provided they ensure that any recipient is bound by the same terms (Article 10).

Box 5.13 *Number of IRRI Breeding Lines Released as Varieties in Different Countries*

Country	No released
Bangladesh	12
Bhutan	2
Brunei	3
Cambodia	10
China	20
Fiji	2
India	36
Indonesia	36
Iran	2
Iraq	2
Malaysia	7
Laos	10
Myanmar	19
Nepal	10
Pakistan	7
Philippines	54
Sri Lanka	2
Vietnam	70
Africa	52
Americas	41

When the CGIAR Centres signed the agreements with the FAO in October 1994, each Centre was free to decide which germplasm from its collections was to be covered by the agreement. Each Centre provided a list of this 'designated' germplasm, and designated different quantities and kinds of germplasm. IRRI designated the entire collection that was held in its International Rice Germplasm Centre in October 1994, including those of its own breeding lines that were deposited in the genebank. IR64, which is accession number 66970, was 'designated' to the FAO on 14 September 1994. Since 1996, IRRI has supplied designated material, including IR64, under a material transfer agreement containing these terms. Non-designated breeding lines are not supplied under MTAs. There are no provisions in either the FAO-CGIAR agreement or the MTAs on benefit-sharing.

Another relevant component of the policy framework is access legislation in the Philippines. However, under the host-country agreement between IRRI and the Philippine government, IRRI has the authority to determine access to its collections (personal communication, Dr Michael Jackson, IRRI, 6 January 1999).

The policy framework for future breeding programmes is likely to be heavily influenced by results of the revision of the IU. In the future, there may well be more conditions for access by breeders to PGRFA from international centres and national genebanks – at both the international and national levels – and practices on access and benefit-sharing adopted by IRRI and the other Centres of the CGIAR.

Benefit-sharing arrangements

Benefits

A number of different benefits have arisen from using the broad range of traditional varieties, pure line selections, varieties produced by hybridisation and the single wild relative involved in the development of IR64. IR64 has a higher yield and more desirable characteristics than the varieties that farmers were growing before it was available, and it requires less pesticides and fertilisers. These benefits, and the governments and farmers with whom they were shared, will be explored in this section.

IR64 is now grown on 10 million hectares of land in Asia, contributing 50 million tonnes of paddy rice (33 million tons of milled rice) per annum throughout the 1990s. With the average person consuming 100 kg of rice per annum, it has therefore fed approximately 330 million rice consumers each year. At US$300 per tonne, the total value of this production is around US$1 billion per annum.

The main benefits of IR64 are:

- *Higher yield, leading to greater food production and food security*: The main benefits of IR64 have been increased production of rice and food security in Asia. The yield potential of IR64 in the tropics is 10 t ha-1 during the dry season and 6.5 t ha-1 during the wet season (Khush, 1995, and personal communication). IR64 can produce 88 kg/day in the dry season and 54 kg/day in the wet season (Khush, 1995). It outperformed IR36, which was the most widely used variety until the release of IR64, by 21per cent in field trials.
- *Desirable traits:* IR64 is resistant to brown planthopper biotypes and green leafhoppers in the Philippines, as well as to one race of bacterial blight, to tungro and grassy stunt virus. It is also moderately resistant to yellow stem borer, and to soil salinity, alkalinity and phosphorus deficiency, Boron toxicity and peaty soils. The grain quality of IR64 is superior to that of IR36.
- *Savings for farmers:* Higher rice production has contributed to lower prices. After adjusting for inflation, rice is now 40 per cent cheaper than it was 30 years ago in most of Asia. This is an important benefit to many poor farmers, who spend over half their income on food. Growing IR64 is also cheaper for the farmer than many other varieties, since its multiple disease resistance and general vigour enable farmers to save the costs they would otherwise have incurred buying pesticides.
- *Environmental and health benefits:* IR64 has contributed to the reduction of the use of highly toxic pesticides throughout Asia, which has greatly reduced pollution and environmental damage. It has also provided health benefits to rice-farming communities. It could also be argued that higher productivity has helped feed growing consumer demand from existing, high quality agricultural land, and reduced the pressures to convert marginal land such as upland areas, steep slopes and tidal wetlands to agriculture. However, the impact on conservation may not be so positive (see below).
- *Access to improved germplasm for breeders:* IR64 has already been used in breeding programmes to develop new and yet more desirable varieties of rice. Two such varieties developed by IRRI and released by the Philippine Seedboard are PSBRc28 (1995) and PSBRc54 (1997). Few private seed companies are involved in rice seed-breeding, and no rice area in Asia is yet planted to rice varieties developed by the private sector. IR64 has been used predominantly by public-sector breeders. However, it is possible that some private companies may have obtained samples of IR64 from sources other than IRRI itself.
- *Information and training:* In addition, a range of applied and basic research and training programmes, projects and student theses have been conducted on IR64 at IRRI and the results have been shared with other scientific institutions world-wide. Some of the data from this work has been published in international journals or IRRI's programme reports. For genetic mapping work involving molecular markers, a permanent mapping population of doubled haploid lines from the cross of IR64 and Azucena has been established at IRRI and is being used to map genes for various traits of agronomic importance, such as root thickness, root length and blast resistance. This mapping population has been shared with scientists in other institutes, and has led to several publications. In addition, a library of bacterial artificial chromosomes (BACs), developed to study molecular genetics, has been prepared to help develop a physical map of rice chromosomes. These BAC clones will be useful for rice genome sequencing work. Deletion mutants of IR64 – special genetic stocks from which certain pieces of DNA have been removed – are being produced at IRRI, and will help determine the functions of genes identified through genome sequencing. The sharing of these benefits has relied on the mutual understanding between scientific and breeding institutions which allows for the free transfer of information and genetic material, and has not necessitated any formal or legal mechanism.

Beneficiaries

Countries

The most desirable lines from IRRI's breeding material are made available to national rice scientists throughout the world via the International Network for Genetic Evaluation of Rice (INGER), and IR64 is no exception. IR64 has been recommended for on-farm production in numerous countries in Asia, Africa and Latin America, including the Philippines, Indonesia, Vietnam, Cambodia, Bhutan, China, India, Burkina Faso, Mozambique and Ecuador. This list includes five of the countries from which the parent varieties and wild relatives originated (Philippines, India, China, Vietnam and Indonesia). IRRI believes that Indonesia, the Philippines, Vietnam and India have benefited the most from growing IR64, in terms of acreage sown to the variety and the population fed, the countries' need for improved germplasm and the suitability of IR64 to their agro-ecological conditions, as well as the contribution of IR64 to their economies.

Farmers

According to IRRI, small farmers, who spend a high proportion of their income on food, are the principal beneficiaries of the savings to be made by growing IR64. Some farmers use it for subsistence purposes, and others sell the rice. IRRI scientists believe that most of the beneficiaries are small, individual farmers, since there are few private seed companies and farmer cooperatives in the countries where it is grown. Government agencies may produce limited amounts of seed for distribution to growers, but this is not a profit-making venture. There have been no significant financial benefits to government agencies, seed companies or farmer cooperatives.

Conservation

There can be no question that the trend towards a small number of highly productive varieties and away from a large number of traditional varieties and landraces will result in the loss of biological diversity. Inevitably, the widespread use of single varieties such as IR64, covering, as it does, 10 million hectares, will contribute to the loss of local varieties and landraces. IRRI believes that in circumstances where farmers will inevitably wish to grow crops with improved characteristics and higher yields, one way to conserve the traditional varieties and even wild relatives for use in future breeding programmes similar to the one that resulted in IR64, is to store samples in genebanks such as its own International Rice Germplasm Centre. Some comfort may also be drawn from the fact that new, high-yielding varieties carry many of the useful genes and characteristics that their parents held, and make some contribution to the conservation of useful genetic traits.

To date, the vast majority of high-yielding varieties have concentrated on favourable agricultural environments. Arguably, raising productivity in these regions has reduced the pressures to convert marginal land such as upland areas, steep slopes and tidal wetlands to agriculture, in order to meet growing consumer demand. Fragile marginal lands, which often have relatively high levels of biodiversity, are under constant pressure from industrialisation and urbanisation. Future rice breeding programmes are likely to focus on increasing rice productivity in unfavourable environments. Increasing the productivity of these areas may serve to prevent the conversion of marginal lands to agriculture.

Conclusions

IR64 shows how modern, high-yielding varieties of agricultural crops can use a wide range of genetic material, both improved and unimproved, from a large number of countries around the world. The wide range of materials used in IR64 was obtained predominantly from government breeding programmes on the basis of free exchange. This approach has facilitated cooperation on breeding programmes between scientific institutions throughout the world, to the lasting benefit of millions of rice consumers and farmers in South and South East Asia and elsewhere. For crops with complex breeding histories, such as rice, it is very difficult to assess the value of the contribution made by the germplasm of individual countries, if this were to be the basis for the allocation of benefits. In these circumstances, the advent of access legislation in many countries, the increasing privatisation of seed-breeding and consolidation of companies, and the growing use of patents not only by companies but by universities and research institutions highlights the importance of concluding the revision of the IU on PGRFA.

(*Sources:* personal communication with Dr Gurdev Khush, IRRI, 1998–9: personal communication with Dr Michael Jackson, IRRI, 1998–9; personal communication with Mary Jean Caleda, Department of Environment and Natural Resources, Manila, 24 February 1998; Khush, 1995; Hawtin and Reeves, 1997; CGIAR Press Release, 11 February 1998; http://www.worldbank.org/html/cgiar/press/germrel.html.)

5.6 Conclusions and recommendations

5.6.1 Conclusions

Policy framework

Some 150 governments negotiating the revision of the IU have established that access to PGRFA and the sharing of benefits arising from the use of PGRFA are a special case. Countries have been, and continue to be, dependent on one another for access to PGRFA for food security. The relatively small profit margins of the seed industry and the high transaction costs involved as germplasm changes hands several times between collection and commercialisation, support the development of an unbureacratic and efficient multilateral system.

Many of the plant breeders interviewed underlined the importance of completing the revision of the IU, and described the continuing policy uncertainty as a barrier to their activities. As Dr Reinhard von Broock of the German company Luchow-Petkus puts it, the problems that some breeders are experiencing in gaining access to PGRFA will not be sorted out 'until the issues of Farmers' Rights are resolved'.

If and when the IU negotiations are finally revised, the new Undertaking is likely to have profound implications for public and private breeders, as well as for farmers and other citizens, as it will set broad terms, including benefit-sharing, upon which germplasm within the multilateral system can be accessed and should be made available.

However, reaching international agreement to establish a multilateral system, which is 'efficient, effective and transparent, to facilitate access to plant genetic resources for food and agriculture' is proving extremely difficult. Many interviewees felt that the trends in science and technology suggest that plant breeders will require more access to PGRFA in the future, but that political and legal considerations concerning access and IPRs mean that access will, in fact, become much harder to obtain.

It is clear from our interviews that there is a common sense of mistrust among countries and between plant breeders and policy-makers. On the one hand, companies fear that countries will 'hoard' germplasm, and believe they have 'unrealistic expectations' about benefits. On the other hand, governments fear that companies are 'unwilling to share benefits'. Such fears and suspicions already operate as barriers to research and to the successful conclusion of partnerships. These issues are discussed at more length in Chapter 10.

Access

Breeders still routinely use germplasm obtained from other countries to develop new plant varieties. Some breeders interviewed noted that patents, and not material transfer agreements and access regulations, are restricting the availability of PGRFA. Others simply pointed to the increasing difficulty of gaining access to PGRFA from various countries.

Because of the extra time and cost involved in working with exotic and primitive materials, few breeders screen these materials randomly; they are more likely to search for genes with a very specific end-use. The most common reasons to use primitive and exotic material are to find resistance to disease and pests, quality traits such as high yield, cold tolerance and other characteristics, to breed sterility into hybrids and to broaden the genetic base of a crop.

Despite this apparent demand for fresh access to PGRFA, a wealth of material – including elite strains, improved lines and primitive material such as landraces and wild relatives – is already available to breeders in the national genebanks and private collections in many countries. While the breeders we interviewed would like to be able to continue to access materials from abroad, many were confident that their breeding programmes could continue with minimal disruption if they were no longer able to do so. In that sense, there is little point in locking the stable door now, as the horse has already bolted. Breeders' comparative lack of motivation to maintain open access in the future, coupled with overexpectations on the part of providers as to what they may obtain from granting access, block agreement both of the IU and individual partnerships.

Benefit-sharing

Despite the increasing privatisation of agriculture, many breeding activities are still conducted in and funded by the public sector. This operates as a heavy subsidy for access to PGRFA, which is often free to companies and public breeders alike, with correspondingly little benefit-sharing. The only benefit available to farmers, who were responsible for all breeding efforts until a couple of centuries ago, is the freedom to purchase improved crop varieties from breeders. Today, demand for access to farmers' germplasm is small, but such access is almost inevitably unrewarded, so that there is no mechanism for sharing benefits with farmers for their role in the conservation of traditional varieties, landraces and wild relatives, and their associated role, however modest, in selection and breeding.

In the seed sector, benefit-sharing is often not directly linked to individual access transactions. Rather, companies may make looser arrangements designed to maintain partnerships with universities and public research institutes that conduct basic research and supply them with improved germplasm. Such arrangements commonly take the form of funding research projects that are disassociated from access. Given the observations above about profit margins, transaction costs and the complex network of actors, the decoupling of access and benefit-sharing in the seed sector may make sense, provided some multilateral mechanism is found that all participating actors regard as fair, and that encourages conservation.

Companies' confidence that adequate collections for the next few decades are easily available without the need for prior informed consent and benefit-sharing, and the disadvantages of working with unadapted and primitive germplasm, mean that they are not prepared to engage in speculative investment in access to new, unadapted or primitive germplasm. This limits their demand for access and their enthusiasm for benefit-sharing. The sad truth is that there is not sufficient, ongoing demand for new materials for there to be the political will to make more commitments on benefit-sharing.

Notwithstanding, our survey found that, in practice, the cost of samples (including downstream benefit-sharing obligations) and their freedom from patent protection, rank low on the list of factors that plant breeders weigh when deciding which materials to acquire and from whom. More important are the calibre of the scientific institution providing genetic resources, the quality of the samples themselves and the simplicity of the permitting procedure or other access and benefit-sharing arrangement required to obtain them. This suggests that organisations supplying genetic resources could charge for access, thus internalising the costs of benefit-sharing, providing they can offer the quality of samples and scientific expertise that companies are seeking. We also found that while it is still the norm for most basic germplasm to be obtained for free, it is becoming gradually more common to pay, and access to improved germplasm is already generally rewarded, often generously. However, these benefits tend to remain in the commercial sector, and are rarely channelled back from private enterprise to those who conserve PGRFA *in* and *ex situ*.

5.6.2 Recommendations

The delegates to the Commission on Genetic Resources still face a considerable challenge in reaching a consensus on the text of a revised IU, but the negotiations have progressed to the stage where its practical implications for individuals and organisations exchanging genetic resources must become clear. Since many of the transactions involving PGRFA are conducted not by the governments who will be party to any revised IU, but by institutions and individuals such as genebanks, universities, companies, farmers and botanic gardens, it will be crucial to involve these actors in the design of the multilateral system, so that the system is feasible to implement. This may require quite lengthy and detailed brainstorming and preparations at the national level.

Issues for consideration include what paperwork (including MTAs) organisations exchanging genetic resources would need to complete in order to comply with any requirements to pass on conditions of use and benefit-sharing under the multilateral system and to track the use of genetic resources. What kind of supporting infrastructure would be required at the national and international level to coordinate and monitor the system, let alone to track the flow of PGRFA, should that be necessary? Organisations that acquire and supply PGRFA both for food and agriculture and for other uses will need to explore material acquisition agreements that integrate the terms of access under the multilateral system for food and agriculture uses, on the one hand, with terms mutually agreed on a bilateral basis with the provider of genetic resources for other uses, on the other hand. It could be complicated and costly for an organisation to use the agreements and run the information management systems necessary to cope concurrently with access to genetic resources and benefit-sharing under the multilateral system and outside it. It would be helpful to have more studies on the administrative requirements and transactions costs of various options under the multilateral system. Assessments of the levels of benefits to which the multilateral system might give rise might help to guide the design of a cost-effective institutional framework.

Any system for access to genetic resources that involves delay and the risk of failing to reach agreement on the terms for the commercialisation of PGRFA would be likely to be unattractive to business, and might jeopardise research investments in PGRFA. This could happen if open-ended, bilateral negotiations between the provider of PGRFA to the system and a company commercialising a derivative of it were triggered at a stage after the company had already conducted some research and development on the materials. Markets for PGRFA are not homogeneous and do not lend themselves to common prices such as those for commodities, but it may ultimately be possible to remove the 'delay and uncertainty' barrier by reaching international agreement on a standard benefit-sharing agreement for access to PGRFA (perhaps defining different ranges of benefit for access to different kinds of PGRFA). At the time of access, companies could commit either to share benefits according to the standard agreement, or to negotiate bilaterally with the provider (or country of origin) when the benefits arose.

As in other sectors, the role of intermediary organisations – those who gain access to genetic resources and pass them on to others for commercialisation – is growing in the seed industry. It is clear from our survey that intermediaries are increasingly setting the terms of access to PGRFA and benefit-sharing. Provider and user institutions alike are now turning more often to intermediaries to make access and benefit-sharing arrangements. It would be useful, therefore, to raise these intermediaries' awareness of the CBD and the IU, and to train them in how to acquire genetic resources and supply them in accordance with the CBD, IU and related national laws.

Organisations that wish to provide PGRFA in return for benefits, and particularly those hoping to participate in the research which adds value to PGRFA, will need to become familiar with the market for PGRFA. Any institution that aims to negotiate an advantageous benefit-sharing package will need to understand what it is that plant breeders are seeking from their partners; the kind of material they are interested in; how they hear about it; what they will pay for it, and the kind of collaborations for value-added research they are willing to contemplate. This chapter has just scratched the surface of a fascinating and complex subject. More information is needed to help policy-makers and institutions working with PGRFA to understand the extremely diverse set of companies and other institutions that seek access to PGRFA and the range of partnerships through which benefits can be shared.

Box 5.14 *What Did the Interviewees Say? Messages from the Breeders Interviewed*

To governments:

- Work to find a win-win solution: design an open, transparent system, preferably without the need for tracking.
- Help the scientists get on with their work. Germplasm is better used than in a genebank. Evaluation and use of genetic diversity is important, so the characterisation, identification and utilisation of germplasm need to be accelerated.
- Do something about the restrictions on collecting and the inadequate funding for maintaining collections. These don't bode well for crops and for people to have enough to eat.
- Without work by plant breeders to improve it, germplasm is not very valuable.
- Remember that the crops in your country are generally not indigenous to you. Reciprocal exchange is helpful.

To companies and other plant breeders:

- Maintain a high ethical standard and do what is required by the law. Specifically, obtain the agreement of the country for collecting. Follow the principle of PIC. Get things clear. Use MAAs and MTAs to define the legal status of collections.
- Be a constructive partner in the policy-making dialogue. Support the multilateral system and be ready to be a participant.
- Get your access now, as you may not be able to in the future!

Notes

1 The distinction between agriculture and horticulture is discussed in Chapter 6.

2 Eight countries – USA, Canada, France, Germany, Japan, New Zealand, Switzerland and the UK – entered reservations, which they subsequently withdrew when the FAO Council adopted the Agreed Interpretation in 1989. Personal communication, Clive Stannard, FAO, 17 March 1999.

3 As well as recognising the enormous contribution that local and indigenous communities and farmers have made to the conservation and sustainable use of PGRFA (Art15(1)), in Article 15.2, 'The Parties agree that the responsibility for realising Farmers' Rights, as they relate to Plant Genetic Resources for Food and Agriculture, rests with national governments. In accordance with their needs and priorities, each Party should, as appropriate, and subject to its national legislation, take measures to protect and promote Farmers' Rights, including (a) protection of traditional knowledge relevant to PGRFA; (b) the right to equitably participate in sharing benefits arising from the utilisation of PGRFA; and (c) the right to participate in making decisions, at the national level, on matters related to the conservation and sustainable use of PGRFA'.

4 Comprising state and territorial agricultural experiment stations, the Agricultural Research Service of USDA, and the Plant Material Centres of USDA.

Kerry ten Kate

Chapter 6
Horticulture

> 'The best way to keep a plant is to give it away.'
> (Old adage of the horticulture industry)

6.1 Introduction

The term 'horticulture' embraces a wide range of activities, from the high-intensity, large-scale commercial production of vegetables, to the hobby of gardening pursued by the enthusiastic (or even the reluctant) amateur. This chapter considers the use of genetic resources to develop new horticultural products. Its main focus is on herbaceous ornamental horticulture.

The first section looks at the major categories of horticultural product; the size and breakdown of the market, and the countries which are the leading producers and consumers for some of these categories. It also identifies the different actors involved in the horticulture industry, and their role in the development and distribution of new horticultural products. The section provides an overview of the different stages of product development, and gives some ballpark figures on the time and investment associated with developing a new horticultural product.

The second section analyses the demand by the different actors in the horticulture trade for access to genetic resources to develop new products. It examines the extent to which, and how, the horticulture industry obtains plant specimens collected from *in situ* conditions, and the extent to which *ex situ* collections, such as national collections and botanic gardens, are used as the source of new materials for the industry. This section also lists some of the important traits sought by plant breeders and explores the source of the genetic resources used in product development.

The third section looks at current practice in benefit-sharing in the horticulture sector, with sections on the monetary and non-monetary benefits that arise from use of genetic resources for horticulture, and some comment on the extent to which they are shared.

A fourth section provides a case study on the registration of new orchid hybrids, and another on a joint venture company established between Nanjing Botanic Garden, in China, and Piroche Plants in Canada. A final section sets out some conclusions and recommendations relevant to this industry sector.

6.1.1 The definition of 'horticulture' and the scope of this chapter

To describe the scope and magnitude of the horticulture industry is difficult because of the paucity of reliable market data, and the unclear boundary between horticulture and agriculture. Section 6.1.2 starts by explaining some of the shortcomings of market data and statistics on horticulture. It goes on to try to draw the distinction between agriculture and horticulture, and then provides some market data on the vegetable seed market, before focusing on ornamental horticulture, which is the main subject of this chapter.

The shortcomings of market data in the field of horticulture

Most of the industry experts we interviewed were extremely sceptical about the accuracy of the available statistics on horticulture markets, which are difficult to compile for a number of reasons. First, definitions of horticulture and the basis for compiling statistics vary from country to country ('The role of horticulture companies in breeding new ornamental varieties', below), so it is difficult to compare like figures and to calculate consistent global values for the products concerned. Second, horticultural products are imported, exported and re-exported, and some major seed companies sell their products to each other, as well as to the distributers and retailers through whose hands products generally pass before reaching the consumer (Goldsmith Seeds, for example, sells seed to Novartis and to Ball, as well as to brokers and distributers).

Figures rarely distinguish between sales of final products and sales of seed to commercial companies for raising into products such as cut flowers and potted plants. Therefore, using either national statistics or available data on the turnover of individual companies, it is difficult to distinguish between different categories of products. The same product may be accounted for twice in a single set of figures. Third, the figures available represent only some sales. National statistics from The Netherlands, for instance, do not include sales of plants grown under contract for export. In some cases, these sales may be significant. For example, contract sales of orchids in The Netherlands are believed to be around 25 per cent of auction sales of orchids, and yet do not appear in the statistics. Furthermore, in some countries, statistics are compiled from data submitted voluntarily by those companies that choose to do so. For example, the USDA conducts a number of surveys of ornamental varieties. Some of these are compiled annually, based on data obtained directly from growers on a voluntary basis, which are inevitably incomplete. Statistics in several countries are estimates calculated on the basis of yield, acreage and value per unit, rather than actual sales. Fourth, vast quantities of horticultural products are produced and consumed by poor or small farmers and by urban families around the world. The value of this production is seldom captured by horticultural statistics. For all these reasons, the figures in this chapter should be regarded as indicative only, at best.

The distinction between agriculture and horticulture

There is no single, accepted definition of 'horticultural' crops, and many possible bases for distinguishing between horticulture and agriculture turn out, on further inspection, to be unsatisfactory. 40 years ago, it might have been possible to distinguish agriculture and horticulture by scale and intensity of production. 'Agricultural' vegetable crops could be thought of as those grown on farms on an industrial scale, often with mechanised production, and 'horticultural' vegetable crops as those grown more intensively on a domestic scale. Today, a single horticultural producer might grow lettuce on some 400 hectares, and use highly mechanised harvesting technology, yet this activity is likely to be classed as 'horticulture' in national statistics and in industry parlance. Conversely, some 'agricultural' production can be extremely intensive.

However, scale remains a reasonable indicator of the distinction between agriculture and horticulture in one respect, since, broadly speaking, horticultural producers are concerned with quality at the level of each individual unit (whether this a cauliflower or a rose stem), whereas agricultural producers of crops such as cereals tend to be concerned with the quality of bulk quantities of produce. Tomatoes for industrial processing into paste or juice are classed as agricultural, whereas tomatoes for use in salads are classed as horticultural. Another indicator used in the horticulture industry is whether or not the producers 'touch' the goods. Although, as ever, there are exceptions, such as cherries shaken straight from the boughs of trees into containers, most horticultural produce – from flowers to fruit – is hand-packed, if not actually hand-picked. This cannot be said of agricultural produce such as corn or soybeans.

Another possible basis for distinction is the use of the harvested produce. Thus mint-growing in the field or glasshouse for sale as a fresh herb is generally considered a horticultural operation, whereas mint-growing in large (or small) fields for harvesting and processing into mint oil is regarded as an agricultural operation. Similarly, in some countries potatoes grown for processing are considered agricultural produce, whereas those grown for fresh market sales are considered horticultural produce. However, this distinction also has problems. In Canada, for example, regardless of final use, potatoes are considered a horticultural crop, whereas in other countries potato production is captured in agricultural crop statistics.

The nature of the crops themselves can also be used to distinguish between agriculture and horticulture. Perishable crops such as fruit, vegetables and flowers are generally regarded as horticultural, and grains, cereals and other crops that can be kept for more than a few days are often regarded as agricultural. However, this basis for the distinction is unreliable in that a single crop can fall into both categories, depending on the method of production. Tomatoes, for example, are considered to be agricultural produce when grown out of doors and uncovered, but are often defined as a horticultural crop when grown in glasshouses.

(*Sources:* personal communication with Bob Bowman, Goldsmith Seeds; Norman Looney, Agriculture and Agri-Food, Canada; Marvin Miller, Ball Seed; and Dr Fordham, Wye College, 16 February 1999.)

The market for vegetable seed

The global market for vegetables is considerably greater than that for ornamental horticultural products, and this is borne out by the relative size of the market for vegetable seed, compared with that for flower seed. In 1996, annual global sales of horticultural seed, including flower and vegetable seed for the commercial and private markets, were approximately US$1.75 billion (Rabobank, 1996). Of this, the commercial market for vegetable seed (ie the market for seed sold to companies which would raise vegetables from it) was estimated at US$1.6 billion. Tomato seed alone accounted for about half of this, and sales of flower seed made up the remaining US$150 million. The markets for vegetable and flower seed have grown in the last two years, and in 1998, sales of horticultural seed by the top seven companies alone were some US$1.8 billion (Table 6.1). Within the market for horticultural seeds, the USA, The Netherlands, France and Japan are the most important suppliers (Rabobank, 1996; FIS, 1998).

While these figures exist for the commercial market for vegetable seed, no reliable data are available on the global market for finished fruit and vegetable produce, due to the complexity of the supply chain (described below), and the vast range of different products, as well as the more general reasons set out above for the shortcomings of market data in the field of horticulture. However, Table 6.2 gives the turnover of the main exporting countires of vegetables and flowers in 1994.

While even broad generalisations can be misleading, the techniques for breeding and propagating vegetable crops, and the nature of the demand for access to genetic resources for vegetable breeding can, in many respects, be compared with those involved in the development of agricultural crops (Chapter 5). For this reason, this chapter concentrates instead on the ornamental horticulture industry, which, with its more modest market and aesthetic rather than nutritional goals, is quite different from vegetable breeding.

Table 6.1 *Global Sales of Horticultural Seed by the Top Seven Companies*

Company	Global sales (US$ m)
Takii	430*
Sakata	390*
Seminis	380
Novartis	225
Ball	180*
Nunhems/Sunseeds	130
Limagrain	100
Total	1.84 bn

*Figure does not only cover seeds

Source: FIS, 1998.

Table 6.2 *Turnover of the Principal Exporting Countries of Vegetables and Flowers in 1994*

Country	Turnover (US$ m)
USA	200
Netherlands	200
France	100
Japan	40
Germany	35
Denmark	20
Belgium/Luxembourg	n/a
Italy	30
Chile	25
New Zealand	6
Others	459
Total	1.12 bn

Source: Assinsel/FIS web page.

The ornamental horticulture industry

Compared with the crops covered in Chapter 5 and the vegetable industry, ornamental horticulture is a very modest market. The ornamental horticulture industry comprises five main areas:

1 herbaceous ornamental horticulture (including annual bedding plants, some potted plants such as impatiens, petunia and geranium);
2 woody ornamental horticulture (a relatively minor component of ornamental horticulture, including shrubs, trees, etc);
3 cut flowers;
4 foliage plants (which includes non-flowering potted plants). Foliage plants are sometimes absorbed into the other categories, with flowering plants included in herbaceous ornamentals, and non-flowering plants in woody ornamentals; and
5 bulbs (which are sold as potted plants and cut flowers, and commercially as starting materials to be grown into potted plants and cut flowers).

The 'big three' producer countries of horticultural produce are The Netherlands (leader of world floriculture production, valued at US$4.7 billion annually); Japan, which has 47,489 hectares under horticultural production (this covers fruit and vegetables, as well as flowers), and the USA, which is the leader in garden flower production with a market of US$1.3 billion (Floriculture Crops Summary, 1997). Other significant exporters include Costa Rica, Guatemala, Singapore, China, Colombia, Ecuador, Kenya, Zimbabwe, and India (Heinrichs and Siegmund, 1998); however, the roster of leading exporting countries is changing constantly.

Throughout history (see Box 6.1 on the 'Tulipmania' and the speculation that threatened the Dutch economy in 1637), markets and prices for horticultural produce have been notoriously fickle and vulnerable to fluctuations. They remain so to this day. Factors such as the globalisation of markets, currency changes, fashion and the weather have a significant impact on companies. Periodically, supply surpluses result in lower prices for producers. For example, in 1996, there was a substantial increase in the volume of floral horticultural products, and thus a corresponding decrease in prices, as the main markets were swamped with imported material. Most developing countries have not earned so much over the last two years, as profitability depends heavily on prices (Pertwee, 1998). In such a climate, the proportion of the market captured by an individual exporting country can fluctuate considerably. In 1997, for example, exports to Europe from Israel, Colombia and Kenya fell, while those from Zimbabwe and Ecuador rose (Heinrichs and Siegmund, 1998).

Some countries which are significant importers of horticultural products, re-export a large proportion of the plants they import. A good example of such a country is The Netherlands, which is not only an important consumer of horticultural produce such as pot plants and cut flowers (Table 6.3, Figure 6.1, Table 6.4 and Table 6.6), but a major centre for international trade. Approximately 25 per cent of floral exports from The Netherlands are re-exports. The Dutch Auction System, which is responsible for 59 per cent of world exports in cut flowers and 48 per cent of world exports of potted plants, is described in Box 6.2.

In 1996, Switzerland led the way in per capita consumption of cut flowers, with each citizen spending an average of US$350 that year. The next highest was Norway, at US$225, Austria (US$190) and Finland (US$180). The Netherlands and Japan stood at US$145 and US$140 respectively, while the USA and the UK lagged with US$86 and US$74 per head. Turning to potted plants, the citizens of Denmark spent most on this category of products (US$210 per capita), closely followed by Norway (US$200), Switzerland (US$190) and Austria (US$130). In The Netherlands, people spent an average of US$93 on potted plants, while in Britain, the figure was just US$25 (Flower Council of Holland, 1997).

Of the four categories of ornamental horticultural product described above, the main markets in herbaceous ornamental horticulture are cut flowers and potted plants.

Box 6.1 *Tulipmania and the First Material Transfer Agreement?*

In 1551, Ogier de Besbeque, the Viennese ambassador to Turkey, wrote of the tulips he had seen in Edirne, and sent seeds or bulbs back to Austria. But the arrival in Antwerp in 1562 of a cargo of tulip bulbs from Constantinople marked the start of an industry that was to make and lose fortunes, and threaten the economy of a nation. By 1576, Charles l'Ecluse, the first monographer of garden tulips and creator of a collection of flowering bulbs at Leiden botanic garden, had already described scores of tulip varieties. The streaked tulips often seen in Dutch old master paintings of the period became the subject of particular extravagance.

In 1623, when the average annual income in The Netherlands was 150 florins, a single tulip bulb could sell for 1,000 florins. The following year, only 12 bulbs of the highly prized tulip Semper Augustus were known to exist, and each sold for 1,200 florins. By 1624, these had doubled in price, and during the height of the *Tulpenwoede* (tulipmania), between 1634 and 1637, each bulb cost 10,000 florins – the price of a substantial house on a central canal in Amsterdam. At the Alkmaar auction in 1637, 180 bulbs were sold for the equivalent of almost US$10 million in today's prices, but in the same year, the bubble burst, prices collapsed, many speculators in tulips faced financial ruin, and the whole country faced economic disaster.

Records from the time offer evidence of an early example of a material transfer agreement. At the start of the tulipmania, a bulb of Semper Augustus was sold for 2,000 florins, under an agreement whose terms prohibited the purchaser from passing the tulip to a third party without the prior consent of the original seller.

Sources: Pavord, 1998; and *Encyclopaedia Britannica*, 1971.

Box 6.2 *The Dutch Auction System*

The Dutch Auction system deals with 59 per cent of world exports in cut flowers and 48 per cent of pot plants. Although 84 per cent of the plants sold through the system originate in Holland, the remaining 16 per cent are sent from all around the world. In 1997, the seven Dutch Auction houses sold US$8.47 billion worth of cut flowers and US$3.84 billion of pot and garden plants. The biggest of these, dealing with about 45 per cent of the market, is Aalsmeer Flower Auction, just ten minutes from Amsterdam's main airport. The Aalsmeer Flower Auction covers an area equivalent to 175 football pitches and maintains its own security force.

An auction is a 24-hour process. Products delivered directly by growers or collected for a fee by the auction's collection service arrive throughout the night before an auction. Lots are sorted and rigorously inspected for quality control in the early hours of the morning. Any lots that do not meet the required standard (flower quality, length, etc) are not allowed to be sold, and the producer may be cautioned. Registered buyers are able to inspect the produce that has been passed for sale. Buyers may work for several companies, or they may specialise in buying a particular product (such as roses). Many larger companies employ their own buyers. Other buyers, known as 'flying dutchmen', purchase products on their own account to sell to smaller markets, shops and dealers. The vast majority of buyers are Dutch, and work as agents for companies around the world. To succeed in his role as an agent, a buyer needs to know details of the grower, of the product and precisely what his customer wants or is prepared to spend.

The auction itself starts at 6.30 am. Lots are circulated throughout the day on trollies. Each buyer sits in the tiered seats in the main auction hall, and is connected by microphone to the auction manager, who introduces each lot briefly. Buyers punch cards into a centralised computer to register interest in particular lots. The auction clock has just one hand, which starts at 12 and moves anticlockwise. The lower the hand falls, the lower the price of the lot. Buyers punch a button when a given lot comes down to a price at which they are willing to buy. The skill and technique in this bidding system lies in predicting the price, which is set purely by supply and demand and which alters every day. Immediately after a purchase is made, the computer prints out an invoice which is attached to the lot. Money is taken directly from the buyer's bank account, and there is no credit system. Most buyers and many of the larger companies occupy designated sites within the auction complex. Lots are taken there by trolley as soon as they are bought. Thus a tulip picked in Holland on Tuesday night could be sold in Aalsmeer at 6.30 am on Wednesday and be sold again on the streets of Chicago the same evening.

Source: personal communication between China Williams and Jonathan Read, Flower Council of Holland, 6 November 1998.

Potted plants

The annual global export market for potted plants is over US$5 billion (Table 6.5). Countries that are major producers of products in this category include The Netherlands, the USA, China, India, Japan and Colombia (Heinrichs and Siegmund, 1998). Since The Netherlands is both a major importer and exporter of potted plants, and is responsible for 48 per cent of global exports of potted plants, Dutch figures (Table 6.3) are a good indicator of those potted plants with the highest sales world-wide. In the USA, the top four flowering pot plants (which comprise about 50 per cent of total US consumption in this category) are poinsettias, azaleas, chrysanthemums and African violets (USDA, 1996).

Cut flowers

As Figure 6.1 shows, The Netherlands is also a major exporter of cut flowers (alone responsible for 59 per cent of global exports of cut flowers). The ten top cut flowers exported from The Netherlands through the auction system are listed in Table 6.4. Other leading exporters of cut flowers include Colombia, Italy, Israel, Spain and Kenya (Heinrichs and Siegmund, 1998).

From our interviews and the available statistics, it seems that the annual global market for ornamental horticultural products (cut flowers, potted plants, foliage and bedding plants) probably lies between US$16 billion and US$19 billion (wholesale value). The lower figure is based on statistics available for markets of certain cut flowers, potted plants, foliage and bedding plants in the USA, Canada, Western Europe, and some figures for Central and South America (but not for Central and Eastern Europe, Asia or Africa). The upper figure is arrived at by allowing for some sales in these other regions (personal communication, Dr Marvin Miller, February 1999, and see Tables 6.5 and 6.6).

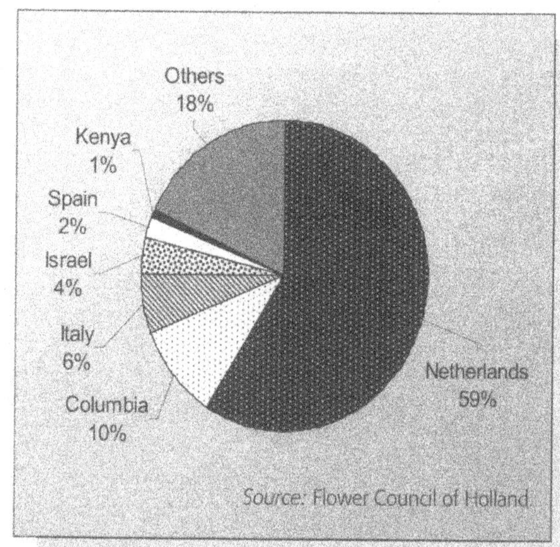

Figure 6.1 *World Exports of Cut Flowers, 1996*

6.1.2 The actors and their role in the development of new ornamental varieties

Overview of horticulture companies

While many horticulture companies – from the single nurseryman to the multinational company – are engaged in growing, distributing and selling ornamental varieties, far fewer are involved in working with genetic resources to develop new commercial ornamental products. Companies breeding ornamental plant varieties fall into three main categories: a small group of

Table 6.3 *Auction Turnover of the Top Ten Pot Plants in The Netherlands, 1997*

	Pot plants	1997 (US$m)	1996–7 %
1	Ficus	222	+3.2
2	Kalanchoe	125	+10.0
3	Dracaena	121	+6.0
4	Dendranthema (Chrys)	109	+20.0
5	Hedera	103	+1.1
6	Begonia	82	+16.6
7	Phalaenopsis	74	+6.2
8	Rhododendron	74	+10.0
9	Spathiphyllum	72	+8.2
10	Hydrangea macrophylla	70	+7.8
	Total	3.04 bn	+9.9

Source: Flower Council of Holland.

Table 6.4 *Auction Turnover of the Top Ten Cut Flowers in The Netherlands, 1997 (US$ m)*

	Cut flowers	1997 $US bn	1996–7 %
1	Rose	1.95	+8.5
2	Dendranthema (Chrys)	1.24	+10.8
3	Tulip	0.6	+6.0
4	Lily	0.54	+20.3
5	Carnation	0.39	+1.1
6	Gerbera	0.38	+16.6
7	Freesia	0.27	+6.2
8	Cymbidum	0.24	+10.0
9	Alstremeria	0.18	+8.2
10	Gypsophila	0.14	+7.8
	Total	5.93	(+9.5)

Source: Flower Council of Holland.

Table 6.5 *World Export Value for Ornamental Horticulture Products (unless otherwise marked, 1996–97, US$ m)*

	UK	Netherlands[1]	EU total	Israel	Kenya	Colombia	USA	World
Nursery stock	1.65	212.82	436.88					
Pot plants	5.44	891.94	1,745.64	19.71	4.6	0.66	68.33	2,430.8
Bedding plants	(both)		(both)					
Cut flowers	13.58	1,793.04	2,086.61	120.23	85.41	384.37	183.96	3,481.79
Bulbs (dormant)	12.41	470.97	507.51					

1 NB Approximately 25% of floral exports from The Netherlands are re-exports

Source: Figures from Heinrichs and Siegmund, 1998.

Table 6.6 *The Top Five Importers of Cut Flowers, Cut Foliage and Indoor Plants, 1996*

	Country	Value of imports (US$ bn)
1	World	5.99
2	Netherlands	3.09
3	Colombia	0.54
4	Italy	0.25
5	Denmark	0.22
6	Israel	0.20

Source: Pertwee, 1998.

multinational companies that account for the majority of sales worldwide; a slightly larger cohort of mainly national companies; and several hundreds of small and medium-sized enterprises, which together account for only a modest share of the global market.

Group one: the major multinationals

Just some ten major multinational seed companies are responsible for roughly 90 per cent of global sales of seed of ornamental varieties, and for breeding the vast majority of the ornamental plants that are subsequently raised, distributed and sold by the chain of actors described in the section below on the time and cost involved in breeding a new ornamental variety.

The major breeders of new ornamental plant varieties – predominantly bedding and garden plant seed – include Novartis (Switzerland), PanAmerican Seed (USA – owned by Ball), Sakata (Japan), Goldsmith Seeds (USA), Sluis & Groot (owned by Novartis), Bodger (USA), Waller (USA – owned by Colgrave), FloraNova (UK), Benari (Germany) and Takii (Japan). The combined annual sales of seed of ornamental horticultural varieties by these companies is probably roughly US$200 million. The five leading flower seed companies alone represent 80 per cent of the global market for seed for cut flowers and bedding plants (rather than for vegetatively propagated plants) (personal conversation with Patrick Heffer, FIS, March 1999).

The horticulture divisions of these companies have annual turnovers of US$60–100 million, employ many thousands of staff world-wide, and market many millions of units of plants each year. They dominate the market for the world's top-selling ornamental varieties. The four top-selling ornamental herbaceous varieties world-wide are impatiens, petunia, geranium and pansy. After these (and in no particular order) come snap-dragons, vinca, begonia, primula, salvia, zinnia, verbena, gerbera and marigold. The popularity of these varieties varies from country to country, and region to region. For example, gerbera is a significant product in Europe, and especially in The Netherlands, but it has a very small market in the USA. Begonia bedding plants are popular on the East coast of the USA, but less so on the West coast.

Group two: middle-ranking companies

The second group of companies, numbering some 25–50, are middle-ranking firms, with annual turnovers of US$5–50 million, and from 100 to several hundred employees, of whom 10 per cent or less are plant breeders. Among these companies are Daenfeldt (Denmark), Sahin (Netherlands), Elidia (France, owned by Limagrain), Hem (Netherlands), Julius Wagner (Germany), Waltz (Germany), Kieft (Netherlands) and David Austin Roses (UK). Most of these companies are based in one country, but market products world-wide.

Generally focusing on a narrower range of products than the major multinationals, these companies make an important contribution to the remaining 10 per cent or so of global sales of ornamental varieties. The German company Waltz, for example, is a major producer of begonias, and Kieft in The Netherlands is a major producer, and also seller, of a broad range of perennials and cut flowers. A small number of companies in Holland and Germany occupy a large share of the market for primulas, and a handful of Dutch companies have an important share of the market for cyclamen.

Over the last decade, several of the companies in this group have been taken over by multinational horticulture companies. For example, the UK company Samuel Yates Seeds, formerly in this group, is now part of the major Japanese Sakata Group, and the multinational Limagrain now owns the French horticulture company Elidia.

Group three: small enterprises

Finally, several hundred small and medium-sized enterprises, from the one-man nursery to the company with 50–100 employees, select and breed ornamental varieties for sale. Their annual turnover may be from as little as US$10,000 or US$20,000, up to a few million dollars each year, and they are likely to operate on a very modest hectareage. According to ADAS, an advisory service for the agricultural and horticulture industry, the British hardy ornamental nursery stock segment, for example, comprises of 2,800 holdings, 80 per cent of which are less than five hectares in size (and account for only a small proportion of total production). Less than 2.5 per cent of the holdings are larger than 20 hectares, but they account for 80 per cent of total hardy ornamental stock production in the UK. While small companies are significant at the local level, and continue to contribute to plant breeding, their combined turnover makes a very small contribution to global sales of ornamental horticultural products.

6.1.3 The role of horticulture companies in breeding new ornamental varieties

The major multinational horticulture companies dedicate substantial research budgets to the development of new products. The combined annual research and development budgets of the companies in the first group described above are likely to lie between US$50 million and US$100 million. The companies in groups two and three, by contrast, invest significantly less in research and development. Middle-ranking companies may invest anything from one to a few million dollars in their breeding programmes. However, the research budget and the costs of breeding a new ornamental variety depend heavily on the extent of the breeding programme, the plant that is the subject of the efforts, and the breeding methods used (see 'Time and cost involved in breeding a new ornamental variety', p165). The two possible 'extremes' for the development of a new variety are, on the one hand, breeding programmes of several years, using genetic engineering and other biotechnologies and costing several million dollars, and, on the other hand, serendipitous selection of a sport of an existing variety which may take just a year or two to test and propagate vegetatively, and little, if any, investment to develop.

Traditional breeding forms the predominant method of research and development, but biotechnologies such as micropropagation, meristem virus elimination, somaclonal hybrids, embryo rescue, and transgenic breeding through genetic engineering are becoming increasingly common in the development of ornamental varieties by multinational companies. Genetic engineering is more common in vegetable breeding than in ornamental horticulture, where it remains the province of the major multinational companies of 'group one' and some smaller, specialist companies such as Florigene Flowers (Box 6.3).

Box 6.3 *The Hunt for the Blue Rose: the Use of Genetic Engineering in Product Development by Florigene Flowers Ltd*

Florigene Flowers Ltd is an Australian-based company with its headquarters in Melbourne, and a fully-owned subsidiary in Rijnsburg, The Netherlands. Research work is carried on in the laboratories in Melbourne, while product development and commercial activities are based in The Netherlands. The company employs a staff of 45, and its shareholders are based in Japan, Australia and Europe.

Florigene was created in 1986, with the aim of producing the world's first 'blue rose' by genetic manipulation. This research is still underway, but in December 1996, Florigene launched 'Moondust', the world's first genetically modified flower (a carnation), and the first flower to be commercialised with Florigene's patented 'Blue Gene', which it isolated from petunias. In Moondust, the Blue Gene expressed itself as a light mauve colour. In 1997, Florigene launched a second product, 'Moonshadow', which achieved a deeper violet colour. The company is preparing to launch the next generation of novel-coloured carnations towards the end of 1999. These will be standard carnations with unique colours ranging from lavender to a dark mauve.

Florigene has also licensed access to another patented gene, the LVL Gene (Long Vase Life), which suppresses the production of ethylene, the naturally-produced chemical that promotes wilting. By preventing production of ethylene the vase life of carnations can be substantially extended. Currently this is achieved by treating the flowers with chemicals, such as silver thiosulfate solution, but this is expensive and there is increasing concern about the environmental impact of such chemicals. Florigene's LVL Gene eliminates the need for preservative treatments. The first products of the LVL range are expected to reach the market early in 2000.

Source: personal communication, Dr Stephen F Chandler, Florigene Europe BV, 17 February 1999.

Some ornamental plants are produced by seed, but most of the big cut flowers are vegetatively propagated from cuttings. In ornamental plants, just as in vegetables and soft fruit, vegetatively-produced material (such as asparagus and strawberry, and, on the ornamental side, lilies, carnations and tulips) is much quicker and cheaper to breed and to propagate than sexually propagated seed (such as tomatoes and potatoes, or, on the ornamental side, pansies and petunias). The cost of raising sexually reproduced plants, not only in terms of the costs of research and development, but also the cost of production, often keeps companies from group two from making the transition into group one.

Until as recently as 20 years ago, most of the ornamental plants developed were open-pollinated (inbred) varieties, but since that time, the vast majority of varieties that have reached the market have been F1 hybrids. The production of F1 hybrids often involves hand-pollination of the female plants, so companies in the first group employ several thousand employees in production sites around the world. While the research and development for breeding is usually conducted in the countries where the company is headquartered, production is often conducted in developing countries, so the 'country of origin' marked on seed packets could be Malaysia, Taiwan, Chile, or any one of several countries in Central America and Africa. The capacity needed to produce seed on this scale means that it is very difficult for companies in group two or three to break into this market.

The main product portfolio of the largest companies lies with new varieties of 'standard' plants such as the top-selling species listed above, but companies in group one are increasingly involved in the development of more unusual 'new' species, such as kangaroo paw and gazanthus. These 'new' species, traditionally the province of companies in groups two and three, may prove to be extremely popular with customers and thus be profitable, or they may not catch on. They thus represent a riskier investment. Small and medium-sized companies work on a broader range of species than the staple, 'top ten' products listed in 'The ornamental horticulture industry', but individual companies tend to work on a smaller number of different product lines, such as Angiozanthus, Clematis, Erica, Ornithogalum, Campanula, Scilla, Tulipa, cacti, Anthurium, etc. They also sell more open-pollinated varieties than the multinationals, whose crops are predominantly F1 hybrids.

As the case study on orchid breeding (p180 below) shows, the pedigree of new ornamental varieties, even those bred by amateurs, can be extremely complicated. However, it is also common for companies, particularly the smaller ones, to market new varieties that have involved very little breeding. The production of a new ornamental variety, for example, often involves more serendipitous observation and selection of plants. According to the British Association Representing Breeders, 'a lot of new plants are, in fact, "discoveries", although there are some breeding programmes. One is always looking for "sports", which are unusual or interesting plants that emerge by natural genetic variation'. Another interviewee concurs, explaining that sports are extremely important in the development of ornamental shrubs. The majority of sports are found in nurseries' stock beds, but some that have been the basis of a new commercial variety have been found in private gardens. The products concerned might simply be propagated vegetatively, selected over a few generations, or result from simple crosses between siblings in order to stabilise their characteristics.

The turnover of the smallest companies may be insignificant compared with the global sales of ornamental varieties, but among the companies in group three can be found some of the most talented horticulturalists involved in the selection, breeding and growing of ornamental plants. These nurserymen often raise plants for niche markets, developing cultivars that are well adapted to local, highly specialised growing conditions. Plants from more mainstream companies may not thrive in these conditions.

Time and cost involved in breeding a new ornamental variety

The cost and time involved in breeding a new ornamental variety varies hugely, according to the plant concerned, the extent of breeding required and the propagation techniques used. The time needed to develop a new variety from scratch can range from one or two years to more than ten, and the cost from virtually nothing to some US$5 million. Several commercial breeders suggested that to develop a new F1 hybrid would take from five to ten years, generally from five to eight years. Breeders' information on costs varied considerably, according to whether the variety was discovered as a sport, or was the subject of an extensive breeding programme, whether it is open pollinated or hybrid, and whether it is vegetatively propagated or produced from seed. Also, many breeders do not develop single products, but rather 'series' of several related varieties (for example, a series of new impatiens in eight different colours, or a series of five matching petunias). To breed a series of new impatiens, petunias or a new chrysanthemum variety would usually take about five years. An additional three years might be required to produce enough 'seed in the bag', ready to sell. The 'average' cost for such an exercise might be US$2–4 million. By contrast, it can cost practically nothing to develop attractive chrysanthemum mutants found in existing varieties and then marketed in their own right (Dr Brian Toulmin, Panamerican Seed, February 1999).

A new rose variety typically takes at least six or seven years, and costs much more than breeding a chrysanthemum (personal communication, Dr Rob Bogers, Research Station For Floriculture and Glasshouse Vegetables (RIKILT-DLO), 1999; Dr Marvin Miller, Ball Horticultural Company, 17 March 1999).

The cost of developing vegetable seed is often higher than for ornamental varieties, and sometimes closer to the considerably higher figures in the agricultural seed market (Chapter 5) (personal communication, Dr Keith Sangster, Thompson and Morgan, 23 February 1999). Only one out of every ten vegetable hybrids developed is a commercial success, compared with perhaps 1 in every 100 ornamental varieties, but the commercial returns are greater for vegetables than for ornamental varieties.

Regulatory requirements and voluntary product standards in the horticulture industry

This section provides a very basic introduction to the regulatory requirements and voluntary product standards that govern the production and marketing of commercial horticultural varieties.

Phytosanitary regulations govern the production, sale and international trade of plants and plant parts such as fruit and seed (Chapter 5). They cover not only fruit and vegetables, but ornamental varieties, too.

Just as there are rules and regulations for agricultural seed, so there are statutory requirements concerning certification, germination, varietal purity, varietal identity and freedom from disease of fruit and vegetables in many countries (Chapter 5). In the EU, for example, there is a category of 'standard seed' for vegetables, which covers vegetable seed tested for varietal purity and identity. The main agricultural and vegetable crops cannot be marketed in the UK, for example, unless the variety is on the UK National List or the EC Common Catalogue. The National List system was introduced in 1973 following the UK's entry into the EC, and is primarily a consumer protection measure. Before a variety can be added to the National List, it must undergo statutory tests to ensure it is distinct from other varieties on the National List or Common Catalogue, and is uniform and stable in its make-up.

Some countries (for example, the Member States of the EU and Japan) have introduced obligations to label ornamental seed, but it is rare for there to be legally binding product approval requirements for ornamental varieties. However, state and commercial laboratories test ornamental horticultural seed on behalf of private companies (who wish to be able to label their products with statements as to their quality), and these laboratories use internationally recognised protocols for testing seeds. For example, the US Association of Official Seed Analysts, the US Society of Commercial Seed Technologists and the International Seed Testing Association have developed standardised tests for seed viability and purity. Viability tests often concern the percentage of seed which germinates under certain standard conditions of temperature, time, light and growth media. Purity tests concern the proportion of inert matter, weeds and other crops' seeds included in samples of the ornamental horticultural produce concerned. Purity tests generally apply to open-pollinated seed grown in fields. For hybrid seed, which is generally hand-pollinated and produced under cover, purity is less of a concern, since pollination with other lines and intrusion by weeds is not a risk.

From seed breeder to market: the interrelationship of private companies

On pages 163–4, we described the small proportion of horticulture companies of all sizes actively involved in breeding new ornamental varieties. As Section 6.2 will show, these companies obtain access to germplasm for the development of new commercial varieties from a range of other actors described in the following sections, namely:

- through purchasing commercially available cultivars on the open market;
- from national genebanks, government agencies and research institutes;
- from university departments and botanic gardens;
- from individuals, societies and associations; and
- through collecting activities by their own staff.

The route from the companies breeding new ornamental varieties to the ultimate consumer quite often involves several stages, as follows:

- *The breeder/producer:* a company in group one, two or three described above conducts research and development, produces seed (or vegetative propagating material) of a new ornamental variety, and supplies it to a broker.
- *The distributor/broker:* distributors and brokers include multinational horticulture companies, such as Novartis and Ball Seed, and smaller companies such as Thompson & Morgan. They may prime the seed (for example, conditioning it for germination and enhanced performance, or coating it so it is easier to sow by machine). They also distribute seed internationally to plug producers or directly to wholesale growers.
- *The plug producer:* the plug producer germinates the seed supplied by brokers or distributors. Early growing has become a highly-mechanised, specialist niche, designed to ensure uniform stand, early germination and good seedling performance. Plug producers often work with trays about twice the size of dinner plates, on which 400 tiny cells

('plugs') each contain a seed which grows into a seedling. When the seedling is some five or six weeks old, the seedlings, in the plugs, are distributed to a wholesale grower.

- *The wholesale grower:* if the seed received by the wholesale grower came direct from a broker, the wholesaler plants the seed and raises little plants. If the seed was already germinated and raised into a young seedling by a plug producer, the wholesale grower simply raises the seedlings for a further month to three months, and sells the young plants to local retail outlets.
- *The retailer:* the retail company, whether a chain store such as KMart in the USA or an individual flower shop, sells the potted plant to the public. From germination of the seed to retail sale usually takes at least eight to ten weeks (eg for impatiens), and often 19–21 weeks (eg for begonias).

This complex chain described above is by no means the only route from the seed producer to the final customer. The advent of 'plug producers' is a recent trend, and many growers still buy seed direct from brokers, and germinate it themselves. Also, several of the largest companies both produce seed and distribute it to wholesale growers. Some multinationals own companies at each stage in the chain, from seed-breeding, through distribution, wholesale growing and sometimes even retailing. Finally, many small family nurseries still buy seed, raise plants and sell them direct to the public, but they face increasingly steep competition from the big retail companies, particularly in the cut flower market. Sales of cut flowers through supermarkets and DIY stores have broadened the market by increasing impulse-buying. Ten years ago, these retail outlets accounted for less than 1 per cent of the cut flower market, but their market share rose to 5–10 per cent in 1994, and is currently 40 per cent. Florists have not lost their share, however, so the size of the cut flower market has increased (personal communication, Dr Philip Bailey, Hamer Seeds).

Not only do plants often pass through many hands en route from the laboratory to the private house or garden, but they often spend time in several countries, as the following illustration concerning the breeding and distribution of Phalaenopis demonstrates. One US company that breeds Phalaenopis sends the improved clones of Phalaenopis seedlings to Japan for further evaluation. There, superior clones are placed into tissue culture for production. The initial cultures are sent to the People's Republic of China for mass propagation, and the resulting plantlets are shipped to The Netherlands to grow. The plants are finally returned to the USA and Japan for finishing and sales.

National genebanks, government agencies and research institutes

Government agencies and agricultural, crop and horticultural research institutes (which may be in the public or private sector) are charged with the conservation of plant varieties and basic and applied research on plant health and crop development. They collect, exchange and maintain genebanks of the genetic resources and conduct the pre-breeding that underpins the horticulture industry. The main focus of these organisations tends to lie with agricultural crops and horticultural vegetables, but some play an important role in the development of commercial ornamental varieties, as the following examples show.

Germplasm collected by government agencies has traditionally been provided for free to horticultural communities, and has formed the basis of a number of commercially successful varieties. For example, in Victorian times, the New Guinea species of impatiens were popular ornamental plants. They were gradually lost from cultivation, perhaps for reasons of disease or insect infestation. In 1970, USDA collectors went to New Guinea to collect a wide range of plants including impatiens. Over 800 impatiens samples, comprising 25 species or subspecies, were successfully collected. While the germplasm brought back to the USA was raised into extremely attractive plants, a poor growth habit prevented it from being commercialised. In 1971, USDA scientists began genetic research on the breeding behaviour of the New Guinea species. From this research, enhanced germplasm was created and distributed to both commercial and public breeders around the world, and was used by private horticultural companies and university researchers to develop cultivars. USDA provided the germplasm for free to the companies and researchers. Today, New Guinea impatiens are among the top ten ornamental plants in sales. Collections such as this one still take place, but now USDA enters into a material transfer agreement (MTA) with the country of origin to enable a benefit-sharing arrangement to be made (personal communication, Dr Robert Griesbach, USDA, 18 February 1999).

In addition to developing plant germplasm, these research organisations conduct another activity that can be just as important to plant breeding, namely the development of production protocols. If a company is to raise and sell several thousand plants, it may be important for them all to bloom together. Horticultural researchers in government agencies such as USDA, and private research institutes such as John Innes Centre in the UK, study the growing conditions of plants and compile extremely detailed protocols describing sequences of conditions of light, temperature, humidity, and post-harvest physiology on an hour-by-hour or day-by-day basis, for the entire growing period of the plant. Such protocols can make the difference between an attractive but unusable new variety, and one with reliable quality that is commercially viable. Manipulation of the growing conditions of plants can also influence the economic value considerably. As Dr Rob Griesbach of USDA explains, 'Take Angiozanthus. As it was developed, it took 18 months to bloom, which would mean it could only be sold to the customer for a price of around US$15. Consumers simply would not pay these prices. By developing production protocols that could induce the plant to bloom in a shorter period of time, it was possible to sell the plant for US$9 and turn the situation around from economic disaster to success story.'

To develop production protocols takes an enormous investment of time, equipment and money. Typically, two scientists might work full time on the production protocol for one variety for two years. Government laboratories in the USA and public and private research institutes in several European countries such as The Netherlands, Denmark, and the UK, maintain the growth chambers necessary for the studies. Private-sector companies generally do not have such facilities. The resulting information has traditionally been made freely available to private companies, although with the gradual privatisation of government research stations and constraints on funding, it is increasingly common for the private sector to pay for the research by the public agency or research institute concerned. On some occasions, several companies join together to sponsor the preparation of a production protocol. Some organisations in The Netherlands, the UK, Denmark and Sweden have exclusive research agreements with individual companies to develop growing specifications for particular plant varieties. As well as research budgets, some organisations are paid royalties on sales of the plant varieties concerned, particularly where they patent the invention involved in developing the protocol. As many thousands of new varieties are launched each year, the decision as to which ones require or merit production protocols may be made by the government agency or research institute concerned, or jointly with the companies who hope to sell a given variety.

University departments and botanic gardens

The historical role of botanic gardens in the introduction of new species and cultivars of vegetable and ornamental varieties is well documented (Field and Semple, 1878; Stearn, 1961; Fletcher and Brown, 1970; Desmond, 1982; McCracken, 1988; Desmond, 1998). An important aspect of the role of botanic gardens is public amenity, so botanic gardens have a particular interest in collecting and raising ornamental plants for public display. Many of the 1,775 botanic gardens in the world are involved in conservation projects, which range from the re-introduction of extinct, endangered or threatened species into the wild, to the curation of *ex situ* collections of living plants and seeds. Research in systematics, conservation techniques, ecology, molecular biology, biochemistry, ethnobotany and many other fields forms a significant part of the work of many botanic gardens, which collect plants in the wild and supply live plants and seed to plant scientists from academia, the public sector and also to companies around the world. The majority of botanic gardens rarely set out deliberately to initiate a breeding programme, but do undertake collecting expeditions around the world, and are known to select specimens of potential horticultural interest from the wild and from their own collections and to pass these on to others who use the material for breeding. Several botanic gardens sell plant material they have acquired for their collections. As Ivor Stokes of the National Botanical Gardens, Wales explains, 'Our horticultural staff will make use of plant material that is bought-in or obtained from other sources and make it available through the shop. It can be assumed that garden-originated sports of various plants will be further developed. This is a common phenomenon as seen in most of the larger gardens.' Botanic gardens still maintain strong links with the horticulture trade, and may select promising wild plants with horticultural potential from their collections for commercial development by others.

The Chicago Botanic Gardens (CBG), for example, has a specific programme, now in its tenth year, which aims to introduce new ornamental plants into the US Midwest horticulture industry. CBG has released nine plants into the industry, and a further 19 are now in the final stages of development and close to being released. The programme is a three-way partnership between CBG, Morton Arboretum and an association of 17 local field nursery growers, under the umbrella of the Ornamental Growers Association (OGA) of Northern Illinois. Material used in the programme is from a variety of sources – from the two participating gardens, the participating nurseries, and from CBG collecting trips which, in the last four years, have included trips to China, Siberia and Russia. All potential plants go through a strict evaluation programme at CBG where they are tested for, on average, 30 different descriptives and characteristics as well as being monitored to check that they are not invasive. This process lasts for a minimum of five years. The plants are then grown up by the members of the OGA and sold on to commercial nurseries. If, after initial trials, the plant shows commercial potential, it is promoted at the various trade and garden shows. CBG protects its interest by taking out a trademark on the name attached to the plant and receives royalties from propagating nurseries on this basis. CBG is currently working on its policy on access and benefit-sharing (personal communication between China Williams and Dr James Ault, CBG, 26 February 1999).

Activities similar to those of botanic gardens are also carried out by staff from university departments of biology, botany, plant science and agriculture, who collect plant specimens, conduct research on them and may provide them to nurserymen and horticulture companies. For example, the Horticulture Department at the University of Georgia (UG) is particularly active in introducing new material into the trade. The Department keeps an 'evaluation garden', in which staff trial a variety of annuals and perennials. Most material is supplied by large seed companies (such as Floranova, Ball and Elidia), which would like UG to evaluate performance ratings on flowering, leaf colour, resistance to insects and diseases, etc. In addition, UG has its own programme of evaluation of new hardy annuals and perennials, particularly those suited to the harsh dry climate of the southern USA. Material trialled originates from a range of sources: sports of existing garden plants, evaluated material, promising material from the gardens of members of the public, and material brought home by staff on holiday. As Dr Armitage puts it, 'a lot of plant work is about good eyeballs'. Plants trialled in the UG garden reach the market in the following ways:

- On the basis of UG evaluations, seed companies may decide to take certain lines to the market.
- UG performs its own trials with material from a range of sources. Plants are evaluated and named and the product is then propagated and marketed by selected growers. The name is usually trademarked, and the packet shows that the product was trialled at UG. An example of a plant that UG has introduced to the market is the verbena 'Homestead Purple' which was marketed in the UK by Blooms of Bressingham. This plant originally came from a private garden.
- One company has recently selected a range of products evaluated by UG and is marketing them under the label 'Athens Select'. The company performs the cleaning and virus indexing, and markets the product. It pays a small royalty to UG and the UG name is on the label. The company is currently preparing a new set of UG plants which it hopes to include in its year 2000 range.
- UG has just brought out its first patented plant, a rosemary called 'Athens Blue Spires'. The patent application is currently before the US Patent and Trademark Office, and UG is preparing other plants to patent. The decision to patent was made for financial reasons: the garden is funded entirely through its collaborations with the private sector. The patented rosemary originated ten years ago from a packet of seeds bought in a KMart store. It was the only seed from the packet that survived, and UG has been evaluating it ever since. All the existing plants are from cuttings from that original plant.

(Personal conversation between China Williams and Dr Allan Armitage, Department of Horticulture, UG, 4 March 1999.)

On a smaller scale, the Herbarium Curator at Reading University in the UK has a specific arrangement with Blooms of Bressingham. In return for funding for collection trips, Blooms have the first right to use any material he collects which has commercial potential. The university generally passes the material to the company just as it is, without further breeding work, on the basis of its appearance in nature. All further research and development work and testing of the plant for commercial potential is conducted by Blooms (personal communication with Dr Alisdair Cullen, School of Plant Science, Reading University, UK, 23 February 1999).

Individuals, societies and associations

Individual horticulturalists often participate in an informal system which acquires and exchanges genetic resources, particularly through societies and associations.

Over several centuries, individual botanists, gardeners and nurserymen have been responsible for a number of important plant introductions. John Tradescant, for example, trained in his family's nursery garden near London and undertook collecting trips to France, The Netherlands and Russia. In 1618, with his friend Sir Dudley Digges, who was creating a garden of his own with Tradescant's help, he sailed to the Arctic Circle. With the use of an Imperial craft lent by the Tsar of Russia, Tradescant botanised on the islands around the Dvina Delta, bringing home roses, hellebore, conifers, red, white and black currants, bilberries, strawberries, at least one geranium, a giant form of sorrel and the larch. On a separate expedition, having heard of a fabulous apricot that grew in Algiers, he sailed to Algeria, a trip which led to the introduction to the UK not only of the apricot, but of gladiolus, lilac, narcisuss, crocus, colchicum, cistus, lychnis, jasmine and many other plants. In 1637, his son, John Tradescant the Younger, sailed for Virginia, where the colony's official register reveals that he was there 'to gather all rarities of flowers, plants, shells, etc'. Subsequent collections by Tradescant the Younger in the West Indies led to the introduction of the false acacia, the yellow lily, the blue-flowered *Tradescantia virginiana*, the yellow primrose, and the tulip tree. Father and son collected *Platanus orientalis* and *Platanus occidentalis*, which they crossed to form *Platanus acerifolia*, the London plane tree (Lyte, 1983).

As Charles Lyte puts it in his book *The Plant Hunters* (1983), 'If John Tradescant the Elder was the founding father of plant collectors, then Sir Joseph Banks was definitely the man who turned plant hunting into a profession'. A trip was planned jointly by the Navy and the Royal Society to seek out Terra Australis Incognita, the Unknown Southern Land, and Banks persuaded the Royal Society, of which he was a Fellow, to allow him to join the crew as the official naturalist. In August 1768, the *Endeavour*, commanded by Captain Cook, sailed for Tahiti. After stopping in Tierra del Fuego, Tahiti and New Zealand, which Cook showed to be an island in its own right, the crew put ashore on what is now known as Botany Bay on the coast of New South Wales. Banks became unofficial Director of the Royal Botanic Gardens at Kew, which he developed into the hub of international plant and seed exchange. He was responsible for the introduction to England of more than 7,000 new plant species, and has many plants named after him, including the Australian genus *Banksia*, which comprises over 40 species of evergreen shrubs and trees.

Individual plantsmen and plantswomen have continued to make a major contribution to the discovery, introduction and maintenance of plants around the world. Some countries, such as the UK, have introduced systems under which individuals maintain 'national collections' of ornamental and other horticultural plants, some of which are not indigenous but are obtained from abroad (Box 6.6).

Box 6.4 *The Horticultural Society of London Commissions a Plant Hunt Expedition to China in 1843*

In 1843, the Horticultural Society of London decided to send Robert Fortune, in charge of the indoor plant section of the Society's gardens, to China. The Society's instructions to Fortune, dated 23 February 1843, were as follows:

You will embark on board board the Emu in which a berth has been secured for you and where you will mess with the captain.

Your salary will be £100 a year from the time of your quitting charge of the hothouse department until you resume it upon your return from China, clear of all deductions and inclusive of the cost of your outfit or such contingent expenses as may be required in carrying out the objects of the Society.

The general objects of your mission are (1) to collect seeds and plants of an ornamental or useful kind, not already cultivated in Great Britain, and (2) to obtain information upon Chinese gardening and agriculture together with the nature of the climate and its apparent influence on vegetation.

Source: Lyte, 1983.

Box 6.5 *The American Orchid Society*

The American Orchid Society (AOS), founded in 1921, is the largest speciality horticultural group in the world, with over 29,000 members in early 1999. Originally founded in the north-eastern USA by a group of wealthy individuals and the nurserymen who supplied them with orchids, today AOS membership is drawn from amateur orchid enthusiasts, as well as the nurserymen and women who continue to supply this hobby-oriented, specialist trade. The society now has about 530 affiliated societies world-wide, approximately 450 of which are in the USA.

AOS aims to support the conservation, display, judging, hybridising and marketing of orchids. In recent years, there has been a rapid growth in the popularity of orchids as flowering potted plants in the USA, and AOS aims to be the primary source of orchid-related information to the world. The AOS Board of Trustees and Officers is drawn from its volunteer leadership, and comprises individuals from many different walks of life, including commercial orchid growing. AOS has a staff of approximately 20, who are responsible for the publication of the Society's monthly and quarterly periodicals and a series of 'how to' books. This staff also communicates with governments and international organisations around the world, and with special interest groups such as IUCN, and the world orchid community. In order to support *in situ* conservation of orchids, AOS supports the concept of only seed-raised or otherwise artificially propagated plants in trade.

Quite distinct from the mass market for the most common varieties, which accounts for some 90 per cent of the orchid trade, many AOS members are specialists who would like to acquire a broad collection of unusual plants. AOS encourages them to exchange plants among themselves as a way of spreading species without increasing the demand for wild-collected material. Formerly, most orchid species originated as wild-collected plants from countries of origin. Orchids are listed in Annexe I of CITES, so that international trade in wild-collected plants has become increasingly difficult, although a great many wild-collected plants are still entering the trade illegally. Because of CITES, the most desirable species (from a collector's point of view) are now propagated either from seed or from stock plants, either in source countries, or elsewhere. Imports by dealers are still largely responsible for the distribution of Asian orchid species, but the burden of paperwork, bureaucracy and associated costs of importing orchids mean that fewer individual dealers are prepared to import orchids, and orchid enthusiasts depend increasingly on importation and distribution by domestic nurseries and the growing number of source-country producers. There is a trend for source-country producers, particularly from Central and South America, to enter the USA with a shipment of a selection of plants for sale directly to consumers at orchid shows or festivals, and for distribution from the particular venues. The plants typically change hands for a price above what the producer would realise at a wholesale level, but below that which a domestic importer would have to ask.

Source: personal communication with Ned Nash, Director of Education and Conservation, American Orchid Society, 4 February 1999.

In addition to the activities of individuals, gardeners' and horticultural societies enjoy a phenomenal membership world-wide. The societies' aims usually entail the conservation, breeding and display of plants, and a wide variety of activities are organised through them, including plant collecting expeditions, the exchange of plants among members and conservation projects. Plant collecting activities commissioned by societies have a long tradition (Box 6.4), which continues today. The members of these societies, such as the Royal Horticultural Society (RHS) and the Alpine Society in the UK, and the North American Rock Garden Society and American Orchid Society (Box 6.5) in the USA, include both amateur and professional individuals and organisations.

6.2 Access to genetic resources

6.2.1 Demand for access to new materials

Our survey asked the horticultural breeders interviewed to explain to what extent they rely on access to new material, and what proportion of their work focuses instead on the manipulation of materials already within their collections. From their responses, it appears that, within all three categories of companies

Box 6.6 *National Collections*

The UK system of National Collections (NC) is coordinated by the National Council for the Conservation of Plants and Gardens (NCCPG), based at the Royal Horticultural Society in Wisley. The system aims to keep endangered garden plants – both wild species and hybrids – in cultivation. The 588 Registered Collections covering 321 genera are held in a range of establishments, including botanical and educational institutions, parks departments, nurseries and private gardens.

Staff from NCCPG visit and inspect, over a period ranging from 2 to 18 months, the collections and records of any horticulturalist or institution wishing to be registered as a NC, and consider the health and state of the plants, the representation of plants in the genus, and evidence that the collection is growing or developing. NC status can be granted to more than one collection of the same genus. In order to expand their collections, holders may go on collecting trips. They also acquire new introductions from other national collections, nurseries and botanic gardens, at home and abroad. Collection holders are encouraged to make their material generally available to the NCCPG, other collections and *bona fide* individuals, commercially, by exchange or as a gift.

An example of a NC is the UK National Collection of *Alstromeria* (Peruvian Lily), held by Parigo Horticultural Co Ltd. in Chichester, West Sussex. This NC maintains wild material from 25 *Alstromeria* 'species', predominantly from South Chile, Peru and Argentina, and 101 hybrids or cultivars. Less than 5 per cent of *Alstromeria* seed acquired by the NC today is wild-collected, and this is mainly acquired from private collectors. As a company, Parigo Horticultural Co Ltd typically charges a few UK pounds for a packet of ten seeds, but, as a NC, they are also obliged to share seeds with other collections or interested parties, either for free or for a small charge.

Another example is the collection of rhododendron species held at Windsor Great Park, which maintains 632 accessions of wild material and some 800 accessions of hybrids/cultivars, the majority of which were collected from the wild, mostly from China, but also from Burma and North-East India, and brought to the UK between 1900 and 1930. The NC obtains surplus stock (under a MTA) from the arboretum of the Royal Botanic Gardens, Kew, from a private nurseryman who collects regularly in China, and from the seed lists of botanic gardens in the USA. It also obtains material from individuals who go on trekking expeditions, whose donations of seed to nurseries and specialist growers become commercially available. The NC releases material to the horticulture industry for a nominal sum where no restrictions (such as the use of MTAs by organisations such as the Royal Botanic Gardens, Kew) prevent this, without further negotiations.

In addition to the conservation of species and cultivars, NCs are involved in:

- **cultivation**: The NC of *Sarracenia* in Preston, Lancashire is made up of authenticated wild material. The species is threatened in the wild and the results of studies conducted on the collection are being used to help *in situ* conservation projects.
- **propagation**: One NC worked with the Micropropagation Unit of London University on the propagation of many of the old cultivars of *Primula auricula*.
- **pest and diseases**: Scientists at Imperial College have been monitoring the disease Firethorn Leaf Miner using the *Pyracantha* NC. RHS Wisley has been working with the *Cornus* NCs, looking at the susceptibility of species and cultivars to Cornus Anthracnose.
- **scientific research**: Collection holders often make important contributions to clarification of nomenclature and taxonomic studies.
- **commercial usage**: Collections have provided *Linum* for cancer research, *Falloia* and *Persicaria* for a Glasgow University study on anti-fungal metabolite production, and *Origanum* for work measuring levels in essential oils. One Herb Garden NC is evaluating the viability of *Echinacea* as a commercial crop in order to alleviate the decline of wild populations through constant harvesting.

Sources: China Williams, 1998, based on personal communication with Graham Pattison, Plant Conservation Officer, National Council for the Conservation of Plants and Gardens (NCCPG), and with Mark Flanagan, Windsor Great Park. Also Pattison and Cook, 1998.

described on page 163 ('Overview of horticulture companies'), there are two distinct groups of horticultural breeders: those for whom access to new materials is an incidental and minor part of their work, and those for whom it is the very essence of their research.

Little demand for access to new materials

For the first group of companies, the hunt for fresh material was just a 'small part of the workload'. This group concentrates on improving their existing material, but the companies are open to injections of new germplasm to introduce attractive new traits, disease resistance or to widen the genetic base, particularly of new plant introductions. As one company put it, 'The overwhelming majority (99.6 per cent) of research focuses on material that is already existing. There is very little search for new material in the wild taking place.'

For this group, it was common for the search for new materials – generally to find new habits in plants, disease resistance or physiological characteristics – to represent no more than 5 per cent of their research activities. As the UK Agricultural Supply Trade Association put it, '95 per cent of the effort is the manipulation of existing material'. According to the interviewees, the majority of material used for breeding is already on the market, and there is enough genetic variation within an individual company's stock that new varieties can be built from this available germplasm, without the need to seek access to more exotic material.

Constant demand

By contrast, several interviewees from companies of all sizes said they were constantly looking for new materials, sometimes through their own collecting trips, but also from amateurs, nurserymen, societies and botanic gardens.

Major multinational companies require access to genetic resources to fuel the development of new varieties, although they can also draw on their own existing, extensive collections of germplasm and genetic manipulation of existing varieties. As Dr Bob Bowman of Goldsmith Seeds explains, 'New genetic resources are our life blood. All the major horticulture companies acquire samples through collecting activities by their staff, and will also acquire specimens in any other way that is legal and ethical.'

Some horticulture companies, particularly those developing new lines of ornamental plants, reported that they rely very heavily on external sources of materials. For a few small companies, the hunt for new material is the main focus of their work. These companies may be able to sell the new materials without the need for substantial improvement in a breeding programme, or they may pass the material to another company to use in its breeding programme. As the representative from one commercial nursery explained, '90 per cent of our time is spent in the search for new germplasm, with 10 per cent spent on breeding.' As another interviewee put it, 'Collecting material and breeding material are two different operations and quite often involve two separate companies or organisations. Therefore [in The Netherlands] there are separate collecting and breeding companies that supply the breeders' market with material.'

While staff in both the largest and particularly in smaller, specialist companies do collect specimens themselves, companies acquire the majority of their new germplasm by receiving seed and plant specimens from genebanks, botanic gardens and other organisations.

6.2.2 Sources of material

Who collects materials for the development of new ornamental varieties?

Many individuals and organisations involved in breeding new horticultural varieties need access to genetic resources, but as their activities are so varied, several breeders described the collection of new material as very much a 'personal thing', driven by an individual's enthusiasm and professional interest. Several interviewees explained that private collectors and plant enthusiasts are an important source for obtaining plant material, and pointed to a small band of roving horticultural collectors, who travel the world and collect new material for use in breeding programmes. 'There are still "planthunters", as in Victorian times,' explained one interviewee. Tony Lord of *Plant Finder* (the RHS directory of plant sources within the British Isles) added that 'Small nurseries, run by *fanatic* individuals are the most important institutions that collect material'.

Horticultural programmes on the radio and the gardening pages of national newspapers in the UK frequently refer to the habit of individuals of bringing back materials from abroad. For example, an article in the *Daily Telegraph* in November 1997 described the international collecting activities of a 'plant explorer' nurseryman with zeal (Lacey, 1997). According to the headline, 'Plant-gathering can have all the drama of a Boy's Own adventure'. The article described the nurseryman's experiences gathering plants around the world, explained that 'the fruits of his expeditions are served up in his nursery', and even gave an address and fax number for his mail-order catalogue.

While their staff do go on field trips, large companies with their own breeding programmes obtain most of their material through acquiring commercial cultivars in the market-place, or from other organisations. As the UK Agricultural Supply Trade Association explained, 'Sometimes seedhouses send out seed-searchers, but this is exceptional. There are no "professional" seed hunters. Instead, botanic gardens and national collections are important sources of new material.' Several interviewees (among them, professional seed hunters) would disagree with part of this assertion, but from the interviews we conducted, it is true that, apart from companies' own collections and commercially available material, the most common source of new material is from national collections and botanic gardens.

Botanic gardens also collect materials around the world, either by mounting collecting expeditions, or by receiving specimens from other organisations. Botanic gardens have traditionally exchanged materials from their collections, using an *Index Seminum* – a list of available seeds – as the basis of an informal seed exchange programme (Box 6.7).

Once companies have obtained a supply of germplasm, they rarely need to return to the source for more material. However, a collecting mission may not gather enough seed to ensure representative samples covering all desirable traits. One specialist collecting company returns to the source country up to three times to collect more samples of promising material. Another interviewee explained that, about 10 per cent of the time, it is important to return to the wild for more material, as collection of limited numbers 'drops' a number of the desired traits.

The use of in-house collections

In-house collections of materials are central to the work of the horticulture industry. All the companies we interviewed hold collections of one kind or another. Most nurseries maintain stock-bed collections (ie many different varieties or breeding materials planted out in fields), plant collections in containers and collections of seed from either commercial sources or in-house collections. Whole, living plants predominate in the collections in most of the market segments of horticulture, although many organisations also hold genebanks containing collections of seeds, and sometimes of vegetative tissues, namely tissues and cuttings. Seedbanks are common among botanic gardens, national genebanks and in companies specialising in vegetable seed, which are often subsidiaries of multinational agricultural seed companies with extremely large collections (Chapter 5).

Most of the materials in the majority of commercial horticultural collections are cultivars. Botanic gardens, by comparison, tend to hold large collections of wild material. In common with agricultural seed companies, horticultural seed companies are reluctant to use unadapted, wild material. The offspring of wild materials often do not run true (ie do not reveal the same physical and other characteristics as their parents), and are

Box 6.7 *Seed Exchange by Botanic Gardens, and the Example of WJ Beal Botanical Garden of Michigan State University*

According to the database maintained by the WJ Beal Botanical Garden in Michigan State University (Beal), over 200 botanic gardens world-wide produce catalogues of seeds which they distribute to fellow botanic gardens and other institutions around the world. In 13 years of computer record-keeping, Beal has received between 36 (in 1985) and 222 (in 1996) catalogues per year. Beal itself sends its catalogue to 424 institutions. Botanic gardens and other recipients of the catalogues can then request samples of the seed listed in them. Each catalogue (also known as a seed list, or *index seminum*) contains varying levels of information about the seed it describes. Some, such as the catalogue distributed by WJ Beal Botanical Garden itself, list not only the taxonomic identification, country of origin of the seed, and the date on which it was collected, but also the exact location where the seed was collected, giving the latitude, longitude and elevation.

The scope of the seed collections listed in catalogues varies enormously from garden to garden. WJ Beal Botanical Garden, for instance, supplies seed harvested from plants it has cultivated in the garden (from seed obtained from all around the world, from the USA to Korea), and also supplies wild-collected seed. In the case of seed from cultivated plants, it lists in its own catalogue the data provided by the garden from which it originally obtained the seed from which the plants were raised. As far as wild-collected seed is concerned, WJ Beal Botanical Garden hopes to improve the diversity of the material listed in its catalogue. Beal has found that wild-collected material is considerably more popular than cultivated material. Over a four-year period from 1994 to 1997, 11 species were offered as both cultivated and wild-collected material. The average ratio of wild-collected to cultivated material requested was 4.5:1. In particular cases, the ratio was much higher: 33:6 for the woody *Nyssa sylvatica* and 28:7 for the herbaceous *Gentiana andrewsii*. Elaine Chittenden, Collections Manager at the Garden, believes that this demonstrates the value of well-documented germplasm over that from unknown sources.

WJ Beal Botanical Garden has itself obtained from 360 (in 1996) to 1,520 (in 1990) new accessions each year from other botanic gardens in this way, and the number of requests by other botanic gardens for seeds that it has met has grown from just over 1,300 in 1994 to just under 3,000 in 1997.

Currently, WJ Beal Botanical Garden does not restrict the use that can be made of the seed it supplies by providing it under MTAs, but Elaine Chittenden has noticed a recent trend for other botanic gardens to supply seed subject to MTAs, some of which specifically mention the CBD. In the future, WJ Beal Botanical Garden will record which gardens use agreements, as it will not be able to place seed supplied subject to certain conditions on its own catalogue for unrestricted use by third parties.

Source: personal communication, Elaine Chittenden, Collections Manager, WJ Beal Botanical Garden, 26 February 1999.

thus too unreliable in quality to be marketed directly. To use wild materials therefore requires time consuming and costly breeding programmes to produce predictable and high quality varieties. This may be a viable option for some non-commercial organisations, but is generally uncompetitive and commercially unattractive for companies.

However, landraces form an important part of some collections, for example in many private collections such as the NCCPG system in the UK. Several institutions interviewed also maintained 'wild' origin material, frequently referred to by horticulturalists as 'species' (as compared with 'cultivars').

Commercial cultivars

With the exception of a small, but growing, minority of ornamental plants protected by patents, most commercial cultivars can be freely used by breeders in the development of new varieties. The most common source of new germplasm for breeders of ornamental varieties is thus commercially available cultivars. Where cultivars are protected by patents, some breeders will license the right to use them in breeding programmes. As one company explained, 'We regularly study the catalogues or information from sales people in other companies in order to find new and exciting cultivars. We produce about 90 per cent of our new cultivars ourselves, and we buy in about 10 per cent of the chrysanthemums from other breeders, under licences to propagate them'. Some interviewees said that it was sometimes difficult to tell whether commercial cultivars were subject to intellectual property rights, and hence the extent to which they could be used in research and breeding. As Dr Itzhak Ayalon, Director of the University of Jerusalem Botanical Garden, explained, his staff are not comfortable using materials received from companies for any purposes other than display. Companies are not always aware of the provenance of their materials, and may not know whether it is wild in origin or the result of a breeding programme. Consequently the garden feels it cannot be sure if it is entitled to use such material for research.

Other sources of material

Other than from their own collections and those of competitors, the horticultural trade receives new materials from a variety of sources, including individuals such as amateur collectors and gardeners; specialised plant societies; botanic gardens; universities and research institutes; public institutions such as national collections and genebanks, and commercial nurseries and other companies.

National collections and botanic gardens

Many horticulture companies obtain seeds or cuttings from national collections and botanic gardens, which form an important *ex situ* repository for materials sourced from the wild.

The majority of national collections and botanic gardens continue to distribute materials without agreements designed to involve source countries prior to commercialisation and to guarantee for them a fair share of benefits, but this practice is gradually changing. Some botanic gardens are among the first organisations involved in horticulture to begin to come to terms with the CBD and with the legal and political implications of suppling materials to companies for commercialisation. Countries are increasingly ensuring, through legislation and MTAs, that collectors – whether individuals or institutions such as universities and botanic gardens – do not collect under 'scientific' or 'research' permits, and yet pass the materials on to third parties who may commercialise them.

In response to, and sometimes in anticipation of, the CBD, several botanic gardens are endeavouring to acquire and supply materials under agreements both to satisfy legal requirements, and to protect their ability to collect in the future. As Dr Itzhak Ayalon put it, 'We [University Botanical Gardens, Jerusalem] have signed agreements that the material will not be used for commercial purposes, and if we commercialise we will need an extra permit'. This is often interpreted by companies as an effective ban on supply of materials by botanic gardens for commercial purposes. As one company explained, 'We tried to get material from Kew, Kirstenbosch and Leiden, but due to the CBD bureaucracy, it was impossible to obtain any'. However, many botanic gardens would be quite content to work in partnership with source countries and companies wishing to commercialise materials from their collections, provided this was consistent with the letter and spirit of the CBD. As Nigel Taylor, Curator of the Living Collections Department of the Royal Botanic Gardens, Kew puts it, 'If the companies are prepared to negotiate terms that provide the source country with a fair share of any monetary and non-monetary benefits that arise and are consistent with the CBD, national law and the terms under which the materials were acquired, that's fine. The trouble is, most companies have not needed to negotiate benefit-sharing arrangements in the past, and it comes as something of a shock to them'.

Amateur individuals and societies

> *'Most breeding is done by amateurs anyway and the nurseries depend on numerous private collections.'*
> *(Tony Lord,* Plant Finder*)*

The British Association Representing Breeders explained that, 'Amateurs are very important suppliers for new material either from their own private gardens or from private expeditions'. Several varieties initially developed by amateurs have led to significant commercial products. For example, Blooms of Bressingham Ltd evaluated and commercialised *Geranium pratense* Summer Skies, which was a seedling from the private garden of Kevin Nicholson, developed privately. Amateur gardeners are often talented growers, but lack the investment and marketing skills needed to launch a new product. Some commercial breeders make 'informal' links to hundreds of amateur growers all around the world. Company breeders may conduct 'site visits' to private gardens and local or national flower shows to look for new varieties that could be of commercial interest. Individual horticulturalists can also be important to companies, as some of them travel the world collecting plants.

Specialised plant societies provide another rich source of material for commercialisation. These groups act as a conduit for introducing new germplasm into the horticulture industry, and some societies organise collecting trips around the world, which may be sponsored by or even involve the participation of professional nurserymen and breeders. In this way, the material collected passes easily to the horticulture industry for potential commercialisation. It is highly doubtful that any of these expeditions involves access and benefit-sharing arrangements with the source country.

Choice of genetic resources

Three main factors drive companies to seek access to new germplasm: the hunt for novelty, for desirable traits such as disease resistance and to broaden the genetic base of existing collections.

Novelty and desirable traits

Many companies are constantly hunting for new, unusual products that will give them a competitive advantage. As one interviewee put it, 'The *"New Factor"* really dominates the market, but is difficult to achieve.' Many plant breeders are constantly on the lookout for new and exciting cultivars. Nurseries specialising in brand new introductions, and also in developing lines that have relatively recently been introduced to the commercial market, such as bromeliads and kangaroo paws, are likely to have a particular interest in acquiring more material, and in introductions from the wild. The popularity of different sources of new materials waxes and wanes over time, in keeping with the fashion. One institution explained that, 'the new Eastern bloc companies are becoming very interesting for nurseries as they might house new cultivars, for example the members of the International Plant Propagators Society in Poland, Bulgaria and Hungary, and institutions in the former Soviet Union'. Tony Lord of *Plant Finder* added, 'There are popular areas in South East Asia, where particular types of Geranium are very common.'

The most common spur for breeders to hunt for novel materials is the desire to find plants with traits such as hardiness in perennials, resistance to disease and pathogens, vigour, 'forcing' and other growth characteristics, fragrance, and ornamental virtues such as long stems, and petals and leaves with attractive patterns, shapes and colour.

Broadening the genetic base

Wild-sourced germplasm obtained many years ago with a narrow genetic base can sometimes form the basis for the development of products with large commercial horticulture markets. Some products have involved little more research and development than several generations of selection of the most promising strains from the progeny of the original wild-collected material. For example, there are thousands of plants of *Phalaenopsis violacea* (now known as *Phalaenopsis bellina*) in cultivation, nearly all of which have the same genetic background. In the early 1960s, a few plants were imported by a US-based horticulture company from the same wild population in Borneo. In 1964, a different US-based company selected two clones ('Country Acres' and 'Borneo') from this population to produce a seed population. This population was recreated and sold to consumers in 1967, 1968, 1971 and 1974. From this gene pool, many horticulturally superior clones with larger, fuller and flatter flowers were selected and used to produce additional populations of the species. Today, most of the clones of *P bellina* in cultivation around the world can be traced back to the same parents (personal communication, Dr Rob Griesbach, USDA, 18 February, 1999).

Another good example of the very narrow genetic base of much of the germplasm in the horticulture industry, this time with more profound conservation implications, is the case of *Cosmas atrosanguineus*. This chocolate-scented dahlia relative, originally introduced to the UK in the mid 1800s, was popular and widely grown in Victorian times, but fell out of fashion for a century. Over the last 20 years it has seen something of a revival, and today the plants are commonly found in nurseries, partly as a result of the efforts of the NCCPG. Endemic to Mexico, the species is now extinct in the wild. Dr Mike Fay and Tim Wilkinson of the Royal Botanic Gardens, Kew, have returned plants of this species to the Botanic Garden of UNAM in Mexico, where Professor Bob Bye and co-workers are endeavouring to secure its future. The plants available in collections in the UK do not set seed, due to incompatibility mechanisms, and so the search was on for other genotypes which would produce seeds. Sadly, molecular tests conducted at Kew have revealed that all the tested samples are genetically identical, making its successful reintroduction a forlorn hope unless other genotypes are found (personal communication, Dr Mike Fay, 26 Feb 1999).

One of the main reasons that companies need to broaden the genetic base of species in cultivation lies not so much in their desire to find new aesthetic characteristics or disease-resistance, but in order to secure fecundity. Hybrid plants propagated through sexual reproduction and the setting of seed (rather than through vegetative propagation) often suffer from a major problem: the fact that they often do not produce many seeds, yet large quantities of seed are required to produce hundreds of thousands of plants needed for the commercial market. Genes from the wild can help to improve fertility.

Local knowledge

Ethnobotanical information is of comparatively little value in the research and development of ornamental varieties, as the industry is looking for interesting aesthetic aspects (eg flower colours, size, etc.), but not for the 'story behind' the plant. In some cases, however, ethnobotany can help to create a marketing advantage. Ian Ashmann of the British Association Representing Breeders tells the story of 'the Robyn Hood Oak', a product developed by Dr Neal Wright, proprietor of Micropopagation Services Ltd, from the very plant on which that great popular hero is said to have stood. The product is sold to consumers on the basis of the folk story behind it.

Local knowledge about plants can be relevant to breeders of fruit and vegetables, who use knowledge derived from 'lifetimes of local experience', including recording data such as climatic conditions during field collections, searching databases to identify potential traits or eliminate materials from research. Ethnobotanical information is also crucial for many collections in botanic gardens. One botanic garden described looking for plants with certain uses by local people, such as resin plants from the Amazon. The garden uses ethnobotanically interesting plants in its collections to stimulate interest in other civilisations. Ivor Stokes of the National Botanical Gardens, Wales, concurred: 'In some cases, the importance of ethnobotanical information will outweigh the ornamental significance of a plant. One will look for plants that are of significance for medicine, food, fibre, building, etc.'

'Raw' genetic resources, or value-added materials

From earlier sections, it will be apparent that companies obtain a wide diversity of materials, ranging from wild-collected, unadapted materials, to commercial cultivars. And while this accounts for a minority of the materials they acquire, companies also acquire pre-bred and pre-selected material from amateurs. One respondent believed that, in the future, there might be a growing market for pre-selected material or pest-resistant varieties for use in horticultural breeding programmes. For several seed companies and firms specialising in fruit and vegetables, improved materials are the main source of material they use from outside their own collections. The UK Agricultural Supply Trade Association explained that, 'The new cultivar has to be a best seller, and it is difficult to justify the cost without using value-added material.'

6.3 Benefit-sharing

'Benefit-sharing' is largely unknown in the horticulture sector, as, indeed, are the implications of the CBD itself. In common with agriculture, there has long been an ethos of 'free exchange of plant material'. While the advent of intellectual property rights has somewhat constrained the 'free' use of plant materials for commercial development, there persists, on the part of plant collectors, a strong feeling that the interests of horticulture and of conservation are best served by spreading new and interesting materials into as many hands as possible, including those of commercial nurserymen. This exchange is frequently conducted without money changing hands, or for a modest fee that simply covers the handling costs. As one interviewee put it, 'There is ample exchange of material with other research institutions world-wide. This exchange is done on a bona fide basis and no money is paid.' The feeling that horticulturalists are engaged in aesthetic work far removed from sordid financial matters still prevails in some quarters. Alain Boggio of Elidia explained, 'I am an artist, not a cheque book.'

However, while specific benefit-sharing arrangements with source countries of germplasm are the rare exception rather than the rule, there is some experience, within the horticulture industry, of the sharing of monetary and non-monetary benefits.

6.3.1 Monetary benefits

Current practice in monetary benefit-sharing

Despite the continuing ethos of free access, the breeders interviewed reported an increasing tendency for companies and other plant breeders to pay for samples. Today, charging for access is common, where ten years ago this was not the case. As one interviewee put it, 'In the past it was common to obtain cuttings or samples from botanic gardens or private collections for free, but in the last 15 years it has became increasingly common to pay for it. There are royalty fees to be paid, especially in the rose industry, and people are prepared to pay. Pirating (stealing material from gardens, collections, etc) is still going on, but is on the decline. Nursery owners can see the benefits of paying the "supplier".'

Even non-commercial organisations such as botanic gardens are accustomed to paying for materials on some occasions. Where commercial materials are used for display, gardens are accustomed to paying commercial rates. Some have established annual budgets for the acquisition of materials, to cover payment for samples and shipping costs. Some public-sector breeders also charge for the use of their genetic resources. From time to time, they announce the availability of new material, and put it up for auction, inviting bids from the commercial horticulture industry. The highest bidder secures the rights to use the material in its breeding programme.

However, it was clear that several interviewees were not accustomed to paying for information. In reponse to our question 'Do you pay for access to genetic resources?', several interviewees understood the question to refer to compensating a collector for his or her activities, and said that they do not pay for such services, other than covering travel expenses. Thus companies pay what is necessary (and competitive) to obtain plant samples from the wild, or to obtain commercial material from a garden centre or nursery, but, as one interviewee put it, 'have not made any payments in relation to the Rio Convention'.

Where companies acquire proprietary materials developed by other breeders, such as strains with proven characteristics or finished cultivars, it is common for compensation to be arranged under a royalty scheme. The British Association Representing Breeders explained that, 'Plant Variety Rights operate for 25 years, and an efficient royalty fee system during that period can ensure that the breeder receives revenues from sales of the protected varieties. Breeders frequently use this to invest in ongoing breeding programmes'. Companies were quick to point out that the money to be made from such royalty schemes is not vast. The 'normal' introduction volume of a new ornamental plant is between 10,000 and 100,000 units. On a unit price of US$7, a royalty fee of 2 per cent would result in returned revenue to the supplier of US$14,000 in the first year. A 'blockbuster' new cultivar would have a maximum turnover of US$3,000,000, which, at the same rate, would result in a royalty payment of US$60,000. An alternative basis for payment to royalties is a flat fee calculated per stem or per bunch sold.

The level of payment is influenced by a number of factors, such as the size of the market, the potential market share of the cultivar concerned, its lifespan on the market, how fashionable the product is, current market rates for payments, and the degree of improvement effected by the ultimate breeder compared with the provider of the material. However, for a number of reasons, quantifying appropriate payments is difficult.

First, the pedigrees of horticultural varieties are frequently complex, and often combine a large number of parents. All of these will have contributed to the genetic make-up of the final variety. While the added value of the resulting variety may sometimes be ascribed to one or a handful of genes from 'new' genetic material (for example, a gene from one plant coding for a new colour may contribute all the novelty when incorporated into an existing variety of another plant), it is more common for genetic material from many parents to contribute towards the desirable features of the new variety ('Orchid breeding and registration: a case study', p180 below). As one director of a seed breeding company put it, 'the other 99 per cent of the genome is crucial to any success'. Second, the horticulture market is well known to be fickle, and highly susceptible to fashion. As the UK Agricultural Supply Trade Association explained, 'Every year there is a new fashion. Especially the bedding plant market suffers from the unpredictability and this makes it difficult to fix a set fee for a product. It is therefore difficult to settle an appropriate market rate that lasts for longer than a year. Products are short lived and if they are not extremely successful, they disappear from the market.' Finally, several interviewees commented that monitoring and enforcing benefit-sharing mechanisms would be difficult, judging from experience in the horticulture industry with royalties, which have proved 'extremely difficult to police'.

When asked if they would be prepared to pay more to obtain new materials for horticultural breeding efforts than they do currently, many respondents said that they might do, but only if the quality of the material justified it. Some of the reasons given by respondents to justify their willingness (or unwillingness) to pay more for access to genetic resources are set out in Box 6.8 p178. Other than bilateral negotiation of monetary benefits through fees and royalties (which currently stay in the commercial sector and do not pass 'up the chain' to the initial providers of genetic resources), there are few partnerships for the development of ornamental varieties that involve the sharing of monetary benefits (but see 'The Pipa Horticultural Company Ltd: a case study', p183 below).

Several horticulturalists interviewed felt that the economics of the horticulture industry had profound implications for the value of benefits that could be shared for the development of a new horticultural variety and appropriate mechanisms for sharing them. As one company put it, 'Compared to the rich industries, like medicine there is little scope to scratch money together. Eventually the vegetable and fruit industry might find the CBD significant, as their products are

Box 6.8 *Would You be Prepared to Pay More to Access Horticultural Materials?*

Yes: some reasons why

- Quality is more important than price. If the quality justifies it, we would be prepared to pay more.
- More financing is available within the large companies that have resulted from recent mergers. Multinationals can afford to pay more for access to genetic resources.
- Novel materials merit higher prices.
- Many institutions continue to provide genetic resources to companies at rates below their market value.

No: some reasons why not

- Competition between companies and even countries will keep the prices down.
- The horticulture industry is 'already squeezed to its limits'.
- The overall budget for buying plant material is squeezed by the enormous shipping costs.
- 'Access to germplasm for research' is not a concrete enough benefit to justify more payment.
- The profit margin in horticulture is getting lower.

economically more important. The overall value of ornamentals is too small in the main. If an economic return could be shown, it might be a different matter.' Yet it is common for just a few cultivars to dominate the ornamental market at any given time. For example, the Reagan 'family' of chrysanthemums alone comprises more than 50 per cent of the Dutch chrysanthemum production (based on auction turnover) – over US$1 billion per annum (Flower Council of Holland, 1997). For such high value products, sufficient sales are generated to justify the transaction costs of establishing some kind of benefit-sharing arrangement, where they are based on genetic resources obtained from source countries. However, some kind of mechanism for benefit-sharing that took into consideration the relatively modest market for ornamental horticulture and the kinds of transactions involved would need to be designed, since none currently exists.

This section will close by describing one mechanism (the Horticultural Development Council levy) used in the UK to raise funds for other purposes, which could possibly inspire new approaches to benefit-sharing, and another voluntary measure proposed (by the Botanic Garden Trading Company Ltd) for the retail flower relay service.

The Horticultural Development Council (HDC) was set up in 1986 by the UK Government following a decision that the horticulture industry, as a beneficiary of research that is publicly funded, through the MAFF, should contribute to the cost of that research. The Council is responsible for the collection of funds through a levy on the turnover of UK horticultural produce, and commissions research on the industry's behalf. Under the terms of the parliamentary order which set up the HDC in 1986, every grower with sales of his own product (excluding haulage and packaging costs) amounting to more than US$40,000 a year is required to register with the HDC. Registered growers are then asked to make an annual return of the value of their sales which, after specified deductions, is used to calculate the levy due. The rate of the levy is set annually, with the approval of the Minister, up to a maximum of 0.5 per cent of net sales turnover. At the maximum rate, the levy generates about US$5 million a year. The levy represents about 20 per cent of the annual budget for horticultural research in the UK, with the remaining 80 per cent provided by the MAFF (HDC, 1998a, p51). There are six sector panels of elected grower members who are responsible for approving research projects. Each panel can coopt or invite experts to advise it; however these experts cannot vote on the proposals. The panels meet regularly to discuss research proposals put forward by universities, research institutes such as Horticulture Research International and ADAS, an advisory service for the agricultural and horticultural industries in the UK, and other bodies involved with horticulture. These proposals are often submitted as a result of problems encountered by growers. Alternatively, they may be for innovative work that will enhance UK horticulture. In some cases, where a piece of work has been identified as being needed, advertisements are placed for appropriate bodies to submit proposals, and these proposals also go forward to the panels for consideration. Once a contract has been placed a grower co-ordinator is nominated to oversee the project with the assistance of one of the HDC technical managers. Effective dissemination of the project results is central to the workings of the HDC (personal communication between China Williams and Fiona Sheppard, HDC, 10 March 1999).

A mechanism that has been proposed for benefit-sharing by the Botanic Garden Trading Company Ltd is to use voluntary royalties on profits from sales of horticultural products to support conservation and also to pay competent national authorities 'licence fees' in return for access to genetic resources in biodiversity 'hotspots' for commercial horticultural development. The Botanic Garden Trading Company Ltd is a start-up company which aims to become a world-wide flower relay and delivery service in competition to existing international delivery services such as Interflora. The global flower relay industry is currently worth over US$7 billion a year. The company sells only commercially available cultivars, and, despite its name, there is no link to the collections held in botanic gardens. However, the company's Chief Executive, Andrew Ross, says that 2.5 per cent of the company's turnover will be paid into the Botanic Garden Foundation, a not-for-profit charity which he is establishing. The Botanic Garden Foundation will fund *in situ* conservation of species of horticultural interest, and Mr Ross also proposes to use it to pay licence fees to governments for access to genetic resources for potential horticultural development. He believes that there is little sign that the companies themselves developing new ornamental varieties are prepared to pay royalties to countries from where the genetic resources were obtained, and proposes instead to gather funds at the retail level, in this case through flower relay and delivery services (personal communication, Andrew Ross, November 1999).

6.3.2 Non-monetary benefits

'Benefit-sharing' is not the norm in the horticulture trade, but a number of 'in kind' benefits are exchanged in the context of collaborative partnerships. Among these, reciprocal access to plant material between non-commercial organisations is important. Another common form of benefit-sharing is the acknowledgement of the name of the provider of genetic resources (whether a grower or amateur) in the name of the cultivar subsequently developed by a breeder, such as *Digitalis purpurea 'Anne Radetzki'*. This practice could be adapted to reflect the country of origin, or even the name of the institution or community which provided the germplasm, although for plants with complex pedigrees involving germplasm from many sources, this may not be workable.

Botanic gardens and other research institutions often acknowledge the provider or country of origin material, through signage in the gardens and through other means of raising public awareness. Some botanic gardens are considering expanding their repertoire for acknowledging support. For example, one plans to offer advertisement and exhibition space to the organisations that contribute to its establishment and displays, and this could include information about countries and institutions providing genetic resources.

Some companies have sponsored PhD programmes for students from countries that have supplied them with germplasm. The PhD is then filed with the company as a source of reference. In terms of sharing the information that arises from breeding programmes with the countries or institutions that provide the genetic resources, larger companies tend to be very private about the information they generate and unwilling to divulge it, whereas smaller companies and amateurs are more willing to share their experiences and results.

Botanic gardens have considerable experience in offering 'benefits in kind' to collaborators providing plant samples for display in their collections. Peter Thoday at the Eden Project describes how the organisation offers student placements and ships books, computers and other equipment to collaborators who face difficulties in obtaining such materials themselves. According to the policy of the Royal Botanic Gardens, Kew, on access to genetic resources and benefit-sharing, what is possible, and thus fair and equitable, varies from case to case, but the different kinds of benefit that it may be able to share include:

- taxonomic, biochemical, ecological and other information, through research results, publications and educational materials;
- benefits in kind, such as augmentation of national collections in the country providing the genetic resources;
- transfer of technology such as hardware, software and know-how;
- training in science, *in situ* and *ex situ* conservation, information technology and management and administration of access and benefit-sharing;
- joint research and development, through collaboration in training and research programmes, participation in product development, joint ventures and co-authorship of publications;
- paid use of local guides, scientists and facilities; and
- in the case of commercialisation, monetary benefits such as royalties.

Several interviewees said that horticulture companies rarely 'outsource' their research in situations in which technology transfer from the company to the researcher might occur, citing the competitiveness of plant breeding for horticulture as the reason for the paucity of collaborative research projects. However, the work of public and private bodies in developing production protocols, and the gradual siting of not only production facilities but a few breeding facilities in developing countries suggest that there might be opportunities for technology transfer and joint research. The following example also shows how joint research can provide the basis for sharing technology and information, as well as royalties.

During the 20th century, wild germplasm of Ornithogalum from South Africa was collected and displayed in botanical gardens throughout the world. In 1988, USDA, the University of California Botanical Garden in Irvine and two US horticulture companies, one of which develops cut flowers and the other potted plants, initiated a cooperative research project to develop hybrid cultivars from the existing germplasm held in US botanical gardens. The role of USDA was to develop enhanced germplasm through the use of embryo rescue technology, and to develop production protocols so that the finished plant varieties could be produced at commercial volumes and quality. The University and the companies made crosses with the species at the botanical garden. The seed capsules were sent to USDA for embryo rescue, and the resulting seedlings evaluated by USDA, the University and the companies. In 1993, the collaboration was expanded to include the Agricultural Research Council of South Africa (ARC), which obtained technology and information through the collaboration. Joint research between the USA and South Africa focused on the development of embryo rescue, molecular taxonomy and genetic engineering technology. The research in the USA focused on developing potted-plant types, while that in South Africa concentrated on the development of types of cut flowers. Funding for the research in the USA was provided by both the private horticultural companies involved and the US Government. Funding for the research in South Africa was provided by both the US and South African Governments. Today, both the USA and South Africa have released new cultivars of both potted plants and cut flowers. Although there was no royalty-sharing arrangement in the 1993 collaboration, because the ARC's involvement began late in the project, a subsequent joint research programme between USDA and the ARC on Erica involves a 50/50 profit-sharing arrangement (personal communication, Dr Robert Griesbach, USDA, 18 February 1999).

6.4 Case studies

6.4.1 Orchid-breeding and registration

Introduction

There are centres of orchid diversity all around the world. In Asia and the Pacific, the main centres of diversity of wild orchid species are the Himalayas, China, Malaysia, Vietnam, Thailand, Laos, the Philippines, Borneo, Sumatra, New Guinea, the Solomon Islands, Fiji, Samoa and Japan. Other important centres of orchid diversity are found in Australia, in Western Australia and Northern Queensland, in Mexico, through Central America, and throughout the tropical regions of South America, including Brazil. Tropical Africa and Southern Africa are home to many native orchid species, and others are found in North Africa, the Mediterranean and the rest of Europe including the UK, and Ireland. There are also many species in the USA and Canada, including several arctic species.

While orchid species have been moved around the world through trade and collection since the 18th century, the first known example of the breeding of a hybrid orchid was in 1854, when John Dominy, the orchid gardener of John Veitch, proprietor of a nursery near Exeter in the UK, crossed two Indian species of *Calanthe* and produced the first 'bi-specific' hybrid. Veitch had been growing the parent plants in his orchid nursery in the UK, since orchid cultivation was a popular pastime among the amateur gardeners of the 19th century. In about 1820, botanists observing orchids in nature had realised that orchids hybridised in the wild, and Dominy's hybrid variety marked the beginning of a series of experiments in inter-specific, and, subsequently, inter-generic, orchid breeding. In 1863, Veitch crossed an orchid of the genus laelia with one of the genus cattleya to make the first bi-generic 'laeliocattleya' hybrid, and thereafter the early experiments focused on crossing similar species in different genera. When amateur breeders realised that this worked, they became more adventurous, and there are now nonageneric hybrids, whose pedigree includes parents from nine separate genera.

The private sector

The commercial development of orchids started in the 19th century, particularly in the UK, where nurseries such as Sander, Low, Armstrong and Brown, McBean and several others started crossing orchids for the small-scale production of flowers for ornamental arrangements. Development of the commercial market at the international level started shortly before the Second World War in Hawaii, Thailand and the USA, and took off in earnest in the USA, Taiwan and Japan after the War. By 1957, the triennial world orchid conferences were attracting substantial interest and participation from private horticultural companies. To this day, these conferences combine lecture series and exhibitions of orchids and are attended by representatives from orchid nurseries and amateur enthusiasts from around the world. During the last 30 years, the number and size of horticultural companies breeding orchids have increased, and there are probably some 800 active commercial orchid breeders around the world today.

The principal countries engaged in orchid breeding are Japan, the USA, Australia, Thailand, and Taiwan, although there are also important groups of breeders in New Zealand, some South American countries, and indeed smaller numbers of orchid breeders in many countries around the world.

The registration of orchid hybrids

Starting in 1893, the *Orchid Review*, a monthly report containing lists of all known hybrid orchids, botanical information and information of interest to orchid growers, breeders and botanists, was produced by Robert Rolse, the orchid botanist at Kew, supported by the major commercial orchid growers of the time. Other botanists and growers followed him, and subsequently the publication was taken over by the Royal Horticultural Society (RHS UK) at the end of the 1980s. The private (UK) Sander company was also keeping records, and in 1946 compiled a consolidated list of orchid hybrids produced artificially between 1856 and 1945.

In the 1950s, Gordon Dillon and Leslie Garay, working at the Oakes-Ames Herbarium of Harvard University, drew up draft rules of nomenclature and also procedures for the registration of orchid hybrids. Independently, the Sander company had developed application forms for breeders to register new orchid hybrids for inclusion in the next edition of Sander's list of orchid hybrids. These two initiatives were then combined and developed into today's system of orchid registration. Under the auspices of various bodies coordinated by the International Society for Horticultural Science established under UNESCO, various international registration authorities for cultivated plants have been established. In 1960, a meeting of the International Orchid Commission at the third world orchid conference, held at the RHS in London, decided to confirm the Dillon-Garay system of nomenclature and registration for orchid hybrids (known as 'grexes'), incorporating relevant sections of the international codes of botanical and horticultural nomenclature. The Commission also decided that the RHS should establish a formal International Registration Authority for Orchid Hybrids (IRAOH), taking over the informal system operated up till then by the Sander orchid nursery. Over the last nine years, the IRAOH has consistently registered the names of some 275 new grexes each month.

An individual who breeds a new hybrid may register the grex name him or herself, or may sell the plant to a commercial horticulture company which might register the name on its own behalf. While there is no legal requirement to register new grexes, the IRAOH believes the great majority are now registered, as there are several important incentives for doing so. The main reason is that no cultivar of an orchid hybrid is eligible for an award of merit (which helps to create market demand) from the major awarding bodies such as the RHS (UK), American Orchid Society (USA), New Zealand Orchid Council, or German Orchid Society, unless it has a full name which includes the registered grex name and a cultivar name. Grex names are like a condensed formula, or shorthand for a vast family tree. The grex names are not translated, but identical around the world.

Of 3,000 grex registrations in 1998, one-third were to multinational horticulture companies, another third to small horticulture companies and nurserymen, and the final third to amateur individuals – which indicates the role of each of these groups in breeding new orchid hybrids. However, the IRAOH is finding it increasingly difficult to distinguish between small firms and multinationals, since small companies may apply to register a grex, even when they are owned by larger corporations. Also, amateur individuals who register grexes may pass them on to commercial growers.

Source material

Of 300 grexes registered in January 1999, 20 (7 per cent) were primary hybrids (ie a species crossed with a species); 100 (33 per cent) were made by crossing one species parent with one hybrid parent, and the balance of 180 (60 per cent) resulted from crossing a hybrid with another hybrid. The latter category is slowly increasing, as the number of potential hybrid parents grows. Mr Peter Hunt, International Orchid Registrar of the IRAOH, believes that, on average, about 20 per cent of grex registrations use 'species' as one or both of the parents. The majority of these 'species' are improved forms of materials originally taken from the wild, often 100 years ago or more, and subsequently selected and bred with siblings and back-crossed with their parents and other ancestors. The remaining 75 per cent (representing the work of about 1,200 people in total in many countries) use proven grexes almost entirely as their raw material. Although registrants of new grexes are invited to state the names of the exact cultivars of the parent species they have used, they are not obliged to do so. If registrants choose to reveal the identity of the cultivar parents used in the breeding programme to the IRAOH, their consent is asked prior to the information being revealed to enquirers.

Mr Hunt estimates that some 99 per cent of the plants of species origin used to create new grexes are taken from existing stocks, rather than from the wild. Amateur individuals and commercial horticulturalists exchange a great deal of material, and can also purchase commercial cultivars, with their grex names, on the open market in Thailand and other countries. Several developing countries, including Thailand, Taiwan, and Brazil, cultivate and sell orchid species and hybrids. The market is fairly international, and the varieties sold in these countries are just as likely to have originated on the other side of the world as to be of domestic origin. For example, many breeders in New Zealand specialise in crossing South American plants; Brazilian breeders sell *Dendrobium* species and hybrids, which are Asiatic in origin; and *Cattleya*, of South American origin, are commonly sold by Taiwanese breeders. Some breeders work with native genera, but the materials used in different parts of the world are governed more by the agroecological conditions favourable to their growth than by their ultimate origin. For example, a large number of Cymbidium species, many of which originate from the Himalayas, require cooler temperatures, and are thus less likely to be grown in the tropics.

The use of orchid species in breeding programmes

Many orchid breeding programmes are complex. Modern day grexes can be the result of from 1 to 20 generations of breeding, during which between 2 and 20 different species from up to nine different genera might enter a plant's ancestry. Some 7,000 species of orchid are held in public or private collections around the world (compared with 25,000 species in the wild), but currently, only about 2,000 species have been used as parents of grexes. Many primary hybrids (both of whose parents are species) continue to be made, and species continue to be incorporated relatively recently in the breeding history of many other hybrids.

Breeding programmes today using species involve a mixture of making new primary hybrids, remaking earlier primary hybrids with 'better' parental cultivars, and introducing species into already existing and often well-established, well proven grexes. However, most primary hybrids are bred just to see what kind of characteristics the progeny possess and do not enter into the horticultural trade in any significant manner. There are a few exceptions, where particular species or primary hybrids make a significant and lasting contribution to the trade. When a plant is collected from the wild, it is grown in a glasshouse for a considerable time to observe its characteristics and see if it has any faults. Use of species material is thus often based on in-breeding and stabilisation of particular traits that result in more desirable products. Once hybrids are made, there is often little recourse to the species parents. The following examples serve to illustrate these points:

- Both *Phalaenopsis amabilis* and *aphrodite* are significant in the potted plant trade serviced by Taiwan. However, the best breeding plants are the result of two or three generations of in-breeding to produce superior lines. While both *Phalaenopsis amabilis* and *aphrodite* are species, not hybrids, each parent will have been bred for at least four generations prior to being used as parents, so some would question whether or not they represent 'true' species. Another example is the species *Paphiopedilum delenatii*, which was originally difficult to grow, but through generations of selection has become widespread in cultivation. *Paphiopedilum delenatii* is also an example of wild-collected material entering the trade through the 'back door'.
- *Paphiopedilum Maudiae*, a primary hybrid initially formed by crossing the species *callosum* and *lawrenceanum* – Charlesworths 1900 – is produced today by crossing the hybrid plants with themselves or their siblings, and not returning to the two species parents.

These cases demonstrate that while access to the species was necessary initially as the source of new parents, 'species' parents are often selected and even crossed with themselves for several generations prior to being used in a breeding programme. Thereafter, hybrids are generally produced by crossing with themselves or siblings, so there is no demand for further access to the species, either for development of the resulting products, or for mass production for the market. As Ned Nash, Director of Education and Conservation at the American Orchid Society, put it, 'The vast majority of plants traded internationally are hybrid orchids of garden origin. They are no closer to species than the roses in your front yard, or the lettuce on your plate. Most "species" material used in breeding has been refined over several years, and hasn't seen a rainforest in several generations.'

Once a new orchid variety has been successfully bred, it takes a considerable time between sowing orchid seed and the flowering of the plant destined for the market as a potted plant or cut flower. An orchid capsule contains up to one million orchid seeds, but each of these may contain as few as two to four cells. These microscopic seeds are placed on agar dishes in sterile conditions with controlled temperature and humidity. It can take up to a year for the cells to start germinating. A second year may pass before the germinated seed produces an identifiable plantlet, and a further two years before the plant grows to any reasonable size. After that, it can take several more years for the plant to flower. Biotechnology has helped to reduce the flowering time of some plants to 18 months.

Box 6.9 shows the first of 16 pages of the family tree for the orchid hybrid (grex) *Paphiopedilum San Dollisco*, which was registered in January 1994. One of *P San Dollisco*'s parents was a cultivar of a species named 'concolor'. However, although one parent was a species, descended from a wild plant, it took 12 generations to breed the other parent (the hybrid 'San Francisco'), so *P San Dollisco* is the result of 13 generations of breeding. Orchid pedigrees can be analysed using an internationally accepted means to measure the percentage contribution of antecedent orchids to each new hybrid, and to count how many times each species was introduced in each generation. Each generation of breeding is counted as 100 per cent, so the total contribution of antecedent orchids to *P San Dollisco* is 1,300 per cent. In total, 455 cultivars of species and hybrids, including seven different species used 228 times, contributed to the genetic composition of *P San Dollisco*. The species 'concolor' was used once in generation two (as one of *P San Dollisco*'s parents) and the six other species were used at various times in earlier generations. *P San Dollisco* does not have a particularly complex pedigree. All the hybrids and species used were from one genus

Box 6.9 *Orchid Pedigree Chart*

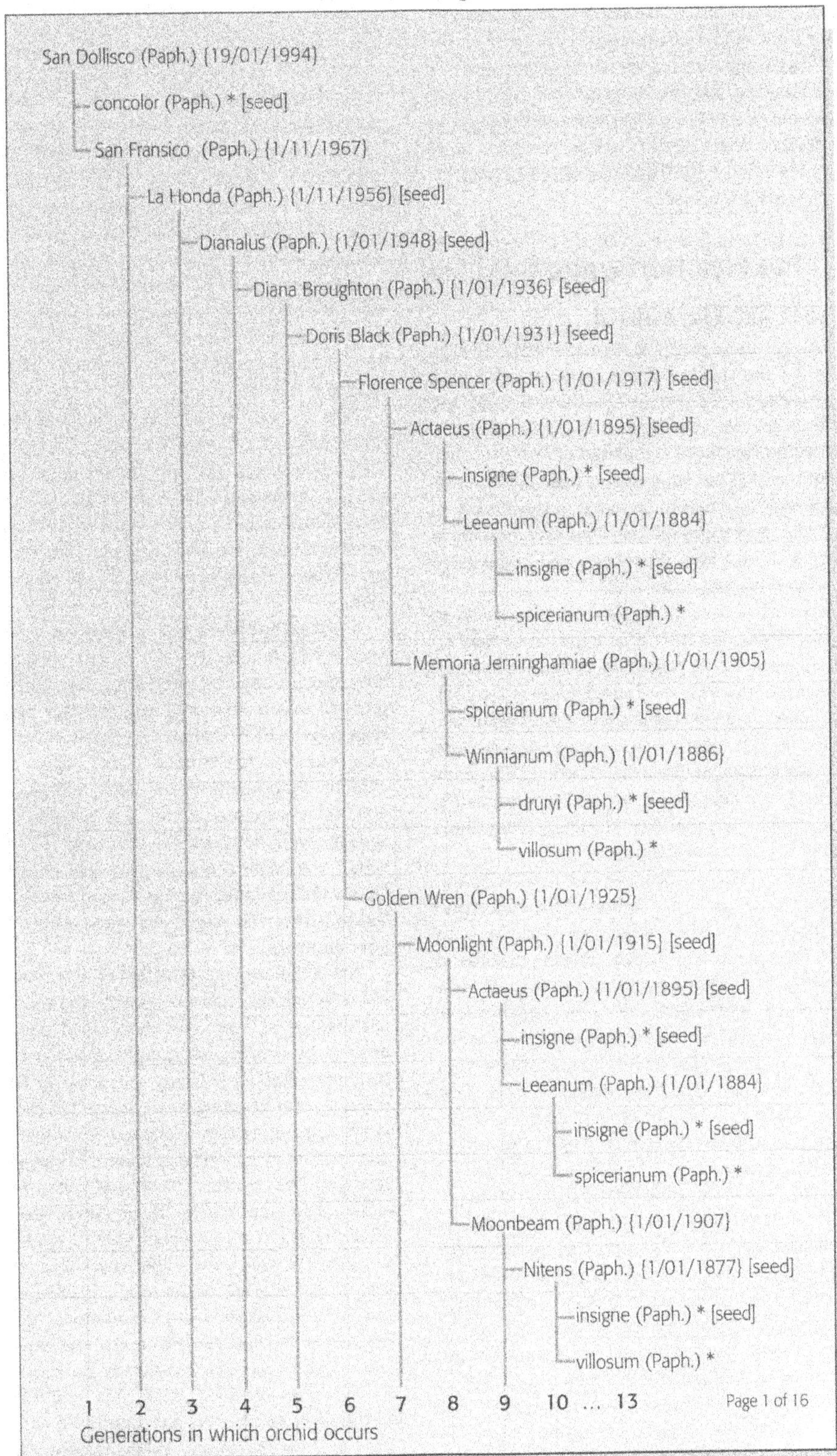

Source: The Orchid Database Company Pte Ltd, 1998.

(*Paphiopedilum*). We also considered using as examples the nonageneric hybrid *Gladysyeeara* Jocelyn, whose pedigree runs to only three pages, but involves nine generations of breeding, incorporating 83 elements, including 22 different species from nine different genera, or Baldan's Sunset Prism (Phal.), whose pedigree is 48 pages long, involving 1,365 elements, which include 105 different hybrids and species.

6.4.2 The Pipa Horticultural Co Ltd

The activities and plant material

This case study concerns a joint venture company, Pipa Horticultural Company Ltd (Pipa), established by the Nanjing Botanical Garden Service Station (NBGSS), and Piroche Plants Inc (Piroche Plants). The major negotiations concerning Pipa were completed before the entry into force of the CBD, so changes have been made subsequently, and will continue to be made in the future, in order that the operations of the company remain in accordance with the new guidelines and regulations established with respect to the CBD.

Piroche Plants is a horticultural company based in Vancouver, Canada, and NBGSS is an organ of Nanjing Botanical Garden, China, relatively independent from the garden itself. Its role is to conduct some of the commercial activities involving the garden's plant material.

In 1992, Piroche Plants first received some plant specimens from Nanjing Botanical Garden. The company's Director then decided he would be interested in receiving a steady supply of tree seedlings from China to sell in Canada and other countries.

Two years of discussions between NBGSS and Piroche Plants led to the identification of a common purpose: the commercial development of Chinese plant resources. The NBGSS and Piroche Plants established Pipa with the aim of developing and introducing new plant species for landscaping, forestry, and horticulture. A board of three directors was immediately established for Pipa, composed of two representatives from Piroche Plants and one from the NBGSS. For the first term of the joint venture, from 1992 to 1996, the President of Pipa was also the director of Nanjing Botanical Garden. For subsequent terms, the President is elected by the Board. Under the terms of the joint venture agreement, Piroche Plants contributes 90 per cent of the US$1 million capital for the joint venture company, and the remaining 10 per cent is contributed by NBGSS. Pipa has used this investment capital to hire eight full-time members of staff, currently from the staff of NBGSS, as well as a number of temporary staff recruited locally in Nanjing, from outside the NBGSS.

In accordance with Chinese laws, regulations and government guidelines, Pipa staff obtain, from the collections maintained in Nanjing Botanical Garden and also from field collections all over China, propagatng materials of interesting plant specimens which are officially allowed to be sent out of the country. Seeds for propagation, including those of wild specimens from the mountainous regions of China, are collected *in situ* by Chinese staff of Pipa or by local people, while other species are purchased in the marketplace at market rates. The plants are taken back to the Pipa premises at Nanjing, where they are propagated, planted, transplanted and seedlings raised from them. The work conducted on the plant materials thus far has been restricted to cutting and grafting. There has been no hybridisation or breeding activity, although this is envisaged for the future.

The number of seedlings sent from Pipa to Piroche Plants and its customers world-wide varies. Recently, the Chinese customs authorities have restricted the export of seedlings following some phytosanitary problems with nematodes. The number of seedlings sent by Pipa to Canada has thus dropped, but staff from Pipa and Piroche Plants hope that shipments will resume in 1999.

Once in the Piroche nurseries in Vancouver, the seedlings are bulked up through further propagation. Some new species and cultivars are subjected to a series of tests to assess how well adapted they are to environments in the USA, Canada and Japan, which are the principal markets for Piroche Plants. Initial tests assess climate adaptation. From 1996 to 1998, more than 100,000 trees were planted out in fields for testing to see how well they perform in all seasons. Piroche Plants sends seedlings to a number of different areas in the USA and Canada for testing so that the company is comfortable that the plants will cope well in a range of environments.

Nevertheless, the majority of the seedlings tested are rejected, for example because the seedlings display low disease or insect resistance, inadequate hardiness, or lack the ornamental quality required for the North American market. It is not unusual, over time, to bring some 10,000 seedlings to Canada for testing, and to select just two or three for further evaluation and commercial use. Piroche Plants takes cuttings from the few seedlings that perform best, and raises them as new cultivars for the market. Some plants are sold in pots after only three or four years. Other trees will take up to ten years to raise to a stage where they are ready for sale on the market in Canada, the USA and Japan for prices ranging from a few Canadian dollars to two hundred Canadian dollars. The company's expenditure and testing on large-scale cultivation and experimentation, including the cost of plants and material, has been between Can$600,000 and Can$1,000,000 each year, for the last few years. While Piroche Plants is thus currently subsidising its research and development of new products, the company anticipates that, in the years to come, more trees from the new, improved selections, will grow to the stage where they can be sold, and the company will see a return on its investment.

Piroche Plants has full rights over the plants brought from China. In the meantime, Pipa is entitled to sell the same plants in the domestic market in China. Although the main business of Pipa so far has been to send Chinese materials to Canada, Piroche Plants also sends landscaping plants from Canada to China, where Pipa propagates and sells them in Chinese markets.

Policy and legal context

Since Premier Li Peng signed the CBD at the Rio Earth Summit in June 1992, China has introduced a number of laws that affect the acquisition, import and export of genetic resources (Xue Dayuan, 1998). The Wild Plants Protection Regulation of 1 January 1997 prohibits illegal collection of wild plants and the destruction of their habitats (Article 9, Wild Plants Protection Regulation, 1997). The State bans the collection of plants 'of the highest category of protection', requiring an individual seeking access to them for 'scientific research, artificial breeding and other special uses' to obtain permission from the local provincial government and then from the responsible Ministry under the State Council. An applicant seeking access to plants with the 'second grade' level of protection must obtain a permit from the authorities in the county local to the proposed collecting activities and then at the provincial level. Applications to import and export wild plants protected by Chinese law or CITES need approval from the local provincial department and then the State Council. The export of 'unnamed or new wild plant species with important values' is banned, and foreigners are prohibited from collecting or purchasing wild plants protected by Chinese law. Field investigations in the habitat of such plants require permission from provincial authorities and approval from central government.

Another law from 1991 subjects the exchange with foreigners of plant reproductive materials (such as seeds, fruit, nuts, roots, stems and seedlings) for use in agriculture and forestry to rules issued by the Ministry of Agriculture and the Ministry of Forestry, which date from June 1991 (Article 10, Regulation for Seeds Management, 1991). Their exchange for research is administered by the Institute of Crop Germplasm Resources of the Chinese Academy of Agricultural Sciences. Any Chinese organisation or individual providing crop germplasm to foreign countries is required to follow certain formalities established in rules that vary according to the nature of the germplasm. Similar provisions exist with respect to tree seeds (Detailed Enforcement Roles of the Regulation for Seeds Management on Tree Seeds, 1995) and domestic animals and poultry and their 'genetic materials' (Article 2, Regulation of Breeding Stock and Poultry Management, 1994).

These laws appear to contain no explicit provisions related to benefit-sharing. However, the Chinese Government is considering reviewing its regulation of access and benefit-sharing, and the existing requirement to obtain permission for access provides the opportunity for the Chinese authorities and an applicant for access to reach 'mutually agreed terms', including those relating to the sharing of benefits, as provided in the Convention.

Members of the Pipa board applied to the Provincial Government Department in charge of establishing private enterprises, which approved the establishment of the company. In addition, Pipa must apply to different authorities in order to obtain the phytosanitary certificates (required for the export of all plants) and the export permits needed for rare and endangered species and those deemed to be of economic importance. The Ministry of Forests, the Environment Bureau and the Ministry of Agriculture all control phytosanitary permits, usually through agents who represent all three agencies. These agents are represented at Customs points in China, and they examine plant and seed sent for export against the relevant rules. The three Ministries and the Chinese Academy of Sciences work together to compile the red data list of rare and endangered species. The export of these species, of CITES-listed species and of other species deemed economically important, is also regulated jointly, or separately, by the three Ministries.

Benefit-sharing

The Pipa joint venture company receives most of its income from two sources: capital contributions and sales of plants. As discussed above, Piroche Plants contributes 90 per cent of the joint venture capital, and NBGSS the remaining 10 per cent. NBGSS has in fact contributed its share of US$100,000 through a combination of cash payments and in-kind contributions of buildings and overheads.

NBGSS has established a pricing scheme under which Piroche Plants pays Pipa twice the cost of collecting seeds and raising seedlings for export to Canada. Thus Pipa's profits arise from a one-off payment by Piroche Plants for each seedling they receive. Once the transaction has been conducted, no further profits will accrue to Pipa, as the price paid per seedling is not linked to the price at which Piroche Plants sells landscaping trees to its international customers, for example, through royalties on sales by Piroche Plants.

Distribution of benefits

Under the terms of the joint venture agreement, profits made by Pipa, after all investment has been finished, are to be distributed in the following proportions:

- 50 per cent for the further development of Pipa;
- 10 per cent to Piroche Plants;
- 10 per cent to NBGSS;
- 10 per cent to Pipa staff; and
- 20 per cent to support conservation activities.

Consequently, Piroche Plants contributes 90 per cent of the initial investment in return for 10 per cent of the profits, while 90 per cent of the profits remain in China. Thus far, the 20 per cent of the profits to be dedicated to conservation have not been distributed. Pipa's board is awaiting project proposals for *in situ* and *ex situ* conservation, as well as others, from which they will select projects for funding.

The joint venture defines only monetary benefits and how they are shared, but the monetary benefits received by the partners have been invested so as to give rise to a range of non-monetary benefits. Just over 70 per cent of the capital invested in Pipa to date has been spent on equipment such as glasshouses, generator, tractor, fencing, water pumps, cooler, materials for propagation, equipment to process ginkgo biloba products, and office buildings and computers.

Conservation

The activities described in this case study have a bearing on conservation in two respects: the sourcing of plants from China, and the intention to distribute 20 per cent of the profits to support conservation projects.

As far as sourcing plant specimens from the wild is concerned, the Chinese staff at Pipa who collect specimens are trained botanists and horticulturalists who remove only seeds, and are careful not to damage the plants themselves, or their environment. It is much harder to ensure that materials collected by local people and sold in local markets were sourced in such a way as not to endanger the plants or cause environmental damage. Pipa can often not tell whether the seed it acquires is from the wild, or from plantations.

Pipa is starting to identify conservation projects to fund with its profits, but has encountered a few difficulties to date. Nevertheless, Pipa hopes to develop reintroduction and *ex situ* conservation or other conservation projects in the near future.

(*Sources:* Xue Dayuan, 1998; personal communication between Kerry ten Kate, Prof He Shan-An of Nanjing Botanical Garden, and Mr Pierre Piroche of Piroche Plants.)

6.5 Conclusions

There are few formal access and benefit-sharing arrangements in horticulture, and few benefits pass to the countries of origin of the germplasm. Most of the work in source countries is not in product development itself, but in cultivation of the finished product which was developed overseas. For this kind of activity, few of the financial benefits support conservation or accrue at the local level. As the International Association of Horticultural Producers (AIPH) puts it, 'Although developing countries such as Kenya and Colombia have substantial areas under cultivation and a large market share of the cut flower industry, control is typically vested in foreign companies (attracted by cheap labour, fast plant growth, lax regulations and a high profit margin) and small farmers rarely benefit significantly from the industry.'

As far as access to germplasm for plant breeding is concerned, most horticulturalists consider only the conservation implications of their work, if that. A common argument is, 'If it's not endangered, why restrict international trade in the germplasm?'

The horticulture industry really operates on two planes: large-scale breeding and marketing for the mass-market; and smaller-scale selection and breeding either for more local or specialised markets, or as part of a chain where products are licensed to bigger companies for production, distribution and sale. Companies working on the first plane do require access to genetic resources for their breeding efforts, although, as in agriculture, this can often be found in their own collections and in those of competitors, collaborators and organisations such as genebanks and botanic gardens.

The 'value' of access to 'new' germplasm by companies working in this way is hard to calculate. However, the companies interviewed appear to be prepared to pay modest fees for access, and even to offer royalties on some occasions, if the economics of product development merit it. The increasing use of IPRs suggests that companies and some research organisations are coming to terms with the privatisation of research and even of germplasm. Companies are becoming more accustomed to using available mechanisms to ensure control over the use of their germplasm and, in some circumstances, to gain bilaterally-negotiated monetary benefits for its use.

The number of companies responsible for the vast majority of global sales of ornamental horticultural products is very small, and their turnover high (compared, at least, with other players in the horticulture industry). A few products are often quite high value. For example, just one cultivar represents one-third of the entire chrysanthemum cut flower market in The Netherlands. For such high value products, sufficient sales are generated to justify the transaction costs of establishing some kind of benefit-sharing arrangement, where they are based on genetic resources obtained from source countries. Finally, a number of 'intermediary' institutions, which acquire genetic resources and may pass them to companies (such as genebanks, botanic gardens and universities), are becoming more aware of the CBD, and increasingly use MTAs that can (at least on paper) trigger benefit-sharing negotiations between companies and source countries. Taken together, these factors might suggest (to an optimist) that the major horticultural companies might be prepared to entertain some kind of benefit-sharing arrangements if simple, cost-effective and 'proportionate' mechanisms were to be found, and the companies could remain competitive.

However, most horticulture companies are quite unaware of the implications of the CBD with respect to access and benefit-sharing. The most important component of the horticulture market is the horticultural seed industry, where open access to germplasm has prevailed for years, and where a multilateral system is still under negotiation (Chapter 5). The links to agriculture, and the historical experience of nurserymen that it is safer not to keep all one's eggs in one basket, and that 'a plant shared is a plant conserved', have contributed to the prevailing ethos of free access, with conservation, not benefit-sharing, uppermost in people's minds. This has been strengthened by more recent experience with CITES, where the wide exchange of cultivars is considered beneficial as it decreases pressure on wild populations. The kind of frequent exchange of germplasm prevalent in many horticulture companies, and in most of the associations, organisations and individuals operating in the more 'informal' second plane – does raise profound questions as to the most appropriate mechanisms for sharing the benefits that arise from access to genetic resources for commercial horticultural development.

With the mass market, the volume of the market through a finite number of dealers and retailers may justify some kind of royalty scheme, despite the transaction costs, but it is difficult to see how such a system would work for low-volume and heavily distributed exchange among the smallest companies, specialists and enthusiasts. It would be difficult to ensure, by means of chains of MTAs, that orchid varieties (for example) were not sold commercially or used for breeding new varieties, and almost impossible to capture and allocate the benefits if they were. In this case, all a source country could really do would be to regulate access, hoping that this would remain controlled, possibly through several hands, until the germplasm reaches the commercial breeder; to charge such up-front fees as the industry would bear, and to lobby in international meetings for voluntary benefit-sharing mechanisms closer to the retail end of the product development and distribution chain.

Kerry ten Kate

Chapter 7
Crop Protection

7.1 Introduction

Agriculture continues to rely heavily on the use of pesticides: chemicals and organisms which are able to kill the weeds, insects and microorganisms which damage crops. Pests and diseases continually evolve and become resistant to pesticides, so that there is a constant need for the development of new crop protection products. The agrochemical industry which produces these products is a US$30 billion a year business (British Agricultural Association, 1998).

Broadly speaking, the three main approaches to crop protection may be described as chemical control, biological control, and genetic improvement of the crop plant itself. The major distinction between these three methods of control is that chemical control methods use chemical compounds to kill pests, biological methods use living organisms, and genetic improvement introduces disease resistance into crops through genetic engineering and traditional crop breeding. These different approaches are presented graphically in Box 7.3, and explained in more detail in Section 7.1.3. All three approaches to developing crop protection products require access to genetic resources, which are defined in the CBD as biological materials of actual or potential value, containing functional units of heredity.

Of the companies and organisations we interviewed, 26 (90 per cent) were developing products based on chemical methods of control, and 9 (31 per cent) were working on biological methods. However, in the coming years, advances in crop protection are likely to integrate chemical control, biological control and the genetic improvement of crops. Integrated Pest Management (IPM) is increasingly acknowledged as the most appropriate approach for the future (Box 7.1). Molecular biology can be used not only to introduce pest- and disease-resistance, and herbicide tolerance to crops, but to improve crop quality and identify new targets for new pesticides.

Box 7.1 *Integrated Pest Management*

Crop	Pesticide to which crop is tolerant
'Roundup Ready' crops	glyphosate ('Roundup')
'Liberty Link' crops	glufosinate-ammonium
'BXN' crops	bromoxynil
'IMI' crops	imidazolinone herbicides
'SU' crops	sulfonylurea herbicides
Poast Protected crops	sethoxydim ('Poast')

Sources: Global Crop Protection Federation brochure, nd; *Integrated Pest Management: the way forward for the crop protection industry*; personal communication with Michael Whitaker, 20 January 1999; personal communication between Adrian Wells and Man-Kwun Chan, IPM Forum Secretariat, 11 August 1998; Meerman et al, 1997.

Research in industry on pesticides and crop development often proceeds hand in hand. One well-known example of linked pesticides and crops is the herbicide glyphosate ('Roundup'), marketed by Monsanto, and the 'Roundup Ready' crops which are genetically engineered for their tolerance to glyphosate. Other examples of products that link pesticides and crops are listed in Box 7.2.

Indications are that the use of transgenic crops bred for resistance to disease and pests is growing rapidly as a method of crop protection. In 1997, an estimated 13.75 million hectares were planted to crops produced by agricultural biotechnology (80 per cent of which were grown in the USA), compared with 5.7 million in 1996 (*European Chemical News*, 1997). In 1998, the total hectarage for all biotechnology crops planted in the world was more than 30 million hectares (personal communication, Diane Herndon, Monsanto, 12 March 1999). Although this approach to crop protection is likely to grow in importance in the future, sales of transgenic crops in 1997 accounted for just US$0.5 billion, some 2 per cent of the agrochemical market (British Agricultural Association, 1998).

Box 7.2 *Some Examples of Linked Pesticides and Crops*

Integrated Pest Management (IPM) is the basis of sustainable crop protection designed to increase agricultural productivity while protecting natural resources and emphasising human resource development. IPM integrates a range of techniques such as the use of biopesticides and biocontrol techniques, culturally appropriate cultivation (eg intercropping, rotations, nursery management and tillage methods) and physical control methods (eg traps and manual weeding), as well as the use of crop protection products. IPM techniques should lead to the appropriate use of crop protection products, maintain pest populations at an acceptable level and encourage natural mechanisms for regulating pests.

IPM developed as the basis for long-term sustainable crop protection and also to overcome some of the negative side effects of the wide use of pesticides (such as the increasing resistance of some pests to certain crop protection products and possible adverse health effects on farmers, farm workers and consumers). The need for improvement is greatest in developing countries, where serious problems still persist. The World Health Organisation (WHO), through the International Programme for Chemical Safety, is currently trying to review the actual incidence of adverse health effects associated with the use of crop protection products, and to identify ways of improving the current situation.

IPM is a knowledge-based system for farmers and pest-control operators. It is the most appropriate combination of biological, physical, chemical and cultural measures that provide the most cost effective, environmentally sound and socially acceptable method of managing the diseases, insects, weeds and other pests under the local circumstances in which they need to be managed. IPM is the crop protection system which best meets the requirements of sustainable agricultural development. It is a component of Integrated Crop Management (ICM) which combines crop protection with farm and natural resource management. IPM has gained widespread support among multilateral and bilateral donors, national governments, the food industry and crop protection companies, all of whom recognise its potential contribution to sustainable agriculture. The wide-scale adoption of IPM is likely to require the active collaboration of public- and private-sector organisations.

Sources: Global Crop Protection Federation brochure, undated. Integrated Pest Management: the way forward for the crop protection industry; personal communication with Michael Whitaker, 20 January 1999; personal communication between Adrian Wells and Man-Kwun Chan, IPM Forum Secretariat, 11 August 1998.

Genetically mediated crop improvement is covered in Chapter 5. This chapter focuses on chemical and biological means for controlling pests. The first section starts with an overview of the different actors involved in the discovery and development of new crop protection products, and the market for them, and then explores which crop protection products comprise of or are derived from genetic resources. It describes the typical stages in product discovery and development for chemical pesticides and biocontrol agents, providing examples of each, with associated costs and timelines, and summarises the regulatory hurdles which a new crop protection product must pass before it is placed on the market.

The second section analyses the demand by scientists from the private and public sectors for access to genetic resources to develop new crop protection products, first exploring which actors require access to genetic resources. It reviews which organisations are involved in collecting genetic resources from *in situ* conditions, and how they can acquire genetic resources held in *ex situ* collections in the private and public sectors. It also explores the kind of genetic resources that are in demand for product development, assessing the balance of 'raw' samples and value-added products, of domestic and foreign material, the relative interest in synthetic compounds and natural products, and the extent to which researchers need to return to the original source for further samples. It concludes with a review of how genetic resources are selected for research, and a prediction of the extent of demand for access to genetic resources for the development of crop protection products in the future.

The third section explores current practice in benefit-sharing in the crop protection sector, with sections on monetary and non-monetary benefits that arise from use of genetic resources for crop protection. A fourth section provides a case study on the development of a specific crop protection product, and a final section offers some conclusions and recommendations.

7.1.1 The roles of actors and markets

The global agrochemical market (including herbicides, insecticides, fungicides, plant growth regulators, rodenticides and molluscicides, but excluding fertilisers) was US$30.2 billion in 1997 (Figure 7.1).

Sales of herbicides accounted for 48 per cent of the market, insecticides for 27 per cent, and fungicides for 20 per cent. Companies from North America and Europe generated almost two-thirds of world sales of agrochemical products in 1997 (Figure 7.2). North America, Europe and Japan are the major centres of the agrochemical industry, and most product discovery and development in crop protection is conducted by enterprises from these regions.

Most commercial production of crop protection products is concentrated in the hands of a small number of multinational companies (see below). However, government agencies, university departments, research institutes, and small and medium-sized enterprises are also engaged in the basic research and discovery work that leads to new crop protection products.

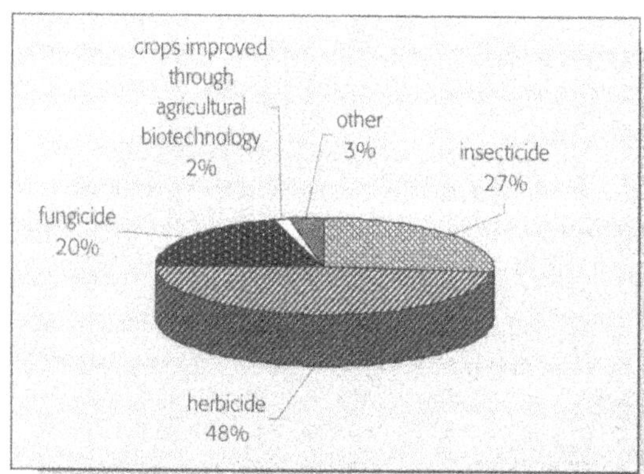

Figure 7.1 *Global Markets for Crop Protection Products, 1997*

Table 7.1 *Sales by the Top Ten Crop Protection Companies, 1997*

Rank	Company	Sales (US$ bn)
1	Novartis	4.2
2	Monsanto	3.13
3	Zeneca	2.67
4	DuPont	2.52
5	AgrEvo	2.35
6	Bayer	2.25
7	Rhône-Poulenc	2.2
8	Cyanamid	2.12
9	Dow Agro Sciences	2.05
10	BASF	1.85
	Total	25.34

Source: Agrow World Crop Protection News, 1998.

The private sector

Since the early 1990s, a flurry of mergers and acquisitions has led to the emergence of a few giant life-science companies which dominate the crop protection sector just as much as they do the seed sector (Chapter 5; and www.css.orst.edu/herbgnl/descr.html, for a 'genealogy' of herbicide companies by Prof Arnold Appleby). Multinationals operating on this scale, such as Novartis, Monsanto, Zeneca, Aventis (to be formed by the merger of Hoechst and Rhône-Poulenc) and Dow AgroSciences, have research and development programmes across the full range of crop protection products. Companies of this size that we interviewed were conducting research on all the chemical, biological and genetically mediated crop protection solutions, and were involved in the entire product development spectrum, including discovery, development, registration of new products, and commercialisation.

In 1996, the top 20 agrochemical discovery companies controlled most of the market for crop protection products. The total sales by these companies were US$28.02 billion, or 91 per cent of the total market of US$30.56 billion (Agrow, 1998a). In 1997, sales of crop protection products by the top ten companies alone were US$25.34 billion, 84 per cent of the global market (Table 7.1 and Figure 7.3). Other than sales by the 20 or so leading companies, many of the remaining sales are accounted for by smaller discovery companies, or generic manufacturers selling existing products once their patent protection has run out. For example, the fastest growing agrochemical companies in 1997 were Fernz-Nufarm and Makhteshim-Agan, both of which own several small, generic manufacturers selling existing products (Agrow, 1998a).

The smaller companies and private research institutions we interviewed tend to specialise in a particular category of crop protection product, such as biocontrol agents, and to focus on a particular stage in product development, typically the discovery phase. In the 1980s, a number of small, start-up biopesticide companies emerged and went public to exploit the market for biocontrol agents, but the market has subsequently consolidated, with many of these (such as Biosys, Entotech and Crop Genetics International) disappearing through mergers or acquisitions (Georgis, 1997).

Smaller companies and private research institutes working on chemical pesticides often conduct bioassays and molecular elucidation, and sell the results to crop protection companies, while others license active compounds from their own screening programmes to clients which include major agrochemical companies. At the other end of the research and development spectrum, some companies are not involved in discovery at all, but focus on the final modification, adaptation and formulation of a product, using compounds obtained from others. Together, such activities account for a very minor proportion of global sales of crop protection

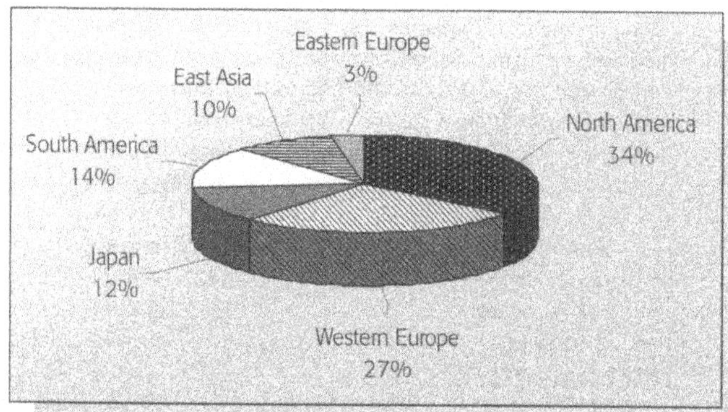

Figure 7.2 *Global Sales of Crop Protection Products by Region, 1997*

Source: British Agricultural Association *Annual Review*, 1998.

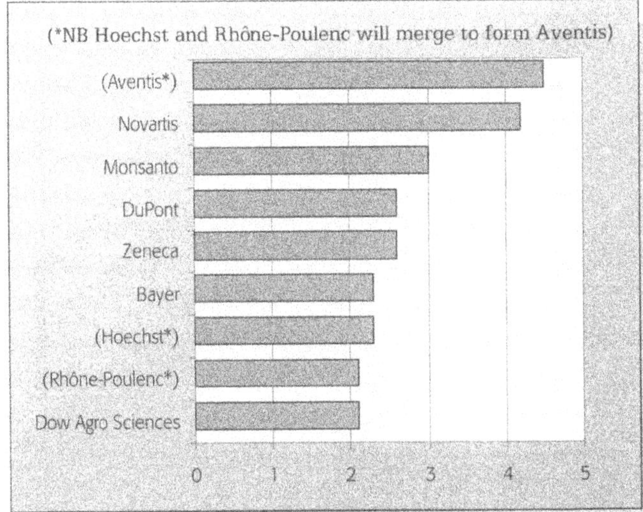

Figure 7.3 *Crop Protection Sales, 1997 (US$ bn) (Excludes Seeds)*

Source: Wood Mackenzie.

products. Research and development investment by the private sector is shown in Figure 7.4

The public sector

Although a number of public research institutes have been privatised over the last decade, public institutions such as universities, botanic gardens, genebanks and research stations affiliated with agriculture ministries continue to play an important role in crop protection research. Most of the research conducted by these institutions is not in product development, but either in 'pure' science, or in 'applied' science at the discovery end of the research and development spectrum. Examples of the former are research in plant, microbial, insect and soil biochemistry, microbiology and ecology, to study organisms' life-cycles, interactions, and diseases. There are also research programmes in studies such as immunodiagnostics and bioinformatics. At the 'applied' end of the scale, public-sector researchers look for new organisms with potential economic applications, and conduct screening programmes and field studies in the search for leads for new products. These institutions hold collections of many different kinds of genetic resources, and frequently enter into collaborations with private companies, who receive access to genetic resources and value-added derivatives, and who sponsor particular research programmes.

In addition, state support for research programmes on crop protection provides an important source of funding for crop protection discovery and development conducted by academics and small and medium-sized enterprises, and the public sector subsidises the use of particular crop protection products, particularly in developing countries. For example, under a scheme initiated by the Federal Government of India, the State Government of Andhra Pradesh subsidises local farmers' application of biological pest control technology, including pheromone traps, by up to 100 per cent for the first two years of use (personal communication between Adrian Wells and Mr Sateesh Kumar, Ecosafe Systems (P) Ltd, Hyderabad, India, 13 August 1998). In another example, the Ministry of Agriculture in Egypt purchases Agrisense's Pink Bollworm pheromone product (Box 7.3) under government tender.

The customers

Chemical pesticides are typically sold to customers in three main sectors: the agricultural sector, through distributors who supply farmers, and government bodies responsible for agricultural advisory services; the industrial sector, generally pest control companies like Rentokil (UK) with markets in domestic and factory pest control; and the consumer products sector, such as DIY stores and garden centres. The customers of biocontrol companies include horticultural glasshouse growers, individual gardeners (through catalogues) and university researchers.

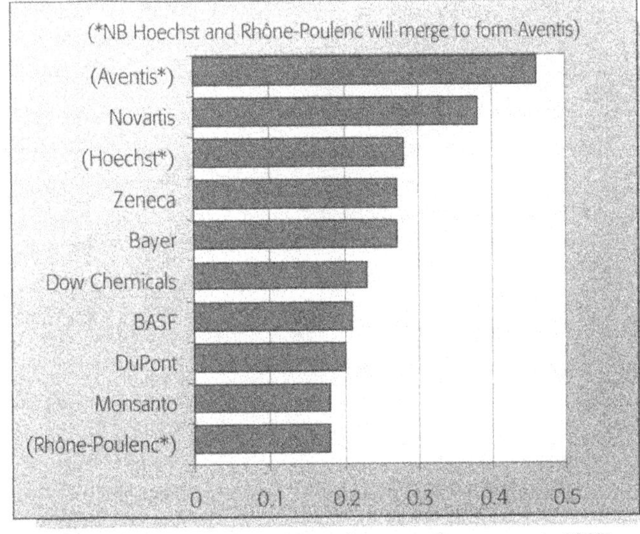

Figure 7.4 *Research and Development Investment, 1997 (US$ bn) (Excludes Seeds)*

Source: Wood Mackenzie.

Box 7.3 *Crop Protection*

Method of control	*Examples**
Chemical control	
A 'Natural products':	
(a) 'Pure' natural products (isolated from nature and not changed chemically).	Glufosinate (Monsanto), Spinosyn (Dow)
(b) Semisynthesised derivatives (modifications of compounds isolated from nature).	Avermectin
B Synthetic compounds:	
(a) Synthetic analogue compounds built from templates originally discovered from natural products.	Azoxystrobin* (Zeneca), Pyrethroids*, Carbamates (see p201)
(b) 'Pure' synthetic compounds (not based on a natural product lead).	2,4-D
C Behaviour-modifying chemicals: Use of naturally-derived and synthetic versions of signalling chemicals from living organisms, such as pheromones, to create insect traps and disrupt mating.	Pink Bollworm pheromone* (Agrisense)
D Growth regulators:	
(a) Insect growth regulators: chemicals which interfere with the growth of pest insects, ultimately killing them.	
(b) Plant growth regulators: chemicals such as gibberillic acid sprayed on to crops to increase the size and quality of fruit, to speed or delay ripening.	
Biological control	
A Toxic protein-producing bacteria: Over 30 recognised subspecies of the naturally-occurring bacterium *Bacillus thuringiensis (Bt)* produce different insecticidal toxins.	XenTari and Dipel* (Abbott)
B Baculoviruses: Naturally-occurring ('wild type') or modified viruses that, once ingested by insects, interfere with their metabolic processes and kill them.	
C Fungal pesticides:	
(a) Fungal insecticides: three species of fungi which are the main source of fungal insecticides: *Verticillium, Metarhizium* and *Beauvaria*.	'Deny'* (Stine), Collego (Ecogen), Devine (Abbott)
(b) Fungal herbicides: fungi which affect the metabolic processes of weed species, killing them.	
(c) Fungal fungicides and microsporidia.	
D Bacterial pesticides: Bacterial bactericides, bacterial fungicides and bacterial insecticides.	
E Beneficial, predatory organisms: Whole, natural predators and parasites, such as insects and nematodes, for insect and mollusc control.	'S. dallatoreanum* (PNGOPRA/Oxford)'
Crop improvement	
A Disease and pest resistance: Breeding disease or pest resistance or tolerance into crop varieties, using either traditional methods or genetic engineering, eg 'biolistics'.	See Chapter 5
B Herbicide tolerance: Breeding herbicide tolerance into crop varieties, using either traditional methods or genetic engineering.	

Companies and public organisations, such as universities and government research institutes, supply genetic resources and value-added derivatives such as compounds, genes and technologies. They sell or license these to each other, and, predominantly, to the multinational companies who develop these 'intermediate' products into final products to be sold to the consumer. Indeed, while most research and development is conducted in-house, there is a trend for the larger multinational bioscience giants to acquire more 'intermediate' products from smaller companies and researchers, which may provide a new and growing market opportunity for institutions conducting research on genetic resources (see p213 below, Section 7.2.4).

7.1.2 Crop protection products based on or derived from genetic resources

Crop protection products based on or derived from genetic resources are represented in both of the methods of crop protection discussed in this chapter, namely chemical control and biological control. The third major approach to crop protection – the genetic modification of the crops themselves – is based entirely on genetic resources, and was covered in Chapter 5. This section will outline briefly the kinds of chemical and biological crop protection product that may involve access to genetic resources.

Chemical methods of crop protection based on genetic resources

The term 'natural products' is used by the crop protection industry to refer to 'pure' natural products – compounds isolated from nature whose chemical composition is not altered by man – and also to compounds isolated from natural sources and then modified through semisynthesis (see p200). Although they are not considered 'natural products' by crop protection companies, another class of compound is originally derived from access to genetic resources. These are synthetic compounds modelled on chemical structures, or 'templates', originally isolated from natural sources, which formed the lead for further discovery and development work. However, new discovery programmes based on compounds of this kind do not involve access to genetic resources. Each of these categories, and some products within them, will be described in this section (Box 7.4).

Two other methods of chemical pest control that may involve access to genetic resources are covered in less detail in this chapter. Insect growth regulators are compounds that kill insect pests by interfering with hormonal control of insect or larval growth. Some hormone mimics are natural products, but the compounds currently used to control larval growth (the benzoylureas) are all synthetic. Another method is to use chemicals to modify the behaviour of pests. Complex, naturally-occurring signalling chemicals such as

Box 7.4 *The Role of Natural Products in Pesticide Discovery and Use*

Use of pesticide	Natural product extracts or semi-synthetic pesticides	Pesticides for which natural products formed the lead for chemically synthesised products
Fungicide	▪ Pyrrolnitrin (from *Pseudomonas*) ▪ Griseofulvin ▪ Kasugamycin and Blasticidin S (against rice blast) ▪ Polyoxins B&D, Validamycin (against powder mildew) ▪ Cyclohexamide	▪ Phenylpyrrole ▪ Azoxystrobin and Kresoxim-methyl (from strobilurin made from a fungus)
Insecticide	▪ Avermectin (from *Streptomyces avermectilis*) ▪ Neem products (against locusts, gypsy moths, aphids) ▪ Spinosyn/Lepicidine (actinomycete fermentation) ▪ Milbemycin (*Streptomyces hygroscopicus*) ▪ Nicotine alkaloids (from tobacco plant) ▪ Ryanodine alkaloid (from *Ryania speciosa* plant)	▪ Carbamates (alkaloidal carbamate from the seeds of an African vine used as arrow poison) ▪ Pyrethroids (from chrysanthemum plant – responsible for 20 per cent of insecticide sales) ▪ Cartap (from Nereistoxin derived from marine annelid, against rice stem borer)
Herbicide	▪ Phosphinothricin-based products ▪ Pelargonic acid ▪ Bialophos/Bialanofos (broad spectrum post emergent)	▪ auxins for 2,4-D ▪ 1,4-cineole for cinmethylin

Sources: Stephen Brewer compiled from Buchel, 1990; Miyakado et al, 1997; Legg, 1998; Cutler, 1998; and personal communication between Kerry ten Kate and Dr David Hunt, Dr Len Copping and Dr Stephen Duke.

pheromones can be developed as the basis for lures to remove predatory insects from crops to traps, where they can be killed. Alternatively, these chemicals can be used to disrupt the mating pattern of insects. Some products use pheromones just as they occur naturally, while others develop synthetic versions through chemistry programmes (Box 7.16 on the natural sex pheromone from the Pink Bollworm).

The 11th edition of the *Pesticide Manual* (1997) lists 759 entries that include all the products registered for use in crop protection today. Of these, 37 are biocontrol agents, eight are natural insect pheromones, and nine are fermentation products or plant extracts (ie 'pure' natural products, or naturally-derived compounds) (personal communication, Dr Len Copping, 5 February 1999). Together, these products, all derived from genetic resources, represent 7 per cent of the products registered for use. Thus 7 per cent of current crop protection products were developed from research programmes involving access to genetic resources (although the resources may have been obtained from within the company's own collections), but these products account for only around 2 per cent of the sales of crop protection products (Box 7.5), since the synthetic products on the market are responsible for higher sales. Crop protection products that, however distantly, owe their origin to nature comprise roughly 10 per cent of the annual global sales of crop protection products. However, discovery of new products from wholly-synthesised analogues once modelled on a template from nature does not require access to genetic resources (although the resources may have been obtained from within the company's own collections).

7.1.3 The process of developing a new crop protection product

This section examines the process for developing a new crop protection product. It distinguishes between two categories of product: chemical pesticides (comprising synthetic chemical pesticides and natural products) and biocontrol agents.

First, we look at chemical pesticides. Since much of the research and development process for synthetic chemicals and natural products is similar, the two are taken together, but distinctions between natural products and chemical compounds are drawn where necessary. The section offers an overview of chemical pesticide research and development.

Research and development for biocontrol agents is quite distinct from research and development for chemical pesticides, so this aspect of crop protection is addressed separately (p205). Finally, the kind of regulations which govern the product approval process for chemical pesticides and biocontrol agents are described at the end of Section 7.1 (p206 et seq).

Box 7.5 *Crop Protection Products Derived from Genetic Resources*

Pesticide	Sales (US$ bn)	Global sales %	Classification of product	Reference
Glufosinate	0.13	0.4	'Pure' natural product	personal communication Steve Brewer, 1998
Spinosyn	–	–	'Pure' natural product: mixture of spinosyns *a* and *d*	(new product; no sales data available)
Avermectin	0.1	0.3	Semi-synthesised derivative	Morrone, 1998
Others[1]	0.3	1.0	'Pure' and semi-synthesised natural products	personal communication Steve Brewer, 1998
Biopesticides	0.08	0.3		Johnson, 1997
Sub-total 'natural products' (including biopesticides)	0.61	2.0		
Pyrethroids	1.2	3.9	Synthetic analogue from natural template	Walsh, 1997
Azoxystrobin[2]	0.2	1.0	Synthetic analogue from natural template	personal communication Mike Legg, 1999
Other strobilurins	0.05–0.1	<0.3	Synthetic analogue from natural template	personal communication Mike Legg, 1999
Others[3]	1.0	3.3	Synthetic analogue from natural template	personal communication Len Copping, 1998
Sub-total 'natural template analogues'	2.59	8.5		
Total products 'derived from' genetic resources	3.0	10.5		

1: eg pyrethrum itself; rotenone; azadirachtin; nicotine; and antibiotic insecticides, acaricides, herbicides and fungicides.
2: Sales of Azoxystrobin by Zeneca alone in 1998. Sales in 1997 by both Zeneca and BASF of all strobilurins were US$17 bn.
3: eg phenylpyrroles; neo-nictotinoids; and nereistoxin analogues.

Source: personal communication with Dr Steve Brewer, Dr Len Copping, and Dr David Hunt, 1998–9.

Box 7.6 *The Stages of Discovery and Development of a Crop Protection Product*

Phase	Activities	Time	Cost
Phase IA	■ Discovery: general screening programme Initial test for activity in 100,000s of compounds (both synthetic chemicals and natural products).	1–2 years (decreasing)	Average cost allocation: 10%.
Phase IB	■ Discovery: fitness evaluation of 5–10 lead compounds Evaluation of characteristics of 'hits' for development potential. Characterisation and optimisation of biological activity by synthesis of analogue compounds, glasshouse and field screening, characterisation of photochemical stability (eg UV-stability).	1–2 years	Average cost allocation: 15%. US$ 8–25 m
Phase II	■ Development of prime lead compound(s) Once a company is satisfied with biology and commercial risk aspects of a potential product, it conducts local 'usage' and registration oriented tests: chemistry; biology and toxicology; and scale-up for production.	4–6 years	Average cost allocation: 65–70%. US$30–70m
Phase III	■ Registration and launch Product approval procedure; once registration looks likely, pre-launch marketing and market support.	2–3 years (registration)	Average cost allocation: 5%. US$1.5–4 m
	Total	8–14 years	US$40–100 m

Sources: interviews, and personal communication with Dr David Hunt.

The discovery and development of chemical pesticides

Overview

To develop a new commercial chemical pesticide can cost between US$40 and US$100 million and take from eight to 14 years (Box 7.6). The most important factors affecting the investment and time involved in product development include the toxicological, ecological and environmental profile of the product; the biological spectrum of its efficacy; the number of target crops; the complexity of the process by which the compound acts on a pest, and how easy it is to formulate the compound into a product. While the costs and times required to develop a pesticide are difficult to generalise and vary significantly from product to product, Box 7.6 shows an approximate chronology and investment profile for the three phases involved in the 'active ingredient life cycle'. The discovery stage (which consists of two parts, the first covering initial screening and the second, development of a promising 'lead'), typically costs US$10–20 million and can last up to four years. The development stage involves developing a lead emerging from the discovery process into a candidate pesticide for product approval. This phase is likely to cost between US$25 and US$60 million, and take a further three years. The greatest costs during the development phase are associated with generating the safety information on the toxicology, environmental impact, and effect on non-target organisms of the potential product. After this, it takes a further 1–3 years for the candidate pesticide to complete the regulatory procedure needed for product approval, which adds a few more million dollars to the cost of product development. In practice, some of the work involved in the first two phases can be conducted concurrently.

Many compounds that initially reveal some biological activity in screens are abandoned because research shows that they are not sufficiently potent, do not have the desired spectrum of activity, operate in the same way and with the same effect as earlier products, are too toxic, or are too expensive to manufacture. Figures vary greatly from company to company, but typically, of 100,000 chemicals tested (Phase IA in Box 7.6), perhaps 100 will enter secondary screening early development (Phase IB in Box 7.6), and about ten will go into full-scale development (Phase II); about two will become products, having successfully completed the regulatory process (Phase III). Some 25 years ago, 1 in 10,000 compounds screened might result in a commercial product. Today, more stringent registration and environmental requirements for candidate pesticides, coupled with the hunt for novelty, mean that a company is more likely to screen 50,000 compounds for each new pesticidal agent approved. Over 99.99 per cent of the compounds screened will thus be rejected, and the cost of research on these abandoned compounds typically amounts to about 10 per cent of sales.

These figures vary, not only from company to company, but from product to product. For example, other figures quoted by scientists interviewed suggest that just ten, rather than 100 compounds proceed to Phase IB, and that for each new chemical pesticide that succeeds to the market, 30,000 (as opposed to 50,000) other candidate chemicals will have been rejected at a cost of US$300 million.

An agrochemical company may market a new product only every year or two. Companies are likely to launch a product containing an entirely new chemical entity (or novel active ingredient) only once every few years, perhaps every five years, although improved formulations or extended uses of existing products are produced fairly regularly, rather like new, improved washing powders.

Discovery

The majority of new commercial pesticides are discovered by testing chemicals for their ability to kill fungi, insects, nematodes and weeds. This exercise consists of two parts: developing a 'screen' to detect biological activity of interest, and finding the chemical compounds (both synthetic and of natural origin) to test in the screens.

Screens

Based on market opportunities and knowledge of the kind of biological activity which may result in a product with good commercial prospects, a company may devise a screen to identify such activity. Developing a screen involves selecting a target such as an enzyme in the weed, insect or pathogen – or the entire organism – upon which it is hoped that a pesticide will take effect. Over the last 20 years, developments in biochemistry have enabled scientists to gain a better understanding of the biological and chemical pathways and reactions in living organisms. This has vastly increased the range of targets available to mechanism-based screens and has improved the specificity of screens available. Recently, biochemical targets on which to base crop protection screens have been identified and isolated through the application of 'genomics' – the sequencing and screening of genes from target pest organisms.

Mechanism-based and whole-organism screens

There are two main kinds of screen. 'Mechanism-based' screens depend on target enzymes or receptors isolated from whole organisms or cells that will reveal specific biological activity (the 'mechanism' concerned) when combined with the chemical to be screened. With this approach, for each possible mechanism of action, a screen with the relevant target must be created. By contrast, 'whole-organism' screens operate *in vivo* and expose an entire plant, microorganism or insect to the chemical being screened, enabling the potential pesticide to operate through a range of different mechanisms during the one test. For both approaches, a collection or 'library', of chemicals to be tested may be built up and maintained (see 'Samples to be screened', p197). Each of these chemicals will be tested many times against an ever-changing array of mechanism-based screens, or new whole-organism screens as these become available. Chemicals that have been tested but do not reveal interesting activity in a particular screen are returned to the company's library for future use in different screens. Companies generally continue to seek new samples to add to the library, containing more chemicals to be screened.

Each approach to screening has its advantages. Mechanism-based screening is the faster, since the results are revealed within a couple of days. Another great benefit of mechanism-based screens is that they give the researcher a good idea of the structure of the compounds screened, which helps to determine the size and shape of the binding site on the enzyme, thus allowing the design of synthetic or semi-synthetic compounds that are more effective inhibitors of the enzyme. However, this method of screening, in which compounds have direct and unencumbered access to the specific enzyme on which they must work, is no indicator that they would function well when applied to an entire plant or pest. If a compound reveals interesting activity in a mechanism-based screen, it is then tested on whole organisms, but it is relatively unlikely to show desirable activity in this much more complex situation.

To show desirable activity against a whole organism, a compound must cross barriers such as insects' shells, maintain its structural integrity as it travels through the plant or insect (rather than being metabolised or degraded), and find its way to the target enzyme within the cellular structure of the plant or insect. Not only are whole organism screens a better indicator of whether a compound is likely to succeed as a product, but they are more versatile, in that they test the full range of mechanisms through which the potential pesticide could operate, rather than a single enzyme responsible for one mode of action. However, whole organism screens take much longer to perform than mechanism-based screens, particularly for large organisms such as entire plants. After spraying the plant with the chemical to be tested, it may be necessary to wait a considerable time in order to tell whether the chemical causes abnormal growth, kills the plant, or has some other effect. For 'pre emergent' pesticides, soil containing seeds is sprayed, and it is necessary to wait until the plant has grown from seed in order to assess the results that the product has on the plant. Whole organism screens are still generally preferred in the crop protection sector, since they identify chemicals which will operate on the actual target pest (rather than on an isolated enzyme).

High-throughput screening

High-throughput screening (described in more detail in Chapter 3) has dramatically increased the annual screening potential of crop protection companies. It has been possible to screen large numbers of fungal specimens for some time. However, 10 or 15 years ago it was only possible to screen for herbicides using whole weed plants grown in pots. At that time, the screening capacity for plants was some 2,000 to 10,000 chemicals a year. Now, crop protection companies can screen some 20,000–100,000 chemicals a year. High-throughput screening for crop protection products uses primarily mechanism-based screens, but companies do also use simple, whole-organism screens based on algae, nematodes, mosquito larvae and other small organisms.

The use of specific enzymes and receptors in mechanism-based screening, a practice which currently dominates the pharmaceutical industry, is evolving in the crop protection sector, but mechanism-based screens involving high-throughput technology have not yet resulted in any new crop protection pesticides (personal communication Dr David Hunt and Dr Len Copping, February 1999).

Samples to be screened

Both mechanism-based and whole-organism screens, whether high-throughput or not, require access to new sources of chemicals to be screened. The chemicals that are tested in screens may be man-made, 'synthetic' chemicals or 'natural products'. Companies may also require access to target organisms to develop into screens, although samples of target pests are generally readily available.

a) Synthetic compounds

A rough estimate (p194 above) suggests that at least 90 per cent of pesticides, and at least this proportion of herbicides currently on the market, have been developed from chemicals created entirely in the laboratory, rather than those derived in any way from nature. Chemists in crop protection companies create and screen large libraries of such synthetic compounds from basic inorganic and organic chemicals, or purchase them from other companies or researchers. Companies running high-throughput screens may need hundreds of thousands of compounds to screen each year, in order to keep this expensive process occupied and productive.

Libraries of synthetic compounds have been built up by companies over the years, as new compounds have been synthesised in small numbers. After a decade of use in the pharmaceutical sector, combinatorial chemistry (Chapter 3) is still in its infancy in the crop protection sector, but gradually becoming more common. Combinatorial synthesis (of small molecules such as peptides) and biosynthesis (of macromolecular molecules such as nucleotides, nucleosides, lipids, fatty acids and some carbohydrates) allows companies to synthesise millions of compounds a year (Balkenhohl et al, 1996). Synthetic chemistry, rather than the search for natural products, is supplying the majority of compounds for the increased screening capacity of high-throughput screening.

b) Natural products

Synthetic chemical pesticides have been in use since the 1890s, when Bordeaux mixture (copper sulphate) was introduced. Subsequently, sulphuric acid, copper sulphate and sodium chlorate were used as herbicides and dinitrophenols as herbicides and insecticides (personal communication, Len Copping, January 1999). Another early chemical pesticide, 2,4-D, was introduced early in the 20th century.

However, until advances in synthetic chemistry in the 1940s and 1950s, most crop protection products were isolated from natural sources, and some natural products, like neem, have an ancient history of more than 1,000 years (Schmutterer, 1995). The term 'natural product' is generally used to describe extracts taken from plants, fungi, microbes, insects or other animals. Scientists in the crop protection industry also consider as 'natural products' naturally-occurring molecules such as phosphinothrycin, which is completely synthesised, but whose structure is identical to the naturally-occurring molecule.

Box 7.7 *Early Fermentation Products for Crop Protection Product*

Product	Year	Application
Streptomycin	1947	control of bacteria
Oxytetracycline	1950	control of fireblight
Blasticidin-S	1959	control of rice blast
Kasugamycin	1965	control of various fungal and bacterial pathogens
Polyoxins	1965	control of fungi
Polynactins	1971	control of mites
Validamycin	1972	control of fungi
Mildiomycin	1979	control of mildew

Natural product pesticides known as 'fermentation products', developed by fermenting microorganisms such as fungi and bacteria were introduced into crop protection from the 1950s. Some of the early fermentation products are listed in Box 7.7. More recent examples are the insecticides and miticides abamectin and milbemectin introduced in 1980, and the insecticide spinosad which was introduced in 1997 (personal commmunication, Dr Len Copping, February 1999).

Today, natural products continue to make a small but important contribution to the development of crop protection products, contributing between 2 and 10 per cent of the US$30 billion world market, depending on how 'natural products' are defined (Box 7.5 and discussion below). As far as insecticides are concerned, a host of insecticidal agents are synthesised by plant and animal species. Approximately 2,000 plant species are known to contain toxic principles effective against insects, but only a small number are used in practice (Fuchs and Schröder, 1990). In the case of fungicides, the 11th edition of the *Pesticide Manual* refers to 167 fungicidal compounds, of which eight (5 per cent) are natural product chemicals and 17 (10 per cent) living systems such as viruses, bacteria, fungi, nematodes, and insect predators and parasites (personal communication, Dr Len Copping, February 1999).

The researchers we interviewed involved in the discovery of novel natural product pesticides were investigating a range of materials, including plant extracts from both well-known sources such as neem and species new to crop protection, as well as insects, nematodes, fungi and viruses. The cost of discovery and development for a natural product is generally much higher than for a synthetic compound, for a number of reasons described in Box 7.8. In summary, it is more labour intensive to source, isolate and elucidate the structure of a natural product than it is a synthetic compound. As Zeneca Agrochemicals puts it, 'the major cost of natural products is "de-convolution" – isolating the active compound out of the natural product mixture', a step which is likely to be unnecessary for synthetic chemicals, whose structure is generally known prior to screening. To show that a natural product compound is new, biologically active and patentable takes much more time and effort than new compound synthesis.

Box 7.8 *Three Reasons Why the Discovery and Development of Natural Product Pesticides are Likely to be More Expensive than for Synthetic Compounds*

1 Obtaining samples

The cost of collecting and extracting natural products – both the cost of the raw material and the labour to source and extract it – is much higher than that of producing chemical compounds in the laboratory. Catalogue vendors sell distinct synthetic molecules for prices that are often in the range US$10–150 per compound.

2 'Deconvolution'

Structural elucidation

In a crude plant extract a large number of compounds are present – probably in excess of 100, each of which could be responsible for signs of biological activity in screens. Fermentation broths of microorganisms also contain many chemicals, the structure and quantity of which vary with the fermentation medium and the time of harvest. The use of separation techniques to fractionate and extract these mixtures with different solvents can reduce the number of compounds in each fraction to roughly ten, and thereafter more work can identify the single active compound from the remaining mixture. This process is costly and time consuming, and one of the most expensive aspects of working with natural products. Improved technologies that can separate mixtures and elucidate structures in tandem (liquid chromatography/mass spectrometry/mass spectrometry and liquid chromatography/nuclear magnetic resonance instruments) make it increasingly possible to identify most of the compounds in an extract before conducting bioassays or molecular screens (personal communication, Dr Stephen Duke, 30 January 1999).

Novelty

Furthermore, after all this work, the natural product may turn out not to be novel, but to be well known. In this case it is extremely likely to be dropped from the research programme, first because it will be difficult to obtain patent protection, and second, because even if a modified form of the compound were sufficiently novel to merit patent protection, the mode of action would probably be familiar, and companies are more interested in seeking products with new modes of action. As one researcher put it, 'It's not like every time you pick up a natural resource, there is something novel there. All the good things that are easy to find have been taken. People are going to more and more remote locations – such as thermal vents – to find organisms with new biology.'

A synthetic chemist creating a compound in a laboratory will know the structure of the chemicals he or she is creating, and can make a compound far faster than a colleague working with natural products could hope to source and 'deconvolute' a natural compound to the stage where its structure is known and novelty established. A chemist could synthesise compounds at the rate of roughly one a day, and thus produce perhaps 200 synthesised, screened and characterised compounds in a working year. With natural products, it is not uncommon for the process of identifying the active constituent from a fermentation broth to take a year, and sometimes as much as three years. A rough estimate, therefore, suggests that it can take 300 times the effort (in terms of man-hours) to produce a screened and characterised natural product, as a synthetic compound.

3 Scale-up

Natural product pesticides – whether 'pure' natural products or semisynthesised derivatives – require constant supplies of wild or cultivated raw materials. Sometimes these can be produced in large quantities through the fermentation of microorganisms, and optimising the efficiency of this process itself represents a substantial cost, but sometimes companies are obliged to cultivate plants, or obtain more supplies from the wild. This, or the alternative of finding an economic synthesis of the compound, adds to the expense and risk of production.

In the USA, in practice, natural products have some regulatory advantage, since natural chemicals such as azadirachtin and spinosad are on the US Food and Drug Administration (FDA) list of Generally Recognised as Safe (GRAS) compounds. Although the GRAS list is maintained by the Environment Protection Agency (EPA), and it is the EPA that registers pesticide products, a product which is on the FDA's GRAS list is usually easier to register as a pesticide.

While certain disadvantages of natural products compared to synthetic compounds are common to both pesticides and pharmaceuticals, several companies interviewed pointed out that the cost structure of the market and the profit margin for crop protection products are very different from those for pharmaceuticals. For research in natural products to be attractive, the cost of research, development and manufacture must be competitive with other approaches. According to several companies, this is rarely achieved for natural products. We found that the majority of most companies' crop protection research budget was dedicated to synthetic chemistry, with natural products taking lower priority, and sometimes even third place after the genetic modification of crops. Several companies cited the high research, development and production costs of using natural products as the reason for the status of natural products as poor cousins to synthetic compounds.

However, natural products remain attractive for three main reasons. First, naturally-occurring chemicals can reveal novel structures that rarely emerge from synthetic chemistry programmes. Second, environmental concerns surrounding the use of chemical pesticides have fuelled the demand for natural products. For both synthetic compounds and natural products, scientists are striving to reduce application rates from kilogrammes of pesticide per hectare to grammes per hectare. To be able to use lower quantities, products must exhibit higher activity. Where natural products are significantly more active than synthetic compounds, the higher costs of developing and producing them may be justified. Third, natural products often reveal novel sites of action (personal communication, Dr Stephen Duke, 30 January 1998).

Selecting compounds for screening

The chemicals to be tested in screens can be chosen at random, in the hope that a small proportion of a sufficiently large number tested will show interesting biological activity (the 'random' approach to screening). Another method of selecting compounds for screening is based on their relationship to families of plants, insects and microorganisms that are known to produce certain classes of compound activity (the 'chemotaxonomic' approach to screening). Yet another method is to choose chemicals because they occur in environments that are likely to cause them to produce certain kinds of compound (the 'ecological' or 'biorational' approach). Samples for screening can also be chosen because they are known to have certain biological activity, from research conducted by reading relevant literature or searching through patents and databases. Where such knowledge is the result of observation by local people, this basis for selecting materials for screening is known as the 'ethnobotanical approach'. A wide variety of literature is used by companies to identify genetic resources for research, including searching patent databases for loopholes in competitors' patents.

From 'hits' to 'leads'

At each stage in the discovery and development process, companies must decide where to invest the time and money needed for the next phase of research, and this may involve choosing between several compounds that reveal biological activity in screens. Companies need to ensure that the one compound, from some 10,000 samples screened which show promise, is a good enough commercial prospect to recoup the costs of its own research and the costs of abortive research on the other candidate compounds rejected during discovery and development. Consequently, scientists look for new products that are likely to generate very high revenues, so new leads are chosen based on a combination of the following, demanding criteria:

- *for pesticides, high activity and selectivity towards the target organism and lower ecotoxicity,* so that the potential pesticide will not damage or be toxic to other organisms and the environment, and so that smaller quantities of pesticide are needed (although the success of non-selective herbicides such as glyphosate and glufosinate means that non-selective herbicides remain of interest to crop protection companies);
- *high potency of the chemical,* so that smaller quantities of pesticide are needed, and to reduce the contamination of food and the environment; and
- *novel chemical structure and new mechanism of action,* so that resources are not wasted developing a product that is similar to other, earlier products, and so that a company can gain a commercial advantage by patenting novel compounds resulting from the research.

As well as avoiding 'me too' products and strengthening the likelihood of obtaining a patent, the novelty of a compound's structure and mechanism of action is increasingly important to overcome the resistance that many pests have developed to existing pesticides (Clough and Godfrey, 1994).

When an extract has revealed potentially interesting biological activity in a screen, the next step for a natural product extract is generally to identify the structure of the compound responsible for the activity, in the process known as 'deconvolution' (Box7.8, p198).

Formulation

Once the compound with pesticidal properties has been identified, researchers need to identify the optimal dose rate and formulation.

Some natural products use the active chemical ingredient in its original, unmodified state. In this case, the remaining development work focuses on the most appropriate and effective formulation for the product. Should it be in liquid or solid form? Solid-form pesticides can be delivered as powders or granules. Liquid formulations can be water-based or in organic solvents. As well as the active principle, several other ingredients or 'adjuvants' are added to help disperse or otherwise deliver the product. Box 7.15 below, and the LUBILOSA case study (p217), both describe research to develop the most appropriate formulation for biocontrol agents.

Semisynthesis

Work to optimise the formulation of a compound as it occurs in nature is often not enough to ensure its efficacy. Compounds as they occur naturally frequently suffer from a range of unattractive characteristics, such as instability (so they break down under field conditions), toxicity, and lack of selectivity (resulting in damage to species other than those targeted by the pesticide). Furthermore, they are often present in low concentrations and are difficult to purify on a large scale. For this reason, many natural products become the subject of chemistry programmes to develop them into more potent and effective compounds by altering their chemical structure. Typically, the fermentation of micro-organisms is used to produce large quantities of the naturally-occurring compound. The structure of the compound is then changed slightly through chemical reactions.

Wholly-synthesised analogues modelled on natural product templates

Some chemical pesticides are entirely synthesised, rather than being extracted from naturally-occurring samples of plant or microbial origin. Compounds are synthesised for three main reasons. First, they may have been synthesised from scratch in the laboratory from inorganic and organic chemicals and not be derived in any way from genetic resources. Second, it may be more cost-effective to synthesise a compound discovered from nature in the laboratory, rather than to obtain sufficient raw material from nature for its scale-up and manufacture, or even to start with natural material and alter it through semi-synthesis. Third, compounds discovered from nature may not only be slightly modified through semi-synthesis, but instead altered almost beyond recognition through lengthy and sophisticated chemistry programmes (Box 7.9). This section will briefly examine synthetic compounds within this third group, since this category is particularly interesting from the point of view of access and benefit-sharing.

Where a compound has demonstrated interesting pesticidal activity but needs to be further developed, attempts are made to improve its activity by modifying its chemical structure. An 'analogue', or compound similar in structure to the original compound, is synthesised. The analogue, or a family of analogues differing to varying degrees but all based on the original structure, forms the lead compound for further screening and chemical synthesis. Computational chemistry, combined with quantitative studies of the relationship between the structure and activity of the compound, is often used to direct chemical synthesis programmes on the lead analogue. Chemists modify the lead compound to find the molecule that is the most potent, the least toxic, and the easiest to synthesise, and which has a novel chemical structure and, preferably, a new mechanism of action. The resulting compound will have been designed for optimal biological, physical and environmental properties, and is often extremely different from the original natural template.

Several compounds that are now wholly-synthesised were discovered from work on analogues with a chemical structure, or 'template' originally derived from the chrysanthemum. The pyrethroids are an example of such a class of compounds. As Box 7.9 describes, 'pyrethrum' is an extract of chrysanthemums long-known for its insecticidal properties, but the naturally-occurring compound suffered from the disadvantage that it

Box 7.9 *Pyrethroids*

Dried and powdered flowers of the chrysanthemum *Tanacetum cinerariaefolium* (previously *Chrysanthemum cinerariaefolium*), which originated in Persia and Dalmatia, have been used since the beginning of the 19th century for the control of household pests. An extract from the flowers known as 'Pyrethrum' has long been used to control flies and mosquitoes, but it is deactivated by air and sunlight. Since the 1940s, chemists have produced a class of compounds derived from the chrysanthemum known as 'pyrethroids', of which esfenvalerate and cypermethrin are perhaps the best examples, given their commercial success as agrochemicals. The structure of the active compound isolated from the natural extract pyrethrum formed the lead for this family of compounds, whose structure has been modified from its natural form in the laboratory. Pyrethroids continue to be an important class of natural product crop protection products, with global sales of over US$1 billion each year.

Source: personal communication with Dr Steve Brewer, August 1998.

ceased to be active when exposed to air and sunlight. Since the 1940s, a whole family of compounds known as the 'pyrethroids' has been developed by chemists through semi-synthesis and total synthesis. If a chemist were to conduct further work on the pyrethroids today, he or she would study the chemical structure of the family of compounds by consulting patent and other literature on the subject, and would then synthesise the molecule in the laboratory, perhaps using existing, related compounds already in the company's collection or purchased from a catalogue as the basis of the synthesis work. The discovery and development programme would therefore have no reason to return to access chrysanthemum plants themselves.

Another example of a wholly-synthesised product that was based on a template originally derived from nature is Azoxystrobin, a broad spectrum fungicide developed by Zeneca (Box 7.10). Azoxystrobin is one of a class of compounds known as the strobilurins. There are now over 500 patents and patent applications on strobilurin compounds. All of these owe their origin to the work of Zeneca with Azoxystrobin, and of Prof Timm and Dr Anke at the University of Kaiserslautern and BASF (personal communication, Dr Len Copping, February 1999). The most likely first source to which a scientist would turn before conducting further work on the strobilurins would be the patent literature. Chemical structures disclosed in patents would offer plenty of leads. A scientist could produce new compounds differing from those described in patents, but modelled on the same basic template, by synthesis in the laboratory, or by semi-synthesis from existing strobilurin compounds in the company's library.

The carbamates provide a third example, and one which many interviewees did not believe was linked in any way to natural products. The family of compounds known as the carbamates is now wholly synthesised, but owes its origin to the compound physostigmine, a naturally-occurring product of the seeds of an African vine known as the 'calabar bean', *P venenosum*. The

Box 7.10 *Azoxystrobin – A Broad-Spectrum Fungicide*

Azoxystrobin (Methyl (*E*)-2-{2-[6-(2-cyanophenoxy)pyrimidin-4-yloxy]phenyl}-3-methoxyacrylate) is one of a new class of fungicides based on a family of natural compounds including strobilurins, oudemansins and myxothiazols, which are secondary metabolites produced by a range of Basidiomycete and Ascomycete fungi. Strobilurin A can be sourced from an edible parasitic mushroom, *Oudemansiella mucida*, found on European beech trees, as well as from the fungi *Strobilurus tenacellus, Bolinea lutea, Xerula melanotricha*, and *Mycena galopoda*. Both Zeneca and BASF have conducted research on these metabolites, and have launched strobilurin products. This box follows the development of one product, azoxystrobin, by Zeneca.

In 1981, Dr Brian Baldwin at Zeneca read a German publication which described the similar function of strobilurin A, oudemansin A and myxothiazol A. Czech and German investigators had discovered that the secondary metabolites concerned kill other species of fungi competing for the same nutrients.

In 1982, Zeneca obtained samples of the metabolites from Prof Timm and Dr Anke, researchers at the University of Kaiserslautern and the Gesellschaft für Biotechnologische Forschung in Germany. Zeneca scientists synthesised other metabolites in their own laboratories, and all the compounds were screened for activity against plant pathogenic fungi at Zeneca's Jealott's Hill Research Station in the UK. It soon became clear that none of the compounds could be used as commercial fungicides in their own right. Oudemansin A, for example, is weaker than conventional fungicides and cannot be produced economically. A programme of chemical synthesis was initiated to create analogues with better physical properties and more cost-effective means of manufacture. Key objectives included stabilisation (to reduce volatility and increase photostability and persistence in field conditions), improved systemicity (to encourage better redistribution of the chemical inside the plant), high fungicidal activity, and low phyto- and ecotoxicity. Zeneca developed a large number of analogues. In 1984, Zeneca applied for its first patent applications on these compounds, and a total of 72 further applications were published up to 9 May 1997. More than 20 other companies and research organisations have investigated strobilurin analogues, mainly as fungicides, and over 500 associated patent applications have been made.

In November 1992, Zeneca announced the successful development of the broad-spectrum, sprayable fungicide azoxystrobin, which emerged from its work on analogues of strobilurins and oudemansins. Azoxystrobin is active against spore germination, mycelial growth and sporulation, and was selected from among 1,400 analogues. Zeneca believes that no other commercial fungicide combines such a broad target spectrum (the product is effective against four major groups of fungal pathogens, some of which have developed resistance to other fungicides) with such high levels of activity. Field trials of azoxystrobin since 1989 have tested the compound against fungal pathogens on more than 60 crops in a wide variety of climatic and environmental conditions. Since commercialisation, Zeneca has continued its research on the product to ensure that targets do not build resistance to it.

Azoxystrobin was first registered in Germany in April 1996 for use on cereal crops. Further registrations for use on 13 crops were received in 20 countries. In June 1997, it also received its first crop registration in the USA under the EPA's Reduced Risk Pesticide Scheme. It is currently being applied to cereals, vines, bananas, tomatoes, peaches, peanuts, pecans and turfgrass, and is sold under a number of trade names such as Amistar®, Quadris®, Abound®, Heritage®, Bankit® and Orvita®.

Source: Adrian Wells, 1998, using Clough and Godfrey, 1998 and Zeneca, 1996.

Box 7.11 *An Illustration of a Different Structure of a Wholly-Synthesised Commercial Herbicide and its Natural Compound Precursor*

Figure 7.5 (a) shows the structure of leptospermone, a naturally-occurring herbicidal component from the bottlebrush plant (*Calistemon* spp). Figure 7.5 (b) shows the structure of a wholly-synthesised derivative, sulcotrione, which has been commercialised in Europe for broadleaf control in corn, after extensive optimisation of the triketone class of compounds.

Figure 7.5 (a) Leptospermone

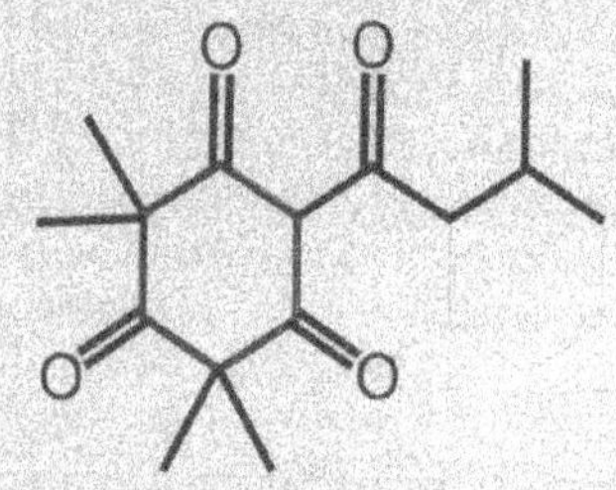

Figure 7.5 (b) Sulcotrione

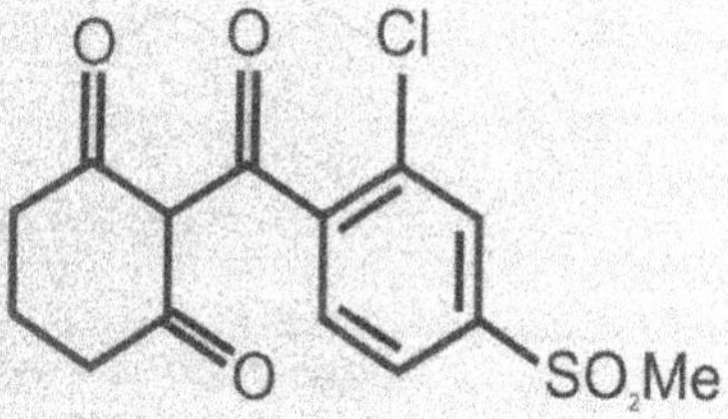

While work on leptospermone led eventually to the development of the commercialised product sulcotrione, it would not be possible to tell from the structure of sulcotrione that it owes its origin to leptospermone.

Source: Lee et al, 1997, and personal communication with Dr Steve Duke, February 1999.

plant alkaloid physostigmine was first isolated from calabar beans in the late 19th century, and characterised in the first half of the 20th century. During the 1930s its methyl carbamate functional group was found to be an inhibitor of acetylcholine esterase. The Union Carbide Corporation manufactured the compound carboryl, sold under the trade mark 'Sevin'. Sevin met with commercial success, and several other companies started to develop and manufacture carbamate compounds. The carbamate compounds currently in use as crop protection products are now totally synthesised, and modelled after one small part of the original, naturally-occurring molecule (*Merck Index*, 12th edn, and personal communication with Dr David Hunt, February 1999).

If you trace far enough back into the development history of wholly-synthesised crop protection compounds, it is possible that you will find a naturally-occurring ancestor on which the work in the field is based, as the examples of the pyrethroids and the carbamates have shown. However, the significance of this fact for access and benefit-sharing is limited. It demonstrates that natural compounds have historically provided the source upon which crop protection products were based (Box 7.11). The natural products currently on the market listed in the *Pesticide Manual* show that they continue to represent a modest, but important, proportion of the crop protection market, and for these there is a continuing demand for access. However, future work on wholly-synthesised analogues patterned on molecules once derived from nature generally does not require access to genetic resources, and thus circumstances will not arise that could trigger negotiations on benefit-sharing. As we will see below, the majority of companies in the crop protection sector do not share benefits arising from sales of wholly-synthesised analogues with the original providers of samples that are not closely linked to the final compound, but became the basis of lengthy chemistry programmes.

Toxicity and efficacy

The successful compound – whether a 'pure' natural product, a semi-synthesised natural product, or a wholly-synthesised analogue, is subjected to a range of further tests in the glasshouse and in the field to determine its phytotoxicity, efficacy, stability in UV light, and all-round development potential.

Development

If a few compounds emerge from the discovery phase and the research team is satisfied that they are strong candidates for commercialisation, they proceed into the lengthy development phase (phase II in Box 7.6). At this stage, the compounds are subjected to a battery of tests designed to assess their chances for product approval.

The compounds are tested in more advanced glasshouse screens which reveal their biological activity against a wider range of related species. In addition, more rigorous toxicity tests are performed to establish that the candidate pesticides are not toxic to non-target organisms, and do not persist in the environment. The testing procedure builds up a picture of the relationship between the chemical structure of the candidate pesticides and their positive and negative features. The synthesis and testing of analogue compounds is repeated and the compounds progressively modified until they achieve the desired characteristics of high efficacy, low mammalian toxicity, low environmental toxicity and reasonable manufacturing costs. If candidate pesticides progress well through this phase, they are tested in field trials to see how they perform in the full range of environmental conditions such as wind, rain, sunlight and drought. Finally, the most promising candidates enter into full-scale field trials.

Another important component of the development phase is to increase the scale of production of the compound, to ensure that, should the product pass the final hurdle of regulatory approval, it could be produced reliably and cost-effectively on a commercial scale.

Biological control agents

Overview

Biological control uses the lifecycle and metabolism of whole natural predators, parasites, parasitoids and pathogens to kill target pests. There are a number of different biocontrol agents (sometimes known as 'biopesticides'), including bacteria and baculoviruses that produce proteins toxic to insects, pathogenic fungi that kill insects and weeds, insects and parasitic nematodes that kill insects and molluscs, and naturally-occurring yeast that kills fungi which infect citrus fruits and grapes.

Biological control agents are often seen not as a substitute for a chemical pesticide, but as a complementary part of an IPM programme (Box 7.1, p188). Their mass introduction can be highly effective in killing pests, especially when the crop protection systems are designed to take into consideration cultural practices and plant selection, thus creating an environment favourable to the predators and to resistance in the host plant. One of the advantages of some biocontrol agents such as those based on fungi or viruses is that, once released, they can reproduce, thus achieving further production of the control organism (in a process known as 'secondary' or 'horizontal' transmission). Under the right conditions, biocontrol agents can kill more pests for a longer period than chemical pesticides. Biocontrol organisms can be formulated as sprayable pesticides and applied either to crops or to the soil on which crops grow. They can be stored in forms such as sachets of dry spores, or liquid concentrates which maintain the predatory organism in a viable form for a sufficient period to give the product a reasonable shelf-life, so that when the biocontrol pesticide is needed, it can be diluted with water and sprayed on to a crop in the same way as a chemical pesticide.

According to Stephen J Linsanki, Director of CPL Scientific, the size of the global sprayable biopesticide market in 1996 was between US$80 and US$85 million, and growing between 10 and 15 per cent annually, with an expected market size of US$150 million by the year 2000 (Johnson, 1997).

Of the different categories of biocontrol agent, the naturally-occurring bacterium *Bacillus thuringiensis* (Bt) has made the biggest commercial contribution. There are over 30 recognised subspecies of the bacterium, each producing different insecticidal toxins, and work continues to isolate strains with new toxins, and to manipulate the Bt genes that encode toxin production using both recombinant and other methods (Georgis, 1997). Annual sales of Bt products as pesticides are now some US$60–70 million (personal communication between Adrian Wells and Dr Jim Haeger, Abbott Laboratory, 29 February 1999; Johnson, 1997).

Box 7.12 *Progress in the Development of Microbial Insecticides*

Issues	1980s	1990s
Products	*Bt*	*Bt*, nematodes, virus, fungi
Cost	Higher than standard insecticides	Comparable to standard insecticides
Shelf-life		
Bacillus thuringiensis	<1 year	2 years
Helicoverpa zea NPV	1–2 months	6 months
Steinernema carpocapsae	1–2 months	5 months
Efficacy	Inferior to standard insecticides	Comparable to standard insecticides
IPM programmes	Minimum usage	Widely used
Market		
Bacillus thuringiensis	Forestry, vegetables	Forestry, vegetables, cotton, corn, fruit trees
Helicoverpa zea NPV	–	Cotton, vegetables, tomato
Spodoptera exigua NPV	–	Vegetables, cotton, grapes
S carpocapsae	Home and garden, cranberries	Home and garden, berries, turfgrass, ornamentals, citrus, mushrooms
S riobravis	–	Citrus, turfgrass
Beauvaria bassiana	–	Cotton, glasshouses

Source: Georgis, 1997.

Table 7.2 *Major Companies Involved in Marketing Microbial Insecticides*

Company	Bt	Nematode	Virus	Fungi
Abbott	+			
AgrEvo	+		+	
Andermatt			+	+
Brickman			+	
BioBest				+
Becker	+			
Calliope	+		+	+
Ecogen	+			
Koppert	+	+		
Microbio		+		
Mycogen	+			
Mycotech				+
Novartis	+	+		
SDS Biotech	+	+		
ThermoTrilogy		+	+	+
Troy BioScience				+
Zeneca Mexicana			+	

Source: Georgis, 1997.

Abbott is the major manufacturer of Bt products, which are used to control larvae of over 55 species of Lepidoptera on more than 200 crops world-wide. The development of Abbott's Bt products XenTari and Dipel is illustrated in Box 7.13 p203.

This chapter contains examples of three other kinds of biocontrol agent: the biological pesticide 'DENY', based on live bacteria *Burkholderia (Pseudomonas) cepacia* type Wisconsin (Box 7.15); the use of the parasitic insect *Stichotrema dallatorreanum* to kill grasshoppers and bush crickets that damage oil palm in Papua New Guinea (PNGOPRA and Oxford University, Box 7.14), and the use of Metarhizium fungi to target locusts and grasshoppers in Southern and West Africa (LUBILOSA: a case study, p217 below).

Biological control is highly dependent on access to genetic resources in the form of living organisms. At US$80 million, biopesticides (of which Bt accounts for the major part) still represent a niche market, since they are often relatively costly, have a narrow environmental window for efficacy, shorter shelf life and special handling requirements. The technology is complex and state support is often needed for biological

Box 7.13 *XenTari and DiPel : Two Biological Control Agents Used in Rotation*

Abbott Laboratories is the world's largest producer of insecticides based on *Bacillus thurengiensis (Bt)*, a naturally-occurring insecticidal microbe. Abbott ferments the microbe itself, and its products are used world-wide to control the larvae of over 55 species of lepidoptera on more than 200 different crops, including vegetables, fruit, cotton and corn. Each subspecies of *Bt* produces a number of specific insecticidal toxins. The quantities in which each toxin is produced vary according to the specific strain of the *Bt* subspecies, and it is these variations which give each strain its target specificity. In order to target a range of pests, combinations of different strains can be used. Abbott's two primary biological insecticides, DiPel® and XenTari®, contain strains of different *Bt* subspecies and, particularly when used in rotation, have been shown to fight resistance to pesticides in a range of pests.

The 'HD1' strain of the *Bt* subspecies *Bacillus thurengiensis* spp *kurstaki* (*Btk*), used in DiPel®, was first discovered and patented by USDA in the 1960s. In 1970, Abbott licensed the strain from USDA and used it in the first formulation of DiPel®. This formulation was subsequently improved over the years for consistency and cost-effectiveness under a variety of crop, weather and pest conditions. Protein from spores of the microbe binds to the gut of larvae, but only those with certain receptor cells, so that DiPel® has the advantage of being highly target specific. Tests have shown that use of the insecticide has a minimal effect on vertebrates, the environment and beneficial arthropods. Several formulations of DiPel® are currently on the market to meet crop and farmer needs. The product is compatible with popular fungicides and insecticides as tank mixes, and is cheap and economical to apply, so it is proving popular for farmers in countries such as Ghana, where it is currently the largest selling biological pesticide. DiPel® has never been patented and Abbott Laboratories does not enjoy exclusive use of the 'HD1' strain of *Btk*. Other *Bt* manufacturers have also isolated and used the strain. According to Abbott, however, DiPel® has retained a significant edge over competing insecticides due to its historical place on the market, as well as the company's emphasis on technical support to growers and strict manufacturing and quality-control procedures inherited from its human pharmaceuticals background.

The potent *Bt* subspecies *Bacillus thurengiensis* spp *aizawa* (*Bta*), used in XenTari®, was discovered in 1960 from a soil sample obtained from fields in Wisconsin by Abbott Laboratories. In 1987, the 'Xentari' strain of *Bta* proved to be the most potent *Bt* strain in screens of more than 100 strains. In January 1990, Xentari was one of four candidates chosen for field trials and further development and, in December 1990, an application was submitted for US EPA Experimental Use Permit (EUP). In February 1991, the strain was submitted for US patent filing and, in July 1991, an EUP was approved. In August 1991, 'Xentari' was submitted to the EPA for product registration, and approval granted in August 1992. As a result of Abbott's patent on the strain, other companies would be able to sell products containing it only under licence. XenTari® is designed for application to vegetable crops and, according to Abbott, can be more effective against pests than other *Bt* products, carbamates and pyrethroids.

Sources: Adapted from Wells, A, 1998, using Abbott *Bt Product Manual*, (1992), pp 2, 3–6.2, 116–17, *Bt History* from Abbott Laboratories, Inc, Abbott Laboratories *Product Sheet 1/98*, DiPel *DF Product Brochure*, Abbott Laboratories, 1997, *Xentari History* from Abbott Laboratories, Inc, and personal communication with Mr BMK Yankah, Managing Director, Jeloise Company, Kumasi, Ghana, 19 August 1998.

Box 7.14 *Use of a Parasitic Insect to Control Defoliation of Oil Palm by Grasshoppers in West New Britain Province, Papua New Guinea*

The Papua New Guinea Oil Palm Research Association (PNGOPRA) is a research body representing private and state-owned oil palm companies, and local smallholders. 50 per cent of its funding is provided by the oil palm industry, and 50 per cent is contributed by the government of Papua New Guinea and project-specific foreign aid – in this case, the EU, which supports Papua New Guinea's oil palm export earnings under the Stabex mechanism of the Fourth Lomé Convention.[1]

This collaboration between PNGOPRA and the Department of Zoology of Oxford University aims to control defoliation of the oil palm in West New Britain by grasshoppers and bush crickets, particularly *Segestidae defoliara defoliara*, using the parasitic insect *Stichotrema dallatorreanum* (strepsiptera). The female form of *S dallatorreanum* lives within its host's abdomen, depriving the host of nutrition and rendering it sterile. This parasite does not occur naturally in West New Britain Province, which supports 70 per cent of Papua New Guinea's oil palm, and control has so far depended on the use of 'Monocrotophos', a highly toxic, trunk-injected, organophosphate insecticide. Monocrotophus is expensive, hard to apply consistently during certain seasons, difficult to monitor in remote areas and frequently applied too late to prevent yield loss.

Previous attempts to introduce *S dallatorreanum* into areas outside its home range have not succeeded due to a failure to account for the parasite's basic biology and ecology. This project seeks a better understanding of *S dallatorreanum* by studying its means of reproduction and feeding behaviour, the host-parasitic relationship, the infection of the target species under captive conditions, releasing infected *Segestidae defoliaria defoliaria* into selected sites in West New Britain Province, and monitoring of the effects on wild populations of the pest.

The Department of Zoology of Oxford University has provided PNGOPRA expertise on the physiology and molecular systematics of strepsipteran parasites, and a formal collaborative agreement was signed in August 1995. Since 1997, the PNGOPRA has conducted trials on captive specimens of the target pest by infecting them with *S dallatorreanum*. These specimens are despatched to Oxford for determination of the success of infection. If sustainable reproduction of the parasites in this target pest species is achieved, experimental releases of captive *Segestidae defoliaria defoliaria*, infected with *S dallatorreanum*, will take place in West New Britain Province in 1999, followed by assessments of the impacts on wild populations of the pest.

PNGOPRA and Oxford University intend to release the resulting pest control technique to oil palm companies and smallholders for free. However, according to the agreement, if PNGOPRA wishes to exercise its exclusive right to market project inventions to other parties, or Oxford University wishes to claim such inventions as its intellectual property, the parties will negotiate further terms, including the development of a royalty-sharing mechanism.

1 The Fourth Lomé Convention is an agreement between 12 members of the EU and 68 African, Caribbean and Pacific (ACP) States, signed on 15 December 1989, which aims, over ten years, to address agriculture and food security, service provision, industrialisation, and cultural, social, regional and environmental concerns through structural adjustment support, debt relief, and investment and development financing. Its trade and commodity provisions include mechanisms such as Stabex, a system designed to stabilise export earnings from certain agricultural commodities by providing ACP countries with financial support aimed at covering shortfalls in earnings due to fluctuations in prices and agricultural production.

Source: Wells, A, 1998, based on personal communication with Dr Kathirithamby, Department of Zoology, Oxford University, 16 February 1998; Kathirithamby, J, Simpson, S Solulu, T, and Caudwell, R, 1998; Solulu, T, Simpson, SJ and Kathirithamby, J, 1998, and Agreement, 'Relating to Research on Oil Palm Insect Pests and their Natural Enemies,' between the Chancellor, Masters and Scholars of the University of Oxford and the Papua New Guinea Oil Palm Research Association, provided by Dr Kathirithamby with kind permission of the University, 18 March 1998.

control programmes to be economically attractive. In this climate, biological control is currently pursued by several research institutes and a few companies. Some of the companies we interviewed were dismissive in their estimation of the likely contribution of biocontrol agents to the gamut of crop protection products in the future. Others, however, including several major multinationals, believed that biocontrol would increase in importance in the future. The quality of microbial insecticides based on Bt, nematodes, viruses and fungi has improved over the last decade, and they are now more widely applied (Box 7.12). It would appear that many major crop protection companies are beginning to divert significant research into biocontrol, and to integrate biological and chemical control as part of their programmes on IPM, and several smaller companies specialise in a growing range of biocontrol products (Table 7.2).

Research and development of biocontrol agents

The research and development process for microbial biocontrol agents can involve the study of receptors and screening of strains of microbes for their biological activity, in much the same way as described in the section on 'Discovery' (p196), and Box 7.6. The major difference is that biocontrol agents apply the living organism itself, while chemical pesticides use compounds extracted or derived from microorganisms and other genetic resources (Boxes 7.13–7.16).

The research and development process for biological control agents that employ insect predators differs significantly from that for chemical pesticides. The approach is largely based on traditional biology, studying the nature of plant disease and the behaviour of natural predators. Researchers find natural enemies of pest species and, using literature searches, ascertain their potential beneficial effects. Glasshouse trials demonstrate how effective the predator is against the pest. Once a predator with proven potential has been identified, a plan to rear it on a commercial scale is formulated. In the case of insect predators, a laboratory colony is established and the possibility of 'mass rearing' is investigated on a small scale. Some insects prove too difficult or uneconomical to rear. In those cases where it appears possible to rear an insect with appropriate predatory behaviour cost-effectively on a commercial scale, the insect is then subjected to more glasshouse trials, for example, on the client company's premises or at a horticultural research station. If successful, the biological control agent is then marketed as a product.

The discovery costs associated with the testing of potential biocontrol agents are relatively modest, as the organisms themselves are collected from the field. The major costs are associated with showing that the biological effect is sufficient to merit commercialisation, and then establishing a suitable production and distribution procedure. Product approval and registration can also be relatively cheap. In the USA, for example, registration of predatory insects by APHIS is far cheaper than the registration of a chemical pesticide by the EPA, and is particularly simple if the organism is already present in North America. It has been suggested that costs of development and approval for a biocontrol product are in the order of US$1 to US$5 million (personal communication, Dr Len Copping, February 1999). These modest costs, however, must be considered in the context of the comparatively small markets for biocontrol products.

Most of the biological-control companies we interviewed focused on work with biocontrol agents, ranging from bacteria, baculoviruses, fungal pesticides and bacterial pesticides to predatory organisms, but a few were also breeding beneficial pollinators for the glasshouse industry, such as bumble bees. Few of these companies were involved in fundamental research such as original observation of predator species, which is usually conducted by universities. Their focus was on the technologies for formulation and application of the final product.

Product approval regulations

In most countries, the registration of crop protection products (CPP), particularly chemical pesticides, is governed by comprehensive regulations that must be satisfied before a product can be marketed and used. While these vary from country to country, and are thus difficult to summarise, some of the more common requirements are described in this section.

Specific regulatory requirements must be satisfied and met prior to the release of a new CPP. Interviews with regulators in the USA, Japan, the EU and Argentina suggest a considerable range in the time and costs associated with product registration. For example, the

Box 7.15 *'DENY' – A Microbial Pesticide Based on a Live, Patented Bacterium*

'DENY' is a biological pesticide based on live bacteria, *Burkholderia (Pseudomonas) cepacia* type Wisconsin. This symbiotic rhizobacterium aggressively colonises crop roots, reducing the sites available for disease pathogens and nematode attack while doing no harm to the plant itself. It also spreads a repelling substance into the surrounding soil and appears to increase the immunity of the crop plants it colonises to diseases of foliage such as white mould. There are four products currently based on *B cepacia* type Wisconsin: Liquid Fungicide, Liquid Nematicide, peat-based Seed Treatment Fungicide and peat-based Seed Treatment Nematicide.

A variety of isolates of *B cepacia* type Wisconsin exist, a range of which were first obtained in Australia between 1985 and 1987 by a PhD student funded by Lubrizol Genetic Inc, a company that later became Agri Genetics. Agri Genetics filed for and obtained a US patent covering the broad-spectrum fungicidal activities of this technology, on 17 January 1989. The patent identifies the specific process of identification: a sequence of seven taxonomic steps through which *B cepacia* must go before it is recognised and classed as type Wisconsin. All subsequent registration actions have used this designation to keep selected type Wisconsin isolates separated from other very similar *B cepacia*, some of which are pathogenic to onions.

In 1989, the Stine Seed Company purchased from Agri Genetics all rights to the *B cepacia* type Wisconsin technology. Stine Microbial Products, a subsidiary of the Stine Seed Company, identified additional *B cepacia* type Wisconsin isolates from corn fields in Wisconsin and surrounding states, and discovered nematicidal properties (in addition to the largely fungicidal isolate obtained from Agri Genetics). Stine Microbial Products was granted a US patent, and in 1992 succeeded in developing dry, peat-based formulations of both the fungicidal and nematicidal isolates of *B cepacia* type Wisconsin as a seed treatment. US EPA registration for this technology, labelled as Blue Circle (BC) Seed Treatment, was completed in 1992. Further formulation work by Stine Microbial Products resulted in the development of liquid concentrate versions of fungicidal and nematicidal *B cepacia* type Wisconsin. These have shelf lives of only six months, but are easier to apply. In 1995, EPA registration of the liquid concentrate formulations was obtained.

Certain varieties of *B cepacia* have been identified as opportunistic human pathogens, raising concerns about the application of *B cepacia* type Wisconsin to food crops and, thereby, its effects on humans. Stine, the EPA and the medical community are collaborating to keep 'type Wisconsin' separate from these other pathogenic varieties. Genotyping and other methodologies have been undertaken on *B cepacia* type Wisconsin, setting a precedent that may become part of the EPA registration system for biological control products.

Sources: personal communication, Mr Vernon Illum, Market VI, LLC, 9.10.98, 14.10.98 and 19.10.98; *'Intercept' Biofungicide*, product information sheet, Soil Technologies Corporation, Fairfield, Iowa.

Box 7.16 *Agrisense BCS Ltd: Biological Control Using a Natural Sex Pheromone from Pectinophora gossypiella, the Pink Bollworm of Cotton*

It has long been known that many pest species use pheromones as a means of communication. Two compounds, Z-7, Z-11-Hexadecadienyl Acetate 50 and Z-7, E-11-Hexadecadienyl Acetate 50, make up the natural sex pheromone produced by the Pink Bollworm, *Pectinophora gossypiella*, a serious pest of cotton throughout the world. Female Bollworms produce this pheromone to attract males for mating. Technology for extracting and identifying such a pheromone did not become available until the 1960s, and it was only in the 1970s that researchers at USDA first identified the two compounds constituting the sex pheromone of the Pink Bollworm.

After this initial discovery by USDA, researchers in the USA, Egypt, Pakistan, India, Europe, Mexico, and Peru began exploring potential application of this knowledge, some independently and some in collaboration, and this resulted in a wealth of knowledge in the public domain. Initially, work focused on synthetically manufacturing the sex pheromone for use in traps as part of Pink Bollworm monitoring programmes. Capture of male Pink Bollworms in such traps provided valuable information on the size and distribution of the pest population and helped farmers make considered judgements about crop protection strategies.

Once chemists had developed economic methods for synthesis of the sex pheromone, its use for mating disruption was evaluated. Under this scenario, the pheromone is released at higher rates than would naturally occur, inundating the air so as to disrupt the ability of male Pink Bollworms to locate individual females. This prevents mating and, as a result, future crop damage. It took many years to discover the optimal method for applying the pheromone in such a manner, requiring development of formulations that degraded less rapidly and that could release the pheromone into the air over an extended period. It also became necessary to come to a better understanding of the Pink Bollworm's ecology to ascertain circumstances for achieving the best results.

Agrisense, a Welsh company that develops and commercialises biological solutions to crop protection problems, contacted researchers in the USA in the late 1980s. The company's subsequent work on the pheromone focused both on trap systems containing the pheromone for use in monitoring as well as controlled-release formulations to disrupt mating. The research and development surrounding these products has enabled Agrisense to build up a body of knowledge on the most economical means of synthesising the pheromone's compounds and the most efficient controlled-release systems. Agrisense has patented a number of synthetic routes to manufacture the two compounds involved, while its controlled release technologies are protected by non-disclosure of the processes and materials involved (trade secret). Such intellectual property protection provides the company with a critical edge over competitors developing equivalent Bollworm pheromone products for the same market.

Agrisense is now marketing Pink Bollworm pheromone technologies to Peruvian, US, Pakistani, Greek and Israeli cotton growers, as well as to the Egyptian Ministry of Agriculture.

Source: Wells, A, 1998, from personal communication with Mr Enzo Casagrande, Agrisense BCS Ltd, Pontypridd, UK, 11.8.89, 27.8.98, 28.8.98, 12.10.98.

registration of a generic product can take as little as three months (in Argentina), and the registration of a new active principle can take as much as 60 months (in the EU).

Other than requirements in some countries relating to the chemistry of the active substance and the description of its characteristics, regulatory requirements for crop protection products can be loosely defined in two categories, namely efficacy and safety.

Efficacy

Efficacy studies on the product, or different variants of the product, are carried out according to predefined protocols designed to assess the overall spectrum of activity of the product; its specificity, crop tolerance and safety, and its general benefits weighed against current commercial standards. Recommendations on the precise use of the individual product arise from such efficacy studies, and are included on the product label and other commercial literature.

Once the basic biological activity of a new crop protection candidate has been defined and a company has moved on to the development of its active substance, it is likely to screen a range of related chemical compounds (analogues). The company may develop a range of different formulations of the active substance, such as wettable powder, emulsifiable concentrate, granules etc. In order for sales of the new product in its various formulations to cover the costs of discovery, development, manufacture and commercial launch, and for the product to be a commercial success, it will need to be registered and used in most major countries of the world. Most companies will conduct trials in a wide variety of countries where the crop on which the product is targeted is grown. This is both to take advantage of the ability to conduct tests over two seasons during a single year (by working in the northern and southern hemispheres), and also to meet specific testing requirements that are still mandatory in most countries.

Safety

Safety requirements are met by extensive testing conducted on both the active substance and the entire crop protection product. Tests may also be conducted on the metabolites of an active substance. The spectrum of safety testing is to determine that the use of the product will cause no short- or long-term adverse effects to users of the pesticide, consumers of the crops it is grown on and other people who may be affected by it. Safety tests also examine the impact on the environment, including soil leaching and persistence and ecological indicators such as its effects on birds, fish and other 'non target' macro- and microorganisms. Safety studies are conducted both in the laboratory and in real-life environments, in order to ensure adequate coverage of

Box 7.17 *Typical Safety Standards for Crop Protection Products Required in Europe and the USA*

Active substance	Crop protection product
Acute toxicity in mammals.	Acute toxicity in mammals.
Sub-chronic toxicity – 90 day.	Acute toxicity to aquatic species.
Chronic toxicity/carcinogenicity 2-year studies in rats and another relevant species.	Crop residues in target crops in accordance with critical good agricultural practice.
In vivo and *in vitro* mutagenicity.	Environmental field studies for degradation, leaching and persistence.
Reproduction and developmental toxicity.	Field biological trials for efficacy and crop safety.
Metabolism and pharmacokinetics in mammals.	Special environmental studies.
Acute studies in birds, aquatic species and micro and macro non-target species.	Risk assessments and modelling to determine appropriate safety margins for operators, consumers and environment. Risk mitigation strategies.
Effects on beneficial organisms.	
Metabolism in plants.	
Environmental studies in the laboratory and field, including leaching, persistence, accumulation, metabolism and uptake.	
Degradation in water and air.	

Source: Dr R Rowe, Dow AgroSciences.

the full spectrum of exposure to the product through its proposed use.

A typical range of safety studies are set out in the appropriate guidelines of the EU (in Directive 91/414/EEC) and the US Environmental Agency (in Section F of the EPA product registration guideline), and are illustrated in Box 7.17.

Once the appropriate studies have been carried out in accordance with national guidelines, and the results collated, risk assessments are conducted to determine the appropriate safety margins to ensure that the product is inherently safe to use as prescribed. Risk assessment criteria and methodologies, as compared with hazard assessments, are still in the early stages of development, but already form the basis of some of the scientific evaluation of crop protection products and will do so increasingly in the future.

Biocontrol products

In many countries, biological control agents which are 'macro-organisms', such as predatory insects, do not require registration at all, and in some countries, the regulatory hurdles for biocontrol agents of other kinds are significantly easier to meet than those for chemical pesticides.

Regulations in the USA distinguish between living organisms that are released to serve as predators, which are controlled by USDA (in order to show that they do not have adverse effects such as preying on desirable species), and biochemical and microbial pesticides which are living entities capable of survival, growth, reproduction and infection, but which are not visible to the naked eye, and which must be approved by the EPA. These biopesticides are easier to register than synthetic compounds because there are fewer data requirements and a special group at US-EPA reviews them. The registration of biopesticides can thus be accomplished in substantially less time than is needed for synthetic chemicals (personal communication between Fiona Mucklow and Dr David Hunt, 17 September 1998; and with Dr Laura Whatley, 12 February 1999).

To take another example, in the UK, the fees for biopesticide registration are cheaper than those for other categories of crop protection product. In addition, many countries, including Australia, Austria, Canada, Finland, France, Germany, Italy, The Netherlands, Sweden and the UK, encourage the use of biopesticides through legislation that promotes IPM (OECD 1996).

International harmonisation

Over the last 40 years, and especially since the early 1980s, a number of treaties and non-binding instruments related to the use and approval of crop protection products have been developed. These address international harmonisation of test requirements, conduct manuals, registration procedures, labelling, classifications, packaging requirements, international trade and transport of crop protection products. Some are indicated in Box 7.18. Perhaps the most universally recognised and accepted is the Codex Committee on Pesticides Residues. This programme operates through the FAO/WHO Joint Meeting on Pesticide Residues (JMPR) and the corresponding, complementary process of the Codex Alimentarius Commission, an intergovernmental group which approves food standards. In the Codex Process, the FAO/WHO JMPR makes recommendations for Maximum Residue Levels (MRLs) and Acceptable Daily Intakes (ADIs), respectively. The review goes through several stages until final approval is given by the Alimentarius Commission, at which point an MRL becomes a Codex Maximum Limit (CXL), and the international standard for products is set (personal communication, Dr Laura Whatley, February 1999).

At the international level, the harmonisation of product registrations and the mutual acceptance of decisions in different jurisdictions still have some way to develop. Numerous initiatives have been explored in the past. More recently, common guidelines and regulatory assessment criteria have been established within the European Union, but the registration of crop protection products is still handled by the competent authority within each Member State. The OECD has initiated a programme of work that pioneers the global harmonisation of regulatory requirements and regulatory assessments together with the sharing of assessing data between countries.

Box 7.18 *International Law Related to Crop Protection Products (CPP)*

About 50 international organisations are concerned with CPPs, including:

- Agencies of the United Nations – FAO; WHO; the International Labour Organisation (ILO); the United Nations Industrial Development Organisation (UNIDO); and the United Nations Development Programme (UNDP);
- The European Plant Protection Organisation (EPPO);
- The Organisation for Economic Cooperation and Development (OECD);
- Non-governmental organisations, including the International Union of Pure and Applied Chemistry (IUPAC); the International Group of National Associations of Manufacturers of Agrochemical Products (GIFAP); and the International Organisation of Consumers' Unions (IOCU).

Elements of international treaty law address chemical pesticides, though not directly their registration as CPPs. The *International Convention on Dangerous Chemicals and Pesticides*, adopted in September 1998, provides for legally binding regulations in the international trade of hazardous chemicals and pesticides, and covers information exchange on national bans and PIC for the import and national production of chemicals for national consumption. The Bamako Convention (1991) bans imports into Africa of hazardous substances which have been banned, cancelled or refused registration by government regulatory authorities in the country of origin on health or environmental grounds. Nine CPPs are persistent organic pollutants (POPs) and are addressed by the United Nations Economic Commission for Europe (UN/ECE) under the *Long-Range Transboundary Air Pollution Convention (LRTAP)*, and a Protocol on POPs is currently under negotiation.

Article 8(h) of the CBD (1992) addresses bio-control agents, though not their registration as CPPs. It obliges parties not to introduce, as well as to control or eradicate, alien species that threaten ecosystems, habitats and species.

Voluntary schemes and codes of conduct relevant to the registration and classification of CPPs include:

- The *International Programme in Chemical Safety (IPCS)*, applied by UNEP, ILO and the WHO;
- The *International Register of Potentially Toxic Chemicals* (1989), applied by UNEP;
- *Recommendations on Tests and Criteria for the Classification of Dangerous Goods* (1991), applied by UN ECOSOC;
- The UNEP *London Guidelines for the Exchange of Information on Chemicals in International Trade* (1987), recommending the development of legislative and regulatory structures, the creation of national toxic chemical registers, and improvement of information exchange;
- The *Consolidated List of Products* replaced the *Codex Alimentarius* Commission (1962), and contains lists of products which have been banned, withdrawn, or severely restricted. It is designed to prevent the uninformed registration of pesticides new to a country's markets but restricted or banned elsewhere;
- Chapter 19 of *Agenda 21, UNCED* 1992, 'Environmentally Sound Management of Toxic Chemicals, Including Prevention of Illegal International Traffic in Toxic and Dangerous Products', contains two programme areas spanning the international assessment of chemical risks, and the prevention of illegal international traffic in toxic and dangerous products;
- Computer Aided Dossier and Data Supply (CADDY) is an electronic dossier interchange and archiving format for plant protection registration procedures recently presented to the OECD Pesticides Forum after agreement by the EU, its Member States, the US-Environmental Protection Agency (US-EPA), the European Crop Protection Association (ECPA), the American Crop Protection Association (ACPA), and the Canadian Crop Protection Institute.

Relevant FAO Codes include the *FAO Guidelines on Environmental Criteria for the Registration of Pesticides* (1985); the *FAO Guidelines for Registration and Control of Pesticides* (1985), *Addenda* (1988); the *FAO Guidelines for Legislation on the Control of Pesticides* (1989); and the *FAO International Code of Conduct on the Distribution and Use of Pesticides* (1985), amended in 1989, which defines the responsibilities of public and private entities in the trade, distribution and use of pesticides. It also includes rules on pesticide management, testing, reducing health hazards and adoption of regulatory and technical requirements, including registration and recording of imports, information exchange and PIC.

Decisions of the OECD Council relevant to pesticide registration include the *OECD Recommendations On Mutual Acceptance of Data (MAD) in the Assessment of Chemicals and Good Laboratory Practices (GLP)* (1988 and 1989); the *OECD Council Decision on Minimum Pre-Marketing Set of Data in Assessment of Chemicals* (1987); the *OECD Decision-Recommendation on the Systematic Investigation of Existing Chemicals,* (1987); and the *OECD Decision-Recommendation on the Cooperative Investigation and Risk Reduction of existing Chemicals* (1991).

The crop protection industry supports the harmonisation of registration processes and regulatory guidelines, since the development costs of bringing a new active substance and associated crop protection products to the market, in a range of countries each with different regulatory regimes, is ever increasing.

7.2 Access to genetic resources

Discovery and development of biopesticides requires access to genetic resources. For chemical pesticides, access to genetic resources is needed to develop screens and to source many of the compounds for these screens. Those developing biological control agents may need access to pests and predators alike. The development of disease-resistant crops may also require access to genetic resources, as described in Chapter 5.

This section explores which organisations require access to genetic resources and from where these genetic resources are obtained, addressing collecting activities and the use of in-house collections. It also reviews the kind of genetic resources in demand by researchers, and how researchers set about selecting them. The section ends with an assessment of the likely future demand for access to genetic resources for the development of crop protection products.

7.2.1 Which organisations require access to genetic resources?

All the organisations conducting research on chemical pesticides and biocontrol agents require access to genetic resources. Thus companies, from the multinationals to the smaller enterprises with one or two staff members run from a university department, public research institutes and universities all seek access to new samples. As the following section shows, some of these scientists collect samples for themselves, while others obtain them from other organisations.

7.2.2 Where are genetic resources obtained?

Collecting activities

Ten out of the 20 crop protection companies and organisations who responded to this question said that their staff occasionally collect samples directly in the field. While collections by their staff accounted for anything between 1 and 70 per cent of their acquisitions, most of those interviewed only rarely acquire resources in this way. The majority of samples are acquired not by company staff, but from other institutions.

As well as sourcing of original samples for discovery, companies obtain materials for industrial-scale manufacture of the final product. Some companies also maintain their own plantations from which to source materials. For example, Murkumbi Bioagro Pvt Ltd, in India, has planted 28,000 neem trees on its own land. These trees supply just a small proportion of the neem the company uses, and the company depends on harvesting by local communities for the bulk of its supplies. According to Murkumbi, 'We buy from markets, and with neem there is a developed auction system based on collection from wild trees on community lands. The auction system has been running from 20 to 30 years.'

In-house collections

Ex situ *collections*

Institutions researching crop protection products often maintain libraries of genetic resources that may contain samples of seeds, soil samples, pure microbial cultures, pathogenic fungi, enzymes, extracts from microbial cultures or plants and compounds such as pheromones. Some companies maintain these collections indefinitely in their libraries, while others maintain only active collections, and discard samples after a certain period if they have not shown promise. Most companies also maintain collections such as colonies of live arthropods, either for research on biocontrol agents, or against which to test potential crop protection products. Compared with the germplasm collections maintained by most seed companies, and the libraries of extracts and compounds maintained by many pharmaceutical companies, collections in the crop protection sector tend to be more focused and modest in scale, although crop protection companies which are part of groups of companies with divisions devoted to pharmaceuticals and other branches of the life sciences can often access the collections held by all companies in the group.

Most companies that develop and market final products use their libraries for in-house purposes alone, although licensing out access to their collections is an option several interviewees mentioned they were considering for the future. Mergers and acquisitions form another source of samples, as the collections of both companies are often consolidated and shared. For companies involved in the production and licensing of precursors rather than final products, the licensing of improved materials from their collections is often an important part of their business.

Some companies concentrate their research efforts on the exploration of genetic resources upon which they have not worked before, whereas for others, the bulk of their work involves the refinement and manipulation of specimens that are well known to them. The companies interviewed ranged in their approaches from working exclusively on new genetic resources – 'We only look for new organisms' – to no expenditure at all on basic research on hitherto unexplored specimens – 'All our work involves known resources. Our academic collaborators do the discovery'.

Others' collections

Some scientists obtain their own specimens for research by collecting them in the field, or by using their own in-house, *ex situ* collections. However, access to collections maintained by others is an important source of materials for crop protection researchers, who access genetic resources from a variety of different sources described in this section.

Public sector

Research institutes and universities tend to maintain something of the 'open access' ethic of exchange that still prevails in much of the seed industry. As one interviewee put it, 'We are in the non-commercial side, which means we are free and open.' For this institution, an important source of genetic resources was exchange between scientists in the research community. Such open exchange depends on the nature of material exchanged. When dealing with potentially dangerous genetic resources such as tropical plant viruses, strict quarantine controls are enforced, operating as a restriction on the exchange of germplasm. Another factor restricting exchange is the increasing use of patents, even by academic institutions, to protect modified genetic resources.

Universities and networks of research scientists form an important source of genetic resources for companies. Organisations conducting research in similar areas – whether in the private or public sector – tend to know each other, and to form networks facilitating access to collections of genetic resources with certain traits. A number of companies interviewed described academics working on agricultural programmes funded by government as an important source of samples. Indeed, as crop protection research is often subsidised by government, public agencies are a common source of genetic resources for companies. Botanic gardens form a fairly important source of genetic resources for several companies. Interviewees also spoke of obtaining material from government research centres, and 'international collections' such as those of the FAO and the CGIAR.

Private sector

Since crop protection companies often run the same kinds of screen, they are generally not interested in acquiring compounds that have already been screened by another agrochemical company. However, they are willing to acquire compounds that have been used in completely different screening programmes, for example in pharmaceutical discovery. As one company said, 'the pharma[ceutical] industry enters into business with us because we are not competition'. Companies engaged in the development of crop protection products involving the genetic modification of crops often exchange genetic resources with seed companies.

Collaboration between companies through research alliances forms another important source of genetic resources, as does the sharing of libraries by subsidiaries within the same corporate family. With the current spate of mergers and acquisitions, the wholesale fusing of two companies' collections can also provide the source of genetic resources.

Finally, an entire category of smaller 'search and discovery companies' specialises in building libraries of genetic resources, which they license to larger companies. Some such companies are spin-offs from university departments, and specialise in the collection and isolation of new and identified compounds, sometimes obtained in the course of the university's academic. These companies license extracts, compounds and cell lines to others.

7.2.3 What kind of genetic resources are in demand?

Raw materials or value-added derivatives?

There is a trend for larger companies to 'outsource' the preparation of intermediate products such as libraries of synthetic chemicals and active compounds isolated from natural products. However, it is still more common for companies involved in discovery and development of crop protection products to acquire raw materials – such as soil samples containing microbes or nematodes and plant specimens – than to purchase value-added materials. Most companies acquire basic extracts and isolated strains of microbes (rather than raw soil samples), but comparatively rarely obtain more value-added materials, such as strains with proven activity. In the case of chemical pesticides, the bulk of the natural products acquired for screening programmes is in the form of extracts. By comparison, in the biocontrol sector, it is more common to acquire 'identified bioactive compounds' – such as pheromones and attractants that have already been identified by academics, and come with associated research data.

Although standard practice is to acquire only partially processed materials, several companies mentioned that they would be open to more value-added collaborations if value-added materials were available. As Dr Dieter Berg of Bayer AG put it, 'If possible, we would like pure compounds. That is the idea that makes us cooperate with suppliers. We would like an extract that we can screen directly, and preferably one that has already been screened revealing interesting activity so that we can re-screen it within the company.' Some interviewees described long-standing collaborations in which they were supplied with value-added materials, some of them from developing countries.

Companies sometimes need to hire external expertise to support their research on genetic resources. As Dr Mike Legg of Zeneca Agrochemicals explained, 'There are so many different experts needed to study natural products that it is impractical to have one of each in-house. We no longer have microbiologists of our own actively isolating microorganisms. Instead, we choose to out-source this activity.' Suppliers of genetic resources may thus be able to find opportunities to provide genetic resources with associated information and accompanying skills in research. Since needs and practices vary from company to company, suppliers need to know which companies are 'outsourcing' which skills in order to take advantage of such opportunities.

One relatively common practice in the crop protection industry is for companies based in different geographical regions with their varied regulatory frameworks, to license each other's final products. Thus, for example, a Japanese company may not enter the US market-place and register crop protection products there, but may instead enter into partnership with a US company, which will be familiar with regulatory requirements in the USA and will have networks for marketing its products. In such an arrangement, the US company would typically license the compounds developed by the Japanese firm, develop them into final products; put the products through the registration procedure in the USA, distribute them through its marketing outlets, and share the profits with the Japanese company.

Domestic or foreign genetic resources?

While the majority of interviewees appeared to know the origin of all the samples used in their research, others said they were often quite unaware of the geographical origin and source of the genetic resources they use in research. As one company explained, 'We often have no idea where a sample came from.'

However, some companies develop a geographical or ecological focus for their collections, such as temperate or tropical regions, often linked to the origin of the crop lines they are working to protect. For companies interviewed, genetic resources acquired from other countries represented from just 10 per cent to over 90 per cent of the materials they acquired. It is noticeable that companies' desire to add to their collections or to source materials from developing countries varies over time. One multinational referred to a subsidiary in India as a major source of genetic resources for several years, but explained that it had subsequently closed down its Indian operations and no longer sourced genetic resources from that country. The changeable nature of the market for genetic resources means that it is difficult for organisations providing them to other countries as a source of income to rely on individual customers for long-term partnerships. A more diversified client-base can help alleviate, but not altogether reduce, this problem.

Synthetic compounds compared with natural products

It is generally a great deal easier for companies to acquire synthetic compounds in the large numbers required for high-throughput screening than it is to acquire samples of natural products on a similar scale. Some of the companies interviewed purchase synthetic compounds on a small scale from academic chemists, but several buy or license whole libraries off the shelf from search and discovery companies. By comparison, collections of natural products are generally not available in such large numbers, so companies tend to purchase natural product extracts through a plethora of testing agreements. One multinational company referred to 'hundreds' of agreements for sourcing samples.

One-off access or return to source

By and large, companies accessing genetic resources for natural product screening programmes only need to obtain an initial sample, and do not need to return for larger volumes of material if a candidate pesticide looks promising. As far as plants are concerned, companies hope to acquire enough material initially to allow the structure of the active compound to be identified without returning to the source for more material. This might typically involve acquiring some 100–500 grammes (personal communication, Dr Stephen Duke, 30 January 1998) of dry material, with accompanying information on whether the material is taken from leaves, stem, roots, flowers or seed. However, when companies acquire plant samples, if the quantities are such that they may need further material, guaranteed re-supply becomes an important factor. Microorganisms pose less of a problem, as they can frequently be bulked up through fermentation processes, without the need to obtain more source materials.

7.2.4 How are genetic resources selected for research?

Methods of selecting genetic resources for research

Interviewees described a range of methods used to select the genetic resources for research and development programmes. Some companies distinguished between factors that are important for selection of samples for the development of chemical pesticides, for biocontrol agents and for new, disease-resistant crop varieties. For example, one company explained the different factors at play when seeking genes (for crop improvement) and chemicals (to screen for potential pesticides). When looking for genes to breed pest-resistant crops, geographical factors are the most important and the company uses a targeted approach, focusing on places where it is likely to find a certain trait, such as heat stability or resistance to drought. However, when the company is looking for natural product samples for screening, the major factor is biological diversity.

As described in the section on the discovery of chemical pesticides (p196 above), the basis for selecting specimens for screening may be random, 'chemotaxonomic', 'ecological' or 'biorational'. Many of the large companies running high-throughput screens still obtain material from as diverse a variety of sources as possible. However, many cited chemotaxonomic and ecological/biorational approaches as the main basis for collection. Directed approaches to collection such as these are often spurred by information that suggests that particular species, genera, ecosystems or families of compound may have interesting properties. Thus literature searches that can provide these tips are often an important first step in a company's research on potential leads. Once a company scientist finds a promising article, he or she may contact the author and ask whether it is possible to obtain the samples described in the article.

Some of the information available to researchers focuses on traditional uses of genetic resources, rather than contemporary biochemical research. The companies we interviewed described using ethnobotanical data in a number of different ways. Most large-scale researchers turn to databases to establish whether historical use of genetic resources may support and speed up the proof of safety and efficacy during the product approval process. Several described using 'knowledge in the public domain', and using any ethnobotanical information supplied with samples. A few use ethnobotanical knowledge to guide the collection or screening of samples, although only to a limited extent. As one company put it, 'We factor ethnobotanical data into our decisions, but it is not the most important aspect of our choice. If suppliers used this to collect samples, for example, for insecticide use, then these samples would be prioritised above others in those screens.' The extent to which ethnobotanical data is used depends on local people's experience with the approach to crop protection concerned. One biocontrol company explained that while traditional farmers 'do not know much about pheromones', they do use certain foodstuffs to attract insects, and the company now sells a biocontrol product using this method, after it identified the active constituents in the material.

For a few companies, often from developing countries, ethnobotanical data is a vital part of product discovery and development. These companies record ethnobotanical information when collecting in the field. Murkumbi Bioagro Pvt Ltd, based in India, has found that ethnobotanical information is highly localised, varying from village to village. The most useful tips came from areas where the company is working on natural pest predators and parasites.

Factors influencing the selection of genetic resources and partners for collaboration

According to market experts (Uttley, 1997), opportunities for suppliers of intermediate products, such as chemical compounds for use in combinatorial chemical research or the production of synthetic or semisynthetic compounds, are on the increase, as new and more complex active ingredients are being developed. The main focus of the major multinational corporations is not the manufacture of these intermediates, so they seek these value-added products from suppliers. Suppliers are chosen for the following qualities:

1 state-of-the-art competence in key technologies and the ability to defend developments by IPRs;
2 readiness to develop new intermediates;
3 ability to deliver intermediates requiring less time to move from discovery to the market;
4 readiness to take a proactive approach and share risks in strategic alliances;
5 ability to develop kilogramme quantities for field trials;
6 transparent communication of costs, to help cost-efficiency exercises;
7 continuous cost improvement, especially when a product approaches patent expiry;
8 validation of the manufacturing process (ie more efficient ways to produce a product at the industrial scale); and
9 environmentally acceptable facilities, with waste reduction initiatives.

Although this list was compiled with reference to factors influencing the selection of suppliers of chemical intermediates, our interviews suggest that, with the exception of the two points specifically linked only to intermediates (points 2 and 3, above), all these factors would be important in companies' choice of suppliers of natural products.

Interviewees cited a number of criteria as important in their selection of the suppliers from whom they would source samples for product discovery and development (Figure 7.6). Some gave the impression that they would go to the ends of the earth for a particularly interesting sample, but the majority appeared to rule out specimens that were not easily obtainable. For the former group, quality of the sample and sometimes the calibre of the collaborating institution that would provide it were paramount. Any extra cost in sourcing the sample, such as following permitting procedures, was seen as 'part of the costs of daily business', and would be integrated into the product price. For the latter group of companies, the ease of the permitting procedure and cost per sample were higher considerations. If it appeared that it would be costly or time consuming to obtain an interesting specimen, this group of companies would abandon the search and focus on other specimens.

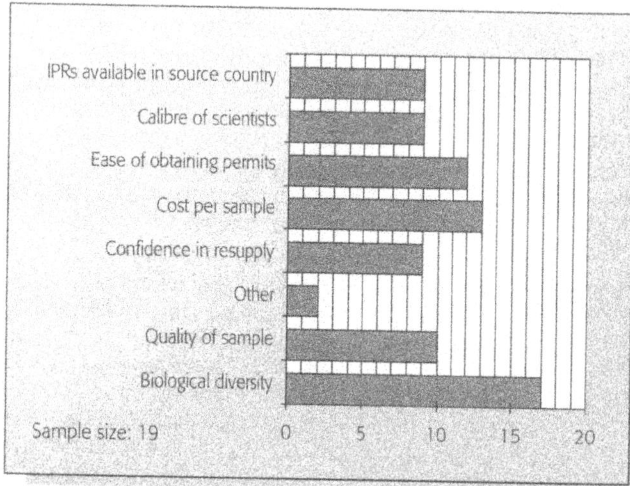

Figure 7.6 *Criteria for Selection of Samples*

Several interviewees stressed the predominance of 'business' considerations, such as the ability to secure a monopoly through the use of IPRs and competitive benefit-sharing arrangements with suppliers, often described as 'sensible licensing and royalty agreements'. Logistical issues can be important, too. In order to preserve interesting secondary metabolites it may be necessary to keep samples clean, fresh and, often, refrigerated. For this reason, reliable electricity supplies and ease of transport are important. These factors, and the reputation of the supplier as an ethical organisation were also cited as criteria (see 'Other' in Figure 7.6). Increasingly, companies need to be confident that issues arising from the CBD have been satisfactorily dealt with. As one company put it, 'We prefer collaborators working in a responsible manner and countries which have signed the CBD.'

7.2.5 Future demand for access to genetic resources

When asked to review their experiences and predict whether, in ten years' time, the demand for access to genetic resources was likely to be greater, the same, or less than today's, 16 out of 19 respondents (84 per cent) believed that the demand for access would increase in the future; two (11 per cent) that it would stay the same and one (5 per cent) that it would decrease. Two reasons were most commonly cited as to why this was likely to be the case. First, the information emerging from research on sequencing plant genomes offers a good source of new biochemical targets for research in crop protection. Second, several interviewees said there was a growing demand for biopesticides in response to the environmental and health concerns associated with chemical pesticides, and felt this would lead to a growing demand for access to genetic resources. Companies were in broad agreement that developments in science and technology would increase the demand for access to genetic resources, although other factors, such as the difficulty of complying with the CBD, might lessen it.

Some crop protection researchers already pay for access to the genetic resources that underpin their work, whereas others obtain them for free, so there was a mixed response to our question as to whether they would be prepared to pay more for access to genetic resources. The majority of interviewees indicated they would be prepared to pay more for access, although some indicated that they would only be prepared to pay for 'value-added' materials, and not 'crude samples'. Companies would be willing to pay in circumstances where they needed the resources for research, if such payment was necessary, and if the price was 'economical' and in line with current market prices (Figure 7.6 above). Some public institutions, however, felt they would be disadvantaged if they needed to pay for access to genetic resources, since they do not have budgets to pay for materials and could not compete with companies. Several interviewees pointed out that research on natural products would need to remain competitive with research on synthetic chemicals, or the former would be abandoned in favour of the latter. As one company explained, 'If the cost gets too high, it shifts the emphasis of our research to [synthetic] chemistry. If a natural product costs ten times the price of a chemical, we would prefer to get ten chemicals in.'

7.3 Benefit-sharing

7.3.1 Monetary benefits

While several institutions, particularly public bodies, universities and research institutions, obtain samples for free, 24 out of 26 interviewees (92 per cent) who answered the question said that they were accustomed to paying for access to samples, generally in the form of a one-off fee and, in some circumstances, royalties or milestone payments. Sometimes an up-front fee allows companies to buy samples outright; on other occasions, the fee secures the right to use material in a defined way, but ownership is not transferred. The price paid for samples varies widely, according to the quality, quantity and nature of the material concerned, so any figures given are not reliable guides. However, some interviewees said that companies would typically pay between US$15 and US$30 for an extract, and US$30 and US$50 for an isolated organism. Biocontrol companies might pay between US$1.5 and US$350 to obtain samples of insects and associated information.

> *'Developing countries overestimate the value of their own resources. Mere provision of genetic resources is not enough. They need to contribute to product discovery and development to get more royalties. We need a fair solution.' (Dr Dieter Berg, Bayer AG)*

Whether researchers pay at all for access to genetic resources, and the scale of payment when they do, are heavily dependent on the degree of improvement of the material and whether knowledge or information is provided with the sample. Companies often pay for 'finished products' but receive 'raw genetic resources' on a 'complementary basis' or for a 'nominal fee'. Similarly, they may pay for a large number of samples from a regular provider, but receive samples under 'one-off' deals for free 'as a good-faith gesture'. Royalty payments are more common for 'value-added' products than for raw samples. However, practices vary from company to company. Some are happy to purchase such value-added products from suppliers, while others wish to maintain quality control and be sure of their title to the final product by conducting their own extraction and analysis. 'It would be risky for us to acquire [derivatives] because it might be the wrong plant or might use others' proprietary methods.'

Payment depends critically on the nature of the institution providing the sample. Public organisations generally do not charge for access, but companies do. Some interviewees said that they may offer royalties to one kind of organisation, but not pay for the same kind of product from another kind of institution. For example, one company explained that, while the outcome depends on the specific agreement, it has agreed to pay royalties to universities, but not, so far, to botanic gardens.

Some companies work out the royalty payments they are prepared to offer at the time the sample is acquired. Others leave the final negotiations until the value of the sample that is being commercialised has become apparent through the research process. One company explained, 'We will determine royalties after we know how much work we had to put in before we could use the compound.' Another said that 'the lawyers do not like this approach, but it keeps the cost down and the speed up in circumstances where you only receive a few samples. Companies are more likely to fix royalty rates at the beginning of a large-scale collaboration involving thousands of samples.'

The magnitude of the royalty payment depends on a number of factors, from level of activity demonstrated by a sample in screens (higher activity leads to higher royalties) to the value of other benefits received. As Mike Legg of Zeneca Agrochemicals explained, 'There may be a trade-off between up-front payments, milestone payments, royalties, technology transfer and other benefits. They will be looked at as a package on a case-by-case basis.' Typical royalty ranges paid by companies are shown in Box 7.19 p216.

7.3.2 Non-monetary benefits

The benefits shared in crop protection agreements tend to be monetary only, unless the provider and receiver of genetic resources are collaborating on a joint programme of research and development. Joint research is generally between scientific organisations, although some companies, particularly those developing biocontrol products, described involving local communities in research programmes, by providing them with free samples of natural parasites and predators in return for information on the success of this 'research and development' at the field level.

In common with the seed industry, many researchers in crop protection saw 'the exchange of materials' as a major benefit of collaboration. The organisations interviewed referred to several other 'in kind' forms of benefit, including travel support. Companies sometimes lend their names to project proposals, thus helping university departments and public research bodies to secure public funding, and are often prepared to cover the costs of taking out intellectual property protection in cases where the IPRs are jointly owned by the company and its collaborator.

Public organisations conducting research on crop protection frequently share information that arises from their research on genetic resources. As one put it, 'There is no question about sharing the results – it happens openly.' Companies, on the other hand, are much more sensitive to competition, and are happier providing information that is not commercially sensitive. Thus they prefer to provide the results of screens if research is not going to be pursued on a given sample than if it is.

Equipment is sometimes sent from companies to their collaborators, especially in developing countries, and some companies allow doctoral students to visit their laboratories, use their equipment, and help with the analysis of the results, in return for a copy of the students' theses.

Where the organisations providing genetic resources have actually protected them with IPRs, companies may 'trade technology for technology'. Cross-licensing of genetic traits is not uncommon in the seed and crop protection industries. Some companies offer concessional licences on certain technologies, particularly in the field of biocontrol. Under such arrangements, a company may allow a collaborating source country to use a product for less than the market price. Other companies collaborate with governments to subsidise the use of crop protection technology by marginal and small farmers.

Many companies provide training opportunities within their laboratories. Staff from the companies' subsidiaries around the world may visit, bringing genetic resources for research, and scientists from external institutions may also be trained. Such training opportunities are rarely in direct exchange for 'one-off' access to genetic resources, and are more commonly in the context of building strategic partnerships with particular institutions.

Box 7.19 *Typical Royalty Ranges for Some Categories of Product*

Raw material/value-added product	Examples of royalty range
Raw samples (eg dried plants, soil samples)	**Range: 0–3%** Examples of specific ranges offered by individual companies: 1) no royalties for raw samples 2) 0.25–0.5% 3) 0.25–2.0% 4) 1.0–3.0%
Extracts (organic or aqueous)	**Range: 0–3%** Examples of specific ranges offered by individual companies: 1) 0.25–0.5% 2) 0.25–2.0% 3) 0–2.0% 4) 1.0–3.0%
Materials with ethno-botanical information	**Range: 0.25–3%** Examples of specific ranges offered by individual companies: 1) 0.25–1.0% 2) 1.0–3.0%
Results of screens provided with materials	**Range: 0.5–3%** Examples of specific ranges offered by individual companies: 1) 0.5–1.5% 2) 1.0–3.0%
Identified, bioactive compound with known structure + activity	**Range: 0.5–2%** Example of specific ranges offered by an individual company: 1) 0.5–2%
Identified pheromones	**Range: 2.5–3%** Example of specific ranges offered by an individual company: 1) 2.5–3.0%
Greenhouse data supplied with biotech compound	**Range: 1–6%** Examples of specific ranges offered by individual companies: 1) 1.0–2.0% 2) 2.0–6.0%
Field-tested and identified bioactive compound	**Range: 2–15%** Examples of specific ranges offered by individual companies: 1) 2.0–5.0% 2) 2.0–5.0% 3) 5.0–5.0%
Commercial product supplied	**Range: 5–50%** Examples of specific ranges offered by individual companies: 1) 10.0–20.0% 2) 5.0–50% 3) 20.0–50.0%

7.4 Case studies

7.4.1 Benefit-sharing by the International Locust Control Programme, *Lutte Biologique contre les Locustes et les Sauteriaux* (LUBILOSA)

Adrian Wells and Kerry ten Kate

Introduction

Locust plagues are an international problem threatening millions of hectares of African farmland. Current locust eradication strategies rely mainly on the monitoring of breeding areas and swarms, followed by treatment with fast-acting chemical insecticides. Effective crop protection, through pre-emptive control of locust swarms early in their development, cannot always be achieved. There are numerous logistical difficulties, such as the remote locations of many breeding grounds, and action therefore comes too late for many farmers. Concerns for user safety and adverse environmental impacts have also made chemical locust control highly controversial. However, the alternatives to chemical insecticides, such as insect growth regulators, suffer from a number of disadvantages, including ineffectiveness against adult locusts.

In 1989, an international, collaborative research programme, Lutte Biologique contre les Locustes et les Sauteriaux (LUBILOSA), consisting of African and European partner institutions, was formed to investigate environmentally-safe, target-specific and effective biological locust control. LUBILOSA's work focuses exclusively on isolates of insect-killing *Metarhizium* fungi, some of which specifically target locusts and grasshoppers. The LUBILOSA programme was provided with US$19.5 million over nine years by four donor agencies: the Canadian International Development Agency (CIDA), the Swiss Development Corporation (SDC), The Netherlands Development Agency (NEDA) and the British Department for International Development (DFID). LUBILOSA is managed by the Programme Management Committee, consisting of representatives from each partner institute, the chairman of the Expert Advisory Committee (a team of independent advisors) and donor representatives.

The Biopesticide Programme of the Commonwealth Agricultural Bureau International (CABI) Bioscience coordinates the LUBILOSA partnership. CABI Bioscience is an intergovernmental, not-for-profit organisation dedicated to the dissemination of scientific knowledge in support of sustainable development. The second partner, the Comité Inter-Etats pour la Lutte contre la Secheresse dans le Sahel (CILSS), also an inter-governmental organisation, is dedicated to reducing the impact of drought in the Sahel and participates through two of its affiliated national programmes: CILSS-AGRHYMET in Niger and CILSS-INSAH in Mali. The third, the Deutsche Gesellschaft für Technische Zusammenarbeit (GTZ), is a German development agency working on behalf of the German Ministry for Economic Cooperation and Development (BMZ). It has been involved in bilateral, and more recently, multilateral, emergency aid measures in support of countries affected by locust plagues. The fourth partner is the Plant Health Management Division of the International Institute of Tropical Agriculture (IITA), an International Agricultural Research Centre based in Benin. These LUBILOSA participants have undertaken fieldwork, laboratory research and field trials of biocontrol agents in close collaboration with African governments, national research institutions, agricultural extension programmes, NGOs and local farmers. In 1998, recognising the need for a private-sector participant to conduct large-scale, commercial production, the LUBILOSA programme developed a partnership with Biological Control Products SA (BCP), a South African company manufacturing and marketing biopesticides. LUBILOSA is currently exploring partnerships with other companies.

The LUBILOSA programme is described below, and summarised in Figure 7.7.

Discovery

180 samples of *Metarhizium* fungi were collected by LUBILOSA's researchers from locusts and grasshoppers (acridoid) across Africa, and bioassayed. 30 of the isolates of *Metarhizium anisopliae* var *acridum* identified were found to be highly virulent in infecting locusts and grasshoppers. One of these strains was collected in Niger in collaboration with staff from CILSS-AGRHYMET (Niger), but under no formal agreement with the Niger government. It was labelled as IMI 330189. This isolate was chosen as the 'standard strain' for further evaluation based on its potential specificity and virulence as a mycoinsecticide (fungal insecticide).

Formulation

The work to develop the fungus for application to locust swarms built on the discovery, in 1988, by Dr Chris Prior of CABI Bioscience that the speed of infection and kill by *Metarhizium* is greater when its conidia (spores) are formulated in oil rather than in water. Oil formulations are suitable for ultra-low volume (ULV) spraying, the most common method of applying locust-control agents in

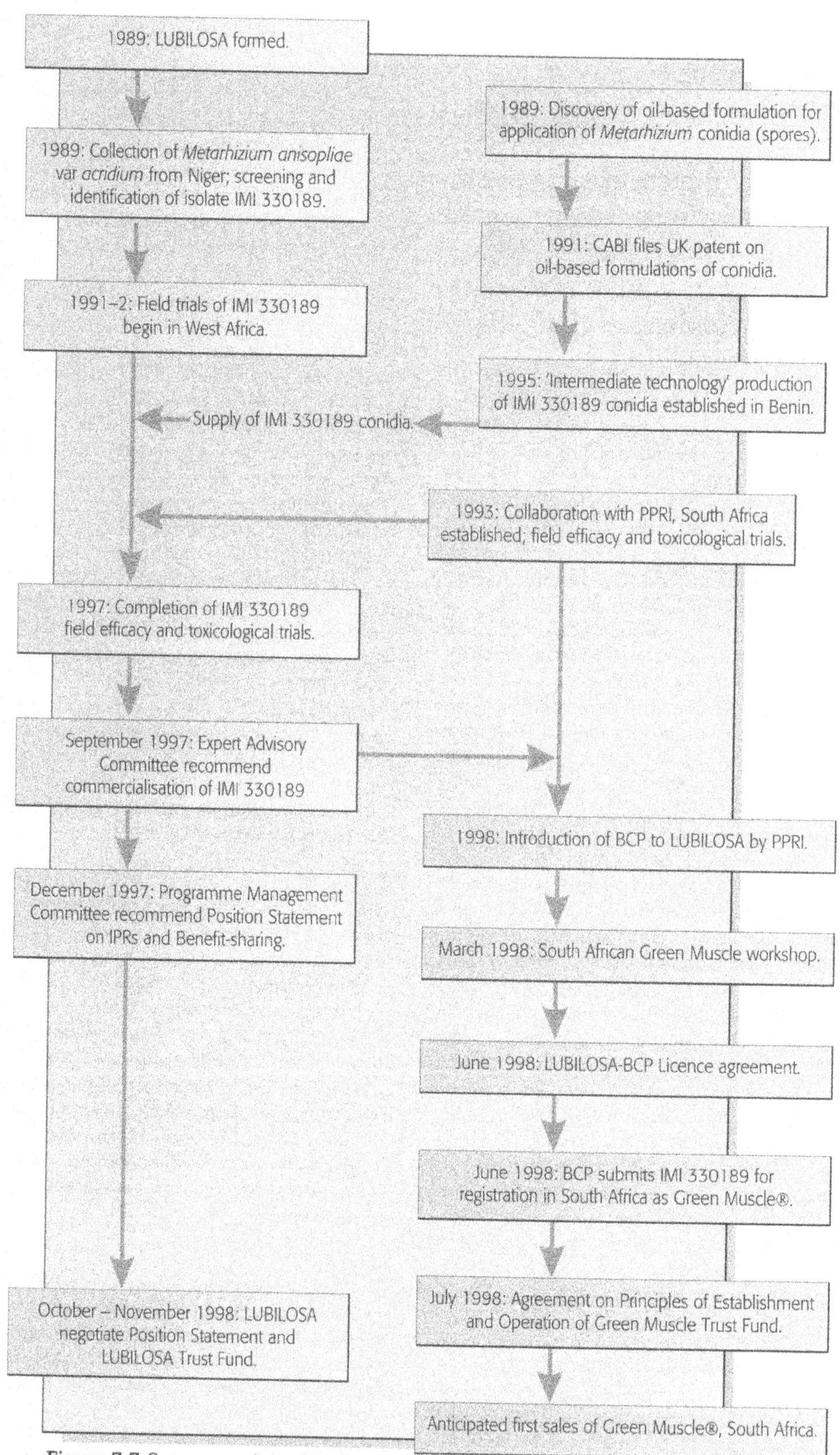

Figure 7.7 *Summary of the Discovery, Development and Commercialisation of IMI 330189*

Africa. Oil formulations have the added advantage of working in semi-arid areas, whereas water-based fungal control has traditionally required humid conditions which do not prevail in many areas that suffer from locust attacks. In 1991, a UK patent was filed on oil-based suspension formulations of deuteromycete fungi, which include IMI 330189, under CABI's name. This ensured the technology remained accessible to LUBILOSA. CABI did not subsequently file patents overseas, because it did not consider it relevant at the time, but will cover the costs of defending the UK patent in the event of infringements. CABI continued its research on formulations up until the end of 1998. During this time, CABI developed an 'SU' formulation for ground-based spraying equipment and suitable for intermediate-technology production, and an 'OF' (oil miscible flowable) formulation for aerial spraying but which can only be manufactured using high-technology equipment.

Efficacy trials

After pilot projects consisting of controlled-droplet applications of IMI 330189, small-scale field trials of the efficacy of the isolate started in 1992 in Niger. Hand-held spraying equipment was used to apply conidia on a few hectares of fallow, peripheral land.

Intermediate-scale field trials on 50-hectare plots, in which formulated IMI 330189 was applied at approximately 2 litres per hectare using ground-based equipment, were subsequently conducted in 1995 in Niger, Mali and Benin by the Crop Protection Training Department (DFPV) of CILSS-AGRHYMET (Niger), CILSS-INSAH (Mali) and IITA (Benin). Scientific evidence of efficacy against a range of target species was obtained.

Meanwhile, a new collaboration was formed with the South African Plant Protection Research Institute (PPRI), part of the ARC, and, during 1994 and 1995, this led to the first large-scale trials in South Africa on the brown locust. Further large-scale field trials, using an aircraft belonging to the Plant Protection Service of Niger, took place on 800-hectare plots in Maine Soroa, Niger, in 1997. A low-volume application rate of 'OF' formulation IMI 330189, at 0.5 litres per hectare, proved successful.

The trials revealed that IMI 330189 has more sustained killing power than alternative treatments, despite lower initial kill, and is effective against other species such as the tree locust, variegated grasshoppers, rice grasshoppers, red locusts and migratory locusts.

The involvement of local farmers has been critical at all stages. The Crop Protection Training Department (DFPV) of CILSS-AGRHYMET (Niger) carried out a training programme in participatory rural appraisal for LUBILOSA staff in July 1995. A socio-economic evaluation of the response of local farmers to IMI 330189 was then conducted in Niger, in collaboration with rural development NGOs. Farmers' problems were assessed and prioritised. Farmers were also informed about the availability and characteristics of IMI 330189, and were encouraged to assist field trials. In 1996, LUBILOSA's NGO partners expressed interest in further trials to encourage farmers to take greater responsibility for crop protection, traditionally a government concern. During 1997, an additional phase of participatory trials was undertaken in which farmers' use of IMI 330189 was compared with standard LUBILOSA applications in Mali, Burkina Faso and Benin.

There are now eight national programmes involved in field testing, spanning West, East and Southern Africa, some of which are the result of countries approaching LUBILOSA and requesting to take part. To date, effective control using IMI 330189 has been demonstrated against locust and grasshopper pests in the Sudan, Benin, Mali, Senegal, Mozambique, Niger and South Africa, among others.

Ecotoxicological trials

In the course of South African field trials, conducted between 1995 and 1998 by CABI Bioscience and PPRI, IMI 330189 was shown to be highly target-specific, with no adverse effect on useful species such as the African bee and no incidence of infections in mammals. Further work by GTZ scientists in Mauritania demonstrated that the fungus has a minimal impact on invertebrates and lizards (which are usually very sensitive to pesticides).

Production of IMI 330189 conidia, commercialisation and marketing

Following submissions of field trial data from LUBILOSA in 1997 and 1998 after the successful completion of efficacy and toxicological testing, the FAO Pesticides Referee Group evaluated and recommended IMI 330189. This gained IMI 330189 international legitimacy, the FAO's approval having been subsequently used by the donor community to gauge the isolate's potential success when used as a mycoinsecticide.

The LUBILOSA Programme has developed a 'two technology' approach to producing conidia in order to meet future demand. One approach is to use locally-available, labour-intensive, 'intermediate' alternatives to imported, capital-intensive equipment technology for small-scale, mass production of IMI 330189 conidia. A pilot project 'Mass Production Team' was established at IITA in Cotonou, Benin, to produce sufficient conidia for field trials. The production process used at the plant consists of two stages, in which a liquid containing IMI 330189 conidia and waste brewer's yeast is first prepared in shake flasks. This is then used to infect sterile rice on which IMI 330189 can grow and produce further conidia in large quantities. The plant can produce sufficient conidia for applications to 2,000–4,000 hectares per year. The other, high technology, and capital-intensive approach, using methods such as down-stream spore extraction and processing, is more appropriate for mass production of high-quality, reliable IMI 330189. Current estimates suggest the market requires sufficient conidia to apply to at least 50,000 hectares each year.

Given these market estimates and the vast areas affected by locusts, LUBILOSA sought private-sector licensees to expand production of conidia beyond the existing output of the Benin plant. Licensees had to be capable of producing conidia of the fungal isolate in an easily handled formulation and on a large enough scale to meet potential demand. A biopesticides company, Biological Control Products SA (Pty) Ltd (BCP), was introduced to CABI Bioscience in 1998 by PPRI. LUBILOSA reviewed BCP's facilities and shareholdings, and CABI Bioscience (signing in the name of LUBILOSA) entered into an exclusive licensing agreement with BCP in June 1998 to produce and market IMI 330189 in South Africa and all other member-states of the Southern African Development Community (SADC). In June 1998, BCP successfully registered IMI 330189 in South Africa, under the trademark 'Green Muscle'. In light of this, BCP is hoping that registration will soon follow in other SADC countries. No sales have yet taken place. To meet demand in other regions of Africa, a second agreement is under consideration with a European-based company and further potential corporate partners are also being considered in Egypt and Kenya. These licensees will also have the option of using the trademark 'Green Muscle' within their designated regions.

Benefit-sharing arrangements

The LUBILOSA Programme has generated a number of benefits, which are shared among partners in Europe and in the African countries involved. These include:

- access to mycoinsectide technology;
- royalties generated from the sale of Green Muscle®;
- capacity building through LUBILOSA's collaborative research and training programme;
- research funding;
- environmental safety; and
- benefits to farmers.

Access to mycoinsectide technology

The LUBILOSA partners and sponsors, all of whom have a development mandate, wish IMI 330189 to be widely available for further research, and also wish Green Muscle and any other products derived from IMI 330189 to be used as widely as possible to control locusts throughout Africa. Conditions for access to the mycoinsectide technology reflect LUBILOSA's 'two technology' approach, described above.

On the one hand, the LUBILOSA participants regard the development of the technology used at the Benin intermediate-technology mass production plant as a public service. The information and expertise generated by the Benin venture is now freely available for replication in other parts of Africa. CABI has therefore given up all IPRs over the simpler 'SU' formulation of IMI 330189, it being better suited to such intermediate-technology production.

On the other hand, CABI continues to maintain its IPRs (on behalf of LUBILOSA) over the 'OF' formulation of IMI 330189. Entities outside the LUBILOSA partnership wishing to commercialise this formulation can gain access to it only if first granted a licence. At the same time, production of 'OF' formulation IMI 330189 requires high-technology equipment, so is only cost effective if performed on an industrial scale by commercial partners. IPRs therefore enable CABI to select only those companies with the necessary manufacturing capabilities, which are willing to share benefits in accordance with LUBILOSA's objectives, as licensees to produce this formulation.

The LUBILOSA partners recognise that the use of IPRs is controversial, given that it restricts access to the protected product. The effect of the CABI patent is that any individual or organisation wishing to commercialise the 'OF' formulation of IMI 330189 must first obtain CABI's permission. The effect of BCP's trade mark on Green Muscle in South Africa is that other companies are not entitled to market a product under the name 'Green Muscle'. Some of the public-sector donors and organisations participating in the LUBILOSA programme are concerned that the exclusive rights granted to private-sector licensees of LUBILOSA's IPRs may conflict with commitments to transparency and public accountability. However, regionally-exclusive licences for production and sales, as well as exclusive rights over trade marks, are important mechanisms by which commercial licensees can recoup their investments in LUBILOSA's technology. They are therefore an incentive for licensees to lay out venture capital and are valuable bargaining chips in encouraging licensees to share benefits.

With these different considerations in mind, the donor agencies supporting the LUBILOSA Programme are currently negotiating a position statement on IPRs and benefit-sharing for LUBILOSA's donors, partners and collaborators. The draft statement covers technology dissemination and transfer; ownership of mycoinsecticide technology (including publicly available information, the origin and distribution of IMI 330189, the patent on oil formulation technology, and donors' positions); benefit-sharing (including trust funds); and action required by LUBILOSA's donors. The position statement is, however, being formulated after having licensed the technology to BCP. LUBILOSA's partners felt that the decision to engage commercial producers could not be delayed until after a position statement had been finalised, given the urgency of ensuring the wide availability of IMI 330189 for application to infested areas.

Royalties generated from the sale of Green Muscle

Green Muscle Trust Fund

With the aim of distributing the benefits which arise from commercialisation and sales of Green Muscle, BCP and CABI Bioscience (acting on behalf of LUBILOSA) are establishing a 'Green Muscle Trust Fund' (GMTF). One-third of the 7.5 per cent royalties (or 2.5 per cent of total proceeds) levied by LUBILOSA on BCP's sales of Green Muscle will be paid into the GMTF and reinvested entirely in South Africa. The funds will be used to build capacity in biopesticides research, development and manufacture by South African institutions, and to increase the availability of environmentally-friendly alternatives to hazardous chemical pesticides. Agreement on the principles of establishment of the GMTF was reached between CABI Bioscience and BCP on 1 July 1998. An account has been opened for the GMTF in a South African bank, and the GMTF has now been approved by LUBILOSA's donors.

Disbursements from the GMTF will be managed by the GMTF Management Board, which comprises representatives from BCP, CABI Bioscience and PPRI (the latter acting as an external body representing the interests of South African agriculture, in its capacity as a member of the ARC). External advisors may also be nominated. The Board will meet within three months of each annual payment of royalties from BCP to consider written proposals for disbursement of funds. Activities for which the GMTF Board may consider funding include:

- studentships to gain academic qualifications based on research relevant to biopesticides, at a South African university, agricultural college or technikon;
- production of educational and promotional materials concerning biopesticides;
- research and development on biopesticides;
- training in the use of biopesticides; and
- purchase of equipment for use by organisations investigating biopesticides in collaboration with PPRI, BCP, CABI Bioscience or other relevant parties.

The term 'biopesticide' may be interpreted as including natural-product based chemicals as well as live biological control agents, at the discretion of the GMTF Board. Allocations will be reviewed annually. Given that no sales of Green Muscle have yet taken place, no disbursements have been made to date.

In order to ensure that benefits are shared equitably across other African nations as well, it has been agreed that the 2.5 per cent of total royalties from sales of Green Muscle in the SADC region outside South Africa may be distributed to a different, international fund, instead of the GMTF.

If research supported by a grant from the GMTF leads to a new commercial product, BCP has the first right to commercialise it. Where this involves BCP commercialising a product arising entirely from the work of a South African reseacher funded by the GMTF, BCP will enjoy sole ownership of any IPRs on that product. This benefit for BCP was negotiated in exchange for the unusually high 7.5 per cent royalty rate. BCP will, however, pay a 5 per cent levy on sales of the product back into the GMTF. If CABI Bioscience (acting on behalf of LUBILOSA) and/or PPRI contribute either research or additional funding to assist a South African researcher in product development, any arising IPR will be owned jointly by those involved, depending on the circumstances. Also, based on the relative contributions of each party, the benefits will be shared by mutual agreement. Here, too, a 5 per cent levy on sales of the product will accrue to the GMTF. The result under each of these scenarios should be an incremental accumulation of funds for further biopesticide research and development in South Africa.

LUBILOSA Trust Fund

In order to ensure the fair sharing of benefits among those African countries which have contributed to the development of Green Muscle, a second trust fund with 'pan-African' scope has been discussed and should be established in 1999. Under the agreement between CABI Bioscience and BCP over the GMTF, the remaining two-thirds of the 7.5 per cent royalties (or 5 per cent of total proceeds) levied on sales of Green Muscle by BCP, will accrue to this proposed 'LUBILOSA Trust Fund'. All royalties from the sales of IMI 330189 in areas other than the Southern Africa region by future commercial licensees will also accrue to this fund; it is not currently intended that further regional trust funds equivalent to the GMTF will be developed. Depending on what they bring to the bargaining table, not all future licensees will pay a 7.5 per cent royalty. LUBILOSA's partners are still debating how disbursements from the LUBILOSA Trust Fund will be directed. However, it is clear that they will in some way benefit the development and use of biopesticides across Africa.

Capacity building through LUBILOSA's collaborative research and training programme

LUBILOSA's European and African partners work closely between themselves, national research institutes, government ministries and other African organisations. The project has therefore provided a unique opportunity for African and foreign researchers to pool resources and data. Benefits, in particular collaborative research and training, have already been shared in this way. For example:

- CILSS-AGRHYMET trained staff from other LUBILOSA partners in participatory rural appraisal for fieldwork in Niger;
- IITA Benin undertook field trials and the development of information packages on IMI 330189 in collaboration with the Benin Crop Protection Service (SPV);
- LUBILOSA undertook collection work and field trials in Niger in collaboration with staff from the Plant Protection Service of Niger;
- CABI Bioscience collaborated with staff from PPRI over field trials in the Karoo region, and;
- a student on study leave from the Niger Plant Protection Service worked with the LUBILOSA team during the 1995 field season in Niger, Mali and Benin, and was trained in insect pathology.

Research funding

Of the US$19.5 million granted to LUBILOSA by donors, a sufficient proportion went to supporting collaborative research in Africa itself to satisfy the donors. The Programme also covered the staff and maintenance costs of LUBILOSA's partner institutions, such as CABI, given the expense and difficulties of coordinating an extensive international network of researchers.

Environmental safety and benefits for conservation

Toxicological trials of IMI 330189, such as those conducted in South Africa by PPRI and CABI Bioscience, and in Mauritania by GTZ, have demonstrated target specificity (affecting only locust and grasshopper species), non-toxicity to vertebrates and user safety.

A priority for LUBILOSA is to ensure that IMI 330189 is a widely-available alternative to environmentally harmful chemicals. BCP's marketing strategy reflects this aspiration in raising public awareness of Green Muscle as an environmentally benign product. In 1998, for example, BCP hosted a workshop in collaboration with CABI Bioscience, bringing together relevant South African interest groups, such as the South African Department of Agriculture, PPRI, locust control officers, farmers' and conservation organisations, LUBILOSA's donor agencies and the media.

BCP intends that initial sales of Green Muscle in South Africa will target environmentally sensitive areas, such as national parks. This is a response to the interest shown in the product by protected area management authorities. While these areas also suffer significant locust damage, they are unable to tolerate environmentally damaging, chemical control techniques and Green Muscle offers them a promising alternative. Furthermore, the application of Green Muscle to these areas may prove an important element in the prevention of future locust damage to neighbouring farmland.

Benefits to agriculture

Farmers will be the principal beneficiaries of Green Muscle, given its safety and efficacy, and the prospects it offers for increased productivity. In South Africa, BCP has liaised closely with the South African Department of Agriculture over the development of a new national locust control policy. BCP is aiming for the inclusion of Green Muscle in a list of Government-recommended locust-control products to be distributed to farmers. Under the new policy, the Government will purchase all products identified on the list. Farmers nominate the products they wish to use, and are supplied them by the Government free of charge. In other parts of Africa, where governments are less able to afford bulk-purchases of locust-control products, they generally obtain them as development assistance from donor organisations for distribution to farmers. Green Muscle may therefore be purchased by both governments and donors.

LUBILOSA has also sought the involvement of farmers in its work through, for example, the dissemination of information on the availability and characteristics of IMI 330189. In Niger and Mali, efforts to obtain the consent of farmers for field trials ensured that they participated both in appraisals of their needs and in fieldwork, through the auspices of CILSS-AGRHYMET (Niger), CILSS-INSAH (Mali) and local NGOs. Farmers in Niger are now engaged in a new scheme designed by CILSS-AGRHYMET to offer a wide variety of entomological collections from the Sahelian region to institutions and researchers. Under it, farmers are paid US$1 for every insect collected.

Benefit-sharing under the BCP-CABI Bioscience licensing agreement is summarised in Figure 7.8, p 223.

Conclusions

Although LUBILOSA's collection work was conducted in collaboration with research and extension personnel from the project's National Programmes, it was not done under formal access and benefit-sharing agreements with source country governments. This was partly due to the fact that LUBILOSA was conceived before requirements for such agreements became international law under the CBD in 1992. Most of LUBILOSA's collection work was done prior to this time as well. As a consequence, the project's benefit-sharing aspirations do not entirely reflect what the CBD would regard as Niger's sovereign rights over of the IMI 330189 isolate of *Metarhizium anisopliae* var *acridum*. Although there is nothing to suggest that LUBILOSA does not recognise such a sovereign right, Niger is not the principal beneficiary under a bilateral access and benefit-sharing agreement with LUBILOSA. Instead, the project's entire work programme, including its benefit-sharing mechanisms, takes a collective, 'pan-African' approach, given that locust control is a 'pan-African' problem requiring 'pan-African' solutions.

LUBILOSA's new position statement does, however, commit the project to upholding the spirit of the CBD. As a post-CBD development, the LUBILOSA Trust Fund will have to reflect this aspiration and LUBILOSA's partners are initiating a process of consultation on appropriate disbursements from the Fund. Formal benefit-sharing agreements with individual source-country governments may have to be brokered if the Fund is to gain legitimacy in light of the position statement. However, the fact that many of LUBILOSA's source countries do not as yet have access laws may significantly delay negotiations and, if so, will conflict with LUBILOSA's targets on progress.

Given such delays, adequate time should be allowed for obtaining the prior informed consent (PIC) of source countries before undertaking collection work. If, however, an international project (such as LUBILOSA) has a number of source countries, obtaining PIC could prove especially time consuming. There will be the need to negotiate separate bilateral access and benefit-sharing agreements with each country in turn.

The development of a single, project-specific, multilateral agreement, involving several source countries together, seems an appropriate solution to this. Such an agreement could be especially relevant where an international project seeks to develop products (such as

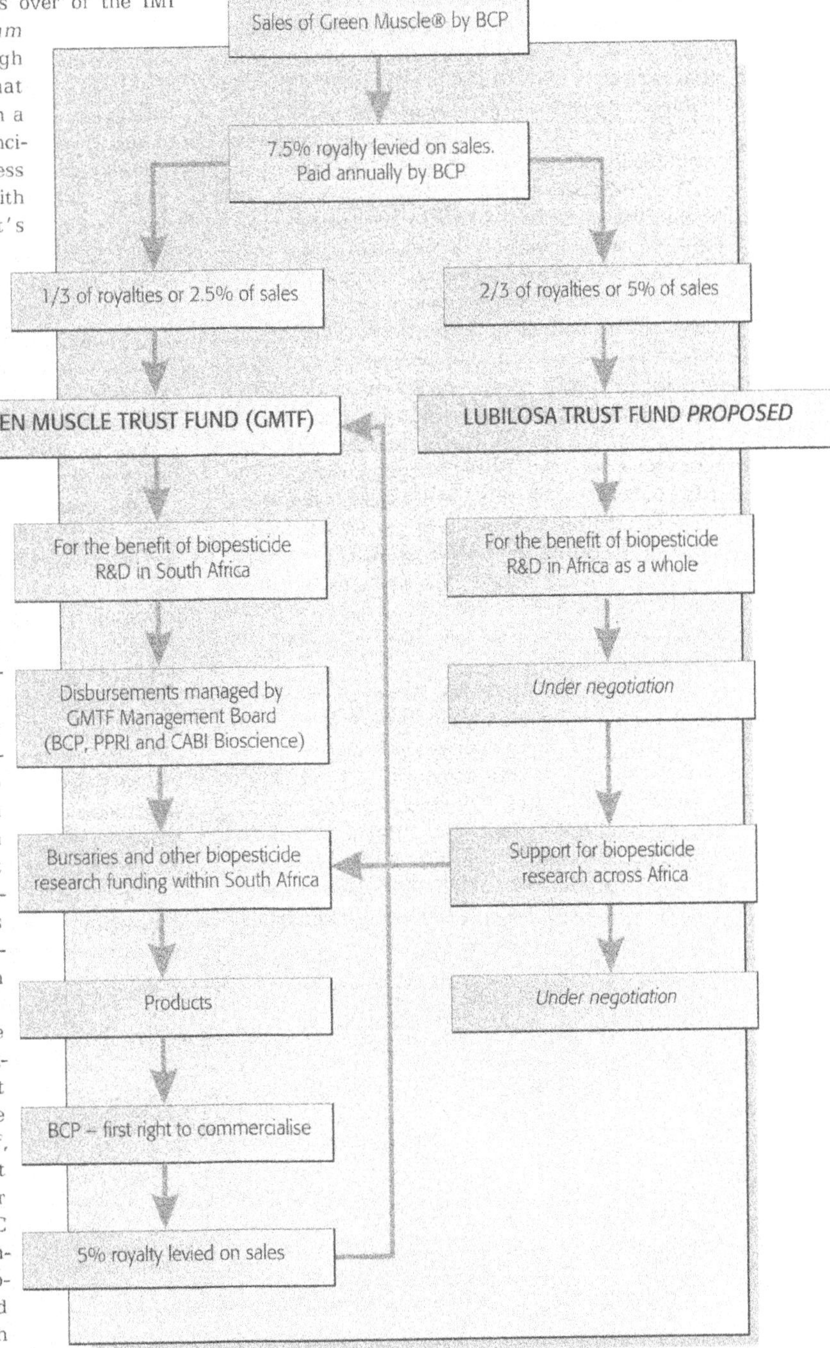

Figure 7.8 *Benefit-Sharing Under the BCP-CABI Bioscience Licensing Agreement*

'Green Muscle') aimed at solving problems of common concern to the source countries involved. A further argument in favour of such a solution is that a bilateral agreement between one source country government and a foreign institution risks redundancy or poor comparative advantage if the same or more promising genetic resources are found in neighbouring countries. As the LUBILOSA project demonstrates, the chances of developing effective products can be greatly increased if access to genetic resources, local research skills, and overseas expertise and funding are pooled between several source countries.

The LUBILOSA experience suggests, however, that benefit-sharing aspirations have to be assessed against other economic considerations, especially when negotiating with prospective commercial licensees. The compromises arrived at include phenomena such as the GMTF. Given LUBILOSA's commitment to encouraging biopesticide research and development across Africa as a whole, the GMTF seems anomalous in retaining a one-third share of the 7.5 per cent royalty levied on BCP for the benefit of South African research, when South Africa is not even IMI 330189's country of origin. The GMTF is, however, the result of a bargain under which BCP's payment of a 7.5 per cent royalty on sales of Green Muscle was made conditional upon its first right to commercialise products developed using GMTF money. This has to be understood in terms of BCP's need, as a commercial partner, to recoup its investments.

(*Sources:* Bateman, R (1997a); Bateman, R (1997b); Biological Control Products and LUBILOSA (1998); *CAB International*, information leaflet; *GTZ Locust Projects;* Jenkins, NE, *et al* (1998); Kooyman, C, *et al* (1997); Kooyman, C and Abdalla, OM (1998); Lomer, C and de Groote, H (1997); LUBILOSA, (December 1997); LUBILOSA, (March 1998a); LUBILOSA, (March 1998b); LUBILOSA, (April 1997); LUBILOSA, (March 1998c); Price, C *et al* (1997); Prior, C and Greathead; Prior, C (1992); Pesticide Referee Group (1998).)

Hansen's Disease Center

Inst de Recursos Biologicos de Argentina

Univ Nacional de Mexico

University of Arizona

Univ Nacional de la Patagonia

Pontifica Univ Catolicia de Chile

American Home Products Corporation (Wyeth-Ayerst & American Cyanamid Co)

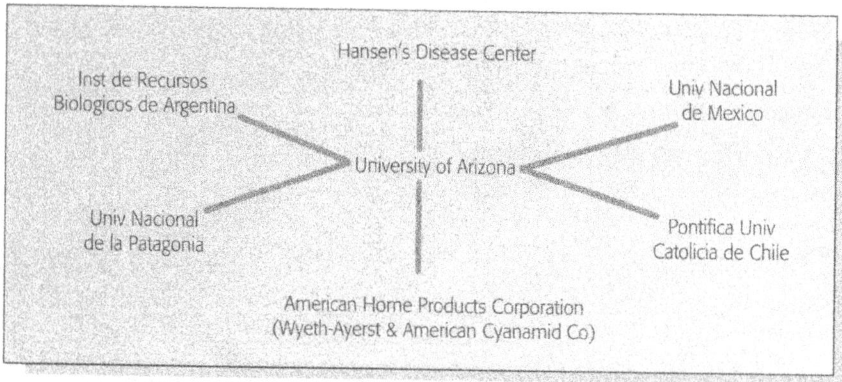

Figure 7.9 *Agreement Structure: Latin American ICBG*

7.4.2 The Latin American International Cooperative Biodiversity Group (ICBG)

Kerry ten Kate, Fiona Mucklow and China Williams

Introduction

The Latin American International Cooperative Biodiversity Group is one of several ICBGs sponsored through a programme initiated in 1991 by three US government agencies: the NIH; the National Science Foundation and the US Agency for International Development (USAID) (Rosenthal, 1998). It consists of a series of bilateral MTAs (Figure 7.9) between the University of Arizona, university institutions in Argentina, Chile and Mexico, and the American Home Products Corporation (AHP), which is the parent company for Wyeth-Ayerst Research Laboratories and American Cyanamid Co. After an initial five-year period (1993–8) involving an overall investment of US$4 million, a grant of US$3 million for a second project phase (1998–2003) has been approved.

The project focuses on the collection of xerophytic plants and associated microorganisms located in the arid and semi-arid ecosystems of Argentina, Chile and Mexico. These are thought to display a wide range of adaptability to extreme conditions such as heat, drought and radiation. Priority is given to arid land plants with rich ethnobotanical information, in the hope of increasing the possible discovery and development of agrochemicals and pharmaceuticals. The project specifically uses collectors with the expertise to choose plants that may have a history of medicinal properties, and which have never been collected or studied before. It aims at funding biodiversity conservation through product discovery and related economic development.

The University of Arizona acts as the coordinator of the project. It receives dried plant samples, and may also receive extracts or fractions of these plants, from the Centro de Investigaciones de Recursos Naturales del Instituto Nacional de Tecnologia Agropecuaria (INTA) in Buenos Aires, Argentina, the Universidad Nacional de la Patagonia (UNP), also in Argentina, and Pontificia Universidad Catolica de Chile. Scientists at UNAM in Mexico City conduct their own chemical studies on Mexican plants. These institutions are responsible for obtaining all the required import and export permits, licences and phytosanitary certificates. Most samples are tested, extracted and fractionated in the laboratories at the University of Arizona, and then sent to be screened for agrochemical leads and activity in the AHP laboratories in the USA, which have the equipment and assays needed for high-throughput screening. The company is not initially given the names of the plant samples that it is screening. Data from their screening process are sent back to Arizona, and if a compound looks promising, AHP is then informed of the name of the plant under a confidentiality agreement. AHP then performs 'de-replication': checking records and databases to see if the compound has already been reported. So far, some promising compounds have been discovered, but it is anticipated that it will take at least another seven years for any product to be commercialised and marketed.

The University of Arizona is responsible for distributing the grants from donor agencies, and also for sharing any royalties that it receives from AHP.

Monetary benefits

The agreements provide for the payment of quarterly salaries to the respective partners in Argentina, Chile and Mexico; the funding of collection trips (including salaries for collectors, technicans, analysts and PhD students); subscriptions for the partners to various publications and purchase of some handbooks; and royalties in the event that a product is commercialised. The MTA between the University of Arizona and AHP provides that the latter pays royalties to the University of Arizona from benefits derived from any agrochemical or pharmaceutical product discovered and developed as a result of the Latin American ICBG bioprospecting arrangements. The precise royalty ranges, calculated on net sales, are confidential and vary in the separate bilateral agreements. The University of Arizona is required to distribute the royalty benefits from future licensed products through the establishment of trust funds for local conservation and development projects in the source countries (yet to be established) as follows:

- 50 per cent to a local conservation fund in the area of collection in Argentina, Chile and Mexico;
- 5 per cent to the collector; and
- 45 per cent to the named inventors of the product.

Non-monetary benefits

A number of non-monetary benefits are shared through the project, as summarised below:

- *Equipment and materials:* The ICBG project funds have been used to purchase equipment for all the participating organisations, including two trucks for Chile and Mexico respectively; computers; herbarium cabinets; and germplasm conservation facilities and laboratory equipment such as HPLCs, rotovaporators, lyophilisers, balances, and chemicals. In addition, AHP donated 12 computers.
- *Know-how and training:* The ICBG involves training and the exchange of know-how through a number of separate initiatives. Staff from the University of Arizona and from the other participating institutions teach natural products chemistry (extraction and identification of biologically active molecules) to faculty members and university students. This involves two-way exchanges between the University of Arizona and the other participating universities. Ten PhD students are currently being trained in bioprospecting and drug discovery in Latin America and the USA. The team is also teaching the knowledge of medicinal plants, their conservation, sustainable collection, planting and propagation to students and teachers at the elementary and high school levels, and to vendors of medicinal plants in communities local to the participating institutions in all three countries.
- *Research collaboration:* Two researchers from Argentina conducted research into tuberculosis (a significant disease in Chile, Mexico and Argentina) for six months at the GWL-Hansen's Disease Center in Baton Rouge, USA. Other participants may have the opportunity for similar research in the future.
- *Database management system:* A database management system developed by the Arid Lands Information Center at the University of Arizona contains information on the research of the ICBG project in both English and Spanish. Available on http://ag.arizona.edu/OALS/oals/oals.htlm, these data are housed on the central server at the University of Arizona and accessible by the participants in Argentina, Chile and Mexico. Staff from the University of Arizona have visited Argentina, Chile and Mexico to train staff from the participating institutions in the use of the database.
- *Conservation:* In collaboration with several municipalities, the ICBG is creating a botanical park in the foothills of the Chilean Andes.
- *Sustainable sourcing:* If an agrochemical product results from this research and requires further sourcing of plant material for its production, AHP is obliged to give preference to the country from which the original sample originated as a supplier of the material.

(*Source:* research by Fiona Mucklow, China Williams and personal conversation with Prof Barbara Timmermann, 8 March 1999.)

7.5 Conclusions

The market for crop protection products derived from genetic resources is estimated at US$0.7 billion or US$3 billion, representing 2 per cent or 10 per cent of the market, respectively, depending on whether only 'pure' and semisynthesised natural compounds and biopesticides, or synthetic analogues modelled on templates from nature are included in the calculations. According to the companies we interviewed, the demand for access to genetic resources is likely to grow in the future. The predominance of high-throughput screens as an approach to the discovery and development of pesticides suggests that companies will continue to need access to genetic resources to fill their research programmes. In addition, there is a trend in the crop protection industry, particularly among the 20 major crop protection companies that control some 90 per cent of the US$30 billion global market, to 'outsource' the preparation of precursor chemicals, or, in other words, to acquire more chemical compounds and derivatives of genetic resources from other organisations rather than preparing them for themselves. Taken together, these factors would suggest that the crop protection industry provides ample opportunity for countries and institutions hoping to supply genetic resources, value-added derivatives and knowledge to the crop protection market.

Despite this up-beat analysis, any hopeful providers of genetic resources need to take a number of factors into consideration if they wish to benefit from partnerships with crop protection companies. First, the costs of research and development for natural product pesticides are greater than those for their synthetic cousins. Costly fees for access, demands for high royalty rates or values of non-monetary benefits, and bureaucratic and lengthy procedures to secure access to genetic resources serve to put companies off this method of research. Instead, companies say that they will focus on materials within their existing collections – which are often comprehensive – or they will concentrate on synthetic compounds to the detriment of natural products. Second, companies wish to acquire only high-quality samples of genetic resources, derivatives and information. They are looking for high levels of genetic diversity to screen, competitively priced samples and samples for which it is simple to obtain collecting permits and access agreements. Third, demand for genetic resources fluctuates, as the fortunes of natural products wax and wane relative to alternatives such as synthetic chemicals and the genetic modification of crops. Institutions that have supplied extracts for years may suddenly learn that an important customer is no longer looking for samples, or is moving out of natural product research. Contracts to provide genetic resources are frequently short-term and precarious. Finally, in the light of the CBD and related national law, companies are increasingly wary of entering into partnerships involving access to genetic resources. They wish to reduce the risk of finding that they have not secured the correct agreements to acquire and use genetic resources, and are dismayed at the lack of certainty and clarity in national law and procedure on access to genetic resources.

This suggests that, in a buyer's market, companies will only look favourably on institutions of high scientific calibre, offering samples of a high quality, based in countries where it is possible to obtain unequivocal prior informed consent. Finally, companies are likely to reduce the number of countries and institutions from which they obtain resources, and endeavour to establish long-term partnerships for the acquisition of large numbers of samples from their chosen few collaborators. The prospects are good for a few well-established organisations around the world, who will be able to provide value-added products, engage in joint research and gain valuable benefits such as technology, training and capacity building. Other organisations are likely to be less favoured partners. Companies will only wish to obtain 'raw' genetic resources or basic extracts from them, for which they will pay modest fees up-front, and are unlikely to transfer proprietary technologies or contemplate joint research.

The crop protection companies interviewed made a number of recommendations to governments regulating access to genetic resources. They recommended that each government should clarify the access procedure and designate a competent national authority. As one company put it, 'It is very difficult to find the right person to talk to. You get bounced around a lot.' They also wish that access regulations would provide legal certainty for researchers once they have followed the correct procedures for access. As another company says, 'We need to have in place an agreement: legal certainty and documentary evidence that we are operating according to the law.'

An overwhelming majority of interviewees stressed the need for governments to keep the access procedure simple, and not to let access regulations operate as non-tariff barriers to trade. The first priority is for countries to set out a simple access procedure at the national level, but several interviewees wished that it would be possible for groups of countries – such as ASEAN or those in Latin America – to harmonise their access laws. One company said it favoured a 'consortium' approach to obtaining materials. Several companies advised governments and institutions providing genetic resources not to be too demanding in their requests for benefit-sharing.

Issues such as these are covered in more depth in Chapter 10 (Companies' responses to the CBD), and in Chapter 11 (Conclusions).

'I am concerned that countries will get greedy and drive away business.'

'[Countries] have to realise that it is a competitive market, or else they will shoot themselves in the foot. If it becomes impractical, then we will go elsewhere. All that does is reveal their naïveté about industry.'

Note

1 The language in Article 2 of the Convention on Biological Diversity defines 'genetic resources' as 'genetic material of actual or potential value' and 'genetic material' as 'any material of plant, animal, microbial or other origin containing functional units of heredity'.

Kerry ten Kate

Chapter 8

Biotechnology in Fields other than Healthcare and Agriculture

8.1 Introduction

'Biotechnology' means the application of biological organisms, systems and processes to the provision of goods and services (OECD, 1994 and 1998). Biotechnology companies apply enzymes and use biologically active compounds derived from genetic resources as an integral part of processes and products in almost every industry sector. Biotechnology techniques can conveniently be divided into two categories: methods used before the discovery of the structure and function of DNA in the late 1950s and early 1960s, such as those used for millennia to make bread, cheese, beer and wine; and recombinant DNA technology developed subsequently. The latter refers to the modification of genes (DNA) within cells and organisms by incorporating genetic material from other sources. Today, techniques from both categories of biotechnology co-exist. Genetic modification, also known as genetic engineering, uses certain enzymes to cut, trim, join and copy DNA in very selective ways. Biotechnologists continue to search for these enzymes, which are generally found in microorganisms but are also produced by plants and animals. One famous example is Taq DNA polymerase, an enzyme used to copy DNA, and the basis of genetic fingerprinting. This enzyme is derived from the thermophilic bacterium *Thermus aquaticus*, originally found in the hot springs of Yellowstone National Park, USA.

The largest market for biotechnology is in the health sector. Some 60 per cent of European biotechnology companies (Ernst and Young, 1998) compared with 68 per cent of US biotechnology companies (Ernst and Young, 1995) work in the field of human pharmaceutical and veterinary products, diagnostics (*in vivo* and *in vitro* tests to detect disease and measure bodily functions), vaccines and other products to administer and deliver drugs. This aspect of biotechnology is dealt with in Chapter 3.

A second important market segment of the biotechnology sector is agricultural biotechnology. 'Ag-bio' companies focus on modifying plant and animal genes to produce organisms with desirable properties such as pest resistance, improved nutritional profiles or the ability to make chemicals more economically. This is dealt with in Chapter 5. Some 'Ag-bio' companies also develop crop protection products to improve the yields of major crops, which are covered in Chapter 7. 5 per cent of European and 8 per cent of US biotechnology companies work in this field (Ernst and Young, 1998 and 1995, respectively).

This chapter will cover some of the remaining major market segments of biotechnology, representing the work of the remaining 33 per cent of biotechnology companies in Europe (Ernst and Young, 1998) and 24 per cent of such companies in the USA (Ernst and Young, 1995).

15 per cent of companies in the USA and almost 27 per cent of those in Europe are 'industry suppliers', selling reagents, biochemicals, enzymes and other tools and services for industrial and food processing. The remaining 9 per cent of the US market and 7 per cent of the European market is served largely by 'environmental' companies, which develop pollution diagnostics and products for pollution prevention and bioremediation; and 'energy' companies looking for fuels from renewable resources. Some of these are also involved in pollution prevention (Ernst and Young, 1998, for Europe, and 1995, for USA).

The first section of this chapter introduces the major categories of biotechnological product and provides an overview of the different actors involved. These actors may be categorised as 'solution providers', who offer a range of products and processes, or as 'problem owners' –companies and other organisations who need biotechnologies to produce their final product. The section gives some examples of how genetic resources are used by these actors, and the size of various market segments. It describes the cost and time involved in a typical product discovery and development process for an enzyme, and summarises the kind of regulatory hurdles which a new biotechnology must clear in order to be placed on the market.

The second section analyses the demand by scientists from the private and public sectors for access to genetic resources to develop new biotechnologies. It looks at which organisations are involved in collecting genetic resources from *in situ* collections, and how they can acquire them from *ex situ* collections in the private and public sectors. It also explores the kind of genetic resources that are in demand for various applications of biotechnology, including the balance of 'raw' samples and value-added products, of domestic and foreign material, and the extent to which researchers need to return to the original source for further samples. The section concludes with a review of how genetic resources are selected for research, and a prediction of the extent of demand for access to genetic resources for the development of biotechnology products in the future.

The third section explores current practice in benefit-sharing in biotechnology, and looks at monetary and non-monetary benefits that arise from the use of genetic resources in biotechnology.

The fourth section provides two case studies on access and benefit-sharing partnerships. The first concerns the partnership between the Diversa Corporation and INBio, which has led to the establishment in Costa Rica of a DNA processing laboratory, and the second looks at New England Biolabs, which has established partner laboratories in China, Vietnam, Portugal, Cameroon, Uganda and Nicaragua.

Finally we set out some conclusions and recommendations relevant to this industry sector.

8.1.1 Introduction to biotechnology products and processes

Biotechnology companies are largely involved in the development of 'products' or 'processes'. Products consist of organisms or the compounds they produce, such as enzymes or metabolites, which either form part of an industrial process or are sold as end products. Biotechnology processes consist of systems such as, for example, a bioremediation system, based on the degrading function of microorganisms, for use *in situ* or in a bioreactor, or a biological process to extract metals, such as copper, from waste tips.

The majority of the institutions and companies we interviewed, especially the larger companies, are involved in all stages of the product lifecycle, from original research, through development to marketing and sales of the final product. Some produce 'intermediate' products, which other companies refine and incorporate into final products. For example, Montana Biotech Corporation isolates and characterises novel microorganisms, from which new compounds are identified and licensed to a variety of commercial markets. A third category of company is not involved in the initial discovery phase, but instead works on processes to formulate, scale-up and manufacture a product obtained from another company. For example, one company obtains genetic resources whose functions are already known, and uses them to develop a final biological system to remove contaminants from waste air.

Some biotechnology products are only a few steps away from the original genetic resource and may still contain DNA. For example, a biotechnology company may supply a bioreactor containing microorganisms to treat industrial waste. Another company might supply freeze-dried microorganisms to the food industry, or for use as cleaning products. When re-hydrated, the microorganisms are re-activated, and start producing the enzymes or other compounds used in the industrial process. Other products are derivatives of genetic resources but no longer contain functional DNA. Enzymes used in many industries fall into this category.

Section 8.1.4 provides more information on the use of genetic resources as the basis for various important categories of biotechnology products.

8.1.2 The actors

Some biotechnology products, such as detergent enzymes and bioethanol fuel, are used by the final consumer, but the predominant role of biotechnology companies described in this chapter (ie companies in fields other than healthcare and agriculture) is to develop products for other companies. Biotechnology companies can thus be pictured as 'solution providers', while the companies and other organisations to whom they supply products can be seen as 'problem owners'. 'Solution providers' offer a range of products such as bioremediation systems, enzymes, biochemical intermediates, and other biological systems. The focus of this chapter is predominantly on the 'solution providers', since it is these companies which access genetic resources and formulate them into products which they sell to the 'problem owners'.

Some 'solution providers' work alone to identify market needs and develop products to meet them, while

Box 8.1 *The Roles of 'Solution Provider' and 'Problem Owner' in One Possible Route from Collection to Commercialisation of a Microorganism*

Material	soil sample →	formulated enzyme preparation →		industrial process
Illustrative relative value in US$	Free	100	10,000	'Multiplier' (see p232)
Timeline	1	2	3	4
Product	sample	strain	enzyme	manufacturing process
Actor	university	culture collection	'solution provider'	'problem owner'
Activity	collects sample deposits strain		develops enzyme supplies preparation to 'problem owner'	uses enzyme in manufacturing process

others work much more symbiotically with their clients, who identify problems and even bring them samples of genetic resources to work on, before the company develops a product or process which it tests in the client's premises. As Box 8.1 shows, there is a limited range of 'solution providers' servicing many different 'problem owners' in virtually all industry sectors. In practice, 'solution providers' can provide solutions in more than one area, and some 'problem owners' find their own solutions.

Figure 8.1 is a highly simplified diagram illustrating one of many possible routes by which a microorganism in a soil sample might ultimately become a commercial product – namely an enzyme used in a manufacturing process. It shows how an academic at a university might collect a soil sample (for free), and deposit a strain in a culture collection. A biotechnology company specialising in developing enzymes for industrial processes (the 'solution provider' in this case) might obtain a sample of the strain from the culture collection for perhaps US$100, and develop from it an enzyme with potential applications in an industrial process, for example, as a catalyst in the manufacture of chemicals. A chemical company (the 'problem owner' which is the client of the biotechnology company) might purchase a formulated enzyme preparation for some US$10,000, and use it as a catalyst in the development of chemicals, food products, or animal feeds. The enzyme would be likely to make the process more efficient, and could save the company large sums of money, many times the price paid for the enzyme preparation. However, this 'multiplier effect' must be considered in the context of the time, cost and risk involved in product discovery and development which is described in Section 8.1.3.

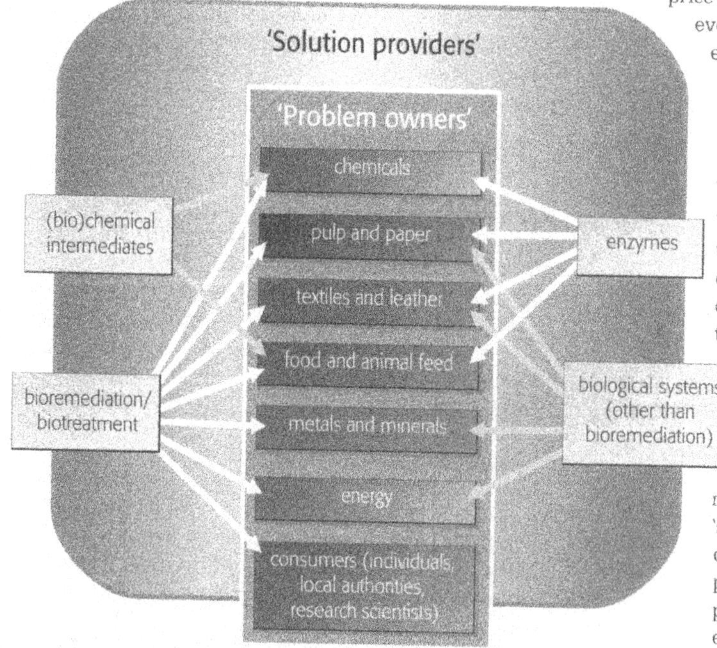

Figure 8.1 *'Solution Providers' and the 'Problem Owners' they Service*

The 'problem owners'

What kinds of individuals and organisations face problems for which biotechnology may provide a solution? 'Problem owners' are generally companies in industries such as chemicals, pulp and paper, textiles and leather, food and animal feeds, metals and minerals, and energy. Sometimes, organisations like local government or real estate developers have problems such as contaminated land, for which they turn to 'solution providers' for help. Individual consumers also purchase biotechnology products, including biologically based products for unblocking drains, and enzyme-based detergents.

The 'solution providers'

There are 'solution providers' in both the public and the private sectors. In 1997, there were roughly 300 biotechnology companies in Europe, and an equivalent number in the USA providing solutions in sectors other than healthcare and agriculture, including industry suppliers, environmental and energy companies. Biotechnology is a young and fast-growing sector. In the last decade, the number of biotechnology companies in the USA has increased by over 50 per cent, and nearly half the European and more than one-third of the US biotechnology companies were founded since 1987 (although these figures include biopharmaceutical and agricultural biotechnology companies) (Ernst and Young, 1997). While companies are being created, the market is consolidating through mergers and acquisitions, so the number of biotechnology companies is likely to stabilise in the next decade and many of the smaller companies will be acquired by larger ones strengthening their 'life sciences' divisions.

Biotechnology companies are so varied in scale and field of operation that they are difficult to categorise. However, it is possible to distinguish three broad groups.

The first group comprises big multinational companies whose main activities lie in chemicals, pharmaceuticals, agriculture and the food sector, and who may be both 'problem owners' and 'solution providers'. These large companies have subsidiaries or departments with biotechnology interests in areas such as materials, bioremediation, biocatalysis and enzymes, some of which are based on traditional fermentation processes (OECD, 1994). Some of these companies purchase biotechnology products, such as clean production technology for their manufacturing facilities, from other companies, and also develop biotechnology products such as biological systems for use by other subsidiaries and members of the group and for sale to third parties on a small scale. They typically allocate between 1 and 10 per cent of their research and development budgets to biotechnology. With the emergence of the new life science giants such as Monsanto, with a turnover in 1997 of US$7.5 billion, and Novartis, this group is of growing significance in the biotechnology market.

A second category of company is the group of small, innovative firms dedicated to a range of biotechnologies, from environmental biotechnology to the discovery and development of specialist enzymes. These companies work with the latest, high technology techniques, and focus on niche markets such as developing specialised processes, enzymes, microbial strains or analyses of specific toxic contaminants. Their turnover is likely to be between US$1 and US$10 million per year, of which anywhere between 10 per cent and 75 per cent may be spent on research and development. Some four-fifths of European biotechnology companies, and three-quarters of US firms, employ less than 50 staff (Ernst and Young, 1995 and personal communication, David Hales, Ernst and Young, February 1999), although, of course, the majority of these companies work in the field of healthcare (page 228 above). Small companies often emerge from universities, where a group of academics discover organisms and processes with potential industrial applications, and establish commercial ventures to exploit them. As with most areas of biotechnology, the majority of start-up companies emerging from universities focus on biopharmaceuticals, with relatively few working in the fields described in this chapter.

The main business of the third group of companies is likely to be in engineering, building and infrastructure, and consultancy. Within construction companies, biotechnology activities are likely to be based in a department or subsidiary which designs, manufactures, constructs and installs bioremediation equipment, and provides related services. The consultancy companies offer biotechnologies as one of the potential solutions to the problems brought to them by clients. For companies in this group, environmental biotechnology may account for less than 1 per cent of their turnover.

Not only companies, but also non-profit organisations are involved in the discovery and development that provides biotechnology solutions to the companies that need them. Universities and research institutes usually do not have the levels of investment, technology and capacity to complete the development of a product, and very rarely to manufacture and market it on a commercial scale, so their involvement tends to be limited to initial discovery and preparation of intermediate products, which they sell on to the private sector for incorporating into final products, and marketing on an industrial scale. For example, INBio collects samples in protected areas, and provides highly purified, clonable, high molecular weight DNA to the Diversa Corporation (see Diversa/INBio case study below). Several universities also supply samples and extracts to biotechnology companies. Some, particularly in Europe, enter into collaborative arrangements with companies. Others, including many in the USA, form spin-off companies to take advantage of opportunities for tax efficiency and venture capital.

These 'intermediary' organisations are very active as collectors and purveyors of genetic resources, and often operate as the link between the collection and commercialisation of genetic resources. As one research institute put it, 'We supply organisms, genes – you name it – to companies, institutions and research laboratories. We give access, swap, license, and have joint ventures in the sense of collaborative research programmes with companies and other institutions which might lead to commercial products such as diagnostic kits, for example. We have also set up spin-off companies.'

In addition to private institutions, many governments have research and development programmes on environmental biotechnologies, and the share of environmental research and development as part of all government-supported research and development is increasing steadily in most OECD countries (Ernst and Young, 1995). Many culture collections are government-funded and maintained.

Governments fund collaborative research between academic institutes and companies through programmes such as LINK in the UK and similar initiatives in the USA, many other countries and regional organisations like the EU, thereby catalysing the development of novel biotechnological products and processes. In general, government bodies do not themselves take up the 'solution provider' role.

8.1.3 Quantifying markets

Estimating the 'market value' or 'global sales' of biotechnology products is extremely difficult. To determine exactly which products have a strong biotechnology component would entail a company-by-company and product-by-product assessment. Not only would these figures be too fragmented and detailed to gather and analyse, but national statistics, figures from trade associations and reports by market analysts do not, as a rule, even estimate them, and may use different definitions when they do.

A much truer value of biotechnologies would be gathered by calculating the value of the biotechnology products to their customers – the 'problem owners'. There is a multiplier effect between the price of the product sold by a biotechnology company and its economic value in industrial processes. Thus an enzyme sold to a food company may make an important contribution to that company's production, effectively earning it many hundreds

Box 8.2 *Rough Estimates of Markets for Various Biotechnology Products*

Product	Annual market (US$ bn)	Source, notes
Environmental biotechnology	40 (early 1990s)	OECD, 1994
	75 (by 2000) or: 56–120	OECD, 1998
... of which soil remediation	10–25	In EC, Japan, USA, OECD, 1998
Industrial enzymes		
Detergent enzymes	0.7	
Starch enzymes	0.16	
Textiles enzymes	0.13	
Baking enzymes	0.09	
Beverages enzymes	0.09	
Dairy enzymes	0.06	
Other (leather, tanning metals, oilfields, etc)	0.24	
Biocatalysts	0.02–0.03	Freedonia Group, 1997
	0.05–0.1	Global catalyst market US$9 bn, CMR
... of which PCR enzymes	0.05–0.1	UNEP/CBD/SBSTTA 2/15
Total industrial enzymes and biocatalysts:	1.5–1.6	Freedonia Group, 1997
	>3 (in 2001)	Freedonia Group, 1997
Diagnostics: DNA polymerase alkaline phosphatase glucose oxidase		
Total (mostly DNA polymerase)	0.15–0.20	Freedonia Group, 1997
... of which 5% in areas other than healthcare	<0.01	Personal communication Dr Ken Murray, Biozyme, 5 March 1999
Biomaterials: forest products and fibres such as cotton produced using biotechnology	Negligible today. Great potential.	
Bioenergy	Negligible today. Great potential.	See Section 8.1.4 below
Rough estimates of total:	Low: 60 High: 120	

of thousands of dollars (Box 8.1.) This 'multiplier' is hard to gauge, and varies for each application of biotechnology. Based on data from the European Association for Bioindustries, the OECD has estimated a multiplier of 6 to 9 (OECD, 1998).

Finally, available statistics for products or market sectors do not reveal what proportion is attributable to biotechnology. For example, global trade in forest products is worth US$140 billion a year, and in cotton US$30 billion a year (FAO *Yearbook*, 1995). There are no data on the proportion of these products produced using biotechnology, which are negligible today, but likely to grow significantly in the coming decades.

Despite all these difficulties in assessing the size of markets, rough estimates of the global market for biotechnology products (other than in healthcare and agriculture) from the information we have been able to find, are set out in Box 8.2, p232. Box 8.3 summarises the biotechnologies covered in this chapter.

Box 8.3 *Biotechnologies Covered in Chapter 8*

Market segment	Examples of use of genetic resources to 'solve problems'
	1 Biotransformations
Waste management	Use of microbes and enzymes to treat residues, wastes and effluents from industrial processes, agriculture and food production.
Bioremediation	The application of biological processes for the clean-up of dispersed pollutants, for example oil spills or contaminated land. Use of genetically modified biosensors to assess land and water contamination and bioremediation potential.
Organic chemicals	Use of enzymes either as catalysts in chemical reactions, to synthesise or break down complex chemicals, or to produce speciality molecules such as certain amino acids that are not naturally occurring.
	2 Materials
Leather	Replacement of polluting, chemical-based technologies, for example 'liming' leather with enzyme treatments, improving production capacity and product quality.
Hydrocarbons, oils and fats	Techniques to modify natural rubber and oil-bearing crops for use in soaps, detergents, paints and lubricants. Production of novel fats and acids. Processes to minimise unwanted by-products.
Mining, minerals and metals	Use of enzymes in bioleaching to extract end-product from ores.
Paper and pulp	Use of enzymes to process wood fibre and lignin, to bleach pulp using less chemicals, to improve paper quality.
Textiles	Use of enzymes in textile processing to produce more fibre, lightweight and superior quality products, and to finish fabrics and garments.
Domestic and industrial detergents	'Biological' (enzymatic) washing agents for industrial and domestic settings. These digest proteins (which cause stains), even at low temperatures.
	3 Energy
Transportation fuels	Ethanol produced from sugar, starch and lignocellulosic feedstocks and methanol from wood can be used as gasoline extenders or substitutes, or to make gasoline additives. Plants and algae produce hydrocarbons that can be refined to form sources of energy. Plant-derived oils can also provide fuel.
Hydrogen and biogas	Hydrogen-producing microorganisms split water molecules to create hydrogen fuel. Microorganisms also manufacture biogas (methane) in landfills, from biomass and as a by-product from the treatment of sewage.
Stationary power	Techniques to convert wood and other biomass into energy.
Oil and gas	Biological processes to remove sulphur.
	4 Food and nutrition
Wine-making and brewing	Wines: improving flavour extraction, clarification and filtration. Beers: improving fermentation of starchy cereals, reducing costs of malt enzymes, improving filtration and producing low-calorie beers.
Animal feeds	Use of enzymes to help animals digest cereals and vegetable proteins in animal feeds; improving taste and processing pet-food meats.
Cheesemaking and dairy	Use of microorganisms and enzymes for processing and production of flavours and ingredients.

8.1.4 Some important categories of biotechnology products

The following paragraphs describe some of the ways in which genetic resources are used by 'solution providers' to meet the needs of 'problem owners'.

Environmental biotechnology, including bioremediation

An important component of the suite of 'environmental technologies' designed to protect the environment by saving energy and materials and by reducing and disposing of waste, involves biotechnology. While some environmental technologies focus on pollution prevention, bioremediation and biotreatment involve the use of biological processes to treat waste. The former generally refers to the *in situ* use of microorganisms, while the latter usually involves the use of bioreactors or *ex situ* treatment (which entails removing contaminated material and treating it off-site, before returning the cleaned material to its original location). Microorganisms are capable of degrading many pollutants and contaminants, and developments in bioreactors and processes have made this both technologically possible and economically attractive. While the use of GMOs may speed the degradation of pollutants, no GMOs have yet been used commercially for bioremediation, although genetically modified microorganisms, especially bacteria, are being applied as biosensors to determine toxicity of environmental samples, assess bioremediation potential, and devise rational bioremediation strategies (Sousa et al, 1998). The bioremediation sector is growing at roughly 10 per cent a year, compared with the overall growth of the environmental technology market of 4–5 per cent a year. The annual market size for soil remediation in the EC, Japan and the USA has been estimated as US$10 billion in 1990 and roughly US$25 billion in 2000 (OECD, 1994). Other estimates are more optimistic. An analysis of contaminated soil treatable by *in situ* methods in Europe has identified a potential market for bioremediation in Europe of US$60 billion (Griffiths, 1997). Box 8.4 shows the results of some bioremediation products.

Box 8.4 *Some Results of Bioremediation Products*

- In the 1980s, the US Air Force used petroleum-degrading bacteria, and a mixture of nutrients and soil moisture conditions to clean up a 27,000-gallon jet fuel spill at Hill Air Force Base, Utah. During the 18-month project, the jet fuel residues in the soil were reduced from an average total petroleum hydrocarbon concentration of 900 mg/kg to less than 10 mg/kg (OECD, 1994).
- A wood-preserving facility in the USA generated waste water contaminated with oily material containing pollutants such as phenols derived from creosote, aromatic hydrocarbons and other aromatic compounds extracted from wood. A small, mobile bioreactor containing microorganisms and nutrients was installed, which removed approximately 99 per cent of the phenolic compounds (OECD, 1994).
- In the aftermath of the 1989 Exxon Valdez oil spill in Prince William Sound, Alaska, an oil spill bioremediation project was conducted by adding nitrogen- and phosphorus-rich nutrients, in the form of different fertilisers, to the contaminated shoreline sites. The nutrients increased the extent and rate of oil degradation significantly, compared with the untreated shoreline (OECD, 1994).
- In the UK, a significant quantity of heating fuel leaked from a burst pipe, sunk through the porous ground on the site in question and polluted a groundwater aquifer, dispersing oil over 3.9 million hectares. Excavation and disposal of contaminated soil was ruled out on account of the large volumes involved and the difficulties of excavation and water removal around buildings and services. Boreholes were installed, and air and nutrients used to encourage the growth of indigenous microorganisms. The site was monitored for six months, after which it was confirmed that the oil concentrations in the soil had been reduced by over 98%, and the concentration of potentially toxic organic components, including benzene, toluene and xylene, was undetectable in both soil and groundwater (OECD, 1994).

The estimated global annual market share of biotechnology for cleaner production in the chemicals, pharmaceuticals, paper and food and feed industries has been estimated as between US$56 billion and US$120 billion (OECD, 1998). According to another estimate, some 20–30 per cent of the total environmental technology market now consists of products or services with a major biotechnology component, representing an annual turnover of US$50–75 billion (OECD, 1994).

Energy

Fossil fuels still provide some 85 per cent of the world's energy, but are non-renewable and may contribute to global warming by the production of carbon dioxide as an end product . Biomass is a renewable source of energy, which offers virtually unlimited supplies and effectively recycles carbon dioxide which is taken up as the plant grows. In contrast, fossil fuels simply release carbon dioxide trapped in plant material over geological time spans. Biomass, either purpose-grown or waste, offers the raw material for a suite of renewable energy sources including heat and electricity, carbon-based liquid fuels such as bioethanol and biodiesel, and hydrogen.

Biotechnology can improve existing processes and open up new plants and microbes to use as energy sources. For example, trees may be genetically modified for the more efficient generation of electrical energy from burning biomass. Several companies are currently developing genetically modified trees for this purpose, but none is yet on the market.

Bioethanol is presently made by the fermentation of simple sugars from sugar-cane or sugar-beet, and corn starch. Converting lignocellulosic biomass to ethanol involves the processing of three major components in biomass: hemicellulose (a polymer of C5 sugars); cellulose (a polymer of glucose) and lignin (a complex polyaromatic material). Following hydrolysis, the hemicellulose and cellulose fractions provide sugars for fermentation to ethanol. Lignin can be used as fuel for steam and electricity production (Sheehan, 1998). In future, much larger quantities of bioethanol may be made available by the breakdown of all three major components from lignocellulosic biomass. Recombinant yeasts (such as improved forms of *Saccharomyces*, or baker's yeast) and bacteria (such as *Bacillus stearothermophilus*, *E coli* and *Zymomonas mobilis*) are under development to aid more efficient fermentation processes and production of ethanol (personal communication, Dr Mike Himmel, March 1999). Genetically modified microorganisms capable of fermenting a broad range of sugars into ethanol are perhaps only one year from commercial production. Genetically modified enzymes to break down cellulose into sugar are closer to four or five years away from commercial production (personal communication, Dr Mike Himmel, March 1999). While lignocellulosic bioethanol technologies are still in the pipeline, conservative estimates suggest that the use of improved enzymes will enable the first commercial, large-scale plants to produce some three to four billion gallons of ethanol in the USA alone in 2010, representing some 3–4 per cent of the US gasoline market. Initially, it is unlikely that bioethanol manufactured in this way will be able to compete with petroleum itself. In the first years of the 21st century bioethanol is likely to be competitive only as a fuel additive. Depending on numerous assumptions about variables, such as the price of gasoline and projections for production capacity, some estimates suggest that bioethanol production will be viable as a profit-making concern (without subsidy) as a fuel additive by 2015, and by 2020 it is quite possible to envisage commercial production of bioethanol from dedicated energy crops at prices competitive with petroleum (Sheehan, 1998). Box 8.5 gives an example of the use of biotechnology in the production of ethanol.

Box 8.5 *The Use of Biotechnology in the Production of Ethanol as an Energy Source*

In 1997, Petro-Canada, one of Canada's largest oil companies, entered into a joint venture with Logen, a biotechnology company in Ottawa which makes enzymes to digest wood waste and crop residues into sugar for refining into ethanol, a renewable energy source. The joint venture involves building a US$14–30 million ethanol test plant which will use bio-engineered enzymes to convert low-cost cellulose into ethanol; co-funding for research and development; and a licensing option to build full-scale ethanol refineries if the venture succeeds.

Source: OECD, 1998.

Biotechnology is also being harnessed to produce biodiesel as crude, or chemically modified, oil from naturally occurring or genetically modified crops such as canola or rapeseed, and photobacteria and other biotechnologies are under development for the production of hydrogen and methane.

At this early stage in the application of biotechnology to energy, it is difficult to assess its economic contribution. However, it seems likely that this market will be a sizeable one in the future, and may match or even outstrip that for environmental technologies.

Materials

Biotechnology applications can improve the quality of materials by changing their composition or structure and can increase the economic and environmental efficiency of their manufacture, use and disposal. For example, the molecule lignin makes up 15–35 per cent of the dry weight of trees, and must be removed during the manufacturing process to make pulp for paper. By manipulating the genes which produce lignin, the lignin content of wood can be reduced, cutting the energy and pollutants involved in papermaking. Even a few per cent reduction in lignin could result in billions of dollars in savings to the paper industry (Podila and Karnosky, 1996), and considerably reduce the environmental impact of papermaking (Box 8.6).

International trade in materials is an important market. For example, forest products are worth some US$140 billion and cotton US$30 billion a year in global trade (FAO *Yearbook*, 1995). Genetically modified trees are under development, but not yet grown on a commercial scale. Similarly, current genetic engineering of cotton is largely to confer pest-resistance, rather than to improve the quality of the material. However, genetically modified materials are likely to make an important contribution to the market for materials in the future.

Biocatalysis

Catalysts are used to speed up the rates and selectivity of chemical reactions. Many inorganic catalysts are robust and work at high temperatures and pressures, but they suffer from the disadvantage that they are generally not very selective in the chemical reactions they facilitate. Enzymes are biological catalysts which are often remarkably selective in the reactions they promote, but are generally not as robust as inorganic catalysts. For example, conventional chemical processes often give rise to so-called 'racemic mixtures of chiral isomers' (molecules which are fundamentally similar in structure, but appear as 'right-handed' and 'left-handed' versions, only one of which may be useful or chemically active). Enzyme-catalysed reactions are so selective that they lead to the production of a single isomer, removing the need for an expensive separation stage later in the process. Chiral molecules are predicted to account for around 80 per cent of the pharmaceutical market by the year 2000, with sales of some US$3 billion per year (personal communication, Steve Brewer, 1998).

In recent years, catalytic molecules have been developed that can bring about reactions that would not occur in nature. Coupled with these developments, the search for new, naturally-occurring enzymes continues. There are two streams of research: hunting for new enzymes, and improving upon currently-known natural proteins or enzymes. The latter approach may be more suitable for creating properties for which natural evolutionary processes are unlikely to have been selected (Zhang et al, 1997; OECD, 1998), such as the removal of chlorine atoms in biodegradation of organic pollutants.

One disadvantage of most known enzymes is that they degrade at the kinds of high temperatures and pressures involved in industrial processes, reducing their utility as catalysts. However, since the 1960s, a number of discoveries of organisms that thrive in extreme conditions, such as high or low temperatures, acidity and alkalinity, pressure, and toxic chemical environments, have led to a growing market for 'extremozymes', or enzymes isolated from such organisms that continue to function in extreme conditions.

The application of the polymerase chain reaction (PCR), using the Taq polymerase enzyme (Box 8.7, p237) for genetic fingerprinting and other applications, relies entirely on the stability of the enzyme at relatively high temperatures. While extremozymes operate in extreme conditions compared with those tolerated by most life-forms, they nonetheless operate at temperatures well below the extremely high temperatures of conventional petrochemical processes. Although enzyme-catalysed processes can thus save money and energy, these gains need to be offset against the cost of removing water from biotechnology product streams. The comparative advantages and disadvantages of petrochemical and biotechnological processes thus need to be compared on a case-by-case basis.European sales of Taq polymerase alone reached US$21 million in 1991. Global sales of PCR enzymes are thought to be in the range US$50–100 million (UNEP/CBD/SBSTTA/2/15). Some observers predict that the market for biotechnology enzymes derived from extremophiles will grow between 15 and 20 per cent a year (New England Biolabs Inc, Beverley, Mass, USA, cited in UNEP/CBD/SBSTTA/2/15), although most of this growth may stem from demand in the diagnostics sector (see below). Indeed, the Freedonia Group predicts that world demand for industrial and speciality enzymes will increase 9.6 per cent per annum to over US$3 billion in the year 2001, led by strong growth in textile processing, pulp and paper, cosmetic and toiletry, and medical and animal feeds markets. Demand growth for speciality enzymes is slated to outpace the growth in demand for industrial enzymes, with particular interest likely in digestive and other therapeutic products. However, industrial applications of enzymes will continue to dominate the market place, due to the size of the detergent and food beverage markets (Freedonia Group, 1997).

Box 8.6 *The Use of Biotechnology in the Paper and Packaging Industries*

The paper and packaging industry

Bacteria are now being used to remove unwanted by-products from the polyaminoamide resins used to strengthen paper and packaging boards. The process for manufacturing these resins results in the formation of unwanted and highly polluting haloalcohols. In the case in question, various ways for removing these chemicals were considered, but all involved other waste products, or were expensive or otherwise unsatisfactory. Biotechnology provided a solution. A 'bioconsortium' of two bacteria – *Arthobacter histidinolovorans* and *Agrobacterium tumefasciens* – acts synergistically to remove the undesirable chemicals when operating in a continuous process. A small unit containing these two bacteria and some resin was attached to the resin stream, and reduced the haloalcohols from around 8,000 ppm to less than 6 ppm.

The pulp and paper industry

The resinous materials known as 'pitch', which constitute 2–8 per cent of the total weight of wood, cause a number of problems in the manufacture of pulp and paper. These include unwanted deposits, the blocking of drains, discoloration, tearing and other faults in paper, and the expense of closing plants for cleaning. In the late 1980s, scientists in Japan discovered that treating pulp with lipase enzymes reduces pitch problems considerably, and enzymes are now the base for several industrial products for pitch treatment. One example is the product of the Clariant Corporation, which sprays a water slurry of the spores of a fungal inoculum of the ascomycete *Ophiostoma piliferum* on to wood chips as they are piled prior to pulping. The fungus invades the wood cells, degrading the pitch.

Source: OECD, 1998.

Food processing

Enzymes, both recombinant and naturally-occurring, play a host of roles in the food and beverages industries, from clean production to product improvement. For example, the milk clotting enzyme chymosin, or rennet, was traditionally extracted from calves' stomachs, but since the 1990s, the gene for the enzyme has been cloned into microbes. Companies now produce the

Box 8.7 *The Discovery of Thermus Aquaticus and the Development of Polymerase Chain Reaction (PCR)*

Yellowstone National Park contains approximately 80 per cent of the geysers and 60 per cent of all terrestrial geothermal features found on Earth. Prokaryotic, or single-celled cyanobacteria, eubacteria and archaea grow in the hot springs, geysers, fumaroles and boiling mud pots of Yellowstone's geothermal ecosystem in temperatures of 40–93° C.

In the summer of 1966, Dr Thomas Brock of Indiana University and one of his undergraduate students, Hudson Freeze, collected pink bacteria and mat samples from the outflow channel of Mushroom Spring at Yellowstone, at a temperature of about 69°C. By October 1966, Freeze had isolated a strain from this first collection, which he designated YT-1 (ATCC 25104). In 1969, Brock and Freeze published a paper based on Brock's taxonomic work and Freeze's study of its DNA, describing the organism which they named *Thermus aquaticus* (Taq). Brock deposited representative cultures of Taq, as the organism is now commonly known, in the ATCC, in Washington DC. The discovery of *T aquaticus* was a scientific milestone, as it demonstrated that life could exist at far higher temperatures than had formerly been believed possible.

Kary Mullis, then a researcher at the American company Cetus Corporation, obtained a sample of YT-1 from the ATCC, for which the company paid US$35. In 1984, he invented a procedure based on the enzymes from *Thermus aquaticus* for which he was to win the Nobel Prize for Chemistry a decade later. Mullis purified from Taq an enzyme known as a thermostable DNA polymerase termed Taq polymerase. DNA polymerases catalyse the synthesis of new DNA molecules, and if this reaction is conducted repeatedly, or 'cycled', it is possible to produce a considerable amount of DNA from what was originally a miniscule sample. This cycling of DNA synthesis has been termed the 'Polymerase Chain Reaction' (PCR), and the technique is used in a wide range of applications, such as the DNA tests used in forensic medicine. During PCR, it is necessary to switch the reaction mixture repeatedly between low and high temperatures. When Mullis first invented the technique, he had been using an enzyme from a bacterium that operated at normal temperatures but that was inactivated during the hot periods of the cycle. When he replaced this enzyme with Taq polymerase, which continues to function during the hot periods of the cycle, this problem vanished.

In 1991, the Swiss pharmaceutical company Hoffmann- LaRoche acquired from Cetus all patent rights to PCR technology, for US$300 million. Annual sales for the licensees of PCR equipment and supplies based on Taq are estimated to be approximately US$200 million and growing, as the technique is applied to a wider range of biotechnology research, medical diagnostics, and environmental and forensic analyses.

Other than the Federal tax dollars paid by the companies in the U.S. who license PCR, which may arguably contribute to the US$20 million annual operating budget from the US Treasury to the Park Service to run Yellowstone National Park, the Park has not received a share of the financial benefits arising from the use of Taq and PCR. The absence of a direct share for the Park in the monetary benefits arising from the use of *Thermus aquaticus* can be largely attributed to two factors: the terms of the research permit issued to Brock by the Park Service in the 1960s, and the access rules of the ATCC where Taq was deposited. The terms of Brock's research permit granted him permission to collect in Yellowstone, but did not contain terms requiring him to share the benefits arising from research on the specimens collected. When Mullis purchased a sample of Taq from the ATCC, he was not obliged to seek the permission of either the ATCC or Yellowstone National Park prior to commercialising PCR, nor to share any of the benefits arising from this use of the enzyme.

This experience with Taq and PCR led the National Park Service to examine the options for benefit-sharing. Accordingly, in July 1996, Yellowstone was asked by the Director of the National Park Service to explore ways to position itself so as to maximise the benefits to resource conservation that could result from research on biological samples acquired from the Park. The process that the Park followed is described below. The first partnership between a commercial company and a US National Park to result from this approach is that between Yellowstone National Park and the biotechnology company Diversa, the subject of the case study on 253.

Source: ten Kate, Touche and Collis, 1998.

enzyme by fermentation, without the need to use calves. The dairy sector also uses a range of lactose-fermenting yeasts as flavouring ingredients. A number of different enzymes are used in the baking industry. Fungal enzymes have replaced the traditional barley malt, and since the 1980s various 'amylase' enzymes have been used to slow down the staling of bread. Enzymes have replaced potassium bromate in bread flour, following concerns about its potential carcinogenic effect, and have improved the processing and softening of whole-grain and high-fibre breads. Biotechnology is also used to convert the waste products of the food industry into raw materials for other processes (OECD, 1998).

Diagnostics

The current global market for diagnostics is in the region of US$150–200 million annually (Freedonia Group, 1997). While 95 per cent of this is probably in the area of healthcare (personal communication between China Williams and Dr Ken Murray, Biozyme, 5 March 1999), with the requirement to detect low levels of pollutants and pathogens in food, drinking water and effluent streams (see Regulations, below), measuring and monitoring systems in areas other than healthcare are growing and are increasingly based on biotechnology. For example, the light-emitting properties of certain microorganisms and fireflies have been used to measure heavy metal and organic pollution (Sousa et al, 1998), while monoclonal antibodies are being developed to detect pathogens such as *E coli* O157 and *Cryptosporidium*. The PCR used in genetic fingerprinting is also being used in a wide range of forensic tests. One company interviewed described developing a diagnostic product to test for the adulteration of cashmere by cheaper fibre.

Box 8.8 *Two Ways to Develop Desirable Enzymes: Protein Engineering and Directed Evolution*

The circumstances in which enzymes occur in nature are often very different from those under which scientists wish them to perform in industrial processes. Industrial chemists are constantly thwarted in their attempts to use enzymes by the fact that enzymes have evolved over billions of years to perform very specific reactions within certain environments. Features that are ideal in the setting of a metabolic pathway may be counterproductive in the man-made circumstances of an industrial process. Conversely, enzymes are not likely to have developed characteristics unnecessary in their natural surroundings but essential in industrial processes. To improve upon undesirable properties of enzymes and to develop new characteristics, molecular biologists and protein chemists use two methods: protein engineering and directed evolution.

Protein engineering consists of studying the manner in which an enzyme is needed to operate and altering its structure using chemistry or genetic engineering to tailor its shape specifically to that of any other molecules with which it must bind during chemical reactions, and to give it new, desirable characteristics such as greater stability in the presence of organic solvents.

Directed evolution, however, enables enzymes to be 'tuned' to operate optimally under specific conditions by mimicking processes of Darwinian evolution in the test tube, so that enzymes evolve and adapt to the desired environment, such as resisting higher temperatures, altering the rate at which the enzyme catalyses reactions, improving its stability in the presence of certain chemicals or its selectivity. One method of directed evolution is 'gene shuffling', in which PCR techniques are used to recombine naturally-occurring genes or gene fragments. The resulting, recombined genes are stored in a 'mutation library' and screened.

One advantage of the directed evolution approach is that it can be used even when little is known of an enzyme's structure or the way in which it catalyses certain reactions. For example, the protease enzyme subtilisin is used as a laundry aid, and is stabilised by calcium. Unfortunately, the laundry process involves metals which react with calcium, destabilising the enzyme so it does not operate. By replicating the environment found in the laundry, scientists were able to 'evolve' an enzyme that catalyses the reactions involved in laundering clothes, but has 1,000-fold increased stability in those conditions.

Sources: OECD, 1998; personal communication with Dr Mike Griffiths and Dr Eric Mathur, March 1999.

8.1.5 The research and development process

Biotechnology embraces so many different products and processes that it is impossible to summarise all possible approaches to research and development. Some biotechnology companies screen compounds, in much the same way as described in Chapter 3 on pharmaceuticals and Chapter 7 on crop protection. Other companies raise crops, using recombinant DNA technology and other biotechnologies, as described in Chapter 5 on major crops. Enzymes are common to much of the remaining biotechnology market, so this section summarises the research and development process for enzymes.

A novel enzyme product may be developed from the starting point of knowledge that a specific type of activity is required. Many raw samples may be screened for this particular activity. Conversely, a sample of soil, for example, which may contain many thousands of species of microorganism, can be randomly screened against a range of assays. Either way, the strains which reveal potentially interesting activities may be isolated as a pure culture of the microorganism, in order to identify it and grow it to produce sufficient quantities of the enzyme for purification. The organism can then be manipulated by conventional mutagenic techniques (such as ultra-violet radiation) to obtain higher growth rates and yields of the enzyme.

It is often difficult and sometimes impossible to isolate a pure culture of certain microorganisms (Box 8.12 below). In these circumstances, an alternative approach is to isolate DNA directly from the soil sample and identify the gene responsible for the enzyme without identifying and culturing the organism itself. The relevant gene may be introduced, by genetic engineering, into a more familiar organism which will express it and produce the enzyme. Analysis of the DNA and gene cloning can now take as little as a few months.

The activity of the enzyme may be further improved by 'protein engineering'. In this process, the three-dimensional structure of the enzyme molecule is obtained by crystallising a pure sample and analysing it using X-ray crystallography and computer modelling. Structural changes that might improve the activity can then be predicted. Box 8.8 describes the technique of protein engineering and another approach, 'directed evolution', in more detail.

Several iterations of these stages may be necessary to optimise performance. Figure 8.2 illustrates the design of the screening process and the steps that may take place. Once a company has identified an enzyme that appears to show the desired activity, it tests it in conditions as similar as possible to those in which the enzyme will be used. An enzyme which is to be used for its ability to break down protein strains when washing clothes, for example, is tested in circumstances that replicate all the possible conditions which might occur in a washing machine.

Once the enzyme itself has been developed, further research and development are required to optimise the manufacturing process (fermentation) and, depending on its final application, the enzyme may have to pass a battery of toxicity studies before the relevant regulatory bodies grant permission for large-scale production and marketing.

Overall, the stages from initial idea to market-place may take from two to five years (Figure 8.3), and cost anywhere from US$2 to US$20 million. This may be compared with the 7–14 years, and up to US$300 million, necessary to develop a new biopharmaceutical product (personal communication, Dr Henrik Dalbøge, Dr Eric Mathur and Dr Michael Griffiths, March 1999).

Screening design
define criteria to enzyme
assay development
primary screening 1–10,000 strains
secondary screening <100 strains
cloning 5–10 strains
DNA sequencing
expression
application
Molecular screening
'Artificial evolution'

Figure 8.2 *Design of the Screening Process*

Source: Novo Nordisk.

8.1.6 The regulatory process

The majority of OECD countries are in the process of developing regulatory systems to cover products comprising of living organisms. These systems include voluntary codes of conduct as well as legislation. In general, industrial **products** consisting of or containing GMOs are covered by specific, horizontal legislation. Legislation tends to be much less uniform or consistent for non-GM organisms (eg naturally occurring microorganisms) and for **processes** involving the use and release of living organisms. Such activities tend to be covered piecemeal by other legislation; either broad legislation concerned with general environmental protection, pollution control, novel chemicals and dangerous substances, or specific product legislation on pesticides, novel food, and animal vaccines. At present, processes involving the release or use of living organisms tend to be covered only in certain countries (eg Norway and, to an extent, the UK) with impact assessment criteria.

In most countries where legislation is in place or is being developed, there are associated guidance, regulations or interpretative documents which outline the issues covered during the assessment of products of biotechnology. These issues fall in to three main areas: basic scientific considerations, health considerations and potential environmental and agricultural implications. Scientific considerations include the characteristics of donor and recipient organisms, such as the source of the organism, factors which might limit its reproduction, growth or survival, and genetic traits. Health considerations include

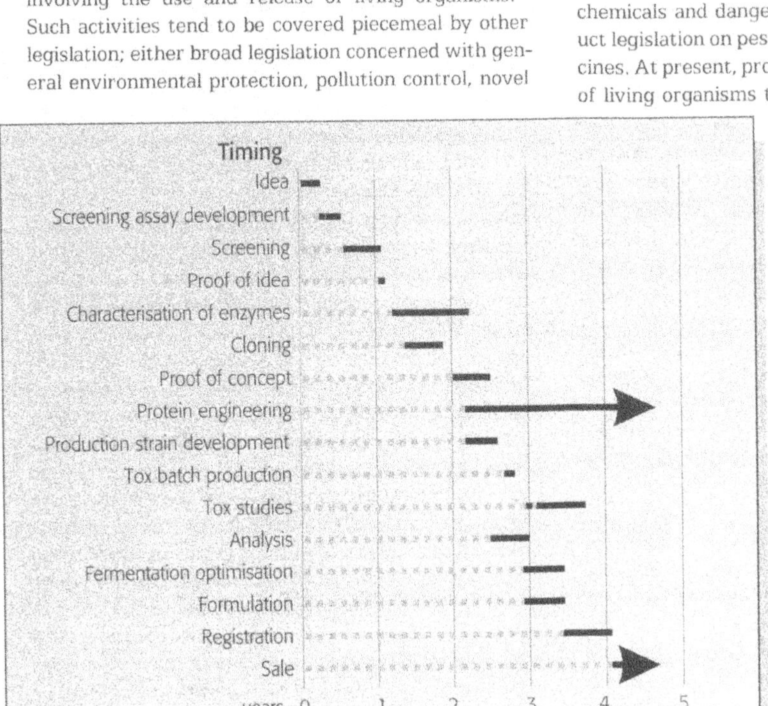

Figure 8.3 *Timing: From Idea to Market-Place*

Source: Novo Nordisk.

the organism's pathogenic properties, its toxigenity, allergenicity, communicability and biological stability. Environmental and agricultural considerations include the ecological traits of the introduced organism, its pathogenity and predicted effect.

The responsibility for assessment usually rests with a country's Environment or Health Ministry, often assisted by advice from a committee of scientific experts.

The EU

The EU is a good example of the piecemeal approach to the regulation of biotechnology. Its regulatory framework is designed to strike a balance between providing adequate protection for health and the environment, while at the same time encouraging the development of and market for biotechnological products.

As far as GMOs are concerned, the EU's regulatory framework includes horizontal legislation specifically covering GMOs (eg (i) **Council Directive 90/219/EEC** which covers any *contained* use of GM micro-organisms, for both research and commercial purposes; and (ii) **Council Directive 90/220/EEC** on experimental and marketing-related aspects of GMOs, which covers any research and development release of these organisms into the environment and contains a specific environmental risk assessment for the placing of any product containing or consisting of such organisms on to the market). Such procedures cover any product containing or involving the deliberate release of a GMO. In the UK, for example, release of such a product requires the approval of the Secretary of State, based on an opinion of the Advisory Committee on Releases to the Environment (ACRE) on the acceptability and environmental safety of the release. In the UK (directly implementing the EU legislation), the rules relate to any release of a GMO or exotic organism, providing it is living and regardless of the kind of product (such as crop, crop protection, or bioremediation product), and of whether it is used in the process stage or present in the marketed product. Processes involving GMOs may also be caught indirectly by legislation implementing Directive 90/219 on contained use of GMOs regulated by the Health and Safety Executive (HSE).

There is little specific legislation (other than sectoral product legislation) on non-GM organisms in the EC that would cover the kind of biotechnology applications discussed in this chapter, such as the use of non-GM organisms in bioreactors, or the use of enzymes in industrial processes. However, **Council Directives 90/679/EEC** and **93/88/EEC** provide for a minimum requirement designed to guarantee a better standard of safety and health as regards the protection of workers from the risks of exposure to biological agents. In the UK, non-GM organisms fall under any legislation covering the release of organisms into the particular environment in which they are being used (eg legislation relating to water, air and soil quality). The exception to this is that if the organism is non-native then it comes within legislation on the release of non-native organisms. This is regulated by the Department of the Environment, Transport and the Regions, which authorises release of non-native organisms following the opinion of a scientific committee.

The USA

The system in the US differs from that in Europe. Both GMO and non-GMO products are covered by sectoral product legislation as well as general horizontal legislation governing, for example, dangerous or toxic substances. The use of plant and animal pathogens is subject to additional, separate legislation designed to prevent the introduction or dissemination of disease.

The **Toxic Substances Control Act** (TSCA), which covers marketed products only, sets out specific guidelines for GMOs which have involved an intergeneric genetic modification. A Biotechnology Programme ('Microbial Products of Biotechnology; Final Regulation Under the Toxic Substances Control Act') was developed in 1997 under TSCA Section 5, authorising the EPA, among other things, to review new chemicals before they are introduced into commerce. Under a 1986 intergovernmental policy statement, intergeneric microorganisms (microorganisms created to contain genetic material from organisms in more than one taxonomic genus) are considered new chemicals under TSCA Section 5. The Act sets forth the manner in which the EPA will review and regulate the use of intergeneric microorganisms in commerce, or commercial research.

Under the Federal **Food, Drug, and Cosmetic Act**, the FDA has authority to ensure the safety and wholesomeness of most foods and food additives, including the enzymes used in the food industry for processing, which are covered in this chapter, whether developed through modern biotechnology or other methods. Pre-market approval is required for food additives, whether or not they are the products of biotechnology. Additives can include spices, flavours, preservatives, thickeners, texturisers and food-processing enzymes. FDA's biotechnology policy treats substances intentionally added to food through genetic engineering as food additives if they are significantly different in structure, function, or amount from substances currently found in food. As many of the food crops being developed using biotechnology do not contain substances that are significantly different from those already in the diet, they do not require pre-market approval. The Act provides for a legal exception to the requirement for pre-market approval for substances which are widely recognised by scientists as being safe (Generally Recognised as Safe (GRAS) substances). GRAS Certification is performed by the marketing company. The FDA does not hold a complete list of GRAS substances. Industry makes its own decision as to whether a substance is GRAS or not. As the FDA has the authority to take legal action against a substance that poses a hazard to the public, in practice companies tend to work closely with the FDA in the development and marketing of new food products, to protect their economic interest.

Regulations on GM food itself, as distinct from the use of enzymes in food processing, are covered more fully in Chapter 5.

(*Sources:* personal communication between China Williams and James Maryanski, FDA, 24 February 1999; Peter Hinchcliffe, DETR, March 1999; Peter Kearns, OECD, 15 March 1999; Michael Schechtman, USDA, March 1999; www.fda.gov; www.epa.gov.)

8.2 Access to genetic resources

With a couple of exceptions, all the companies and organisations interviewed – ranging from the one-man band to the multinational with several thousands of employees – access genetic resources in their research. Two of the companies do not work directly with genetic resources, however, but instead purchase ready-made enzymes from suppliers who isolated them from genetic resources. Access to genetic resources is thus a crucial component of the work of many biotechnology companies, but usually represents just a small part of their running costs – from less than 0.5 per cent to a few per cent.

Genetic resources are either collected or acquired from other organisations. We discuss each method in turn.

8.2.1 Who collects genetic resources for biotechnology research and development?

A large proportion of those interviewed do collect samples themselves, although they may be divided into two groups. For roughly half the companies and organisations whose staff collected samples, this represented a comparatively unimportant method of acquisition, sometimes described as 'rare', and often accounting for less than 10 per cent of the materials acquired by the company. For the other half of the organisations, this was the predominant manner in which they acquired materials, often accounting for over 90 per cent of their acquisitions. Perhaps the group that collects the most genetic resources in the field of biotechnology is the university sector. However, within the private sector there is a group of companies, generally those with less than 100 employees, and quite frequently spun off from universities, that can be described as the 'hunter gatherers' of the biotechnology world. These companies identify a problem, either through their own research or as brought to them by a client, then seek genetic resources that can fix the problem.

Some companies collect quite literally 'in their own back yard', while others favour the extreme conditions in which extremophilic microorganisms are found. As Dr Phil Caunt of Biotal put it, 'We believe that the bugs are out there if you just go out and look for them.' This company tends to go to local 'exposure' sites as it is easier and cheaper than collecting in more distant venues. An example of a company collecting biological samples further afield is Montana Biotech Company. Collections by members of its staff account for 95 per cent of its library of microorganisms. The library consists primarily of microorganisms that can withstand conventionally extreme conditions such as high temperatures and low pH. These extremophiles are isolated from thermal features found all over the world including Iceland, Europe, deep sea vents, and Yellowstone National Park. Microbial samples in the Montana Biotech library not collected by its staff are purchased from commercial bacterial collections. Another company which collects extremophiles obtains only 20 per cent of its collections from members of staff; 20 per cent are obtained from universities, and the remaining 60 per cent from external collectors based in the countries from where it collects, several of whom it has trained in collecting techniques.

While many of the collectors we spoke to were used to obtaining collecting permits in protected areas, and some even to negotiating access and benefit-sharing agreements with collaborators, overseen by the providing government, others were accustomed to a more informal approach, such as taking soil samples when on vacation. There is a growing realisation among biotechnology companies and academics that this is no longer an appropriate approach, but there has been some (justifiable) confusion about whether microorganisms fall within the scope of national access and permitting regulations. While microorganisms are often not explicitly listed in wildlife protection laws and other regulations that have a bearing on access, in many jurisdictions they are already covered by access laws, and recent access regulations and those under development tend to cover microorganisms in their scope, avoiding this possible 'loophole' common in the past (Chapter 2).

While the quantities of samples needed for biotechnological research are usually small – typically a few grams of soil or millilitres of water – it is nevertheless possible to cause ecological damage when obtaining them. Some interviewees raised the importance of taking care to ensure that the collecting activities do not cause any such damage. As one company explained, Yellowstone National Park is important to the scientific community because of its wide diversity and high concentration of extreme environments. Biological diversity is maintained through the strict control of human intervention into the natural resource. In such an environment, it is important not to contaminate different ponds, nor to damage microbial mats, some of which may have taken hundreds of years to develop. One company described how activities tapping the thermal features in Iceland as a source of heat have unfortunately resulted in the loss of biological diversity in some of its hot springs.

8.2.2 *Ex situ* collections

Companies' own collections

The vast majority of companies and organisations involved in biotechnology research maintain collections of genetic resources or their derivatives. Companies tend to fall into two categories:

1. companies for whom an important part of their business is to build and improve their collections, and to license access to samples to other companies who conduct research and product development on them; and
2. companies that use their collections as the basis for their own product discovery and development, and are less likely to grant access to other companies.

Many kinds of genetic resources are represented in company collections, including plants, insects, human genetic material, animals, fungi, bacteria, streptomycetes, actinomycetes, viruses and other microorganisms. Microbial cultures are the most common form of collections of genetic resources in the companies interviewed. In many cases, companies maintain collections of the derivatives of genetic resources – such as enzymes, purified compounds and some extracts – in addition or as an alternative to samples of the genetic resources themselves. These derivatives may be prepared by the company itself from the original sample, purchased from other companies, or obtained from university researchers or company partners.

Box 8.9 *Some Collections of Genetic Resources and Derivatives Held by Biotechnology Companies and Organisations Interviewed (Illustrative Examples Only)*

'General' category *(including derivatives)*
- extracts
- enzymes
- compounds
- cultures and strains of anaerobic, thermostable and soil-borne microorganisms
- the chymosin gene from different microorganisms
- sequence data of the genetic information obtained from microorganisms
- extremophiles
- soil samples
- DNA
- RNA

Mammals
- mammalian cells as tissue cultures
- human DNA
- leukaemia cells

Insects
- lepidopteran species
- ticks
- mosquitoes
- dried nematodes

Marine organisms
- fish DNA
- marine organisms
- sea worms

Algae

Plants
- dried plants
- living trees
- plant extracts
- plant tissue cultures
- seed
- transgenic plants (including trees)

Fungi
- cultures of fungi
- gene libraries of fungi
- penicillins
- yeast strains

Archaea

Bacteria
- actinomycetes
- bacterial plasmids
- eubacteria
- freeze-dried samples of bacteria
- gene libraries of bacteria
- lactic bacteria
- streptomycetes

Viruses
- arthropod-borne viruses (stored in liquid nitrogen)
- virus collection

Box 8.10 *Examples of Ex Situ Collections Used for Biotechnology*

A university collection

The National Collection of Industrial and Marine Bacteria (NCIMB), Aberdeen, UK, is a private company wholly owned by Aberdeen University. Approximately 7,000 organisms are held in Open Deposit and sold on a predominantly small-order level. The majority of organisms are from overseas, obtained from other culture collections and scientists. Two-thirds of sales are to academic institutions, one-third of sales to clients overseas. The NCIMB also acts as an International Deposit for Patent Purposes, and approximately 1,500 strains are currently lodged for patent purposes, available subject to patent regulations only. (*Source:* personal communication, Dr Nick Green, NCIMB, March 1999.)

National culture collections

Belgium: The Belgian Coordinated Collections of Microorganisms (BCCM) is a consortium of four culture collections financed by the Belgian government. The collection contains:

- over 16,000 strains of bacteria;
- over 1,500 plasmids;
- over 6,500 strains in the biomedical fungi and yeast collection; and
- over 25,000 strains in the agro-industrial fungi and yeast collection.

Strains are sold and delivered on a world-wide basis to clients in the scientific and industrial community. (*Source:* BCCM website, www.belspo.be/bccm/bccm.html.)

China: The Centre for Culture Collection of Microorganisms is in charge of the collection and preservation of type cultures of microorganisms in China. At present the Collection has 5,500 strains of fungi; 1,900 strains of yeast; 2,200 strains of bacteria, and 1,400 strains of actinomycetes. The Centre is entrusted by the Patent Bureau of China to preserve organisms for the purposes of patent procedure. 340 such strains are currently being preserved. The Centre provides strains across the board to education, research and industry.
(*Source:* http://www.im.ac.cn/imcas/junbao.html.)

USA: The American Type Culture Collection (ATCC) is the largest culture collection in the world. It houses over 92,000 different strains, the majority of which are non-hazardous cultures and derived materials preserved without charge for scientists all over the world. It was established in 1925 to fill the need for a central collection of microorganisms to serve scientists world-wide. Since its foundation, it has distributed over 2.5 million cultures, and in 1997 alone shipped over 133,000 cultures. It also acts as a Patent Depository under the Budapest Treaty. ATCC is a non-profit organisation and relies on funding from Federal government (10 per cent), but more importantly from fees from the distribution of cultures and from life-science companies, institutions and organisations.
(*Sources:* personal communication with Dr Ray Cypess, ATCC, March 1999; ATCC 1997 *Annual Report and Mission Statement.*)

An international collection

The International Mycological Institute (IMI) houses a collection of 21,000 cultures or preserved strains, of which 2,000 are bacteria and 19,000 fungi. It is currently in the process of acquiring a collection of nematodes. The IMI became an international collection in 1982, and is funded by CABI. Its collection is drawn from 130 source countries, half from the 42 CABI member countries. The collection was established mainly for CABI's own research purposes, but is made available to depositor countries (free) and to academics (55 per cent of users) and industry (45 per cent) for a small supply charge. Some 4–5000 strains are supplied annually, the majority to users in the UK (80–85 per cent).
(*Source:* personal communication with Dr David Smith, March 1999.)

The World Data Center for Microorganisms

The World Data Center for Microorganisms, in Saitama, Japan, has information on 500 collections holding between 250,000 and 300,000 cultures of all kinds, principally those of not-for-profit organisations, but including information on some culture collections of private companies. (*Source:* http://www.wdcm.nig.ac.jp/wfcc.)

The size of these collections varies enormously. Generally speaking, collections held by biotechnology companies are not as great, in terms of numbers, as those held by pharmaceutical companies engaged in high-throughput screening, or by certain seed companies, although the range of different kinds of genetic resources contained in their collections may be greater. Some biotechnology companies hold very small and constantly changing collections. One, for example, explained that, 'We maintain just those enzymes we are currently evaluating'. Multinational companies, however, build and maintain large libraries of genetic resources and derivatives for their own use or to license to customers (Box 8.9).

Some companies have built up large collections and are currently not engaged in further collecting activities. As one company put it, 'Most of the cultures [we] use today go back a very long time. For example, the cultures which produce penicillin originate back to the 2nd World War. A certain cellulase-producing strain was isolated from the rotten tent of an American sergeant during World War 2. He had packed his tent away wet.'

It is costly and time consuming to maintain culture collections, so each company weighs a number of factors when deciding whether or not to maintain certain strains in its collections, or to expand its collections. While libraries of strains form the basis for one traditional method of product discovery, namely the screening of microbial isolates, alternative technologies such as protein engineering, directed evolution and gene shuffling enable companies to make greater use of existing resources and to streamline their collections. These alternative approaches also mean that, in some cases, it is no longer such a priority for companies to gain access to large collections of new samples. The companies interviewed in this survey maintained collections of the various kinds described in Box 8.10.

8.2.3 Acquisition of materials from other organisations

The vast majority of interviewees acquired materials from *ex situ* collections. The biotechnology community relies heavily upon culture collections for the provision of samples, and several interviewees obtain their samples exclusively from such collections. 'Search and discovery' companies whose business is to establish and maintain collections for licensing to others are another common source of samples. Companies also obtained specimens from their clients and from university staff who frequently collect genetic resources from *in situ* conditions. On rare occasions, the companies interviewed also obtained samples from genebanks. Box 8.10 gives examples of *ex situ* collections used in biotechnology.

The heavy reliance by the biotechnology sector on culture collections for access to genetic resources is significant from the perspective of benefit-sharing, as most cultures held in these establishments were acquired prior to the CBD. These and post-CBD collections continue, in most cases, to be supplied to companies and other researchers without any restrictions on commercialisation or requirement to share benefits with source countries (unless specifically requested by the depositor).

Within non-profit organisations such as university departments and research organisations, the prevailing ethos is still one of 'free exchange' among academics. While some culture collections charge for access to samples (Section 8.3, below), few impose conditions on recipients when they obtain them. However, this practice is changing fast. Culture collections sometimes check organisations applying for access to strains with potentially dangerous applications before supplying them, and increasingly provide specimens under MTAs. According to the German culture collection DSMZ, 'We check the purchaser first (for example, to check that they are not are involved in biological warfare). We also use material transfer agreements, when requested to do so by the depositor, to prevent the use of the sample for commercial processes without the permission of the depositor.' Universities, too, increasingly use MTAs.

The use of MTAs is also common in the private sector, with companies using 'Research Sample Agreements' when exchanging samples for research. Common terms in such agreements include prohibitions from transfer to third parties; restriction of use of the material to non-commercial activity; obligations to use samples according to good laboratory practice; and disclaimers for any damage or liability arising from use by the recipient of the strain.

Box 8.11 *Examples of MTAs Used by Culture Collections*

International Mycological Institute (IMI)

Acquisition: Since 1992, IMI has requested its suppliers to provide proof that all relevant PIC has been granted to collect and supply the material. IMI accepts material only upon proof that suppliers have 'taken reasonable action' to obtain the relevant permission to collect and deposit the material.

Supply: All new customers are required to sign and return a 'Culture Supply Form' before any cultures are sent to them, and this acts as an umbrella agreement for all subsequent supply. The one-page form states that material is supplied on the condition that it is not commercialised. If the recipient subsequently wishes to commercialise the material, then it needs to enter into a separate and further agreement with IMI. These agreements vary but include conditions to share benefits with the country of origin and take into account any 'added value' by CABI. Material deposited by IMI member countries is re-supplied to them without restriction.

Deutsche Sammlung von Mikroorganismen und Zeikulturen GmbH (DSMZ)

Acquisition: The 'Accession Form' to be completed by individuals depositing organisms in the German culture collection DSMZ contains the clause 'NOTE: I understand that subcultures of the deposited strain will be distributed at the discretion of the DSMZ (for a reasonable fee to cover actual expenses).' It requests 'as much information as possible' about the origin of the strain, including the 'geographical area', but does not mention PIC.

Supply: Under the heading 'Restrictions on the Use and Distribution of Certain Strains', the DSMZ website (http://www.dsmz.de/restrict.html) recommends that its customers should, prior to using strains from the DSMZ catalogue that bear a patent number, consult the original depositor or the appropriate patent documents to avoid possible infringement of the patent. It also points out that some strains are marked 'for research only', and that these may be used for research purposes only and not distributed to third parties.

American Type Culture Collection (ATCC)

In 1997, the ATCC introduced a series of new generic MTAs for investigators seeking access to materials within the ATTC (personal communication with Dr Leslie Platt, ATCC, 20 February 1998). Under this regime, in order to access materials from the ATTC, an individual or institution was obliged first to sign a MTA. As of April 1998, ATCC used six different kinds of MTA, designed to meet the various circumstances and terms under which specimens are deposited in the Collection and supplied from it (ten Kate, Touche and Collis, 1998). As of 12 March 1999, model supply agreements were still available on ATTC's website (http://www.atcc.org/mmf/mmf23.html), but these are currently under review (personal communication, Dr Ray Cypress, 11 March, 1999).

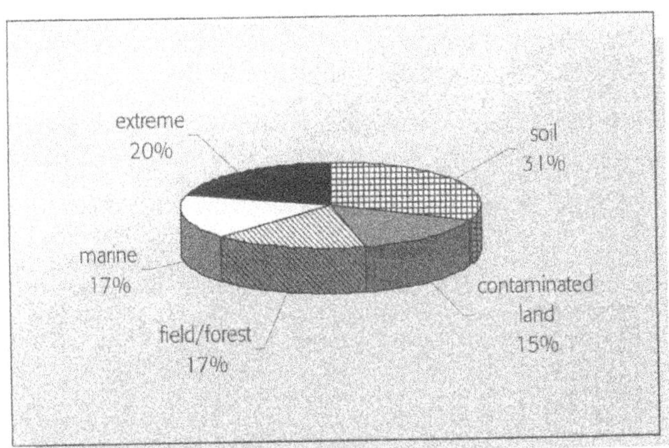

Figure 8.4 *Ecosystems from which Interviewees Acquire Samples (percentages derived from a sample of 20 companies)*

Several culture collections and other intermediary institutions such as seed banks and botanic gardens now supply samples of plants and microorganisms under MTAs. MTAs are contracts obliging recipients of the samples to which they refer to treat them and their derivatives in a manner consistent with the terms of the MTA. Common terms might prevent a recipient from commercialising a strain or derived product without the prior permission of the culture collection. MTAs offer the providing institution using them a mechanism to trigger benefit-sharing negotiations, and to involve the source country in the negotiation. Currently, the MTAs used by culture collections do not explicitly mention benefit-sharing, and rarely mention the rights of the depositor of a strain. Box 8.11 provides examples of the use of MTAs by culture collections.

8.2.4 How are genetic resources selected for research and development?

Biorational approach

Several of the organisations interviewed use a range of different methods to select materials on which to conduct research. However, the 'ecological' or 'biorational' approach (ie selecting samples from environments which suggest that their genes and metabolites enable them to survive there) proved to be by far the most popular. It was the most commonly used, and often the first choice of method of selection. As the Diversa Corporation said, 'We collect from ecosystems that are likely to reveal considerable diversity: rainforests, areas that show a great diversity of plant life, deserts, volcanoes, Antarctic snow pacts.' Ecological 'niches' of interest need not be the more extreme and exotic ecosystems of the world, but may be food processing plants, garbage dumps or other environments where organisms may evolve that are able to survive in these conditions, or to conduct useful functions, such as degrading materials (Figure 8.4).

Application oriented selection

A common approach to product discovery in biotechnology is 'application orientated selection'. This involves identifying the problem one wants to fix (for example, the need to find an enzyme that breaks down cellulose, but does not stop doing so at comparatively high temperatures), then hunting for genetic resources to meet that need. One option is to search for genetic resources with the necessary characteristics through conducting field collections using the 'biorational' approach, or asking external organisations such as university research groups for samples with the desired activity. In another kind of application-oriented selection, collections are

Table 8.1 *The Number of Known and Estimated Microbial Species in Relation to Material Held in Service Culture Collections*

Group[1]	Number of species		Material held in culture collection		
	Known	Estimated	Total number	% of known species	% of estimated species
Algae	40,000	60,000	1,600	4.0	2.6
Fungi	69,000	1,500,000	11,500	16.7	0.8
Prokaryotes[2]	4,204	40,000 to 3,500,000	2,300	54.7	5.8 to 0.07
Viruses	5,000	130,000	2,200	44.0	1.7

1 Group names are used in the colloquial and not in the formal taxonomic sense. Data compiled from Wilson (1988); Hawksworth (1991b); Nisbet and Fox (1991); Bull et al (1992); and World Conservation Monitoring Centre (1992).

2 Including *Archaea* and *Bacteria*. Upper limit for predictions includes uncultivated bacteria.

Source: personal communication with Dr Gilson P Manfio, March 1999.

made on a random basis, studied closely so that their properties and functions are known, and then held so they can be put to use should a related problem arise. Table 8.1 shows the number of microbial species in relation to material held in service culture collections, and Table 8.2 shows the locations and numbers of such collections world-wide.

Given the ubiquity and diversity of microorganisms (Box 8.12), several companies still collect on a random basis. One company explained that it looks across a range of different environments and, in effect, conducts random searches. It hopes to identify as many restriction enzymes as possible, and has an open mind as to where it may find them. Another company that uses random methods of collection seeks thermophiles but, instead of using the source location to determine research and seeking them from extreme environments, it depends on its isolation techniques to find thermopiles in samples taken from more mundane environments.

Table 8.2 *About Culture Collections*

Region	Number of collections	Number of cultures
Asia	147	118,034
South America	54	22,175
North America	70	241,434
Oceania	60	78,323
Africa	13	8,361
Europe	140	347,241
Total	484	815,568

Note: It is important to note that cultures held in specific countries may not necessarily have originated from those countries. In developing countries, samples are more likely to be standard international reference and type strains. Conversely, a significant number of cultures held in collections in developed countries originated from samplings in developing countries. For this reason, and for reasons of quality control, companies seeking samples from developing countries are currently likely to obtain them from collections held in industrialised countries.

Source: Sugawara and Kirsop, 1994.

Box 8.12 *About Microorganisms*

- Our knowledge of microorganisms (archaea, bacteria, filamentous fungi, yeasts, microalgae, protozoa and viruses) is limited compared with our knowledge of animals and plants. It is difficult to estimate the extent of microbial genetic diversity using conventional approaches, since only a relatively small portion (eg 0.1–1 per cent for bacteria) can be cultivated under laboratory conditions. One difficulty lies in applying the species concept to microorganisms such as archaea and bacteria (*Current Opinion in Microbiology*, 1998). The number of validly-named and accepted microorganisms is currently around 150,000 species, of which fungi account for almost 50 per cent, and algae and protozoa contribute between 20 and 25 per cent. Bacterial and virus species together contribute less than 5 per cent.
- Microbial genetic resources (MGR) replicate frequently (sometimes every 30 minutes), leading to changing populations both in the environment and during cultivation in the laboratory. If the conditions in which they are stored and propagated are not expertly regulated, replication can lead to 'strain drift' (ie genetic modification from the original sample) and changes to original strain properties.
- Because of their microscopic size, vast numbers and the difficulty of isolating and maintaining pure cultures in the laboratory, microbial genetic resources cannot be accurately enumerated. There can be no 'base line' for inventorying purposes. They may be transferred across borders by wind, water, animals or human intervention.
- In the case of fungi, the collections registered in 1990 with the World Data Centre for Microorganisms (WDCM) held some 11,500 species, which constituted about 17 per cent of the described species, and less than 1 per cent of the estimated number in nature (Hawksworth, 1991). In studies of tropical regions and hitherto unexplored habitats, it is not unusual for 15–30 per cent of fungi found in some investigations to be unknown to science, and about 1,700 new species of fungi are described each year (Stackebrandt, 1994). 8 per cent of fungal species are endangered (World Federation of Culture Collections (WFCC), 1996).
- The vast majority of microorganisms has not been cultured or characterised in detail. Obtaining pure cultures for characterisation requires appropriate growth conditions, which can be difficult to find and maintain. However, one of the main reasons for the inability to assess microbial diversity is the failure to recover so-called 'uncultured' and 'resting' forms from environmental samples (Morita, 1975; Hugenholtz et al, 1998). These organisms can be detected microscopically but cannot be brought into culture using the known cultivation methods. It is estimated that more than 99 per cent of microorganisms observable in nature typically cannot be cultivated by using standard techniques (Amann et al, 1995). These forms include many mutualistic symbionts and parasites of other organisms. Only recently, through the use of molecular methods involving the direct extraction of nucleic acids from the environment, application of the PCR technique and sequence analysis of ribosomal genes (16S rRNA), have bacteriologists begun to assess the biodiversity of prokaryotes in environmental samples (Stackebrandt, 1994; Hugenholtz et al, 1998).
- Because of their microscopic nature, MGRs cannot be tracked and monitored conventionally. They cannot be readily fingerprinted for authentication purposes, although tools for doing so may become available in the future. According to WFCC, 'Scope for piracy exists unless universally accepted procedures are developed and adopted' (WFCC, 1998).
- Within a population of microorganisms in the environment, isolates of the same species may show slight genetic variation when sampled in different locations or even at different times. Since they may display different characteristics, different isolates of the same species can thus be held usefully in a collection for taxonomic or screening purposes. However, many organisms have currently been isolated only once (WFCC, 1998).

Sources: references cited above; UNEP, 1995; personal communication with Dr Gilson P Manfio, March 1999 and Dr Vanderlei P Canhos, February 1999.

Information

Genetic resources themselves are not the only source of discoveries. Information – whether in the form of a gene sequence or a paper describing a new discovery – often provides the stimulus for research on a particular sample. Many companies search databases to pinpoint gene sequences and other information that may be of interest. For example, Brewing Research International – an organisation working on discovery and development of microorganisms for brewing – searches the yeast and barley genome databases. However, the work of some companies is so specialised – for example, the absorption of metals by plants – that relevant databases do not exist. While genomic databases, for example, are a relatively new phenomenon of the last decade, there is a tremendous volume of literature, from reference works that are decades or even hundreds of years old, to specialist journals containing information relevant to product discovery and development, ranging from biochemical abstracts to satellite maps. Scientists frequently conduct a 'library search' to see what other specialists have published on the subject in question, as this is a fertile source of useful leads. Many companies have on-line search facilities for patents and literature. If a literature search turns up a strain of potential interest, the company approaches the culture collection or other organisation that holds the strain to obtain a sample. One company described how it looks through chemical abstracts, searching for a particular reaction or a particular class of microbes. As all papers give the corporate source for the genetic resources, it is just a matter of contacting the person and asking them if it is possible to use their organism.

8.2.5 The nature of genetic resources sought

Microorganisms are the category of genetic resources most commonly used by biotechnology companies working in fields other than pharmaceuticals and agriculture. As can be seen from Box 8.12, they are also one of the least well known groups of organism.

> *'Microbes don't recognise frontiers, they recognise niches, and these niches may occur all over the world.'*

In conducting our interviews, two approaches to sampling microorganisms emerged, which, at first sight, appear contradictory. Some interviewees took the view that 'many microbes are ubiquitous, so I don't need to collect outside my back yard', while others explained that 'microorganisms exhibit interesting characteristics in particular environments, and that is where I look'. In fact, both arguments are valid, and reveal different ways of approaching the search for novelty (in the form of organisms, properties, lead compounds or even raw genetic material). Culture-dependent studies reveal that representatives of some bacterial divisions, for example, are cosmopolitan in the environment, whereas others appear restricted to certain habitats (Schlegel and Jannasch, 1992, cited in Hugenholtz et al, 1998). It is a question of which strategy (or maybe a mixture of both) a company chooses to pursue to find interesting biological materials to put through its screening programmes and assays.

Endemic microorganisms or those associated with specific environments

For a number of reasons, some microorganisms may only be found in certain environments, may exist in greater numbers in such environments, or may only reveal certain infra-specific properties (such as the production of secondary metabolites) under particular conditions. Each of these cases can be illustrated as follows. Some microorganisms such as endophytic bacteria and mycorrhizal fungi are host-specific, existing only in specific associations with the host plants. These microorganisms are therefore endemic to the regions that the plants inhabit. In extreme environments such as the Arctic and Antarctic, sulphur pools and thermal vents, highly specialised organisms predominate, and these may represent a different population of microorganisms from those found in more equable conditions. All cultivated representatives of *Aquificales*, thermophilic bacteria that metabolise hydrogen, have been obtained only from high-temperature environments, which suggests a specialised habitat niche for this group (Hugenholtz et al, 1998). In addition, a comprehensive study of fungi showed that *Fusarium compactum* isolated from different geographic locations differed in their capacity to produce secondary metabolites (Talbot *et al*, 1996).

Ubiquitous microorganisms

Using modern molecular techniques which do not rely on culturing, it is apparent that there is a huge diversity of organisms in natural habitats. For example, representative samples of several bacterial divisions have been identified in a wide range of habitats, suggesting the cosmopolitan or ubiquitous distribution of the corresponding organisms in the environment and, potentially, their broad metabolic capabilities (Hugenholtz et al, 1998). Organisms found in extreme conditions may also be found in many milder environments. For example, thermophiles (heat-loving organisms) have also been found in the cold water of the Mariana Trench, and *Thermus aquaticus*, the organism sampled from the thermal pools of Yellowstone National Park, from which

Taq DNA polymerase was developed (Box 8.7), may be found in other thermal habitats all over the world. Again, exchange with other researchers of organisms discovered by Genencor in alkaline soda lakes has revealed that similar microorganisms can be found in other alkaline environments (including food processing waste) world-wide (personal communication, Professor Brian Jones, Genencor, 31 January 1998).

Thus some companies are happy to source their samples in their own back yards, or, indeed, anywhere, while others focus their search in extreme environments and specific locations. While a few companies we interviewed obtain all their samples from within the country where they are based, many obtain samples from culture collections, which are likely to maintain and supply accessions from all around the world. Several companies interviewed obtain specimens exclusively from abroad; for others, the proportion of foreign materials obtained lies somewhere between these two extremes.

Many companies rely heavily on a network of collaborating experts with specialist knowledge in a number of fields, whether taxonomists, microbiologists, or even geologists. As Montana Biotech explained, complete interpretation of the company's research results requires the integration of information from numerous resources, such as experts in the field, literature and internet sources and nucleic acid and protein databases.

In some cases, well-known species form the basis of companies' research. For example, Slater UK uses only common agricultural crops to develop phytoremediation products which are likely to be acceptable in the EU market. At the other extreme, 99 per cent of the organisms with which Montana Biotech works are novel and have not been previously identified. Many of these organisms are from the kingdom archaea, a genetically distant and distinct form of life.

In addition to conducting their own research upon the raw samples and derivatives such as basic extracts that they acquire from their collaborators, many companies acquire more value-added products, such as active isolates, taxonomically identified strains of microbial cultures, purified enzymes, DNA and genes, and commercial products and processes incorporating these. Some outsource a great deal of their research, under contract, to other groups, thereby removing the need to maintain broad in-house expertise and specialist equipment. Others form opportunistic partnerships as circumstances arise. Montana Biotech Corporation described how once, when it found a microorganism which produced a unique polymer, it collaborated with a university that had the facilities to test it. On another occasion, it found a microbe which could metabolise methane, but did not have the facilities to test its action on an industrial scale, so it collaborated with a large-scale methane production plant. Companies tend to out-source only the less commercially sensitive work. Thus the early stages of discovery, or the later stages of formulation and product development, may be contracted out, but the finer details of discovery and development are likely to be researched exclusively in-house.

While many companies obtain these value-added products and services from abroad, few, if any, were obtained from developing countries, which more commonly supply raw samples, basic extracts or occasionally cultures. However, in some cases, companies deliberately seek out value-added partnerships in countries where they do not have a distribution network.

8.2.6 Resupply

Generally speaking, once a biotechnology company has a sample in its collections on which it can base its research, it does not need to return to the source for a further supply. This is particularly true of microbial cultures, which can often be bulked up by fermentation. Diversa explained that, 'we sub-clone the gene of interest into an over-expression host [ie an organism that produces more of the desired enzyme or compound than normal], and then ferment up the culture to obtain large quantities of a specific enzyme or compound'. As Dr Paul Hamlyn of the British Textile Technology Group said, once a company has obtained a culture, it is likely to maintain it in-house, but may occasionally need to re-source material from a culture collection, as there is the possibility that some material may mutate. In the case of plants, it is more common for companies to need to return to the source to obtain larger quantities of the original sample, if initial tests show promising results.

8.2.7 Choice of suppliers and collaborators

Companies are able to obtain supplies of genetic resources from a bewildering range of countries and organisations, so it is interesting to examine which factors affect their choice of supplier. The calibre of the scientists in the organisations supplying the genetic resources was a consistently high criterion. One typical response was from Dr Peter Innes of Micro-Bio: 'The calibre and expertise of the research group in the area in which Micro-Bio wants to develop a product is most important when selecting an institution [with which to collaborate]'. Some companies acknowledge the importance of the calibre of collaborating institutions, but recognise that this varies enormously around the world, and do not allow this factor alone to determine the selection of partners, when access to a diversity of genetic resources is also an important consideration. 'In the countries in which we collect, molecular biology skills vary widely from world-class to almost non-existent,' explained Diversa. Dr Peter Grant of Celsis International noted that companies can usually tell the calibre of the scientists with whom they are considering collaborating from the published papers.

While securing access to as wide a diversity of genetic resources as possible is an important factor for some companies, for others it is less so. Biotal, for example, has 'a very small requirement for diversity, as we are working with one or two taxonomic groups of bacteria only.'

Several companies stressed that the simplicity of obtaining permits for collection and export was an important factor in deciding where to work. 'The primary consideration is the quality of the sample, but if permits are difficult we might be inclined to go elsewhere,' was a typical observation. When permitting procedures are not simple and streamlined, much time, money and effort may be expended, thus delaying the research cycle. However, other companies felt less compunction about dealing with the permitting process. As one company put it, 'In most countries it is quite straightforward to get the material out, and on many occasions we have not bothered with permits. If there were problems accessing material, we would just go elsewhere. The company has a firm belief that similar microbes occur in the same environments all over the world. For example, you could probably obtain the same bug from the Atlantic as you do from the Pacific, so we do not feel restricted by [the need to work in any particular] country.'

Several of the companies interviewed described problems sending their products – particularly those containing pathogenic organisms – to other countries, whereas they faced few problems in acquiring samples. One explained that the simplicity of the process for obtaining permits is very important to it, but more for export than for import, since its formulations have to pass the quarantine inspection service.

Box 8.13 *Respondents' Reasons as to Why Demand for Access to Genetic Resources for Biotechnology will Grow*

- Demand for alternatives to synthetic chemicals is growing.
- Greater use of gene expression libraries is likely to lead to greater demand for access to new genes.
- There are many new categories of wastes to treat, so there will be a growing demand for access to new sources of microbes.
- There is increased interest in plant oils, as biotransformation can improve their quality.
- Industry is more comfortable with genetic modification, and therefore there is greater flexibility with respect to the sources the DNA may come from.
- Biodiversity is very rich and may contain organisms which are applicable to industry but as yet unexplored.
- Enzymes with new properties may come from a new genetic resource.
- The amount of information at present is limited and may be opened up by the generation of new molecular biology databases, leading to a demand for access to a wider range of genetic resources.
- The biological part of waste treatment is slowly becoming more important.
- Diversification into new forestry products will increase demand.
- More information on bioinformatics is available, leading to better experimental design and an ever-increasing use of biological materials.
- There is potential to use 'wild flower' species which grow on contaminated sites.
- We can explore the synergies between phytoremediation and bioremediation.

Cost

While the quality of samples appears to be companies' prime concern, the cost of samples, and of collaborations overall, is another important consideration when they decide from whom to obtain genetic resources. Companies are well aware that the huge microbial diversity world-wide creates something of a buyer's market, and they are determined to source materials as economically as possible. As one company put it, 'It is important to remember that the microbes we are working with are not unique. For example, if a microbe is found in a Yellowstone hot spring and is then also found in a deep sea vent, we will go to Yellowstone as it is the most economic source. Alternatives would be to go to other companies or countries.'

Infrastructure

Another factor singled out by some companies as being important when they choose where to work is the availability of infrastructure that facilitates arrangements. 'Ease of communication and infrastructure issues such as ease in getting samples in and out of countries are also important to us.'

From respondents' replies, it is clear that they take all these factors into account when deciding whether to mount their own collecting activities, or to acquire material from a culture collection or other collectors. Of particular importance are the quality of the sample and the calibre of collaborating institutions.

Predicted interest in access to genetic resources in the future

Demand for access to genetic resources depends upon a number of factors, particularly the relative attraction of alternative methods of product development, and the economic competitiveness of working with 'new' genetic resources compared to the alternatives such as synthetic chemistry and the use of existing collections. Interviewees cited a number of reasons as to why demand for access might increase (Box 8.13), and why it might decrease. Despite the wealth of existing collections held by biotechnology companies, the majority of interviewees (75 per cent of 27 who answered this question) felt that the demand for access to genetic resources would grow in the future; 19 per cent thought it would remain the same, and 7 per cent thought it would decrease. It should be noted that an important method of obtaining access to genetic resources in the biotechnology sector involves obtaining samples already held in culture collections, in transactions which only exceptionally involve benefit-sharing with the depositor. In the biotechnology sector, companies were particularly reluctant to answer the question, 'Would you be prepared to pay more for access to samples than you do currently?', although several interviewees from universities and other public-sector organisations felt that companies frequently get a 'good deal' in terms of sample price, and many samples are provided for free.

8.3 Benefit-sharing

In the course of their commercial transactions, biotechnology companies share a number of monetary and non-monetary benefits with the organisations that supply their genetic resources and derivatives. These transactions are generally not considered in the light of 'benefit-sharing', but rather as an inevitable part of the bargain in order to maintain access to quality samples, and collaboration with high-calibre scientists, and to remain competitive in the future. The transactions themselves take several forms. One of the most common is the supply of genetic resources through licensing agreements, which initially allow the recipient to screen the materials, and, if the results are positive, to conduct further research and commercialise the results. This kind of agreement is often used to provide companies with access to libraries of samples, extracts, compounds or other value-added derivatives. Where there is more collaborative research, joint ventures may be used as the framework for sharing the risks and rewards of research and development, including access and benefit-sharing.

Many of the smaller biotechnology companies are spin-offs from universities, or in the 'start up' phase. They do not yet have products on the market, and are relatively inexperienced in negotiating licensing and benefit-sharing arrangements. Their calculations for benefit-sharing often seem surprisingly vague. As one interviewee put it, 'I have no feel for what is a good price. It doesn't influence the end price of our products. I trust [our suppliers] to charge us cost, plus a fair profit. We do not even have a contract with them, just a gentleman's agreement.'

Since resources may have passed through many hands (and, notably, been deposited in and accessed from culture collections) before reaching the company that ultimately commercialises a product, benefit-sharing with 'source countries' is relatively rare. Several interviewees believed that the issue of benefit-sharing was not relevant to them, as they did not collect themselves, but obtained their materials from intermediary institutions such as culture collections.

Nevertheless, developments on two fronts suggest that, in future, there may be more mechanisms to channel benefits to the governments and stakeholders conserving genetic resources *in situ* and ex situ. First, the benefit-sharing commitments of some companies – particularly the smaller ones – appear to be led by the requirements of providers of genetic resources. Progress in benefit-sharing is thus often at the instigation of intermediary organisations such as culture collections, which are increasingly using MTAs containing some kind of benefit-sharing provisions, as well as charging fees for access to samples. Second, those companies whose staff collect genetic resources or establish access arrangements with intermediary institutions overseas are aware that access legislation increasingly requires collectors to define benefits as part of a collecting permit (see the Diversa/INBio agreement: a case study, p253 below) – although they may not consider working in such countries.

8.3.1 Monetary benefits

Monetary benefit-sharing in biotechnology development takes many forms, from the payment of a fee per sample, through sponsorship of a research grant (effectively payment for sub-contracted research activities), to milestone payments and royalties. Although it is rare to do so, several companies occasionally used milestone payments, joint ventures, and stakes in the equity of a company in exchange for providing access to genetic resources.

The value of benefits arising from the discovery and development of enzymes is considerably lower than for pharmaceuticals, largely due to the considerable difference in their relative market value. A technical enzyme produced by microbial fermentation costs several orders of magnitude less per gram of pure protein than a similar weight of a pharmaceutical product.

The majority of companies and institutions interviewed pay for access to samples, although payment may simply be a nominal fee to purchase a culture from a collection, or a DNA sequence. However, a common price for an 'ordinary' strain is typically between US$80 and US$350 for a vial containing a few milligrams of culture. If a company requires a more unusual organism, it usually seeks a quote from the provider, and may consider paying several thousands of dollars.

Many culture collections operate different price schemes for for-profit and not-for-profit organisations. For example, the prices listed on the website of the Belgian Coordinated Collections of Microorganisms (BCCM) range from US$50 for universities and non-profit organisations to obtain plasmids, to US$160 for access to cDNA libraries for commercial entities (www.belspo.be/bccm.htm). The German Collection of Microorganisms and Cell Cultures (DSMZ) website lists prices ranging from US$14 for non-profit groups to access certain viruses, to more than US$100 for access to plant cell cultures and strains deposited under patent regulations (www.gbf.de/dsmz/dsmzhome.html). The American Type Culture Collection (ATCC) website shows prices ranging from US$60 to US$275 per strain, and ATCC offers discounts on some items to non-profit organisations (www.atcc.org/codes.html). The International Mycology Institute has a standard supply charge of US$80 to cover maintenance and handling costs, although there is a discount price of US$50 for universities using strains for teaching purposes (personal communication, David Smith, IMI, 3 March 1999).

Companies sub-contracting the sourcing of specimens and research work may pay either for specific results, or for an agreed work programme, the outcome of which is uncertain. These arrangements generally involve a confidentiality agreement, and clarification of the ownership of any IPRs over the results of the research. One rule of thumb adopted by some companies is that any IPRs arising from this research belong to the company, unless the university or research institution conducting the research makes a significant advance outside the area envisaged by the original research agreement, in which case the royalties would be shared.

Royalties are not always offered to companies acquiring genetic resources. In particular, a flat fee is more common than the offer of royalties when the samples supplied are 'raw materials', and come with no indication of activity or application. Some interviewees did not feel that royalties were the most attractive form of benefits for suppliers to receive, as it takes such a long time before the collaborator may see any 'benefit'. For example, it may be five to 15 years between the identification of an enzyme and money coming in. Although universities often ask the company in question for royalties, after discussion, the partners frequently agree that 'a four-year post-doctoral research grant may be of more immediate benefit'.

Table 8.3 *Royalties Offered to Collaborators in Enzyme Discovery*

Input	Activity of collaborator	Royalty range %
Sample	Pretreatment	0.05–0.1
Strain (either identified or not)	Isolation of organism	0.1–0.2
Active strain	Detection of activity	0.2–0.5

Source: Novo Nordisk, personal communication.

However, as Tables 8.3 and 8.4 reveal, the payment of royalties is common in the biotechnology sector. The range for royalty rates for the development of enzymes is generally lower than the comparable range for pharmaceuticals, lying between 0.05 per cent and a few per cent, depending on the value added by the organisation providing the genetic resource. Companies have fairly idiosyncratic approaches to royalty payments, so there is really no 'industry average'; however, the range of royalties offered by Novo Nordisk is fairly typical (Table 8.3). In Table 8.4, which collates the answers provided by several interviewees, values for royalties suggested by several companies are coloured darker.

A number of factors affect the magnitude of the royalties offered in the event that genetic resources or derivatives supplied contribute to a successfully commercialised product. As Dr Lene Lange, Senior Principal Scientist in Enzyme Research at Novo Nordisk, puts it, 'The main point is the quality of the overall collaboration: the scientific level, validity and credibility; the mutual interests between the collaborating partners; scientist to scientist cooperation; and communication. We do not simply buy samples like groceries.' The factors mentioned by interviewees as relevant to their decisions on royalty rates include the current market rate for royalties and the market share that the final product enjoys, but the key to a higher royalty rates is to provide value-added genetic resources, such as strains with proven activity and clear commercial applications. Where the provider has obtained IPRs over the product he or she is providing, the amount of royalties will be correspondingly higher. In this sector, the provision of information with a sample, in general, and ethnobotanical information in particular, appears to be of little relevance to companies and does not contribute to a higher share of monetary benefits.

As Professor Mike Turner of Ensynthase Engineering explains, royalty rates are not always negotiated up-front at the time of access. Typically, a company may pay for access to 50 cultures, and then negotiate royalties if and when any desired activity is found.

Table 8.4 *Royalty Rates for Genetic Resources and Derivatives for Biotechnology*

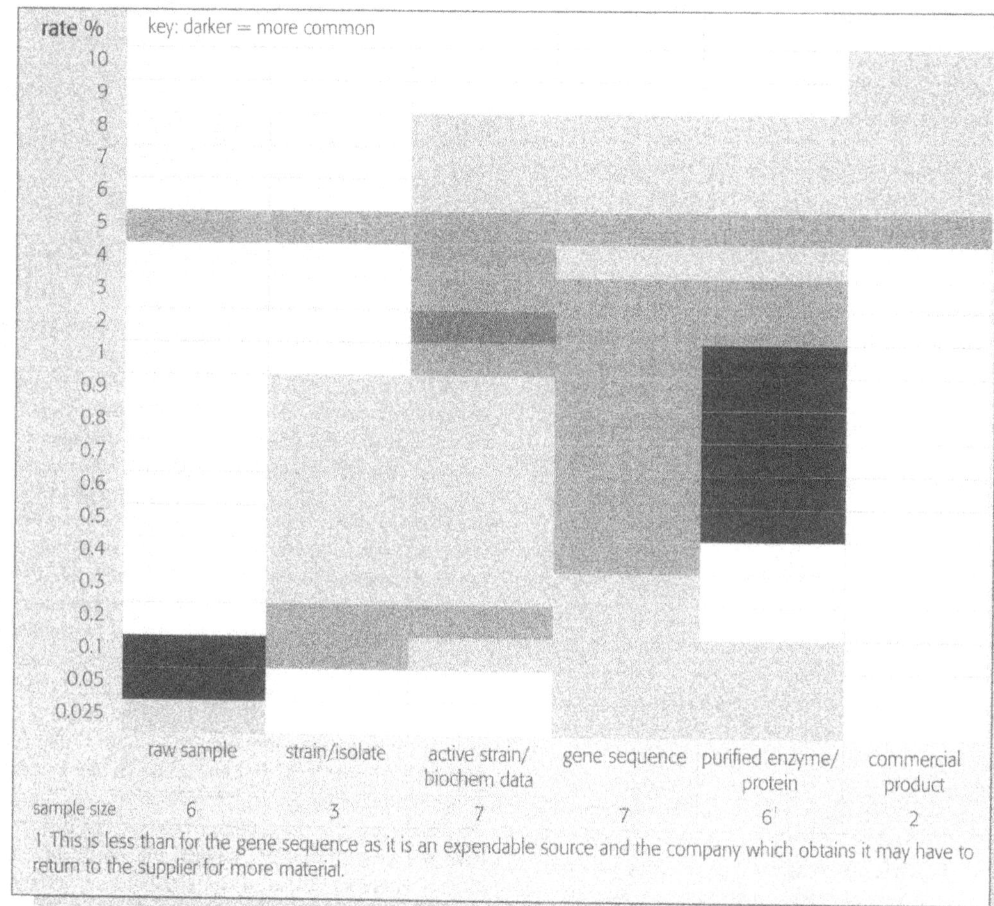

When asked whether they would be willing to pay more for access to genetic resources, our interviewees demurred. A number pointed out that while they would look for some microbial genetic resources in certain locations (for example, certain extremophiles), many are ubiquitous, for example, Pseudonomade. The possibility of sourcing ubiquitous microorganisms locally, or even of sourcing extremophiles from a number of different sites, reduces the amount companies are willing to pay to collect in any one site. If the prices asked by one supplier went up, they would simply turn to another.

8.3.2 Non-monetary benefits

It is relatively common for biotechnology companies to share non-monetary forms of benefit. They share information and research results, transfer technology, train their collaborators and contribute to capacity building in the institutions from which they obtain supplies. Indeed, one company emphasises non-monetary benefits, feeling that money is not always the best way to support research and to obtain genetic resources from some of the countries it works in. Another common form of non-monetary benefit-sharing among research institutes and non-profit organisations is the exchange of samples, in a reciprocal system of access to genetic resources. Culture collections, for example, may provide institutions that deposit strains with them with three or four cultures in exchange.

Companies are prepared to share data and information, provided they can protect confidentiality and the opportunity to patent discoveries. As one company put it, 'We provide information, advice and research results when it is appropriate. The collaborator may only be working on one small part of the process that leads to a new product, in which case the company would only share results relevant to this. We would not share information such as production cost and other proprietary information.'

Several of the companies we spoke to train staff from collaborating organisations, either when they visit their collaborators – for instance, working alongside them in a university – or by inviting them to work in their own corporate laboratories, providing them with access to larger and better facilities than may be available in the collaborator's own organisation, and to expensive technology such as nuclear magnetic resonance equipment. For example, in 1998, Novo Nordisk began a collaboration with the National Centre for Genetic Engineering and Biotechnology (BIOTEC) in Bangkok, Thailand. The programme envisages training and technology transfer involving a Thai student, who will spend some of her time in Denmark in the laboratories of Novo Nordisk (Box 8.14). The PhD student and her advisor in Thailand collect samples from various ecological niches. Isolation and preliminary enzyme testing is done in Bangkok (through liquid fermentation and testing for extracellular enzymes). Based on these results, a selection of strains is transferred to Denmark for further molecular cloning and expression work. This is also done by the PhD student herself, but under guidance of the Novo Nordisk scientists (personal communication, Dr Lene Lange, Novo Nordisk, 10 March 1999). Other examples of training and staff exchanges are described in the two case studies below.

Box 8.14 *Examples of Non-Monetary Benefits Shared with Providers of Genetic Resources for the Development of a New Enzyme by Novo Nordisk*

- Support to establish and maintain culture collections
- Introduction to isolation and preservation techniques
- Contact with taxonomists to identify microbial genetic resources
- Contact with expert scientists
- Training in:
  - safety procedures
  - sterile techniques
  - enzyme assay techniques
  - molecular biology
- Technology
- Assistance with and co-authorship of scientific papers.

Source: Novo Nordisk, personal communication.

In addition, companies can help to build the capacity of their collaborators. Several interviewees had experience in establishing laboratories with collaborators in developing countries. While the calibre of scientific counterparts is one of the prime criteria by which companies select collaborators, they are also used to working with institutions with a wide range of abilities, 'from world-class to almost non-existent', as one company put it, and are happy to work with them to build their skills, provided there is some advantage to the company in the collaboration. In exchange for access to genetic resources, one interviewee had built the capacity of a collaborator to the extent that it has now become a competitor. Again, the following case studies offer some insight into the kinds of technologies and skills that organisations can acquire in the context of access and benefit-sharing partnerships.

8.4 Case studies

8.4.1 The Diversa/INBio agreement

Kerry ten Kate

This case study concerns an agreement between the Costa Rican National Biodiversity Institute (INBio) and the Diversa Corporation.

INBio

INBio was established as a private, not-for-profit organisation in 1989 by the Ministry of Environment and Energy (MINAE) of the Government of Costa Rica, with the aim of strengthening and reviving the value of biological diversity in Costa Rica. Comprising of a central research facility at San José and 28 research stations and offices in major conservation areas of Costa Rica, the organisation now employs 165 people.

INBio's strategy concerning biodiversity is 'save, know, use', and its mandate reflects these priorities, embracing programmes in all three areas. National priorities and key tasks of INBio include the National Biodiversity Inventory, disseminating findings and results, promoting non-destructive uses of biodiversity with the private and public sectors, linking with international networks of other biodiversity management institutions, and responding to planning and fund-raising needs. As far as knowledge of biodiversity is concerned, INBio's major contribution is its inventorying activities. Costa Rica is a 'megadiversity' country, and the current estimate for the number of species is greater than 500,000, of which some 18 per cent have been described. INBio has conducted inventories of this biodiversity since its inception, and has generated over 14,000 web-pages of information on Costa Rican biodiversity. An important component of INBio's work to promote the sustainable use of Costa Rican biodiversity is its programme on Bioprospecting.

INBio's first research agreement with a company was the much-publicised agreement with Merck and Co (signed in November 1991). Since then, INBio has signed a number of other commercial research agreements, including agreements with the US pharmaceutical company Bristol-Myers Squibb (1993); the Swiss-US company Givaudane Roure for the development of new fragrances (1994); the Italian manufacturer INDENA for the development of new phytochemicals and phytomedicine (1996); the German service and contract research company Analyticon (1996); the Costa Rican company La Pacifica and the British Technology Group for the development and possible commercialisation of a bio-nematicide from a tropical legume (1994), and the US-biotech company Phytera for the development of cell cultures from plants (1998). To date, INBio's bioprospecting agreements have contributed more than US$390,000 to MINAE; US$710,000 to conservation areas; US$710,000 to public universities, and US$740,000 to other groups at INBio, particularly the Inventory Programme.

Diversa

Diversa is a company specialising in the discovery, modification, and manufacture of novel enzymes and bioactive compounds resulting from research on biological samples. Since 1994, Diversa has discovered more than 1,000 enzymes and now serves customers world-wide. With headquarters in San Diego, California, the company was initially established as 'Recombinant Biocatalysis, Inc', until its name was changed to 'Diversa Corporation' in August 1997. Diversa bases its research on molecular biology and functional genomics, and is particularly interested in sourcing materials from all extreme environments represented on earth, including the thermal pools of Yellowstone National Park and Iceland, and the volcanoes of Indonesia and Mexico. The company works in three principal areas:

1 *biocatalysis:* exploration of biocatalytic functions for application in the synthesis of pharmaceuticals and speciality chemicals enzymes;
2 *transgenic plants:* expression of enzymes and bioactive molecules in plants, for uses such as insect and disease control; and
3 *discovery of recombinant natural products:* the discovery of bioactive products from recombinant sources for the generation of lead compounds for new pharmaceuticals.

Rather than culturing organisms that express genes and compounds of special interest, Diversa captures nucleic acids by sampling DNA directly from the environment and cloning the genes from these uncultured microorganisms. The company uses ultra-high throughput robotic screening systems to identify novel enzymes and bioactive compounds expressed from both single genes and gene pathways. Once Diversa has identified enzymes and bioactive compounds demonstrating interesting characteristics, it licenses them to client companies in the pharmaceutical, seed, crop protection and biotechnology sectors.

The partnership

In 1995, Diversa and INBio signed a three-year agreement, which was renewed and expanded in 1998. The first three years of the partnership involved biannual collections of soil and water by Diversa and INBio. The samples were all processed at the Diversa laboratories in San Diego. During this collaboration, the relationship between the two parties deepened, and in October 1998, Diversa set up a DNA processing laboratory at INBio's bioprospecting division in San José. Now that the biotechnology laboratory has been established at INBio and a molecular biologist could be hired with Diversa funds, INBio is able to do more value-added work. Interactions between the two institutions are more frequent and there is a greater exchange of ideas and more collaborative research.

The 1991 agreement between INBio and Merck was the first benefit-sharing agreement between a drug discovery company and an institution in a biodiversity-rich country. The agreement between INBio and Diversa is the first of its kind in the field of gene prospecting (Nader and Rojas, 1996), and was the model for the well-known Yellowstone-Diversa Cooperative Research and Development Agreement (CRADA) (ten Kate, Touche and Collis, 1998).

Both partners feel they have learned from comparing their experiences in this partnership with others. According to Diversa's former Chief Executive, Dr Terrance Bruggeman, it takes a great deal of energy, cost, managerial time and personal effort to make sure that agreements of this kind are negotiated properly and fairly and that all the necessary consents are obtained along the way. This can be quite a burden for small companies. INBio's experience, gained through other partnerships, was a great help in establishing the partnership. The agreement with Diversa was concluded after just two months of negotiations, compared with several of Diversa's other agreements, which have taken between three and a half and four years to negotiate. Since both INBio and Diversa are experienced in developing such partnerships, both organisations, unusually, already had bioprospecting agreements ready to use as the basis for their negotiations. On this occasion, Diversa produced the first draft of the agreement.

Diversa and INBio are collecting samples of microorganisms associated with insects, nematodes, epiphytes and other organisms from a broad range of biotopes, including mangrove swamps, coral reefs, forest soils and tropical leaf litter. Diversa is looking for enzymes, structural proteins and other secondary metabolites for a variety of applications including biotechnology, crop protection and pharmaceuticals. Within the arena of biotechnology, enzymes derived from microorganisms can have many potential applications, of which the following are examples:

- *cellulose degradation:* microorganisms which live in the guts of insects are well adapted to degrading cellulose fibres, so Diversa is sampling microorganisms from the guts of termites and other insects;

- *biomass conversion:* cellulose-degrading enzymes may also have applications in the conversion of biomass to ethanol; and
- *chemical processes:* Diversa is interested in the discovery of enzymes that are effective in separating racemic mixtures of compounds, and which will permit the chiral synthesis of pharmaceutical intermediates. Enzymes also have applications in polymer chemistry, and facilitate the production of longer, more uniform polymers than those produced by chemical processes.

Diversa is also interested in finding organisms that live symbiotically with insects and microorganisms, and which may produce secondary metabolites with potential pharmaceutical and agricultural applications as anti-fungal, anti-infective or antimicrobial agents. Thus the work is interdisciplinary and Diversa could well discover leads with applications in several different industry sectors.

Under the terms of both the 1995 and 1998 agreements, INBio collects samples using both its own techniques and proprietary technology provided by Diversa. INBio guarantees that it will not collect and process samples for other companies using Diversa's proprietary technology. Otherwise INBio is free to provide other companies with DNA from the same environments. All DNA sequences, isolated by INBio for Diversa, become Diversa property, and all micro-organisms, isolated from the sites, remain the property of Costa Rica.

In return, Diversa pays the salary and overheads of at least one INBio staff member, and commits to pay undisclosed royalties to INBio, in the event that Diversa licenses a product based on samples obtained from INBio to a client company. In addition to these monetary benefits, Diversa provides INBio with non-monetary benefits such as technology, equipment, capacity building and access to Diversa's high-throughput DNA sequencing facility.

Access

All bioprospecting activities and programmes of investigation within Costa Rica on genetic or biochemical material require an access permit. The Costa Rican Biodiversity Law ('Ley de Biodiversidad 7788'), which came into force in April 1998, regulates access to genetic resources. However, pending implementing regulations and a court challenge as to the constitutionality of two articles concerning the powers vested in the bodies established under the new Law, INBio's bioprospecting work continues under the earlier law, the Costa Rican Wildlife Law ('Ley de la Vida Silvestre') which authorised MINAE to grant concessions, contracts, rights of use and licences for conservation and sustainable use, and other relevant laws related to natural resources, such as those pertaining to forests and protected areas. Applications for access are handled by a Technical Office within MINAE. Both the 1995 and 1998 agreements between INBio and Diversa were obtained pursuant to the Wildlife Law.

The Biodiversity Law authorises MINAE to manage and conserve biodiversity, and to grant access to genetic resources. A decentralised and inter-departmental body under MINAE, known as the National Commission for the Management of Biodiversity, is to formulate and coordinate policies on access. Access applications will be handled by a new Technical Office to be established under the National Commission.

The work of INBio is based on a cooperative research agreement with MINAE which specifically sets the terms and conditions for INBio's biodiversity inventory and bioprospecting activities. According to this agreement, INBio will donate 10 per cent of all bioprospecting budgets and 50 per cent of all income from royalties to the MINAE, for conservation purposes.

The Wildlife Law requires that every entity intending to investigate Costa Rican Biodiversity must seek prior authorisation from the General Directorate for Wildlife at MINAE. The applicant must submit a description of each project, and each person who will collect specimens needs individual permission to do so. The export of biological material requires further permission, which is only granted if the material was collected within an authorised and registered research project. INBio applies for an access permit from the Technical Office for the whole, three-year period of the collaboration. The technical office at MINAE studies each access application from INBio and has the right to reject the application on reasonable grounds. Should any unusual circumstances arise, or the plans change in any way from those described in the project proposal, INBio will contact MINAE for their approval. Under the agreement, INBio has the right and duty to impose limitations on collection activities if at any point the habitats are deemed too fragile to support the activities concerned. The terms of INBio's agreement with MINAE require it to ensure that its own collecting activities, and those of collaborators with whom it works in the field, cause no undue environmental impact.

Before each collecting trip, INBio informs the Conservation Area staff about who will be going to the area to collect, and requests permission to do so. The Conservation Area staff can ask for more information on the proposed activities, and for a report of the work once it is done. Most of the samples that are the basis of the agreement between INBio and Diversa are collected by INBio staff, but staff from Diversa visit Costa Rica on average twice a year, and on these occasions, they usually go on a collecting trip, accompanied by their hosts from INBio.

Research and development

One member of INBio's staff, Ms Myriam Hernandez, a molecular biologist, is supported full-time by Diversa. Since the agreement was renewed in 1998, Diversa and INBio jointly direct her research programme. Currently INBio staff isolate DNA from the samples collected, which requires the proprietary technology supplied by Diversa. Diversa process results in highly purified, clonable, high molecular weight DNA suitable for the production of complex gene libraries, some of which not only express proteins, but also contain clusters of up to 100 genes for the synthesis of small molecules.

The purified DNA is transferred to Diversa's San Diego laboratories for the construction of gene expression libraries. While INBio does have the capacity to conduct some secondary screening in Costa Rica, the terms of Diversa's agreements with some of its client companies mean that Diversa is obliged to carry out certain screening activities itself, and not to transfer proprietary screens provided by its clients outside the company. Indeed, Diversa also wishes to keep some of its own proprietary technology in-house. However, the partnership is very interactive and constantly evolving. Both parties are committed to investigating options for more value-added work to be conducted at INBio, provided this meets the requirements of Diversa's clients.

Currently, Diversa passes purified proteins and other derivatives of genetic resources that do not contain nucleic acids to its clients. The companies only obtain derivatives and not the Costa Rican organisms themselves, and are thus unable to reproduce the genetic resources. Should they wish to do so, they would need to negotiate with Diversa, which would involve INBio in the decision.

According to Diversa's former Chief Executive Officer, Dr Terrance Bruggeman, to produce a fully developed product from the leads emerging from initial discovery and screening of an enzyme for biotechnology applications would generally take the company between two and three years and some tens of millions of dollars. The process could involve screening tens or even hundreds of thousands of organisms before a desired gene trait is revealed.

Since the partnership between INBio and Diversa was only established in 1995, and the product discovery and development process for an enzyme can take several years, no specific product development programme has arisen yet from the collaboration. However, Diversa is already investigating some leads. Some of the samples received from Costa Rica have been used to create a series of secondary metabolite extracts that are presently being screened by about ten pharmaceutical companies world-wide who are customers of Diversa. In the biotechnology field, a thermophilic enzyme is currently being tested by one of Diversa's client companies in the oil industry. The enzyme is pumped at high temperatures and pressures into disused oil fields, to help reclaim remaining oil. Another enzyme is being used by a company in the food industry to improve the nutritional quality of animal feed. While there have currently been no biotechnology leads from Costa Rican microorganisms, Diversa is obliged to keep INBio scientists fully informed of the results of its screening efforts.

Benefit-sharing

Monetary benefits

Diversa pays the costs of collecting activities, the salary of Ms Myriam Hernandez and an overhead for the use of the infrastructure at INBio for the three-year period of the agreement. The products that may emerge from the partnership could be of many kinds, involving different levels of investment in research and development, and giving rise to different profit margins. Consequently, in this, as in similar agreements, royalty levels are defined by the product category. The general distinction is between pharmaceutical products in one schedule, containing a higher royalty range, and other potential applications, such as biotechnology and agriculture, in another schedule with a lower royalty range.

Technology transfer and capacity building

Diversa has provided INBio with equipment for a molecular biology laboratory worth over US$30,000, including fermentation kits, centrifuges and PCR equipment. This enabled INBio to complement its existing laboratory capacity. INBio's ultimate goal is to establish its own functional molecular biology laboratory. Diversa has also transferred to INBio a series of techniques that should help INBio to select samples more effectively and to isolate DNA of interest.

INBio believes the techniques in molecular biology that it has acquired through its partnership with Diversa are important to the organisation for a number of reasons:

- Molecular taxonomy is becoming an important tool for the taxonomist and is thus essential for the inventory approach that INBio uses as part of its mandate from MINAE.
- The survival of threatened populations depends on the conservation of their genetic diversity. The extent of this diversity must be determined in order to initiate rational conservation measures. This requires infrastructure, knowledge and equipment for molecular biology.
- The search for new genes is becoming the major preoccupation in bioprospecting, and skills in biotechnology and genetic engineering are thus critical for the long-term success of INBio's bioprospecting activities.

The method of accessing genetic information directly from nature is complex and requires state-of-the-art technology which was only recently developed and is only available in a few companies, such as Diversa. The collaboration with Diversa is, for INBio, an important step that can help the institution build a new level of capacity in biotechnology and genetic engineering. According to Dr Werner Nader, INBio's business development manager, INBio's biotechnology laboratory could not have been developed at this juncture without the cooperation of Diversa. INBio's experience is that countries such as Costa Rica can only acquire this kind of high-technology through step-by-step investments in infrastructure, equipment and human resources. INBio believes that a small company like Diversa offers the advantage of openness and flexibility in transferring technology and know-how, and is thus a particularly attractive partner.

Training

Prior to taking up her position with the INBio-Diversa partnership, Ms Hernandez trained for eight weeks at Diversa's US laboratory in order to prepare for the work in Costa Rica. In addition, she and other INBio staff have been invited to spend one month at the US laboratories of Diversa at the company's expense, to be trained in molecular biology techniques. Visiting researchers from Diversa train INBio staff on the job and are helping to set up a functional molecular biology laboratory. Year-round, a member of INBio staff supervises and supports Ms Hernandez.

Further training is likely to go beyond the skills which it is necessary for INBio to perform to meet its obligations under the agreement with Diversa. The exact nature of the training is decided jointly by INBio and Diversa during the course of the project.

Joint research and development

INBio's expertise in conservation, taxonomy and inventories helps Diversa to understand the nature of biodiversity in Costa Rica, and Diversa's skills and technologies improve INBio's knowledge of the depth of what is contained in the rainforest canopy. Staff in both organisations have established regular reporting systems, which operate in both directions. Diversa uses its database to keep track of the results of work on INBio samples and informs INBio on a regular basis of the results. All INBio's agreements oblige its industrial partners to inform INBio of the results of their work on a regular and strictly defined basis.

Both INBio and Diversa feel that the close collaboration between them reflects the level of scientific expertise of both parties, which allows for joint research by peers in each organisation. According to Eric J Mathur, Diversa's Director of Molecular Diversity, 'Biodiversity prospecting agreements require far more than signed contracts; they only become productive through close personal work and frequent contacts between the scientists of the collaborating institutions. These delicate relationships require hands-on management and real-time feedback on the status of research and product development. Most countries are not interested in simply selling their soil. They want to take part in the discovery effort. At Diversa, we spend considerable time working with the scientists from the source countries to build capacity and fully integrated discovery programmes.'

(*Sources:* Nader and Rojas,1996; Tamayo, Nader and Sittenfeld, 1997; Nader and Mateo, 1998; ten Kate, Touche and Collis, 1998; personal communication between Adrian Wells and Dr Nicolas Mateo, INBio, August 1998; personal communication between Kerry ten Kate and Dr Terrance Bruggeman, Diversa, November 1998; personal communication between Kerry ten Kate and Dr Werner Nader, January and February 1999; personal communication between Kerry ten Kate and Dr Eric Mathur, Diversa, February 1999; presentation by Dr Nicolas Mateo 16 October 1998, at the Workshop on Access to Genetic Resources, Technical, Legal and Ethical Aspects, 15–16 October 1998, São Paulo, Brazil.)

8.4.2 New England Biolabs

China Williams and Kerry ten Kate

New England Biolabs (NEB), Inc, a biotechnology company with an annual turnover of US$50 million and a staff of over 200, was established in 1975 in Beverly, Massachusetts. The company's particular expertise lies in discovering, developing and cloning restriction enzymes. Restriction enzymes protect bacterial cells from infection by foreign DNA (viruses), and are used for the study and manipulation of genetic material. Restriction enzymes were first purified from bacteria over 20 years ago. Subsequently, nearly 3,000 restriction enzymes exhibiting over 200 different recognition specificities have been found, by screening over 10,000 bacteria and archaea. Approximately 30 per cent of these recognition specificities were first discovered at NEB.

NEB maintains a database of restriction enzymes and related literature, called REBASE, which can be accessed from its website (www.neb.com), as can its annual product catalogues, listing the enzymes available to its customers. The company currently supplies over 180 restriction enzymes, half of which are produced by recombinant DNA technology. As well as selling restriction enzymes, NEB also provides technical support regarding restriction enzymes, including the REBASE database and a wide variety of other products used in molecular biology.

Partnerships to establish laboratories in developing countries have formed an important part of NEB's business. These laboratories provide the company with access to microbial genetic resources and intensive screening programmes for new restriction endonucleases. NEB has established restriction enzyme screening laboratories in China (1987), Vietnam (1989), Portugal (1996), Cameroon (1996), Uganda (1996) and, most recently, in Nicaragua (1998). About 10 per cent of NEB's restriction enzyme products were discovered through the screening collaborations.

The basis for collection

Restriction enzymes produced by bacteria and archaea have been isolated from many different environments around the world. Dr Donald Comb, NEB's President, explains that, in fact, restriction enzymes are often found from ubiquitous organisms. 'If you take a sample of water from a lake or an ocean or some dirt from any place in the world and try to culture the bacteria or archaea from that sample, only about 2–3 per cent of the different organisms present can be cultured. These tend to be the same organisms all over the world ... Our experience with restriction enzymes is that finding different specificities does not depend on where in the world you look but how hard you look. For example, the widely used enzyme *Xho I* was found in an organism from my back yard as well as in an organism that grows at 100°C on black smokers found in thermal vents at a depth of 3000 meters.' Thus NEB's strategy for sourcing genetic resources is simply to obtain samples from a wide range of environments. NEB hopes that the royalties earned by the partner laboratories from the discovery of new restriction enzymes will partially support self-directed science activities in these laboratories.

The company has little experience of entering into formal access agreements, and has not yet negotiated directly with the government in any of the countries where it has established partnerships, relying instead on its partner laboratories to obtain collecting permits and ensure that they are entitled to provide bacterial strains to NEB for commercial purposes.

The respective roles of NEB and its partner laboratories overseas

At present, NEB's partner laboratories are involved primarily in the collection, cultivation and initial screening of micro-organisms. Some have the expertise to clone the restriction enzymes they discover. All the laboratories are involved in the isolation of pure cultures of bacteria from a variety of environmental locations, growing these pure cultures in small amounts, and then breaking open the bacterial cells to determine if they contain restriction endonucleases, and, if so, to find the specificity of the enzyme. This is done by mixing tiny amounts of bacterial lysate with DNA and looking at the pattern of DNA fragments produced by electrophoresis on agar gels. Scientists and students learn the very basic techniques of molecular biology by conducting these assays. The equipment is not expensive and a grant of US$30,000 can usually launch the career of a few scientists.

When a new specificity is found, a drop of bacterial solution is dried in a small tube and shipped to NEB for large-scale production and sale to the scientific community. In most cases, the amount of enzyme produced by the bacterial species is low and the final purified protein is very expensive. Therefore, either the screening laboratory, or a scientist at NEB, will attempt to clone the enzyme into laboratory strains of *Eschericia coli* – a bacterial species that has been used for over 60 years in thousands of laboratories world-wide to study and understand the molecular mechanisms of life. If the cloning is successful, and hundreds of times more enzyme is produced through expression in *e coli* rather than in the original bacterium, the price to the scientific community is reduced 10 to 20-fold.

Work at NEB to develop strains received from partner laboratories may take anywhere from a couple of months to a year or more, depending on the amount of enzyme produced by the native strain. While this development work is predominantly conducted at NEB, on some occasions the company and its partner laboratories conduct joint research. For example, on a recent visit to the Vietnam laboratory, NEB scientists worked together with their partners to analyse some especially difficult enzymatic activities they had discovered. Staff from NEB and the partner laboratory also discussed how to conduct the new screening methods being developed at NEB, and made plans to transfer the technology involved from NEB to Vietnam. Thus it is possible that, in the future, partner laboratories will become more involved in the further research and development of products. To date, 20 restriction endonucleases found by partner laboratories have been produced and marketed by NEB.

Benefit-sharing

Establishing and maintaining partner laboratories: equipment and salaries

NEB provides its partners with laboratory equipment and money to pay the salaries of the staff. The equipment transferred to a partner laboratory typically includes a centrifuge, gel electrophoresis apparatus, a sonicator, pH meter, stirrer, simple chromatography equipment, an incubator, plastic and glassware, and chemicals and the microbiological media needed to grow bacteria in. It costs approximately US$30,000 for NEB to establish a laboratory from scratch, provided that the space is freely available and that there is a regular supply of electricity. In addition to this figure for equipment, NEB pays the salaries of the laboratory staff. In some countries, such as Vietnam, NEB works with a laboratory that has already been established within a university or similar institute. In these cases, equipment may already be available, so the costs of equipping the laboratory are likely to be lower.

Once the partner laboratory has been established, NEB provides expertise, reagents, and a yearly credit towards the purchase of NEB products which are used in molecular biology research, and the costs of two shipments each year from the USA to the laboratory. NEB provides yearly funding to the laboratories and allows the director of each laboratory to allocate salaries according to its own programme. In addition, two scientists at NEB, Carol Polisson and Richard Morgan, help support the general activities of the partner laboratories. They are in regular contact by phone, e-mail and fax, offering technical assistance and trouble-shooting any problems with the equipment. They aim to visit each laboratory every other year for approximately one week to review activities and needs.

Royalties

In its agreement with each partner laboratory, NEB undertakes to pay a 5 per cent royalty on sales of enzymes found by partner laboratories and, to date, these enzymes have generated over US$400,000 a year in sales, of which over US$20,000 in royalties have been returned to partner laboratories. The royalty rate remains the same, whether the enzyme is produced directly from the original strain or NEB contributes additional research and development. Royalties are calculated and returned annually, and deposited according to the directions of the partner laboratory.

Joint research and training

NEB and its partner laboratories conduct joint research, and training is another important element of the partnership. NEB staff train each of the scientists involved in the partner laboratories, both at the company's facility in Beverly, MA, US, and when NEB scientists visit the partner laboratories. The training covers a range of skills, including techniques for the collection of samples and cultivation of bacteria, and methods of protein purification and characterisation. Newly-employed scientists at partner laboratories usually visit NEB for a 3–6 month training period. NEB covers all the expenses for these visits, including air fares. The company organises lessons in the English language for its visitors, when these are necessary. Training activities continue periodically for the duration of the partnership. For example, the director of the Vietnam laboratory will return to NEB in the spring of 1999 for a few months to update her skills and learn new techniques.

National research priorities and capacity-building

A 'knock-on' benefit that results from NEB's training programme is that the skills learned by the staff of partner laboratories are often transferred to others in the country, particularly when the partner laboratories are established within institutions such as universities. NEB partner laboratories in academic institutions in China, Cameroon and Vietnam have shared expertise and equipment with other researchers and students in those countries.

Partner laboratories can also use the royalties received from NEB to expand research in areas of their own countries' interest. For instance, staff from NEB's partners in Vietnam are also conducting research into polio and tuberculosis in their own country. NEB's partnership with a laboratory in Cameroon was established through joint research with a Cameroonian PhD student. The student came to NEB to work on river blindness and was co-supervised by Dr Francine Perler. On completing her PhD, she worked independently from NEB. After several years she visited NEB again, where she trained for six months in NEB's restriction enzyme laboratory, and returned to Cameroon to establish a partner laboratory. This was a benefit to NEB, too, since the company wished to expand its training programme to cover research into new enzymes.

Acknowledgement on patents and other publications

As yet, NEB has not patented any of the strains developed from samples received from its partner laboratories. However, if the company does so in the future, NEB will provide appropriate credit to staff from partner laboratories. Partner scientists would be named as inventors in the same way as scientists at NEB. All benefits are shared directly between NEB and the partner laboratory, except for the partnership in Vietnam, where a portion of the NEB funding is paid directly to the Ministry of Health. Directors of the partner laboratories decide how the benefits received from NEB should be allocated.

For its part, NEB has benefited from these arrangements by gaining access to interesting restriction enzymes which have become commercial products and by developing good screening programmes.

(*Sources:* personal communication between China Williams and Dr Donald G Comb, Dr Fran Perler, Carol Polisson and Rick Morgan, all of NEB, November 1998–February 1999.

Wider conservation implications

None of NEB's partnerships contain provisions dedicating specific benefits to conservation of the ecosystems where the samples were sourced, but NEB has established and partly funds an affiliated private foundation, New England Biolabs Foundation (NEBF). NEBF primarily supports scientific research for environmental projects in developing countries. This has ranged from whitefly eradication research in Dominica, West Indies, to research into solar energy, and funding for a Biotechnology Working Group report on patents and private ownership of life forms. In addition, NEB provides financial support for environmental research undertaken by the Marine Conservation Biology Institute (MCBI) and the Sustainable Ecosystems Institute (SEI).

8.5 Conclusions

The biotechnology industry is poised to make a growing contribution to product development, not only in the health and agriculture sectors, but also in environmental management, energy, materials, biocatalysis, food processing and diagnostics. Three-quarters of the companies and institutions we interviewed who replied to this question felt that the demand for access to genetic resources on the part of biotechnology companies working in these areas would inevitably grow. They attributed this to the fact that science is uncovering more potential applications for genetic resources in areas such as waste treatment, and to the availability of new tools that will facilitate the use of genetic resources, such as gene expression libraries, molecular biology databases and bioinformatics.

Microorganisms are among the most important genetic resources to the biotechnology industry, and they have a number of characteristics with important implications for access and benefit-sharing. Many taxa are poorly known. For example, it is not unusual for 15–30 per cent of fungi found in some investigations in tropical regions and unexplored habitats to be unknown to science (Stackebrandt, 1994). The vast majority of microorganisms have been neither cultured nor characterised in detail. Once microorganisms are held *ex situ*, if the conditions in which they are stored and propagated are not expertly regulated, there is likely to be genetic modification from the original sample. Studies have shown that while some microorganisms appear restricted to certain habitats (which may nonetheless occur in a number of countries), others are cosmopolitan in the environment, and are likely to be found in many different countries. Because of their microscopic size, it is difficult to track and monitor microorganisms. Their ability to replicate means that once companies have obtained a sample, they are unlikely to need to return to the source for further supplies.

Together, these factors suggest a number of conclusions for access and benefit-sharing with respect to microorganisms. First, it is hard to determine from where a strain originated, and to track access to and exchange of microorganisms, so that monitoring and enforcement of access and benefit-sharing legislation and material transfer agreements (MTAs) will be difficult, unless the parties cooperate. Second, even for extremophiles, companies may be able to obtain samples from more than one country, thus limiting their willingness to pay for access to a particular habitat in one country. Companies are well aware that the huge microbial diversity world-wide creates something of a buyer's market, and they are determined to source materials in the most economically competitive manner possible.

Companies pursue a mixture of strategies for finding interesting biological materials to put through their screening programmes and assays. Some take samples from their own back yards, while others seek out more extreme environments. Some companies work on known resources, while others seek novelty. It is common for biotechnology companies to collect their own samples, but they also obtain materials from a range of other institutions. The materials they acquire range from 'raw' biological samples to derivatives such as enzymes and isolated genes. Factors which affect companies' selection of materials and partners include the quality of samples, the calibre of collaborators, the cost of samples, the availability of infrastructure such as transport and electricity, and, importantly, the ease of obtaining permits for samples.

Since resources may have passed through several hands before reaching the company that ultimately commercialises a product, benefit-sharing with 'source countries' is relatively rare. Culture collections are an important source of genetic resources for the biotechnology industry. They are beginning to use MTAs for the supply of genetic resources, and these may prohibit the commercialisation of products derived from the genetic resources without separate agreements. None of these agreements yet refers explicitly to benefit-sharing, but they could act as a trigger for benefit-sharing negotiations. A few culture collections are now using MTAs to clarify the terms on which they acquire specimens, knowing that this will affect the terms under which they can supply them, and acknowledging that source countries may wish to control the supply of genetic resources to third parties and to negotiate benefit-sharing arrangements.

Universities are another important source of samples for biotechnology companies, but they rarely use MTAs to supply samples, so any link between access to genetic resources and benefit-sharing with the source country is likely to be broken.

Monetary benefit-sharing in biotechnology development takes many forms, from the payment of a fee per sample and sponsorship of research grants (effectively payment for sub-contracted research activities), to milestone payments and royalties. A flat fee is more common than the offer of royalties when the samples supplied are 'raw materials' and come with no indication of activity or application. Prices for access to samples from culture collections are readily available, but there appears to be less of a clear 'going rate' for access to samples and value-added products compared with those in pharmaceuticals and agriculture. Many of the smaller biotechnology companies are spin-offs from universities or in the 'start-up' phase, so they do not yet have products on the market, and are relatively inexperienced in negotiating licensing and benefit-sharing arrangements.

Companies have fairly idiosyncratic approaches to royalty payments, so there is really no 'industry average'. The range for royalty rates for the development of enzymes is generally lower than the comparable range for pharmaceuticals, reflecting the considerable difference in their relative market value, and lies between 0.05 per cent and a few per cent, depending on the value added by the organisation providing the genetic resource. The most important factor affecting the magnitude of royalty payments is whether the provider supplies raw samples or value-added materials such as strains with proven activity and clear commercial applications. Where the provider has obtained IPRs over the product he or she is providing, the amount of royalties is correspondingly higher.

It is relatively common for biotechnology companies to share non-monetary forms of benefit. Companies share information and research results, transfer technology, train their collaborators and contribute to capacity building in the institutions from which they obtain supplies, although this often grows informally during a relationship with a supplier, rather than being prescribed up-front. Companies are prepared to share data and information, provided they can protect confidentiality and the opportunity to patent discoveries.

Sarah A Laird

Chapter 9
The Natural Personal Care and Cosmetics Industry

9.1 Introduction

The 'natural' personal care and cosmetics market is growing rapidly, with estimated average annual growth of 8–25 per cent. In contrast, the mainstream, largely synthetic or petrochemical ingredient-based market growth averages 3–10 per cent per year. The 'natural' segment of this industry is anticipated to make up 10 per cent of total sales by the year 2000 (*NBJ*, 1998).

The 'natural' segment of the personal care and cosmetics industry is extremely difficult to characterise, however. The size and approach of companies vary dramatically. Many companies – small and large alike – include a tiny amount of botanical ingredients in their products primarily for marketing benefits, and with no intended contribution to the product's efficacy. Other companies, usually small to medium in size, work to make 'natural' 100 per cent of the ingredients used in their products, replacing those of artificial or petrochemical origin. These companies usually operate under corporate policies that prioritise 'naturalness', and sometimes wider environmental and social concerns. Other, usually large, companies emphasise efficacy over 'naturalness', but in an effort to develop more effective cosmeceuticals, or therapeutic products, screen natural alongside synthetic compounds in search of new leads for product development. The latter strategy is much closer to that of the pharmaceutical industry than it is to that of many companies in the natural personal care and cosmetic sector, who do little more than re-formulate known agents.

For the most part, company research and development programmes in this industry are small and conservative, and are not characterised by substantial risk-taking and innovation. The lion's share of company investment in new product development is spent on marketing and packaging. However, increasing calls for therapeutic products, and stricter regulations in Europe and elsewhere, have directed more attention to research and development demonstrating safety and efficacy. Although the majority of finished-product manufacturing companies conduct little research on new leads for product development, many speciality ingredient supply companies do so in a search for products that meet manufacturers' specifications, or in response to wider market demand for active and 'performance enhancing' natural ingredients.

Because of this tremendous variation in the role of natural ingredients in product development, the ways in which companies seek access to materials and share benefits are also diverse. Some companies subcontract with collectors to gather large numbers of samples for mechanism-based screens; others send staff researchers to the field for ethnobotanical and other collections, including prospecting for marine organisms. Most companies make use of the suite of species already well-known to the market, and re-formulate existing ingredients, rather than prospect for new species. Only a very few companies conduct or commission field collections of samples. However, all companies review literature, databases, and material promoted at trade shows and by speciality chemical suppliers and other intermediaries, as a source of leads for new product or ingredient development.

Benefit-sharing in the natural personal care and cosmetics industry has largely taken the form of charitable donations, sometimes as an established percentage of sales. In a few cases, charitable donations are targeted at countries or communities from which the company acquires raw materials, and – more rarely – leads for product development. Technology transfer, training, and capacity-building have also resulted from commercial partnerships associated with the sourcing of raw materials, but generally in limited form. Still uncommon is benefit-sharing tied to access to genetic resources or traditional knowledge, and within the industry there exists significantly little awareness of any legal or ethical obligations on the part of companies.

9.2 The natural personal care and cosmetics market

9.2.1 Definition of category

The personal care and cosmetics sector encompasses a wide range of products and companies. The *Nutrition Business Journal* (*NBJ*) (1998) includes in 'Natural Personal Care' the following categories of products: cosmetics, feminine hygiene, hair products, baby care, nail care, oral hygiene, bath items, deodorants, shaving, skin care, bath/toilet soap, and fragrances. Many of these products would also fall under the 'cosmetics, toiletries, and fragrance' market which has been defined to include: oral care, soaps and bath products, hair care products, and beauty products (skin/body care, perfumes and fragrance, and make-up) (Kalorama, 1997). 'Beauty sales', on the other hand, include fragrances, make-up, skin care, hair care products, cellulite creams, deodorant and shaving creams. Excluded from 'beauty sales' are bar soap, razors, toothpaste, foods, diet supplements, medicine, vitamins, and detergents (*Women's Wear Daily*, 1997). This chapter addresses the broad category of 'cosmetic, toiletries and fragrances' – including skin care, oral hygiene, shaving products, fragrance, hair care, and make-up – grouping all under the heading 'personal care and cosmetics'.

9.2.2 Global markets for personal care and cosmetic products

The global market for personal care and cosmetic products was US$55 billion in 1997; however estimates vary depending upon product categories included in calculations (*NBJ*, 1998). The largest personal care and cosmetic markets are found in Europe, North America, and Asia. The 1997 US market for personal care and cosmetic products was US$36.4 billion, with a compounded annual growth rate of 3.4 per cent (Bucalo, 1998; and Table 9.1).

Retail sales of personal care products in the EU in 1997 were DM 70.1 billion, up 3.6 per cent over 1996. Germany topped the market with 24.1 per cent of sales, followed by France (19.8 per cent), Italy (14.1 per cent), the UK (13.8 per cent), and Spain (9.8 per cent), although France leads in per capita spending (IKW, 1998). The European Cosmetic, Toiletry, and Perfumery Association reported 1995 European (France, Germany, Italy, Spain, and the UK) retail sales of cosmetics, fragrances and toiletries as US$45 billion. Toiletries accounted for 27 per cent of this market, followed by hair care (26 per cent), skin care (21 per cent), perfumes and fragrances (16 per cent), and cosmetics (10 per cent). European retail sales of cosmetics have experienced only moderate to slow growth since the late 1980s (Gain, 1997).

Table 9.1 *The US Cosmetics and Toiletries Market, 1997*

Market	Total sales %	US$ (bn)
Make-up	23.7	8.6
Fragrance	18.4	6.7
Hair care	14.5	5.3
Skin care	13.7	5.0
Personal hygiene	13.5	4.9
Oral hygiene	10.5	3.8
Shaving products	5.6	2.0

Source: Bucalo, 1998.

Prior to the recent economic crisis, the Asian market for cosmetics was estimated to grow an average of 15–20 per cent a year through the end of the century, with lower growth expected in mature markets like Japan (<3 per cent/year) and rapid expansion anticipated in countries like China (>35 per cent/year) (Fost, 1997). In 1996, China's skin care market was estimated at US$1.6 billion, with make-up sales at US$355 million and fragrances at US$207 million (Fost, 1997). Total 1996 sales of personal care products in South East Asia were US$5.8 billion, up 10.7 per cent from 1995 sales of US$5.2 billion, and US$3.6 billion in 1991 sales. Hair care products led sales with 31.4 per cent of the market, followed by beauty products (30.2 per cent), oral care (22.7 per cent) and toilet soaps (15.6 per cent). The region's largest market was in Korea, with sales of US$1.4 billion, representing 24.8 per cent of the South East Asian market. Thailand and Indonesia each had more than a 20 per cent share of the 1995 regional market, followed by Taiwan, Malaysia, the Philippines, Singapore, and Vietnam (Kalorama , 1997b).

Brazil is the largest market in Latin America, making up 39 per cent of the cosmetics and toiletries market. Brazil is tipped to be the second largest market in the world for cosmetics and toiletries by the year 2010. The Mexican market makes up 31 per cent of the total Latin American market, and is more than double those of Argentina (11 per cent), Chile (5 per cent), Colombia (6 per cent), Peru (2 per cent), and Venezuela (6 per cent) (Egginton, 1997).

9.2.3 Types of markets and outlets for products

Personal care and cosmetic products are sold in a variety of retail outlets targeting different segments of the market. **Prestige markets** are characterised by higher priced products with limited distribution, often sold as multiple-item regimens in department and speciality stores. Prestige products usually contain distinctive ingredients and are expensively packaged. Most are sold by trained salespeople or skin care specialists (Kalorama, 1998).

Mass markets feature lower-priced products with broad distribution, sold through drugstores, discount stores, supermarkets, and other mass-merchandise outlets. Mass market products are usually single-item products, often considered 'fast movers' because of their low cost, wide availability, and broad applications (Kalorama, 1998). The distinction between prestige and mass markets is blurring, however, as companies once geared solely to prestige markets sell their products at a wider range of outlets, including those traditionally served only by 'mass market' products (Winter, 1994).

In addition to prestige and mass markets, **alternative markets** have grown up in recent years. Alternative market products are sold through direct marketing (eg NuSkin, Freeman Cosmetics, Avon, and QVC); speciality retailers (eg Bath and Body Works, The Body Shop, Garden Botanika, and Estee Lauder's Origins); salons (eg Aveda and Nature's Fresh!Northwest), and health food stores (eg Jason's Natural Cosmetics, Aubrey Organics, Nature's Elements, and Albert-Culver Co) (Kalorama, 1998; *NBJ*, 1998).

Many of the companies emphasising natural personal care and cosmetics fall within the alternative market category, although prestige and mass marketers increasingly make use of natural ingredients. A breakdown of 1997 sales of natural personal care products was topped by health food store sales of US$660 million, followed by health and beauty speciality stores (US$340 million), the mass market (US$130 million), department stores (US$110 million), mail order/direct advertising (US$100 million), multi-level marketing (US$1.35 billion), and practitioner sales (US$100 million) (*NBJ*, 1998).

Speciality retail channels are the fastest growing outlet for natural personal care products (Bucalo, 1998). Direct or multi-level marketing is also expanding, and some retail companies have opened direct selling to reach new customers. For example, The Body Shop has had tremendous success with its Body Shop Direct division, which sells products door to door in areas lacking retail outlets (J. Morris, personal communication, 1999). Within Brazil, direct sales are common. It has the largest number of Avon sellers (400,000) after the USA. The Brazilian-based Natura company has 60,000 direct sales representatives. These sellers approach rural customers living far from traditional shopping channels (Egginton, 1997).

9.2.4 The natural component of the personal care and cosmetics market

Natural personal care products accounted for US$1.4 billNon of the 1996 US$45 billion personal care market (*Natural Pharmacy*, 1997). By 1997 this figure had grown to US$2.8 billion of a total personal care market of US$55 billion (*NBJ*, 1998). Average annual growth in the natural segment of the market is estimated at 8–10 per cent (*NBJ*, 1998). The investment bank Adams Harkness and Hill, Inc (AH&H, Boston) places growth rates at a higher 20–25 per cent per year, well above average industry growth projections of 5–10 per cent per year. AH&H estimate natural personal care and cosmetics markets of US$6.5 billion by the year 2000, comprising more than 10 per cent of the world-wide personal care and cosmetics market (*NBJ*, 1998).

The natural component is more significant in some product categories than others. 23 per cent of all 1997 hair product sales in the US health and beauty care market were labelled 'natural'. 38 per cent of skin care products were labelled 'natural', and 12 per cent of bath items used 'natural' in product descriptions. However, only 2 per cent of cosmetics, 3 per cent of oral hygiene, 2 per cent of deodorants, 1 per cent of fragrances, and 0 per cent of nail care products were marketed as 'natural' personal care products (*NBJ*, 1998).

Growth in natural personal care and cosmetics markets is global, with Asian, Latin American, European, Australian, US and other consumers seeking out therapeutic and natural products. For example, in South East Asia, several local manufacturers have successfully introduced new products with plant extracts like cucumber, apricot, ginseng, iris, and aloe, and are marketing brands in competition with overseas companies like The Body Shop and Australia's Red Earth (Kalorama, 1997b).

Top prestige, mass, and alternative marketers of natural personal care products in the USA include: Aubrey Organics; Aveda Corporation; Avon Products; Bath and Body Works; The Body Shop; Carme Inc; Clairol; Clarins; Cosmair; Del Laboratories Inc; Freeman Cosmetic Corporation; H2O Plus; Jason Natural Cosmetics; Leiner Health Products Inc; Levlad Inc; Origins Natural Resources Inc (Estee Lauder); SC Johnson and Son Inc, and Tom's of Maine Inc (Kalorama, 1996). The number of small and large companies entering this market is on the rise, and during the last few years there has been a massive entry into this arena by the large mainstream manufacturers. As Nancy Rosenzweig of Tom's of Maine said, 'Anybody who wants to be a viable player in the personal care industry wants a stake in the natural products business' (*NBJ*, 1998).

9.2.5 Structure of the industry

Companies active in the natural personal care and cosmetics segment can be broadly grouped into **supply companies**, including wholesalers of raw botanical material and speciality chemical ingredient producers and formulators, and **manufacturers and marketers** of finished products.

Botanical raw material suppliers

Botanical raw material is supplied to the personal care and cosmetics industry through the same channels as those supplying the botanical medicine industry. Most of the wholesalers of raw plant materials – including exporters, traders, brokers, and agents – sell to a range of industries that include botanical medicines, personal care and cosmetics, pharmaceutical, nutrition and food, dyes, pet products, and household products. In addition to commodity and bulk-level trade in botanicals, in a small minority of cases speciality suppliers source 'green' (eg certified organic or 'sustainable') or 'fair trade' (certified ethically or socially-sound) material on behalf of companies. In a few cases, companies undertake direct sourcing partnerships with suppliers or local communities (9.6 below; for more detail on the trade in bulk raw botanical materials, see Chapter 4).

Speciality chemical ingredient manufacturers

Speciality chemical ingredient manufacturers formulate and supply ingredients and blended products to manufacturers of finished consumer products. Thousands of compounds are supplied to the personal care and cosmetics industry as primary ingredients (with unique marketing or functional characteristics) or additives. 'Performance-enhancing' compounds (eg emulsifiers, viscosity modifiers, speciality surfactants) are of increasing importance in the industry. These compounds result from intensive research and development and specialised manufacturing (Fost, 1997).

Speciality ingredient companies conduct a large portion of personal care and cosmetics industry research. They research and develop new compounds to market to finished product manufacturers, or in response to a manufacturing company's request for particular performance enhancing properties or activity. As Ron Joseph, Director of Research and Development at H2O+, said, 'We leave it to people who supply raw materials to us to do the screening and identify active ingredients ... H2O+ asks for a certain activity and type of product, and the ingredient supplier gives it to us. We will then play around with levels in the finished formulation.'

Increased demand from manufacturers for natural ingredients in stable and useful form has created a supplier market of chemical companies that provide popular natural ingredients and invent new ones by processing and recombining biological products. These companies may also help marketers with the formulation and design of new natural personal care products. Supply companies working with natural ingredients include: Aloecorp; Amerchol Corporation; Arista Industries Inc; Bio-Botanica; Carubba Inc; Centerchem Inc; Croda Inc; Florida Food Products; Haarmann and Reimer; Henkel of America Inc; Hoffmann-LaRoche Inc; Indena; International Sourcing Inc; Koster Keunen Inc; Nurture Inc; Oils of Aloha; Purac of America; RITA Corporation; Scher Chemicals Inc; Technichem Inc; Terry Laboratories, and Tri-K Industries Inc (Kalorama, 1998).

Speciality chemical suppliers also increasingly test new ingredients, as well, supplying finished product manufacturers with a product dossier substantiating all claims. Peter Cade, Director of Technologies at Croda Inc, explained, 'Whether products are natural or

Table 9.2 *The World's 15 Largest Beauty Companies*

Rank	Company	Country	Sales 1996 (US$ bn)	% change over 1995
1	L'Oreal Group	France	9.587	+13
2	The Procter & Gamble Co	USA	7.1	+9
3	Unilever	The Netherlands & UK	6.969	+18
4	Shiseido Co Ltd	Japan	4.874	+5
5	The Estee Lauder Co	USA	3.29	+10
6	Johnson & Johnson	USA	3.03	+9
7	Avon Products	USA	2.9	+4
8	Wella Group	Germany	2.262	+9
9	Sanofi SA	France	2.211	+3
10	KAO Corp	Japan	2.2	+4
11	Revlon Inc	USA	2.17	+12
12	Beiersdorf Ag	Germany	2.096	+4
13	LVMH	France	1.753	-3
14	Henkel KGAA	Germany	1.716	+94
15	Joh. A Benckiser	Germany	1.603	+17

Source: Women's Wear Daily, 1997.

synthetic, we need to come up with a package that demonstrates the function ... this is something we must do as a chemical specialty manufacturer – we demonstrate what an ingredient does.' Hossein Janshekar, health and performance chemicals consultant at SRI Consulting (Zurich), echoes this: 'Chemicals are now being packaged with guidelines and advice about how to use a product and achieve a particular type of performance ... Tailor-making ingredients is also a trend, and advice must come along with the chemicals ... This started in the late 1980s and early 1990s and will become more prevalent' (Gain, 1996).

Finished product manufacturing and marketing companies

Manufacturing companies marketing natural personal care and cosmetic products vary in size and approach. They may be small companies, marketing primarily to health food stores, with a turnover of less than US$10 million, or large multinational companies selling mass and prestige products with sales in the billions. Companies may conduct little research on product ingredients, primarily formulating their products, or they may operate advanced in-house research and development, including screening compounds for chemical activity.

The world's 75 largest finished product manufacturing cosmetic companies in 1996 accounted for just over US$76 billion in wholesale volume of beauty sales (including fragrances, make-up, skin care, hair care products, cellulite creams, deodorant and shaving creams). The USA has the largest number of companies in the top 75 (28), followed by France (11), Germany (7), Japan (7), Italy (7), UK (5), South Korea (4), Brazil (3), Spain (2) and Sweden (1). The US cosmetics industry also leads in terms of sales volume, with 1996 sales of more than US$29 billion, representing more than one-third of the world-wide total. Of the top 15 companies, however, eight are based in Europe, five in the USA, and two in Japan. As Table 9.2 shows, all of world's top 15 companies had 1996 sales in excess of US$1.5 billion.

Today, most multinational companies feature a natural product line, and for some this is a great deal more than a marketing ploy. Companies like Estee Lauder and Elizabeth Arden (of Unilever's Home and Personal Care division) conduct advanced research on natural products, including screening for active ingredients. While the natural trend has caught on with mass and prestige market manufacturers, alternative marketers are also expanding sales of natural personal care and cosmetic products. In contrast to the larger companies entering the natural labelling field, however, the majority of alternative finished product marketing companies have annual sales of less than US$10 million (Box 9.1).

Box 9.1 *The Largest Natural Personal Care Manufacturers Supplying Natural Food Stores in the USA*

Company	*HQ city*
US$30–40 m	
Aubrey Organics	Tampa, FL
US$25–20 m	
Levlad Inc (Nature's Gate)	Chatsworth, CA
US$20–US$25 m	
Tom's of Maine	Kennebunk, ME
Jason's Natural Cosmetics	Culver City, CA
US$10–15 m	
Kiss My Face	Gardiner, NY
US$5–10 m	
Abkit Inc (Camocare)	New York, NY
Burt's Bees	Raleigh, NC
Earth Science Inc	Coronoa, CA
Zia Cosmetics	San Francisco, CA
US$3–5 m	
Avalon Natural Cosmetics Inc	Petaluma, CA
Rachel Perry Inc	Chatsworth, CA
Reviva Labs Inc	Haddonfield, NJ
Scandinavian Natural H&B	Perkasie, PA
Desert Essence (Consac Ind)	Chatsworth, CA
US$1–3 m	
Orjene Company Inc	Long Island City, NY
Shikai	Santa Rosa, CA
Geremy Rose	Brattleboro, VT
Weleda Inc	Congers, NY
Alba Naturals Inc	Santa Rosa, CA
Borlind of Germany Inc	New London, NH

Company size US$ m	Number of companies	Wholesale revenue US$ m
>20	5	135
10–20	3	35
5–10	10	60
1–5	20	56
<1	180	76
Total	218	363

Source: NBJ, 1998.[1]

9.3 What is 'natural'?

Natural personal care and cosmetic products are considered a 'super-trend'. The term 'natural' is an important marketing claim used widely by personal care and cosmetic companies, but there is little widespread agreement on its meaning. For example, many claimed natural ingredients, like animal products, petrochemicals, and alcohol, are not what consumers imagine when purchasing a natural labelled product. In fact, very few products on the market today could be claimed 100 per cent natural, with most incorporating synthetic preservatives, and many artificial colours, fragrances, and petrochemicals.

In many cases, natural ingredients form a trivial amount of the total formula, and do not play an active role in product efficacy. Despite the prevalence of botanicals in product names, labels, and marketing, the activity and efficacy of a product is often based on a common core of synthetic and petrochemical ingredients. As one company representative interviewed as part of this study said, 'Many companies are in the labeling and marketing business – they want products to look like complex mixtures of many natural ingredients, but in fact most don't have anything in the way of research departments, and the natural element is usually a tiny fraction added on to the product at the end for marketing purposes.' Exaggerated claims of 'naturalness' are found in large and small companies alike, including many of the companies that market products as 'green' through alternative channels.

Companies that misrepresent the naturalness of their products obviously confuse and crowd the market. As Nonie Faggat, owner of Nonie Beverly Hills, which includes a line of alpha hydroxy acid (AHA) products said, '[Companies] put pictures of fruits [on a label] and call it natural. Sometimes I am so disgusted. At the end of the label you'll see the parabens, urea, sodium laureth sulfate, all those chemicals that are not really good for you. What you put on your skin goes directly into your bloodstream. They'll say parabens are sourced from oils in the earth. Give me a break! It's a petrochemical!' (*NBJ*, 1998). Horst Rechelbacher, founder of Aveda Corporation, echoes this: 'I call it greenwashing, or brain washing' (Chittum, 1996).

Aubrey Hampton, of Aubrey Organics, has gone as far as publishing a list of the 'ten most commonly used synthetic ingredients in cosmetics and how to avoid them'. These are the 'ten chemicals I most want to see off the labels of so-called natural hair and skin care products', he says. They include (Hampton, 1994):

1 *imidazolidinyl urea* and *diazolidinyl urea* (commonly used preservatives);
2 *methyl-, propyl-, butyl-*, and *ethyl-paraben* (inhibit growth of microorganisms and extend shelf life of products);
3 *petrolatum* (mineral oil jelly used in lip balms);
4 *propylene glycol* (can be a vegetable glycerine mixed with grain alcohol, both of which are natural, but are usually a synthetic petrochemical mix used as a humectant);
5 *PVP/VA copolymer* (petroleum-derived chemical);
6 *sodium lauryl sulfate* (synthetic substance used in shampoos for its detergent and foam-building abilities; 'it is frequently disguised in pseudo-natural cosmetics with the parenthetic explanation "comes from coconut"');
7 *stearalkonium chloride* (chemical used in hair conditioners and creams);
8 synthetic colours;
9 synthetic fragrances; and
10 *triethanolamine* (used in cosmetics to adjust the pH).

Natural ingredients are generally more costly to incorporate into products, and in some categories – such as preservatives and fragrances – can be difficult to work with, compared with their synthetic or non-botanical counterparts. One company researcher interviewed as part of this study explained, 'Most fragrances are 90 per cent artificial. No one is really asking for completely natural fragrances, and if they do they would be highly disappointed. The fragrance industry is a chemical industry. We use only natural ingredients, but we don't make perfumes – it would be very difficult to make an all natural perfume. A fragrance lab might work with 2–4,000 different compounds. We have 80 natural compounds to work with. The nicest scents – like rose – are very expensive if all natural.' Companies marketing products of 100 per cent botanical origin may also experience problems with shelf life when shipping products around the world, or with unexpected reactions between natural ingredients. The division between natural and 'synthetic' ingredients is further confused when natural substances are processed, often using synthetic chemicals, in order to make them functional and stable (Kalorama, 1998).

Misrepresentation of natural products also extends to the manner in which ingredients are sourced. In a study conducted by the Organic Trade Association, for example, it was found that over the last five years the use of the word 'organic' in cosmetic advertising increased by 60 per cent; however, many labelled 'organic' products do not contain organic ingredients. Aubrey Hampton, of Aubrey Organics, has filed a petition with the FDA asking that anyone using the term 'organic' on a personal care product be required to meet the same standards applied to food – that is, 95 per cent of the product must actually be organic. Although 'organic' is a well-established term in the food industry, there is no legal definition for organic personal care and cosmetic products. Aubrey Organics is concerned that the word 'organic' could become as watered down as the word 'natural' (Chittum, 1996; *NBJ*, 1998).

While some companies make excessive and inaccurate claims about the role of botanical ingredients in the efficacy of a product, numerous others include active ingredients of botanical origin, but do not draw attention to this fact on their labels. These companies may list an ingredient under the active compound name only. This is particularly true for heavily researched and tested therapeutic products, for which an ingredient's natural origin is less important than claims associated with its beauty-enhancing properties. Refined components of plants such as centella, liquorice, rosemary, horseradish, and green tea form the basis of many skin care products today, but are not always highlighted as of botanical origin on labels (Brown, 1996).

Regulatory agencies are similarly perplexed by the term 'natural'. The US FDA has no legal definition for the use of the word, and concedes that a company can wave a leaf over a vat of material, and then sell the resulting product as 'natural'. 'We haven't developed a regulatory guideline on what constitutes natural cosmetics ingredients,' said John Bailey of the FDA. The Cosmetic, Toiletry, and Fragrance Association (CTFA) similarly does not provide guidance on the use of the term 'natural'. The lack of substantial regulatory requirements means that claims relating to natural or 'organic' are effectively left to the manufacturer. While industry 'inches' towards self-regulation, it will largely be left to consumers to determine the 'definitional barrier' between natural and non-natural (*NBJ*, 1998; Ouellette, 1997).

In summary, then, a natural personal care and cosmetic product might have one of three broad types of relationships to the natural world:

- *Marketing product:* This product is marketed as natural with inclusion of negligible quantities of natural material solely for marketing purposes. These products characterise the bulk of sales in mass, prestige, and alternative markets;
- *Active ingredient product:* This product contains scientifically validated active ingredients of natural origin, most likely combined with synthetic processing agents, colourants, fragrance, and other ingredients. These products tend to emerge from the sophisticated research and development programmes of mass and prestige companies. Marketing may not highlight the natural origin of the ingredients, and is likely to emphasise efficacy and therapeutic potential;
- *Whole product:* This product contains scientifically or traditionally validated active ingredients of natural origin. All other ingredients are also of natural (botanical, marine, etc) origin, and do not include petrochemicals. Marketing and company identity are likely to be tied closely to 100 per cent natural ingredient claims.

9.4 Trends that impact demand for natural personal care and cosmetic products

Natural products have been identified as a 'super trend' within the personal care industry, due to a variety of factors, including increased consumer interest in all things 'natural'; demand for therapeutic products; demographic shifts, and the entry of large companies into the market.

1 **Increasing consumer sophistication and interest in all things natural.**

Across sectors, consumers are calling for healthier and more natural products. Increased consumer sophistication and awareness of ingredients, performance, and health benefits are changing the personal care and cosmetics industry. Higher quality, more effective products are now available at more affordable prices than hitherto, and the market is increasingly segmented in an effort to appeal to niches.

Consumers perceive natural ingredients as 'good' and 'pure'. As founder and president of Forever Living Products International, Rex Maughan, said, 'We're looking at products from the rain forests in the South Pacific and South America. There are so many products that are wonderful we've found and we're going to be coming out with more of them ... . We're finding all over the world that more people are looking for natural things instead of taking synthetics' (*NBJ*, 1998). Eva Frederichs, founder of Eva's Esthetics, predicts: 'Botanical research will be the next trend in skin care – bigger than AHAs' (Dunn, January 1998).

In an FDA poll conducted in 1996 in the USA, 72 per cent of all respondents felt the word natural on a cosmetic product was important. In the same year, sales of natural skin care products increased by 30 per cent (*Natural Pharmacy*, 1997). As Dylan Reinhardt, Manager of Corporate Communications at Tom's of Maine, put it: 'There has been a significant increase in people's awareness of and interest in natural products, but also of the environmental impacts of the products they purchase. They often want products that are not only efficacious, but are not harmful to the environment, and have some social benefits, like creating a market for indigenous peoples' materials.'

2 **Stagnant markets and new ingredients.**

In many parts of the world, the personal care and cosmetic market is crowded. Companies' market shares are likely to stagnate unless they re-formulate their products to address the needs of niche markets, incorporate new ingredients, and heighten the performance of products (Gain, 1997). Natural ingredients and lines have attracted consumer interest and have helped boost sales in some product categories. For example, in skin care, companies anticipate that products containing beta hydroxy acids (BHAs) and other therapeutic additives will generate the kind of interest alpha hydroxy acids (AHAs) did a few years ago (Bucalo, 1998). Ingredient and product innovation over the last four years has responded primarily to consumer demands for natural products (Egginton, 1997).

3 **The entry of mass and prestige market companies and their large advertising budgets.**
The personal care industry is dominated by multi-billion dollar, multinational companies. In 1996, for example, Procter and Gamble, Bristol-Myers Squibb, Helene Curtis, Cosmair, and Alberto Culver accounted for more than 50 per cent of all hair care sales in the USA, and ten companies produced 73 per cent of the fragrance segment's dollar value (Ouellette, 1997). The top five mass market skin care companies – Cosmair, Unilever, Scott's Liquid Gold, Beiersdorf, and Johnson & Johnson – claimed 85 per cent of 1997 USA mass market sales (Kalorama, 1998).

Large multinational companies have recently entered the natural personal care and cosmetic segment, long dominated by small, alternative manufacturing and marketing companies. 'In the last three or four years, you can look at the mass market and see a very fast evolution of mass marketers trying to jump on the natural personal care bandwagon,' said Mark Egide, president and co-owner of Avalon Natural Cosmetics (*NBJ*, 1998). The entry of multinational companies into natural personal care and cosmetics has meant, in some cases, expanded interest in natural products within companies' internal research and development divisions. In others, large companies have acquired smaller companies with a stronger natural base. For example, Bristol-Myers recently acquired Matrix, and Estee Lauder has acquired nearly half a dozen small independent companies in recent years, including Aveda Corporation (*NBJ*, 1998; Dunn, May 1998).

As mass and prestige marketing companies enter the natural labelling field, advertising revenues spent proclaiming the benefits of natural products are increasing dramatically. Although many of these mass and prestige products contain small and sometimes insignificant amounts of natural ingredients, the message to consumers that 'natural is better' is gaining ground. One representative of a natural personal care company interviewed as part of this study said that: 'In this industry there is no idea of "you get what you pay for". Prices are random and unrelated to costs, and there is often little difference in quality between the most expensive, and the cheapest product.' Images projected in marketing campaigns are therefore critical to consumer perceptions of product value, and as large companies market natural as the latest best approach, the already significant consumer interest in these products will continue to increase.

4 **Changing demographics.**
Changing consumer demographics mean that the personal care and cosmetics industry is focused on a greater range of products for an aging population. Older consumers are demanding multi-functional, therapeutic products that moisturise, provide UV protection, and are mild. Baby boomer generation demand for hair colour to hide grey hair, for example, has resulted in rapid growth in this segment, and was the driving force behind recent nominal growth in the hair care market (Bucalo, 1998). Another recent development is larger numbers of men using facial preparations and hand and body moisturisers (Wilck, 1997). Younger female consumers are also considered particularly interested in things 'natural', and less attracted to expensive packaging ... 'they don't want big names. They want something plain-packaged, handwritten,' said Rebecca Gadbury of Youth Glow Private Label Skin & Body Care, which is introducing 80 products, including a new line of 'high tech naturals' containing 'natural ingredients scientifically proven to work' (*NBJ*, 1998).

5 **Therapeutic products: cosmeceuticals.**
Demand for 'sophisticated' personal care and cosmetic products is on the rise around the world. The trend is away from products that superficially enhance beauty but have no biological effect, to 'therapeutic' products that might, for example, repair damaged tissues, smooth, protect from the sun, and moisturise. This has led to increased use of new, active ingredients, including natural products with defined constituents and specific biologic effect (Iwu, 1996).

Therapeutic personal care and cosmetic products are called 'cosmeceuticals' because of the grey area they occupy between pharmaceuticals and cosmetics (Ouellette, 1997). As one director of research at a company interviewed as part of this study said, 'There is a whole new category of products that put hard science behind historical claims ... . Consumers want something that works, but they don't want to get a prescription.' Jean Julien Baronnet, Rhône-Poulenc's president of personal care, detergents, and surfactants world-wide, described this trend: 'For example, we'll launch a sunscreen cream that moisturises as well as protects and a shampoo that cleans and offers sun protection ... This cross over between cosmetics and pharmaceuticals is a trend that started two years ago, and since you need about four years to develop them, some products have not been launched ... Cross-technology is in' (Gain, 1996).

Sales of cosmeceutical products in 1997 were an estimated US$769 million, a 400 per cent increase over 1993 levels. Annual growth in this segment is anticipated to average 15–18 per cent over the next few years, growing into a market of at least US$1.6 billion by 2002 (Kalorama, 1998; Ouellette, 1997).

Although natural ingredients are now part of a super-trend, they have long formed part of industry research and product development efforts. As one industry representative at a large company that conducts screening of natural products said, 'Cosmetic companies have been interested in natural products forever. It pre-dated current fashion, and will continue afterwards. The early Egyptians used plants in their cosmetics. It is not a fad/marketing thing, and is very much a long term phenomenon. Natural-based materials are central components or ingredients in our products and are not there just for show – natural botanicals are integral to the psyche of the industry.'

9.5 Scientific and technological trends

In response to consumer demand for natural and therapeutic products, the personal care and cosmetics industry increasingly seeks leads and raw materials for product development in the natural world. Botanicals, marine organisms, vitamins, and other natural products provide active compounds that contribute to product efficacy, assist with improved delivery systems, and replace petrochemicals, artificial preservatives, surfactants, and other synthetic ingredients.

Karl Raabe, Product Management Director of Cosmetic Ingredients at Henkel, the 14th largest personal care and cosmetics company in the world, described the situation in Germany, where 'there's a strong trend toward natural ingredients such as plant extracts used as active components, or surfactants based on natural raw materials such as coconut or palm kernel oil. Petrochemical-free ingredients are also in vogue ... . In general the industry is moving towards animal-free, vegetable based ingredients. There is also a trend to use ocean-derived products in ingredients such as seaweed, chitin from shrimp shells, or fish oils' (Gain, 1996).

Classes of natural products of interest to product development teams include: bio-saponins (steroids and triterpenoids); flavonoids (bioflavonoids and biflavonoids); amino acids; non-protein biocomplexes; proteins and phytoamines; anti-oxidants; alpha- and beta-hydroxy acids; formulation aides; and vitamins (Iwu, 1996). Many products contain ingredients that yield these compounds, including green tea, marigold, camomile, ginger, rosemary, and aloe (Box 9.2).

Box 9.2 *Examples of Plant Activity with Cosmetic Applications*

Plants rich in:

- *gallic* or *catechic tannins, flavenoids* or *anthocyanin* derivatives will have astringent activity – eg butcher's broom, colt's foot, hawthorn, horse chestnut, horsetail;
- *saponins* will have soothing properties – eg lesser celandine;
- *flavonoids* will have firming properties – eg wild thyme;
- *essential oils* will have antiseptic properties – eg cinnamon;
- *phenols* will have free radical scavenging properties – eg green tea;
- high molecular weight *proteins* will have tightening properties – eg lupin.

Source: Brun, 1998.

Therapeutic products are developed as anti-inflammatories, anti-irritants, anti-allergens, ultra-violet screens, modulators of collagen and elastin, and for their firming, smoothing, soothing, moisturising, or rejuvenating properties. These products might be based on re-formulations of known agents, or may result from sophisticated research and development programmes that look for new agents, including those derived from plants, marine organisms, and other natural products. For example, Chanel developed a skin care product designed to correct dry skin by answering the question: 'Why do some plants remain lush in even the most bitter cold and dry climates?' Called 'Source Extreme Dual Benefit', the product's key ingredient is cryocytol, which is found in the larch tree (*Larix* spp). Larches grow at altitudes of over 2,000 metres. 'We looked at nature and found the larch tree ... . We extracted a number of liquids from the bark and studied its defense mechanisms,' said Dr Mausner, VP of Research and Development. Aloe and camomile were added to soothe the skin, vitamin A to reduce lines and wrinkles, and C and E to fight free radicals (Dunn, January 1998).

Large cosmetics companies such as Estee Lauder and Elizabeth Arden run natural product samples against mechanism-based assays similar to those employed by pharmaceutical companies (Chapter 3). Promising active compounds are isolated and identified, and the mode of action studied. As Jon Anderson, of Estee Lauder Companies, wrote: 'The research trend for Skin Care Products is moving toward the development of highly refined raw materials of natural origin with defined constituents imparting a specific biological effect to benefit healthy skin. Ingredients target mechanism-based systems to modulate enzymes or receptors present in the skin to prevent and protect skin from damage or to repair damaged skin' (Anderson, 1996).

The current trend in research and development towards better definition of natural product extracts is due to a combination of factors, including increasingly stringent regulations in the EU, scientific advances that allow improved study of natural compounds, and consumer demand for more sophisticated products. As Ron Joseph, Director of Research and Development at H2O+, said, 'Natural extracts can vary in active ingredient content, colour, and odor. It depends on how it was grown and where it was grown (climate, weather, variety, etc), as well as other variables. Standardizing extracts can help limit the variation'.

There are three types of natural product extracts present in the market-place for personal care and cosmetic manufacturers (Brun, 1998):

- *Extracts of well-documented plants with little scientific confirmation.* No scientific claims are made for extracts in this category, which come from plants well-known for their useful properties, often as documented in the literature. Their inclusion in products is primarily for marketing purposes, so little or no testing has been performed, even though extracts of good quality, used at sufficiently high levels, could provide some degree of efficacy. For this group there is usually no guaranteed content of any active molecule, only an indication of plant material concentration.
- *Extracts that make claims based on scientific confirmation.* For these extracts, certain properties may be claimed, substantiated *in vitro* or *ex vivo*. The extracts have been scientifically evaluated, through either laboratory testing (enzyme inhibition or activation) or human cell cultures (keratinocytes, fibroblasts, malanocytes, etc). They may offer a guaranteed content of the active ingredient, and the price is usually higher.
- *Clinically-tested extracts.* These extracts are offered as *in vivo* substantiated active ingredients. This group is the most scientifically reliable, because extracts have been clinically tested on humans. Included here are the most highly purified fractions, requiring advanced purification technologies.

9.5.1 Product and ingredient trends

A 'cocktail' approach to personal care and cosmetic products, in which products contain multiple ingredients and serve multiple functions, is increasingly common. These cocktails often involve vitamin (particularly A, C and E) and botanical combinations. Many of the botanical ingredients are those with established markets and name recognition in the botanical medicine industry, including ginseng, ginkgo, green tea, and echinacea. The company La Prarie, for example, formulated a product with vitamin C, AHA, and extracts of ginseng, horsetail, liquorice, green tea, heather, apple, comfrey, mallow, and seaweed. The company's latest moisturiser includes vitamins, green tea, camomile, mint, fennel, hops, yarrow, allantoin, sweet almond milk, liquorice root, alpha bisabolol, and vitamin B complex. The development of multiple function products requires what one industry representative called a 'cross-pollination of technology', in which formulae incorporate technologies from a range of product categories like sun, face, and body care (Dunn, January, 1998; May 1998; Kalorama, 1998).

Introduced world-wide into the mass market in 1994, AHAs have driven much of the growth in the skin care market in subsequent years. AHAs are fruit acids used to treat dry and severely dry skin, unblocking and cleaning pores, improving skin texture and tone, managing acne problems, and protecting against harsh environments (Wilck, 1997; Kalorama, 1998). AHAs include: glycolic acid (derived from sugar-cane juice.); lactic acid (from sour milk and tomato juice); malic acid (from apples); tartaric acid (from grapes and wines); and citric acid (from citrus fruits and pineapples).

Although AHAs are still an important ingredient in many products, they are now partially eclipsed by products touting the antioxidant benefits of vitamin C, and its ability to increase collagen production in the skin. Lancome's major launch for 1998 was the vitamin-C based Vitabolic Deep Radiance Booster, which is said to increase the skin's radiance. Another hydroxy acid has also gained favour with manufacturers in recent years: BHA, derived from plant cells and from the sugar derivative luconamide, is considered less irritating to the skin than AHAs. Significant new growth ingredients include enzymes, antioxidants, vitamins A, C and E, marine organisms, and botanicals (Dunn, May 1998; Brun, 1998; Kalorama, 1998).

Technological developments are also influencing industry demand for natural products. Recent developments in delivery systems designed to time-release higher concentrations of active ingredients at a rate that will not irritate the skin include microsponges, liposomes, phospholipids, oxygen emulsifiers, and silicone emulsions (Kalorama, 1998). In January 1998 Estee

Lauder launched a new product called 'Diminish', a retinol treatment that includes vitamin A in a pure form to reduce the appearance of wrinkles, age spots and discolorations. However, because many consumers find pure vitamin A irritating, Estee Lauder formulated the product with thalaspheres, developed from collagen derived from fish. Thalaspheres isolate the retinol from oxygen and other ingredients, so it remains stable and pure; they also act as a slow delivery system. 'Diminish' also contains green tea and marine extracts along with vitamins C and E (Dunn, January 1998).

Demand for natural ingredients has also spurred research on flavourings, colourings, preservatives and processing aids used in personal care and cosmetic products. Processing aids include surfactants, texturisers, clarifying and chelating agents, tannins, higher alcohols, and foaming agents – all of which combine to improve consistency, delivery, and appearance of products (Winter, 1994). Many suppliers, for example, are looking for so-called natural preservatives to complement naturally-derived active ingredients. Commonly used natural preservatives include grapefruit seed oil, and Vitamins A, C and E. The chemical company Brooks Industries recently developed a natural preservative based on willow bark (*NBJ*, 1998). But some in industry are not convinced: 'The problem with natural preservatives is that there is nothing on the market right now that meets the commercial standards of preservatives,' said Mr Hinden of ISP Sutton Laboratories (Ouellette, 1997). But Aubrey Hampton of Aubrey Organics disagrees. Having tracked the shelf life of products preserved with grapefruit seed oil for three years, he has concluded that 'it's probably superior to [the commonly used synthetic] methylpropylparaben'.

Interest in mild natural surfactants has created demand for alkyl polyglucosides and other natural options (Van Arkum, 1997). Of the 2,000 flavourings used in food and cosmetics, and by the drug industry, 500 are natural, derived from a wide variety of plant extracts and essential oils. Demand for natural flavourings for lipsticks, dentrifices, and mouthwashes is on the rise. Natural texturisers, which give products a desired feel and appearance, include acacia, which is used to thicken hairdressings and hold unruly hair in place (Winter, 1994). Some compounds can be derived from both natural and synthetic or petrochemical sources. For example, sodium laurel sulphate can be derived from coconut oil or from a petroleum source (*NBJ*, 1998).

As Peter Matravers, VP of Research and Development at Aveda Corporation, said, 'We are a botanicals-based company, and are constantly looking for plant ingredients to replace synthetic compounds. One strategy is to look at many plant chemicals and select those with a chemistry similar to synthetic compounds – we can then build on the scientific literature on synthetic chemistry, which is much larger than the plant chemistry literature. Structure-activity relationships are one of our strategies. We also try to build more complicated natural plant chemistry, by combining two or three naturally-derived ingredients, to get the benefits and functions of synthetic ingredients. This is all in an effort to replace all petrochemical ingredients in our products.'

However, this process is not easy. Dylan Reinhardt, Manager of Corporate Communications at Tom's of Maine, explained: 'It takes quite a bit of effort to make a product that is similar to well-established products but made with natural ingredients – just doing that is very difficult, and there is often quite a challenge to introduce new ingredients that people don't know a lot about. We have stuck with more conventional ingredients, things that are well-documented, because they are what Tom's as a small company, with limited research capacity, can manage.'

9.5.2 The use of traditional knowledge

Many natural ingredients in personal care and cosmetic products derive from traditional use of species. Commercial applications grow from widespread and common traditional use, as well as highly specialised knowledge of narrow ethnic and geographic provenance. Widespread traditional use of bixa as a skin and food dye highlighted its potential as a colouring agent in make-up (see The Yawanawa and Aveda Corporation *Bixa orellana* Project case study, p281). NuSkin's 'Ava puhi moni' products cite 'indigenous peoples throughout Polynesia [who] have known the secret to soft, shiny hair for generations' as their inspiration. The flower clusters of the ava puhi plant are used in NuSkin's 'ethnobotanical shampoo and light conditioner'. Aubrey Organics' 'Rosa Mosqueta' line was inspired by the use of this mountain rose 'by South American Indians for hundreds of years for its healing and moisturising properties to the hair and skin' (Hampton, 1994).

However, the flow of traditional knowledge is not only from the South to Northern research laboratories. The Indonesian personal care company Martina Berto, for example, searches the globe for traditional knowledge as a lead for product development; in addition to work with local communities in Indonesia, and a review of literature from China and India, the company consults the German pharmacopeia. Staff travel to Europe on a regular basis to seek out new natural ingredients for their products, like lavender, rose, and grape seed.

A number of the new botanical ingredients in personal care and cosmetic products are drawn from their traditional use as medicines, and subsequent incorporation into the botanical medicine industry. For example, Amway recently launched a body firming gel, which uses echinacea to firm skin (Dunn, January 1998). The commercial applications for many other natural personal care ingredients – like cohune oil from Guatemala (see the Cohune oil case study, p287) – do not grow directly from traditional use. Cohune oil is used locally as a cooking oil, but prior to its commercial development by Conservation International and Croda Inc was not found in use for personal care.

Areas of traditional use with relevance to personal care and cosmetic product development include: wound healing, antiseptic, anti-irritant, anti-inflammatory, anti-infective, body decoratives, toning (mud packs), and mouth and teeth cleaning (Iwu, 1996). Traditional knowledge might be more appropriate for some product lines than others, however. For example, Peter Cade, Director of Technologies at Croda Inc, said, 'We wouldn't use traditional knowledge for, say, a supermarket anti-dandruff shampoo, for which it would have no importance whatsoever. But for natural lines, or products making direct claims relating to mildness and naturalness, traditional knowledge could be useful.' Another company representative interviewed said, 'Traditional knowledge would be most useful in developing therapeutic products, with desired activity.'

This point was echoed by a spokesperson from a large company, who said, 'Traditional knowledge is very useful. It is not "folklore". It is far more serious than that – more "ethnomedicine". There is a long history of people using plant remedies, and there exists a tremendous literature. Traditional knowledge is particularly good for some product areas, such as anti-oxidants, fragrances, soothing agents, and astringents.'

NuSkin's Epoch line includes 'ethnobotanical' bath and foot care products containing botanicals derived from the study of indigenous cultures. Firewalker Foot Cream includes an ingredient used by Polynesians. Darren Bartels, marketing specialist said, 'Research indicates that people who walk on hot coals have actually used this leaf to wrap on their feet to soothe and cool those areas.' But traditional knowledge is more than a marketing tool for NuSkin, according to Bartels: 'We don't just include ethnobotanicals because they are natural. We also want to create a nice product.' To secure proprietary position, NuSkin sometimes obtains patents, and asks suppliers to provide herbarium voucher specimens to make sure plants have been correctly identified by ethnobotanists (*NBJ*, 1998).

Raintree Nutrition Inc bases its product development on traditional knowledge of Amazonian species, and describes its development of new product leads as follows:

> *Research begins by working with indigenous tribal healers and shamans as well as rain forest community herbal healers called curanderos and other natural health practitioners, herbalists, and researchers in South American cities, to target potential plants which have important medical values and benefits. Worldwide document gathering and literature searches compile and compare what other ethnobotanists, botanists, and researchers have discovered and tested about the properties, uses, and effectiveness of the plants.*

Most press coverage and public relations materials released by companies marketing natural personal care and cosmetic products include reference to traditional use of species. Aveda's 1994 *Source Book*, for example, includes the following: 'Our most valuable resource is indigenous peoples, for they are living libraries of ancient wisdom and ways. Working with them as partners, we have combined their botanical knowledge with today's technology to create a wealth of flower and plant products'. NuSkin's Force for Good Campaign was launched because 'We have received invaluable knowledge from indigenous peoples around the world about the ways they use botanicals. This knowledge is integral to the formation of the Epoch line of products'.

Of the companies interviewed, all make use of traditional knowledge in some form. Every company consults databases and literature on traditional knowledge as a guide to a species' activity, and many access information through the internet. A majority of companies reported receiving information on traditional uses from vendors, brokers or distributors, or other intermediaries who market potential ingredients to manufacturers of finished products. Aubrey Organics' 'Rosa Mosqueta' line, for example, was developed when Chilean researchers brought the oil to the company owner and 'asked him to work with it' (Hampton, 1994).

Field ethnobotanical collections are not common in this industry, but some companies undertake collections of samples in a systematic manner in order to feed high throughput screens. Ethnobotanical information often accompanies samples. A number of owners and upper level management in personal care and cosmetics companies also conduct ad hoc field 'collections' or 'explorations' to gather ideas for product development. Although general information on these activities is provided to the public and press, a great deal of secrecy surrounds the elements of field collections. Companies consider this information proprietary, and were reluctant to share much in the way of detail, primarily, it appeared, for fear competitors might follow in their footsteps.

9.5.3 The acquisition of natural product samples

Most of the companies that use natural ingredients acquire information on a species' traditional use and scientific validity through literature, databases, intermediary suppliers, trade shows, and other outlets in their home countries. Raw material and bulk ingredient suppliers might promote new natural ingredients to finished product manufacturers, or supply ingredients or formulae that manufacturers have identified through the literature as of possible interest. A long and complex chain of raw material exchange, involving numerous parties, is involved in new natural ingredient development. It is generally a targeted approach, in which species of interest are first identified, and then a source of raw materials sought. Random and larger scale 'prospecting' for new natural leads remains uncommon.

Exceptions do exist, however. A number of personal care and cosmetic company executives and researchers actively seek out traditional knowledge and species from high biodiversity regions and might undertake collecting trips in search of new leads. For example, in its 1997 annual report, Yves Rocher described the following prospecting activities:

> *In order to encourage the discovery of new active principles and knowledge of the sea, Dabiel Jouvance [a staff researcher] is an active partner in several ocean exploration projects. In 1997, the scientific and humanitarian mission 'Auracea' visited the Socotra Archipelago in the Indian Ocean, a little known site housing extraordinarily diverse flora and fauna. The objective of the research was to bring back new substances with potential cosmetics applications.*

Rainforest Nutrition Inc sources samples through joint ventures and contract agreements with established extractive reserves, indigenous tribal associations and other non-profit groups in the Amazon.

Many natural personal care companies have grown from strong personal interest in natural ingredients on the part of the founders (eg Ales Group, Aveda, The Body Shop, Neal's Yard, Tom's of Maine, Yves Rocher, and Rainforest Nutrition). As described, this often translates into continued interest and involvement in new product development, including field trips to collect samples for further study in the company's laboratories. Some of the companies interviewed mentioned that staff pick up materials while on trips, or that people from around the world send them ideas and samples in the mail to research.

More systematically, larger companies with mechanism-based screening programmes sub-contract with for-profit brokers (eg Estee Lauder contracted with the for-profit broker Biotics for the supply of extracts from a number of countries), research institutions, and other intermediaries. These intermediaries collect samples in much the same way as for the pharmaceutical industry (Chapter 3), but with greater emphasis on traditional use, and an eye towards raw material sourcing strategies, which are of immediate concern to companies in the personal care and cosmetics industry.

9.6 Sourcing raw materials for manufacture

Personal care and cosmetic companies seek reliable, affordable, and consistently high quality raw materials for product manufacture. These factors will determine the viability of a new commercial product or ingredient, much as in the botanical medicine industry. As the representative of one large company said, 'Uncertainty surrounding the supply of raw materials is one of the most difficult aspects of natural products. Even if a plant has really interesting ingredients, it is useless to us unless we have figured out a reasonable, reliable supply.'

Companies in this sector customarily purchase raw materials at low prices and in bulk as commodities, again much as in the botanical medicine industry. Packaging, presentation, and marketing absorb the majority of most company budgets, and sourcing of raw materials, and in some cases research and development, a relatively small fraction. Additionally, most manufacturing and marketing companies are many steps removed from the original sources of their raw material.

Most companies source raw materials from dozens of countries. The material has usually passed through many hands before it reaches a manufacturing company, and most companies find they cannot get satisfactory details on its origin. Many do not consider this important, however, as long as the material meets their specifications and price requirements. As one company representative put it: 'The supply industry involves a very complex and lengthy distribution system of brokers. Each will claim its sources are good, but there are about 100 people between them and the source, and they have no idea. Price is everything to most purchasers, so the larger issues get lost along the way.'

To date, there is limited consumer pressure on industry to change raw material sourcing relationships, and few calls for certified organic, fair trade, or otherwise environmentally or socially sound sources of raw material. While opposition to animal testing for personal care and cosmetic products has taken widespread hold, concerns relating to environmental and social impacts have done so only to a limited extent, and more so in some

European countries than the rest of the world. There is therefore little awareness within most companies of benefit-sharing and equitable partnerships related to the sourcing of raw materials. Again in parallel to the botanical medicine industry, we are confronted with the paradox of consumers asking for things more 'natural', but paying little attention to the impact that the delivery of these natural products has on the environment and local communities living in areas from which they are sourced.

Companies with a stake in the organic or socially-sound nature of their materials tend to be more closely connected to their sources. Aubrey Organics works directly with farmers in a number of countries, sourcing aloe vera from Honduras (where it is hand filleted by farmers and shipped to Florida), shea butter from Africa, and eucalyptus from Australia. For companies developing 100 per cent natural personal care products, the quality of raw material is critical. As Marc Wethmar of Weleda Ag said, 'Because Weleda works with virtually no synthetic substances, if raw materials differ in quality it can dramatically alter the quality of the product – we are much more vulnerable to inconsistency in the quality of raw materials than companies that use larger proportions of synthetic ingredients.'

Other examples of such companies are Neal's Yard, which initiated direct sourcing relationships with communities in Brazil and Ladakh, and The Body Shop, which developed 'Ethical Guidelines/Criteria for Buyers' that include requirements for sustainability, state a preference for sustainable management systems, and prohibit the harvest of endangered species. The Body Shop Community Trade Programme established a number of relationships with local communities for the supply of fairly traded and environmentally-sound raw materials, although these supply only a small fraction of the company's total raw material consumption. In an effort to systematically conduct rapid environmental and social monitoring of small-scale suppliers, The Body Shop has also entered into a partnership with the certifying agency Société Générale de Surveillance Group (SGS).

Aveda is similarly seeking larger portions of environmentally- and socially-sound raw materials, as part of its wider corporate strategy. As Peter Matravers, VP of Research and Development, said, 'Aveda's product development strategy is based on a commitment to replace petrochemical and synthetic ingredients by using renewable resources – organically grown or wild harvested, free of petrochemicals, synthetic fertilizers, and herbicides, and assisting in the maintenance of the world's biodiversity. This means that the company does not want to develop monocultures of hybrids that replace the natural environment. We try to form partnerships with sources to develop organic material grown in conditions that conserve biodiversity.'

Weleda is actively working to bring 100 per cent of the plant material it consumes under its direct control. 75 per cent of its material is grown on Weleda farms, or through organic or otherwise environmentally-sound sources; but 25 per cent is still wild harvested, because these species are 'hard to get any other way'. Weleda would like to have more control over these sources, but achieving 100 per cent sustainable and fair trade sources is a major undertaking. As one small company representative said, 'We have to chose our issues. We have barely been able to establish a system in which our suppliers do not test on animals. We have not yet gotten to other areas.'

A large majority of companies interviewed preferred cultivated sources of raw material to wild harvested. Many companies have no idea how the raw materials they consume break down into cultivated versus wildcrafted material, however. Most think their raw materials are primarily of cultivated origin, but few could confirm this. Some companies – like Weleda and Yves Rocher – run their own farms, but these supply only a portion of their total demand for raw material.

Larger companies often invest to ensure that sources of their key raw materials are in cultivation. As one company representative interviewed explained:

> *We much prefer species that are easily accessible, commonly-known, and widely-available, such as green tea, which we will use in a lot of our products. If something is not currently in cultivation, we will try to get it there, because we need to rely on the source. We cannot rely on wild-harvested material. Even with cultivated materials there are dangers – you can have a monsoon somewhere. Or you can grow a plant in one field one year and it will come out fine, and the next year it will not meet specifications (colour, odor, desired components, etc). We tell suppliers that if the material does not meet specifications to not bother bringing it to us.*

9.7 Regulatory environment

The regulatory framework for personal care and cosmetic ingredients is increasingly conservative, as led by recent EU legislation. For a number of years, the regulatory requirements in Japan were the world's strictest, and acquiring approval in Japan became the standard by which products for the global market were developed. Europe now sets the regulatory 'floor' for international personal care and cosmetic products, as part of a trend towards increased regulation.

The Cosmetics Directive, the key piece of EC regulation, was modified by a sixth amendment governing the composition, manufacture, packaging, and labelling of cosmetics. According to the new regulations, any new extract must have either guaranteed content of an active molecule detectable in the finished product (either directly measured or indicated by the presence of a tracer), or substantiated efficacy, with a clinical dossier (Brun, 1998). As a result, companies are changing the way they represent ingredients on labels, and require testing and proof of product authenticity from chemical suppliers, who now package chemicals with guidelines and advice on how to use a product and achieve a particular performance (Gain, 1996).

These regulations mean that new extracts are unlikely to be developed and issued without tests and guaranteed content of active molecules (Brun, 1998). So demand for ingredients 'new' to the market, including natural products, is diminished. As Peter Cade, Director of Technologies at Croda Inc, said,

> *In Europe the system seems to be designed to restrict the development of new compounds. If you develop something really new, with nothing like it around, it could cost several million dollars and take several years to get approval. If there are several analogous substances, and parallels on toxicity and environmental effects, it might cost a few hundred thousand dollars. In this environment, it is easier to get approval for natural products like botanicals with a long history of traditional use, but if the compound is completely novel, then it will be difficult and expensive to get approval.*

Meanwhile, in the USA the FDA does not require pre-market approval for cosmetics. If a safety problem arises after a cosmetic is on the market, the FDA can take action, but in general the FDA pays little attention to cosmetics, the surveillance of which accounts for less than 1 per cent of its budget. This lack of attention grows from an historical belief that the skin is close to a perfect barrier that prevents applied chemicals from penetrating the body. We now know this is not the case, amply demonstrated by the use of nicotine patches, but regulations have yet to catch up (Winter, 1994).

The FDA draws a clear line between drugs and cosmetics. While cosmetics companies cannot make the therapeutic claims allowed for drugs, they also receive little scrutiny. The only category of personal care and cosmetic ingredients that requires FDA approval is colourants, for which the FDA has an approved list. Cosmeceuticals blur the distinction between drugs and cosmetics, but to date companies market their products as cosmetics and get around expensive and time-consuming pharmaceutical testing. According to John Bailey, director of cosmetics and colours for the FDA, many companies phrase product claims in such a way as to imply therapeutic effects, but still word them carefully enough to avoid having the products categorised as drugs (Ouellette, 1997). On the other end of the spectrum, many of the smaller natural personal care companies slip under the FDA radar screen, and may not even use FDA-approved language on their labels. Product safety and quality ultimately come down to the individual companies, most of which build checks into their research and production systems, and place a premium on reputations for high quality, effective products (Kalorama, 1996).

The regulatory environment is currently putting pressure on companies to steer a conservative course. According to Peter Cade at Croda Inc, 'The big personal care and cosmetic manufacturers tend to ask first about a new ingredient: Is it approved for use in Europe, Japan, Asia, and Australia? and then might, as a secondary issue, show interest in it as a naturally-derived or botanical product. Croda must invest the time necessary to get approval for ingredients or demonstrate that the material is within classes of approved material before companies will be interested – this is increasingly important for multinational companies that want to market in many countries.'

The US regulatory system is proving that most receptive to the introduction of new ingredients. Companies selling products primarily to domestic US markets are allowed greater room for innovation. However, multinational companies dependent upon global sales follow the lead of European regulatory frameworks. Although loopholes in regulations still exist, requirements for clinical dossiers, substantiated efficacy, and guaranteed content of an active molecule are likely to create a short-term disincentive for incorporation of species and compounds new to the market.

9.8 Demand for access to 'new' species

As discussed above, companies incorporate natural ingredients into products in response to consumer demand for both natural and therapeutic products. However, this is primarily a conservative industry, not characterised by risk-taking and innovation. Although scientific and technological capacity to analyse and understand the function of natural products is increasing in effectiveness and decreasing in cost, most companies' research and development is directed towards re-examining and re-formulating species known to the market. As Karl Raabe at Henkel said, 'In general, there are no completely new ingredients in our products and there is more of a tendency to use variations of existing ingredients' (Gain, 1996). Bob Rumsbey, VP of Operations and Manufacturing at Liz Claiborne Inc, echoes this: 'We are not out there using state of the art materials ..We want something on the FDA and CTFA approved lists, and the lists approved in Europe. It can be extremely expensive to bring a completely new product to market, potentially many millions of dollars, so we prefer to stick with ingredients on the approved lists.'

A glance at most of the natural products coming on to the market confirms that the majority of natural ingredients are already well known. However, because this industry has long incorporated natural ingredients into its products, the base from which it works is very wide. Marc Wethmar of Weleda Ag explained: 'At least US$1 million must be spent on clinical trials to find new medicaments, so it is better to modify existing ones. In cosmetics, sometimes we can develop products based on new raw materials – eg we now use shea butter – but it is rare. We don't do a great deal of research on new materials because we have an existing background with a large number of materials – about 260 plant species with which we regularly work. So we don't really need to work on exotic or tropical plants. We have this base of plant material and knowledge from which to work.'

Concerns relating to available supplies of raw materials also impact companies' demand for new ingredients. As one director of research at a medium-sized company put it: 'We stick mainly with what is known and available – we piggyback on others' product lines, and don't want to be the only people out there with an ingredient. We need to ensure a sustainable supply, which means that there has to be enough demand for suppliers to keep supplying, so enough buyers on the scene. A new species is not attractive if others are not using it as well.'

Demand for access to new natural products follows from companies' research and development strategies. Most small and medium-sized companies re-formulate known ingredients, but some prospect for new leads in high biodiversity countries and others purchase scientifically-validated natural extracts with guaranteed active ingredient content. In some cases these extracts are derived from 'new' natural ingredients. A few companies seek out new leads and access to resources and traditional knowledge in order to enhance the marketing potential of products. Larger companies engaged in mechanism-based screening demand access to a range of natural product samples, often with an emphasis on novelty and diversity. Company decisions relating to development of commercial products from new natural ingredients involve consideration of regulatory requirements, availability of reliable and high quality raw material sources, and cost.

9.9 Benefit-sharing

While natural personal care and cosmetic products are part of a 'supertrend', benefit-sharing associated with the commercial use of traditional knowledge and resources is developing more slowly. Across the sector there exists little awareness of the CBD and access and benefit-sharing provisions (Chapter 10). Access to 'new' species and traditional knowledge rarely involve direct contact between companies and source countries. Information and samples travel through multiple institutional, corporate, print, or electronic channels before reaching a company. However, even when companies actively prospect for new leads, there appears little awareness of the need to link these activities to benefit-sharing in a way more formal than charitable donations.

The personal care and cosmetics industry is characterised by secrecy at the research and product development levels, with proprietary new ingredients carefully guarded, including their sources. As we have noted, few companies are willing to elaborate on their field collecting and raw material acquisition practices, although some describe community-based partnerships and global collecting efforts in general terms as a way of demonstrating corporate commitment to natural ingredients, the environment, and in some cases social equity. The following is a discussion of some of the monetary and non-monetary benefits that have featured in commercial partnerships to date.

9.9.1 Monetary benefits

Personal care and cosmetic companies purchase botanical material used in the manufacture of commercial products in bulk, paying a price per kg for material of specified quality and consistency. When companies purchase samples for screening, they pay fees for samples, usually following pricing practices common to sample collection in the pharmaceutical industry (Chapter 3). Royalties are not commonplace in this industry, but might be tied to the supply of samples, as in the pharmaceutical industry. A single personal care and cosmetic ingredient has the potential to yield large revenues, and isolated, patented compounds are increasingly the norm in this industry. There exists the potential for companies to capture an exclusive right to market what might be the next 'hot' product ingredient and source countries and communities can benefit financially from these developments through royalties and advance payments.

Advance payments, intended to cover collaborations and an agreed upon workplan, are uncommon in the research and development phase of product development, although they could become common practice should companies begin to work more closely with source-country research institutions (Chapter 3). Advance payments are more common for the sourcing of raw materials. For example, The Body Shop, Weleda Ag, Aveda Corporation and others invest funds in local partnerships in order to develop sustainable, fair trade sources of raw materials.

Charitable donations

Many of the companies marketing natural personal care and cosmetic products have established charitable foundations or programmes to support humanitarian and environmental causes. Yves Rocher launched a widespread tree-planting programme in France in 1991, called 'An Arboretum for Every School'. The Aveda Corporation supports community development projects in some of the areas it sources raw material. Some companies 'tithe', or set aside a percentage of sales for internal foundations which support social and environmental projects. Tom's of Maine provides 10 per cent of pre-tax profits to non-profit organisations 'benefiting the environment, human need, arts, and education'. The company has also taken steps to support community-based programmes in areas in which it conducts research. Following attendance at the Rio de Janeiro UN Conference on Environment and Development (UNCED) in 1992, Tom's of Maine, in conjunction with the Rainforest Alliance, supported a range of ethnobotanical research and community development projects through its charitable wing, while collaborating on a commercial basis with local universities and research institutions.

NuSkin donates 25 cents from each Epoch line product sold to its 'Force for Good' charitable campaign in order 'to assist in the preservation and continuation of an indigenous culture or humanitarian cause'. Although NuSkin secures patents on products in its 'ethnobotanical' Epoch line, it does not appear to link these to royalties or financial benefit-sharing in source countries. Rather, the company provides monetary benefits in the form of grants to communities, as well as non-monetary benefits through its consulting ethnobotanists.

9.9.2 Non-monetary benefits

Non-monetary benefits take a variety of forms, but are primarily associated with the sourcing of raw materials. They range from guaranteed markets for products at preferential prices, to the development of infrastructure and capacity for domestic industries based on local raw materials. In some cases, companies have also collaborated with research institutions, and have transferred technology and built capacity. A range of in-kind benefits and donations to community projects also result from commercial collaborations.

Raw material sourcing partnerships

A number of partnerships have been created based on the sourcing of raw materials, often with the express purpose of contributing to environmental and social objectives, and sharing commercial benefits. Partnerships of this kind are increasingly common for alternative marketers, and are promoted by many companies in their marketing campaigns. Many companies interviewed – including AE Hobbs, the Ales Group, Yves Rocher, Aveda, Weleda, Raintree Nutrition, and The Body Shop – link the sourcing of raw materials to benefits for local communities or conservation programmes.

In the case of bixa (see the case study, p281), Aveda Corporation made a multi-year investment to develop sustainable and socially responsible sourcing of a key raw material for its cosmetics and hair product lines. Although bixa is only one of hundreds of botanical ingredients used by Aveda, the company hopes to increase the proportion of material it sources through this approach. It sees sourcing partnerships as a key form of benefit-sharing for local communities. As the Aveda *Source Book* (1994) says: 'Through contract farming with indigenous cultures around the globe, Aveda is helping to improve their economies and preserve their way of life'.

The Body Shop also employs a community-based approach to sourcing some of its raw materials. Through its Community Trade Programme, the company is trying to achieve 'long-term sustainable relationships, which are one way of using trade as a mechanism for people in these communities in need to benefit through employment income, skill development, and social initiatives'. The Body Shop's 'fair price' for material is intended to cover production costs and fair wages, and 'enables an investment in the community and the future'. The Body Shop's criteria for these partnerships (The Body Shop, 1997) are:

- working with community organisations that already exist to represent the interests of the social group concerned;
- working with groups with limited access to resources, education, healthcare, and outlets for their goods;
- trade must benefit the producers, so benefits must be re-distributed throughout the community in ways that promote development;
- the partnership must be commercially viable – price, quality, capacity, and accessibility must be carefully considered; and

- the activity must meet Body Shop standards for environmental and animal protection.

By 1997, The Body Shop had entered into more than 25 community trade partnerships with suppliers world-wide, purchasing through these avenues a total of £2.1 million of raw materials and accessories in 1995–6. Partners in this programme include (The Body Shop, 1997):

- *Brazil:* The Kayapo villages of A-Ukre and Pukanev supply Brazil nut oil for use in the Brazil Nut conditioner and five other products; The Body Shop supplied around US$80,000 worth of start-up infrastructure investment to the Brazil nut oil processing business, and a commitment to an agreed upon price per kg of oil well above market rate. The Body Shop also buys babassu oil from COPPALJ, a cooperative in the state of Maranhao in north-east Brazil; in 1996–7 The Body Shop purchased approximately 38 tonnes of refined babassu oil as an ingredient for lotions and make-up; again, The Body Shop agrees to provide a secure market, at a preferential price.
- *Ghana:* Women's groups in the Tamale region supply shea butter.
- *Nicaragua:* The Juan Francisco Paz Silva Cooperative supplies sesame seed oil (2.4 tonnes in 1995–6) for use in ten Body Shop products.
- *USA:* The Native Americans of the Santa Ana Pueblo of New Mexico supply six tonnes of blue corn per year for use in the Blue Corn Scrub Mask .

Weleda concentrates on environmental sourcing, actively working to increase company control over its raw material sources. The species used in product manufacture are primarily sourced from Europe, and company efforts include working with small farmers in Germany to train them in organic cultivation techniques. Marc Wethmar of Weleda Ag explained that: 'Farmers are invited to Weleda for training in medicinal plant cultivation, and are then offered a purchase contract. In this way the company helps to develop knowledge in medicinal plant cultivation and better approaches to agriculture like organic and biodynamic.'

In sourcing its raw materials, Raintree Nutrition Inc works with private landowners, established extractive reserves trying to preserve critical sections of rain forest, and government agencies, in order to wildharvest materials. Harvesting is undertaken by 'local rain forest communities and villages as well as Indigenous Tribal groups and communities'. According to Leslie Taylor, founder of Raintree, 'This provides more profits to local communities, and creates an incentive for conservation of the forest – people earn eight to ten times more from wildharvesting than slash and burn.'

A number of companies work with small farmer cooperatives around the world. Yves Rocher works with cooperatives in Burkina Faso to source sesame, and Aubrey Organics has direct sourcing relationships with a number of small cooperative suppliers around the world. While it was not initiated by Croda Inc, its sourcing partnership with Conservation International for the supply of cohune oil and other ingredients was established to assist community development and conservation initiatives in a number of countries in South America and elsewhere (see Cohune oil case study, p287).

Raw material sourcing partnerships usually guarantee markets at preferential prices. This has allowed some communities and farmers to invest in sustainability and capacity building, as well as a range of non-monetary benefits associated with capacity building like equipment and training. Companies often supply in-kind donations, as well, like medical supplies and road improvements. As the case studies make clear, however, community-based sourcing strategies take many years and numerous points of refinement to reach a comfortable trading partnership, and many large companies are unwilling to invest time and resources in this approach to raw material sourcing.

It should also be noted that many consider community-based sourcing partnerships detrimental to local communities in the long run. Critics claim that commercial marketing of community products emphasises consumption instead of changing broader problems in the way commercial companies and governments work with communities. Entry into the market of isolated communities is also considered a dangerous approach to 'conservation' and 'development' (Corry, 1993).

We must be cautious in generalising about benefits that can accrue to local communities from sourcing partnerships, since there are many variations by region and community. For example, what might seem 'obvious' steps towards benefit-sharing may in fact not increase a community's bottom line at all. Clay (in press) describes the case of PRONATUS, a company based in the Amazon that manufactures personal care products and dietary supplements for local markets. Merely adding value to products did not improve their bottom line, and it was not until they lowered production costs, and made the operation more efficient that increased benefits were realised. International markets were also not a guarantee to greater benefit-sharing, and the company came to realise that local markets afforded the best opportunities for their products – based largely on local and traditional preparations, and geared towards local consumers. Finally, the company realised the benefits in a raw material sourcing strategy based on 'a steady supply of higher priced raw materials rather than a massive, inexpensive supply which might discourage producers from supplying more in the future.'

Research and training

In some cases, personal care and cosmetics companies have entered into research collaborations with institutions in high biodiversity countries. These collaborations can result in a number of spin-off benefits, including laboratory equipment, access to scientific literature, training, and networking assistance. The Body Shop, for example, collaborated with the Institute of Ethnobiology of the Amazon between 1991 and 1994, and then with the Foundation for the Federal University of Para in Brazil, and the University's Natural Products Chemistry Department on research into new species for product development. The collaboration included establishment of a natural products laboratory at the University. In 1994, Tom's of Maine established a partnership with the Federal University of Ceara, in Brazil, to conduct research on species of potential use as substitutes for synthetic ingredients.

In-kind benefits

The bixa case (see p281) illustrates some of the in-kind benefits possible in commercial partnerships with source countries and communities. Infrastructure building, roads, schools, healthcare, and other benefits can result, albeit usually on a very localised basis. For example, NuSkin's Force for Good campaign provided benefits like sources of clean water, schools, and health facilities to villages it works with in the South Pacific. More than US$40,000 was donated to the village of Falaelup, Western Samoa, to build an aerial walkway through a 12,150-hectare forest reserve for education and tourism; US$10,000 was provided to build a visitor's centre and ecological library at the 4,050-hectare Tafua Rainforest Reserve in Western Samoa; and through the Seacology Foundation, US$15,000 was provided for a solar-powered well in Haiti.

Benefits associated with the use of traditional knowledge

Traditional knowledge is usually accessed by companies through databases and literature, and in some cases through multiple intermediaries. As a result, access to knowledge and the sharing of benefits have been de-linked, and examples of benefit-sharing associated with the use of traditional knowledge are few. Although some companies provide grants to communities with which they work, commercial agreements and compensation tied to the commercial value of the knowledge supplied are rare.

In the case of bixa, Aveda Corporation sought a species with widespread and common use, in order to avoid the implications of commercialising a species whose use grows from specialised or restricted traditional knowledge. Aveda is developing a manifesto to guide its work with indigenous communities, which recognises traditional resource rights, sets out terms for PIC, benefit-sharing, and sustainable sourcing of raw materials, and provides for technology transfer in the form of supply industries.

In 1993 the Body Shop launched a project to develop an 'Intellectual Property Rights' policy that recognises the 'accumulated wisdom of centuries of tribal lore as a valuable commodity, and compensates accordingly'. According to its press release on the project, as part of its on-going relationship with the Kayapo Indians in Brazil, The Body Shop planned to implement an 'IPR agreement' to ensure that any commercial development of products benefits the Kayapo. 'The issue of fair trade is vital to The Body Shop. It means an end to exploitation of humans, animals, and environments. It means guaranteeing full intellectual property rights to indigenous peoples whose knowledge and resources people have, until now, been exploited with impunity,' said Anita Roddick, Managing Director of The Body Shop.

The use of indigenous peoples' names and images in the marketing of personal care and cosmetic products is widespread and on the rise. Steps should be taken to ensure communities not only provide PIC for this use, but share in the financial benefits resulting from product sales which their images promote. In some cases this might take the form of a grant to the community for healthcare, community development, or other projects, and in others a negotiated royalty on sales might be more appropriate (see the Yawanawa and Aveda Corporation *Bixa orellana* Project case study, p281).

Benefits for conservation and development

Many of the companies marketing natural personal care and cosmetic products are committed to things natural and wider environmental concerns. The founders/CEOs of the Ales Group, Aveda, Aubrey Organics, The Body Shop, Raintree Nutrition, Tom's of Maine, and Yves Rocher, for example, publicly support natural products, and environmental issues feature prominently in annual reports and company public relations materials. As a result, company charitable giving, and in some cases business practices, work to minimise the company's impact on the environment, and promote conservation and social equity.

The main environmental considerations addressed by companies in this sector are waste reduction, packaging, recycling, and other issues associated with the manufacturing process, and the company's role in its home country. Companies are also working to avoid petrochemical ingredients and animal testing, and in some cases to develop progressive conditions for workers in their facilities. Some of the companies in this sector are actively involved in The Social Venture Network, Businesses for Social Responsibility, and other groups working to set standards of best practice for industry. A few have signed on to the CERES Principles (Coalition for Environmentally Responsible Economies), which – in addition to numerous principles that primarily influence manufacturing and distribution practices – commit companies to 'sustainable use of renewable natural resources, such as water, soils, and forests' (Chapter 10).

9.9.3 Benefit-sharing conclusions

For the most part, companies in the personal care and cosmetics industry are poorly informed of the environmental and social impact of their sourcing strategies around the world. Likewise, few have considered the implications of sample collection, and development of new leads from traditional knowledge or species 'new' to the market. Equitable benefit-sharing linked to access to resources and knowledge is a concept unfamiliar to most, only a handful of whom have heard of the CBD. As a result, benefit-sharing is often viewed as a charitable contribution, a gesture of corporate goodwill, rather than an obligation. Pilot-level projects of the kind we discuss in the case studies, and which have been tried on a very small scale around the world, provide some indication of the types of benefits that can result from raw material sourcing collaborations. Fewer examples exist of partnerships associated with the discovery and product development phase. Should demand for new natural products continue, it is hoped that benefit-sharing in the future will come to include a wider range of benefits packaged to transfer technology and build capacity in the research and development, as well as raw material sourcing, phases.

9.10 Case studies

9.10.1 The production and marketing of a species in the 'public domain': the Yawanawa and Aveda Corporation *Bixa orellana* Project, Brazil

May Waddington[2] and Sarah A Laird

Overview

In 1993 the Yawanawa Community of the Gregorio River area of Acre, Brazil, established a cooperation agreement with the US personal care and cosmetics company Aveda Corporation. The agreement addressed the commercial farming of 30 hectares of *Bixa orellana* ('bixa') trees, in multi-cropping systems that also included the economic species pupunha and Brazil nut. Bixa is a common and widespread species throughout the neotropics, with a long history of traditional use – it is therefore a species that the Yawanawa and Aveda consider in the 'public domain'. This case was the first such collaboration between a commercial company and an indigenous community in the Acre region. The Aveda Corporation uses bixa as a colourant in its lipstick and other personal care and cosmetics products.

The partnership between the Yawanawa and Aveda involved a package of benefits over a number of years, including the supply of technical assistance; start-up financing to cover initial production, local processing, transportation and structural investments such as warehouses and dryers; facilitation of certification of the product as organic; and a guaranteed market for a portion of the product, at an agreed-upon price. The company sought to avoid what might be a patronising distribution of benefits unconnected to the agreed workplan, but the first three years of the collaboration convinced staff that investments in social benefits were vital to the success of the project. These included a health centre with dental and medical equipment; malaria screening instruments and medication; purchase of an equipped office in the town of Tarauaca; and a solar energy system in the village.

The main players

The Yawanawa

The Yawanawa tribe consists of approximately 330 people inhabiting the upper banks of the Gregorio River in the county of Tarauaca, Acre. They share 92,859 hectares of land (99.95 per cent in forest) with the 175 members of the Katukina do Sete Estrelas village, who arrived in the area approximately 60 years ago. In 1983, the reserve was the first demarcated in the State of Acre, under Decree no. 89.257/83.

To assure greater control over and coordination of their internal and external market relations, the Yawanawa created the Organizacao dos Agricultores e Extractivistas Yawanawa do Rio Gregorio (OAEYRG) in 1992. The organisation incorporates existing leadership structures in which Assemblies elect individuals to oversee activities. OAEYRG legally represents the community in partnership agreements and external financing matters, overseeing the sale of the community's products and the purchase of basic goods.

Aveda Corporation

Aveda is a personal care and cosmetics company with a mission to produce high-quality products; it works to substitute petrochemical and synthetic ingredients with botanical ingredients, and to operate in an environmentally-friendly manner. In 1998 Aveda was bought out by the larger US cosmetics firm Estee Lauder. Most of the research and development conducted by Aveda is formulation chemistry, testing for safety and efficacy, and 'creative' research, rather than screening. Aveda trademarks many of its formulations, but does not pursue patenting.

FUNAI

FUNAI is a Brazilian federal government agency under the Ministry of Justice, in charge of indigenous peoples' affairs. It provides official authorisation to enter the Yawanawa territory, and has sent teams to supervise the Yawanawa-Aveda collaboration. It has also assisted the collaboration by sending dentists to the area, and clearing customs for medical equipment.

Institutions that have acted as partners or consultants to the Yawanawa-Aveda collaboration, or that have contributed medical and other assistance, include the following:

- *ITAL/CAMPINAS (Instituto de Técnologia de Alimentos de Campinas)* – consultants who have tested for the percentage of bixine in seeds, advising on processing, storage and manufacturing;
- *Institute Biodinamico* – an organic certification agency, whose consultants have advised on agronomic techniques;
- *Project RECA (Reflorestamento economico adensado)* – based in Acre, this group has volunteered advice on reforestation and has supplied seed of pupunha and Brazil nut;
- *EMBRAPA (Empresa Brasileira de Pesquisa Agropecuaria)* – the Brazilian federal agriculture agency, based in Acre, that supplied bixa seeds and technical advice in the early stages of the collaboration;
- *CEPATUR* – a branch of EMBRAPA in Para that volunteered technical advice on solar drying of bixa seeds;
- *Polydrier*– a private company made up of staff from the University of Viscosa, Minas Gerais that contributed technical advice and mechanical bixa seed dryers and storage facilities;
- *COSAI/FNS* – the Indian Health Agency of the Ministry of Health that assisted in the transportation of medical supplies; trained a Yawanawa Health Agent in malaria screening, and provided malaria medication;
- *Comissao Pro-Indio* (CPI-Acre) – worked for many years in a range of ways to assist the Yawanawa, including with the school programme and the demarcation of the area in 1983, and have provided a home base in Rio Branco for the community. Although not formally part of the Aveda-Yawanawa collaboration, the foundation they laid was critical to the success of this project;
- *Associacao dos Dentistas de Campinas* – together with the Prefeitura de Campinas, has donated dental equipment, including dental chair, motor, tools, etc;
- *SHARE* (Supporting Hospitals Abroad with Resources and Equipment) – based at Johns Hopkins Medical School, donated medical equipment worth US$15,000 to the community;
- *Direct Relief International* – donated medical equipment worth US$30,000, due to arrive in the area in 1999;
- *VARIG Airlines* – has provided free or discounted transportation for medical equipment; and
- *Other suppliers* – including Siemens Solar (California); Thin Light (California); Solar Electric Specialities (California), and Amazonas Solar (Manaus) – provided discounts on equipment.

The species/product

Bixa orellana (*urucu* in Portuguese; *achiote* in Spanish; *rocouyer* in French; and annatto in English) is native to and widespread throughout the neotropics. It is a shrub, often cultivated around villages and in indoor yard gardens. A red dye, bixin, obtained from the aril of the seed is used in the Amazon for decorative body painting, to ward off evil, and to go to war. Bixa paint is also reputed to repel insects. Rope is sometimes made from the fibrous bark, and bixa is also employed in a variety of folk remedies including those for epilepsy, fever, dysentery, and venereal diseases (Smith et al, 1992).

Bixa has many traditional uses across the region, and the species and knowledge regarding its use are common and widespread, particularly with regards to its dyeing and food colouring properties. For example:

- the Waimiri Atroari use the red dye obtained from the aril of the seed for body painting (Milliken et al 1992);
- the Chacabo of Bolivia grow bixa not only for a red dye, but also for its edible seeds that are cooked in butter. The leaves are reported to alleviate body pains by pressing them against affected areas (Boom, 1987);
- in Colombia, the red dye, if ingested, is said to be an effective antidote for cyanide poisoning (as from untreated manioc) (Schultes and Raffauf, 1990);
- the Awa Indians of south-western Colombia include bixa in their tree gardens and use it as a food colouring (Orejuela, 1992); and
- the Kayapo in Brazil use *urucu* for body paint and the husks are a favoured mulch in swidden fields (Hecht and Posey, 1989).

Commercial use

The almost tasteless colouring properties of bixa are used in a range of industries, including food, beverages, cosmetics and pharmaceuticals. The food industry uses the pigment (bixine) primarily in the dried meats industry (eg salami, sausages), but also in dairy products (cheese and butter), pasta, breakfast cereals, desserts and drinks. Intermediary companies process seeds to extract bixine in different ways depending upon the end use. Oil (generally soy) is used to extract bixine for use in the food industry. Water-soluble norbixine (bixa salt), obtained through the inclusion of soda in a chemical process, is used in the beverages and cosmetics industry. Local markets in Brazil trade in 'colourau' – a combination of corn or manioc flour, bixa seeds, and oil. In the north-east of Brazil, a family's average consumption of colourau is 1 kg per month.

Bixa has a long history in international trade. Bixine was first identified in 1825, but has not been synthesized (Smith et al, 1992). During the 19th century, the Brazilian Amazon exported significant quantities of bixa powder to Europe. In the mid-1900s, bixa was replaced by Red Dye #3 as the preferred colourant of food and cosmetics; however in the 1970s Red Dye #3 was shown to be carcinogenic. In 1990 the FDA banned the use of Red Dye #3, which sparked renewed interest in bixa on the part of food processors and cosmetic companies, because bixine is safe for consumption and for the skin.

The commercial production of bixa in Brazil

In 1998, 12,000 hectares in Brazil were under bixa cultivation, producing 7,000 tons of seeds (Dr Abel Reboucas, personal communication 1999). The largest farms are located in the states of Bahia, Paraiba, São Paulo and Para. The major bixa processor – the multinational company Christian Hansen – has a large farm in the state of Para and produces bixa for use in colour agents that it sells internationally.

The proportion of bixa seeds to bixine or norbixine is approximately 10:1. Norbixine and bixine are both used in water soluble cosmetics like soaps and conditioners, but only bixine is used in oil soluble products like lipsticks. In 1991 1,000 tons of bixa (14.3 per cent of production) were transformed into bixine. 20–30 per cent of the bixine, earning US$1.4 million, was exported to the USA. 2,000 tons (28.6 per cent of production) were transformed into norbixine and sold in powder form or water solution, earning US$4.8 million. The remaining 4,000 tons were transformed into colourau, with a turnover of US$12 million.

The price paid for bixa varies depending upon bixine content. In 1993, participants of the II National Congress of Natural Colors estimated that the 'break even' price for bixa was US$0.70 per kg. In 1994, prices reached US$2.50–4.00 per kg, but this fell to US$1.60–2.20 per kg in 1995. The marked drop in price in 1995 can be explained by major buyers forcing down prices by dumping bixa on the market. The 1998 price was US$0.80–1.20. On the whole, bixa is a lucrative crop, even in smallholdings; in 1998 prices as low as US$0.75 for bixa in São Paulo state still led to higher earnings for smallholders than they could realise from oranges.

The development of a commercial partnership: the Yawanawa and Aveda Corporation

Some 70 years ago, the Yawanawa became involved in the extraction of rubber, which was their main cash product until the collapse of the price of rubber in 1991. During this time, demand for external goods – such as fuel, ammunition, clothes, and medicines – was established. Prior to the explusion of the rubber 'barons' from the area in 1984 these goods were purchased from the stores of the Sete Estrelas Seringal (rubber extraction area) and the Kashinawa Seringal in exchange for balls of rubber, or services rendered to the rubber barons (hunting, opening trails, felling forest, transportation of cargo, etc). By the time the Yawanawa-Aveda project was initiated in 1992, the Yawanawa were suffering from the fall in the price of rubber, which had forced community members farther into the forest in search of larger quantities of rubber, and to nearby towns. For these reasons, the tribe was looking for a new economic alternative that would allow them to remain in the area and would favour cohesion of the community.

For the Yawanawa, bixa was a positive species for economic development because:

- it is native to the region;
- it is a perennial shrub that could help recuperate degraded lands, as part of multi-cropping systems that included other economic species like Brazil nut and pupunha;
- it is adapted to traditional farming practices, and there is local know-how on cultivation techniques;
- it is a product familiar to and used in the local culture, without being considered sacred or otherwise having a restricted use;
- it is a product with widespread use and distribution, thereby being of the 'public domain';
- the product is widely used as a food colouring and so has substantial local markets upon which the community could depend if international demand proved unstable;
- the prospects for an increase in commercial value appear good.

Aveda's first contact with the indigenous communities in Brazil was in 1992 at the UNCED in Rio de Janeiro, and the Earth Parliament Conference, organised by the International Society of Ethnobiology. Recognising that a partnership between a company and indigenous communities would be complex, the company drafted general guidelines for its first few years of investigating this type of collaboration in Brazil. According to the guidelines, Aveda would:

- not prospect for new botanical or chemical products; if communities wanted to work with Aveda, they would select a 'public domain' species (ie a widely used and distributed species);
- collaborate with national research institutions (rather than foreign) and domestic companies, which had received clearance and proper permits for any material under study;
- acquire the approval of FUNAI before entering into collaborations with indigenous communities.

Aveda was interested in the natural food colouring and dyeing properties of bixa for use in its cosmetics line, and as produced from organic sources. The company also wanted to ensure it collaborated on the basis of a widely used and known species, in order to avoid IPR issues associated with species that have restricted distribution and use. The company also sought to incorporate raw material into its cosmetics line that had applications in a range of industries – such as food and cosmetics – and thereby avoid establishing a dependency relationship between community and company. Staff at Aveda are researching expanded applications for bixa in their personal care and cosmetics lines, particularly as an ingredient in colour shampoos and conditioners, as well as permanent hair dyes. Aveda has found that this relationship allows them to improve company spirit and educate their network of personnel, professionals and clients, by involving them in the collaboration, which has included written materials and publications, workshops, lectures, and visits from Yawanawa.

The Bixa Project began with the planting of bixa on 30 hectares of land, in conjunction with other economic species, primarily pupunha, Brazil nut, and guarana. Processing machinery was purchased, and technical, marketing and administrative support provided to develop capacity in packaging, transport, and processing of colourau, and to set standards for quality control. Aveda staff worked directly with the Yawanawa Community Association in the development of this research and cultivation programme for bixa. Technical support was also provided by EMBRAPA (Empresa Brasileira de Pesquisa Agropecuaria), ITAL Campinas, Instituto Biodinamico, Projeto RECA, CEPATUR, and Polydrier. In return, Aveda has the right to be first buyer of preference if the price they offer meets any other price. Aveda is not required to buy all the bixa produced, but has made a commitment to assist in placing the product in the market should Aveda not require all of the material.

Aveda has agreed to pay the Yawanawa a price of US$2.40 per kg for bixa which is of high quality. This is part of an agreed price structure that is calculated as: 2 x market rate (US$1.50 for 1998) minus 20 per cent (which goes towards covering the original US$50,000 Aveda investment) = US$2.40. Once the original investment is paid off, the 20 per cent will no longer be deducted from the price Aveda pays.

At present, Aveda consumes the equivalent of 12 tons of bixa seeds per year, and is likely to demand increased amounts in the future as new products under development come on the market. The Yawanawa and Aveda are also negotiating an agreement with the São Paulo-based company Formil/Flora Brasil to process and standardise the pigment before it is exported to Aveda. Although Aveda does not benefit financially from this arrangement, it guarantees organically processed seeds derived from the Yawanawa bixa crop, which allows for marketing benefits that make possible the social and production investments.

This partnership grew from a confluence of factors: increased international attention on environmental issues in the Amazon; the Yawanawa's desire to establish an equitable economic partnership; a consumer market for products from ecologically-sound sources; and the resulting interest of companies like Aveda, whose products are directed to this public and whose philosophy is based on principles of environmental sustainability. For Aveda, the commercial benefits of marketing such partnerships are significant, and allow them to make this type of investment – and to share benefits. The closer a company is to the final consumer, and the more superfluous its product to survival, the more a company requires a positive public image, and the less it can afford negative press.

Benefit-sharing arrangements

Benefit-sharing has taken a range of forms throughout the course of this collaboration, and has developed alongside the relationship between Aveda and the Yawanawa. The 'package' of benefits developed was intended, on the one hand, to improve the community's ownership over the means of production, through the development of a bixa supply industry; and, on the other hand, to diminish their level of dependence on external goods (eg fuel and medical care). Box 9.3 provides a general summary of the benefit-sharing package.

Key issues related to benefit-sharing raised during this collaboration are examined below.

Intellectual property rights

Bixa is widely known and used throughout the neotropics, and is available in any market-place in the region. The agreement between Aveda and the Yawanawa was therefore not based on the use of restricted traditional knowledge, and does not involve commercialisation of a product previously used only on a local or subsistence level. The collaboration was based on the Aveda company's desire to avoid 'prospecting' for new species. The second product under consideration by Aveda for commercialisation is pupunha, which was planted alongside the bixa trees. It was decided that ethnobotanical research would not be carried out in order to identify species new to the market until legislation was in place to protect indigenous peoples' knowledge within Brazil (such as Senator Marinas' PL 364 'Access to Biogenetic Resources' Bill). All products under development had to be widely used and known. This was a central point for the company. As May Waddington said, 'This project was intended to strengthen the community and its ability to preserve the forest and their style of life; it was not intended to uncover a "miracle" plant.'

Commercial benefit-sharing in a community

In an effort to avoid a patronising approach, Aveda initially tried to tie all benefits to the commercial partnership, such as guaranteed price, a promise to seek alternative buyers, and provision of equipment directly linked to bixa production. After a few years, however, it became clear to the company that the project would only be successful if other infrastructure assistance and support was provided. For example, malaria was killing on average six members of the community per year. Medical facilities were installed in a joint effort between Aveda, government institutions, and other companies which discounted or donated equipment. The installed system includes a microscope and training for malaria screening in the village; since it was put in place there have been no casualties due to malaria.

A solar electric system established for bixa processing facilities was extended to homes in the village. This was critical to the community, not only to avoid the purchase of fuel, but also to respond to community concerns at 'being in the dark', which is seen as the main indication of *atraso*, or delay, for rural communities. The office established in Tarauaca to serve the Bixa Project is also used by elders who must go to town to pick up retirement cheques from the government. By having a place to stay in town, more dignified relations between the Yawanawa and town people are possible.

Although Aveda has covered some emergency flights and health crises, where possible the company seeks to stimulate action on the part of government agencies, and facilitate access of the Yawanawa leaders to government officials. The Yawanawa have consistently improved their relationships with government agencies, which has led to improved services. For example, six teachers were hired by the state government; a cholera crisis that at one time would not have received attention was promptly dealt with by Federal government agencies, who also extended assistance to other villages in the region.

The ultimate objective of the benefit-sharing component of this project is to allow the Yawanawa to control their economy more effectively. Not only are they providing a product – bixa – to external markets, but they have strengthened their internal economy with crops like pupunha, and proposed sugar production. The project appears to be functioning well within the traditional structures – benefits are determined and distributed according to traditional systems, and bixa production has been integrated into other village economic and cultural activities. But this took many years to achieve. To move on to more value-added processing and manufacturing of bixa products in the villages is likely to take a great deal more time, if it is to be done in a way that does not damage their subsistence economic activities like fishing, hunting, and planting. Learning to manage the initial stages (harvest, storage, processing, packaging, and transportation) without 'ceasing to be Indians' has been a great deal of work over seven years.

A community will reasonably stop work harvesting bixa if, for example, they are needed to build the house of a member, or cultivate a neighbour's field – these are practices that no commercial partnership should alter. Reasonable time horizons, expectations, and incremental steps, are the most effective way to achieve success in commercial-community collaborations.

Use of the community's image in marketing

Aveda uses the community's image in its marketing of bixa products, as agreed with the Yawanawa. Originally, Aveda considered paying a royalty for this use, but upon closer examination found that more significant benefits could result from the wider partnership than a small royalty on net sales. Aveda calculated that its annual contribution to the Yawanawa averaged the same as a 5 per cent royalty on gross sales of products derived from bixa (1994 gross sales figures were approximately US$550,000). Since royalties on the use of indigenous peoples' images in marketing would be calculated based on a smaller percentage of net sales, Aveda considered it more advisable to fold the benefits that should accrue from use of the community's image into the whole benefit-sharing package. Aveda considered this a critical issue to address, however, because communities are often not consulted, and are rarely compensated for the use of their images in marketing. Any company working with communities is likely to

Box 9.3 *Benefit-Sharing Arrangements of the* Bixa Orellana *Project*

1993–4 (Project start-up costs):

- US$50,000 loan to OAEYRG to cover planting and production costs for bixa, pupunha, and Brazil nut; this sum is being paid back through the deduction of 20 per cent from the price paid by Aveda to the Yawanawa for the seeds.

1993–7:

- US$25,000 provided per year, to cover technical studies and consultancy input from ITAL, Instituto Biodinamico, and others.
- Construction of a warehouse, subsequently used as a community school.
- Construction of new warehouse, storage, and processing facilities.
- Purchase of an office and equipment in Tarauaca, including one year maintenance costs.
- Repairing all boat engines and the nine-ton boat in the village (US$5,000); purchase of a 9hp diesel engine.
- Provision of machinery, including de-podder, mechanical dryer, motors, solar dryers, scales, grinding mill, suspended silo for 7.5 tons of seeds.
- Purchase of a 500 square metre lot of land at the conjunction of the river and the BR364 for warehousing and travel facilities.
- Development of a water pasturisation unit.
- 1997 health project costs covered by Aveda: US$37,000.

1998:

- Equipment provided by Aveda (US$23,600).
- Mechanical dryer (US$15,600).
- English language training in the USA (US$24,000).
- Extension of solar electricity system (US$16,500).
- Final inspection of bixa for organic certification (US$5,000).

On-going benefits:

- Provision of machinery and equipment.
- Provision of supplies, including work clothing and boots, tools, silos.
- The salary and expenses of the project coordinator.
- Payment of double the market price.
- Guaranteed purchase.

In-kind benefits:

- Construction of a health centre, including medical and dental equipment and solar energy system.
- Road improvements (Nova Esperanca to Sete Estrelas).
- Installation of a solar energy system for 40 houses; this saves the Yawanawa from purchasing fuel for home use, and diminishes external cash expenditure.
- Facilitation of donations from international relief agencies.
- Facilitation of government agencies' provision of health and other services, such as malaria and cholera relief, and well-building.
- Assisting with commercial networking on behalf of the Yawanawa, eg the agreement between the Yawanawa and the São Paulo company Formil/FloraBrasil.
- Construction of houses and other works to assist in the re-unification of the Yawanawa who were dispersed along the Gergorio River and in cities.
- Training of Yawanawa students in English, marketing, healthcare, law, and administration.
- Travel costs for leaders and members of the council to attend meetings with most of the parties involved in the collaboration, including researchers, consultants (eg Instituto Biodinamico), Food Technology Institute, Aveda research and development personnel, Formil/FloraBrasil, suppliers, and congresses on natural colouring agents.

Collateral benefits:

- Improved contact and standing with government officials, which leads to improved services, quicker response to health threats, etc.
- Improved commercial capacity in the Yawanawa: for example, the Yawanawa are beginning to market their own colorau in two forms – powder for the regional rural market, and another, trade-marked product produced through a collaboration with Formil/FloraBrasil for urban markets like São Paulo.
- Strengthening of the community's economy, and empowering the community to control their trade with the external market.

make use of this relationship in marketing materials, but the ways in which images are used, and the terms under which this occurs, must be clear and agreed upon by the community.

Conservation impacts

The production and processing of bixa is conducted in an environmentally sound manner: the farming and processing systems are organic, as certified by Instituto Biodinamico; farming also involves multi-cropping of a range of species on degraded land as a first step towards reforestation. One of the species included in this system – pupunha – enriches the soil. Perhaps most critically, construction of the Brazilian highway BR364 reached the banks of the Gregorio River in 1997. The Yawanawa will be exposed to pressures to log their land for cash which they might be able to reject with alternative income sources established through the Bixa Project, as well as through awareness raised on the implications of deforestation on the part of community members who have travelled and visited other communities as part of this collaboration.

9.10.2 Cohune oil: marketing a personal care product for community development and conservation in Guatemala – an overview of the Conservation International and Croda Inc partnership

Jennifer Morris[3] and Sarah A Laird

Overview

In 1992, Conservation International (CI) and Croda Inc began discussions on the supply of sustainably-produced rain forest products for the development of a Croda line of natural raw materials. Croda is a bulk ingredient supplier to a number of markets, primarily personal care and cosmetics. Through its CroNatural™ line, it currently sells cohune oil derived from a species native to Central America, the oil of which is used by some communities in cooking and candle-making, but not for personal care. Since 1992, CI and its local partners have developed a marketing programme for cohune oil that includes a range of benefits for local communities, institutions, and conservation objectives.

The species

The cohune palm (*Orbignya cohune*) is a monoecious palm native to the wet Atlantic lowlands of Central America. It is cultivated south to Panama, and in northern South America. Known as *corozo* locally in Central America, it grows abundantly in lowland tropical forest, appearing on all types of soils up to 600m altitude (Duke, 1989). Cohune grows at an average density of 15 palms per hectare. It prospers best with direct sunlight, and has been successfully integrated into reforestation projects (*Informa*, 1979).

Traditional uses

Cohune is a versatile species, with multiple traditional uses. In Guatemala, the leaves and branches are used as roof thatching; the shell is used as fuel, to make handicrafts, and as animal feed; the oil is used for cooking and candle-making. The bulk of cohune use is for subsistence; however, there is also limited small-scale sale of the oil for local use as a cooking oil. There appears to be no history of traditional use of cohune as an ingredient in hair or skin care products.

The commercial product

Cohune oil is extracted from cohune palm nuts collected from the forest floor in northern Guatemala. The nuts are de-shelled, pressed into oil by local communities and then exported to a US-based ingredient manufacturer (Croda) that cleans the oil and sells it to finished product manufacturers. Cohune oil is a non-drying oil with an emollience that is considered finer than coconut oil. It therefore leaves skin and hair soft, and is a useful additive to shampoos, conditioners, soaps and lotions. Companies purchasing cohune oil from Croda to date include: Levlad/Nature's Gate, Equinox, and California Tan. The oil is relatively inexpensive, and coupled with its unique marketing story, its prospects remain positive. In 1998, cohune oil was certified organic by IFOAM-accredited Oregon Tilth. It is intended that certified organic cohune will become a part of Croda's CroNatural™ line in 1999.

The players

- *Conservation International – Guatemala (CIG):* the EcoEmpresas department of CIG seeks to protect northern Guatemala's 2 million hectare Mayan Biosphere Reserve through conservation-based enterprises which economically sustain the local population still dependent upon the forest for survival. Its primary role in this case is to provide on-going technical and business development assistance to Industria Petenera de Corozo (INPECO).
- *Industria Petenera de Corozo (INPECO):* is a registered cooperative based in the community of La Má quina in northern Guatemala. Its primary role in this case is to purchase de-shelled corozo nuts from local collectors, and press them into oil.
- *EcoMaya, SA:* a private marketing and trading company based near the CIG office in Flores, Guatemala, EcoMaya helps cooperatives such as INPECO to package, ship and deliver goods to the market-place. Established by CI in 1996, EcoMaya is currently owned by eight local cooperatives and CIG, which maintains shares in the company. Its primary role in this case is to facilitate the delivery of goods from INPECO to Croda; to receive and distribute payment, and generate new sales.
- *Croda Incorporated (New Jersey):* the US manufacturing division of Croda International PLC, Croda Inc is a leader in the provision of high-quality raw materials and speciality chemicals to many markets, particularly the personal care and cosmetic markets in the USA. Its primary role in this case is to clean and process cohune oil received from EcoMaya; deliver it to customers, and generate new sales.
- *Conservation International – Conservation Enterprise Department (CED):* based in Washington, DC, CED works with local partners to promote ecologically sound businesses and industry practices that contribute to CI's mission of biodiversity conservation. Its primary role in this case is to provide marketing assistance to

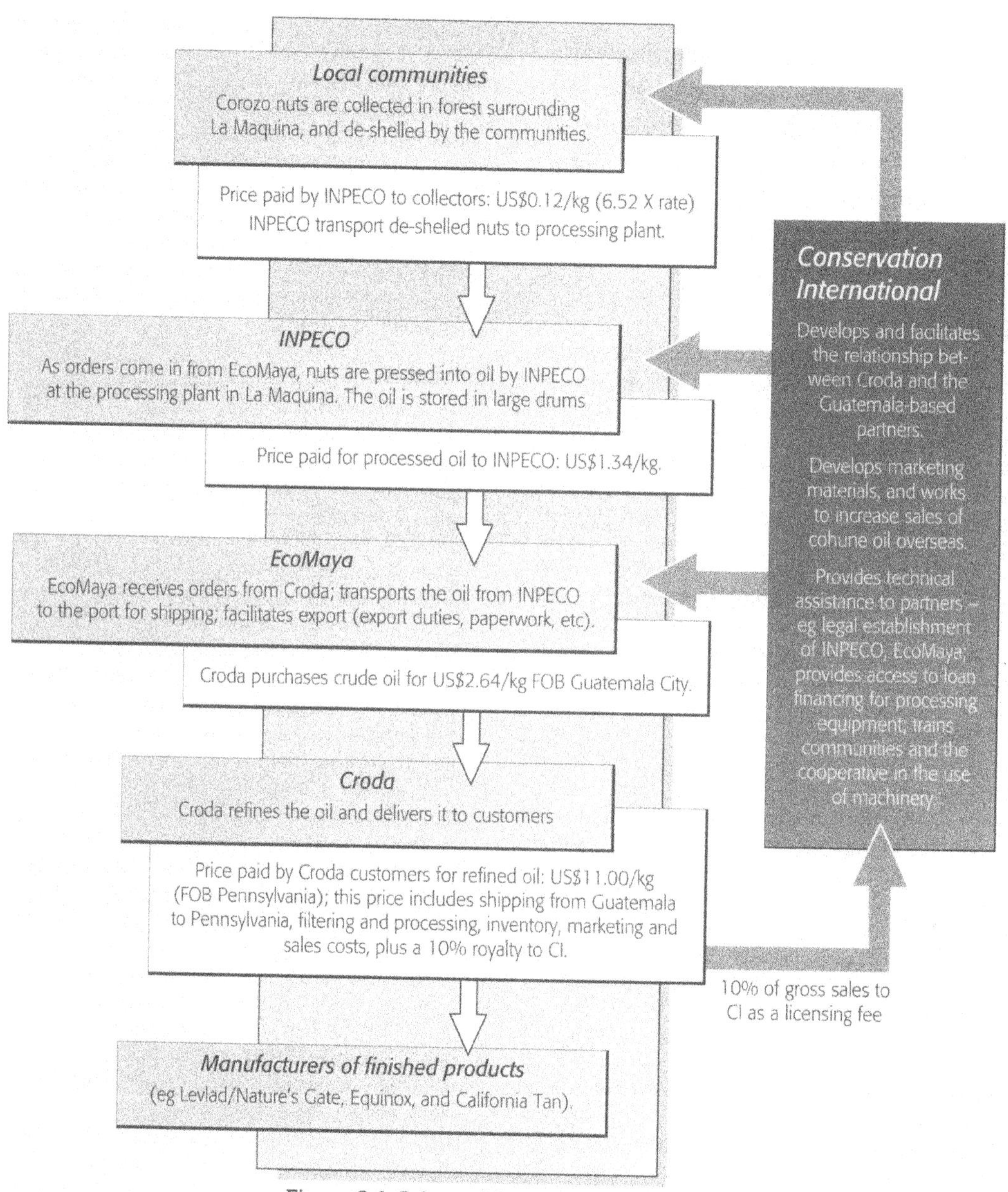

Figure 9.1 *Cohune Oil: Production and Marketing Chain*

EcoMaya and CRODA, and facilitate communications when needed.

The roles of these players are shown in Figure 9.1.

Cohune oil: the development of a commercial product

In 1991, realising the conservation and commercial potential for personal care ingredients found in the rain forests, CI began to search for a private-sector partner who could evaluate and market raw materials deemed by in-country botanists, and local communities, to be viable for commercialisation. Croda Inc was chosen due to its commitment to rain forest-derived ingredients and community development, as well as its strong position in the global market-place which gives it access to leading personal care companies.

At the same time, CIG identified cohune nuts as a possible basis for an enterprise which would give local people the economic incentive to protect their natural resources. Since oil-press technology and experience existed locally, cohune oil production was attractive as a potential source of local employment based on a sustainable enterprise. Initial resource surveys determined that individual trees were sufficiently abundant in the wild to support expanded use of cohune products. As a result, samples of the oil were sent to Croda laboratories for testing. Test results showed that the properties of the oil, with a similar fatty-acid distribution to coconut oil, demonstrated positive potential as an ingredient for use in personal care products.

Croda Inc first became interested in cohune oil as a potential personal care product based on scientific studies undertaken in the laboratory, rather than through local traditional use. As a result, the collaboration with local communities generates benefits associated with production of the oil, rather than those that might be tied to commercial use of traditional knowledge connected to the product.

Croda-CI agreements

Between 1992 and 1994, CI entered into a series of agreements with Croda Inc detailing a relationship in which CI and affiliated local partners would supply Croda with raw materials for testing, manufacture and resale. In addition to promoting the market for cohune, Croda also agreed to provide a royalty of 10 per cent of gross sales to CI for the use of CI's name in marketing materials and to contribute to CI's conservation activities.

In the first Product Assessment Agreement (1992), Croda agreed to test all new raw materials within three months to determine commercial viability. During this three-month period, Croda retains exclusivity over the samples. However, if samples are not found to be commercially viable, CI regains all usage rights over them. Samples for the personal care industry (eg essential oils, resins, waxes, and botanicals) can degrade relatively quickly, so timing is critical. In 1992, Croda tested seven new raw materials, of which five, including cohune, were selected as commercially viable. These five ingredients became known as the CroNatural™ line. Three product ingredients are from Guatemala (cohune, allspice and jaboncillo), one from Peru (Brazil nut oil), and one from the Solomon Islands (ngali nut oil).

The 1992 agreement was revised in 1998, with Croda being held responsible for reimbursing CI for sample collection time and expense should Croda fail to test the samples within the agreed time. This change occurred due to sample spoilage under the first Product Assessment Agreement. In 1998, eight new products were tested by Croda's recently-acquired French company, Sederma, with preliminary test results not yet completed (early 1999).

In 1992, after Croda indicated its desire to commercialise cohune oil, purchase agreements were developed to solidify expectations of both Croda and CI-affiliated suppliers. During this early stage, INPECO and EcoMaya did not yet exist. The machinery for oil production was located near the CIG office with future INPECO members and CIG staff involved in the oil's production. Since INPECO was not yet legally-established, the supply agreement was developed between CIG as the supplier and Croda as the buyer. In these early agreements, Croda agreed to purchase a specified amount of product each year, based on strong sales projections indicating that Croda would have little difficulty selling its inventory of cohune.

The third agreement between Croda and CI signed in these early years dealt with promotion, marketing and the use of the jointly-owned logo for the rain forest line, called Renewable Rainforest Resources™ (RRR). For Croda's use of CI's name under the promotion of the RRR logo, Croda paid CI 10 per cent of gross sales of the CroNatural™ ingredients to end-product manufacturers. The RRR mark remains the only logo jointly-owned by a non-profit and private-sector ingredient manufacturer which can be used by personal care companies with minimum purchases of CroNatural™ ingredients. The initial idea behind the use of the RRR logo was for personal care companies interested in demonstrating their use of sustainably-harvested ingredients to display this commitment visually by using the logo on product packaging. The logo is currently being used by several companies, including Equinox (in their Rain Forest Resources™ line) and the Rain Forest Company (in their Source™ line).

Benefit-sharing

Benefit-sharing in this case is largely intended to promote a sustainable and alternative source of income based on the forest resource, for communities living around the Maya Biosphere Reserve in Guatemala. As a result, most of the benefits are channelled either directly (ie fees and employment associated with local collection and processing of nuts) or indirectly (ie royalty fees paid by Croda to CI which contribute to technical assistance and general programming) to serve conservation objectives.

The development of the commercial potential of cohune oil to date has included strong institutional and capacity-building on the marketing front, particularly in the establishment and development of INPECO and EcoMaya. Improvement in business skills has been significant, and results in part from the shared nature of the risk and investment made by NGOs, local communities, and private-sector partners.

Capacity building in the scientific and higher technical levels is also underway. A local Guatemalan university has collaborated with CIG to develop a Material Safety Data Sheet (MSDS) for cohune, and has determined the basic fatty-acid chain of cohune oil. This is an important step towards greater independence of INPECO/EcoMaya marketing capacity. It is intended that in future there will be greater involvement of local research institutions and universities as collaborators in INPECO/EcoMaya marketing initiatives.

The benefit-sharing aspects of the Croda-CI partnership are outlined in Box 9.4, p291.

Conclusions

Green marketing: 'Green' marketing is a valuable tool that can more closely tie the commercialisation of natural products to benefits for local communities, and the conservation of biodiversity. In addition, it can improve a company's market position by appealing to customers concerned with 'green' issues. However, the key to long-term success for the international marketing of forest products is to not rely solely on green consumerism.

Unique properties and active ingredients: The products most likely to generate benefits over time for local communities and conservation programmes have unique properties or active ingredients that can sustain the vagaries of demand for green products, and will be valued by end-product manufacturers irrespective of the community or environmental story. These kinds of products are few and far between, however, so commercialisation activities linked to conservation programmes should explore a package of products, with varying market – as well as social, cultural, and ecological – profiles.

Diversification of end uses of raw material: Due to the risks involved and the often short life-cycle of raw materials in the personal care market, it is vital for local suppliers with large resource and monetary investments in one product to diversify the potential end-product applications for this raw material. For example, a Brazil nut supplier in Peru – Candela, SA – is using every part of the Brazil nut in an attempt to minimise reliance on one application. Candela sells Brazil nuts as a food product, presses the nuts into oil for the personal care industry, and fills the nut pods with scented wax to make aromatherapy candles. In 1998, Croda developed a protein using Brazil nut meal which is used by Aveda to replace quaternised hair keratin protein in five Aveda products.

The importance of local partnerships: While many companies source raw materials from developing countries, few source directly from community-based enterprises, which can be challenging due to poor communications, limited transportation networks, language barriers, and other factors. EcoMaya is an example of an invaluable link between isolated communities and private partners. Based in a town, and equipped with phone, fax, email and bilingual staff, EcoMaya plays a crucial role in the facilitation between supplier and purchaser, providing private-sector companies with a greater level of comfort than if they were to purchase direct from the communities. However, before an intermediary relationship such as this is pursued, careful analysis must be made to determine if the intermediary's margin allows for maintenance of a competitive price for the product. Cohune's price remains positive, though cost controls are consistently monitored to ensure long-term competitiveness with the inexpensive substitutes prevalent in the market.

In addition to the importance of marketing intermediaries to the long-term success of a commercial partnership of this kind, local research institutions should also be actively involved. Universities and research institutes can conduct analyses of raw materials important for commercial customers. These institutions can build domestic technical and scientific capacity to work directly with companies, and will thereby enhance a community or local-level group's independence. A wider base of domestic capacity upon which local suppliers can draw allows them to market products more effectively through a number of different channels.

9.11 Conclusion

The personal care and cosmetics industry incorporates increasing quantities and varieties of natural ingredients into products. This is primarily a response to consumer demand for natural and therapeutic products, but is also in part an expression of increasing scientific and technological capacity to analyse and understand natural products. While consumer demand has led to expanded use of natural ingredients, scientific, technological and regulatory trends have in many countries converged to ensure that natural ingredients are better defined and tested.

As we have seen, there is an enormous range in research, development and marketing approaches within this industry, with resulting dramatic differences in the ways that companies seek access to natural products and share benefits. Some companies 'prospect' for new leads to run against mechanism-based

Box 9.4 *Benefit-Sharing: the Case of Cohune*

	Inputs	Benefits – monetary	Benefits – non-monetary
INPECO	Labour Equipment	Payment per supply of processed oil (US$1.27/kg). Employment (approximately 30–40 collectors, plus 24 members of the cooperative manage the cracking and pressing of cohune nuts). Shareholders in EcoMaya. Asset-generation through equipment.	Conservation of corozo palms, considered valuable for forest management in the Maya Biosphere reserve. Development of local industry. Technical assistance provided by CIG. Training in machinery use and oil production. Building of local cooperative.
EcoMaya	Labour (staff of three). Marketing of cohune oil (attendance at trade shows; production of marketing materials).	Receives a margin from sale of the oil to Croda.	Building of local capacity in trade and marketing of non-timber forest products. Developing expertise in supply-side relationships with international importers.
Conservation International	Two-year pilot project which included purchase of machinery, employment of local communities to process oil; working capital to buy nuts; facilitation of transport; development of raw material storage and processing facilities. Assisted with legal establishment of INPECO and EcoMaya. Provided technical assistance, including training in machinery use and business development. Developed FONDOMAYA as a source of micro-financing for INPECO. Obtained grant financing for facility upgrade.	10 per cent of gross sales from Croda's CroNatural™ line.	Helps to achieve a conservation objective of CIG, to protect areas within the Maya Biosphere Reserve containing corozo palms. Provides eco-economic alternatives to Peten communities – another primary objective of the CIG programme. The project has yielded numerous lessons and insight that might be applied to other projects in Guatemala and elsewhere. Capacity building for staff in appropriate technologies for oil production.
Croda Inc	Testing a range of new materials for efficacy and potential use. Marketing and storing products. Provision of technical assistance (eg how to ship oil without spoilage). Royalty contribution – 10 per cent of gross sales – for use of the CI name, logo, and story in conjunction with sale of products.	Receives a margin on sale of products to end product manufacturers.	Marketing benefits of association with a non-profit environmental organisation.

screens, as part of their search for therapeutic products. Others seek out new ingredients or re-formulate known natural ingredients as part of a strategy to achieve products as close to 100 per cent natural as possible, including replacement of synthetic and petrochemical-based preservatives, fragrances, and processing agents. Many companies merely source raw materials to add – purely for marketing ends – a natural sheen to what are otherwise primarily synthetic or petrochemical-based products.

Demand for natural ingredients is likely to continue in this industry, particularly for the more serious research and development programmes, but widespread use of natural ingredients could be tempered by regulatory forces, and should consumers lose interest in things 'natural'. Demand for access to 'new' natural leads from geographically diverse sources is not significant and, if necessary, most companies can acquire what they need at home or through traditional distribution channels.

Benefit-sharing associated with natural product use is not widely practised in this industry. The link between access to new samples – whether for screening, formulating into products as active ingredients, or as marketing 'fluff' – and benefit-sharing is almost always severed in practice, and generally in the minds of company staff, as well. A few companies selling to alternative markets have established fair trade, organic, sustainable, or other sourcing partnerships that involve a package of benefits for local communities. But these partnerships usually make up only a tiny fraction of the company's raw material sources. Many companies provide grants or donations through their charitable arms, some to community or conservation projects in areas from which they acquire traditional knowledge or raw materials for product development.

Companies openly publicise their innovative 'prospecting' strategies for natural products. Photos of company founders meeting with indigenous peoples to talk about new plant ingredients, annual reports detailing research staff's scientific expeditions on the high seas, labels extolling the virtues of a traditional use that led the company to a favoured ingredient, are common in this industry. To a greater extent than most of the sectors covered in this book, the natural personal care and cosmetics industry makes prominent use in its product marketing of the diverse sources of its raw materials, and traditional use associated with ingredients. However, few of these companies have heard of the CBD. Steps should be taken to increase awareness of companies' obligations under the Convention, and to develop industry standards for benefit-sharing 'best practice' associated with commercial product development.

Notes

1 'This list is the result of information provided by company executives, analysts, and reputable published materials. *NBJ* has made every effort to be accurate, but figures are not the result of audits and are therefore not guaranteed to be accurate. Errors and omissions are unintentional.' (*NBJ*, 1998.)

2 May Waddington is a staff member of the Aveda Corporation, based in Rio de Janeiro, Brazil; she has been involved in the collaboration between the Yawanawa and Aveda since its inception in 1993.

3 Jennifer Morris is the Manager of Wild Harvest Products in CI's Conservation Enterprise Department. Her product area includes personal care and cosmetic products. In collaboration with EcoMaya and Croda Inc, she markets cohune oil to finished-product manufacturers.

Kerry ten Kate and Sarah A Laird

Chapter 10
Industry and the CBD

10.1 Introduction

The Convention on Biological Diversity (CBD) is a treaty between governments, but it is of central importance to business. Provisions of the Convention and national laws that implement them set the scene for any company or individual seeking access to samples of plant, animal, or microbial origin for scientific research or as the starting point for commercial development. And just as the CBD is important to business, so the involvement of the private sector is essential for the successful implementation of the treaty.

The CBD seeks to integrate conservation and sustainable use, and calls for the fair and equitable sharing of the benefits arising from the use of genetic resources. As we have seen in the chapters covering specific sectors of industry, the private sector is a major user of genetic resources. Furthermore, participation in scientific research conducted by industry and the transfer of technology developed in the private sector are among the most effective ways to share benefits. The views of the private sector on the CBD, and the nature of commercial partnerships will inevitably inform the manner in which benefits are shared in practice. They will also influence the extent to which biological resources are used sustainably and whether this use will create incentives for conservation.

Finally, motivation on the part of business to comply will be essential for the success of access legislation for three main reasons. First, the monitoring and enforcement of access legislation and agreements is difficult, since it involves tracking and identifying the source and date of collection of specimens and also a product's movement through the discovery and development pipeline. Second, law and procedure on access is unclear in the vast majority of countries. Third, user countries show little inclination to introduce laws to support enforcement of access agreements in the countries where companies conduct their research and development. Consequently, voluntary compliance by industry will be essential, if the principles of prior informed consent (PIC) and the fair and equitable sharing of benefits are to be realised.

The business community would like access regulations to be more practical to implement, and governments and NGOs would like companies to adhere to the objectives of the CBD. In order to improve the current situation for both sides, it is important to assess, as this chapter sets out to do, the extent to which companies are aware of the CBD and its objectives, and the manner in which it has influenced business practices to date.

Gradually, the CBD is becoming better known in the business community, and is beginning to influence the nature of commercial partnerships involving the use of genetic resources. At the same time, governments regulating access are learning more about the commercial use of biodiversity and the needs of business. However, progress in finding practical solutions on access and benefit-sharing has been stymied by a number of factors.

First, the debate on 'bioprospecting' at the international level has become highly polarised and rhetorical. A quick sketch in exaggerated and oversimplified language would present two entrenched positions as follows. Source countries and NGOs view all companies and academic collectors as 'biopirates' and all source countries and communities as their vulnerable victims. On the other side, a caricature of business' perspective is that the CBD is driven by grasping politicians who over-estimate the value of their genetic resources. Greedy governments will hold companies to 'ransom' by withholding their PIC for commercialisation once companies have invested hundreds of millions of dollars in research and development, and not a penny of the 'benefits' shared would help conservation or the people of the country.

Looking on the bright side, this polarised debate has raised awareness of the historical inequities in the genetic resources trade, which may help to prevent the same situation arising in the future, and it has led to the publicising and castigation of some contemporary inequitable activities, some of which have subsequently been rectified. However, it has principally created an atmosphere of distrust, and through exaggeration and ill-founded propaganda, has led some governments, companies and NGOs to disengage from a debate which would benefit from their involvement.

A significant minority of companies still does not agree with the principle of sovereignty over genetic resources enshrined in the CBD. As one company put it, typically confusing access to genetic resources with patent rights over novel innovations, 'What's in nature should be available freely. You shouldn't have patent-type laws which apply to natural organisms. You can't patent an organism from the wild.' Several agricultural companies felt that individual governments should not have 'exclusive' rights over plant genetic resources. As another company put it, 'Countries are practising "genomic imperialism". They operate on the basis that one gene in a plant is yours, so the whole plant is yours.'

Another major barrier to progress has been the lack of factual information on which to base negotiations. First, there is not enough information on the practical challenges of regulating access, the manner in which industry uses genetic resources, or the diversity and nature of partnerships leading to the commercial use of biodiversity. Second, industry is ill-informed about the CBD and access legislation, and, as our interviews have revealed, industry's opinion of the CBD and companies' strategies for complying with or avoiding it, are often based on second-hand, inaccurate information. Many interviewees referred to the low awareness and understanding of the CBD in their respective sectors. They attributed this, in part, to the sparse coverage of the CBD – and, specifically, of its provisions on access and benefit-sharing – in professional publications; the lack of information explaining the steps they need to take, and the striking uncertainty surrounding the procedural and substantive requirements on access in most countries of the world.

The obvious potential impact of the climate change negotiations on the energy sector ensures the involvement of industry bodies in those negotiations, but the CBD is more indirect and subtle in its repercussions, and companies frequently do not see the relevance of the CBD to their work. As Edgar Asebey, President of Andes Pharmaceuticals, puts it: 'There has been a striking lack of private sector involvement in the CBD policy process. The CBD has potentially huge implications for the private sector, as well as for the environment, but the private sector is all but absent from the discussion. The national measures [on access] have teeth on them, so companies should be involved. But there has to be an incentive for the private sector to participate in these discussions.'

The lack of involvement of industry is a problem for several reasons. First, access and benefit-sharing laws have been drafted without the benefit of suggestions from industry as to how they might work in practice. Consequently, the laws that have resulted simply put companies off working in the countries where they have been introduced. For example, the Philippines Executive Order and Implementing Rules and Regulations is currently under revision, since many scientists in the Philippines and foreign companies found it unworkable. One reason was the bureaucracy and cost involved in complying with the procedures it introduced. Another was the concern by companies that the Order would require them to license technology to the Philippines. But the experience in the Philippines, often cited as the bête noire of access laws, does have a positive message. The authorities in the Philippines have now actively sought feedback from interested parties. They have explained that the Order sets a framework that is intended to be flexible, and that benefit-sharing requirements (including commitments to license technologies) will be negotiated on a case-by-case basis, as appropriate. They are also redesigning the administration of the access procedure, for example by decentralising some aspects of access regulation. Philippines universities can now qualify for academic research agreements by negotiating a Code of Conduct with the authorities, and committing to abide by it.

This chapter sets out to assess the level of understanding of the CBD by the private sector, and its commitment to the provisions of the CBD and national laws on access and benefit-sharing. It starts by comparing the level of awareness and understanding of the CBD in different industry sectors, and considers whether the CBD is seen as a positive or negative development from the private sector's perspective, and why. It then relates companies' experiences when obtaining collecting permits and complying with access legislation, and examines the different approaches that companies adopt to ensure legal title to the materials from which their products are developed. The chapter concludes with a discussion of how companies are responding to the access and benefit-sharing provisions of the CBD and access legislation; how corporate behaviour is changing in its wake, and what kind of informal and formal strategies and policies companies have adopted.

10.2 Companies' awareness and understanding of the CBD

Overall, 74 per cent of the 185 interviewees who answered this question said they had heard of the CBD (Table 10.1). Awareness was significantly lower in companies in botanical medicine, personal care and cosmetics and horticulture than in pharmaceuticals, biotechnology, the seed industry and crop protection. One reason for this may be that early discussions on equitable relationships involving the exchange of genetic resources started well before the CBD, but focused on the activities of companies in the pharmaceutical, biotechnology, and agriculture industries. These industries are now relatively familiar with the concepts of prior informed consent (PIC) and benefit-sharing. As one representative from a pharmaceutical company put it, some companies feel they were 'probably exceeding the mandate of the Convention on Biodiversity before it was written'.

limited that it would not enable them to understand the implications of the CBD for their daily business. For a few companies, this was a source of concern. As the UK Agricultural Supply Trade Association put it, 'We are worried. We do not know the details of [the CBD], nor how it will affect us.' Another company in the seed industry said that the CBD was causing trepidation, as companies believed it would become important, but did not understand why.

Most of the people we interviewed are aware that the CBD promotes conservation, but few could articulate its objectives. Several, therefore, could not see the link between their business and the CBD. For example, a natural products researcher at a large pharmaceutical company said: 'Pharmaceutical research is not connected to conservation ... Natural products research decisions have nothing to do with the CBD – there is no link whatsoever.' Many companies believe the CBD concerns phytosanitary and export permits, or – a common misunderstanding in companies in the botanical medicine, and personal care and cosmetics sectors – that the

Table 10.1 *Companies' Views on the CBD*

	Proportion of interviewees who had heard of the CBD			Proportion of interviewees who thought the CBD was a positive, negative or neutral influence			
Industry sector	Sample size (answered question)	% who had heard of CBD	% who had not heard of CBD	Sample size (answered question)	% who felt CBD was 'positive'	% who felt CBD was 'negative'	% who felt CBD was 'neutral'
Pharmaceuticals	26	96	4	24	83	8	8
Botanical medicines	21	62	38	12	92	0	8
Major crops	32	94	6	27	44	26	15
Horticulture	30	63	37	24	46	25	29
Crop protection	24	92	8	18	67	44	22
Biotechnology	31	90	10	24	58	21	33
Personal care	22	36	64	8	100	0	0
Total	186	(ave) 74	(ave) 26	137	(ave) 63	(ave) 19	(ave) 18

General awareness of the CBD, however, does not signify understanding. Judging by answers to our questions on: how the CBD has influenced the activities of each company; how a company can be sure that it has good legal title to samples of genetic resources acquired for research; and what policies the company has developed concerning the CBD, it is clear that, six years after the entry into force of the CBD, a large proportion of companies have only a very rudimentary grasp of its objectives and articles. Many interviewees heard about the CBD indirectly, frequently from organisations providing them with genetic resources. The information they received was often inaccurate, inconsistent, and so

CBD is exclusively concerned with the collection of endangered species and the overexploitation of species in the wild. Several companies could not distinguish between the CBD and CITES, and were unable to elaborate about either treaty. Many more were unaware that the CBD covers access to traditional knowledge, as well as to plants and animals.

With the exception of the pharmaceutical companies we interviewed and some better-informed companies in other sectors, few interviewees were aware of the third objective of benefit-sharing. They were rarely familiar with the language in Article 15 of the Convention, which balances sovereign rights over genetic resources, PIC and benefit-sharing with a positive obligation to facilitate access, on mutually agreed terms. Many companies do not distinguish between access to genetic resources and IPRs, which, while related, are two distinct issues.

Several interviewees repeated a number of fairly common misunderstandings about the scope of the CBD and its access and benefit-sharing provisions, including which activities on the part of companies triggered its access provisions. Some of the more common 'myths' are as follows:

Scope:

- The CBD does not apply to our industry if we do not use endangered species or over-exploit raw materials.
- The CBD does not apply to materials with no known value, derivatives such as proteins or compounds, microcorganisms, or *ex situ* collections.
- The CBD does not apply to common species with wide distribution.

Activities regulated under the CBD:

- The CBD has no implications for organisations that do not collect genetic resources themselves.
- The CBD does not apply to academic research. Only commercial benefits need be shared.

10.3 Companies' opinions of the CBD

Predictably, corporate opinion on the CBD and national access and benefit-sharing measures is mixed, ranging from the uninformed and indifferent, through the negative, to the positive. However, even companies that are positive about the CBD are concerned about the direction that national access and benefit-sharing measures are taking, and the feasibility of their implementation. For example, although 83 per cent of pharmaceutical industry respondents felt that the CBD was 'positive', this was generally qualified to mean 'in principle'. And it is noticeable that, over the two-year course of this study, many companies have come to believe that the implementation of the CBD has gone badly wrong. As Reimar Bruening, Director of Natural Products at Millennium Pharmaceuticals, said,

> *The CBD has had an overall counterproductive effect on natural products research. Expectations are unrealistic, administrative hassles and red tape overwhelming, and if you can get to drug leads in other ways, why still bother to do inherently expensive natural products research in such a politicised environment? From the beginning, the big companies were seen as thieves that come at night and take things away. There exists the totally unrealistic notion that pharmaceutical companies routinely make billions of dollars from natural sources.*

Some of the main reasons why companies viewed the CBD as a 'positive' or 'negative' influence are summarised below.

10.3.1 Positive views on the CBD

As outlined in Table 10.1 above, the majority of individuals interviewed in every sector who had heard of the CBD believed it was a positive development, in that it:

- promotes equitable relationships;
- promotes the conservation of valuable *in situ* and *ex situ* resources;
- potentially clarifies issues relating to ownership and procedures to obtain access to genetic resources; and
- supports best practice in industry.

However, most companies do not believe that the CBD has much of an impact on their work, so most of their positive sentiments relate to the broad goals of the Convention and the contribution they feel it could make in the future. A few interviewees said they could already notice benefits brought about by the CBD in their work, although, as we will discuss below, the majority felt that the current impact is negative.

The companies who responded positively to the CBD tended to be those most engaged in the policy debate, and those involved in collaborative research partnerships with providers of genetic resources. Some in this category found that their willingness to embrace the CBD conferred a competitive advantage, since it enabled them to secure access to genetic resources and removed any risk surrounding their right to develop commercial products from these resources. Some companies operate under a 'highest common denominator' principle, offering the same kinds of benefit-sharing arrangements in countries that do not have access legislation as in those whose laws introduce stringent benefit-sharing requirements.

Many felt the CBD could assist in the development of best practice for industry. One crop protection company representative looked forward to guidelines for companies, so they could 'know how to share with developing countries'. Others felt the CBD created a more level playing field, balancing inequalities in bargaining positions. As one company based in an OECD country put it, 'Unfortunately, the multinationals exploit the exchange between the developing and developed countries. The CBD is valuable that way'. Several companies expressed concern at the increasingly common charges of biopiracy, and feel that the CBD could help clarify the terms for relationships, prevent them from being wrongly accused of 'ripping off' providers of genetic resources, and help them to gain a positive public image.

Some interviewees felt that the CBD requirement for PIC would help companies avoid the risk of law suits concerning the ownership of genetic resources or the right to exploit them. Intermediary companies and organisations report that customers increasingly require some guarantee of clear ownership or rights of exploitation, which will be easier in the post-CBD world. Some companies also felt that the CBD process provided an important opportunity to shape the legal and policy framework within which business must operate.

A number of interviewees felt the CBD offered the hope of finally removing a great deal of the past confusion surrounding permitting procedures and ownership of samples, enabling them to be quite sure, at last, that they were complying with relevant laws. Looked at in one way, the CBD has created new legal obstacles for companies, but in another way, it reflects broad international consensus on the need for equitable partnerships that companies would inevitably face in their dealings with business partners around the world. A minority of the companies and organisations interviewed believe that the CBD and access legislation are helping to clarify the situation. They suggested that access legislation is getting a worse press than it deserves. As Genencor put it, abiding by the access regulations introduced by countries may 'sound alarming in the abstract, but need not be so in practice'.

Finally, a few of the companies interviewed believe that the CBD is helping to raise awareness of the importance of conservation, of the dangers of the loss of genetic diversity, and consequently of the need to promote the conservation of biodiversity *in situ* and to enhance *ex situ* collections of genetic resources. Some interviewees believed that regulating access to genetic resources would promote conservation, and it would help to stop uncontrolled and unsustainable collection. Others felt the CBD offers opportunities to seek funding for projects contributing to conservation.

Some of the key positive influences that the CBD could have on business are summarised by David Corley, Natural Products Chemist at Monsanto:

> *The CBD is a very positive development. I think that in the long-term it is the only way companies can maintain their freedom to operate. It allows us an understanding of the rules – a level playing field – and provides us with the grounds of credibility. The CBD means that we will not be challenged in world court for 'stealing' something – it allows us to source materials in a legal way ... . Countries have something companies want, and there must be a fair exchange in order to acquire access.*

10.3.2 Negative views on the CBD

The majority of company representatives interviewed expressed serious concerns about the CBD and the national access and benefit-sharing legislation it has spawned. Most of the positive comments about the CBD related to the results it might achieve broadly for society in the future, but the negative comments from companies were largely based on their experiences to date, and fears that the CBD would impact on their daily business.

Central concerns focused on:

- lack of clarity in the CBD and national access and benefit-sharing measures;
- difficulty in keeping up with the plethora of different measures adopted by countries;
- bureaucracy and delay involved in following access procedures;
- unrealistic expectations on the part of governments and provider country institutions;
- a perception that the CBD rejects scientific traditions of research collaboration and exchange of specimens, and that this will damage research on biodiversity and new product development;
- belief that the CBD and access legislation create an additional disincentive to conduct natural products research, which is already under pressure from new technologies and market trends; and
- unreasonable transaction costs.

The three major groups of concerns expressed by many companies are first, the lack of clarity concerning access rules; second, the bureaucracy and transaction costs involved in following them; and third, the lack of understanding of the role of business on the part of regulators and institutions providing access to genetic resources.

Lack of clarity, a policy vacuum and the resulting impasse

By far the most common complaint voiced by interviewees about the CBD and its implementation is the lack of clarity in individual countries as to procedures to obtain access. There is a strongly held view that countries do not make it easy even for well-intentioned companies to do the right thing, and huge frustration that countries are making demands of scientists and companies which they cannot fulfill, since it is impossible to find out who is the access authority in many countries. Companies from all around the world feel that the CBD has created obligations in circumstances where the procedures do not exist to comply with them, and that while the CBD and access laws are fine in theory, there is no practical way to implement, monitor and enforce them. Several spoke of not seeing how the principles of PIC and benefit-sharing could be translated into reality.

As one executive at a small pharmaceutical company explained: 'Countries we tried to work with in Latin America did not have a permitting process, no rules to follow, basically no way for them to say "Yes". It makes it very hard for companies to work there ... the absence of any means to implement the CBD creates the problems. I would like to think we are beyond debating the basic ideas – now we need protocols.'

Another researcher from a small pharmaceutical company said, 'The CBD gives national governments sovereign rights. They should be figuring this stuff out, or should delegate the responsibilities ... . Right now it is not clear. The access and benefit-sharing discussion at a national level is not really moving, and companies and collectors are operating in a vacuum ... . We don't know where to go for permission to collect ... . We are trying to push the government, because we want guidelines to direct us in our collections.'

The same problem is faced by companies hoping to obtain specimens from the country where they are based. For example, Kodzo Gbewonyo, President of Bioresources International, a Ghana-US company, said, 'Governments should specify what they want done. Companies do not know where we stand, and it makes us nervous. It makes it difficult for us to reassure our collaborators ... . We need a national level policy to guide agreements, including where benefits should go to with regards to healers, conservation, and so on ... . Companies and collectors are operating in a vacuum.'

Many companies said that they simply did not know where to turn to for accurate information on the rules pertaining to access in the vast majority of countries. Some rely entirely on their collaborators in the countries concerned, and choose not to ask too many questions. Others felt they simply could not rely on partner institutions, who were often unaware of the issue, poorly informed, and anxious to secure research collaborations at any cost. As Dwight Baker, Senior Project Manager in Drug Discovery Services at MDS Panlabs put it: 'Many of the in-country institutions are so poor that they will say anything to get some cash, and it is difficult for companies to know what is required.'

Bureaucracy and transaction costs

In countries where clear access rules do exist, interviewees often complained that the rules were too cumbersome for them to wish to work there. The vast majority of companies and other organisations interviewed revealed a 'low pain threshold' for the procedures needed to secure legitimate access to genetic resources and complained about the bureaucracy, inefficiency, delay and transaction costs involved in following national requirements on access and benefit-sharing. Companies are particularly troubled by what they perceive as a growing divergence between access measures introduced by policy-makers and what is feasible on the ground. As one representative of a biotechnology company said, 'I wish the politicians and diplomats who write these policies would go out in the field and try and get prior informed consent.'

Several companies pointed out that the costs involved in following national access procedures – in terms of benefit-sharing commitments, travel, communications and staff time – made working in those countries uncompetitive. However, others were unmoved by the prospect of higher costs for access, saying that they would simply pass on any extra cost to the consumer. Some companies involved in the better publicised access and benefit-sharing agreements (SCBD, 1998) reported that these negotiations had taken between two and three years to conclude. This was fine for a 'one-off' experimental project, but could not be the norm for the company. They thus believed that these partnerships could not be used as benchmarks that industry should replicate elsewhere.

Several companies referred to 'a quagmire of government approval' and many are concerned that this will get worse, rather than better, as growing numbers of countries introduce different measures on access and benefit-sharing. Access to agricultural genetic resources might prove an exception, if countries are ultimately able to agree a revised International Undertaking on Plant Genetic Resources for Food and Agriculture. A number of interviewees stated that it was beyond a company's capacity to familiarise itself with the plethora of laws. This posed a particular problem for those working in several locations, and for smaller companies, for whom access regulations 'will be a nightmare'. Several expressed the wish – a wish which currently appears to be politically unrealistic – for countries to harmonise their laws on access. As Dwight Baker, Senior Project Manager of MDS Panlabs suggested, 'Rather than country by country laws, a unified set of standards should be developed which apply across the board. It will be difficult to get countries to agree to this, but companies need to know what the rules are.'

Instead of encouraging equitable partnerships, many interviewees feel that the effect of the CBD has been to restrict science, and that countries' access laws run counter both to the obligation in Article 15(2) for countries to facilitate access to their genetic resources, and to their obligations under other articles to promote research, sustainable use and conservation. This effect is felt not only by institutions hoping to collect genetic resources on field expeditions, but by those who receive materials sent from around the world by institutions with whom they have collaborated for years. A seed company representative felt that the effect of the CBD would be to disrupt and interfere with companies' relationships with many universities around the world, and feared this could affect academic research.

The overriding view is that the effect of permitting procedures and access regulations is to restrict access to, or exchange of, genetic resources. Several company representatives said that, since entry into force of the CBD, certain groups, such as botanic gardens, are no longer providing access to their collections, or only doing so under very strict terms. They felt that the CBD could only slow down access and 'require more people to be paid'. To compensate for this expense, products would become more expensive, damaging the competitiveness of industry. In these circumstances, some researchers might be encouraged to break the rules. As Reimar Bruening, Director of Natural Products at Millennium Pharmaceuticals, said,

> *Had the CBD been done in a different way, it could have worked well – positive publicity, foundations set up to provide funds to villages where things were discovered, research collaborations with local educational centers. Now funds go to opaque government channels and other bureaucratic places where companies have very little influence over the effectiveness and whereabouts of their investment.*

Lack of understanding of business

The third common concern shared by companies is that regulators and providers of genetic resources do not understand the interests, activities and proper role of business. According to our interviewees, the most common misperceptions harboured by governments and institutions providing genetic resources are:

- the conviction that 'raw' genetic resources are more valuable than companies believe them to be, leading to unrealistic demands for up-front payments;
- the feeling that business should be involved in or pay for conservation and strive for social equity, when many companies do not see this as their role; and
- the belief that companies would be prepared to engage in lengthy procedures at the national and local level to obtain access and to share benefits, when in fact it is not cost effective for them to do so, and their interest in access is limited.

There is widespread concern within business that continuing negotiations under the auspices of the CBD and FAO are based on a misunderstanding of the costs, risks and timelines involved in product development, and a lack of awareness of the range of different markets and approaches to product development. As a result, many respondents felt that providers of genetic resources overvalued the 'raw materials' they sought to provide to companies. Consequently, countries demanded monetary and non-monetary benefits that companies considered unrealistic, and misinterpreted aspects of past partnerships, citing them wrongly as precedents. Several interviewees were keen to explain that the much cited US$1 million payment to INBio by Merck covered a specific, long-term workplan, and was not, as it is commonly believed to be, the equivalent of an 'access fee'. As one interviewee from the horticulture sector put it: 'In some ways, the CBD is quite naïve. It suggests that the key to countries' wealth and prosperity lies in the [genetic resources of] their rainforests, but in reality this is not true.'

This misunderstanding of markets and the role of business has led to poorly designed access laws and unreasonable expectations for partnerships. Frustrated by this lack of understanding, some companies have withdrawn from international and national discussions on these issues, including the CBD and FAO policy processes. One natural products researcher from a large pharmaceutical company said, 'I went to these meetings on the Convention on Biodiversity, and I am clear on the issues – prior informed consent, benefit-sharing, and so on – but not at all on how they are supposed to work in practice. There are no guidelines, no standardised procedures.' Companies have also found that a large element of the transaction costs associated with access can be attributed to the time involved in communicating basic information about product development to potential partners, and dispelling an atmosphere of hostility and suspicion.

Companies frequently warned that this atmosphere is creating a disincentive for the very research which can create benefits to be shared. As one executive at a small pharmaceutical company said,

> *Prior to the CBD, there existed difficulties with natural products research – re-supply problems, difficulties in isolating chemicals, others areas that are more 'sexy' like genetics and combinatorial chemistry – but the CBD adds another reason for companies to cut back, or not expand on natural products research .... Implementation of the CBD seems to deal more with allocation of benefits, rather than also promoting the research that creates benefits .... It would be nice to give some thought to how the CBD could help developing countries attract equitable and more forward-looking natural products research, rather than discouraging research altogether.*

Finally, some companies stated that it is not the business of the private sector to tackle conservation and equity issues in access and benefit-sharing partnerships. Companies will operate according to the laws in a country, but the motive of the private sector is to make profit, and if the laws of a country hinder economic competitiveness, companies will go elsewhere. Several interviewees said that they do not believe that it is the role of business to achieve social goals such as 'fairness and equity'. Other companies, however, expressed distaste for negotiations with governments, because they felt it is naïve to believe that deals struck with governments will benefit conservation or local people. Several said they thought that money exchanged for access to genetic resources would be swallowed by bureaucracy, or find its way into the pockets of corrupt officials.

10.4 The impact of the CBD on business, and companies' response

Despite the strong feelings both for and against the CBD, almost every interviewee felt that, at present, the treaty makes only a negligible impact on his or her company in its daily business. As one industry representative put it: 'Companies do not think the Convention on Biological Diversity has any impact on them, so there is no need to comply. Once it affects the bottom line they will notice.'

Several companies – particularly from the botanical medicines, personal care and cosmetics, and horticulture industries – feel the CBD is largely irrelevant to their sector. As one company said, 'The CBD is not a subject that concerns the horticulture industry.' Indeed, many companies in these sectors do not seek access to species new to the trade, and the regulatory environment often encourages them to be conservative, and to focus on better-known species. However, many other companies in these sectors seek market differentiation with 'new' products, some of which require access to genetic resources (see, for example, the case of *Panax vietnamiensis* and Nature's Way in Chapter 4). Furthermore, companies in the botanical medicines, personal care and cosmetics sectors regularly seek access to traditional knowledge as a guide to species use, formulation, and confirmation of safety and efficacy, although this information is almost always accessed through databases and literature. As a result of these factors, for many companies in these sectors the CBD is, in fact, far from irrelevant.

Access and benefit-sharing are still not clearly regulated in the majority of countries, and monitoring and enforcement of such access regulations as do exist are scanty. In these circumstances, it is easy for companies to obtain specimens (for example, by receiving seeds or soil samples through the post) as they have done for years, not knowing whether this is legal or illegal. There are few mechanisms for bringing new rules and their significance to the attention of companies, many of whom continue 'business as usual'.

While the majority of companies report that the CBD does not affect their business practices today, many of those interviewed believed that it would in the future. As one industry representative put it, 'the CBD is a mosquito with the possibility of becoming an alligator'. Even today, a number of companies reported changes in business practices already visible across their industries.

Companies who are aware of the CBD usually respond in one of four ways: to ignore it, to work around it, to work with it, or to try and shape its further development and implementation. The few companies with a relatively good understanding of the CBD are fairly evenly distributed in the first three categories. Very few have engaged with the CBD and national policy-making processes to endeavour to shape them to their advantage.

Current changes in business practice, which many interviewees reported as a response to the CBD include:

- a decrease in corporate collecting activities, consolidation of collecting programmes into fewer countries, and in some cases a concentration on domestic collections;
- greater recourse to material from *ex situ* collections such as culture collections and compound libraries, rather than samples acquired through collecting activities;
- an increased role for intermediaries as brokers of access and benefit-sharing relationships, as well as suppliers of samples; and
- the increasing use of MTAs.

10.4.1 Decrease in and consolidation of collecting activities

In the face of legal uncertainty in the absence of access legislation, and high transaction costs where legislation does exist, companies are finding it increasingly difficult to collect genetic resource samples.

Many companies are now taking the CBD objectives – in some cases as manifested in permitting procedures and access legislation – more seriously. Fewer company staff collect, and fewer do so 'informally' without permits, for example when attending a conference or on holiday. Dr Rajaram, Head of the Wheat Programme at CIMMYT, felt that collecting would now have to be done legally, where, in the past, much was done illegally with scientists on holiday taking samples for themselves. However, such practices continue to be widespread in less aware companies, particularly in the biotechnology and horticulture sectors. One company, for example, turns a blind eye to materials sent without permits by a collaborator in a certain area of Brazil, and said that it does not check the paperwork as it is quite aware that it is illegal to remove the material. In the personal care and cosmetics industry, some companies receive plant materials from staff on trips and collaborators around the world, including information on traditional uses. This fact is often included in marketing and promotional materials.

Several companies mentioned that one result of the CBD is that they now collect in fewer countries. As one natural products researcher in a large pharmaceutical company put it,

> *We used to have 30–40 countries where we collected. But you can no longer collect without some kind of agreement. Our legal department says we have to pursue this country by country. It would be too costly to do this for so many countries. We will end up working with a couple of countries at a time .... We have changed our collecting strategy, and no longer conduct widespread collecting campaigns in many countries. Previously, we could work in 40 countries in a few months, now it takes a year to get permission to work in one – it is just too much.*

There is considerable nostalgia for the old days. As one interviewee reminisced, 'Before the CBD, we sent people all round the world to collect and they came in with thousands and thousands of species. It's all in our seed bank. Now, each year we investigate a few species.' Although the majority of companies reported that access to genetic resources is becoming increasingly difficult, a number pointed out that once materials have been obtained, there are very few checks in place to prevent them leaving the country. As one collector put it, 'It is sometimes so easy getting material that it is frightening.'

Many companies are becoming more selective about the countries in which they collect. Several said they would avoid working in countries that have adopted stringent access regimes. As one company explains, 'We are steering away from countries that have enacted very strict national legislation. There are places you can't work any more. Basically, you can't work in Australia, Brazil, the Andean Pact countries ... you can't get permits. Without permits, we do not work.'

The decision of Bristol-Myers Squibb not to work in the Philippines following the Philippines Executive Order is described in Box 10.1. This perspective was shared by a number of the companies interviewed.

A common response is to collect more at home, or at least within national territory. One pharmaceutical company makes more extensive use of the tropical biological diversity found in the country's overseas territories. Another envisages a similar strategy: 'If the lead of the more restrictive countries, such as the Philippines, Australia, Andean countries, and Brazil, were followed by everyone, then natural products research as we know it would change dramatically ... . But this is very unlikely to happen. If push comes to shove, there is always our own flora, which is very poorly studied.'

For one US-based company, the CBD made work at home more attractive: 'The CBD influenced our choice to work in the United States first. It was not planned that way, but the costs and benefits of different approaches proved it to be the best first step.' However, companies are increasingly facing access and benefit-sharing measures within their own countries, including the USA. Hearing about the agreement between Diversa and Yellowstone National Park (which is currently the subject of a lawsuit brought by several NGOs), another US-based company re-evaluated its practice. It now avoids collecting on government property.

Box 10.1 *One Company's Response to the Philippines Executive Order*

Excerpt from letter by Bruce S Manheim, Jr, Fox, Bennett and Turner, on behalf of Bristol-Myers Squibb, to the Hon Timothy Wirth, Under Secretary for Global Affairs, Department of State, USA, 3 November 1995

'A recent Executive Order from the Philippines implementing these provisions of the Convention creates a severe impediment to any natural products research by Bristol-Myers Squibb and its collaborators in that country ... Indeed it must be emphasised to the Convention parties that Bristol-Myers Squibb will not pursue natural products research in those countries that impose requirements similar to those contained in the Philippine Executive Order ... [G]overnment initiatives that place onerous restrictions on those seeking access to genetic resources or do not afford appropriate protection to intellectual property rights will result in fewer efforts to survey natural sources for pharmaceuticals and that will ultimately work to the detriment of environmental protection ... Among other things, one requirement of such an agreement is that, for endemic species, "the technology must be made available to a designated Philippine institution and can be used commercially and locally without paying a royalty to a Collector or Principal" ... Although the Philippine EO goes on to provide that other arrangements may be negotiated "where appropriate and applicable", this requirement amounts to nothing more than a compulsory licensing scheme. Moreover, while somewhat ambiguous, the provision would appear to allow the designated Philippine Institution to use the technology commercially both within and outside of the country. This scheme is flatly inconsistent with the Agreement on Trade-Related Aspects of IPR (TRIPS) negotiated in the Uruguay Round of the GATT. It also contravenes the interpretive statement accompanying the instrument of ratification for the Convention submitted by the administration to the Senate ... [Consultations] to the best of our knowledge, ... did not include pharmaceutical firms, such as Bristol-Myers Squibb, whose natural products research efforts will be seriously undermined by the Philippines EO and other similar initiatives.'

10.4.2 Greater recourse to *ex situ* collections

Some 2.5 billion specimens are held in biosystematic reference collections world-wide; an estimated 1,600 botanic gardens world-wide hold over 4 million living material accessions which represent about 80,000 species of plants (Hawksworth, 1995), and over 6.1 million accessions are stored world-wide in *ex situ* germplasm collections (FAO, 1996). As the interviews with companies from each of the industry sectors covered in this book have shown, companies commonly turn either to their own, existing collections, or to these institutions as the source of materials. The vast majority of materials held in *ex situ* collections were acquired before the CBD entered into force, and are thus not subject to its provisions on access and benefit-sharing. Although managers of *ex situ* collections are increasingly using MTAs to regulate access to materials within their collections, the vast majority of those materials can still be accessed without any commitments to share benefits on the part of the recipients. Several interviewees, particularly in the seed and horticulture industries, as well as in the pharmaceuticals and biotechnology sectors, said that, should access laws make the acquisition of new materials difficult to achieve, they would concentrate their efforts on existing *ex situ* collections. Some companies were determined to make 'the best of a bad situation'. One company observed that if the CBD discouraged access, it would at least encourage the study and use of *ex situ* collections, leading to the better characterisation of the genetic resources already held in collections.

10.4.3 The increasing importance of intermediaries

Several interviewees mentioned that their main response to the CBD has been to transfer the onus for obtaining collecting permits and negotiating access and benefit-sharing arrangements to the organisations from whom they acquire genetic resources, whether institutions with whom they collaborate in their own or source countries, specialist collectors, or *ex situ* collections.

In most cases, the companies who found no difficulty in obtaining materials relied upon counterpart organisations in the country in question to obtain collecting permits for them. One company explained that it felt the advantage of working with collaborators was that the company itself did not have to deal with permits. Another described how, if it proves difficult or impossible to secure collecting permits, the company abandons the prospect of working in the country concerned unless it knows of a local collaborator who will help it obtain material.

Some companies rely on intermediaries to bypass or ignore complex and difficult local permitting procedures. As one biotechnology company said, 'Generally collaborators do the collecting and we don't ask.' It is worth observing at this point that most of the countries that have introduced laws on access govern access by nationals as well as by foreigners, so this route does not necessarily exonerate the company or the national collector from access and benefit-sharing obligations. Others find other methods of obtaining samples without the bother of dealing with permits. One company described how it had regularly used the postal system to send samples through the mail, until there were problems with Customs, whereupon the company took to using courier services. Another company described 'turning a blind eye' to permits in countries such as China and Russia where there was 'such chaos' that the efficient regulation of access appeared a distant prospect.

However, the use of intermediaries is not universally a step taken to get around, or avoid dealing with, permitting procedures. Many companies employ intermediaries to ensure that they are adhering strictly to local laws, and are operating according to high standards. While some companies take steps to avoid encountering the CBD (for example, by collecting in countries without access legislation, or focusing on existing *ex situ* collections), a few others go to great lengths, often at considerable expense, to secure access that is authorised by government. These companies believe this will bolster their reputations as reliable partners for the supply of material. In some cases, companies wary of their obligations under the CBD said they had turned down collections of unclear provenance and without supporting documentation.

Some intermediaries felt that the increasing reliance placed on them to obtain PIC and to negotiate benefit-sharing is a major new responsibility, and one for which they are poorly equipped, in terms of resources and expertise. The difficulties and expense inherent in such a role are poorly understood by either those from whom they obtain genetic resources or those to whom they supply them. Also, their role is only partially monitored, which some felt left them open to pressure from the companies to whom they provide derivatives of genetic resources acquired from source countries. Some intermediaries felt that recipient companies would need to share the administrative burden faced by intermediaries in negotiating access and benefit-sharing arrangements, and that some level of scrutiny is needed not only at the stage where intermediaries obtain access, but also where they supply derivatives to companies. This might assist a level playing field and fair negotiations.

10.4.4 Increased use of MTAs and emphasis on legal acquisition of samples

Several companies referred to the increasing use of material transfer agreements (MTAs), both when collecting in the wild and when receiving materials from other organisations such as *ex situ* collections. Many interviewees see this as inevitable, but also consider it a burden. One representative from a horticulture company, for example, cited a botanic garden that he believed would rather compost plant material than deal with the hassle of producing MTAs under which to supply it.

However, although many intermediary institutions use MTAs reluctantly, several companies are now insisting that specimens are supplied to them under agreements, as articulated in some corporate policies (10.4.5). This practice is currently most common in the pharmaceutical industry, but also exists in the larger crop protection and biotechnology companies. Several interviewees believed the practice would spread.

Companies go to different lengths to confirm that intermediaries have acquired the samples they supply legally; that they have valid title to them, and that they are entitled to supply them to the company for potential commercial application.

Some companies make no effort to do so at all. In interviews, these companies said that there was no way that one could be sure that materials had been legally obtained, so they simply did not worry about the question. One explained that they do not know the history of the materials they receive, and believed there was no way of checking. Another summed up the feelings of this group: 'How can we know?'

A second group of companies trusts suppliers to have obtained materials legally, relying on suppliers who have a good reputation and are thus likely to have obtained all the necessary permits and agreements. Word of mouth and a small community allow companies to determine the reliability and reputation of suppliers. As one company put it, 'we know most of the sources, and whether they are *bona fides*'. Even with reliable suppliers, however, some companies know that full authorisation might not be sought in all cases, or that the company should oversee suppliers' practices and demand proof of proper permits. As one seed company explained: 'The company does not ensure that government has authorised a certain institution to supply genetic resources, but we should.'

A third group is not content to leave title to the genetic resources provided by suppliers unclear. This group requires evidence of the legal acquisition of specimens in writing, or incorporates terms in its contracts with provider institutions stating that the provider has acquired materials legally and is passing full title to the recipient. One company described how it insisted on seeing 'evidence of the proper paperwork'. Dr Michael Pauly of Novartis Seeds explains that 'our company requires suppliers to provide a legal document demonstrating that they have title to the material, and documenting their ability to pass the material on to us'. Another requires assurances in the contract for the acquisition of the materials that 'access has been handled correctly'. This approach is relatively common in the large seed and crop protection companies, and in some pharmaceutical companies.

> *'We don't acquire samples from people with dark sunglasses.' (Brian Foody, Iogen Corporation)*

One company involved in drug discovery described how, with the current lack of clear access regulations, it is not always possible to be certain that provider institutions have supplied sufficient evidence that they have authority from government to provide access to genetic resources: 'If an organisation that is part of or close to government says that it is authorised to provide you with genetic resources, is that enough? Do you still have to go to another government department to check whether this is true? Sometimes the best you can say is that, based on your reasonable efforts to find out, access was authorised by government.'

10.5 Corporate policies on access to genetic resources, traditional knowledge and benefit-sharing

Very few companies have developed policies in response to the CBD, let alone clear and detailed public documents designed to ensure that the acquisition of materials complies with the CBD and national laws on access. Several of the staff interviewed responded that their company did have a policy, but that it was either not in written form, for internal use only and not accessible to the public, or under development. When asked what kind of issues these unpublished and nascent policies addressed, interviewees' answers were usually vague. Several companies said it was their policy 'to obey the law', others that they 'always obtain all necessary permits' and several said 'we support fair partnerships', but did not enlarge upon what a fair partnership might mean in practice.

Box 10.2 *Comparison of Five Corporate Policies Introduced in Response to the CBD*

	Glaxo Wellcome	Novo Nordisk	Xenova
Scope and standard	▪ *Scope*: Pharmaceutical research and development ▪ *Standard*: 'Policy is to ….', 'Practice is to …', 'Agreements will be …'	▪ *Scope*: (All activities) healthcare and enzyme discovery. ▪ *Standard*: Requirements on the use of and access to genetic resources. 'Guiding principles, which we will do our utmost to live up to'.	▪ *Scope*: Natural product drug discovery. ▪ *Standard*: Reasonable efforts to comply with the following principles.
PIC and legal title	▪ Collaborate with organisations that possess the expertise and the authority to obtain materials from whatever source. ▪ Collaborate with bona fide suppliers. ▪ Conclude agreements with prospective sample suppliers only when they can provide documentary evidence that they have permission from appropriate government authorities to collect such samples.	▪ Acknowledge and respect that access to genetic resources requires PIC. ▪ No microbial strain or natural material obtained without proper PIC from the country of origin will be included in screening. ▪ All materials screened should be covered by contracts and/or MTAs. ▪ Contracts should be cleared by the proper authority in the country of origin.	▪ Deal only with suppliers acting in good faith and which have undertaken in writing to comply with the CBD and applicable national laws on access to genetic resources. ▪ Requires reasonable documentary evidence that suppliers are duly authorised by the government of the source country to participate in sample supply for the purposes of natural product discovery. ▪ Requires suppliers to certify in writing that they have obtained the PIC of any relevant local interests in the source country to obtain and use the samples for the purposes of natural product discovery.
Benefit-sharing	▪ Reimburse suppliers for the costs incurred in collecting natural product source samples. ▪ Reward suppliers for their expertise (eg in taxonomic classification). ▪ Cover freight costs of natural materials. ▪ Use MTAs which may refer to intermediate forms of compensation and involve a financial benefit payable to the supplier if a commercial product results from screening the natural product supplied. ▪ Require a significant proportion of this payment to the supplier to be returned to the source country to support science training and education at the community level.	▪ Acknowledge and respect benefits arising from the utilisation of genetic resources should be shared fairly and equitably with the country of origin, reflecting the contribution made. ▪ Conditions should be on mutually agreed terms and should include benefit-sharing, IPRs and technology transfer arrangements where appropriate. ▪ The country of origin will be mentioned in relevant publications and patent applications.	▪ Will seek to include in MTA: payments for the supply of materials. ▪ May involve additional short-term and/or long-term compensation to an appropriate, government-approved organisation in the source country (or to others with the agreement of such organisation). ▪ May include a financial benefit if a product is commercialised from the source material. ▪ Require a significant proportion of any longer-term payments to be returned to the source country to support conservation, education and scientific training at community level.
Conservation and sustainable use	▪ Acquire only taxonomically classified samples whose supply is reproducible and sustainable. ▪ Neither seek nor knowingly support the collection of endangered species.	▪ See Position Paper on Biodiversity.	▪ Will neither seek, nor knowingly support, the collection of endangered species.
Process and indicators		▪ 1999 target: patent applications and publications submitted in 1999 onwards to state the country of origin of genetic material. ▪ Filing of patent applications to be reported to donor country authorities. ▪ 1999–2000 target: procedures for monitoring the implementation of the requirements.	▪ As XDL gains experience of implementing the policy and dealing with organisations in source countries, it will review the policy from time to time and make such modifications as it considers necessary.

Bristol-Myers Squibb	Shaman	
■ *Scope*: Prospecting of biological diversity. ■ *Standard*: Collectors must comply with applicable laws and regulations. Agreements will vary, but 'follow certain basic principles'.	■ *Scope*: Use of plants and traditional medicine to develop modern therapeutic agents. ■ *Standard*: Terms and conditions under which Shaman may conduct research and development; consistent with the international codes of scientific conduct and ethics; conduct its business with integrity.	Scope and standard
■ Collectors must, at all times, comply with applicable laws and regulations and obtain PIC from the host government. ■ Where such collecting is to be based on ethnobotanical knowledge, BMS has agreed to collaborate in such efforts only where those providing such information have given their written consent (after being informed as to how such information will be used and what their rights could be to potential benefits).	■ Committed to respect and support for the rights of indigenous peoples and communities. ■ Respects community's rights to say 'no'. ■ Respects confidentiality of information provided by community. ■ Conducts research in accordance with regional and national laws, customs, and cultures. ■ Explains thoroughly the purposes and potential commercial outcome of the research to all parties and will not proceed without their informed consent. ■ Conducts business in a manner which promotes stability within communities. ■ Conducts business in accordance with international professional society codes of conduct.	PIC and legal title
■ Provide, when appropriate, advance benefits to the country whose biological resources are being investigated. These benefits may include: assistance in screening natural products for diseases of that country; financial support for patent protection; royalties to research organisations on resulting commercial products (expected to be used in part for conservation and protection of local biodiversity). ■ Consider utilising the country as a source of supply and/or cultivation of necessary raw natural product materials for any commercially developed product. ■ When BMS does not directly enter into agreements, it expects its collaborators to do so, and will return a share of royalties to 'the country and its indigenous peoples'.	■ Provides three stages of benefits, the immediate-, mid-, and long-term. ■ Immediate benefits: a portion of Shaman's research budget will be allocated to communities to provide cultural and biological conservation programmes; facilitate and enhance community's research efforts; and improve healthcare, education, land and water use and conservation. ■ Mid-term benefits: assisting local communities to create new or expanded supply industries for natural product pharmaceutical products. ■ Long-term benefits: a percentage of all profits is to be shared with all the 'Culture Groups' around the world with which Shaman has previously worked.	Benefit-sharing
■ The requirement that the company's research partners (and, in turn, their agents) acquire biological samples only in an environmentally responsible manner.	■ Committed to sustainable resource utilisation that does not harm species harvested or commercialised. Has developed sustainable harvesting and management programme in several Andean countries.	Conservation and sustainable use
	■ Shaman will continue to inform itself about local, regional, and national laws, customs, and cultures, and conduct its business with integrity.	Process and indicators

Several staff in research who were responsible for concluding access agreements were not sure whether there was a corporate policy on this issue and referred us to public or external relations departments. Some company representatives felt that public policies are no more than public relations gimmicks. As one company representative put it: 'the implementation of any of this stuff will always be left to staff working with natural products – so what is the point of drafting a public document?' But corporate policies have the important advantages of providing a tool for transparency and good corporate citizenship, enabling companies to communicate their positions and commitments to source countries and others.

Five companies that have introduced public policies, or have drafted documents outlining their 'best practice' in response to the CBD, are Glaxo Wellcome, Novo Nordisk, Xenova, Shaman Pharmaceuticals, and Bristol-Myers Squibb. We analysed the policies introduced by these companies against a list of five elements: the scope of the policies and the standard of behaviour to which the company concerned subscribes; whether and how the policy addresses PIC, benefit-sharing, conservation and sustainable use of biodiversity; and whether there is a process for continual improvement of the policy and mention of verifiable indicators of its implementation. This section summarises the results of the analysis (Box 10.2).

Glaxo Wellcome

Discovering New Medicines from Nature: Policy for the Acquisition of Natural Product Source Materials, is a publicly-available policy introduced by Glaxo in 1992 and now sent to all Glaxo Wellcome's suppliers of natural materials. It acknowledges the need 'to conserve rare species and not to imperil biodiversity' and states that the company considered the issues involved in seeking leads for new medicines from nature, and that the policy document sets out the company's views and mode of conduct. The policy states that the company 'understands the impact that unauthorised and/or unrestrained removal of natural materials from their indigenous habitats can have on the ecology and economy of a country', and addresses the quantities of material sourced; the kind of institutions with which the company will collaborate; the requirement of supplier organisations to provide written evidence of proof of PIC and government permission to supply samples; and benefit-sharing. Other than reimbursement of costs, reward for expertise and 'financial benefits' whose magnitude 'will recognise the relative contribution of the discovery of the bio-active principle to the subsequent development of the commercial products', the nature of monetary or non-monetary benefits is not discussed. Other than covering costs and sharing benefits in the event of commercialisation, the timing of benefit-sharing is not addressed in the policy, although in practice, through its partnerships, the company shares benefits in the short, medium and long term (personal communication, Dr Melanie O'Neill, 1998; Box 3).

The document addresses conservation by stating that the company will acquire only taxonomically classified samples whose supply is reproducible and sustainable, and will neither seek nor knowingly support the collection of endangered species. The policy distinguishes between the terms for the purchase of samples of natural products, described in Box 10.2, and 'broader philanthropic efforts to conserve resources of which these materials are a part'. Financial support for conservation and environmental projects is a matter for the Glaxo Wellcome Appeals Committee. The policy makes no reference to future development, and does not provide indicators, nor mention mechanisms for scrutiny.

Novo Nordisk

In 1995, Novo Nordisk Health Care Discovery developed a document entitled *Acquisition of Natural Resources for the Development of New Pharmaceuticals*, which stated the company's respect for the sovereign rights of nations to their own natural resources, and acknowledged 'that benefits arising from the utilisation of natural resources by any other party should be shared fairly and equitably with the donor country'. The policy affirmed the company's intention to establish research agreements with organisations supplying material in compliance with national legislation and international law; its requirement for supplying organisations to provide documentary evidence of 'all necessary authorisations and permits' to dispose and make use of the material; and benefit-sharing. The company's approach to access and benefit-sharing evolved in the next two years, and, in its 1997 report on environment and bioethics, it set out 'Guiding principles for Novo Nordisk's implementation of the Convention', which applied to both its healthcare and enzyme businesses (Box 10.2 p304).

By 1998, the company's *Environmental and Bioethics Report* reported on progress in implementing the principles in its partnerships in several countries, and highlighted some of the difficulties of doing so. The report identifies two main requirements if providers and users of genetic resources are to cooperate successfully on the implementation of the CBD. The first is an effective system for establishing prior informed consent without too much bureaucracy: 'We fully support the objectives of the Convention on Biological Diversity (CBD). However, more speedy and simpler procedures for obtaining access to genetic resources under the CBD provisions are urgently needed, including streamlined processes for obtaining prior informed consent from relevant authorities and communities on mutually agreed terms.' The second is for users to be able to identify whose prior consent for access is needed: 'For our researchers and for our collaborating scientists, it is critical to know who to contact before entering into collaborations on biodiversity'. The report concludes that a 'limiting factor for Novo Nordisk to be in full compliance with our guiding principles has been lack of implementation of national CBD procedures'.

The 'Requirements' do not explicitly distinguish between monetary and non-monetary benefit-sharing, nor state when benefits will be shared, but the *1998 Environmental and Bioethics Report* sets out examples of 'collaborations with organisations from different regions of the world involving both monetary and non-monetary benefits to the providing country'. While the requirements do not explicitly rule out benefit-sharing for products developed from samples obtained prior to the entry into force of the CBD, their scope appears to be restricted to post-CBD materials, since the company 'will do our utmost to live up to [the guiding principles set out in the 'Requirements'] for all material covered by the CBD. (The CBD came into force in December 1993.)' Beyond endorsing the CBD's objective of conservation, the document makes no mention of conservation or sustainable sourcing.

The company's use of published targets demonstrates its commitment to the continual improvement of its policy on the implementation of the CBD. The target for 1998 was to develop formal corporate requirements on the use of and access to genetic resources in keeping with the CBD. This done, the target for 1999 is that all relevant patent applications and publications submitted from 1999 onwards should state the country of origin of genetic material, and the company's target for 1999–2000 is to develop procedures for monitoring the implementation of the company requirements on the use of and access to genetic resources. The report explains that external contracts aimed at gaining access to genetic resources are being evaluated to ensure compliance with the new company requirements, and that the company is engaged in negotiations with several governments in order to obtain final PIC for material obtained by Novo Nordisk through 'third party arrangements'.

Xenova

The Board of Xenova Discovery Ltd (XDL) approved a *Policy for the Acquisition of Natural Product Source Materials* in November 1998. The policy provides a framework within which the company will seek to negotiate formal agreements and collaborate with suppliers of natural product source materials. It addresses the requirement for its suppliers to act in good faith and comply with the CBD; to obtain PIC from government and 'any relevant local interests'; the use of MTAs with each supplier; and the sharing of monetary and non-monetary benefits with 'government-approved organisations in the source country'. The policy mentions not only payment for the supply of material, but the fact that agreements 'may involve additional short-term and/or long-term' benefits (Box 10.2). As the company gains experience of implementing the policy and dealing with organisations in source countries, it will review the policy and make such modifications as it considers necessary. The document states that 'the implementation of the policy must depend, in part, on there being in place appropriate organisations within source countries with which agreements can be reached'. As it stands, this statement is ambiguous, and could mean either that the company will not work in countries where it is not possible to reach agreements consistent with its policy, or that it will not apply its policy in such circumstances. The commitment in the policy is to make 'reasonable efforts to comply' with the principles it contains.

Bristol-Myers Squibb

Bristol-Myers Squibb's corporate policy takes the form of a statement of the principles that guide the company's approach (Baker et al, 1995). It addresses requirements that the company will make of its research partners, in terms of environmentally responsible sampling, compliance with applicable laws and regulations in source countries and the need to obtain PIC from the host government and providers of ethnobotanical knowledge. It also states that, 'when appropriate', the company will provide 'advance' benefits including support for research on source-country diseases, financial support for patent protection and royalties on products eventually commercialised (Box 10.2; Baker et al, 1995).

Shaman Pharmaceuticals

Shaman Pharmaceuticals has developed an *Agreement of Principles* in order to establish the terms and conditions under which Shaman conducts research and development programmes on the use of plants and traditional medicine in order to develop modern therapeutic agents. The agreement outlines Shaman's philosophy, including its commitment to respect the rights of indigenous peoples and local communities to choose not to sell or commoditise their knowledge and skills, flora, fauna, crafts, objects, and artifacts. Communities' rights to privacy, confidentiality, and to fair compensation are also supported. The purposes and potential commercial outcome of Shaman's research are to be explained to all parties, and will not proceed without their PIC.

The *Agreement of Principles* requires that sample collection of plant materials of between 3 and 40 kg dry weight will only be undertaken in cases where the conservation, management, and ecology of a given species will not be endangered, threatened, or harmed by such collections. Supplies of raw materials on a large scale are also intended to be 'sustainable' and developed in conjunction with in-country collaborators.

The *Agreement of Principles* also elaborates upon the company's commitment to benefit-sharing in the immediate-, mid-, and long-term. Immediate benefits are intended to provide support for cultural and biological conservation programmes, to facilitate the community's own drug development and research efforts, and/or improve the community's health, education, land and water use, and conservation programmes. Mid-term benefits include assisting local communities in the development of supply industries in cases in which Shaman has commercialised a molecule for drug development that cannot be synthesised economically on a large scale. Long-term considerations include a share in the profits from products, which will be distributed in part through The Healing Forest Conservancy, to all the 'Culture Groups' Shaman has worked with. Shaman also seeks to provide technical assistance and capacity building on a case by case basis, including training, laboratories, the provision of scientific materials, literature, and other tools for the development of traditional natural products in provider countries.

None of the five policies and sets of principles described here addresses whether the companies will endeavour to share benefits arising from the commercialisation of samples acquired before the CBD entered into force (although the Novo Nordisk policy appears to rule this out), nor whether they will ensure that any benefit-sharing or other commitments made as a result of the policies will be passed on to any third parties licensing products from them. This is particularly relevant in the case of smaller companies which may not carry a product the whole way to the market-place, but may license it on to another company for this purpose.

While few companies have policies that directly respond to the access and benefit-sharing provisions of the CBD, several interviewees mentioned that they were developing them. For example, soon after the Convention came into force, both the Aveda Corporation and The Body Shop began processes to address the 'IPRs' of indigenous peoples. The Body Shop initiated a process in 1994 to develop an *Intellectual Property Rights* agreement to guide their collaboration with the Kayapo Indians in Brazil. The Aveda Corporation has declined to research any species not in the 'public domain' until Brazil has enacted access and benefit-sharing laws that provide a basis for commercial best practice. The company is developing a *Manifesto* that outlines a framework for collaborations with communities, including requirements of PIC, and benefit-sharing.

In addition to these specific corporate policies introduced in response to the CBD, corporate literature of other kinds sometimes mentions access and benefit-sharing. For example, Monsanto mentions access and benefit-sharing in its 32-page 1997 annual report on *Sustainable Development Including Environmental, Safety, and Health Performance*. A half-page section of this report, headed 'Bioprospecting Contract Seeks Fairness', states that 'countries rich in biodiversity are working with Monsanto and other companies to identify commercially valuable plant life and ensure proper compensation and resource protection', and lists a number of issues that must be worked out when contracts are developed, including

> *who can claim the genes from these plant products and who deserves compensation? The host country? The research institution that collected the samples? The company that ran thousands of tests to derive the compound? Or indigenous people who already understand the plant's properties? ...*
> *How can a company engaged in bioprospecting secure patent protection for any discoveries made? How can Monsanto help preserve biodiversity while encouraging host countries to develop their economies?*

The section does not seek to provide answers to these questions, nor to explain in any detail the company's approach, but states that Monsanto supports the CBD and has been involved in 'the negotiation process to resolve outstanding concerns' (which could mean either the attendance as observers at the CBD Conference of the Parties of representatives from Monsanto, or negotiation of individual access agreements). The section concludes by summarising the benefit-sharing arrangements of the ICBG project involving Monsanto, Washington University, and the Aguaruna people in Peru (Chapter 3), concluding that 'Monsanto also ensures that royalties and other compensations are returned to Peru. In exchange, Peru provides legal certainty that Monsanto has the patent rights to any compounds discovered under this contract'.

10.5.1 Corporate policies addressing related environmental and social issues

A number of companies in the botanical medicines and personal care and cosmetics sectors, participate in industry-wide efforts to promote environmental and socially responsible business practices. These include industry groups like the Social Venture Network, and Businesses for Social Responsibility. Some companies have adopted the CERES (Coalition for Environmentally Responsible Economies) Principles. These Principles are divided into ten categories: protection of the biosphere; sustainable use of natural resources; reduction and disposal of wastes; energy conservation; risk reduction; safe products and services; environmental restoration; informing the public; management commitment; and audits and reports. This suite of issues is indicative of those customarily highlighted in company policies in these sectors. Additionally, some companies have made a commitment not to test products or ingredients on animals, and to purchase only organically-sourced raw material.

A number of corporate and institutional policies and guidelines developed for these sectors are relevant to the conservation and sustainable use objectives of the CBD. In response to growing concern on the part of consumers and some companies that species are over-exploited in the wild, a number of industry and conservation groups have drafted policies to promote sustainable and high-quality sources of raw materials. Non-profit organisations, like the Rocky Mountain Herbalists' Coalition, the United Plant Savers, TRAFFIC, and the IUCN Medicinal Plant Specialist Group, conduct research and promote education relating to threatened medicinal plant species in trade, and produce guidelines for ethical wildcrafting of species. These usually involve sustainable harvesting, but can also include the fair sharing of benefits with harvesters.

One interviewee in the seed sector said his company subscribed to the European Association for Bioindustries' 'Core ethical values' which mention health, safety, environmental protection, ethical research, animal rights and informed consent of patients, but which do not address PIC for access to genetic resources other than human genetic resources, nor benefit-sharing.

10.5.2 Professional codes of conduct related to access to genetic resources, traditional knowledge and benefit-sharing

Although few companies have corporate policies, several industry researchers have participated in the drafting of professional society codes of conduct, such as the Guidelines for Members of the American Society of Pharmacognosy. These Guidelines promote 'the facilitation of technology transfer, and the fair and equitable sharing of results and benefits arising from the utilisation and commercialisation of source country natural resources'. Staff from the NCI, Merck Research Laboratories, Glaxo Group Research, and numerous academic research institutions which collaborate with industry, assisted in the drafting of this document (Cragg et al, 1997). Other professional societies and international organisations that have developed Codes of Conduct intended to encourage members to obtain PIC and to share benefits with source country institutions and communities include the Society of Economic Botany, the International Society of Ethnobiology, and the Society for Applied Anthropology.

10.6 Institutional policies in response to the CBD

Perhaps because they are more heavily involved in the 'cutting edge' of collecting samples around the world and distributing them to the scientific community and to companies, a growing number of organisations such as botanic gardens, culture collections and even some genebanks are beginning to develop institutional policies and MTAs in response to the CBD. Very few universities have yet considered doing so. Groups such as the NCI (USA); the SIDR (UK); the National Botanical Institute (South Africa); and the botanic gardens of Kew (UK), Rio (Brazil), Bonn (Germany), Missouri and New York (USA) already operate policies on the CBD that cover some areas of their activities. Under the umbrella of a pilot project funded by the UK Department for International Development (DFID) and coordinated by the Royal Botanic Gardens, Kew, these and a number of other botanic gardens from around the world have developed a harmonised and integrated approach, comprising policy guidelines and material acquisition and transfer agreements. These are set out in 'Common Policy Guidelines for Participating Gardens on Access to Genetic Resources and Benefit-Sharing' (www.rbg.ca/cbcn/cpg_index.html). A similar pilot project (MOSAICC – Microorganisms Sustainable Use and Access Regulation – International Code of Conduct) sponsored by the EC is developing a policy and MTA for use by culture collections and other users of microbial genetic resources. In the field of agriculture, the International Agricultural Research Centres of the CGIAR use MTAs and abide by policies that focus on IPRs, but currently do not mention benefit-sharing.

Strathclyde Institute for Drug Research (SIDR)

SIDR has developed a *Heads of Agreement* between SIDR and 'providers' of plant material for screening. This agreement commits SIDR to pass on 60 per cent of any fees obtained from third parties in respect of access

Box 10.3 *NCI Policy on Benefit-Sharing: A Comparison of the Letter of Collection and the Memorandum of Understanding*

Benefit	Letter of Collection (LoC)	Memorandum of Understanding (MoU)
Monetary benefits /royalties	▪ Depends upon the negotiation of benefits between a licensee of a patented product and a source country, as required by the LoC.	▪ Depends upon the negotiation of benefits between a licensee of a patented product and a source country, as required by the MoU.
IPRs	▪ Joint patent protection is sought for all inventions developed collaboratively by NCI and source country organisation employees. ▪ All licences on patents arising out of the collaboration refer to the agreement and all licensees are apprised of it.	▪ As for LoC.
Joint research	▪ During the course of the contract, the NCI (in collaboration with the source country organisation/government), assists the appropriate source country institute with capacity building for drug discovery and research (including screening capabilities). A senior source country scientist/technician is invited to NCI's labs for one year or may use NCI technology useful in furthering work under the agreement. ▪ Once an agent has been approved by the NCI for preclinical development, the basis on which source country scientists can participate in such development is negotiated. 'Sincere efforts' are made to transfer knowledge and expertise to the source country organisation. ▪ Further development of compounds submitted by source country scientists to NCI for screening (as separate initiatives from submissions by the collection contractor) is conducted by NCI in consultation with the relevant source country organisation.	▪ Facilities being available, the source country organisation undertakes in-house primary anti-cancer and anti-viral screening of synthetic compounds and natural extracts, for later submission to NCI screens along with data sheets. ▪ Once NCI's advanced anti-AIDS and anti-cancer screens are completed, the source country organisation undertakes bioassay-guided fractionation to isolate pure active compounds. ▪ If fractionation facilities cannot be established at the source country organisation, suitably qualified source country scientists are sent to NCI labs for isolation studies. ▪ Otherwise, as for the LoC.
Technology transfer	▪ With regard to source country participation in the further development of specific agents, 'sincere efforts' made to transfer technology to the source country organisation, subject to IPRs.	▪ The NCI assists in providing necessary bio-assays for the source country organisation to undertake fractionation, subject to available resources.
Information:	▪ NCI provides the results from screens of extracts to the source country (subject to confidentiality until the DTP has a chance to file patents).	▪ The NCI must repatriate the results of its advanced anti-cancer and anti-HIV screens within 90 days.
Rights to supply further material	▪ NCI requires licensees to seek resupply of source material from source countries. ▪ The Collection Contractor collaborates with the source country organisation over possibilities for mass propagation.	▪ As for the LoC.
Transfer to third parties/licensing	▪ Third-party recipients of material sent by the NCI must compensate the source country as appropriate.	▪ The NCI will not distribute material to third parties without prior consent from source countries. If a source country organisation wishes to collaborate with such third parties, the NCI will put them in touch with one another.
Publications	▪ Permission of a traditional healer is sought prior to publication of any information he/she has contributed, and he/she is acknowledged.	▪ Publication of data arising out of the MoU takes place at a time agreed upon by the source country organisation and the NCI.

Source: adapted from ten Kate and Wells, 1998.

to extracts for evaluation purposes. Results of all such evaluations will be communicated to the provider, and should species prove of further interest, SIDR will negotiate a re-supply contract, again passing on 60 per cent of any fees to providers. Any agreements negotiated by SIDR with a third party will commit SIDR to sharing 60 per cent of any net income gained by SIDR with providers of the original samples. The provider of the plant material agrees not to supply the material to any third party for a period of two years.

The *Heads of Agreement* are not intended to be legally binding, but nevertheless constitute the basis agreed by the provider and SIDR for negotiations of a legally binding agreement. The provider supplies SIDR with the scientific names of plants provided, but SIDR depends upon the provider to provide correct identification, security of re-collection, and to obtain legal title to the material. SIDR will not work with individuals, however, preferring instead to work with governments, research institutions, universities, and recognised NGOs. As Professor Alan Harvey, Director of SIDR, says, 'Our ability to work with external companies in this area relies entirely on the quality of all the participants in our natural product network.'

The National Cancer Institute (NCI)

Over the last 40 years, the NCI has facilitated the preclinical anti-tumour screening of more than 400,000 compounds and materials submitted by a wide range of grantees, contractors, pharmaceutical and chemical companies, and other private and public scientific institutions world-wide. Contracts awarded for the collection of plants in tropical and subtropical regions world-wide include those to the Missouri Botanical Garden (MBG) (collecting in Africa), the New York Botanical Garden (NYBG) (collecting in Central and South America) and the University of Illinois at Chicago (UIC), assisted by the Arnold Arboretum and the Bishop Museum in Honolulu (collecting in South-East Asia). Since 1988, the NCI has developed three standard agreements that contain terms relating to the sharing of both monetary and non-monetary benefits: the Letter of Intent (first issued in 1990); the Letter of Collection (introduced in 1992 and still used in some countries); and the Memorandum of Understanding (introduced in 1995). Each successive model has marked an evolution in terms of the degree of research collaboration contemplated between NCI and the source country, the number of non-monetary benefits such as technology transfer offered to the source country, the role of source countries in negotiating benefits arising out of commercialisation, the scope of the benefits to be shared, and the contracting parties involved.

Box 10.4 *The Main Elements of RBG, Kew Policy on Access to Genetic Resources and Benefit-Sharing*

The *'Policy on Access to Genetic Resources and Benefit-Sharing of the Royal Botanic Gardens, Kew (RBG, Kew)'*, which came into effect in January 1998, addresses the supply of genetic resources; the fair and equitable sharing of the benefits arising from their use; the commercial use of genetic resources; and the further development of RBG, Kew's strategy on access and benefit-sharing.

- **Acquisition**: The policy outlines requirements for obtaining the PIC of source countries and stakeholders and states that RBG, Kew will attempt to clarify the terms of acquisition in written material acquisition agreements (MAAs) with its partner suppliers.
- **Supply**: The policy provides that RBG, Kew will supply genetic resources to recipients under MTAs which will be consistent with any terms under which those resources were supplied to RBG, Kew. These agreements will contain terms which oblige recipients to share fairly and equitably the benefits of their utilisation of those genetic resources. The MTAs will also prohibit recipients from commercialising those genetic resources without the prior written agreement of RBG, Kew and from passing those genetic resources or their derivatives on to third parties without ensuring that the third parties enter into similar agreements. The policy specifies that RBG, Kew may choose not to supply genetic resources to recipients if it is not satisfied that they are acting in accordance with the CBD, relevant national laws and agreements.
- **Benefit-sharing**: RBG, Kew will share benefits arising from the utilisation of genetic resources and will do so consistent with the terms under which those genetic resources were acquired. The kinds of benefits to be shared may include both monetary and non-monetary benefits. Monetary benefits will typically be appropriate in the case of commercialisation. Non-monetary benefits can arise from any use of genetic resources – from purely taxonomic research to the development of new pharmaceutical or horticultural products – and may include training, information, technology transfer and joint research and development programmes. In addition, RBG, Kew commits to making efforts to share benefits fairly and equitably in situations which are not covered by explicitly agreed terms, such as the use of materials from 'pre-CBD' collections.
- **Commercialisation**: In the special case of commercialisation of materials held within RBG, Kew collections, the policy provides that any commercialisation of genetic resources from 'post-CBD' collections will be subject to the agreement, and PIC, of the source country of those genetic resources and, where appropriate, the supplier and relevant stakeholders. With respect to commercialisation of 'pre-CBD' genetic resources, RBG, Kew's policy is to negotiate with commercial partners to share benefits with source countries in the short-, medium- and long-terms.
- **Further development**: RBG, Kew recognises that its policy will need to be revised periodically in order to reflect the wishes of the UK government, source country governments and partners overseas, based on changes in law and acknowledged best practice. RBG, Kew is also working with other botanic gardens to develop, harmonise and spread best practice.

Source: CBD Unit, RBG, Kew, March 1999.

Each of the three prototype agreements contains provisions addressing:

- IPRs (involving the payment of royalties and possibilities for joint ownership of patents);
- technology transfer, training and capacity building (involving the training of source country scientists in NCI laboratories);
- confidentiality of ethnobotanical data, including PIC from traditional healers prior to publication and adequate acknowledgement of their contributions;
- joint research;
- the communication of research results to source country institutions;
- resupply (collaboration over the resupply of additional material for discovery, development and scale-up for manufacture); and
- obligations of third-party licensees to share benefits with the source country.

The evolution of the approach by the NCI to access and benefit-sharing is set out in Box 10.3.

The Royal Botanic Gardens, Kew (RBG, Kew)

In order to honour the letter and spirit of the CBD, RBG, Kew has developed a policy on access to genetic resources and benefit-sharing (Box 10.4). In April 1992, RBG, Kew introduced a benefit-sharing policy with respect to the screening of its plant accessions for novel chemicals. In 1995, RBG, Kew began a broad consultative process involving staff throughout the gardens, the UK government, collaborators in the UK and overseas, and international experts. The consultative process culminated in the Kew 'Policy on access to genetic resources and benefit-sharing', which came into force in January 1998. RBG, Kew is currently gaining experience in implementing the 1998 policy and building the capacity to do so. The policy document will be reviewed in the coming months to reflect the experiences of Kew staff and external collaborators in working with it; developments in policy and law related to access and benefit-sharing; and the results of initiatives such as the botanic gardens' pilot project on access and benefit-sharing described in 10.6 above. A summary of the policy of the Royal Botanic Gardens, Kew on access to genetic resources and benefit-sharing is included in Box 10.4 p311.

10.7 Conclusions

A high proportion of the companies and other organisations interviewed had heard of the CBD, but many interviewees did not have a good understanding of its scope. Most companies and other organisations involved in access to genetic resources are still ill-informed about the implications for their business of the CBD. There is little easily-accessible and 'business-friendly' information on access and benefit-sharing. The pharmaceutical, biotechnology, crop protection and seed industries appear to be more familiar with the provisions of the CBD and national access laws than horticulture, botanical medicines and personal care companies.

Many companies were positive about the objectives of the CBD, but the vast majority of those interviewed had major concerns about its implementation. Concerns included the lack of clarity surrounding access rules, the bureaucracy and transaction costs involved in following them, and the lack of understanding of the role of business on the part of regulators and institutions providing access to genetic resources.

Almost every interviewee felt that, at present, the treaty makes only a negligible impact on his or her company in its daily business. Gradually, however, the CBD's influence over the nature of commercial partnerships involving the use of genetic resources is growing. Several interviewees mentioned current changes in business practice as a result of the CBD, including: a decrease in and consolidation of corporate collecting activities; greater recourse to material from *ex situ* collections; an increased role for intermediaries as brokers of access and benefit-sharing relationships in addition to suppliers of samples; and the increasing use of MTAs.

A number of voluntary guidelines for professional scientists such as botanists, ethnobotanists and pharmacologists have been developed over the last decade. Several intermediary organisations involved in collecting genetic resources, and sometimes in supplying them to industry, have developed institutional policies and MTAs on access and benefit-sharing. A number of companies have developed policies on environmental and social issues, and while some even refer to the CBD in corporate literature such as annual reports, only a handful of companies have developed and adopted specific policies on the acquisition of genetic resources and benefit-sharing. Such of the latter policies as exist, predominantly in the pharmaceutical sphere, mark a good first step, and it appears that others in industry are following suit.

Kerry ten Kate and Sarah A Laird

Chapter 11
Conclusions

11.1 Overview

The CBD defines 'genetic resources' broadly, as biological materials of actual or potential value, containing functional units of heredity. Through its provisions on access to genetic resources and benefit-sharing, the CBD has an important bearing on the work of any person or company requiring access to genetic resources of any kind, whether for academic study or for commercial research and development. The CBD marks an important watershed in the international understanding of countries' rights and responsibilities on these issues. However, its practical implementation poses an enormous challenge for the 175 parties that have ratified it, and for the many sectors of industry that need access for product discovery and development. Together, they must find workable rules and procedures that reflect the rights of sovereign states, communities, individuals and companies, but deliver partnerships that are 'fair and equitable' in the context of the risks and rewards of product development. The rules and procedures need to be speedy, simple and efficient.

This chapter will start with some opening observations about policy and practice in access and benefit-sharing, before turning to markets for natural products, the extent of commercial demand for access to genetic resources and traditional knowledge, and the nature of benefit-sharing arrangements. It will also address the role of intermediary institutions, companies' views of the CBD, and corporate and institutional policies developed in response to the CBD and access legislation. Finally, it will consider briefly the contribution that access and benefit-sharing partnerships can make to sustainable development and conservation of biological diversity.

11.2 Introduction

The 'access and benefit-sharing' provisions of the Convention are particularly complex to implement. They interact with a number of other areas of law and policy, and apply to a broad range of academic and commercial activities by an extremely diverse group of individuals and organisations. Key stakeholders in the transfer of genetic resources include governments, research and conservation institutions, private companies, and local and indigenous communities. Relevant legal and policy frameworks include not only access and benefit-sharing measures drafted to implement the CBD, but also law and policy on trade and investment, environment, natural resources, private and intellectual property, and constitutional law (Glowka, 1998; Gollin and Laird, 1996; ten Kate, 1995). The right of self-determination and the territorial and resource rights of local people are critical to the effective distribution of benefits to local communities, and to the conservation of biodiversity (UN *Draft Declaration on the Rights of Indigenous Peoples*, 1993; International Working Group for Indigenous Affairs, 1997).

The Convention's objectives reflect a commitment by governments to facilitate access to genetic resources in return for a fair and equitable sharing of benefits such as technology transfer (CBD Article 1); an exchange that has been described as a 'grand bargain' (Gollin, 1993).

This 'bargain' is often portrayed as one between the technologically advanced 'North' and the biologically rich 'South', based upon the 'realisation that the world's biological resources are distributed in approximately inverse proportion to its material wealth' (*Nature*, 1998). A large number of products have been, and continue to be, developed in the North from genetic resources obtained from the South, but 'South-South' transfers remain significant among developing countries (eg genetic resources for agricultural commodities such as cocoa, coffee, oil palm and bananas), as do exchanges between industrialised countries (eg DNA polymerase isolated from Yellowstone National Park and used in biotechnology laboratories around the world) (Chapter 8). Every country, even the most biologically diverse, is dependent upon genetic resources from outside its borders for food (Chapter 5). Botanical medicines are also often pan-tropical in distribution and use (eg senna species), or are traded between developing countries (eg *Prunus africana*, which is exported from Cameroon to South America, as well as to Europe, Japan and the USA) (Chapter 4). There are also 'North-South' transfers of genetic resources, not only in agriculture. Ivermectin, a treatment developed by Merck and Co for the river blindness which afflicts millions of people and livestock in tropical countries, was developed from a microorganism collected on a Japanese golf course.

Additionally, access to genetic resources is not always the subject of international trade. Local access to genetic resources is often sought by domestic companies and institutions, as we have seen in the case of Jeevni, developed from a plant used traditionally by the Kani tribe by the Arya Vaidya Pharmacy, and obtained from the Tropical Botanic Gardens and Research Institute in Kerala (Chapter 4). In these cases, it is companies, scientists and communities within a country who access genetic resources, innovate, transfer technology and conserve biological diversity. Thus the 'grand bargain' is fundamentally between different actors in civil society, rather than between states.

Just as 'access' cannot be pictured as a simple flow of genetic resources from South to North, so it is increasingly untenable to picture 'benefits' such as technology and finance inevitably generated in the North and shared (or not, as the case may be) with the South. Many countries in the South have built extensive capacity to innovate and add value to their genetic resources, rather than supplying raw materials alone for research in the North. As we have seen across the industry sectors covered in this book, the number of joint ventures and partnerships based on research and development conducted in source countries is gradually growing. As a result of this increased capacity, the NCI in the USA is able to build partnerships involving more drug discovery in high biodiversity countries than was possible just five years ago. In some cases, such as the joint venture between the Government of Sarawak and the US pharmaceutical company Medichem Research, high biodiversity countries are participating in research on endemic species. In others, such as the Nigerian ICBG, partnerships create the opportunity for research on diseases prevalent in the source country that do not otherwise attract much research attention within multinational pharmaceutical companies (Chapter 3).

Benefit-sharing itself is extremely complex, involving a package of benefits that go well beyond financial benefits such as royalties. Technology transfer, capacity-building, and in-kind benefits, delivered over the short-, intermediate-, and long-term, characterise the most effective benefit-sharing packages, but it is difficult to draw more detailed conclusions at this general level, since successful benefit-sharing arrangements differ from sector to sector and case to case. Best practice in benefit-sharing evolves in tandem with technological and scientific developments, and also reflects changes in market and regulatory environments. The most effective form of benefit-sharing appears to result from well-developed partnerships between the private sector and source country institutions. Technology transfer and capacity building within these partnerships frequently become more valuable to the organisation supplying genetic resources over time, involving a more extensive and a wider range of benefits. For example, during the initial three-year term of the agreement between INBio and the Diversa Corporation, all samples were sent to the company's laboratories in the USA for analysis, but when the agreement was renewed in 1998, Diversa set up a DNA processing laboratory at INBio's bioprospecting division in Costa Rica (Chapter 8).

The policy and practice of the commercial use of biodiversity is moving very fast. The number of countries that have introduced or are exploring different options for law and policy on access has grown from just a handful in 1996 to over 40 by the beginning of 1999. Scientific developments over this period in the fields of biology, chemistry, genomics and information technology have been equally rapid, revealing a vast range of new targets for the development of medicines and agricultural products, and transforming the process of discovery and development. Biological discoveries which would once have taken years can now be completed in days, thanks to new technologies such as combinatorial chemistry, ultra-high-throughput screening and laboratories-on-a-chip (Chapter 3).

In response to these scientific and technological developments, the constellation of companies, nearly as diverse as the genetic resources on which they work, form a shifting pattern of partnerships in an increasingly globalised economy. Discernible among this complex pattern are trends towards – on the one hand – consolidation through mergers and acquisitions, and – on the other hand – a proliferation of small companies that specialise in aspects of discovery or development. Thus 'life science titans' such as Monsanto, Novartis and Aventis evolve alongside a host of small research biotechnology companies such as Millennium, AMRAD,

MDS Panlabs, and Tularik, to whom, particularly in the area of healthcare, the larger companies 'outsource' an increasing proportion of their research.

The richness and complexity of the legal, political, scientific and socio-economic framework for the commercial use of biodiversity does not lend itself to generalities and simple conclusions. However, this chapter will endeavour to draw some broad conclusions from the study as a whole.

11.3 Markets for natural products

A question central to access and benefit-sharing discussions, and one which we have set out to answer in this book, is 'What is the market for genetic resources in each sector?'

For some industries – such as the seed industry, horticulture, and the botanical medicines industry – all products sold are derived from genetic resources (Table 11.1).

Table 11.1 *Global Sales for Products Derived from Genetic Resources, 1997 (US$ bn)*

Product	US$ bn
Botanical medicines	20+
Ornamental horticultural produce	16–19
Agricultural produce	300–450+
Commercial agricultural seed	30

The challenge involved in estimating global sales of finished agricultural products and horticultural fruit and vegetables cannot be overestimated. Global sales of commercial seed amount to some US$30 billion a year, but this seed is sold to farmers, who multiply it and sell it to their customers, which include distributors and industrial and food processing companies. These companies either sell produce (such as potatoes) directly to the consumer, or they manufacture products (such as bread) which are then sold to the consumer, often via other retail outlets. This complex chain makes any estimation of the global market for the products arising from global agricultural and horticultural vegetable crops extremely complex. If one assumes, for the sake of argument, that once the different products have passed through all these hands, the final value of the produce reaching the consumer is ten times the commercial sales of seed to farmers, this would suggest that the global market for agricultural produce is some US$300 billion a year. If the multiple is closer to 15, then the value of the global market for agricultural produce is some US$450 billion a year. While such assumptions and estimates are highly speculative, it seems safe to assume that the global market for agricultural produce is well in excess of the US$300 billion a year global sales of pharmaceuticals.

The position in the other sectors explored in this book is rather different. Pharmaceuticals, personal care and cosmetic products and crop protection products, can be discovered, developed and manufactured in a number of different ways, not all of which involve the use of genetic resources. Thus only a proportion of the products in each of these sectors can be regarded as derived from genetic resources. Estimates for total global gross sales in these industries, and the component of product sales that are of natural origin, are as shown in Table 11.2.

Table 11.2 *Estimated Global Sales in the Pharmaceutical, Personal Care and Cosmetics, and Crop Protection Industries, and Component of Products Derived from Genetic Resources, 1997 (US$ bn)*

Product	US$ bn	Sales of natural origin (US$ bn)
Pharmaceuticals	300	75 (using a conservative 25%)
Personal care and cosmetics	55	2.8+
Crop protection	30	0.6–3

Genetic resources and derivatives such as biologically active compounds and enzymes form an integral part of all applications of biotechnology. Estimating the value of applications of biotechnology in fields other than healthcare and agriculture, such as bioremediation, chemical processes, materials and energy is extremely difficult, since figures do not distinguish between sales of products produced by biotechnology and those produced in other ways. Also, there is little data on the final value to industry of the use of biotechnology in industrial processes. Depending on the assessment of the contribution of biotechnology to environmental biotechnologies (Chapter 8), upper and lower estimates for the value of global sales of biotechnology products for 1997, other than in the fields of healthcare and agriculture, are between US$60 and US$120 billion.

A very crude estimate of the combined market value of products derived from genetic resources in 1997 for these seven sectors, therefore, might fall anywhere between US$500 billion and US$800 billion. (The market data and rough calculations by which these figures were reached are set out in Chapters 3–9.) This figure does not account for subsistence or locally-traded products, which are not included in national and international statistics; nor does it incorporate other industry sectors that use genetic resources in other ways. It is also important to note that this figure reflects gross market sales – ie it does not distinguish between the costs associated with access to genetic resources, and those involved in product research and development, marketing, and manufacture. It offers little indication, therefore, of what companies might be willing to pay for access, nor the nature and value of the many different benefits which arise from access to genetic resources. The figures simply indicate the scale and scope of some different markets for products derived from access to genetic resources. As the following section will show, while markets for products developed from genetic resources are extremely valuable, the genetic resources required for research and development can be obtained in a number of ways that do not necessarily require payment or trigger benefit-sharing negotiations.

11.4 The nature and extent of private-sector demand for access to genetic resources

This book has also sought to answer the questions, 'What is the extent and nature of private-sector demand for access to genetic resources and traditional knowledge?' and 'Will this demand increase or decrease in the future?'

While the figures cited above suggest that there is some demand for access to genetic resources, they do not indicate the extent of the demand for access to genetic resources other than those already available to researchers, nor do they offer any indication of likely future demand. As we have shown, a number of factors influence the nature and extent of demand for genetic resources, but two in particular are likely to determine whether commercial demand for access will increase or decrease in the future: developments in science and technology, and trends in law and policy. Across all the sectors reviewed, we found broad agreement that developments in science and technology will change the nature of the demand for genetic resources, and that private-sector demand is likely to continue at current levels, and in some sectors may even increase. At the same time, however, many companies believe that the legal and policy climate could lead to a decrease in demand.

11.4.1 Scientific and technological advances

Some of the scientists we interviewed believed that current developments in science and technology would lead to a decline in demand for access to genetic resources, but many felt they would open up further possibilities for research and development and increase demand. Both sets of arguments are presented here.

Decline in demand for access

Alternative approaches to product discovery and development

Some interviewees reported that developments in science and technology are leading them away from the use of genetic resources such as plants and microbes, as they invest more in alternative approaches to product discovery and development. For example, in the pharmaceutical sector, combinatorial and synthetic chemistry, and biotechnology involving human genes offer routes to product discovery and development that many companies find more attractive than 'natural products' which rely on access to plants and microorganisms. Developments in science and technology that underpin these alternative approaches have made them economically attractive. As a result, financial and management pressures, coupled with the belief that these approaches offer more promise for the discovery of new products in the future, have led some pharmaceutical companies to either close or drastically scale down their natural products research (eg SmithKline Beecham). In the seed sector, tools such as genomics may enable companies to make greater use of existing collections, and reduce their need for new samples of genetic resources.

Selective acquisition of samples

Companies that do still seek access to genetic resources for their research and development programmes are rarely involved in amassing large new collections of natural products, as several did in the 1980s. Companies in the pharmaceutical, biotechnology, botanical medicine and personal care and cosmetics sectors commonly pursue a strategy that could be termed 'cherry picking'. This involves a focused and targeted selection of a relatively small number of samples, based on specific chemotaxonomic or biorational leads, in order to fill gaps in existing collections. Companies are often more selective in terms of the quality of samples, accepting only those for which there is adequate accompanying taxonomic, geographical, ecological and other information.

Increasing use of existing collections

Another major factor that suggests demand for access to 'new' genetic resources may decline in the future is the increasing use of materials from existing, *ex situ* collections (whether from in-house corporate collections, or from external institutions such as botanic gardens or culture collections). Although it is now more common for managers of *ex situ* collections to use MTAs to regulate access to materials within their collections, the vast majority of those materials can still be accessed without any commitments to share benefits on the part of the recipients. Seed and horticulture companies largely work from *ex situ* collections, and the use of such collections by pharmaceutical and biotechnology companies is growing.

Growing demand for access

The reasons set out above suggest that demand for access to genetic resources will decline in the future, but other arguments suggest it will grow.

New tools to explore and use genetic resources will increase demand

Many of the individuals interviewed in our survey were of the firm belief that trends in science and technology are creating opportunities for the further exploration of genetic resources, which will inevitably lead to higher demand for access. For example, developments in genomics and bioinformatics promise better experimental design and a host of new molecular targets to explore. A better understanding of biological systems will open avenues for their future exploitation, whether in pharmaceuticals or crop protection, and many believe that there will be a growing demand for access to genetic resources to support and provide new leads for this research. Recombinant DNA techniques mean that there is greater flexibility to use a wider range of genetic resources in combinations that were inconceivable just a few years ago. Developments in science and technology offer the ability to study plants, animals and microorganisms more quickly and cost effectively. This means that there may well be a growing demand for access to a wider range of species in sectors such as pharmaceuticals, cosmetics and botanical medicines.

Consumer demand for natural products

According to our interviewees, another trend in fields as diverse as cosmetics, medicine, crop protection and waste treatment, is the growing attraction of natural alternatives to synthetic chemicals. This interest is spurring the investigation of unexplored organisms which may offer new, interesting products.

11.4.2 Legal and policy environment

A large majority of the companies and other organisations interviewed report that law and policy on access is currently deterring partnerships with industry. If future developments continue along the same vein as recent laws, the demand for access to genetic resources is likely to decrease. Companies and intermediaries interviewed for this study reported a range of difficulties with the legal and policy frameworks for access and benefit-sharing, which can be summarised as, first, lack of clarity, legal uncertainty and excessive bureaucracy and, second, unrealistic expectations for benefit-sharing (see Chapter 10).

Lack of clarity, legal uncertainty and excessive bureaucracy

Lack of clear procedure or authority for providing access creates considerable uncertainty and risk, not only for companies and intermediaries, but for governments and institutions in provider countries. Companies cannot be sure that they will secure clear and free title to develop products, nor can they assess the time implications and costs associated with seeking access before they apply. Companies have also found that the 'goal-posts' keep moving, as provider-country governments and institutions change the procedures and demands for benefits during access and benefit-sharing negotiations.

Such uncertainty, and the cost and time currently required to secure access will be sufficient to deter most companies that would be prepared to negotiate in good faith. This problem is presented by companies as the fundamental reason why demand for access to genetic resources is likely to decrease in the future. Interviewees felt that conscientious companies may decide not to seek access to genetic resources in these circumstances, but others may respond by endeavouring to evade legal obligations. This is often at the instigation of in-country partners who are anxious to pursue or maintain collaborations but fear that this may be impossible if they 'play by the rules'. Difficulties of this kind reported by companies include:

- problems identifying government officials empowered to negotiate and grant access to genetic resources;
- the large number of officials from different departments and different levels of government with authority to determine access, who are thus involved in each access negotiation;
- the lack of clear guidance on whose consent is needed to ensure that 'mutually agreed' activities have satisfied all legal agreements exhaustively;

- difficulty in reaching a level of confidence that, once an agreement has been successfully negotiated with the appropriate institutions and authorities, the need to obtain consent from others will not arise in the future; and
- the time needed to negotiate and reach agreement with all the officials concerned, as well as other stakeholders such as local and indigenous communities.

Unrealistic expectations for benefit-sharing

Another factor that dampens companies' enthusiasm for access to genetic resources is their common perception that the benefits demanded in return for access are disproportionate, considering the relative contributions to the final product made by the genetic resources and the investment in research and development.

Many interviewees felt that there has been unwarranted 'hype' about the value to industry of access to genetic resources and the contribution that 'bioprospecting' could make to a national economy. They believe this has raised unrealistic expectations on the part of developing countries as to the value of the benefits that can be gained in exchange for access to genetic resources, and that this misapprehension is a barrier to partnerships which could be mutually beneficial.

A large number of companies also felt unable to agree to provider countries' requests to publish the nature of their agreements, or to share research results or transfer technology before patent protection has been obtained. Accidental or deliberate disclosure could threaten companies' ability to obtain the patent protection which is often all-important to industry as well as affecting the likelihood of significant benefits that could be shared with countries of origin.

11.4.3 Demand for access to traditional knowledge

Private-sector demand for access to genetic resources is often accompanied by demand for information and knowledge associated with the traditional use, formulation, and management of the material in question. The extent of private-sector demand for traditional knowledge varies significantly across sectors, and from company to company. Several of the chapters in this book have documented how traditional knowledge has played an important historical role in the development of pharmaceutical and botanical drugs, new varieties of agricultural and horticultural crops, crop protection products, and personal care and cosmetic products. This information continues to play a role in industry research and development programmes today. Traditional knowledge is still used as follows:

- *Pharmaceuticals:* In a very small minority of industry discovery programmes, traditional knowledge is collected alongside samples, and is used to guide research strategies. However, in most pharmaceutical companies, traditional knowledge is accessed by consulting databases and academic literature once a lead has shown promise.
- *Botanical medicines:* Traditional knowledge is used as the basis for identification of potential leads, and to assist in safety and efficacy determinations and in product marketing. In almost all cases, traditional knowledge is acquired through consultation of databases, literature, and through middlemen and brokers, rather than directly through field collections on the part of companies.
- *Crop breeding:* Private seed companies make little use of traditional knowledge concerning crops in their breeding programmes, but they do use germplasm that has been pre-bred by other organisations to which genes from traditional varieties may have made an important contribution.
- *Horticulture:* Many popular ornamental varieties and many horticultural vegetable crops owe their existence to domestication and selection over long periods of time. However, traditional knowledge is rarely used in the selection and breeding of new horticultural varieties today.
- *Crop protection:* A very small proportion of the companies interviewed use ethnobotanical knowledge to guide the collection or screening of samples, but many companies consult databases containing information on local uses of plants to help them decide whether to pursue or to abandon research in certain areas. While a minority of interviewees said they would prioritise the screening of samples obtained with ethnobotanical data, the royalty ranges offered for materials with this information do not appear to be materially higher than those offered for samples or extracts supplied without it.
- *Biotechnology (other than healthcare and agriculture):* While many biotechology applications – such as brewing and bread-making – are based on traditional knowledge dating back millennia, contemporary biotechnological research appears to make little use of ethnobotanical or traditional knowledge. All interviewees were questioned and just one interviewee involved in breeding trees for the paper and pulp industry described an interest in ethnobotanical data. Traditional knowledge was not mentioned by any of the interviewees in this sector as a factor that influenced benefit-sharing.
- *Personal care and cosmetics:* Most personal care and cosmetic companies include a natural line today, but the role of natural products in industry research and development remains minor;

however, this industry has its roots in traditional uses of natural products and marketing materials often make reference to traditional uses. In some of the larger companies, more serious screening and research and development efforts are underway which build in part upon traditional knowledge. Traditional knowledge is acquired in most cases via databases, literature, and industry middlemen.

11.5 The nature and extent of benefit-sharing across industry sectors

Benefit-sharing varies dramatically across and within sectors, and recent trends respond in varying degrees to international policy developments such as the CBD. Common practices in the sharing of monetary and non-monetary benefits in each industry sector are discussed in the respective chapters, as are some of the more progressive projects involving benefit-sharing packages designed by companies, governments, research institutions and NGOs. Here we will provide general conclusions for each sector, and examine some differences between sectors that have implications for the development of policy and practice in access and benefit-sharing. There are striking differences between sectors, which suggests that it will be necessary for access and benefit-sharing measures to be flexible enough to reflect this diversity.

Many companies from all sectors are willing to pay fees for samples, and, in some cases, royalties on net sales. Willingness to share non-monetary benefits is mainly confined to collaborative research relationships; however, these embrace not only the more value-added activities of discovery and development, but also fairly basic activities like the processing of extracts. Across all sectors, access is often de-coupled from benefit-sharing. In many cases, materials existing in *ex situ* collections are available without the obligation to share benefits. In others, access and benefit-sharing are severed when benefits arise many decades after the original access, and in a completely different part of the world. A prime example of this is SmithKline Beecham's anti-cancer drug Topotecan, based on samples of *Camptotheca acuminata* sent from China to the USA in 1911, and put on the market in the USA in 1996, more than 80 years later (Chapter 3). Access to traditional knowledge is often de-coupled from benefit-sharing because it is most commonly sought by companies through literature and databases.

Overall, there is a gradual but palpable trend towards more creative benefit-sharing, involving monetary and non-monetary benefits in the short, medium and long terms. This trend has been driven by a confluence of interests, and reflects the evolution of public opinion, NGO advocacy, private-sector responses to requests from source country collaborators, and the initiative of intermediary institutions that broker interests of provider country authorities and commercial users. Widely publicised cases such as the partnership between INBio and Merck have served to some extent as a template for the development of subsequent benefit-sharing arrangements. However, there has been a growing appreciation that what is 'fair and equitable' is likely to differ substantially according to industry sector, product area and individual research and development programme, and that successful benefit-sharing arrangements are those tailored to the specific circumstances of an individual case. Experience and 'best practice' in benefit-sharing have progressed quite significantly on a number of fronts in the decade since the concept first emerged.

Non-monetary benefit-sharing – whether in the form of research collaborations, supply of literature, or in-kind benefits such as medical assistance – is viewed by some companies as a form of charitable contribution rather than as part of the cost of research and development. Other companies consider these forms of benefit-sharing as integral to any collaborative partnership with providers of genetic resources, and an essential requirement for business that allows them to secure access to high quality samples and to work with high calibre collaborators.

11.5.1 Pharmaceuticals

The global market for pharmaceuticals and the research budgets and profit margins of companies in this sector are relatively high compared with those of other industry sectors using genetic resources. This fact, and the extensive collecting programmes to fuel the high-throughput screens developed in the 1980s, have led to greater experience in establishing benefit-sharing agreements in this sector than in any other. The partnerships of pharmaceutical companies were among the first to attract international attention before the principles of PIC and benefit-sharing were articulated in the CBD. Today it is usual for a pharmaceutical company to pay royalties on net sales of commercial products, which was not the case ten years ago. Milestone payments at key stages in the development process, in addition to the initial fees for samples or grants to cover research, are also common. The factors which affect the magnitude of royalties are fairly consistent, and tend to be reflected by specific clauses in supply contracts. They include:

- the current market rate for royalties;
- the likely market share of a given product;
- the relative contribution of the partners to the inventive step and development;
- the degree of derivation of the final product from the genetic resource supplied; and, for some companies,
- the provision of ethnobotanical data with the sample.

As far as non-monetary benefits are concerned, the approach of many pharmaceutical companies and their suppliers has evolved over the last ten years, so that there is a discernible trend towards a 'package' of monetary and non-monetary benefits delivered across time. Furthermore, the capacity of many countries harbouring biological diversity to engage in value-adding research has grown over the last decade, and companies are increasingly open to collaborations with provider-country institutions if they are confident of the quality and cost effectiveness of their work.

Botanical medicines, personal care products and cosmetics

Benefit-sharing in the botanical medicine and personal care and cosmetics sectors has developed primarily in the context of the supply of raw materials for the manufacture of products. These industries do not depend upon large numbers of samples for screening programmes (the exception being the cosmetics industry's recent incorporation of high-throughput screening as a discovery tool), and development of new products tends to be targeted and based on information derived from literature and databases. In some cases, however, companies have established partnerships for the sourcing of raw materials for manufacture that include the transfer of technology and capacity building.

Access and benefit-sharing is often severed at both the level of traditional knowledge of a species use, which is sourced from literature and databases, and the genetic resource itself, which might be grown outside its original habitat, and treated as a commodity. *Panax vietnamensis* and kava (Chapter 4), are exceptions to this rule. Demand for access to new species is increasing, alongside developments in scientific and technological capacity that allow for more efficient and affordable study and testing of natural products. It is therefore likely that these industries will require closer relationships with source countries and communities in the future. In these circumstances, benefit-sharing packages can be designed to link not only raw material supply, but also commercial research and product development activities, to local capacity- and institution-building.

Biotechnology applications other than in healthcare and agriculture

Biotechnology companies often obtain for free samples that were collected by university researchers. Licensing agreements for access to value-added genetic resources and biotechnologies are rarely seen by companies in this sector as 'benefit-sharing', but rather as an inevitable part of the bargain in order to maintain access to quality samples, to enjoy the advantages of collaboration with high calibre scientists, and to remain competitive in the future. Many of the smaller, start-up companies, often spun off from universities, are unfamiliar with the CBD. Several believe that the microorganisms which form the mainstay of so much of the biotechnology sector are not 'covered' by the CBD, and most are inexperienced in negotiating licensing and benefit-sharing arrangements. Rather than initiating benefit-sharing agreements of their own, these companies follow the lead of intermediary organisations such as culture collections, which are increasingly supplying materials under MTAs. Genetic resources may have passed through many hands (and, in the case of microorganisms, are often deposited in and accessed from culture collections) before reaching the biotechnology company that ultimately commercialises a product. The link between access and benefit-sharing is thus often broken. Benefit-sharing with 'source countries' is relatively rare, and usually confined to occasions when companies collect genetic resources themselves, or establish access arrangements with intermediary institutions overseas.

Comparatively few companies are accustomed to developing formal benefit-sharing agreements to comply with access legislation, but such agreements as do exist typically involve technology transfer and training as well as commitments to pay royalties. However, the sharing of information and capacity building often arise informally in business relationships. Those who obtain their materials, especially derivatives such as enzymes, exclusively from other suppliers, are generally unfamiliar with the CBD. Their 'benefit-sharing' extends only as far as the purchase price or licence fee for the derivative concerned.

Major crops

The seed industry involved in the development of major crops approaches benefit-sharing in a very different way from the pharmaceutical industry. The exchange and commercialisation of plant genetic resources for food and agriculture have also been the subject of public scrutiny and intergovernmental negotiations for more than a decade, but benefits are shared in a much more indirect fashion in the seed industry than in the pharmaceutical industry. Our interviews with seed breeders revealed that it is still common for many seed companies to obtain genetic resources for free or for a nominal handling fee, particularly if the germplasm acquired is 'unimproved', although licensing agreements are common for access to elite germplasm. Many actors are involved in the chain from initial access, through pre-breeding and commercial development, to sale of the final product to the farmer or consumer. The gradual privatisation of the seed industry in many parts of the world, and the growing use of licences as more seed is patented, mean that sophisticated agreements do occur towards the end of this chain, but benefits do not pass back directly along the chain to each contributor, particularly as the vast majority of the materials used have been obtained from collections maintained by seed companies themselves, or by national governments. Several seed breeders we interviewed from both the public and private sectors felt that the increasing use of IPRs, and not developments in law and policy on access and benefit-sharing, were driving change in partnerships in the seed industry. The majority of researchers in agriculture view unrestricted, reciprocal access to genetic resources as the major benefit 'shared' through the current informal system of exchange, although the sharing of research results, access to technology and capacity building also take place, predominantly in public-sector, but also in private relationships. These benefits are often not limited to access to specific germplasm, but flow between institutions engaged in collaborative research that involves access to germplasm. At the national level, tax revenue from the seed industry supports research and pre-breeding in the public sector, including some extension services to farmers. Industry also supports research in academia directly, by endowing chairs and funding basic research programmes in universities. At the international level, donor governments support the work of the CGIAR.

Horticulture

'Benefit-sharing' is almost unheard of in the field of commercial ornamental horticultural development, particularly among the many small companies and amateurs involved in collecting and breeding ornamental horticultural plants. Several of the ornamental horticultural companies and institutions interviewed had never heard of the CBD. Few partnerships have yet been structured with the CBD in mind. However, commercial arrangements do exist that involve royalties and payments of fees. A number of 'in kind' benefits may also be shared. Among these, reciprocal access to plant material between non-commercial organisations is important. Another common form of benefit-sharing is the acknowledgement of the name of the provider of genetic resources (whether a grower or an amateur) in the name of the cultivar subsequently developed by a breeder. Some of the horticulture companies interviewed have sponsored student placements within the company or enrollment on courses of higher education. Others have provided books, computers and other equipment to collaborators who face difficulties in obtaining such materials themselves.

Pharmaceutical and agricultural seed companies may 'outsource' work and enter into joint ventures and collaborative research agreements that provide a framework within which technology transfer can take place. By contrast, horticulture companies that breed new ornamental varieties tend to conduct the majority of their research in-house. They are open to acquiring germplasm – ranging from wild-collected material to the result of sophisticated breeding programmes – from other breeders, but it is comparatively rare for them to engage in joint research with other companies on the development of new ornamental germplasm. Collaborative research involving technology transfer and capacity building does take place within the ornamental horticulture industry, but tends to centre around the development of production protocols for the commercial production of new varieties of ornamental plants once the breeding programme has been completed.

Crop protection products

Crop protection products are often developed by departments or subsidiaries of companies that are involved in pharmaceutical development. Libraries of compounds and gene-based materials may be tested for use in both areas, and the technologies for sourcing and screening materials are also broadly similar, although the economics of product development in the sectors are extremely different. Where such links with pharmaceutical companies exist, crop protection companies are more likely to be familiar with the CBD, with the kinds of partnerships available in the pharmaceutical sector, and more prepared to enter into contractual arrangements with suppliers that involve benefit-sharing. However, the crop protection industry has some similarities with the seed industry in that basic research into crop protection is most often conducted in the public sector (from where genetic resources are passed, often for free, to industry). Furthermore, most crop protection companies have synthesis programmes which use strategies involving model compounds based on templates originally discovered from nature as natural products. To date, some crop protection companies have seen little rationale for sharing the benefits that arise from the development of such products with the provider of the genetic resources that led to the discovery of the template. This is in contrast to pharmaceuticals, where royalty arrangements generally guarantee benefit-sharing, however modest, even for wholly synthesised analogues.

11.6 The role of intermediaries

In the vast majority of cases, and across sectors, companies gain access to samples through intermediaries such as botanic gardens, universities, research institutions, genebanks, and for-profit brokers. In addition to providing collection and scientific services, these intermediaries – many based in the North – tend to broker access and benefit-sharing relationships with source countries, sometimes as agents of companies, and sometimes independently. In some instances, a number of different intermediaries may be involved between the initial collection stage and ultimate commercialisation. Consequently, intermediaries play a key role in determining benefit-sharing relationships.

Companies rarely collect on their own behalf, and may be reluctant to negotiate directly with source countries. They frequently turn to intermediaries to obtain genetic resources, either by collecting in the field or by providing samples held in *ex situ* collections. Many are anxious for intermediaries to shoulder the responsibility of satisfying legal and administrative requirements for access, and in particular are unwilling to become involved in direct negotiations with governments or local stakeholders, such as local authorities, institutions, and communities. Intermediaries are thus increasingly relied upon not only to collect, identify, and guarantee re-supply of promising materials, but to acquire government approval for collections, to broker benefit-sharing agreements, and even to take responsibility for seeing that any benefits that arise are shared fairly and equitably in the source country. This may be compared with the trend in access and benefit-sharing legislation, which seeks to establish links between companies conducting research on genetic resources and institutions in the countries that provided them.

As a rule, the agreements between intermediaries and companies deal only with the rights and benefits of the immediate parties, such as royalties for the intermediaries, should a commercial product develop. The agreements rarely touch on the rights of third parties such as source country institutions. In the majority of cases, the resources change hands under conditions that do not check that PIC has been obtained, and that neither require benefits to be shared with the original provider nor restrict transfer to third parties.

Companies rarely seek written representations from intermediaries that the intermediaries obtained the genetic resources lawfully, or that they will share their benefits within the country of origin. When interviewed, most companies used language such as 'we trust' or 'we assume' when referring to the authority of intermediaries to provide them with genetic resources. Others stated that they did not check that intermediaries were able to pass them clear title to the genetic resources, but rather relied on working with 'high quality' or 'reputable' institutions. As few intermediary institutions currently have policies on access or benefit-sharing, and are often unfamiliar with the provisions of the CBD or national measures on access, this trust may be misplaced. A few companies interviewed do require intermediaries to give a positive undertaking in the agreement between them that the intermediary is entitled to provide the genetic resources to the company. Other companies insist on seeing permits and other paperwork to satisfy themselves that the materials were lawfully obtained and that there is thus unlikely to be any challenge to the company's title in the future.

For these and other reasons, access to raw materials for product discovery in most industry sectors is usually obtained on terms that do not trigger the sharing of benefits through partnerships with provider countries or institutions. The connection between countries providing genetic resources and product development has been severed, and there is no mechanism by which benefit-sharing can take place. The critical role of intermediaries in brokering access and benefit-sharing relationships, and the tendency for benefit-sharing to be 'de-coupled' from access, as discussed above, are important considerations for those drafting access and benefit-sharing regulations.

11.7 Views on the CBD, and corporate and institutional policies

As we have seen in the previous chapter, companies' responses to the CBD are mixed, and the development of best practice in industry varies accordingly. For the most part, awareness of the CBD was highest within the pharmaceutical, crop protection, seed, and biotechnology sectors, although direct impacts on corporate business practices are greatest in the pharmaceutical sector. Horticulture, botanical medicines, and personal care and cosmetics sectors tend to be largely unaware of the content of the CBD. Companies in the pharmaceutical sector reported the most experience with access and benefit-sharing measures and the CBD policy process, and a number of companies have drafted policies on the acquisition of natural products.

Awareness and experience of partnerships that reflect the CBD vary enormously between sectors, from company to company even within single companies, where individual researchers and management staff may differ in their perspectives. However, awareness of the CBD is growing rapidly, and more and more companies report that they are changing their business practices in response to it. The most common changes in business practice as a result of the CBD include:

- a decrease in corporate collecting activities; consolidation of collecting programmes into fewer countries, or even to domestic collecting activitites;
- greater recourse to material from *ex situ* collections, such as culture collections and compound libraries, in place of samples acquired through collecting activities;
- an increased role for intermediaries as brokers of access and benefit-sharing relationships, as well as suppliers of samples; and
- the increasing use of MTAs.

11.8 Access and benefit-sharing, sustainable development, and the conservation of biodiversity

The CBD emerged as a multi-faceted treaty encompassing the link between conservation, sustainable development, and equity. It reflects the complexity of biodiversity conservation, its close ties to commercial activities associated with genetic resources and growing awareness of the need for a fair and equitable sharing of the benefits arising from access to genetic resources. As a result, the private-sector activities discussed in this book are integral to implementation of the treaty, although forming only one part of the whole.

This book has focused on the nature of demand for access to genetic resources, and benefit-sharing practices in seven industry sectors, and did not set out to examine the cumulative effect of access and benefit-sharing agreements on either the conservation of biological diversity or the development paths of countries. However, the practices we have described lead to the following conclusions on the potential contribution of the commercial use of biodiversity to sustainable development and conservation.

Commercial activities involving genetic resources can provide direct benefits for conservation programmes and protected areas in the form of financial benefits for park systems, projects, and government departments involved in biodiversity conservation (eg the dedication of 10 per cent of research funds and 50 per cent of royalties by INBio to the Ministry of Natural Resources in Costa Rica; the dedication of funds from the CALM-AMRAD partnership to conservation). They can also provide a source of funds to support activities related to conservation such as resource surveys, taxonomic research and inventories, and other activities integral to ecosystem and species management (eg the taxonomic work of the Sarawak State Department of Forests and UIC under a collecting agreement with the NCI, and INBio's national inventories funded by biodiversity prospecting collaborations).

Sustainable economic activities based on biodiversity can boost sustainable development, providing an incentive for conservation and an alternative to more destructive income-generating schemes. For example, within the pharmaceutical, botanical medicines, personal care and cosmetics industries, the basis for employment and income-generation associated with the sustainable supply of raw materials has been created in some cases (eg the work of Shaman Pharmaceuticals with *Croton*; the work of NCI on *Ancistrocladus*; the Aveda Corporation's bixa programme in Brazil, and The Body Shop community trade programme). The provision of value-added derivatives of genetic resources such as extracts for screening; pre-bred materials for crop development, or the processing of material into finished products for local, regional, or international markets, also provides jobs. Markets to supply such value-added materials can promote sustainable development in several ways, including by creating employment, by supporting trade in higher value products, by generating export revenues, and by substituting for imports.

Sustainable development is also served by the capacity building and technology transfer that result from commercial research collaborations. For instance, efforts are currently underway in a number of countries, many with the support of the WHO, to study and standardise traditional medical systems. The skills and capacity needed to undertake these studies and integrate the results into national and local healthcare, are important benefits that can be developed through partnerships involving access to genetic resources, and that can result in improved and affordable local healthcare (eg Balick et al, 1996; Akerele, 1991). Scientific capacity to study tropical diseases and locally-important health conditions and agricultural problems can be improved by biodiversity prospecting research collaborations (eg the African and Suriname ICBG cases, Chapter 3; and the LUBILOSA case study, 7.4.1). Partnerships with companies can also provide local institutions with training, technology, access to market information and other forms of capacity building that will allow them to develop relationships and work more effectively with the private sector. They can also help to build local programmes and domestic or joint venture companies (eg Sarawak-Medichem Pharmaceuticals, Chapter 3; the *Panax vietnamensis* case, Chapter 4; the Pipa horticultural joint venture company, Chapter 6; the partnership between INBio and Diversa, Chapter 8; the Aveda-Yawanawa collaboration, Chapter 9).

An obvious, if often overlooked, conclusion is that long-term partnerships involving more than a one-off transfer of 'raw' genetic resources provide the best context for benefit-sharing. These can involve a range of benefits, some of which might have immediate application to conservation objectives, others of which could incrementally build domestic capacity to conduct research on tropical diseases, for example, or to develop products for local and regional markets. Benefit-sharing partnerships that involve a range of stakeholders, including local communities, research institutions and governments also help to spread benefits so as to make the maximum contribution to conservation and sustainable development. However, simple partnerships may be more efficient and cost-effective, and companies will only be interested in working with stakeholders who can contribute to profitable – and sustainable – business activities and positive public relations. Companies cannot be compelled to enter into benefit-sharing arrangements, but only encouraged to include various organisations in their activities. However, governments can set policy frameworks that protect basic rights and create incentives for partnerships that serve wider national interests such as conservation and sustainable development, in addition to supporting profitable enterprise.

Note

1 Article 2 of the CBD defines 'genetic resources' as 'genetic material of actual or potential value', and 'genetic material' as 'any material of plant, animal, microbial or other origin containing functional units of heredity'.

Kerry ten Kate and Sarah A Laird

Chapter 12 Recommendations

12.1 Overview

This chapter is divided into recommendations on access regulations, which are mainly directed to governments; recommendations concerning the role of intermediary organisations, which are directed both to governments and to the intermediary organisations themselves; and recommendations on corporate practice, which are directed to companies and industry associations. It also proposes a number of indicators which governments, companies and other institutions could use to assess whether access legislation and individual partnerships are fair and equitable, and some elements that companies and other organisations might consider for use in institutional policies on access and benefit-sharing.

12.2 Recommendations to governments on access regulations

The recommendations to governments regulating access centre around the need for simple, unbureaucratic, streamlined and flexible access regulations. Some of the recommendations concern the process by which access regulations are developed and administered, and others concern the substantive provisions contained in access regulations.

- **Participation**. Governments will need to consult a wide range of stakeholders in order to draft effective access and benefit-sharing regulations. Stakeholders include local communities, scientists, research institutions and *ex situ* collections, as well as a wide range of governmental departments including science and technology, health, industry, finance, justice, and environment. Primary responsibility for the CBD at the national level often rests with the Ministry of the Environment, but its access and benefit-sharing provisions require a good knowledge of recent developments in science, technology and industry, which expertise may be found in other departments and in the private sector.

  The development of access regulations requires a two-way process: not only do regulators need to understand more about product development, but companies need to know more about the CBD and the opportunities to participate in the policy-making process.

- **A team of industry advisors.** This book has illustrated the enormous diversity among companies in the same industry sector, let alone the differences between industry sectors. Before developing rules or procedures on access, governments need to be familiar with these, since access and benefit-sharing measures are likely to be counter-productive if they are not grounded in the practical realities of the businesses affected by them. Regulators need more information on the size and nature of the different markets for genetic resources and products derived from them, as well as the manner in which different industry sectors undertake research and development.

  If source country institutions know what companies are looking for in partnerships, they can develop the appropriate skills and form partnerships that will bring valuable benefits to the country. Regulators could consider forming a team of industry advisors by establishing links with their Ministry of Trade and Industry, or even with a local business school. The team could follow developments in each industry sector that requires access to genetic resources, and keep abreast of technological developments, companies' research strategies and what companies are seeking from partnerships. This kind of information could help provider countries set realistic objectives and demands for benefits when negotiating access arrangements.

Once the framework of access laws has been developed, scientists negotiating access and benefit-sharing arrangements need to be well informed on the CBD, access legislation and best practice in benefit-sharing, in order to establish what is fair and equitable in individual circumstances. Regulators could consider asking their 'industry advisors' to prepare briefing documents for institutions that will negotiate access and benefit-sharing arrangements, including a list of questions for them to ask potential commercial partners in order to elicit the information needed to design an appropriate access and benefit-sharing agreement.

- **National bioprospecting strategy**. Understanding the needs of users such as industry is just one part of the information required to develop successful access legislation. Prior to drafting access legislation, regulators should involve a range of stakeholders in developing a national access and benefit-sharing strategy (ten Kate, 1995). This strategy would involve an assessment of the opportunities for partnerships (perhaps in conjunction with 'industry advisors', as above). It would also identify the needs and priorities of the country in terms of technology transfer, capacity building, local diseases and requirements for other products, national biological resources and human and institutional capacities. The strategy would then identify particular targets, such as viable national industries based on genetic resources, a trained workforce with skills in particular areas, or national research programmes geared to curing significant diseases in the country not addressed by international pharmaceutical companies. Specific programmes would set out how the country could move towards the agreed target, for example by developing a 'critical mass' of institutional capacity and trained staff.
- **The focal point**. The biggest frustration and difficulty experienced by scientists and companies world-wide is the problem of identifying which government agency is authorised to grant prior informed consent for access. Governments should nominate a focal point on access to genetic resources. Even if several departments and levels of government are involved in authorising access to genetic resources, the focal point could coordinate applications for access so applicants would know exactly whom to contact. It would be extremely helpful if each focal point would prepare a short paper explaining when and from whom PIC is needed in the country concerned, and describing the procedure involved.

  Focal points must have the knowledge and experience necessary to process permits, which is why the access legislation of several countries establishes technical advisory bodies to assess the merits of access applications and proposed benefit-sharing arrangements. Well-intentioned companies seeking PIC have expressed concern at dealing with government staff without the necessary knowledge of commercial, scientific, market, and regulatory trends.
- **Simplicity and cost-effectiveness**. Regulators of access need to coordinate and streamline the procedure for government representatives from different departments and national, regional and local government to review access applications and negotiate access and benefit-sharing agreements. The substantive and procedural requirements for access and benefit-sharing need to be proportionate to the costs and time-frames of research and development. Almost all the companies interviewed pointed out that if access regulations were too bureaucratic and expensive to follow, this would make product discovery and development involving genetic resources uncompetitive. They would no longer seek access and would instead pursue alternative approaches, such as synthetic chemistry, or make more extensive use of their own collections.
- **Flexibility.** The procedures required of applicants for access and the nature and value of benefits demanded of them in access laws and in individual negotiations will need to be flexible in order to accommodate the many different potential uses of genetic resources, and thus of corresponding benefit-sharing arrangements. If access legislation is to be flexible enough to encompass all uses of genetic resources, for example taxonomic study, pharmaceutical development and horticulture, it will need to set basic procedural and substantive requirements, and then allow for the individual negotiation of benefit-sharing terms, since the transaction costs and the benefits that arise from different uses vary so greatly. Alternatively, different regimes for access will be needed for different uses of genetic resources.

  Measures introduced by countries on access and benefit-sharing should leave room for the development, on a case-by-case basis, of benefit-sharing partnerships which yield more significant returns for source countries than rigidly defined requirements for monetary benefit-sharing would allow.

A national strategy for access and benefit-sharing could help to create a broad framework that would identify national priorities for benefits but be open and flexible enough to stimulate a broad range of benefit-sharing partnerships. Within this framework, governments could allow collaborating institutions within their countries the freedom to identify their own priorities for benefit-sharing and develop innovative ways to maximise upon collaborations with industry. Indicators for the fair and equitable sharing of benefits, such as those proposed below, could help governments to oversee these decentralised activities.

- **Clarity of scope and terms.** Government authorities or other institutions providing access to genetic resources should ensure that any access and benefit-sharing agreement states clearly the uses to which the provider is willing for the genetic resources and information concerned to be put. Access agreements should clarify the rights of the recipient to pass materials to third parties, typically by obliging him or her not to pass the materials on to any third parties, or at least to ensure that the third parties are bound by similar terms. These terms should include the obligation to return to negotiate the conditions for any subsequent use different from that with the provider government.

  The scope of access regulations should be clear. Sometimes the position with respect to certain categories of genetic resources, such as microorganisms or human genetic material, is not clarified, and the geograpical scope of the regulations is also often ambiguous. Several interviewees were concerned about various 'loopholes', particularly the role of academics from domestic universities and research organisations, who collect without access restrictions, but then pass material to companies. This is not popular with companies who adhere to access requirements, since it is seen to undermine the entire access system, as well as their competitiveness. They recommend that academic and government institutions be regulated in the same way as companies. By contrast, several intermediaries interviewed felt there should be a clear distinction between collection for commercial research and collection for purely scientific research, as has been tried in the Philippines. A solution to this apparent dilemma is to provide for a 'fast-track' access procedure for scientific and taxonomic institutions, but to ensure that they understand that the terms of the access agreement limit their use to purely scientific purposes and restrict their freedom to transfer specimens to third parties.
- **Legal certainty**. Access legislation should be designed in coordination with other national law (such as a country's constitution, legislative provisions on the rights of indigenous communities, and law on property rights). This should ensure that all rights and interests are taken into consideration, and that once an applicant for access has secured the necessary consents, he or she has good legal title to use the genetic resources in the manner described in the access agreement, and is not subject to legal challenge in the future, provided he or she honours the terms of the agreement.
- **Confidentiality.** Access legislation needs to allow for certain information to be kept confidential. MTAs can contain clauses stipulating the extent to which the various partners are entitled to use, or prohibited from using, information and research results that arise during a collaboration.
- **Conservation.** Access legislation can incorporate components that not only require collectors to source materials sustainably, but specify that a share of benefits arising in access agreements – whether these benefits are monetary or in-kind – should be dedicated to conservation activities.
- **Enforcement**. Companies committed to honouring the letter and spirit of the CBD are keen for governments to take steps to enforce access regulations. Monitoring and enforcing access laws and agreements is notoriously difficult, since it is so easy to exchange genetic resources. However, countries can take a number of steps. One obvious step is an aggressive education campaign within academic and corporate circles in the country, so that these institutions are aware that they should not simply post genetic resources out of the country, or provide them to other institutions without access and benefit-sharing agreements. Another step is to train the staff at Customs and to coordinate with them, so that they can monitor the entry and exit of individuals from organisations that are likely to collect genetic resources and check when they leave the country with samples whether they have the correct documentation. Another option might be to train Customs staff to work with the postal services to check mail from institutions involved in the handling of genetic resources on a random basis. These activities might build upon infrastructure already in place to implement CITES.

12.3 Recommendations on the role of intermediary organisations

- **Awareness-raising.** Intermediaries include for-profit brokers, universities, research institutes, culture collections and botanic gardens. All these kinds of organisation collect or otherwise acquire genetic resources, and pass them on to other organisations, including to industry. Increasingly, companies are relying on intermediary organisations not only to collect genetic resources, but also to broker access and benefit-sharing arrangements. It is critical that intermediary organisations are aware of the responsibilities incurred by the collection and supply of material to companies, and the pivotal importance of collectors' activities to the successful implementation of the CBD. Since intermediaries are often completely unaware of the CBD and the requirements of national laws, awareness-raising programmes will be needed, at both the international level (for example, through professional societies or umbrella organisations representing botanic gardens, universities, genebanks, zoos, etc) and the national level.
- **Clarity of access regulations.** As explained above, the application of access regulations to domestic intermediary organisations needs to be clear, as does any distinction between academic and commercial access agreements, and any associated restriction on the intermediary's use of the genetic resources it acquires.
- **Proof of PIC and title.** Companies are increasingly requiring proof by intermediaries of their authority to provide genetic resources. Since access rules and procedures are unclear in most countries, this poses a considerable challenge to intermediaries. Government authorities will need to work with intermediary organisations to develop appropriate access regulations and to train intermediaries in how to obtain the requisite consents and how to use MTAs.

  It is important that organisations' budgets for collecting expeditions include an element to cover the time and expense of seeking PIC from governments and other stakeholders. Some general guidance or detail on these activities could be included in agreements between companies and intermediaries.
- **Position of *ex situ* collections.** Policy makers and intermediaries need to clarify the situation for access to *ex situ* collections, particularly those acquired before the entry into force of the CBD. In the field of agriculture, this topic is under consideration during the revision of the IU. In the meantime, managers of *ex situ* collections and other intermediaries can develop voluntary approaches, for example by clarifying in a written policy the basis on which they will obtain access to genetic resources, and on which they will provide access to the genetic resources within their collections.
- **Clear institutional policies.** Because intermediaries commonly broker access and benefit-sharing arrangements and exchange genetic resources in countries where there are no clear rules on access, it is important that intermediary organisations develop policies to articulate their approach to acquiring PIC, reaching mutually agreed terms, and sharing benefits. To the extent possible, these policies should be harmonised within the collecting community, to promote a shared understanding of 'best practice', to ensure that the rules are workable for different kinds of intermediary institutions, and to avoid 'undercutting' by unethical collectors.

12.4 Recommendations to industry

- **Raise awareness.** A process of awareness-raising and consultation with industry is badly needed. Many companies are unaware of the ways in which the CBD may affect their research and development activities.

  Companies and industry associations should draft briefing documents for their staff and members which address the challenges posed by the Convention, and should identify necessary changes in company practices in response to the CBD and access legislation. Existing documents offer scant information on the provisions of the CBD on access and benefit-sharing, despite the fact that this aspect of the CBD will directly affect business practices.
- **Honour international and national law.** Several interviewees stressed how important it was for the private sector to honour the letter and spirit of the CBD and access legislation. As one company put it, 'Do the right thing. Don't flout the regulations – it makes life difficult for everyone'.
- **Develop a corporate policy.** Compliance with the CBD is an inter-disciplinary and cross-cutting issue in much the same way as is corporate strategic planning or environmental management. In order to train all staff in the necessary procedures, to

ensure a consistent approach within the company, and to be able to communicate the company's approach on access and benefit-sharing, companies should institute a process involving all relevant staff in the development of a coordinated, corporate policy on access and benefit-sharing. Possible elements of such a policy are set out below.

- **Rationalise collecting activities**. Since it is increasingly hard to obtain access to new samples from many countries, several companies interviewed recommended that their peers should focus on collecting in their own back yard, and exchanging more samples with other companies. However, many interviewees believed that it was both possible and important to secure access to more diverse samples. Companies could achieve this by concentrating their collecting efforts in just a few countries, with whom adequate access and benefit-sharing agreements would be negotiated.

  Companies should also ensure that all collecting activities are undertaken in a sustainable way. There have been a few cases of over-collection of material from the wild, particularly of species that have already shown some commercial promise. Corporate policies should include language committing the company to sustainable supplies of wild-harvested samples and raw materials for discovery, development and subsequent manufacture.

- **Manage partnerships**. Several interviewees recommended that companies should use agreements to record rights and responsibilities, and to demonstrate full compliance with the CBD and access legislation. As one company put it, 'Create a paper trail in case you are challenged, and be clear on the terms and conditions under which you received the resources.'

  Others recommended that it was important to form links with appropriate government authorities. While business relationships are generally between companies or scientific institutions, clearance from government is now the only way to ensure risk-free title to genetic resources.

  Another recommendation from interviewees was to offer fair deals in partnerships, and only to employ reputable people on whose honesty a company can rely. Several interviewees felt that the best way to secure access to genetic resources in the future was to find an intermediary institution in each source country from where the company wishes to obtain materials, and give them guidelines under which to obtain specimens. Such an approach would require the intermediary to obtain all the necessary consents prior to obtaining the material and passing it to the company for potential commercial use. Intermediary organisations will need to be well-prepared to assume this role (12.3 above). In addition, companies will need to give credit for the extra time, skills and resources that such a role requires of intermediaries, and to share the financial and administrative burden with them.

- **Influence international negotiations**. Participation in UN negotiating sessions is a particularly unrewarding experience for the private sector. Companies find the inevitable delays, the loosely structured debate, the erratic agendas, the emphasis on negotiated text rather than concrete results, and the limited opportunities for participation extremely frustrating. The private sector does, however, need to be more in tune with developments in intergovernmental negotiations, and more closely involved in the formulation of national negotiating positions. This may best be achieved by nominating representatives from industry associations to attend the meetings. Continuity is important, since familiarity with the process, the debate and the individuals involved in the negotiations takes some time to build and is easier to maintain once achieved. Translating the rather abstract results from intergovernmental negotiations into their specific implications for business is a time consuming process. So, too, is proactive communication of this information to individual companies. Such communication is important, since it would appear from this survey that the vast majority of employees are oblivious to the implications for their business of the CBD. Industry associations may wish to consider dedicating more staff time to the issue of access and benefit-sharing. The rapidity of developments in this field and their significance for companies suggest that it is not possible for a single staff member to cover other issues such as biosafety and climate change as well as access and benefit-sharing.

- **Participate in national policy-making processes.** Companies should contact government authorities in countries where they are based or from where they obtain significant quantities of genetic resources in order to become involved in national processes to design measures on access and benefit-sharing. As one company put it, 'We need to be in there discussing. Working from the inside allows one to influence matters.'

12.5 Indicators of fair and equitable benefit-sharing

A question commonly raised concerning the fairness and equity of benefit-sharing arrangements is 'fair and equitable according to whom?' The CBD itself does not define 'benefit', 'sharing', 'fair' or 'equitable', nor does it offer much guidance on who is the arbiter of the standard of 'fairness and equity'. What parties themselves believe to be satisfactory when they agree on the terms for access will be the primary determinant of fairness and equity, although this will depend upon the bargaining position of each party, which in turn depends upon the information and negotiating skills available to each. The views of the government of the provider country, expressed in the access legislation regulating any such partnerships, will also contribute to the determination of what is fair and equitable, by setting a framework for individual agreements, and, depending on the degree of intervention by the state, by becoming involved in individual agreements as a party or reviewer. Should a dispute arise, the question of what is 'fair and equitable' may come to be determined by a tribunal such as a judge and jury or an expert panel. Such a tribunal would be likely to draw on existing best practice in access and benefit-sharing, and thus to compare the agreement under scrutiny against some kind of industry bench-mark in order to determine what a reasonable practitioner in the field would be likely to consider fair.

Any objective assessment of whether a particular access agreement is fair and equitable is thus difficult for two main reasons: first, because the primary arbiters of fairness and equity are the parties to the individual agreement, so that the test is more subjective than objective, and second, because what is fair and equitable will depend upon the individual circumstances of an agreement. The circumstances of specific agreements are likely to involve several complex factors, ranging from the type and nature of samples supplied to the nature of a collaborating institution or community. Consequently, it is difficult to generate any indicators that are sufficiently general to apply to a range of potential partnerships, but that nonetheless offer a tool that is precise enough to assist prospective partners to access and benefit-sharing agreements.

Despite the difficulty of making an objective assessment of what is fair and equitable, practical demands for help by governments, companies and others suggest the effort is worth making. This study aims to help by contributing to the body of information on current and best practice in the field. Finally, we offer the following questions that parties considering entering into an access and benefit-sharing arrangement could ask themselves in order to determine whether a given or proposed arrangement is fair and equitable. As it is impossible for each partner to determine whether an arrangement is fair without participating in meaningful discussions on its terms, these indicators relate not only to the contents of an arrangement, but to the process by which it was made.

12.5.1 Process indicators

- Were the benefits identified and defined jointly by the provider of genetic resources and the user?
- Was there PIC for access?
- Were all affected parties (eg government, research institutions, local communities) represented in the provider's granting of consent?
- Are provider and user clear which variables affect the type and value of benefits agreed?
- Is it clear from the agreement which benefits were precisely defined at the time that the agreement was made, and which benefits must be defined later in the partnership once the use of the genetic resources becomes clear?
- If some of the benefits are to be defined after the initial agreement is made, is there a process stipulated in the initial agreement for reaching agreement during discovery and development on the type and value of benefits?
- Was the agreement based on full disclosure by the user of how it intends initially to use the genetic resources, and a process determined by which other uses might be approved by the provider?
- Did both the provider and the user of genetic resources have available to them the information enabling them to assess the likely value of the results of access (including the probability of success of a commercial product and the likely size and value of the market for the product)?
- Did both the provider and the user of genetic resources have available to them the negotiating skills and legal assistance needed to reach agreement?

12.5.2 Content indicators

- Are both monetary and non-monetary benefits included in the agreement?
- Are benefits shared at different points in time, from initial access, through discovery and development, and for the duration of sale of a product?
- Are benefits distributed to a range of stakeholders?

- Does the agreement include a 'package' of different benefits?
- Is the agreement based on the standard terms of either the provider or the user of genetic resources, or was it tailored to the specific needs of both parties?
- Does the magnitude/value of benefits vary according to degree of exclusivity of access?
- Does the magnitude/value of benefits vary according to the value added to the genetic resources by the provider (whether by supplying derivatives of the raw genetic resources, such as purified compounds, or by providing information concerning the raw genetic resources, such as ethnobotanical information or data on traits)?
- Is a mechanism established for the distribution of benefits within the provider country over time?
- Is benefit-sharing linked to a set of objectives or principles (eg conservation of biodiversity, sustainable development) that address wider national, as well as local and institutional priorities?

12.6 Corporate and institutional policies on access and benefit-sharing

Development of a corporate or institutional policy on access and benefit-sharing offers several advantages to a company or other organisation. Preparation of a policy provides an opportunity and a mechanism for a company to familiarise itself with the letter and spirit of the Convention and access legislation. It will result in a management tool that can protect the company from liability by ensuring compliance with required standards and procedures. It can enable more proactive companies to design tools for continuous improvement in their supplier and user chains. It can contribute to the development of a company's research and development strategy, since the process of developing such a policy will help the company to identify parameters such as the number of countries it is likely to work in, its main suppliers and collaborators, and the monetary and 'non-monetary' costs of partnerships. It provides a tool for 'transparency' and good corporate citizenship, enabling companies to communicate their positions and commitments to suppliers and other outside collaborators. Finally, a good policy is a vehicle for positive public relations.

A corporate policy on access and benefit-sharing might contain:

- a statement of principle;
- commitments by the company (eg compliance with local legislation, etc);
- a list of practical tools for implementation of these commitments, for example: MTAs, binding procedures for staff, incentives for compliance, disciplinary measures in the case of non-compliance;
- internal and external objectively verifiable indicators for compliance;
- measures for enforcement;
- a process for monitoring and evaluation; and
- a process for evolution/continuous improvement.

Possible elements for inclusion in corporate and institutional policies developed in response to the CBD are outlined in Box 12.1

Box 12.1 *Possible Elements for Inclusion in Corporate and Institutional Policies Developed in Response to the CBD*

1 Scope and standard

- scope of activities and resources covered by the policy;
- reference to the CBD, its access and benefit-sharing provisions, and to national law; and
- standard of effort.

2 PIC and legal title

- prior informed consent for collection from all interested parties (eg government, research institutions, local communities);
- requirement of proof that an organisation supplying genetic resources has title to the materials and is authorised to supply them to the company for product discovery and development; and
- respect for traditional resource rights of local communities.

3 Benefit-sharing

- monetary and non-monetary benefits;
- timing of benefit-sharing;
- sharing of benefits arising from use of pre-CBD materials; and
- sharing benefits with a range of stakeholders (eg government, research institutions, local communities).

4 Conservation and sustainable use

- sustainable sourcing for manufacture;
- dedication of benefits to conservation; and
- commitment to not over-harvest species during collections.

5 Process and indicators

- objectively verifiable indicators that the policy has been applied;
- options for scrutiny; and
- continual improvement.

In addition to the policy, it will be essential to develop, in parallel, means of implementation to ensure that it is effective, respected and not just seen as corporate PR. Such means of implementation might include material transfer (or 'bioprospecting') agreements; procedures for staff; information management tools to track genetic resources and their derivatives through research and development processes, and to restrict their supply to third parties; and indicators and measures for monitoring, evaluation and improvement.

12.7 Summary of recommendations

12.7.1 Recommendations for governments regulating access

- Understand the different user industries, and the differences among them:
  – their demand for access to genetic resources and traditional knowledge;
  – the use made of the resources in product discovery and development;
  – the costs and risks involved and the magnitude and nature of the benefits generated.
- Understand different kinds of possible partnerships and mechanisms for sharing benefits.
- Be realistic.
- Design access measures flexible enough to deal with different genetic resources and uses.
- Keep access procedures and conditions simple, speedy and efficient.
- Establish a national focal point or competent national authority for access which has the competence to process unambiguous collecting permits and access and benefit-sharing agreements.
- Keep the benefit-sharing requirements in laws simple. Allow parties to reach mutually agreed terms within this framework.
- Assume the administrative burden of local level PIC and benefit-sharing arrangements.
- Build capacity to attract beneficial partnerships.
- Don't legislate without a strategy or the capacity to implement the laws introduced.
- Cooperate with other governments to harmonise access regulations around the world.

12.7.2 Recommendations for companies and other organisations seeking access

- Develop an accurate understanding of the CBD and national access legislation.
- Understand the priorities of provider countries and their openness to flexibility.
- Engage in policy formulation at the international and national levels, whether at CBD meetings or through involvement in national processes (such as the formulation of access legislation), and whether as an individual company or through industry associations.
- Develop a company or institutional policy that will give governments, research partners and others a clear understanding of the company's principles and practice on access and benefit-sharing.
- Develop tools – such as MTAs and guidelines for employees – to ensure all staff implement this policy and that it is enforced.
- Involve professional societies and industry associations in discussions on the CBD and its relevance to the private sector; develop standards of 'best practice'.

David J Newman and Sarah A Laird

Appendix A

The Influence of Natural Products on 1997 Pharmaceutical Sales Figures

Studies undertaken in recent years to assess the contribution of natural products to pharmaceutical sales have focused on the number of prescriptions dispensed, usually in the USA, and in some cases for selected therapeutic categories (Grifo et al, 1997; Cragg et al, 1997a; 1997b; Farnsworth and Morris, 1976). While these studies have provided a valuable indication of the importance of natural products to medicine, they do not shed significant light on the value of natural products to pharmaceutical companies' bottom line.

In an effort to overcome some of these intrinsic biases, and to offer a different perspective on the role of natural products in the pharmaceutical industry, we examined the ten top-selling drugs of the world's top ten companies, for the latest year that figures are available: 1997. There are obvious limitations to this approach: for example, the study does not include data on Viagra (sildenafil), which was not approved for sale in the United States until 27 March 1998. We also acknowledge that sales of the top ten drugs of any one company – while making up a majority of the company's earnings – do not represent a full picture of the company's products.

The world's top ten pharmaceutical companies, according to the *Financial Times* (1998) are as follows: Merck & Co (USA); Glaxo Wellcome PLC (UK); Novartis Pharmaceuticals Corp (Switzerland); Bristol-Myers Squibb Co (USA); Pfizer Inc (USA); Hoffman La Roche Inc (Switzerland); American Home Products (USA); Johnson & Johnson (USA); SmithKline Beecham (UK); and Hoechst Marion Roussel (Germany). In order to provide greater breadth to the study, we also included three other top US pharmaceutical companies – Eli Lilly Co, Schering Plough, and Pharmacia & Upjohn, as well as the largest Japanese pharmaceutical company, Sankyo.

Companies are ranked by ethical pharmaceutical sales, not by gross sales of the corporate group to which the pharmaceutical divisions or subsidiaries belong (*MedAd News*, May 1998; September 1998). For each company, we examined the ten top-selling products, grouping each into one of four categories based on the ways natural products contribute to the discovery of a drug:

Biologicals An entity that is a protein or polypeptide either isolated directly from the natural source or more usually made by recombinant DNA techniques followed by production using fermentation. Denoted in the following tables by 'B'. An example is insulin.

Natural product An entity that though occasionally manufactured by semi-synthesis or even total synthesis, is chemically identical to the pure natural product. Denoted in the following tables by 'N'. Examples are Vitamin C, paclitaxel, and cyclosporine.

Derived from a natural product An entity that starts with a natural product that is then chemically modified to produce the drug. An example is Amoxicillin, where the starting material is a penicillin produced by fermentation which is then chemically modified. Another example is simivastatin (Zocor). Denoted in the following tables by 'ND'.

Structural class from a natural product This is a material where the parent structure came from nature and then materials were synthesised *de novo* but following the natural template. At times, there is a very subtle 'chemical decision' that has to be made as to whether or not a compound is an 'ND' or 'S*'. An example is Acyclovir. Denoted in the following tables by 'S*'.

Synthetic Denoted in the following tables by 'S'.

As demonstrated in Table A.1, the contribution of natural products to the sales, and ultimately the profits, of an individual company's pharmaceutical products ranges from greater than 50 per cent to less than 10 per cent. In the case of Merck & Co, 'N'- and 'ND'-assigned drugs account for 50.6 per cent of 1997 sales, rising to 74 per cent if the class of drugs derived from a natural product structural type ('S*') is included. In contrast, the Johnson & Johnson Group shows only 8.6 per cent of sales from 'N' or 'ND' drugs, although 15.2 per cent of its product sales are biologicals ('B').

In addition to analysing the sources of the ten top sellers of the top pharmaceutical companies, we looked at the 25 top-selling drugs world-wide – all of which are 'blockbuster' drugs, with sales greater than US$1 billion per year. We grouped these drugs by disease type to allow for further comparisons across therapeutic categories (Table A.3).

The 25 top-selling drugs world-wide are ranked according to sales by trade name. Most rank tables in commercial, as distinct

Table A.1 *Proportion of Top Ten Drug Sales, by Source*

Company	1997 global sales of pharmaceuticals, US$ bn	Biological B	Natural/ or Derived From (N or ND)	Synthetic (S)	Derived Synthetic (S*)
Merck	13.28	nil	6.7190 (50.59%)	1.8610 (14.02%)	3.0950 (23.32%)
Glaxo Wellcome	13.09	nil	2.1352 (16.30%)	3.9606 (30.26%)	2.4550 (18.75%)
American Home Products	11.08	nil	1.6743 (15.12%)	1.2438 (11.22%)	0.4479 (4.04%)
Novartis	9.73	0.2694 (2.76%)	1.5344 (15.77%)	3.7068 (38.10%)	nil
Bristol-Myers Squibb	9.93	nil	2.9060 (29.26%)	1.4590 (14.69%)	1.5210 (15.31%)
Pfizer	9.24	nil	1.1670 (12.63%)	6.7340 (72.88%)	nil
Hoffman La Roche	8.32	0.2268 (2.73%)	1.7121 (20.58%)	0.8929 (10.73%)	0.6938 (8.34%)
Johnson & Johnson	7.7	1.1690 (15.19%)	0.6580 (8.55%)	3.8210 (49.62%)	nil
Hoechst Marion Roussel	7.68	nil	0.9570 (12.46%)	1.6856 (21.95%)	0.3402 (4.43%)
SmithKline Beecham	7.5	1.1186 (14.92%)	2.3790 (31.72%)	2.0810 (27.75%)	nil
Eli Lilly	7.38	1.4500 (19.65%)	0.7790 (10.56%)	3.8155 (51.70%)	nil
Schering Plough	6.1	0.5980 (9.80%)	0.7990 (13.09%)	2.3080 (37.84%)	0.2830 (4.64%)
Pharmacia Upjohn	5.03	0.3492 (6.94%)	1.5859 (31.53%)	0.2935 (5.83%)	nil
Sankyo	3.6	0.3486 (9.69%)	1.6363 (45.49%)	0.4266 (11.86%)	0.0925 (2.57%)

from scientific publications, use the trade name of the drug rather than the generic name of the compound. Use of the trade name shows that Zocor (simivastin) is the leading seller. However, if the generic compound names are used, the best-selling compound emerges as omeprazole – which is sold under the trade names Losec and Prilosec. Omeprazole outsells simivastin by US$1.34 billion.

Table A.3 demonstrates the distinctive influence of drug sources by disease area. For example, the best-selling drugs in 1997 were the anti-ulcer medications, totalling US$8.5 billion. One drug – omeprazole – a synthetic agent developed by Astra, and sold both by Astra and through a joint venture with Merck, accounted for US$5.06 billion, or 59.6 per cent of sales in this area. All of the top four anti-ulcer drugs are of synthetic origin.

Cholesterol-lowering drugs followed the anti-ulcer in top sales. All of the top four sellers in this category are 'statins' – natural products or slight modifications thereof, grossing US$7.52 billion. Simivastin (ND) was the second-highest selling drug compound in 1997, and the top-selling product, marketed under the trade name Zocor by Merck and Co. The hematologic agents are all biologicals, and have sales of US$3.4 billion. The top two sellers in this category – Procrit and Epogen – are the same compound (epotein alfa), sold by Johnson & Johnson and Amgen Inc, respectively. If sales for the compound epotein alfa were tallied, they would come to US$2.33 billion, placing them above number five Zantac (ranitidine), at number four best seller overall.

Approximately 67 per cent of antibacterial sales are from compounds from naturally-derived sources. In antihypertensives, the true synthetics (both are calcium channel blockers) account for 57 per cent. All antidepressive agents are synthetic, as are the sole examples of conventional antihistamines, NSAIDs, antimigraine and heartburn drugs included in the top 25 best-selling drugs.

In conclusion, sales of the 25 best-selling drugs world-wide, apportioned according to source type, suggest the economic value of non-synthetic agents is significant. 42 per cent of sales of the top 25 best sellers is made up of biologicals, natural products, or entities derived from natural products, with a total value in 1997 of US$17.5 billion. Natural products and their derivatives (N and ND, above) contribute US$11.6 billion – some 28 per cent of the sales (Table A.2). Natural sources continue to be a major player in the sales of ethical pharmaceutical agents approved for use in 1997.

Table A.2 *Top 25 Drugs: Sales by Source*

Source of agent	Gross sales (US$ bn)	% of top 25 drug sales
B	3.39	8.2
N/ND	11.59	28.1
S	23.75	57.6
S*	2.5	6.1

Table A.3 *Top 25 Drugs by Worldwide Sales, Arranged by Disease Type*

Ranking	US brand type name	Generic name	Marketer	1997 global sales (US$ bn)	US patent expiration
Anti-ulcer					
2	Losec (S)	omeprazole	Astra AB	2.82	2001
6	Prilosec (S)	omeprazole	Astra Merck	2.24	2001
5	Zantac (S)	ranitidine	Glaxo Wellcome	2.26	1997
17	Pepcid (S)	famotidine	Merck & Co	1.18	2000
Cholesterol-lowering					
1	Zocor (ND)	simvastatin	Merck & Co	3.58	2005
13	Pravachol (N)	pravastatin	Bristol-Myers Squibb	1.44	2005
14	Mevalotin (N)	pravastatin	Sankyo Co	1.41	2002
22	Mevacor (N)	lovastatin	Merck & Co	1.1	2001
Hypertension					
4	Vasotec (S*)	enalapril maleate	Merck & Co	2.5	2000
7	Norvasc (S)	amlodipine besylate	Pfizer	2.22	2007
21	Adalat CC (S)	nifedipine	Bayer Corp	1.1	1991/2007
Antidepressant					
3	Prozac (S)	fluoxetine hydrochloride	Eli Lilly & Co	2.56	2001
10	Zoloft (S)	sertraline hydrochloride	Pfizer	1.51	2005
11	Paxil (S)	paroxetine hydrochloride	SmithKline Beecham	1.47	2005
Hematologic					
18	Procrit (B)	epoetin alfa	Johnson & Johnson	1.17	2004
19	Epogen (B)	epoetin alfa	Amgen Inc	1.16	2013
24	Neupogen (B)	filigrastim	Amgen Inc	1.06	2013
Antibacterial					
9	Augmentin (ND)	amoxicillin	SmithKline Beecham	1.52	2002
15	Biaxin (ND)	clarithromycin	Abbott Labs	1.3	2003
12	Cipro (S)	ciprofloxacin	Bayer Corp	1.44	2004
Antihistamine (H_1)					
8	Claritin (S)	loratadine	Schering-Plough	1.73	2012
Immunosuppressant					
16	Sandimmune (N) & Neoral	cyclosporine	Novartis	1.25	1995/2010
NSAID					
20	Voltaren-XR (S)	diclofenac sodium	Novartis	1.11	1993/2007
Antimigraine					
23	Imitrex (S)	sumatriptan succinate	Glaxo Wellcome	1.09	2006/2008
Heartburn					
25	Propulsid (S)	cisapride	Johnson & Johnson	1.05	2007

Note: B = biological; N = natural product; ND = derived from natural product; S = synthetic; S* = synthetic derived

Sources: MedAd News, May 1998; *Med Ad News*, September 1998; Cragg et al 1997a.

Appendix B

Scientific and Common Names of Selected Botanical Medicines

Scientific name	Common name
Aesculus hipposcastanum	horse-chestnut
Allium sativum	garlic
Aloe barbadensis; A vera; A ferox; Aloe spp	aloes
Arctium minus; A lappa	burdock
Astragalus membranaceus	astragalus
Calendula officinalis	calendula
Camellia sinensis	green tea
Capsicum spp	red and cayenne pepper
Cassia/Senna spp	senna
Catharanthus roseus	rosy periwinkle
Caulophyllum thalictroides	blue cohosh
Cimicifuga racemosa	black cohosh
Crataegus spp	hawthorn
Dioscorea villosa	wild yam
Echinacea purpurea; E augustifolia	echinacea/coneflower
Eleutherococcus senticosus	Siberian ginseng
Ephedra sinica	Ma Huang
Glycyrrhiza glabra	liquorice
Ginkgo biloba	ginkgo
Glycine max	soy
Harpagophytum procumbens	devil's claw
Hedera helix	ivy
Hypericum perforatum	St John's wort
Hydrastis canadensis	golden seal
Mangifera indica	mango
Matricaria chamomilla; M recutita	camomile
Mentha x piperita	peppermint
Myrtus communis	myrtle
Oenothera biennis	evening primrose
Panax ginseng	Asian ginseng
Panax quinquefolius	American ginseng
Panax vietnamensis	Vietnam ginseng
Pausinystalia johimbe	yohimbe
Piper methysticum	kava (kava kava)
Plantago afra; P major	psyllium/plantain
Psidium guayaba	guava
Prunus africana	pygeum
Rhamnus purshiana	cascara buckthorn/ cascara sagrada
Sambucus nigra	elderberry
Serenoa repens	saw palmetto
Silybum marianum	milk thistle
Symphytum officinale	comfrey
Tabebuia spp	pao d'arco
Tanacetum parthenium	feverfew
Trichopus zelanicus travancoricus	arogyapacha
Turnera diffusa	damiana
Ulmus fulva	slippery elm
Uncaria guianensis; U tomentosa	cat's claw
Urtica dioica	stinging nettle
Vacinnium macrocarpon	cranberry
Vacinnium myrtillis	bilberry
Valerian officinalis	valerian
Viscum album	mistletoe
Vitus spp	grape seed
Zingiber officinale	ginger

Sarah A Laird

Appendix C

Regulatory Frameworks for Botanical Medicines

The following is a brief review of the regulatory framework for botanical medicines in a few key countries, regions, and international bodies: the EU, Germany, Japan, the USA, and the WHO.

C.1 The EU

The European Union regulatory system recognises phytomedicines as therapeutic agents derived from plants or plant parts, including preparations. Phytomedicines do not include isolated, chemically defined substances, but are considered an active entity, although they may contain hundreds of chemical constituents (Foster, 1995). EU Directives have created a comprehensive legislative framework for pharmaceutical products, which is also used for botanical medicines (Directives 65/65/EEC and 75/318/EEC). Requirements for the documentation of quality, safety, and efficacy are laid down in Directives 91/507/EEC. The procedures are decentralised, and as a general rule provide that acceptance by one national authority, based on a monograph called a 'Summary of Product Characteristics' (SPC), is sufficient for subsequent registration in other Member States. If differences of evaluation between Members occur, agreement is reached through arbitration (Steinhoff, 1998).

In 1989, the European Scientific Cooperative for Phytotherapy (ESCOP) was formed under the auspices of the EC to advance the state of botanical medicines. The main objectives of ESCOP are to establish harmonised criteria for the assessment of botanical medicine products, to give support to scientific research, and to contribute to the acceptance of phytotherapy on a European level. ESCOP published 50 plant species monographs (SPCs) in 1996–7. An SPC describes the species, preparation, quality, active constituents contributing to the claimed effect, therapeutic indications, dosage, and pharmacological properties. Pharmacodynamic and pharmacokinetic properties, and preclinical safety data are also included, as is a complete reference list. The Ad Hoc Working Group on Herbal Medicine Products, founded by the EMEA in early 1997, reviews draft SPCs (Steinhoff, 1998).

European producers have joined together to form the European American Phytomedicine Coalition (EAPC) to improve representation of their interests to the FDA and other bodies in the USA. This has included petitioning the FDA for use of European product data in the registration of existing European botanical medicines as OTC drugs (Gruenwald and Goldberg, 1998).

C.2 Germany

On 1 January 1978, the Second Medicines Act came into force, setting new standards for marketing authorisation for botanical medicines. These standards were in accordance with the European framework for the handling of medicines, and require proof of quality, safety, and efficacy. To meet the requirements of this Act, the German Government carried out a review process to establish clear criteria for active ingredients, and to make transparent to industry the kinds of products that might be authorised. A multi-disciplinary commission of experts –Commission E – was responsible for the review and evaluation of more than 300 medicinal plants (Steinhoff, 1998).

Commission E reviewed species monographs and research summaries for each new botanical medicine, and decided whether the product could be labelled for the proposed use. Monographs for botanical medicines include information on the drug's constituents, indications (including those for the crude drug or preparation), contraindications (if known), side effects (if known), interactions with other drugs or agents (if known), details on dosage or preparations, the method of administration, and the general properties or therapeutic value of the herb. The review process also yielded guidelines for manufacturers relating to quality control, safety, and labelling. Manufacturers must supply pharmacological, toxicological, and clinical data, which for well-known plants can be drawn from literature. Under the monograph system, the German market includes about 60,000 phytomedicine products (Foster, 1995).

The Federal Institute for Drugs and Medical Devices is responsible for the assessment of medicines and the verification of submitted dosiers with respect to quality, safety and efficacy. Criteria for registration are set out by European directives and guidelines, such as the *Note for Guidance on Quality of Herbal Remedies*, the *European Pharmacopoeia*, and national guidelines and directives. Criteria developed by Commission E and positive monographs are widely used to document the safety and efficacy of botanical medicines. In August 1994, the Fifth Amendment of the German Medicines Act provided a new procedure for proof of quality, safety and efficacy, widening the scope of exiting legislation to include traditional use. Traditional usage as proof of efficacy is accepted for certain products, mainly those sold outside pharmacies. These products must be labelled 'as traditionally used'. This new system might allow legal proof for a large number of preparations lacking sufficient scientific documentation. In contrast to other botanical medicines, the quality dossiers of 'traditional' products are not checked by the health authority. This regulation contrasts with EU requirements for marketing medicinal products (Steinhoff, 1998).

C.3 Japan

Japanese traditional medicines include 'folk' medicines, and TCMs, adapted into Japanese Kampo medicine (McCaleb, 1996; Steinhoff, 1998). Kampo drugs are formulae consisting of from five to ten different plant species, many formulated in industrialised granular, powdered or other forms based on classical decoction. The Ministry of Health and Welfare oversees the manufacture and marketing of botanical medicines. Approval processes for Japanese botanical medicines and Kampo medicine are quite different, due to differences in methods for evaluation. In the evaluation of Kampo medicine, importance is given to the 'empirical facts or experience', such as reference data and clinical testing. For traditional Japanese medicine, the pharmacological activity of each ingredient contained in a raw herb is evaluated. A monograph system is employed, or in cases where the monograph is not yet completed, regulators are guided by claims in the *Japanese Pharmacopoeia* (Steinhoff, 1998).

Based on the experience of doctors practising Kampo medicine, the Government approved 210 formulae as OTC drugs. New Kampo drugs, however, are regulated in the same way as Western drugs, and must undergo time-consuming and expensive testing. In the 1980s quality control of Kampo medicine increased, when regulations for Good Manufacturing Practice were extended to Kampo drugs, and in 1988 the Japan Kampo Medicine Manufacturers' Association drew up self-imposed guidelines (Steinhoff, 1998).

C.4 The USA

Until 1994, botanical medicines were considered food additives under the NLEA (1990). In order for medical claims to be made, either the active ingredient had to be subjected to an OTC monograph, or the manufacturer had to go through the NDA process for pharmaceuticals. Foods were considered safe unless proven otherwise, but food additives – unless they appeared on the FDA *GRAS* (Generally Regarded as Safe) list (which features around 250 species) (Box C.1) – were considered unsafe, unless the manufacturer proved them otherwise and received FDA approval (Foster, 1995).

Consumers and industry found this regulatory regime excessively restrictive, and lobbied the US Congress until in 1994 it passed the Dietary Supplement and Health Education Act (DSHEA). DSHEA was enacted to provide consumers with current and accurate information on supplements, and created the new legal food class of 'dietary supplements' (defined as ' a vitamin, a mineral, an herb, or other botanical (or) an amino acid') (McCaleb, 1996; DSHEA, 1994). DSHEA defined dietary supplements; established the manner in which literature may be used in connection with sales; specified statements of nutritional support that may be made on labels, and certain labelling requirements; provided for the establishment of regulations for GMP; and placed responsibility for ensuring product safety on manufacturers (Commission on Dietary Supplement Labels, 1997).

Under DSHEA, dietary supplements are presumed safe, and the burden of proving otherwise rests with the FDA. The FDA reserves the right to withdraw a product from the shelves should it make false or misleading claims, or be suspected of adulteration, but it must conduct testing to prove the product represents a health risk (Datamonitor, 1997; Gruenwald and Goldberg, 1997). As Brad Stone of the FDA said, 'Any product that makes specific claims to treat, cure, or prevent a specific condition or ailment is technically classified as a drug.' But if the FDA has not received any complaints about a botanical medicine product, and in the absence of a public health hazard, it falls into a legal grey area (New Hope, 1998). 'The result is that there are a lot of products on the market that little is known about,' says FDA deputy commissioner for policy, William Schultz (Greenwald, 1998).

DSHEA allows companies to sell botanical medicines for their health-enhancing properties, without going through lengthy and costly drug-approval procedures. It also allows companies to make brand-specific claims, provided they pay for the research to back such claims. But marketers are explicitly forbidden from using an important marketing tool – the 'bold claim'. In the OTC drug market, for example, this includes 'proven effective' or 'doctor recommended'. DSHEA allows marketers to make only 'structure and function' claims, which are relatively vague, and requires the disclaimer that 'this product is not intended to diagnose, treat, cure, or prevent any disease' (Kalorama, 1997). For example, the label on Celestial Seasonings GinkgoSharp Herb Tea states that: 'GingkoSharp gives you the synergy of Ginkgo and Siberian Ginseng to help you stay at the top of your game mentally'. This is followed by the required: 'These statements have not been evaluated by the Food and Drug Administration. This product is not intended to diagnose, treat, cure, or prevent any disease'.

Box C.1 *GRAS – Generally Regarded as Safe*

GRAS lists include species regarded by a consensus of scientific opinion as safe for use. Countries compile *GRAS* lists to give guidance on and provide legal protection for a wide range of species used by product formulators. The FDA *GRAS* list for food additives includes around half the herbs sold by the botanical medicines industry. They are included on the *GRAS* list as ingredients for alcoholic beverages, or as natural flavourings. In Australia, the 1989 Therapeutic Goods Act created the Australian Register of Therapeutic Goods. The Traditional Medicines Evaluation Committee (TMEC) evaluates non-prescription traditional medicines, and advises on their registration (Steinhoff, 1998). The Australian register, or *GRAS* list, includes 1,300 species of plants from around the world that are considered acceptable as ingredients for the formulation of medicines (Iwu, 1996).

Most *GRAS* lists grow from European traditions and species use. High biodiversity countries tend to refer to these lists, but they inadequately reflect local species diversity and use. Companies, universities, professional associations, and regulatory authorities in high biodiversity countries can develop their own *GRAS* lists that incorporate carefully evaluated native plant species (Iwu, 1996).

The FDA officially regulates botanical medicines under DSHEA through its food and drug divisions. It sponsors scientific conferences on botanical medicines, develops analytical methods to assess the safety risks of a plant preparation, and – if it suspects a dietary supplement is unsafe – must carry out tests to prove it represents a health risk. The FDA is currently undertaking research on ephedra, kava (isolating kavalactones), and blue cohosh (tetratogenic alkaloids). While the FDA is responsible for packaging claims, the Federal Trade Commission (FTC) regulates advertising standards for herbal products, and has begun to monitor the trade more closely (Gruenwald and Goldberg, 1997). In 1998 the FTC took legal action against seven manufacturers that had broken rules requiring advertising to be truthful and verifiable (Brody, 1999).

DSHEA has eased the production and marketing of botanical medicines, and has allowed the botanical medicines market to expand rapidly. Entry of large pharmaceutical and OTC companies is likely to create further pressure to 'rationalise' the regulatory environment, while increasing efforts at quality control and standardisation. Two trends appear to be converging in the USA today. On the one hand, there is a trend towards liberalisation of Federal regulations, as manifested in DSHEA. On the other hand, there are increasing calls within and outside the industry for quality control, and scientific validation of products – along the lines of the German regulatory model.

In June 1997, the Commission on Dietary Supplement Labels drafted a public comment that included the following: 'The Commission believes it would be logical and desirable for the US over-the-counter (OTC) drug system to include therapeutic claims for botanicals, at least for those having a long tradition of use and general recognition of safety and efficacy based on adequate studies ... In many other industrialised countries ... claims for traditional botanical remedies and medicines are permitted, often with specific disclaimers, as a unique category of nonprescription products within the drug regulatory system'.

The rapid increase in the use of botanical medicines over the last decade has not resulted in increased reports of consumer harm. This is probably due to the fact that most herbs have been consumed by humans for hundreds, or even thousands, of years. WHO's *Guidelines for the Assessment of Herbal Medicines*, for example, find historical use of a substance a valid form of safety and efficacy documentation, in the absence of scientific evidence to the contrary:

> *A guiding principle should be that if the product has been traditionally used without demonstrated harm, no specific restrictive regulatory action should be undertaken unless new evidence demands a revised risk-benefit assessment ... Prolonged and apparently uneventful use of a substance usually offers testimony of its safety.*

For this approach to work, however, commercial botanical medicine use and formulation must correlate with the traditional (Herb Research Foundation, 1998).

Although DSHEA is the most widely used regulatory framework for botanical medicines in the USA, botanicals can also be regulated as approved OTC drugs (instead of as 'dietary supplements'), which can carry more specific therapeutic claims. In 1992, the EAPC made a formal application to the FDA OTC drug division that European phytomedicines be reviewed as old drugs based on their extensive use in Europe. The FDA has not yet responded. Botanicals approved as OTC drugs in the USA today include aloe (*Aloe ferox*), cascara sagrada (*Rhamnus purshiana*), psyllium (*Plantago afra*) and senna (*Cassia senna*), as laxatives; peppermint oil (*Mentha x piperita*) as an antitussive; red pepper (*Capsicum* spp) as a counter-irritant; slippery elm (*Ulmus rubra*) as a demulcent; and witch hazel (*Hamamelis virginiana*) as an astringent. In addition, 50 botanicals, including Chinese herbal formulae, gingko, and saw palmetto, have been submitted under the IND/NDA (Investigational New Drug/New Drug Application) process (Brevoort, 1998).

The US regulatory environment for botanical medicines remains in flux, with a number of areas, including GMP, labelling rules, and possible development of an FDA OTC review for botanicals, currently on the table. It is not unlikely, given increasing awareness of German and other regulatory systems, that a rationalised approached to botanical medicines regulation along these lines will arrive in the not too distant future in the USA.

C5 The WHO

The World Health Organisation (WHO) has been a major force in stimulating collaborative research on traditional, largely plant-based, medicines (Foster, 1995). In 1978, the Declaration of Alma-Alta recommended that proven traditional remedies be incorporated into national drug policies and regulatory measures. Following this, in 1991 the WHO established the Traditional Medicine Programme to facilitate the integration of traditional medicines into national healthcare systems; promote the rational use of traditional medicines through the development of technical guidelines and international standards in the fields of botanical medicines and acupuncture; and to act as a clearing house for the dissemination of information on various forms of traditional medicines (ICDRA) (Steinhoff, 1998).

In 1991, the WHO drafted *Guidelines for the Assessment of Herbal Medicines*, which contain basic criteria for the assessment of safety, efficacy, and quality assurance. These *Guidelines* were the first comprehensive and uniform model guidelines issued for national health authorities, industry, academics, and others, and were intended to facilitate the assessment and registration of botanical medicines based on scientific results. The *Guidelines* are also intended to accommodate cross-cultural transfer of traditional medicinal plant knowledge (Steinhoff, 1998; WHO, 1991; Foster, 1995).

To develop criteria and general principles to guide evaluation of botanical medicines, the WHO Regional Office for the Western Pacific organised a meeting of experts in 1992 to develop guidelines for research on botanical medicines. In 1994, the WHO Regional Office for the Eastern Mediterranean published *Guidelines for Formulation of National Policy on Herbal Medicines*. In 1996, a WHO consultation on 'WHO Monographs on Selected Medicinal Plants' took place in Germany; at this consultation 28 monographs were adopted. WHO monographs are intended to provide scientific information on the safety, efficacy, and quality control of widely used medicinal plants; to provide models for Member States to develop their own monographs; and to facilitate information exchange. 32 monographs are in preparation, which contain information on: (part 1): botanical characteristics, major active chemical constituents and quality control of each plant; and (part 2): summaries of clinical applications, pharmacology, posology, possible contraindications, and precautions, and potential adverse reactions (Steinhoff, 1998).

Appendix D

Tentative List of Crops Under Negotiation as the Scope of the Multilateral System in the Revised International Undertaking

Common name	Genus[1]
Rice	*Oryza*
Oats	*Avena*
Rye	*Secale*
Barley	*Hordeum*
Millets	*Pennisetum*
	Setaria
	Panicum
	Eleusine
	Digitaria
Maize	*Zea*
Sorghum	*Sorghum*
Wheat	*Triticum*
Peanut	*Arachis*
Cowpea	*Vigna*
Pea	*Pisum*
Beans	*Phaseolus*
Lentils	*Lens*
Soybean	*Glycine*
Potato	*Solanum*
Sweet potato	*Ipomoea*
Yams	*Discorea*
Cassava	*Manihot*
Bananas, plantains	*Musa*
Citrus	*Citrus*
Sugar-cane	*Saccharum*
Beet	*Beta*
Pumpkins, squashes	*Cucurbita*
Tomato	*Lycopersicon*
Coconut	*Cocos*
Tannier	*Xanthosoma*
Taro	*Colocasia*
Cabbages, rape, mustards	*Brassica*
Onion, leek, garlic	*Allium*
Chickpea	*Cicer*
Faba bean	*Vicia*
Pigeon pea	*Cajanus*
Melons	*Cucumis*
Flax	*Linum*
Sunflower	*Helianthus*
Cotton	*Gossypium*
Oil palm	*Elais*
Forages	
Grasses (Graminaceae)	*Agropyron*
	Agrostis
	Alopecurus
	Andropogon
	Arrhenatherum
	Axonopus
	Brachiaria
	Bromus
	Bothriochloa
	Cenchrus
	Chloris
	Cynodon
	Dactylis
	Elymus
	Festuca
	Hyparrhenia
	Ischaemum
	Lolium
	Melinis
	Panicum
	Paspalum
	Phalaris
	Phleum
	Poa
	Schizachyrium
	Setaria
	Themeda
Legumes (Leguminosae)	*Aeschinomene*
	Alysicarpus
	Arachis
	Bauhinia
	Calopogonium
	Canvalia
	Centrosema
	Clitoria
	Coronilla
	Desmodium
	Dioclea
	Galactia
	Indigofera
	Lablab
	Lathyrus
	Lespedeza
	Leucaena
	Lotus
	Lupinus
	Macroptilium
	Medicago
	Melilotus
	Neonotonia
	Onobrychis
	Pueraria
	Stizolobium
	Stylosanthes
	Teramnus
	Trifolium
	Trigonella
	Vetiveria
	Zornia

Note

1 Genera are indicated only to clarify to which genus a particular crop belongs.

Kerry ten Kate

Appendix E

Regulatory Frameworks Relevant to the Release of Crop Varieties

E.1 Seed certification

Seed certification schemes vary from country to country. In Japan, the UK, and Argentina, for example, the government controls seed certification. The EU prescribes EU-wide minimum quality standards. In some countries, seed registration is the prerequisite for certification. The EU has equivalence arrangements with most OECD member countries, such that if they can guarantee that the production of seed is to the same standards as required in EU Member States, their seed can be sold in the EU under their own labels.

In the USA, seed certification is voluntary, and is administered by independent authorities. Various international certification and seed testing schemes, including those of the OECD, will be described in this section, followed by a few illustrative examples from specific regions (the EU) and countries (the USA, Japan and Argentina).

E.1.1 International certification and seed testing schemes

OECD

The OECD Schemes for Varietal Certification of Seed Moving in International Trade (OECD Seed Certification Schemes) are voluntary schemes designed to ensure an equivalence in certification standards between countries, ensuring that international trade is maintained at minimum standards acceptable to the importing countries. Once an individual country has committed to participate in a particular Scheme, its National Designated Authority is obliged to ensure that there is full compliance with the rules for all seed which carries the country's OECD label. The OECD publishes the 'List of Cultivars Eligible for Certification' – an annually revised official list of cultivars accepted by National Designated Authorities as eligible for certification in accordance with Rules of the OECD Seed Schemes.

The OECD Seed Certification Schemes use two procedures to check the progress of a cultivar at different stages in the seed production process. The first is to test plants grown in control plots and seeds and seedlings grown in the laboratory, using samples of seed from a seedlot. The second procedure entails field inspection of growing seed crops, on one or more occasions, to report on their condition.

The rules for listing of the following cultivars are addressed under the OECD Seed Certification Scheme: Herbage and Oil Seed; Cereal Seed; Beet Seed; Seed of Subterranean Clover and Similar Species; Maize and Sorghum Seed; and Vegetable Seed. There are voluntary 'Purity Standards' for each of these schemes. The country of registration is the country where the cultivar is registered on the National List following satisfactory tests for distictness, uniformity and stability. (website: www.worldseed.org/~assinsel/igos.htm)

FAO

The 1982 Quality Declared Seed System was introduced under the auspices of the FAO, for use in countries where there are insufficient resources for a fully developed seed quality control scheme such as seed certification.

ISTA, AOSCA and FIS/ASSINSEL

The International Seed Testing Association (ISTA) is the only world-wide organisation dedicated to seed testing on an international scale. The ISTA 'Rules for Seed Testing' constitute a standard reference manual used by the seed industry throughout the world. Other seed testing standards are those of the Association of Official Seed Certifying Agencies (AOSCA); or those of the International Seed Trade Federation (FIS/ASSINSEL: International Rules for Seed Testing and the International Seed Analysis Certificate).

The EU

EC regulation lays down minimum quality requirements for the marketing of agricultural and vegetable seeds. The Common Catalogue Directive sets out uniform rules for the acceptance of cultivars and requires Member States to compile national lists of cultivars that have been accepted for certification and marketing within their jurisdiction. Certification, which requires labelling of the certified seed, is a prerequisite for marketing seed in any EU Member State. The scope of the regulations for certification and production covers 'Breeder's Seed', 'Basic Seed' and 'Certified Seed' for crop and seed production. For 'Standard Seed' of vegetable plants, there is limited official checking, which involves 'post-control tests' that demonstrate that the prescribed standards have been met.

The Common Catalogue Directive also introduces procedures for the EU-wide production of a Common Catalogue of Varieties comprising all the cultivars on the national lists. Each Member

State must implement the directives through statutory provisions.

In order to allow inclusion of a cultivar of an agricultural crop on a Member State's national lists, the cultivar concerned must fulfill conditions of being distinct, uniform and stable (DUS), and have, in the case of agricultural varieties, satisfactory value for cultivation and use (VCU). The DUS requirement for National Listing in an EU Member State is usually confined to comparisons between varieties on the EU Common Catalogue. For Plant Breeder's Rights, comparisons may be extended to include other varieties in 'common knowledge' as required by the UPOV Convention. 'VCU' requires a new cultivar to show a clear improvement over other cultivars on a national list.

Once a cultivar is registered, it is accepted for certification and marketing by each Member State. The DUS tests will include the analysis of varietal and analytical purity and germination, and will result in the cultivar's description, which is then used to determine identity and purity during seed certification.

The UK

As an example of the seed certification regulations of one EU Member State, let us take the UK. A new seed variety (of the main agricultural and vegetable species) may not be marketed in the UK unless the variety is on the UK National List, administered by the Ministry of Agriculture, Fisheries and Food (MAFF), or on the EU Common Catalogue, and certified. In order to be awarded National Listing, the applicant must conduct official tests and trials designed to assess whether the variety is DUS and, in the case of agricultural crops, whether it has VCU. The duration of most DUS and VCU tests is a minimum of two years and the trial protocols differ for each crop, although ISTA rules are followed in general.

After addition to the National List, the seed must be certified by one of the three certification authorities – MAFF, Scottish Agricultural Science Agency (SASA), and Department of Agriculture for Northern Ireland (DANI) – in order to be marketed in the UK. UK certification functions according to a pedigree (or generation) system, and distinguishes between different categories of seed ('breeder's seed', prebasic through to the C3 generation of commercial seed). On entry to the scheme, an applicant must apply to one of the three certification authorities and submit a seed sample to ensure it complies with minimum purity and germination standards.

The USA

In the USA, seed legislation exists at the Federal and State levels. For seeds to be marketed in the USA, they must be labelled for quality (1939 Federal Seed Act). To determine seed quality, several tests are undertaken to establish the seed's purity, germination and quality.

While there are different standards and procedures for several 'national' and 'interstatal' certification programmes for 'foundation', 'registered', or 'certified' seed, there are no registration requirements for the commercialisation of new seed varieties. Under the OECD Seed Schemes, the USA provides a mechanism for seed certification which is administered by the USDA-Agricultural Research Service (ARS). The USA participates in the OECD Seed Scheme and also adheres to the AOSCA minimum genetic certification standards in order to certify the quality of seed, to underpin confidence on the part of consumers and other customers for seed, and to facilitate trade. Ordinarily, the procedures required by these schemes are followed when the seed is to be shipped to another OECD Member Country.

Apart from OECD certification for varietal purity, the US provides a seed certfication system that is voluntary and which is not under Federal or State authority. Rather, the US system is run by independant agencies and organisations in each State, usually the State Crop Improvement Agency. The great majority of commercialised seed is not certified; however, certified seed is labelled with a special colour labelling tag, demonstrating that the breeder has completed tests described by AOSCA rules.

Japan

In Japan, the registration of 'conventionally bred and transgenic plant' seeds is authorised by the Ministry of Agriculture, Forestry and Fisheries (MAFF). Each variety must be tested by way of an on-site inspection and/or a growing test in Japan. Japan participates in the OECD Seed Schemes, and seed certification is undertaken at MAFF for herbage and oil seeds, as well as beet seeds. Labelling requirements exist in Japan, and seeds are accorded the 'Japan Agricultural Standard' (JAS) label by MAFF. The Quality Control and Consumer Service supervises the plants which are approved or permitted by MAFF as JAS Approved or Permitted plants. JAS also monitors JAS-labelled products for quality and properness.

Argentina

According to Argentina's regulatory system, the National Institute of Seeds (INASE) is the authority charged with the registration and certification of cultivars. Registration comes in to effect when INASE places the new cultivar on the National Cultivar Registry, and thereafter, certification may be granted following the Official Certification procedures. Seed may not be sold in Argentina until the conditions of registration and certification are satisfied. In order to market a new seed variety in Argentina, the breeder must first undertake trials in Argentina: such as for two years in three locations or for three years in one location. Phytosanitary Certficates may be obtained from the Directorate of Seed Certification and Trade at INASE. Once the applicant is awarded a Property Certificate (in effect PBRs) and registration on the National Cultivar Registry, the seed must be labelled in order to be sold. This condition applies especially for imported seed. The seed must be labelled in a certificate class according to the OECD Seed Scheme.

E2 Regulation of the release of GMOs

Regulation of the release of GMOs varies from country to country, but they still need to undergo variety registration, in addition to satisfying human health, environmental protection and biosafety issues before being eligible for the market. Harmonisation of biotechnological regulation is underway as efforts such as standardising interpretations of key terms are negotiated by the OECD, too. Prior to the release and/or marketing of a GMO in the UK, the consent of the Secretary of State must be obtained. In Japan, the guidelines in respect of the production and sale of rDNA organisms are focused on biosafety and risk assessment criteria. Under Argentinian regulations, GMOs must be registered and certified. The cost of a GMO consent can range from zero (USA); to US$3,260 (UK); and US$326 (Mexico). The time periods of approval of the release into the environment and/or its commercialisation vary. Typically, the procedures allow for some two to three months, but the period for approval can be over two years where more data have been requested. Public registers or notification of approved GMOs exist in the USA, Japan, the EU/UK, and Argentina. Labelling of GMOs is required in EU Member States, and in Argentina. The USA, however, requires labelling only when there is a material concern about health or safety. Japan has no apparant GMO Labelling schemes in place, however the Japanese government requires labelling for transgenic crops (as for seeds). A list of approved GMOs exists in all these markets.

Europe

The European system is a process-based one, whereas the American system is product-based. There is a comprehensive EU-wide framework covering all stages of work with GMOs (including GMO crops) designed to protect human health and the environment. The relevant EC Directive 90/220 (as amended) covers deliberate release into the environment, and marketing of GMOs.

A breeder wishing to introduce genetically modified seed in the EU must approach a Member State to act as rapporteur, and the rapporteur must be satisfied that the application (or 'notification') meets European health and safety requirements. Consent to market products containing or consisting of GMOs is conditional upon prior written consent by the EC, via the EC, once Member States have made a decision based on qualified majority voting. If a risk analysis was carried out, any consent will be based on its results, and also on compliance with the relevant Community product legislation; and satisfaction of requirements concerning the environmental risk assessment. The notification dossier, with a favourable opinion from the rapporteur, is forwarded to the Commission within 90 days of receipt of the notification, or the applicant is notified that its application was rejected. The Commission forwards the dossier to the competent authorities of the other 14 Member States. Within 60 days, these should give their written consent to the notification so that the product can be placed on the market, and should inform the other Member States and the Commission. Once all national committees present their findings, a vote is taken. If a qualified majority votes in favour, the product may be commercialised in the EU. However, before this system comes into effect, Member States must set up regulations or pass laws implementing the Commission's decision. So far, Austria and Luxembourg have refused to do so for genetically modified foods. The Commission publishes in the *Official Journal of the European Communities* a list of all the products receiving final consent under EC Dir 90/220, together with a specification of the GMO or GMOs contained therein, and their use or uses. The procedures described above are free of cost.

In the EU, mandatory labelling will be necessary for all genetically modified foods. The relevant Regulations provide that label must indicte the characteristics or properties modified and the method by which these were obtained; it must inform the end consumer of the presence in the novel food, or ingredient, of material which is not present in an existing equivalent foodstuff and which may have implications for the health of certain sections of the population, or give rise to ethical concerns; and it must inform the end consumer of the presence of GMO. The Compulsory Labelling Directive relates specifically to genetically modified soya, and genetically modified maize placed on the market prior to the Novel Foods Regulation, making it compulsory for all European food manufacturers to label products containing genetically modified soya and maize, and setting a precedent for future labelling of all genetically modified foods.

The USA

The import, interstate movement and environmental release of certain genetically engineered plants and micro-organisms is regulated by the Biotechnology and Biological Analysis and the Permit Sciences Divisions of Scientific Service (BSS) of the Animal and Plant Health Inspection Service (APHIS) of USDA. New pesticidal substances introduced into transgenic plants are regulated by the EPA as 'biological pesticides' and thereby require registration prior to their commercialisation. Before such authorisation, the EPA must however obtain clearance from APHIS in respect of its non-regulated status.

When a GMO is to be used and commercialised in the USA, a petition for a determination of non-regulated status is generally submitted to APHIS. The complete submission (but without confidential business information) is made available to the public for their comment for a period of 60 days. The average time for processing a petition is a maximum of 180 days (mandatory maximum review time). However, as in other countries, the 'clock' can be stopped to request more data. There are no costs associated with the submission of petitions. When a GMO is afforded deregulated (non-regulated) status it may be used like any other variety of crop. The list of non-regulated GMOs is available on the APHIS website. The general trend in US regulation is to modify the regulations to allow for more expeditious and efficient review and processing, based on knowledge and insight gained in viewing the behaviour of different groups and classes of GMOs. In order to conduct GMO field tests in the USA, a permit or notification is required by APHIS.

For transgenic plants modified to contain novel pesticidal substances, the EPA requires small-scale field testing which consists of terrestrial studies (10 acres or less of land) and aquatic studies (1 surface-acre or less of water). Such GMO testing prior to registration with the EPA also comes under the authority of APHIS. All EPA registrations are listed in the Federal Register which is open to public comment.

In the USA, there is no requirement to label GMOs, unless they pose a material health or safety risk, or the GMO-product presents different characteristics. The US rationale is that there is no common characteristic of GMOs that would justify their being labelled as a group, so that no product is labelled as containing genetically modified material. If the group commercialising the GMO wishes to label it voluntarily, the label must be truthful, provide useful information and not be misleading. Under general labelling requirements for seeds, transgenic seeds must be labelled accordingly.

Japan

Before the production or sale of rDNA organisms for use in agriculture in Japan, a total safety evaluation of the rDNA organism must be undertaken, on the basis of the characteristics of the

Box E.1 *Comparison of Plant Breeders' Rights under UPOV 1978 and 1991 and Patent Law*

Provisions	Plant Breeders' Rights		Patent law
	UPOV 1978	UPOV 1991	
Protection coverage	Varieties; undertaking to protect the largest possible number of botanical genera and species; minimum: 24 after eight years.	Varieties; obligation to protect all botanical genera and species; phasing-in over 5 or 10 years possible.	Inventions
Protection term	Minimum 15 years from grant (18 years for trees and vines)	Minimum 20 years from grant (25 for trees and vines)	Minimum 20 years from filing date (Article 33 TRIPS)
Protection scope	Minimum scope: producing for purposes of commercial marketing, offering for sale and marketing of propagating material of the variety.	Minimum scope: producing, conditioning, offering for sale, selling or other marketing, exporting, importing, stocking for above purposes of propagating material of the variety.	In respect of a product: making, importing, offering for sale, selling and using the product; stocking for purposes of offering for sale, etc. In respect of a process: using the process; doing any of the above-mentioned acts in respect of a product obtained directly by means of the process.
Breeders' exemption	Yes, but hybrids (and like varieties) cannot be exploited without permission from the holder of rights in the protected inbred line(s).	Yes, but hybrids (and like varieties) cannot be exploited without permission from the holder of rights in the protected inbred line(s). Essentially derived varieties cannot be exploited in certain circumstances without permission from the holder of rights in the protected initial variety.	There is some question over the applicability of the research exemption to patented inventions. Exploitation of a new variety incorporating a patented invention would require permission of the patentee.
Farmers' privilege	In practice: Yes. The implicit result of definition of minimum protection scope. Usually implicit under national law.	Must be spelled out in national law; subject to reasonable limits.	No.
Prohibition of double protection	Any species eligible for PBR cannot be patented. Special provision (Article 2(1)).	No provision.	Most countries exclude plant varieties from patentability on the basis of express provision in patent law, case law or administrative practice.

Sources: Derived from Dutfield, 1999 and van Wijk et al, 1993. Personal communication with André Heitz, UPOV, 22 February 1999 and 10 June 1999.

host, the rDNA molecules and the vectors involved. The risk assessments of 'GMOs', or rDNA organisms, are carried out independently of seed registration by MAFF. The 'Guidelines for Application of Recombinant DNA Organisms in Agriculture, Forestry, Fisheries, the Food Industry and Other Related Industries' set out the relevant procedures for the Japanese regulation of GMOs (including transgenic crops) and follow closely the recommendations of the OECD.

Japan has a multi-Ministerial approach to regulating GMOs. Although it is the process used to create the GMO (rDNA technology) that is the trigger for regulatory requirements, the biosafety of the end product and intended use of the end product are the basis for the risk assessments, which are undertaken by different organisations. MAFF conducts the environmental risk assessment of GMOs in Japan and the MHW has responsibility for food safety assessment of GMOs.

For rDNA experimentation at the confined testing stage, the Science and Technology Agency performs an environmental risk assessment, while MAFF performs environmental biosafety assessments and feed additive assessments for ag-biotech products intended for commercialisation. MAFF requires field trials in Japan for the environmental biosafety assessement, even if the product is not intended for cultivation in Japan. The Society for Techno-Innovation of Agriculture, Forestry and Fisheries (STAFF) coordinates the required field trials with the petitioner and a national agricultural institution. The trials involved are closed greenhouse, semi-closed greenhouse, isolated field, and open field trials. An rDNA Application Special Committee undertakes biosafety reviews and advises MAFF and the Director-General of Agriculture, Forestry and Fisheries Research. So far, 90 entries have been approved by MAFF, of which 48 are transgenic plants. There are no apparent labelling requirements for GMOs in Japan; whereas for transgenic crop seeds, the same labelling requirements will apply as for seeds in general under JAS.

Argentina

Applications to license experimentation on, and releases into the environment of genetically modified organisms are administered by the Argentinian Ministry of Agriculture, Fisheries and Food, subject to the favourable decision by the National Advisory Commission for Agriculture Biotechnology (CONABIA). Applications are initially processed by the INASE. A breeder may apply for various licences, for manipulation of organisms through recombinant DNA techniques at official or other institutions, laboratory-hothouse testing, first and repeat small-scale field testing, first and repeat large-scale field testing, or pre-commercial breeding.

Before commercialisation, transgenic seeds and plants must be registered on the National Cultivar Registry and be certified. If the GMO is an 'agrochemical product' (crop protection product), compliance is necessary with requirements set forth for commercialisation of agrochemical products by the National Service for Health and AgriFood Qualtiy (SENASA). If this is so, the Technical Coordination of CONABIA will request review by the National Bureau of AgriFood Markets regarding the convenience of commercialising the transgenic material. Upon completion of all the described steps, CONABIA's Technical Coordination will compile the pertinent information for the purposes of preparing a final report to the Secretary of Agriculture, Livestock and Food for its final decision.

Upon deciding whether to grant a licence, the following considerations are taken into account: risks to the environment, human or animal health; characteristics of the material (organism subject to control/donor-receptor-vector system); purpose of the trial and operational scheduling proposed; biosafety methods and procedures to be applied; intended fate; final disposition of the GMO; and transportation method. Argentina follows the UNEP Technical Guidelines on Biosafety and also has a fast-track procedure in place for the supervision of GMO field trials. Finally, as is the case for seeds in general, transgenic plant (seeds) must be labelled.

Box E.2 *Countries with Laws or Draft Laws on Plant Variety Protection*

At least 100 countries now have laws or draft laws on plant variety protection.

There are 43 UPOV members, as follows:

1 On the basis of the 1961/1972 Act, and a list of protected species: 2 – Belgium, Spain.

2 On the basis of the 1978 Act: 20 –
(a) protecting all genera and species: Argentina, Canada, Chile, France, Hungary, Mexico, New Zealand, Norway;
(b) list of protected species: Austria, Brazil, China, Czech Republic, Kenya, Panama, Paraguay, Portugal, Switzerland (the list includes over 100 families), Trinidad and Tobago (two families), Ukraine, Uruguay.

3 On the basis of the 1978 Act, but with a law essentially conforming to the 1991 Act: 10 –
(a) Australia, Bolivia, Colombia, Ecuador, Finland, Ireland, Slovakia;
(b) Italy, Poland, South Africa.

4 On the basis of the 1991 Act: 11 –
(a) Denmark, Germany, Israel, Japan, Netherlands, Sweden, UK, USA
(b) Bulgaria, Republic of Moldova, Russian Federation.

States and Organisations which have started the accession procedure to UPOV: 13 –

Belarus, Costa Rica, Croatia, Estonia, EC, Georgia, Kyrgyzstan, Morocco, Nicaragua, Romania, Slovenia, Venezuela, Zimbabwe.

States with other laws or drafts on plant variety protection: 47 –

Africa: Algeria, Egypt, Ethiopia, Organisation Africaine de la Propiété Intellectuelle (OAPI) Member States (Benin, Burkina Faso, Cameroon, Central African Republic, Chad, Congo, Côte d'Ivoire, Gabon, Guinea, Guinea-Bissau, Mali, Mauritania, Niger, Senegal, Togo), Tanzania, Tunisia, Zambia;
Americas and the Caribbean: Cuba, Dominica, El Salvador, Guatemala, Honduras, Peru;
Asia: Bangladesh, India, Indonesia, Iraq, Malaysia, Pakistan, Philippines, Republic of Korea, Sri Lanka, Thailand, Vietnam;
Europe and Central Asia: Cyprus, Greece, Kazakhstan, Latvia, Lithuania, Tajikistan, Turkey, Uzbekistan, Yugoslavia.

Source: André Heitz, UPOV, 10 June 1999.

E.3 Plant variety rights

Box E.1 compares plant breeders' rights under UPOV 1978 and 1991 and patent law, and Box E.2 provides information on countries with laws or draft laws on plant variety protection.

The European Union

Under the Council Regulation on Community Plant Variety Rights (EC 1994), plant breeders are granted cultivar protection throughout the EU. The Regulation is based on UPOV 1991. The European Plant Varieties Office is authorised to issue EU Plant Variety Certificates. Examination usually takes a minimum of two years, extending to about three years in total for later testing and evaluation. An applicant must furnish information that supports the distinct, uniform and stable 'DUS' criteria.

In the EU, patents are granted under the European Patent Convention (EPC). The EPC operates in a similar fashion to the Patent Coooperation Treaty: if a patent application is filed under the EPC, it will be effective (if granted) in each country that the applicant lists on the application; the applicant may make a single application for patent protection in any one or more of the EPC Contracting States by designating the countries of their choice on the application and paying a fee for each (which may be cheaper than making separate applications in each country). The EPC provides protection subject to the national law of each country.

The UK

The UK is a member of UPOV, and grants PBRs to breeders if the new variety is DUS and meets the criteria for novelty set out in the *Plant Breeders' Rights Handbook*. The Plant Variety Rights Office (PVRO) is authorised to administer PBRs at the national level, while the Community Plant Variety Office (in France) administers EU PBRs. EU and UK PBRs cannot operate simultaneously; however, the UK allows UK PBRs to be suspended whilst EU rights are exercised, which allows UK rights to be re-invoked if EU rights are terminated. Once granted, the UK provides PBRs for a term of 25 years for all species except trees, vines and potatoes, which are granted 30 years of protection.

There are two possible routes by which an applicant may obtain patent protection in the UK: by international application, nominating the UK as a designated country, or by applying directly to the UK Patent Office. Under UK patent law, patents are concerned with the technical and functional aspects of products and processes. The applicant must submit a patent specification to the UK Patent Office, the invention must be new, involve an inventive step, be capable of industrial application, and not be 'excluded'. Furthermore, it is not possible to obtain a patent for an animal or plant variety. International patent application allows for protection of a patent in countries other than the UK. In order to seek such international patent protection, an applicant must first file for a UK patent. Thereafter, the applicant has 12 months to file any foreign, European or world patent (just as with the Patent Cooperation Treaty) application. An applicant may file for a European or a world patent directly, or file in the UK first.

The USA

The 1970 US Plant Variety Protection Act, and its subsequent amendments, allows for protection of the production and sale of seed by owners or developers of new varieties of plants that are sexually reproduced (by seed) or tuber-propagated. Variety protection – which results in a Plant Variety Protection Certificate for the breeder or owner – is available to US citizens and citizens of countries that are members of UPOV, and may be obtained from the Plant Variety Protection Office (PVPO) within the USDA. The candidate variety is evaluated according to criteria such as novelty, uniformity, and stability. The time for consideration of an application to the PVPO for a Plant Variety Protection Certificate is generally 18 to 24 months, but this time can be longer. The applicant must provide data related to the variety's genealogy, uniformity and stability. An additional requirement is data from field trials related to the novelty and objective description of the variety from a minimum of two years or from two locations. The PVPO maintains databases for crops of both public and private varieties in order to allow the applicant to determine the distinctness of the candidate variety. Upon issuance of a Plant Variety Protection Certificate, the holder is granted 20 years of protection, unless the variety protected is a tree or vine, which is provided 25 years of protection. The Plant Variety Protection Certificate gives the holder the right to sue infringers of the rights accorded to him in the certificate, under the Plant Variety Protection Act, in the US courts. In addition, some rights attach immediately at the time of application.

A researcher can also protect a new variety by applying for a utility patent or a plant patent. For instance, an importer of an asexually reproduced plant (such as citrus fruit), can secure protection by way of a plant patent or a utility patent, obtained from the Patent and Trademark Office, within the US Department of Commerce. Both types of patents grant the the patent holder protection for 20 years on the invention, and prevent others from making, selling or using the patented invention without permission of the patent holder.

Japan

Japan is a member of UPOV and hence provides the breeder with the possibility of obtaining Plant Breeders' Rights on new varieties. The Plant Varieties Protection Certificate may be obtained from the Seeds and Seedlings Division of MAFF. In order to obtain patent protection in Japan, an applicant must file his application with the Japan Patent Office. The protection of a Japanese patent extends for 25 years.

Argentina

Argentina is a member of UPOV, and grants PBRs to breeders according to the 1978 Act. Property Certificates may be obtained from INASE, the Instituto Nacional de Semillas (Ministerio de Economia, Buenos Aires) (Section A above). Argentinian patent law is under development.

Useful Contacts and Sources of Information

General

Biodiversity-related secretariats

Convention on Biological Diversity (CBD)

World Trade Centre
393 St Jacques Street Suite 300
Montreal
Quebec H2Y 1N9
Canada
Tel: 1 518 288 2220
Fax: 1 514 288 6588
Web: www.biodiv.org

Convention to Combat Desertification

Secretariat of the Convention to Combat Desertification
PO Box 260129
Haus Carstanjen
D 53175
Bonn
Germany
Tel: 49 228 815 2800
Fax: 49 228 815 2899
Email: secretariat@unccd.de
Web: www.unccd.ch/

Convention on International Trade in Endangered Species (CITES)

15 chemin des Anémones
CH-1219
Châtelaine-Genève
Switzerland
Tel: 41 22 979 9139/40
Fax: 41 22 797 3417
Email: cites@unep.ch
Web: www.wcmc.org.uk/CITES/

Convention on Migratory Species (CMS)

Martin Luther King Street 8
D-53175 Bonn
Germany
Tel: 49 228 815 2401/02
Fax: 49 228 815 2449
Email: cms@unep.de
Web: www.wcmc.org.uk/cms/

Ramsar Convention on Wetlands

Rue Mauverney 28
CH-1196
Gland
Switzerland
Tel: 41 22 999 0001
Fax: 41 22 999 0002
Web: www.ramsar.org

United Nations Framework Convention on Climate Change (UNFCCC)

PO Box 260124
D-53153
Bonn
Germany
Tel: 81 75 705 2701
Fax: 81 75 705 2702
Email: secretariat@unfccc.de
Web: www.unfccc.or.jp/unfccc

World Heritage Convention

The World Heritage Centre UNESCO
7 place de Fontenoy
75352 Paris
07 SP
France
Tel: 33 145 68 1889
Fax: 33 145 68 5570
Email: wh-info@unesco.org
Web: www.unesco.org/whc/

Inter-governmental organisations

Organisation for Economic Cooperation and Development (OECD)

2 rue André Pascal
75775 Paris
Cedex 16
France
Tel: 33 1 4524 8200
Web: www.oecd.org/

United Nations Convention on Trade and Development (UNCTAD)

Palais des Nations
1211 Geneva
Switzerland
Tel: 41 22 907 1234
Fax: 41 22 907 0043
Email: ers@unctad.org
Web: www.unctad.org/
UNCTAD Biotrade Initiative
Web: www.biotrade.org/

United Nations Environment Programme (UNEP)

PO Box 30552
Nairobi
Kenya
Tel: 254 262 1234/3292
Fax: 254 262 3927/3692
Email: ipainfo@unep.org
Web: www.unep.org/

United Nations Educational, Social and Cultural Organisation (UNESCO)

7 place de Fontenoy
75352 Paris
07 SP
France
Tel: 33 14 568 1000
Fax: 33 14 567 1690
Web: www.unesco.org/

United Nations Food and Agriculture Organisation (FAO)

Viale delle Terme di Caracalla
00100 Rome
Italy
Tel: 39 65 7051
Fax: 39 65 705 3152
Web: www.fao.org

FAO Agriculture Department

Web: www.fao.org/ag/

FAO Commission on Genetic Resources for Food and Agriculture

Web: www.fao.org/ag/agp/agps/pgr/cgrfa.html

FAO Plant Genetic Resources for Food and Agriculture (PGRFA)

Web: web.icppgr.fao.org/

United Nations Industrial Development Organisation (UNIDO)

Vienna International Centre
PO Box 300
A-1400 Vienna
Austria
Tel: 43 12 6026
Fax: 43 12 69 2669
Web: www.unido.org/

World Trade Organisation

Web: www.wto.org/

National government contacts

These contacts have been selected because, to the best of the authors' knowledge, they are the offices responsible for coordinating access to genetic resources in the countries concerned.

Australia

Biodiversity Convention and Strategy Section
Biodiversity Conservation Branch
Biodiversity Group
Environment Australia
Fax: 61 26 250 0723
Email: veronica.blazely@ea.gov.au

Bolivia

Ministry of Sustainable Development and the Environment
National Secretariat of Natural Resources and the Environment
Fax: 591 231 6230
Email: recgenet@dncb.rds.org.bo
Web: coord.rds.org.bo/

Brazil

Divisao do Meio Ambiente (DEMA)
Ministerio das Relaçoes Exteriores
Esplanada dos Ministerios
Anexo Administrativo I
Sala 635
Brasília DF
70170.900
Brazil
Fax: 55.61.411.6012/224.1079

Colombia

Ministerio de Medio Ambiente
Calle 37 No 8–40
Santafé de Bogotá
Colombia
Tel: 571 338 3900
Fax: 571 340 6210
Email: info@mma.rds.org.co
Web: www.minambiente.gov.co/

Costa Rica

Ministerio de Ambiente y Energia
Email: root@ns.minae.go.cr
Web: www.minae.go.cr/

Ecuador

Instituto Ecuatoriano Forestal y de Areas Naturales y Vida Silvestre
Eloy Alfaro y Av Amazonas
8 piso
Quito
Ecuador
Tel: 593 254 8924
Fax: 593 256 4037

Eritrea

Department of Environment
Ministry of Land Water and Environment
PO Box 976
Asmara
Eritrea
Tel: 2911 120 311/125 887
Fax: 2911 126 095
Email: env@env.col.com.er

Fiji

Ministry of Local Government and Environment
PO Box 2131
Suva
Fiji
Fax: 679 30 3515

India

Ministry of Environment and Forests
Paryavaran
Bhawan
CGO Complex
Lodi Estate
New Delhi 110003
India
Telefax: 91 11 436 1712

Malawi

The Secretary to the Genetic Resources
and Biotechnology Committee
National Research Council of Malawi
PO Box 30745
Lilongwe 3
Malawi

Peru

Consejo Nacional del Ambiente
Av San Borja Norte 226
San Borja
Lima
Peru
Fax: 511 225 5369
Email: ddda@conam.gob.pe
Web: www.conam.gob.pe/

Seychelles

Ministry of Foreign Affairs Planning and Environment
PO Box 656
Victoria, Mahé
Seychelles
Tel: 248 22 46 88
Fax: 248 22 48 45

South Africa

Department of Environmental Affairs and Tourism
Private Bag X447
Pretoria 0001
South Africa
Fax: 271 2322 2682
Email: omb_pb@ozone.pwv.gov.za

Venezuela

Ministerio del Ambiente y de los Recursos Naturales Renovables
La Dirección de Vegetación
Email: info@marnr.gov.ve
Web: www.marnr.gov.ve/

Selected NGOs, research institutes and networks

Aberdeen University Molecular Biology Department UK

Web: mcb1.ims.abdn.ac.uk/

African Centre for Technology Studies

Web: www.anaserve.com/~acts/home.html

Biodiversity Action Network (BIONET)

Web: www.igc.org/bionet

Biowatch South Africa

Web: www.oneworld.org/saep/forDB/ Biowatch.html

Birdlife International

Web: www.kt.rim.or.jp/~birdinfo/birdlife/

CABI Bioscience

Web: www.cabi.org/institut/biosci.html

Center for International Environmental Law (CIEL)

Web: www.igc.apc.org/ciel/

Center for International Forestry Research (CIFOR)

Web: www.cgiar.org/cifor/

Centre for Science and Environment

Web: www.oneworld.org/cse/

Centro Internacional de Agricultura Tropical (CIAT)

Web: www.ciat.cgiar.org/

Centro Internacional de Mejoramiento de Maiz y Trigo (CIMMYT)

Web: www.cgiar.org/cimmyt/

Centro Internacional de la Papa (CIP)

Web: www.cgiar.org/cip/

Ciencia y Tecnología para el Desarrollo (CYTED)

Web: www.cicyt.es/ivpm/cyted.htm

Commonwealth Scientific and Industrial Research Organisation (CSIRO)

Web: www.csiro.au/

Conservation International

Web: www.conservation.org/

Consultative Group on International Agricultural Research (CGIAR)

Web: www.cgiar.org

Coordinadora de las Organizaciones Indigenas de la Cuenca Amazonica (COICA)
Web: www3.satnet.net/coica/

Council for Scientific and Industrial Research (CSIR)
Web: www.csir.co.za

Eden Foundation
Web: www.eden-foundation.org/

Environmental Resources Information Network (ERIN)
Web: kaos.erin.gov.au/erin.html

European Working Group on Research and Biodiversity
Web: www.oden.se/~ewgrb

Foundation for International Environmental Law and Development (FIELD)
Web: www.field.org.uk/

Friends of the Earth International Secretariat
Web: www.xs4all.nl/~foeint/

Genetic Resources Action International (GRAIN)
Web: www.grain.org

Greenpeace International
Web: www.greenpeace.org/

Horticulture Research International (HRI) (UK)
Web: www.hri.ac.uk/

Indian Institute of Management
Web: csf.Colorado.EDU/sristi/

Indian Institute of Public Administration
Web: www.res.bbsrc.ac.uk/

Indigenous Peoples Biodiversity Information Network (IBIN)
Web: www.ibin.org/

Indigenous Peoples Coalition Against Biopiracy
Web: www.niec.net/ipcb

Inter-American Biodiversity Information Network (IABIN)
Web: www.nbii.gov/iabin/

International Center for Agricultural Research in the Dry Areas (ICARDA)
Web: www.cgiar.org/icarda/

International Center for Living Aquatic Resources Management (ICLARM)
Web: www.cgiar.org/iclarm/

International Centre for Genetic Engineering and Biotechnology (ICGEB)
Web: www.icgeb.trieste.it/icgebhom.htm

International Centre for Research in Agroforestry (ICRAF)
Web: www.cgiar.org/icraf/

International Chamber of Commerce (ICC)
Web: www.iccwbo.org/

International Crops Research Institute for the Semi-Arid Tropics (ICRISAT)
Web: www.cgiar.org/icrisat/

International Development Research Centre
Web: http;//www.idrc.ca

International Food Policy Research Institute (IFPRI)
Web: www.cgiar.org/ifpri/

International Institute for Sustainable Development (IISD)
Web: www.iisd.ca/

IISD, Environmental Negotiations Bulletin
Web: www.iisd.ca/linkages

International Livestock Research Institute (ILRI)
Web: www.cgiar.org/ilri/

International Plant Genetic Resources Institute (IPGRI)
Web: www.cgiar.org/ipgri/

International Rice Research Institute (IRRI)
Web: www.cgiar.org/irri/

International Service for National Agricultural Research (ISNAR)
Web: www.cgiar.org/isnar/

International Water Management Institute (IWMI)
Web: www.cgiar.org/iimi/

IUCN Environmental Law Centre
Web: www.iucn.org/themes/law/index.html

IUCN The World Conservation Union
Web: www.iucn.org

Maryland Biotechnology Institute
Web: www.umbi.umd.edu/

National Biological Information Infrastructure (NBII)
Web: www.nbii.gov/

National Institute of Health (botanical medicine)
Web: www.nal.usda.gov/fnic/IBIOS

Natural Resources Defense Council
Web: www.igc.apc.org/nrdc/

The Nature Conservancy
Web: www.tnc.org/

Nuffic-Centre for International Research and Advisory Networks, Indigenous Knowledge
Web: www.nuffic.nl/ciran/ik.html

Organisation for Economic Cooperation and Development (OECD) Megascience Forum
Web: www.oecd.org/dsti/mega/

Rainforest Alliance
Web: www.rainforest-alliance.org

The Royal Botanic Gardens, Kew
Web: www.rbgkew.org.uk/

Rural Advancement Foundation International (RAFI)
Web: www.rafi.org/

Third World Network
Web: www.upm.edu.my/~webworks/twn/

TRAFFIC International
Web: www.traffic.org

Tufts Center for the Study of Drug Development
Web: www.tufts.edu/med/research/csdd/

United Plant Savers
Web: www.plantsavers.org

Via Campesina – UNORCA
Web: laneta.apc.org/unorca/

von Humboldt Biological Resources Research Institute
Web: www.humboldt.org.co/

West Africa Rice Development Association (WARDA)
Web: www.cgiar.org/warda/

Working Group on Traditional Resource Rights
Web: users.ox.ac.uk/~wgtrr

World Business Council for Sustainable Development (WBCSD)
Web: www.wbcsd.ch/

World Conservation Monitoring Centre (WCMC)
Web: www.wcmc.org.uk

World Economic Forum (Davos Forum)
Web: www.weforum.org/

World Foundation for Environment and Development (WFED)
Web: www.wfed.org/

World Rainforest Movement
Web: www.wrm.org.uy

World Resources Institute
Web: www.wri.org/wri/

Worldwatch Institute
Web: www.worldwatch.org/

World Wide Fund for Nature International
Web: www.panda.org

On-line listservers and information networks

BIODIV-CONV: CBD and related information List-Server. To subscribe, send message "subscribe biodiv-conv" to:
majordomo@igc.org

BINAS: Information Network and Advisory Service, service of the United Nations Industrial Development Organisation (UNIDO). Monitors global developments in regulatory issues in biotechnology:
www.binas.unido.org/binas/home.html

BIO-IPR: Intellectual property rights related to biodiversity and associated knowledge. To subscribe, send message "subscribe" to:
bio-ipr-request@cuenet.com

BIODIV-L: List-server on biological diversity. To subscribe, send message "subscribe biodiv-l" to:
majordomo@ns.bdt.org.br

BIOSAFETY: CBD Working Group Policy and Science Updates. To subscribe, send message "subscribe Biosafety listserver" to:
acfgenet@peg.apc.org

BioSafety Journal, Online Journal
www.bdt.org.br/bioline/by

BIOWATCH: To subscribe, send message "subscribe BIOWATCH listserver" to:
majordomo@sunsite.wits.ac.za

CHM – The Convention on Biological Diversity's Clearing House Mechanism
Web: www.wcmc.org.uk/~dynamic/chm /eng_main.html

ENV-BIOTECH: Bi-weekly news bulletin, Intellectual Property and Biodiversity News. To subscribe, send message "subscribe env-biotech" to:
Majordomo@igc.apc.org

G7ENRM: The G7 Environment and Natural Resources Management Project
Web: enrm.ceo.org/

HERB: Medicinal and aromatic plants; cross-cultural medicine and/folk/herbal medicine (Anadolu University Medicinal Plants Research Center). To subscribe send message to:
listserv@vm3090.ege.edu.tr

International Institute for Sustainable Development (IISD): To subscribe, send message to:
chadc@iisd.org
Web: www.iisd.ca/linkages/journal/IISD Sustainable Developments
Web: www.iisd.ca/sd

IPR-SCIENCE: Intellectual property in science, academic-industry links, sociological/ethical/legal analyses, inventiveness and exploitability. To subscribe email
mailbase@mailbase.ac.uk

NTFP-BIOCULTURAL-DIGEST: non-timber forest products and ethnobotany. To subscribe send message to:
majordomo@igc.org and in the body put subscribe ntfp-biocultural-digest (your e-mail address).

On-line directories

Agrilaunch
Web: www.agrilaunch.com/

Seed Quest
Web: www.seedqueStcom/

Industry sector section: Associations, NGOs and IGOs

Pharmaceutical industry selected resources

Industry organisations, trade associations, regulatory bodies

American Chemical Society
1155 16th Street NW
Washington DC
USA
Tel: 1 202 872 4600
Web: www.acs.org/

American Association of Pharmaceutical Scientists (AAPS)
1650 King Street
Alexandria
VA 22314-2747
USA
Tel: 1 703 548 3000
Fax: 1 703 684 7349
Email: aaps@aaps.org
Web: www.aaps.org/

American Pharmaceutical Association
2215 Constitution Avenue NW
Washington
DC 20037 USA
Tel: 1 202 628 4410
Fax: 1 202 783 2351
Web: www.aphanet.org/

Association of the British Pharmaceutical Industry
12 Whitehall
London SW1A 2DY UK
Tel: 44 171 930 3477
Web: www.abpi.org.uk/_private/welcome/ default.htm

Association of Information Officers in the Pharmaceutical Industry
Glaxo Group Research Ltd
Greenford Road
Middlesex, Greater London
UB6 0HE, UK
Tel: 44 181 577 1576
Web: www.aiopi.org.uk/

Australian Therapeutic Goods Administration
PO Box 100
Woden ACT 2606
Australia
Tel: 612 6232 8644
Web: www.health.gov.au/tga/

Bundesverband des pharmazeutischen Großhandels - PHAGRO e V
Savignystr 42
60325 Frankfurt
Germany

Chemical Specialties Manufacturers Association (CSMA)
1913 Eye Street NW
Washington DC 20006
USA
Tel: 1 202 872 8110
Web: www.csma.org/

Chinese Pharmaceutical Association
A38 N Lishi Road
Beijing 100810
People's Republic of China
Tel: 86 10 831 6576
Web: info3.bta.net.cn/tcm/chinacpa.htm

Committee of Experts on Pharmaceutical Questions
Council of Europe
F-67075 Strasbourg Cedex
France
Tel: 33 88 412 000

European Agency for the Evaluation of Medicinal Products (EMEA)
7 Westferry Circus
Canary Wharf
London E14 4HB UK
Tel: 44 171 418 8400
Fax: 44 171 418 8416
Web: www.eudra.org/

European Commission Directorate General III (Industry)
Pharmaceuticals and Cosmetics Unit
Web: dg3.eudra.org/

European Federation of Associations of Health Product Manufacturers (EHPM)
c/o European Advisory Services SPRL/BVBA
50 rue de l'Association
B-1000 Brussels
Belgium
Tel: 32 2209 1145
Fax: 32 2223 3064
Web: www.ehpm.org/

European Federation of Pharmaceutical Industries' Associations (EFPIA)
250 Avenue Louise bte 91
B-1050 Brussels
Belgium
Tel: 32 2626 2555

European Pharmaceutical Marketing Research Association (EPhRMA)
2 Woodburns Rd
Brooklands
Sale
Cheshire MB3 3SY
UK
Web: www.ePhRMA.org/

German Federal Institute for Drugs and Medical Devices
Seestrasse 10
D-13353 Berlin
Germany
Tel: 49 3045 4830
Fax: 49 3045 4832 07
Web: www.bfarm.de/gb_ver/

International Federation of Pharmaceutical Manufacturers Associations (IFPMA)
30 rue de St-Jean
PO Box 9
1211 Geneva 18
Switzerland
Tel: 41 22 340 1200
Fax: 41 22 340 1380
Web: www.ifpma.org

International Conference on Harmonisation
C/o IFPMA
Web: www.ifpma.org

International Federation of Pharmaceutical Wholesalers
Web: www.ifpw.com/

International Pharmaceutical Federation
Andries Bickerweg 5
NL-2517 JP
The Hague
Netherlands
Tel: 31 70 363 1925
Web: metalab.unc.edu/pwmirror/pharm web92.html

Japanese BioIndustry Association (JBA)
26-9 Hatchobori 2-chome
Chuo-ku
Tokyo 104
Japan
Tel: 813 5541 2731
Fax: 813 5541 2737

Japanese Pharmaceutical Manufacturers Association (JPMA)
Web: www.jpma.com

Latin American Association of Pharmaceutical Industries (LAAPI)
Acoyte 520
1405 Buenos Aires
Argentina
Tel: 541 903 4440
Web: www.pjbpubs.co.uk/scrip/scrhome html

Malaysian Organisation of Pharmaceutical Industries (MOPI)
5B Lorong Rahim Kajai 13
Taman Tun
60000 Kuala Lumpur
Malaysia
Tel: 603 717 3486

Medicines Control Agency
Market Towers
1 Nine Elms Lane
London SW8 5NQ
UK
Tel: 44 171 273 0000

Ministry of Health and Welfare (MHW) Japan
Web: www.ncc.go.jp/mhw/index.html

National Conference of Standards Laboratories (NCSL)
1800 30th Street Suite 305B
Boulder
CO 80301
USA
Tel: 1 303 440 3339
Web: www.ncsl-hq.org/

National Pharmaceutical Industries Syndicate
88 rue de la Faisanderie
F-75016 Paris
France
Tel: 33 14 503 8888

Organisation for Economic Cooperation and Development (OECD) Environmental Health and Safety
Web: www.oecd.org/ehs/

Organisation of Pharmaceutical Producers of India
Cook's Building
1st Floor
324 Dr Dadabhoy Naoroji Road
Mumbai 400 001
Maharashtra
India
Tel: 91 22 204 5509

Pharmaceutical Manufacturers Association of Canada (PMAC)
302-1111 Prince of Wales Drive
Ottawa ON
Canada K2C 3T2
Tel: 1 613 727 1380
Web: www.pmac-acim.org/

Pharmaceutical Research and Manufacturers of America (PhRMA)
1100 15th Street NW
Suite 900
Washington
DC 20005
USA
Tel: 1 202 835 3400
Web: www.Phrma.com

United States Food and Drug Administration
Web: www.fda.gov/

Botanical medicines selected resources

Industry organisations, trade associations, regulatory bodies

Alternative Medical Association (AMA)
7909 SE Stark Street
Portland
OR 97215
USA
Tel: 1 503 253 4301

American Botanical Council
PO Box 201660
Austin
TX 78720
USA
Tel: 1 512 331 8868
Web: www.herbalgram.org/

American Herb Association (AHA)
PO Box 1673
Nevada City
CA 95959
USA
Tel: 1 916 265 9552

American Herbal Products Association (AHPA)
4733 Bethesda Avenue
Suite 345
Bethesda
MD 20814
USA
Tel: 1 301 951 3204
Web: www.ahpa.com

British Herbal Medicine Association
Sun House
Church Street
Stroud GL5 1JL
UK
Tel: 44 1453 751389

Bundesvereinigung Deutscher Apothekerverbande (ABDA)
Deutsches Apothekerhaus
Carl-Mannich-Strasse 26
65760 Eschborn
Germany
Web: www.abda.de/

Council For Responsible Nutrition
1300 19th Street NW
Suite 310
Washington
DC 20036-1609
USA
Tel: 1 202 872 1488
Fax: 1 202 872 9594
Web: www.crnusa.org/

European Federation of Associations of Health Product Manufacturers (EHPM)
c/o European Advisory Services
SPRL/BVBA
50 rue de l'Association
B-1000 Brussels
Belgium
Tel: 322 209 1145
Fax: 322 223 3064
Web: www.ehpm.org/

European Herbal Infusions Association (EHIA)
Gotenstrasse 21
20097 Hamburg
Germany
Tel: 49 40 2360160

European Scientific Cooperative on Phytotherapy (ESCOP)
Argyle House
Gandy Street
Exeter
Devon EX4 3LS
UK
Tel: 44 1392 424 626

Herb Research Foundation (HRF)
1007 Pearl Street
Suite 200
Boulder
CO 80302
USA
Tel: 1 303 449 2265
Web: www.hrf.org/ or
www.herbs.org/index.html

International Herb Association
1202 Allanson Road
Mundelein
IL 60060
USA
Tel: 1 847 949 4372
Web: www.herb-pros.com/

Natural Products Marketing Council (NPMC)
PO Box 500
Truro NS
Canada
B2N 5E3
Web: www.gov.ns.ca/govt/
foi/Natural.htm

Society for Medicinal Plant Research
Am Grundbach 5
97271 Kleinrinderfeld
Germany
Tel: 49 931 6102271
Web: : www.rz.uni-uesseldorf.de/
WWW/GA/

US Pharmacopeia (USP)
12601 Twinbrook Parkway
Rockville
MD 20852
USA
Tel: 1 301 881 0666
Fax: 1 301 998 6806
Web: www.usp.org/

United States Food and Drug Administration (USFDA)
Web: www.fda.gov/

Seed industry selected resources

Industry organisations, trade associations, regulatory bodies

American Seed Trade Association (ASTA)
601 13th Street
#570 South
Washington
DC 20005-3807
USA
Web: www.amseed.com/

Argentine Chamber of Seed Traders
Argentina (CSBC)
Web: www.argenseeds.com.ar/

Asia and Pacific Seed Association
PO Box 1030
Bangkok 10903
Thailand
Tel: 662 940 5464
Fax: 662 940 5467
Email: apsa@apsaseed.com
Web: apsaseed.com/

Association Internationale des Selectionneurs (ASSINSEL)
Chemin de Reposoir 7
CH-1260
Nyon
Switzerland
Tel: 41 22 361 9977
Fax: 41 22 361 9219
Email: assinsel@iprolink.ch
Web: www.worldseed.org

Federation Internationale du Commerce Semences (FIS)
Chemin du Reposoir 7
CH-1260
Nyon
Switzerland
Tel: 41 22 361 9977
Fax: 41 22 361 9219
Email: fis@iprolink.ch
Web: www.worldseed.org

France Seed Industry Association
Web: www.amsol.asso.fr/

United Kingdom Agricultural Supply Trade Association (UKASTA)
3 Whitehall Court
London SW1A 2EQ
UK
Tel: 44 171 930 3611
Fax: 44 171 930 3952
Email: enquiries@ukasta.org.uk
Web: www.ukasta.org.uk/

University of Nebraska, Agriculture Information Network
Web: www.unl.edu/agnicpls/assoc.html

Horticulture industry selected resources

Industry organisations, trade associations, regulatory bodies

American Horticultural Society (AHS)
Web: www.ahs.org/

American Nursery and Landscape Association
Web: www.anla.org/

American Society of Horticultural Science
113 S West Street
Suite 400
Alexandria
VA 22314-2824
USA
Tel: 1 703 836 4606
Web: www.ashs.org/

Commercial Horticultural Association
Global CHA Secretariat and Export Promoter
National Agricultural Centre
Stoneleigh Park
Kenilworth
Warwickshire
CV8 2LG
UK
Tel: 44 120 369 0330
Fax: 44 120 369 0334
Email: Pgrimbly@aol.com
Web: www.ukexnet.co.uk/hort/cha/index. html

International Federation of Agricultural Producers (AIPH)
60 rue St Lazare
Paris 75009
France
Tel: 33 14 526 0533
Fax: 33 14 574 7212
Email: ifap@club-internet.fr
Web: www.ifap.org

Nursery Industry Association of Australia
Web: www.niaa.org.au/
Ontario Horticultural Association
Web: www.interlog.com/~onthort/

The Royal Horticultural Society
Web: www.rhs.org.uk/

Crop protection industry selected resources

Industry organisations, trade associations, regulatory bodies

American Crop Protection Association (ACPA)
1156 Fifteenth Street NW
Suite 400
Washington
DC 20005
USA
Tel: 1 202 296 1585
Fax: 1 202 463 0474
Web: www.acpa.org

Argentina Crop Protection Association (CASAFE)
Rivadavia 1367
piso 7 "B"
1033 Buenos Aires
Argentina
Tel: 541 381 2742
Fax: 541 383 1562
Email: casafe@casafe.org
Web: www.casafe.org/

Asia-Pacific Crop Protection Association (APCPA)
1405 Rasa Tower
555 Pahonyothin Road
Chatuchak
Bangkok
Thailand 10900
Tel: 662 937 0487
Fax: 662 937 0491
Web: www.apcpa.org

British Agrochemicals Association (BAA)
Tel: 44 17 3334 9225
Fax: 44 17 3336 2523
Web: 195.50.78.12/baa/

Crop Protection Institute
21 Four Seasons Place
Suite 627
Etobicoke
Ontario
Canada M9B 6J8
Tel: 416 622 9771
Fax: 416 622 6764
cpic@croppro.org

European and Mediterranean Plant Protection Organisation
Web: www.eppo.org/

European Crop Protection Association (ECPA)
6 Avenue E Van Nieuwenhuyse
1160 Brussels
Belgium
Tel: 322 663 1550
Fax: 322 663 1560

Global Crop Protection Federation (GCPF)
Avenue Louise 143
B1050 Brussels Belgium
Tel: 322 542 0410
Fax: 322 542 04 19
Email: norma@gcpf.org
Web: www.gcpf.org/

Japan Crop Protection Association (JCPA)
5-8 1-Chrome Muromachi
Nihonbashi
Chuo-Ku
Tokyo 103
Japan
Tel: 813 3241 0215
Fax: 813 3241 3149

Biotechnology industry selected resources

Industry Organisations, Trade Associations, and Regulatory Bodies

Belgian BioIndustries Association
490 Avenue Louise B9
B 1050 Brussels
Belgium
Tel: 322 646 0564
Fax: 322 640 3759
Email: jviseur@bba-bio.be
Web: www.bba-bio.be/

BioIndustry Association
14/15 Belgrave Square
London SW1X 8PS
UK
Tel: 44 171 565 7190
Fax: 44 171 565 7191
Email: admin@bioindustry.org
Web: www.bioindustry.org/

Biotechnology Industry Association
1625 K Street
Suite 1100
Washington
DC 20006
USA
Tel: 1 202 857 0244
Fax: 1 202 857 0237
Web: www.bio.org/welcome.html

Confederation Mondiale de l'Industrie de la Sante Animale (COMISA)
1 rue Defacqz
Bte 6
B-1000 Brussels
Belgium
Tel: 322 541 0111
Fax: 322 541 0119
Email: comisa@fedesa.be

European Association for BioIndustries
Avenue de l'Armée 6
1040 Brussels
Belgium
Tel: 322 735 0313
Fax: 322 735 4960
Web: www.europa-bio.be/

Green Industry Biotechnology Platform
Web: www.gibip.org/

Japan BioIndustry Association
26-9 Hatchobori 2-chome
Chuo-ku
Tokyo 104
Japan
Tel: 813 5541 2731
Fax: 813 5541 2737
Email: nozaki@jba.or.jp
Web: www.jba.or.jp/

National Center for Biotechnology Information
Web: www.ncbi.nlm.nih.gov/

National Institute for Biological Standards and Control
Blanche Lane
South Mimms
Potters Bar
Herts
EN6 3QG
UK
Tel: 44 17 0765 4753
Fax: 44 17 0764 6730
Email: enquiries@nibsc.ac.uk
Web: www.nibsc.ac.uk/

The Netherlands BioIndustries Association
PO Box 2260 AK
Leidschendam
Netherlands
Tel: 31 70 327 0464
Fax: 31 70 317 6215
Email: niaba@xs4all.nl
Web: www.xs4all.nl/~niaba/
Postbus

UNIDO: National Biotechnology links can be found at:
Web: www.binas.unido.org/binas/link.html

Personal Care and Cosmetic Industry

Industry organisations, trade associations and regulatory bodies

American Society for Quality Control (ASQC)
611 E Wisconsin Avenue
PO Box 3005
Milwaukee
WI 53201-3005
USA
Tel: 1 414 272 8575

Australian Therapeutic Goods Administration
PO Box 100
Woden ACT 2606
Australia
Tel: 612 6232 8644
Web: www.health.gov.au/tga/

Coalition for Environmentally Responsible Economies (CERES)
711 Atlantic Avenue
Boston
MA 02111
USA
Tel: 1 617 451 0927

Committee of Experts on Cosmetic Products
Council of Europe
F-67075 Strasbourg Cedex
France
Tel: 33 88 412 000

Cosmetic and Detergent Industry Association Germany
Karlstr 21
60329 Frankfurt
Germany
Tel: 49 69 2556 1323

Cosmetic Association – Asia
1765 Ramkhamhaeng Rd
Huamark Bangkapi
Bangkok 10240
Thailand
Tel: 662 314 1415
Web: www.ctfas.com/aca.html

Cosmetic Industry Buyers and Suppliers (CIBS)
36 Lakeville Road
New Hyde Park
NY 11040
USA
Tel: 1 516 775 0220

Cosmetic, Toiletry and Fragrance Association (CTFA)
1101 17th Street NW
Suite 300
Washington
DC 20036
USA
Tel: 1 202 331 1770
Web: www.ctfa.org/

Cosmetic, Toiletry, and Perfumery Association Belgium (COLIPA)
rue du Congres S-7
B-1000 Brussels
Belgium
Tel: 322 227 6610

Cosmetic, Toiletry and Perfumery Association England
Josaron House
5–7 John Princes Street
London W1M 9HD
UK
Tel: 44 171 491 8891

European Commission Directorate General III (Industry) Pharmaceuticals and Cosmetics Unit
Web: dg3.eudra.org/

Fragrance Foundation
145 E 32nd Street
New York
NY 10016 USA
Tel: 1 212 725 2755
Web: www.fragrance.org

Independent Cosmetic Manufacturers and Distributors (ICMAD)
1220 W Northwest Highway
Palatine
IL 60067
USA
Tel: 1 847 991 4499
Web: www.icmad.org/

Industrieverband Koerperpflege- und Waschmittel eV (IKW)
Karlstraße 21
D-60329 Frankfurt-am-Main
Germany
Tel: 49 692 556 1331
Fax: 49 692 37 631

International Federation of Societies of Cosmetic Chemists (IFSCC)
24/26 Rothesay Road
Luton
Beds LU1 1QX
UK
Tel: 44 158 226 661
Web: www.ifscc.org/index.htm

Ministry of Health and Welfare (MHW) Japan
Web: www.ncc.go.jp/mhw/index.html

OECD Environmental Health and Safety
Web: www.oecd.org/ehs/

Society for Cosmetic Chemists (SCC)
120 Wall Street
Suite 2400
New York
NY 10005
Tel: 212 668 1500
Web: www.scconline.org/

Society of Cosmetic Chemists of South Africa (SCCSA)
c/o SA Association for Food Science and Technology
PO Box 91182
Auckland Park 2006
Republic of South Africa
Tel: 27 11 726 2376

United States Food and Drug and Administration
Web: www.fda.gov/

Selected marketing and research publications and data information companies

Advances in Phytochemistry
Web: www.fiu.edu/orgs/psna/

Advanstar Communications Inc
Web: www.advanstar.com/

AgPlus Network
www.agplus.net/

Agribiz
Web: www.agribiz.com/

Argus Research Corporation
Web: www.argusresearch.com/

Biotechnology and Development Monitor
Web: www.pscw.uva.nl/monitor/

Ceres Corporation
Web: www.ceresgroup.com/

Chemical and Engineering News
Web: pubs.acs.org/cen/

Chemical Market Reporter
Web: www.chemexpo.com/schnell/cmr.html

Chemical Week
Web: www.chemweek.com

Data Monitor
Web: www.datamonitor.com/

Deloitte and Touche
Web: www.dttus.com/

Drug and Cosmetic Industry (DCI)
Web: www.cosmeticindex.com/

Drug Topics
Web: www.drugtopics.com

The Economist
Web: www.economistcom/

Happi Magazine (Household and Personal Products Industry)
Web: www.happi.com

Hartman and New Hope
Web: www.hartman-newhope.com

Health Foods Business
Web:www.healthfoodsbiz.com/INDEX.html

Health Supplement Insider
Web:
www.hsrmagazine.com/articles/871edlet.html

Herbalgram
Web: www.herbalgram.org/

Herbs for Health
Web: www.consciouschoice.com/herbs/

Information Resources Inc
Web: www.infores.com/

Inverizon International Inc
Web: www.inverizon.com/

Journal of Natural Products
Web: acsinfo.acs.org/journals/jnprdf/

Journal of Ethnopharmacology
Web:
www.elsevier.nl/inca/publications/store/5/0/6/0/3/5/index.htt

Kalorama Information Market Intelligence Reports
Web: directory.findexonline.com/findex/about.html

Natural Business Communications
Web: www.naturalbiz.com/index.html

Natural Foods Merchandiser
Web: www.nfm-online.com/

NPD Beauty Trends
Web: www.npd.com

Nutrition Business Journal (NBJ)
Web: www.nutritionbusiness.com

SPINS – Spence Information Services
Web: www.spenceinfo.com/

Stat Publishing
www.statpub.com/

Supplement Industry Insider
Web: www.vpico.com/

The Tan Sheet
Web: www.fdcreports.com/abouttan.html

Wall Street Journal
Web: www.wsj.com/

Whole Foods
Web: www.wfcinc.com/

Glossary

AHAs (alpha hydroxy acids): Fruit acids that include glycolic acid, lactic acid, malic acid, tartaric acid, and citric acid. AHAs flake off old skin and are used to treat dry and severely dry skin, unblocking and cleaning pores. They drove much of the growth in skin care during the early 1990s.

agrobiotechnology: The research on and development of agricultural products such as crop varieties and crop protection products by modifying genes to confer desirable properties such as pest resistance or improved nutritional profiles.

allele: One of two or more forms of a gene arising by mutation and occupying the same relative position (locus) on homologous chromosomes.

antibiotic: An antimicrobial compound produced by living **microorganisms**, used therapeutically or sometimes prophylactically in the control of infectious diseases. Over 4,000 antibiotics have been isolated, but only about 50 have achieved wide use.

aromatherapy: The therapeutic use of pure essential oils and other substances obtained from flowers, plants, and aromatic shrubs, through inhalation and application to the skin. Generally based on traditional practices from around the world.

asexual reproduction: Reproduction of a plant or animal without fusion of male and female **gametes**. It includes vegetative propagation, cell and **tissue culture**.

assay: A technique that measures a biological response; the determination of the activity or concentration of a chemical. (See **bioassay**.)

Ayurveda: A philosophy and healing system developed over thousands of years in India, in which patients are characterised by the elements of earth, water, fire, air, and ether. Employs botanical preparations, usually combinations of a number of herbs.

***Bacillus thuringiensis* (Bt):** A natural enemy of insects which was isolated from dead silk worms. This bacterium kills insects with the help of a protein, the so-called Bt-toxin. More than 50 Bt-toxins have been detected, each with its own characteristics.

backcross: The cross of a hybrid with either of its parents (or a genetically equivalent individual).

bacteria: Members of a group of diverse single-celled organisms; organisms lacking a nucleus.

bacteriophages: A group of viruses whose hosts are specifically bacteria.

beauty products: Includes *skin/body care products; perfumes and fragrances; make-up* (lipstick, lip cream, face powder, foundation, eye shadow, eyeliner, eyebrow pencils, blush, and nail care products).

BHAs (beta hydroxy acids): BHAs are the latest hydroxy acid offered by manufacturers. Derived from plant cells and from the sugar derivative luconamide, BHAs are supposed to be less irritating to skin than the **AHAs**.

bioassay: The determination of the activity or concentration of a chemical by its effect on the growth of an organism under experimental conditions.

bioavailability (biotechnology): The degree of availability to **biodegradation** of pollutants in contaminated soil or land.

bioavailability studies (pharmaceuticals): *In vivo* studies conducted in healthy human volunteers required to ensure that the final marketed dosage formulation is bioequivalent to the dosage formulation used in Phase I, II, and III clinical tests. **Bioavailability** is the rate and extent to which an active drug ingredient or therapeutic moiety is absorbed from the drug product and becomes available at the site of drug action.

biocatalyst: An **enzyme**, used to catalyse a chemical reaction.

biochemical: A product produced by chemical reactions in living organisms.

biodegradation: The microbially mediated process of chemical breakdown of a substance to smaller products caused by **microorganisms** or their **enzymes**.

biodiversity: (See **biological diversity**.)

bioenergy: Energy made available by the combustion of materials derived from biological sources.

bioinformatics: A scientific discipline that comprises all aspects of the gathering, storing, handling, analysing, interpreting and spreading of biological information. Involves powerful computers and innovative programmes which handle vast amounts of coding information on **genes** and **proteins** from **genomics** programmes. Comprises the development and application of computational algorithms for the purpose of analysis, interpretation, and prediction of data for the design of experiments in the biosciences.

biolistics: In molecular biology, a method developed to inject **DNA** into cells by mixing the DNA with small metal particles and then firing the particles into the host cell at very high speeds.

biologics: Vaccines, therapeutic serums, toxoids, antitoxins and analagous biological products used to induce immunity to infectious diseases or harmful substances of biological origin.
(See **biopharmaceuticals**.)

biological control agent: The use of living organisms to control **pests** or disease. May be a single organism or a combination of a number of different ones.

biological diversity (biodiversity): The variability among all living organisms, including diversity within **species**, between species, and of the **ecosystem**. The variability among living organisms from all sources including, *inter alia*, terrestrial, marine and other aquatic **ecosystems** and the ecological complexes of which they are part; this includes diversity within species, between species and of **ecosystems**.

biological resources: These include genetic resources, organisms or parts thereof, populations, or any other biotic component of **ecosystems** with actual or potential use or value for humanity.

biomass: All organic matter that derives from the photosynthetic conversion of solar energy: the total mass of living organisms in an **ecosystem**.

biopesticide: Naturally occurring biological agents used to kill **pests** by causing specific biological effects rather than by inducing chemical poisoning. The idea is based on mimicking processes that arise naturally (eg protecting the coffee bean by its caffeine content), and is argued to be favourable to conventional chemical pesticides as it is more easily biodegradable and more target specific.A pesticide in which the active ingredient is a **virus**, fungus, or **bacteria**, or a natural product derived from a plant source. A biopesticide's mechanism of action is based on specific biological effects and not on chemical poisons.

biopharmaceutical: Recombinant protein drugs, recombinant vaccines and **monoclonal antibodies** (for therapeutic roles). Biopharmaceuticals are still only a small part of the pharmaceutical industry, but of increasing importance. (See **biologics**.)

bioprocess: Any process that uses complete living cells or their components (eg **enzymes**, chloroplasts) to effect desired physical or chemical changes.

bioreactor: A contained vessel or other structure in which chemical reactions are carried out (usually on an industrial scale), mediated by a biological system, **enzymes** or cells. A bioreactor can range in size from a small container to an entire building.

bioremediation: The use of biological agents to reclaim soils and waters polluted by substances hazardous to human health and/or the environment; it is an extension of biological treatment processes that have been used traditionally to treat wastes in which **microorganisms** typically are used to biodegrade environmental pollutants.

biosynthesis: The synthesis of molecules by living organisms or their components.

biotechnology: Any technological application that uses biological systems, living organisms, or derivatives thereof, to make or modify products or processes for specific use.

biotransformation (bioconversion): The conversion of a compound from one form to another by the actions of organisms or **enzymes**.

botanical: A substance derived from plants; a vegetable drug, especially in its crude state.

botanical medicine: A medicine of plant origin, in crude or processed form; used in this book to represent herbal, or plant-based, medicines that are not consumed as isolated compounds (as are pharmaceuticals). Includes single herb, herb combination, and herb combined with non-herbal ingredient products. Delivery formats include capsules, tablets, herbal teas, extracts, tinctures, and bulk herbs.

broad spectrum: A pesticide which is active towards a wide variety of weeds or other **pests**. Often used to describe an **antibiotic** that is effective against a wide range of **microorganisms**.

brown bag sales: Sales by farmers to other farmers of seed they have saved.

cell fusion: A technique of fusing two cells from different species to create one **hybrid** cell for the purpose of combining some of the genetic characteristics of each original.

clone: Cell or organism identical to an ancestor with respect to **genotype** and **phenotype**.

combinatorial chemistry: Automated parallel synthesis of hundreds or thousands of compounds at a time; can be directed to produce 'drug-like' molecules and molecules compatible with molecular-based screens.

cosmeceuticals: Products that straddle the boundary between **cosmetics** and drugs, with increasingly sophisticated bioactive properties. They include anti-ageing, slimming, moisturising, sunscreen, and hair-growth preparations. Unlike cosmetics or general **skin care** products which claim only to mask or retard skin ageing, cosmeceuticals change, or claim to change, the structure of the skin.

cosmetic (FDA definition): '(1) Articles intended to be rubbed, poured, sprinkled or sprayed on, introduced into, or otherwise applied to the human body or any part thereof for cleansing, beautifying, promoting attractiveness, or altering the appearance, and (2) Articles intended for use as a component of any such articles, except that such terms shall not include soap'.

cosmetics and toiletries: Products in the following categories: **oral care**; **soaps and bath**; **hair care**; and **beauty** (**skin/body care**, perfumes and fragrances, and make-up).

country of origin of genetic resources: The country which possesses **genetic resources** in *in situ* **conditions**.

country providing genetic resources: The country supplying **genetic resources** collected from *in situ* sources, including populations of both wild and domesticated **species**, or taken from *ex situ* sources, which may or may not have originated in that country.

cross-breeding: The breeding of distinct and genotypic types or forms in plants. This may entail the transfer of pollen from one individual to the stigma of another of different **genotype**.

cross-pollination: The transfer of pollen from the stamen of a flower to the stigma of a flower of different **genotype**, but usually of the same **species**.

cross: The act or product of cross-fertilisation between different individuals.

cryopreservation: The storage of plant material at very low temperatures (-196°C) in liquid nitrogen.

cultivar: Distinct form or **variety** of domesticated plant derived through breeding and selection and maintained through cultivation.

Decision of the Conference of the Parties: A formal agreement of the Conference of the Parties to the CBD that leads to binding actions. It becomes part of the agreed body of decisions by the Conference of the Parties that direct the future work of the Conference of the Parties and guide action at the national level.

deconvolution: Isolating the active compound out of a natural product mixture.

diagnostics: Measuring and monitoring systems. **Healthcare**: Human diagnostics refer to *in vivo* and *in vitro* tests to detect disease and measure bodily functions. Often **monoclonal antibodies**, used in a variety of diagnostic applications, including pregnancy and fertility testing, drug testing and the diagnosis of infectious diseases. Some monoclonal antibodies are also designed for *in vivo* cancer detection. **Biotechnology:** Non-medical diagnostics are increasingly based on biotechnology, and are designed to detect pollutants in the environment and effluent streams, and **pathogens** and contaminants in food and drinking water.

dietary supplements: An umbrella term used in the USA for a group of products including vitamins, minerals, **herbs**, and natural medicines. Also known as 'nutritional supplements' This book focuses on the botanical and natural medicine component of the dietary supplement market.

DSHEA: The Dietary Supplement and Health Education Act, passed in 1994 in the USA. Created the new legal class ' **dietary supplement**', and resulted in expanded market opportunities for the **botanical medicines** industry.

diploid: Having a pair of **homologous chromosomes** with the exception of the sex chromosome, the total number of chromosomes being twice that of a **gamete**.

deoxyribonucleic acid (DNA): The molecule that generaly encodes all genetic information. It consists of two strands or chains of sub-units (known as nucleotides).

DNA: see **deoxyribonucleic acid**.

domesticated or cultivated species: A species in which the evolutionary process has been influenced by humans to meet their needs.

drug development: Includes chemical improvements to a drug molecule; animal pharmacology studies; pharmacokinetic and safety studies in animals, followed by Phases I, II and III clinical studies in humans.

drug discovery: The process by which a lead is found, including the acquisition of materials for screening; identification of a disease and therapeutic target of interest; methodology and assay development; advanced screening; and identification of active agents and chemical structure.

ecosystem: A dynamic complex of plant, animal and **microorganism** communities and their non-living environment interacting as a functional unit.

elite: Advanced **germplasm** in a breeding or crop improvement programme. Synonymous with superior performance characters, adaptation or quality.

emulsifiers: These product ingredients help maintain a mixture and ensure consistency. They also influence smoothness, volume, and uniformity.

entry into force: **Protocols** and any amendments to them are not binding in international law until they have been ratified by an agreed number of countries. In the case of the CBD, **ratification** by 30 countries was needed for the treaty to enter into force. The CBD entered into force for the first 30 **Parties** on 29 December 1993. It enters into force for other Parties 90 days after each ratifies.

enzyme: A **protein** which catalyses the conversion of a substrate to a product. Other than a few well-established enzymes such as papein and trypsin, most enzyme names can be recognised by the suffix *-ase*, eg cellulase, protease, etc.

ethical pharmaceutical: Products, including biological and medicinal chemicals, used for the cure, alleviation, mitigation, treatment, prevention, or diagnosis of disease in humans or animals and promoted primarily to the medical, pharmacy, and allied professions. Ethical drugs include products available only by prescription as well as some **over-the-counter (OTC)** drugs.

excipients: Binders and fillers without activity or nutritive value, used in the manufacturing of capsules and tablets.

exotic: (1) Not native to a given area; either intentionally transplanted from another region or introduced accidentally; (2) In plant breeding, it refers to plants types that are from outside a breeding region or exhibit traits that are uncommon to the prevalent crop plant type.

extracts (Botanical medicines): Greater concentrations of original **herbs** produced by solvent extraction. Extracts come in the form of **tinctures**, fluid extracts, solid extracts, and powdered extracts. Their quality is normally expressed either as *strengths*, or they are *standardised* against a chemical or group of chemical compounds.

***ex situ* conservation:** The conservation of components of **biological diversity** outside their natural habitats.

extreme environments: Environments characterised by extremes in growth conditions, including temperature, salinity, pH, and water availability, among others.

extremophile: A **microorganism** whose optimum growth is under extreme conditions of temperature, etc.

financial mechanism: The Convention establishes a mechanism for providing financial resources to help developing countries implement the Convention; the **GEF** is operating the mechanism on an interim basis.

functional foods: Foods that are considered to have a positive beneficial effect on health, by the addition of **active ingredients**, or by making bioavailable existing ingredients. Includes ' functional' modified or fortified soft drinks and other beverages, bread, dairy products, cereals, and snacks.

gamete: Specialised **haploid** cell (sometimes called a sex cell) whose nucleus and often cytoplasm fuses with that of another **gamete** in the process of fertilisation.

gene: The units of heredity transmitted from generation to generation. Each gene is a segment of nucleic acid carried in the **DNA** encoded for a specific protein. More generally, the term 'gene' may be used in relation to the transmission and inheritance of particular identifiable traits. The basic unit of heredity, a gene is an ordered sequence of nucleotide bases comprising a segment of DNA. A gene contains the sequence of DNA which encodes one polypeptide chain. The sum of an organism's genes is known as its **genome**.

gene mapping: Determination of the relative positions of **genes** on a **DNA** molecule (chromosome or plasmid) and of the distance, in linkage units or physical units, between them.

generics: Copies of well-known drugs for which patent protection has expired. Companies specialising in generics invest little on research, or only on research in manufacturing procedures. The average price of a generic is 30 per cent below that for patented products.

genetic engineering: Manipulation of **DNA** to form a **hybrid** molecule,a new combination of non-homologous DNA (so-called **recombinant DNA**). The technique allows the bypassing of all the biological constraints to genetic exchange and mixing and may even permit the combination of genes from widely differing species. Genetic engineering developed in the early 1970s.

genetic marker: A **gene** with a clear, unambiguous **phenotype** used in genetic analysis to identify individuals that carry it or other linked genes. May act as a probe to mark a nucleus, chromosome or locus.

genetic material: Material of plant, animal, microbial or other origin containing functional units of heredity.

genetic resources: **Genetic material** of actual or potential value.

genome: The genetic endowment of an organism. When expressed, this will result in the observable characteristics or **phenotype**.

genomics: The study of **genomes** including genome mapping, **gene** sequencing and gene function. The use of this information in the development of therapeutics.

genotype: (1) The genetic constitution of an organism or group of organisms; (2) A group of organisms that share the same genetic composition.

germ cell: A small organic structure or cell from which a new organism may develop.

germplasm: The genetic material which forms the physical basis of heredity and which is transmitted from one generation to the next by means of germ cells.

Global Environment Facility (GEF): The multi-billion-dollar GEF was estabished by the World Bank, UNDP and UNEP in 1990. It operates the Convention's ' financial mechanism' on an interim basis and funds developing-country projects that have global biodiveresity benefits.

GMO (genetically modified organism): The modification of the genetic characteristics of a **microorganism**, plant or animal by inserting a modified **gene** or a gene from another variety or species. GMOs may be microorganisms designed for use as **biopesticides** or seeds that have been altered genetically to give a plant better disease resistance or growth.

Good Manufacturing Practices (GMP): Government or industry set standards for the production of safe, efficacious, and high-quality ingredients and products.

Group of 77 (G-77) and China: The G-77 was founded in 1967 under the auspices of the UNCTAD. It seeks to harmonise the negotiating positions of its 132 developing-country members.

habitat: The place or type of site where an organism or population naturally occurs.

hair care products: Shampoos, conditioners, hair dye, salon preparations, hair oil, gels, brilliantine, pomade, lacquers, wax, rinses, and hair treatments.

haploid: Having the number of chromosomes present in the normal **germ cell** equal to half the number in the normal **somatic cell**.

herbs: Plants or plant parts valued for medicinal, savoury, cosmetic, flavouring, or aromatic qualities. This study focuses on herbs with medicinal properties, but has chosen the term ' **botanical medicines**' to describe them.

herbal supplements: Botanical or herbal medicines, also known as herbal **dietary supplements**, and herbal nutritional supplements. A sub-set of the broader category ' dietary supplement' in the USA.

homeopathy: A system that attempts to stimulate the body to heal itself by activating defence and healing mechanisms with a medicine typically derived from natural sources (80 per cent of which are botanical).

homologous chromosomes: Contain identical linear sequences of **genes** and pair during **meiosis**.

homozygous: An individual having two identical **alleles** of a particular **gene** or genes, and so breeding true for the corresponding characterisitcs.

horticulture: The cultivation of ornamental and vegetable plants in gardens or smallholdings (market gardens). *Hortus* = garden (Latin).

hybrid: Individual plant resulting from a **cross** between parents of differing **genotypes**. Hybrids may be fertile or sterile, depending on qualitative and/or quantitative differences in the **genomes** of the two parents. Hybrids are most commonly formed by sexual cross-fertilisation between compatible organisms, but techniques for the production of hybrids from widely differing plants are being developed by **cell fusion** and **tissue culture**.

hybridisation: (1) The act of crossing two different individual organisms of differing genetic constitution from different populations or different **species**; (2) A molecular procedure in which single strands of **DNA** and/or **RNA** are mixed and subsequently bind to one another. The degree of binding is a measure of the relatedness of the strands. The procedure is used to detect RNA or DNA using suitable probes.

IARC: International Agricultural Research Centre.

impact assessment: An evaluation of the likely impact on **biological diversity** of proposed programmes, policies, or projects.

in-bred: A cross between parents of similar genetic constitution which could be from the same blood line (shared ancestry).

in-bred line: In plants, a line produced by repeated selfing and selection. Results in truebreeding and a fixed **genotype**.

integrated pest management (IPM): The challenging or control of **pests** through a tailored programme of different strategies including **biological control agents** and agrochemicals.

***in situ* conditions:** The conditions where **genetic resources** exist within **ecosystems** and natural **habitats**, and, in the case of domesticated or cultivated **species**, in the surroundings where they have developed their distinctive properties.

***in situ* conservation:** The conservation of **ecosystems** and natural **habitats**, and the maintenance and recovery of viable populations of **species** in their natural surroundings and, in the case of domesticated or cultivated species, in the surroundings where they have developed their distinctive properties.

***in vitro*:** Literally 'in glass'. Experimental reproduction of biological processes in isolation from a living organism, eg **tissue culture**.

***in vivo*:** Taking place in a living organism.

landrace: Farmer-developed **cultivars** of crop plants which are adapted to local environmental conditions.

life sciences companies: Companies which combine businesses in pharmaceutical, agricultural chemicals and products, and food and nutrition.

line: A **homozygous**, pure breeding group of individuals **phenotypically** distinct from other members of the same **species**. Broader than **strain**.

markers (botanical medicines): Chemical compound markers against which botanical medicine extracts are standardised; these markers are not necessarily the ' active' ingredient.

marker assisted selection (MAS): The use of **molecular markers** to follow the inheritance of **genes**, particularly those genes which cannot be readily identified. Selection of a **marker** flanking a gene of interest allows selection for the presence (or absence) of a gene in a new progeny.

mass marketers: Companies that market products through drugstores, discount stores, supermarkets, and other mass-merchandise outlets. Mass market products are usually single-item, low cost products, with wide availability, and broad applications. (See **prestige marketers**.)

mass selection: Breeding method whereby seed from a number of individuals is selected to form the next generation. Selection criteria are relaxed until later generations and **crosses** are performed at random.

me-too drugs: A variation on existing drugs that are already under patent protection.

mechanism-based screening: A **receptor**- or **enzyme**-based screen against which a range of materials can be run, including natural products such as plants, marine organisms, fungi, and **microorganisms**, but also synthetic compounds.

medicinal and aromatic plant material: Whole plants and plant parts (including seeds and fruits) used primarily in perfumery and pharmacy. Includes fresh, dried, uncut, cut, crushed, and powdered material.

meiosis: The process of division of sexual cells in which the number of chromosomes in each nucleus is reduced to half the normal number found in normal **somatic cells**. When two sexual cells fuse, each contributes its half of the chromosomes. The resulting embryo contains the full chromosome complement. Cells with half the chromosomes are called **haploids**: those with the normal chromosomal complement, diploids.

meristem tip culture: A cell culture developed from a small portion of the **meristem** tissue of a plant.

meristem: The tip of a growing plant shoot or root.

metabolites: Chemical products of metabolism; the biological synthesis or breakdown carried out by cells or their components.

microbe: Synonymous with **microorganism**.

microorganisms: Groups of microscopic organisms, some of which cannot be detected without the aid of a light or electron microscope, including the **viruses**, the **prokaryotes (bacteria** and archaea), and eukaryotic life forms, such as protozoa, filamentous fungi, yeasts and microalgae.

micropropagation: The use of biotechnological methods to grow large numbers of plants from very small pieces of plants, often from single cells using **tissue culture** methods.

molecular marker: A molecular selection technique of **DNA** signposts which allows the identification of differences in the nucleotide sequences of the DNA in different individuals. *Agriculture:* A tool which allows crop geneticists and breeders to locate on a plant chromosome the **genes** for a trait of interest. It is considered more efficient than conventional breeding as it has the potential to greatly reduce development times and substitutes laboratory selection for much of the fieldwork.

monoclonal antibody: A homogeneous antibody population derived from one specific B-lymphocyte or **hybrid** cell.

monograph: A document that summarises the botanical characteristics, historical use, and major active constituents and quality control of a plant. Monographs also include summaries of clinical applications, pharmacology, posology, possible contraindications and precautions, and potential adverse reactions.

mutagen: Agent that induces a mutation within an organism, such as X-rays, gamma rays, neutrons, and certain chemicals such as carcinogens. KP is an agent capable of inducing a mutation (a change that alters the sequence or chemistry of bases in the **DNA** molecule) in the **genetic material** of an organism.

NARS: National Agricultural Research System.

national delegation: One or more officials who are empowered to represent and negotiate on behalf of their government.

natural foods industry: An umbrella term for the store-based sale of health food, groceries, botanical **personal care products**, and all **dietary supplements**.

natural product drugs: Drugs of natural origin classified as original natural products, products derived semi-synthetically from natural products, or synthetic products based on natural product models.

nematode: Roundworms or threadworms, often internal parasites of animals and plants. They are significant economic **pests** on foodcrops as few crops are immune to attacks of these creatures which inhabit the soil about the roots of plants. The development of **nematode**-resistant varieties of crop plants is important to food growth economics.

new chemical entities (NCEs): Newly developed drugs, characterised by their therapeutic innovation. An innovation may be substantial when it completely changes a therapeutic field. Companies specialising in NCEs invest heavily in product research. The discovery and development of NCEs is at the heart of pharmaceutical research and development, and accounts for most of the money spent on pharmaceutical research and development (eg an estimated 83 per cent in the USA). NCEs are discovered either through screening existing compounds or designing new molecules.

non-governmental organisation (NGO): In the context of the CBD, NGOs include environmental groups, indigenous peoples' organisations, research institutions, business groups, and representatives of city and local government.

non-papers: Issued informally to facilitate negotiations, these do not have an official document symbol although they may have an identifying number.

non-party: A state that has not ratified the Convention and may attend as an **observer**.

nurseryman: A person who grows plants for a living (for ornamental and amenity use).

nutraceuticals: An umbrella term for any substance considered a food or part of a food and that provides medical and health benefits, including the prevention and/or treatment of disease. Products in this category include **dietary supplements**, entire diets (eg macrobiotic), isolated nutrients, medical foods, ' designer' **biotechnology**-enhanced foods, and processed ' **functional' foods** such as cereals, soups and beverages.

observer: The Conference of the Parties and its subsidiary bodies normally permit accredited observers to attend their meetings. Observers include the United Nations and its specialised agencies, the International Atomic Energy Agency, non-Party states, and other qualified governmental and non-governmental organisations.

OECD (The Organisation for Economic Cooperation and Development): comprises Australia, Austria, Belgium, Canada, Czech Republic, Denmark, Finland, France, Germany, Greece, Hungary, Iceland, Ireland, Italy, Japan, Republic of Korea, Luxembourg, Mexico, The Netherlands, New Zealnd, Norway, Portugal, Spain, Sweden, Switzerland, Turkey, the UK, and the USA.

oral care: Toothpastes, toothbrushes, mouthwashes, breath fresheners, dental floss, and other products used in the mouth;

orthodox seed: Seed that can be dried and stored for long periods at reduced temperatures and under low humidity.

over-expression host: An organism that produces more of the desired **enzyme** or compound than normal.

over-the-counter (OTC): A drug product sold over the counter in pharmacies, and available without a prescription.

Party: A state (or regional economic integration organisation such as the EU) that ratifies or accedes to the CBD becomes a Party 90 days later and is then legally bound by its provisions.

pathogen: A disease-producing agent, usually restricted to to a living agent such as a bacterium, fungus or **virus**.

pathway: a sequence of reactions undergone in a living organism.

personal care products: A broad category that includes **cosmetics**, feminine hygiene, **hair care products**, baby care, nail care, oral hygiene, bath items, deodorants, shaving, **skin care**, bath/toilet **soap**, and fragrances.

pests: Any **species**, **strain** or biotype of plant, animal or pathogenic agent injurious to plants and plant products.

pharmaceutical dosage formulation and stability testing: Those operations involved in incorporating an active compound into a dosage form suitable for administration to the patient; the development of analytical methodology to measure the active compound; and the techniques used to assure that the compound is stable over the shelf-life of the dosage form.

phenotype (-typic): The characteristics of an organism that result from the interaction of its genetic constitution with the environment.

pheromone: A volatile hormone or behaviour-modifying agent. Normally used to describe sex attractants – for example bombesin for the moth Bombyx – but includes volatile aggression-stimulating agents (eg isoamyl acetate in honey bees). A hormone-like substance secreted into the environment by certain animals, especially insects.

phytomedicine: Medicinal products based on standardised **active ingredients** within a herbal base. This term is sometimes used more broadly to include all plant-based medicines, but in this book is used more narrowly as defined.

phytonutrients: Naturally-occurring compounds found in fruits and vegetables, such as beta carotene, capsaicin, and flavonoids.

plenary: An open session of the entire Conference of the **Parties** to the CBD where all formal decisions are taken.

pollen culture: A culture of plant cells derived from pollen in a synthetic medium: the progeny generated will have a single set of chromosomes.

polyketide: A large family of structurally diverse natural products synthesised by repeated cycles of condensation of thioesters; they include **antibiotics**, antiparasitics and anti-cancer drugs.

post-emergent: A herbicide which acts after the seed has germinated.

pre-breeding: The development of **germplasm** to a state where it is viable for breeders' use. Primarily involves the evaluation of traits from **exotic** material and their introduction into more cultivated backgrounds.

pre-clinical studies: The various tests conducted in whole animals and other test systems, such as cell cultures, to determine the relative toxicity of a compound to living systems. These are referred to as pre-clinical studies – tests conducted and evaluated prior to the first administration of the compound to humans. Also included in this category are two-year carcinogenicity assays which typically overlap with the chemical testing phase. Pre-clinical tests include: toxicity (how poisonous it is and what side effects might be expected); **bioavailability** (how effectively it is taken up into the body and delivered to the tissue where it is needed); pharmacokinetics (how it is metabolised, and therefore how long it stays in the body); and whether it has the desired physiological effect.

preservatives: As an ingredient in cosmetics, preservatives offer protection against infection under conditions of use, and prevent the decomposition of the product due to microbial multiplication. A **cosmetic** preservative is ideally active at a low concentration, against a wide range of **microorganisms**, and over a wide acidity/alkalinity range. It should also be compatible with other ingredients in the formulation, as well as nontoxic, non-irritating, non-sensitising, colourless, odourless, and stable.

prestige marketers: Companies that market multiple-item regimens with limited distribution, primarily in department and speciality stores. Prestige products usually contain special (sometimes ' therapeutic') ingredients, are expensively packaged and higher priced.

primary metabolites: compounds ubiquitous in living organisms and essential for life, such as carbohydrates, the essential amino acids and polymers derived from them.

primitive cultivar: Crop forms developed from **landraces**. Improvement through selection restricted to a few specific characteristics and often more uniform in nature than a landrace.

process development for manufacturing and quality control: The scale-up of production of bulk chemical and finished-dosage forms from small research quantities to large quantities required for marketing. This includes the attendant quality control procedures required to maintain the safety and efficacy of the product for the duration of its shelf-life.

progeny testing: Procedure to establish the **genotype** of a parent by recording the genetic status of off-spring/progeny.

prokaryote: An organism (eg bacterium) having a cell lacking a nucleus.

protease: An **enzyme** that catalyses the hydrolytic breakdown of **proteins**.

protected area: A geographically defined area which is designated or regulated and managed to achieve specific conservation objectives.

protein: Any of a class of nitrogenous compounds forming an essential part of living organisms and having large molecules consisting of one or more chains of amino acids linked together.

protein engineering: The generation of **proteins** (specifically **enzymes**) with subtly modified structures, thus conferring new properties such as changed catalytic specificity or thermal stability.

protocol: Can be used to expand or strengthen the terms and commitments of a Convention, but must be signed and ratified anew before it enters into force.

protoplast fusion: Any induced or spontaneous union between two or more **protoplasts** to produce a single bi- or multi-nucleate cell. Fusion of nuclei may or may not occur subsequent to the initial protoplast fusion.

protoplast: A plant cell from which the cell wall has been removed by mechanical or enzymatic means. Protoplasts can be prepared from primary tissues of most plant organs as well as from cultured plant cells.

random screening: This form of screening treats all samples equally, and works through extract and compound libraries. Compounds may be screened singly or in mixtures.

ratification: After signing the Convention, a country must ratify it, for which it often needs the approval of the parliament or other designated body. The instrument of ratification is submitted to the UN Secretary-General, who acts as the Depositary. 90 days later the country becomes a Party. As of mid March 1999 the CBD had 175 parties.

rational drug design: Lead compounds are identified based on a molecular understanding of the drug and its **receptor**, often by using computer technology to aid in the determination of the 3-D structures of molecular targets. Molecular modelling can be used to design new structures from scratch, or to look at a database of existing compounds, to select classes of compounds for screening, or to manipulate naturally occurring molecules.

reagent: A substance used to cause a reaction, especially to detect another substance.

recalcitrant: A term applied to pollutants which are not biodegradable or are only biodegradable with difficulty.

recalcitrant seed: Seed that cannot withstand either drying or temperatures of less than 10°C and cannot therefore be stored for long periods as orthodox seeds can.

receptor: A molecular structure within a cell or on the surface characterised by selective binding of a specific substance and a specific physiologic effect that accompanies the binding; for example, cell surface receptors for peptide hormones, neurotransmitters, antigens, complement fragments and immunoglobulins and cytoplasmic receptors for steroid hormones.

recombinant DNA (r-DNA): A strand of **DNA** synthesised in the laboratory by splicing together selected parts of DNA strands from different organic species, or by adding a selected part to an existing DNA strand.

recombinant DNA technology: The process of excising segments of **DNA** from one **species** of organism and inserting them into the DNA of another species.

recombinant (micro) organisms: Organisms whose **phenotype** has arisen as a result of recombination.

Recommendation of the SBSTTA: The SBSTTA delivers its advice to the COP in the form of recommendations that are numbered sequentially.

recurrent selection: A breeding method aimed at increasing the frequency of favourable **genes** through repeated cycles of selection and selected crossing of individuals.

regional economic integration organisation: An organisation constituted by sovereign States of a given region, to which its member States have transferred competence in respect of matters governed by this Convention, and which has been duly authorised, in accordance with its internal procedures, to sign, ratify, accept, approve or accede to it.

RNA: A molecule with similar structure to **DNA** that is involved in a number of cell activities, especially **protein** synthesis. Some **viruses** have RNA as their **genetic material**.

SBSTTA: The Subsidiary Body for Scientific, Technical and Technological Advice.

secondary metabolites: Compounds such as phenolics, turpenoids and alcaloids that are of restricted occurrence in living organisms. Based on primary metabolites, they are responsible for a range of functions such as defence, recognition and storage.

Secretariat to the Convention on Biological Diversity: Staffed by international civil servants and responsible for servicing the Conference of the Parties and ensuring its smooth operation, the Secretariat to the Convention on Biological Diversity makes arrangements for meetings; compiles and prepares reports, and coordinates with other relevant international bodies. The Secretariat of the Convention on Biological Diversity is administered by UNEP and located in Montreal, Canada.

self-pollinated: See **in-bred**.

skin/body care products: Facial cream; facial cleansing cream; cold cream; face packs; moisturisers; skin whitening products; body creams and milks; talcum powder; anti-cellulite creams; slimming gels; body contour creams; shaving products, and sun-screen products.

soaps and bath products: Toilet soaps (bars, cakes, moulded shapes, medicated), liquid soaps, shower gels, and perfumed bath salts.

somatic cell: Any cell other than a **germ cell**.

speciality biotechnology products: Include **enzymes**, fine chemicals, speciality food products and food ingredients. Non-medical diagnostics for detection of pesticides in the environment and contaminants in food.

species: A taxonomic rank below a genus, consisting of similar individuals capable of exchanging **genes** or interbreeding.

sport: An aberrant form of a plant that arises vegetatively out of a somatic mutation, usually of a bud or a shoot.

square brackets: Used during negotiations in UN meetings to indicate that a section of text is being discussed but has as not yet been agreed.

standardised extracts: Contain a specified amount of a presumably 'active' chemical compound or chemical group. **Extracts** containing high concentrations of pharmacologically active chemicals are produced based on the physical and chemical properties of these compounds, using appropriate solvents (eg water, ethanol, methanol, and hydroalcoholic mixtures) and temperatures.

Subsidiary Body of the Convention on Biological Diversity: A committee that assists the Conference of the Parties. The Convention defines one permanent committee: the Subsidiary Body for Scientific, Technical and Technological Advice (**SBSTTA**). The Conference of the Parties may establish additional subsidiary bodies as needed; for example in 1996 it set up the Open-ended Ad Hoc Working Group on Biosafety.

surfactants: Ingredients in **personal care products**, these compounds reduce the friction between two surfaces, and include **emulsifiers** and emulsion stabilisers, solubisers, and texturisers. Emulsifiers help maintain a mixture and ensure consistency. Solubisers mix oil and water. Texturisers give products a desired feel and appearance, eg 'body'.

sustainable use: The use of components of **biological diversity** in a way and at a rate that do not lead to the long-term decline of biological diversity, thereby maintaining its potential to meet the needs and aspirations of present and future generations.

strain: A population of cells all descended from a single cell; also called a **clone**. A group of organisms within a **species** or **variety** distinguished by one or more minor characteristics; a variety of bacterium or fungus used for culturing. The term is mostly associated with cells, **bacteria**, fungi and **viruses**, but is sometimes applied to plants.

technology transfer: The transfer of knowledge or equipment to enable the manufacture of a product, the application of a process, or the rendering of a service.

tinctures: Produced through a simple process of infusing **herbs** in alcohol.

tissue culture: *In vitro* methods of propagating cells from animal or plant tissue.

total extracts: Contain the whole spectrum of ingredients present in the original **herb**, plus any new active compounds formed during processing.

traditional Chinese medicine (TCM): The primary healthcare for 20 per cent of the world's population, the system of medicine developed over thousands of years in China, which treats the patient holistically, and includes herbal preparations – usually combinations of between five and ten **species**.

traditional knowledge: The knowledge, innovations and practices of local and indigenous communities. As used in the CBD, those elements of traditional knowledge that are relevant to the conservation and **sustainable use** of **biodiversity**.

transformation: Uptake of naked **DNA** by a competent recipient **strain**.

transgenic: Organisms into which **DNA** from another **species** are introduced by, for example, microinjection or retroviral infection.

ultra-high-throughput screening: Fully automated, around-the-clock, screening of compound libraries in a variety of molecular-based **assays**. The result is the capability to merge the increasing capacity for the development of new screening targets and the production of chemical diversity to reduce cycle times in **drug discovery**.

unadapted: Material lacking the characteristics or agronomic performance for cropping in a specific area.

variety: A taxonomic rank below subspecies in botany, varieties are usually the result of selective breeding and diverge from the parent **species** or subspecies in distinct but relatively minor ways. Usage varies in different countries.

vector: 'A carrier'. In genetic manipulation the vehicle by which **DNA** is transferred from one cell to another. An agent of transmission; for example, a DNA vector is a self-replicating segment of DNA that transmits genetic information from one cell or organism to another.

virus: The smallest known type of organism. A noncellular entity that consists minimally of **protein** and nucleic acid, and that can replicate only after entry into specific types of living cells, and then only by usurping the cell's own systems.

wide crossing: In plant breeding this refers to the process of undertaking a cross where one parent is from outside the immediate genepool of the other, ie **landrace** or primitive line crossed with a modern **cultivar**.

zygote: Fertilised ovum of animal or plant formed from the fusion of male and female **gametes**.

Bibliography

General

Note: All bibliographies for chapters that are not sector-specific are included in the general bibliography.

Akerele, O, Heywood, V and Synge, H (1991), *Conservation of Medicinal Plants*, Columbia University Press, New York

Anuradha, R V (1997), 'In Search of Knowledge and Resources: Who Sows? Who Reaps?', *RECIEL* (RECIEL), vol 6, no 3, 263–73

Balick, M J, Elisabetsky, E and Laird, S A (eds) (1996), *Medicinal Resources of the Tropical Forest*, Columbia University Press, New York

Bridson, D and Forman, L (1998), *The Herbarium Handbook*, Royal Botanic Gardens, Kew, London

Brush, S B (1996), 'Is Common Heritage Outmoded?', in S B Brush and D Stabinsky, *Valuing Local Knowledge: Indigenous Peoples and Intellectual Property Rights*, Island Press, Covelo CA

Burrell, I (1998), 'Animal Smuggling is the Most Lucrative Crime After Drugs', *The Independent*, 4 August

Carew-Reid, J (1994), *Strategies for National Sustainable Development*, Earthscan, London

Carr, G (1998), 'The Pharmaceutical Industry Survey', *The Economist*, 21 February

Comisión del Acuerdo de Cartagena (1996), *Decisión 391: Regimen Comú n sobre Acceso a los Recursos Genéticos Gaceta Oficial*, añ o XII, no 213, Lima, 17 July

Costa e Silva, E da (1996), 'The Protection of Intellectual Property for Local and Indigenous Communities', *European Intellectual Property Review*, vol 17, no 11, 546–9

Cragg et al (1997), 'Interactions with Source Countries, Guidelines for Members of the American Society of Pharmacognosy', *Journal of Natural Products*, 60, 654–5

Diversity (1998), 'Access Scenarios', *Diversity Magazine*, vol 14, nos 1 and 2, Bethesda

Downes, D (1997), *Using Intellectual Property as a Tool to Protect Traditional Knowledge: Recommendations for Next Steps*, Center for International Environmental Law, Washington DC

Dun and Bradstreet (1997), *Key British Enterprises: Britain's Top 50,000 Companies*, vols 1–4, Dun and Bradstreet, High Wycombe

Dutfield, G (1997), *Can the TRIPS Agreement Protect Biological and Cultural Diversity?*, ACTS, Nairobi

Economist, The (1998), 'The Pharmaceutical Industry', 21 February; 'The World in 1998'

Environmental Law Institute (1996), *Legal Mechanisims Concerning Access to and Compensation for the Use of Genetic Resources in the United States of America*, Washington DC

FAO (1996), *World List of Seed Sources*, FAO Plant Production and Protection Division, Rome

FAO (1999), *Revision of the International Undertaking on Plant Genetic Resources*, Consolidated Negotiating Text Resulting from the Deliberations During the Fifth Extraordinary Session of the Commission on Genetic Resources for Food and Agriculture, CGRFA/IUND/CNT/Rev.1 FAO, Rome, web: www.fao.org.ag/cgrfa/docs8.htm

Glowka, L (1994), *A Guide to the Convention on Biological Diversity*, IUCN, Switzerland

Glowka, L (1995), *The Deepest of Ironies: Genetic Resources, Marine Scientific Research and the International Deep Sea-Bed Area*, a paper distributed for comment and discussion at the first meeting of the SBSTTA of the CBD, Paris, 4 September

Glowka, L (1998a), *A Guide to Designing Legal Frameworks to Determine Access to Genetic Resources*, Environmental Policy and Law Paper no 34, IUCN Environmental Law Centre, Bonn

Glowka, L (1998b), *A Guide to Undertaking Biodiversity Legal and Institutional Profiles*, Environmental Policy and Law Paper no 25, IUCN Environmental Law Centre, Bonn

Glowka, L, Plän, T and Stoll, P (1998), *Best Practices for Access to Genetic Resources*, information paper distributed at the fourth meeting of the Conference of the Parties of the CBD, Institute for Biodiversity and Nature Conservation, Regensburg, May

Gollin, M (1993), 'An Intellectual Property Rights Framework for Biodiversity Prospecting', in Reid et al, *Biodiversity Prospecting*, WRI, Washington DC

Gollin, M and Laird, S A (1996), 'Global Policies, Local Actions: The Role of National Legislation in Sustainable Biodiversity Prospecting', in *University Journal of Science and Technology Law*, 2, L 16, Boston

Grifo, F and Rosenthal, J (1997), *Biodiversity and Human Health*, Island Press, Washington DC

Gruenwald, J (1997), 'Europe Leads in Herbal Remedies', *NBJ*, January/February, vol 12

Gupta, A K (1996), *Getting Creative Individuals and Communities Their Due: Framework for Operationalizing Article 8j and 10c*, SRISTI, Ahmedabad

Hawksworth (1995),

Hoagland, K E and Rossman, A Y (1996), *Global Genetic Resources: Access, Ownership and Intellectual Property Rights*, Association of Systematics Collections, Washington DC

Janestos, A C (1997), 'Do We Still Need Nature? The Importance of Biological Diversity', *Consequences*, vol 3, no 1

Juma, C (1989), *The Gene Hunters*, Zed Books, London

ten Kate, K (1994), *Improving Policy Cooperation Between Governments and Industry: a Report by the WICE Working Group on Policy Partnerships*, World Industry Council for the Environment, International Chamber of Commerce, Paris

ten Kate, K (1995), *Biopiracy or Green Petroleum? Expectations and Best Practice in Bioprospecting*, ODA, Environmental Policy Dept, London ten Kate, K and Laird, S (1997), findings presented at workshops at the third meeting of the SBSTTA of the CBD, Montreal

ten Kate, K and Strong, M F (1993), 'Environment and Development: Future Prospects in the Light of the United Nations Conference on Environment and Development' in M Nazim and P Polunin (eds) *Environmental Challenges from Stockholm to Rio and Beyond*, Foundation for Environmental Conservation, Geneva, and Energy and Environment Society of Pakistan, Lahore

La Vina, A G M, Caleda, M J A and Baylon, L L (eds) (1997), *Regulating Access to Biological and Genetic Resources in the Philippines: A Manual on the Implementation of Executive Order No 247*, Department of Environment and Natural Resources of the Philippines, Quezon City

Lee, W E, Bell, B M and Sutton, J K (1982), *Guidelines for Acquisitions and Management of Biological Specimens*, Association of Systematics Collections, Kansas, Washington DC

Martin, G (1995), *Ethnobotany*, Chapman and Hall, Cambridge

McConnell, K (1996), *The Biodiversity Convention – A Negotiating History*, International Environmental Law and Policy Series, Kluwer Law International, London

Monsanto (1998), 'Monsanto Releases Seed Piracy Case Settlement Details', press release by Karen K Marshall and Jennifer O'Brien, Monsanto Company, St Louis, Missouri, web: www.monsanto.com

Moran, D and Pearce, D W (1997), 'The Economics of Biodiversity', in T Tietenberg and H Folmer (eds), *International Yearbook of Environmental and Resource Economics*, Edward Elgar, Cheltenham

Mugabe, J et al (1997), *Access to Genetic Resources*, IUCN, Nairobi

OECD (1996), *Intellectual Property, Technology Transfer and Genetic Resources*, OECD, Paris

OECD (1996), *Saving Biological Diversity*, OECD, Paris

Oldfield, M L and Alcorn, J B (eds) (1991), *Biodiversity Culture, Conservation and Ecodevelopment*, Westview Press, Boulder CO

Pearce, D and Puroshothaman, S (1993), 'Protecting Biological Diversity: The Economic Value of Pharmaceutical Plants' in T Swanson (ed) *Biodiversity and Botany: The Value of Medicinal Plants*, CUP, Cambridge

Pearce, D W, Moran, D and Krug, W (1999) 'The Global Value of Biodiversity', report prepared for UNEP, Nairobi (forthcoming)

Plimmer, J R and Parry Jr, R M (1994), 'Registration of Biopesticides', in P A Hedin, J J Menn and R M Hollingworth (eds), *Natural and Engineered Pest Management Agents*, Ch 33, American Chemical Society, Washington DC

Posey, D A (1996), *Traditional Resource Rights: International Instruments for Protection and Compensation for Indigenous Peoples and Local Communities*, IUCN, Gland

Posey, D A and Dutfield, G (1996), *Beyond Intellectual Property*, IDRC, Canada

Prescott-Allen, C and Prescott-Allen, R (1986), *The First Resource*, Yale University Press, New Haven

Richman, A and Witkowski, J P (1998), 'Herb Sales Still Strong', *Whole Foods*, South Plainfield NJ, October, 19-26

Secretariat of the Convention on Biological Diversity (1996a), *The Impact of Intellectual Property Rights Systems on the Conservation and Sustainable Use of Biological Diversity and on the Equitable Sharing of Benefits from its Use: A Preliminary Study*, UNEP/CBD/COP/3/22, CBD Secretariat, Montreal

Secretariat of the Convention on Biological Diversity (1996b), *Knowledge, Innovations and Practices of Indigenous and Local Communities: Implementation of Article 8(j)* Montreal, UNEP/CBD/COP/3/19, CBD Secretariat, Montreal

Secretariat of the Convention on Biological Diversity (1997a), *A Compilation of Case Studies Submitted by Governments and Indigenous and Local Communities Organisations*, UNEP/CBD/TKBD/1/Inf.1, CBD Secretariat, Montreal

Secretariat of the Convention on Biological Diversity (1997b), *Survey of Activities Undertaken by Relevant International Organizations and Their Possible Contribution to Article 8(j) and Related Articles*, UNEP/CBD/TKBD/1/Inf.2, CBD Secretariat, Montreal

Secretariat of the Convention on Biological Diversity (1998), *Implementation of Article 8(j) and Related Provisions*, note by Executive Secretary, UNEP/CBD/COP/4/10, CBD Secretariat, Montreal

Spencer, N, (1998), 'Mellow Yellow', *The Independent*, Saturday 11 April

United Nations Commission on Sustainable Development (1997), *Overall Progress Achieved Since the United Nations Conference on Environment and Development*, Report of the Secretary-General, Distr GENERAL E/CN.17/1997/2/Addn 15, 21 January, 7–25 April

United Nations Draft Declaration on the Rights of Indigenous Peoples, 1993

Wilson, E O (1988), *Biodiversity*, National Academy Press, Washington DC

Wilson, E O (1992), *The Diversity of Life*, Penguin Books, London

World Business Council for Sustainable Development and IUCN (1997), *Business and Biodiversity*, WBSCD and IUCN, Cambridge

WRI (1999), *Biodiversity*, www.wri.com

WTO – Committee on Trade and Environment (1996a), *Excerpt from the Report of the Meeting Held on 21–2 June 1995: Record of the Discussion on Item 8 of the Committee on Trade and Environment's Work Programme*, WTO, Geneva

WTO – Committee on Trade and Environment (1996b), *Report of the WTO Committee on Trade and Environment*, WTO, Geneva

Zakri, A H (1995), *Prospects in Biodiversity Prospecting*, Genetics Society of Malaysia, Kuala Lumpur

Chapter 3: Natural Products and the Pharmaceutical Industry

ABPI (1997), 'Significant Agreement Hailed by Pharmaceutical Industry', ABPI, London, 24 June

ABPI (1998), 'Pharmaceutical Facts and Statistics', ABPI, London

Akerele, O, Haywood, V and Synge, H (eds) (1991), *Conservation of Medicinal Plants*, CUP, Cambridge

Albers-Schonberg, G (1996), 'Is It Too Late for Natural Products?' in *Science in Africa: Utilising Africa's Genetic Affluence through Natural Products Research and Development*, Sub-Saharan Program, American Association for the Advancement of Science, Washington DC

Andrews, P, Borris, R, Dagne, E, Gupta, M P, Mitscher, L A, Monge, A, de Souza, N J and Topliss, J G (1996), 'Preservation and Utilisation of Natural Biodiversity in Context of Search for Economically Valuable Medicinal Biota,' *Pure and Applied Chemistry*, vol 68, no 12, 2325–32

Anon (1997), 'Natural Products', *Chemical Marketing Reporter*, Focus

Report, Schnell, New York, 15 September, 12ff

Argus Research Corporation (1998), *Top 100 Pharmaceutical Companies Sorted by 1997 Revenue*, Argus, New York

Aylward, B A, Echeverria, J, Fendt, L and Barbier, E B (1993), *The Economic Value of Species Information and its Role in Biodiversity Conservation: Case Studies of Costa Rica's National Biodiversity Institute and Pharmaceutical Prospecting – A Report to the Swedish International Development Authority*, London Environmental Economis Centre and the Tropical Science Center, in collaboration with the National Biodiversity Institute, Costa Rica

Baba, S, Akerele, O and Kawaguchi,Y (eds) (1991), 'Natural Resources for Human Health: Plants of Medicinal and Nutritional Value', Proceedings of the first WHO Symposium on Plants and Health for All: Scientific Advancement, Kobe, 26–8 August

Baker, J, et al (1995), 'Natural Product Drug Discovery and Development: New Perspectives on International Collarboration', *Journal of Natural Products*, vol 58, no 9, September, 1325-57

Balick, M B, Elisabetsky, E and Laird, S A (eds) (1996), *Medicinal Resources of the Tropical Forest: Biodiversity and its Importance for Human Health*, Columbia University Press, New York

Baltazar, E M, et al (1998), *El Manejo Sostenible de Sangre de Drago o Sangre de Grado: Material Educativo*, Shaman Pharmaceuticals and the Healing Forest Conservancy, San Francisco CA

Baumann, M, Bell, J, Koechlin, F and Pimbert, M (1996), *The Life Industry: Biodiversity, People and Profits*, Intermediate Technology, London

Borris, R P (1996), 'Natural Products Research: Perspectives from a Major Pharmaceutical Company', *Journal of Ethnopharmacology*, vol 51, 29–38

Bossong-Martines, E (ed) (1988), *Industry Surveys: Health Care and Pharmaceuticals*, Standard and Poors Industry Surveys, New York

Bowles, I A, Clark, D, Downes, D and Guerin-McManus, M (1996), 'Encouraging Private Sector Support for Biodiversity Conservation: The Use of Economic Incentives and Legal Tools', *CI Policy Paper*, vol 1, Conservation International, Washington DC

Bowman, W C and Harvey, A L (1995), 'The Discovery of Drug', *Proceedings of the Royal College of Physicians, Edinburgh*, vol 25, 5–24

Bravo, E and Gallardo, L (1998), 'Biopiracy of Epipedobates Tricolor: Biodiversity Campaign, Accion Ecologia', Bio-IPR Listserver, Quito, November

Brush, S B and Stabinsky, D (eds) (1996), *Valuing Local Knowledge: Indigenous People and Intellectual Property Rights*, Island Press, Washington DC

Business Communications Inc (1998), 'Plant Derived Drugs to Reach $30 Billion Worldwide in 2002', Business Communications Inc, Norwalk CT

Business Wire: HealthWire (1998), 'Diversa Signs Bioprospecting Agreement with the Institute of Biotechnology at the National Autonomous University of Mexico', 6 November

Capson, T (1999), 'Biological Prospecting in Tropical Rainforests: Strategies for Finding Novel Pharmaceutical Agents, Maximizing the Benefits for the Host Country and Promoting Biodiversity Conservation', Smithsonian Tropical Research Institute, Panama (forthcoming)

Carlson, T J, Iwu, M M, King, S R, Obialor, C and Ozioko, A (1997), 'Medicinal Plant Research in Nigeria: An Approach for Compliance with the Convention on Biological Diversity', *Diversity*, vol 13, no 1, 1997

Carlson, T J et al (1999), *Case Study on Medicinal Plant Research in Guinea: Prior Informed Consent, Benefit Sharing and Compliance with the Convention on Biological Diversity*, Shaman Pharmaceuticals, San Francisco CA (forthcoming)

Carrizosa, S (1996), 'Prospecting for Biodiversity: The Search for Legal and Institutional Frameworks', PhD Dissertation, School of Renewable Natural Resources, University of Arizona

Carte, B (1997), 'Natural Products and the Changing Paradigm in Drug Discovery: An Industry Perspective', paper given at UNIDO seminar, Davao, Philippines, July 1997 (unpublished)

Chemical Marketing Report (1997), 'Natural Products', 15 September

Chemical Week (1995), 'China Sets Goals for Chemicals: Ambitious Ideas', 29 November

Chetley, A (1990), *A Healthy Business? World Health and the Pharmaceutical Industry*, Zed Books, London

Chinnock, J A, Balick, M J and Sanchez, S C (1997), 'Traditional Healers and Modern Science – Bridging the Gap: Belize, a Case Study', in A Paul, D Wigston and C Peters (eds), *Building Bridges with Traditional Knowledge*, Botanical Garden Press, New York

Christoffersen, R E and Marr, J J (1995), 'The Management of Drug Discovery', in M E Wolff (ed) *Burger's Medicinal Chemistry and Drug Discovery*, 5th edn, vol 1, *Principles and Practice*, Wiley, New York

Ciba Foundation (1994), *Ethnobotany and the Search for New Drugs*, Ciba Foundation Symposium 185, Wiley, New York

Coghlan, A (1997), 'Shark Chokes Human Cancers', *New Scientist Archive*, 26 April 1997, www.last-word.ns/970426/shark.html

Congressional Research Service (CRS) (1993), 'Biotechnology, Indigenous Peoples and Intellectual Property Rights', *Report for Congress*, Library of Congress, Washington DC,16 April

Cowell, A (1998), 'Zeneca Buying Astra as Europe Consolidates', *The New York Times*, 10 December

Cox, P A (1994), 'The Ethnobotanical Approach to Drug Discovery: Strengths and Limitations', in *Ethnobotany and the Search for New Drug*, Ciba Foundation Symposium 185, Wiley, New York

Cragg, G M, Boyd, M R, Khanna, R, Newman, D J and Sausville, E A (1999), 'Recent Advances in Phytochemistry', vol 33 (forthcoming)

Cragg, G M, Newman, D J and Snader, K M (1997), 'Natural Products in Drug Discovery and Development', *Journal of Natural Products*, vol 60, no 1, 52–60

Cragg, G M, Newman, D J and Weiss, R B (1997), 'Coral Reefs, Forests and Thermal Vents: The Worldwide Exploration of Nature for Novel Antitumor Agents', *Seminars in Oncology*, vol 24, no 2, 156–63

Cragg, G M, Schepartz, S A, Suffness, M and Grever, M R (1993), 'The Taxol Supply Crisis – New NCI Policies for Handling the Large-Scale Production of Novel Natural Product Anticancer and Anti-HIV Agents', *Journal of Natural Products*, vol 56, no 10, 1657–68

Crooke, S (1998), 'Optimising the Impact of Genomics on Drug Discovery and Development', *Nature Biotechnology*, Supplement, vol 16

Crucible Group (1994), *People, Plants and Patents: The Impact of Intellectual Property on Biodiversity, Conservation, Trade and Rural Society*, IDRC, Ottawa

Datamonitor (1998), 'Estimates of Prescription-Only Medicine Sales', www.datamonitor.com

Davis, S (1993), 'Pathways to Economic Development Through Intellectual Property Rights', World Bank Environment Department, Washington DC

Demain, A L (1998), 'Microbial Natural Products: Alive and Well in 1998', *Nature Biotechnology*, January, 3–4

DiMasi, J A, Hansen, R W, Gabowski, H G, et al (1991), 'The Cost of Innovation in the Pharmaceutical Industry', *Journal of Health Economics*, vol 10, 107–421

Dow Jones Newswires (1999), 'EU Approves Zeneca-Astra Merger, if Firm Gives Up Rights to Drugs', Dow Jones, New York, 1 March

Drews, J (1998), 'Biotechnology's Metamorphosis into a Drug Discovery Industry', *Nature Biotechnology*, Supplement, vol 16

Drug Topics (1998), 'Top 200 Brand-Name Rx Drugs', *Drug Topics Red Book*, Medical Economics Company, New Jersey

Duke, J A (1993), 'Medicinal Plants and the Pharmaceutical Industry', in J Janick and J E Simon (eds), *New Crops: Exploration, Research and Commercialization*, Proceedings of the Second National Symposium, Indianapolis, 6–9 October 1991, Wiley: New York

Dumoulin, J (1998), 'Pharmaceuticals: The Role of Biotechnology and Patents', *Biotechnology and Development Monitor*, June, no 35

Duncan, D E (1998), 'A Shaman's Cure: Can a West African Plant Slow the Scourge of Diabetes?', *LIFE*, Fall

Dunn-Coleman, N and Prade, R (1998), 'Toward a Global Filamentous Fungus Genome Sequencing Effort', *Nature Biotechnology*, January, 5

Economist, The (1997), 'Genetic Engineering: Building to Order', 1 March

Economist, The (1998a), 'The Pharmaceutical Industry Survey', 21–7 February

Economist, The (1998b), 'Beyond the Behemoths', 21-7 February

Economist, The (1998c), 'Horn of Plenty', 21–7 February

Economist, The (1998d), 'From Blunderbuss to Magic Bullet', 21–7 February

Economist, The (1998e), 'Trials and Tribulations', 21–7 February

Economist, The (1998f), 'Piggy in the Middle', 21–7 February

Economist, The (1998g), 'Drug Industry: European Unions', 12 December

Economist, The (1999), 'Ethnobotany: Shaman Loses its Magic', 20 February

Edwards, R (1996), 'Biotech Firm "Embarrassed" by Leaked Plant Deal', *New Scientist*, 29 June

Elliot, S and Brimacombe, J (1986), 'Pharmacy Needs Tropical Forests', *Manufacturing Chemist*, October, 31–4

EMEA (European Agency for the Evaluation of Medicinal Products) (1998), 'Pre- Submission Guidance for Users of the Centralised Procedure', London, November

Ernst and Young (1998)

EuropaBio (1997), *Benchmarking the Competiveness of Biotechnology in Europe*, EuropaBio Independent Report, Brussels

Farnsworth, N R (1994), 'Ethnopharmacology and Drug Development', in *Ethnobotany and the Search for New Drugs*, Ciba Foundation Symposium 185, Wiley, New York

Farnsworth, N R (1988), 'Screening Plants for New Medicines', in E O Wilson (ed) *Biodiversity*, National Academy Press, Washington DC

Farnsworth, N R, Akerele, O, Bingel AS, Soejarto D D, Guo Z (1985), 'Medicinal Plants in Therapy', *Bulletin of the World Health Organisation*, vol 63, 965–81

Farnsworth, N and Morris, R W (1976), *American Journal of Pharmacology*, vol 148, 46–52

Faulkner, D J (1992), 'Biomedical Uses for Natural, Marine Chemicals', *Oceanus*, Spring

Finch, J (1998), 'Last of the Billion Dollar Drugs', *Guardian*, 31 January

Gbewonyo, K (1995), 'Biodiversity Prospecting', *World Journal of Microbiology and Biotechnology*, vol 11, 251–2

Gbewonyo, K (1997), 'The Case for Commercial Biotechnology in Sub-Saharan Africa', *Nature Biotechnology*, vol 15, April

Geuze, M (1998), 'Patent Rights in the Pharmaceutical Area and their Enforcement: Experience in the WTO Framework with the Implementation of the TRIPS Agreement', *Journal of World Intellectual Property*, vol 1, no 4

Glaxo Wellcome (1997), *Discovering New Medicines from Nature: Policy for the Acquisition of Natural Product Source Materials*, Hertfordshire

Glowka, L, Plan, T and Stoll, P T (1997), 'Best Practices for Access to Genetic Resources: Background', paper for the correspondent workshop on 'Best Practices for Access to Genetic Resources', co-sponsored by the DGXI European Commission and the German Federal Ministry of the Environment, Nature Conservation and Nuclear Safety, Cordoba 14–17 January

Gollin, M A (1994), 'Patenting Recipes from Nature's Kitchen: How Can a Naturally Occurring Chemical like Taxol be Patented?', *Biotechnology*, April, vol 12

Gower, J (1998), 'The Evolving Role of Collaboration in Biotech', *Nature Biotechnology*, Supplement, vol 16

Grabowski, H and Vernon, J (1994), 'Returns to R&D on New Drug Introductions in the 1980s', *Journal of Health Economics*, vol 13

Green, D and Cookson, C (1998), 'SmithKline-Glaxo 100bn Merger Plan: World's Biggest Corporate Deal Outflanks US Rival and is Set to Create Largest Pharmaceuticals Group', *Financial Times*, 31 January–1 February

Greaves, T (ed) (1994), *Intellectual Property Rights for Indigenous Peoples: A Sourcebook Society for Applied Anthropology*, Oklahoma City OK

Grifo, F (1994), 'Chemical Prospecting: An Overview of the International Cooperative Biodiversity Groups Program', in J Feinsilver (ed) *Emerging Connections: Biodiversity, Biotechnology and Sustainable Development in Health and Agriculture*, Proceedings of Pan American Health Organization Conference, San Jose

Grifo, F and Rosenthal, J (eds) (1996), *Biodiversity and Human Health*, Island Press, Washington DC

Grifo, F, Newman, D, Fairfield, A S, Bhattacharya, B and Grupenhoff, J T (1996), 'The Origins of Prescription Drugs', in F Grifo and J Rosenthal (eds), *Biodiversity and Human Health*, Island Press, Washington DC

Gruber, A B and John, W (1996), 'Back to the Future: Traditional Medicinals Revisited', *Laboratory Medicine*, vol 27, no 2, February, 100–108

Guerin-McManus, M, Famolare, L, Bowles, I, Stanley, A J, Mittermeier, R A and Rosenfeld, A B (1998), 'Bioprospecting in Practice: A Case Study of the Suriname ICBG Project and Benefit-Sharing under the Convention on Biological Diversity', in *Case Studies on Benefit-Sharing Arrangements*, Conference of the Parties to the Convention on Biological Diversity, 4th meeting, Bratislava, May

Gullo, V P (ed) (1994), *The Discovery of Natural Products with Therapeutic*

Potential, Butterworth-Heinemann, Boston

Haseltine, W (1998), 'The Power of Genomics to Transform the Biotechnology Industry', *Nature Biotechnology*, Supplement, vol 16

Hawksworth, D L (1992), 'Microorganisms', in B Groombridge (ed) *Global Biodiversity*, World Conservation Monitoring Centre, London

Haycock, P (1998), 'Europe Discovers Bioentrepreneurship', *Nature Biotechnology*, Supplement, vol 16

Holmstedt, B, Wassen, SH and Schultes, RE (1979), 'Jaborandi: An Interdisciplinary Appraisal', *Journal of Ethnopharmacology*, vol 1, 3–21

Horrobin, D F and Lapinskas, P L (1993), 'Opportunities and Markets in the Pharmaceutical and Health Food Industry', in K R M Anthony, J Meadley and G Robbelen (eds), *New Crops for Temperate Regions*, Chapman and Hall, London

ICH (International Conference on Harmonisation) (1998), 'ICH Structure', Geneva

Iwu, M M (1994), 'African Medicinal Plants in the Search for New Drugs Based on Ethnobotanical Leads', in *Ethnobotany and the Search for New Drugs*, Ciba Foundation Symposium 185, Wiley, New York

Iwu, M M and Laird, S A (1998), 'The International Cooperative Biodiversity Group: Drug Development and Biodiversity Conservation in Africa: A Case Study of a Benefit-Sharing Plan', in *Case Studies on Benefit Sharing Arrangements*, Conference of the Parties to the Convention on Biological Diversity, 4th meeting, Bratislava, May

Joffe, S and Thomas, R (1989), 'Phytochemicals: A Renewable Global Resource', *AgBiotech News and Information*, vol 1, no 5, 697–700

JPMA (Japanese Pharmaceutical Manufacturers Association) (1998), 'Approval and Licensing of Drugs', Tokyo

Kalorama Information Market Intelligence Reports (1996), *The Chinese Pharmaceutical Market*, Find/SVP Publishing, New York

Kalorama Information Market Intelligence Reports (1997), *The Pharmaceutical Products Market* (Pacific Rim), Find/SVP Publishing, New York

Kapoor, P, Trotz, U O'D and Simon, O (eds) (1992), 'The Use of Medicinal Plants in the Pharmaceutical Industry', in *Summary of Discussions on a Standardized Methodology and Guidelines for Use*, Guyana, June 1988

ten Kate, K (1995), *Biopiracy or Green Petroleum? Expectations and Best Practice in Bioprospecting*, ODA, London

ten Kate, K and Wells, A (1998), 'The Access and Benefit-Sharing Policies of the US National Cancer Institute: A Comparative Account of the Discovery and Development of the Drugs Calanolide and Topotecan', in *Case Studies on Benefit Sharing Arrangements*, Conference of the Parties to the Convention on Biological Diversity, 4th meeting, Bratislava, May

ten Kate, K, Touche, L and Collis, A (1998), 'Yellowstone National Park and the Diversa Corporation Inc', in *Case Studies on Benefit Sharing Arrangements*, Conference of the Parties to the Convention on Biological Diversity, 4th meeting, Bratislava, May

King, S R (1994), 'Establishing Reciprocity: Biodiversity, Conservation and New Models for Cooperation Between Forest-Dwelling Peoples and the Pharmaceutical Industry', in T Greaves (ed) *Intellectual Property Rights for Indigenous Peoples: A Sourcebook Society for Applied Anthropology*, Oklahoma City OK

King, S R (1996), 'Conservation and Tropical Medicinal Plant Research', in M B Balick, E Elisabetsky and S A Laird (eds), *Medicinal Resources of the Tropical Forest: Biodiversity and its Importance for Human Health*, Columbia University Press, New York

King, S R and Carlson, T J (1995), 'Biocultural Diversity, Biomedicine and Ethnobotany: The Experience of Shaman Pharmaceuticals', *Interciencia*, May–June, vol 20, no 3, 134–9

King, S R et al (1997), '*Croton lechleri* and the Sustainable Harvest and Management of Plants' in *Pharmaceuticals, Phytomedicines and Cosmetics Industries*, paper given at the International Symposium on Herbal Medicine, Workshop III: Environmental Protection Concerns, Honolulu

Kinghorn, A D and Balandrin, M F (eds), *Human Medicinal Agents from Plants*, ACS Symposium, series 534, American Chemical Society, Washington DC

Kleiner, K (1995), 'Billion-Dollar Drugs are Disappearing in the Forest', *New Scientist*, 8 July

Kozlowski, R (1998), 'Chemical Wizardry', *Oxford Today*, vol 10, no 3

Krattiger, A F, McNeely, J A, Lesser, W H, Miller, K R, St Hill, Y and Senanayake, R (1994), *Widening Perspective on Biodiversity*, IUCN and the International Academy of the Environment, Gland, Switzerland

Laird, S A (1993), 'Contracts for Biodiversity Prospecting', in W V Reid, S A Laird, C A Meyer, R Gamez, A Sittenfeld, D H Janzen, M A Gollin and C Juma (eds), *Biodiversity Prospecting: Using Genetic Resources for Sustainable Development*, WRI, Washington DC

Laird, S A (1995a), *Access Controls for Genetic Resources*, WWF–International Discussion Paper, WWF, Gland

Laird, S A (1995b), *Sustainable Sourcing of Raw Materials: Weighing the Benefits*, WWF People and Plants Program Background Paper, WWF, Gland, Switzerland

Laird, S A (ed) (2000), *Equitable Partnerships in Practice: The Tools of the Trade in Biodiversity and Traditional Knowledge*, a People and Plants Programme Conservation Manual, Earthscan, London (forthcoming)

Laird, S A (1999), 'One in Ten Thousand? The Cameroon Case of *Ancistrocladus korupensis*', in C Zerner (ed) *People, Plants and Justice: Case Studies of Resource Extraction in Tropical Countries*, Columbia University Press, New York

Laird, S A and Lisinge, E E (1998), 'Benefit-Sharing Case Studies from Cameroon: *Ancistrocladus korupensis* and *Prunus africana*', in *Case Studies on Benefit Sharing Arrangements*, Conference of the Parties to the Convention on Biological Diversity, 4th meeting, Bratislava, May

Laird, S A and Wynberg, R (1997), 'Biodiversity Prospecting in South Africa: Towards the Development of Equitable Partnerships', in J Mugabe, C Barber, G Henne, L Glowka and A La Vina (eds) (1997), *Managing Access to Genetic Resources: Towards Strategies for Benefit-Sharing* ACTS, Nairobi and WRI, Washington DC

Larvol, B and Wilkerson, J (1998), 'In Silico Drug Discovery: Tools for Bridging the NCE Gap', *Nature Biotechnology*, Supplement, vol 16

Lesser, W (1994), *Institutional Mechanisms Supporting Trade in Genetic Materials: Issues under the Biodiversity Convention and GATT/TRIPS*, Environment and Trade 4, UNEP, Geneva

Macilwain, C (1993), 'MIT's Amazon Outpost', *Nature*, vol 365, 9 September

Macilwain, C (1998), 'When Rhetoric Hits Reality in Debate on Bioprospecting', *Nature*, vol 392, 9 April, 535–41

Magainin Pharmaceutical (1998), www.squalamine.com/magaininnews 2.html

McFarling, U (1998), 'The Code War: Biotech Firms Engage in High-Stakes Fight Over Rights to the Human Blueprint', *San Jose Mercury News*, 17 November

McGinn, D (1998), 'Viagra's Hothouse', *Newsweek*, 21 December, 1998

MedAd News (1998a), 'Brands Rule Supreme', Eagle Publishing Partners, New Jersey, December

MedAd News (1998b), 'Marketers and the Value of their Blockbuster Medicines', vol 17, no 5, Eagle Publishing Partners, New Jersey, May

MedAd News (1998c), 'The top 100 Prescription Drugs by Worldwide Sales', vol 17, no 5, Eagle Publishing Partners, New Jersey, May

MedAd News (1998d), 'Top 50 Companies', Eagle Publishing Partners, New Jersey, September

Medical Sciences Bulletin (1994), 'Oral Pilocarpine Hydrochloride for Radiation-Induced Dry Mouth', Pharmaceutical Information Associates

Mestel, R (1999), 'Drugs from the Sea', *Discover*, vol 20, no 3, March

Meza Baltazar, E et al (1999), *Conservando la Biodiversidad Cultural 'Sangre de Grado' y el Reto de la Produccion Sustentable en el Peru*, Propaceb, Lima

Meza Baltazar, E M, et al (1998), *El Manejo Sostenible de Sangre de Drago o Sangre de Grado: Material Educativo*, Shaman Pharmaceuticals and the Healing Forest Conservancy, San Francisco CA

Monsanto (1997), *Report on Sustainable Development, Including Environmental, Safety and Health Performance*, St Louis, Missouri

Moore, S (1998), 'Phytopharm Grants Pfizer Rights to Obesity Drug', *Wall Street Journal*, 25 August

Moore, S (1999), 'Hoechst Faces Wobbly Support for the Rhône-Poulenc Accord', *Wall Street Journal*, 25 February

Moo-Young, M (1983) (ed in chief), *Comprehensive Biotechnology: The Principles, Applications and Regulations of Biotechnology in Industry, Agriculture and Medicine*, vol 4, 'The Practice of Biotechnology: Speciality Products and Service Activities', Robinson, C and Howell, J (eds), Pergamon Press, Oxford, New York

Moran, K (1995), 'Returning Benefits from Ethnobotanical Drug Discovery to Native Communities', in F Grifo and J Rosenthal (eds), *Biodiversity and Human Health*, Island Press, Washington DC

Moran, K (1998), 'Mechanisms for Benefit-Sharing: Nigerian Case Study for the Convention on Biological Diversity', in *Case Studies on Benefit Sharing Arrangements*, Conference of the Parties to the Convention on Biological Diversity, 4th meeting, Bratislava, May

Mugabe, J, Barber, C, Henne, G, Glowka, L and La Vina, A (eds), *Managing Access to Genetic Resources: Towards Strategies for Benefit-Sharing*, ACTS, Nairobi, and WRI, Washington DC

Novo Nordisk (1996), *Environmental Report: Summary of Novo Nordisk's Environmental Performance*, Denmark

Novo Nordisk (1997a), *Acquisition of Natural Resources for the Development of New Pharmaceuticals: A Policy Statement from Novo Nordisk Health-Care Discovery*, Denmark

Novo Nordisk (1997b), *Extracts for Screening: Biodiversity and the Search for New Microbes*, Denmark

OECD (1997), *Issues in the Sharing of Benefits Arising out of the Utilisation of Genetic Resources*, co-authored by Kerry ten Kate and Jan Keppler, ENV/EPOC/GEEI/BIO(97)4, Group on Economic and Environment Policy Integration; Expert Group on Economic Aspects of Biodiversity, OECD, Paris

Office of Technology Assessment, US Congress (1993), *Pharmaceutical R&D: Costs, Risks and Rewards*, OTA-H-522 Government Printing Office, Washington DC

O'Neill, M J and Lewis, J A (1993), in A D Kinghorn and M F Balandrin (eds), *Human Medicinal Agents from Plants*, ACS Symposium Series 534, American Chemical Society, Washington DC

Paul, A, Wigston, D and Peters, C (eds) (1997), *Building Bridges with Traditional Knowledge*, Botanical Garden Press, New York

PhRMA (1998), *PhRMA Facts*, Washington DC

Pinheiro, C (1997), 'Jaborandi (Pilocarpus Sp. Rutaceae): A Wild Species and its Rapid Transformation into a Crop', *Economic Botany*, vol 51, no 1, 49–58

Pollack, A (1999), 'Shaman Says it is Exiting Drug Business', *The New York Times*, 2 February

Popovich, B (1997), 'Prescriptions for Health', *Chemical Market Reporter*, CMR Focus Report, Schnell, New York, 15 September, 3–6ff

Poste, G (1998), 'Molecular Medicine and Information-Based Targeted Healthcare', *Nature Biotechnology*, Supplement, vol 16

Purcell, D (1998), 'Navigating Biotechnology's New Fiscal Opportunities', *Nature Biotechnology*, Supplement, vol 16

RAFI (1994), *Conserving Indigenous Knowledge: Integrating Two Systems of Innovation*, commissioned by UNDP, Ottawa, Canada

RAFI (1997), *Communique*, RAFI, November/December, Ottawa, Canada

RAFI and IPBN (Indigenous Peoples' Biodiversity Network) (1994), *COPS ... and Robbers: Transfer-Sourcing Indigenous Knowledge: Pirating Medicinal Plants*, Occasional Paper Series, vol 1, no 4, November

Reed, N (1998), 'Notes of Harmony', *GlaxoWellcome World*, September

Reid, W V, Laird, S A, Meyer, C A, Gamez, R, Sittenfeld, A, Janzen, D H, Gollin, M A and Juma, C (eds), *Biodiversity Prospecting: Using Genetic Resources for Sustainable Development*, WRI, Washington DC

Richter, R and Carlson, T (1998), 'Letter to the Editors: Reporting Biological Assay Results on Tropical Medicinal Plants to Host Country Collaborators', *Journal of Ethnopharmacology*, vol 62, 85–8

Richter, R, et al (1999), 'Mutualism: An Ethnomedical Research Collaboration between Shaman Pharmaceutical and the Institute of Traditional Medicine in Tanzania', *Journal of Ethnobiology*, (forthcoming)

Rosenthal, J (1998), 'The International Cooperative Biodiversity Groups (ICBG) Program', in *Case Studies on Benefit Sharing Arrangements*, Conference of the Parties to the Convention on Biological Diversity, 4th meeting, Bratislava, May

Rouhi, M (1997), 'Seeking Drugs in Natural Products', *Chemical and Engineering News*, American Chemical Society, 7 April, 14–29

Royal Society of Chemistry (1996), 'Phytochemical Diversity: A Source of New Industrial Products', Abstracts for the Royal Society of Chemistry Industrial Affairs Division, Biotechnology Group, University of Sussex, meeting 14–17 April

Scrip's (1997), *Scrip's 1997 Yearbook*, January, PJB , Richmond

Secretariat of the Convention on Biological Diversity: Case Studies on Benefit-Sharing Arrangements, Conference of the Parties to the Convention on Biological Diversity,

Bratislava, May 1998, www.biodiv.org

Shaman Pharmaceuticals (1999), 'Company Press Release: Shaman Pharmaceuticals Restructures to Focus on Value of Shaman Botanicals', San Francisco CA

Shearson Lehman Brothers (1991), *PharmaPipelines: Pharmaceutical Winners and Losers in the 1990s*, Lehman Brothers Securities, London

Simpson, R D, Sedjo, R A and Reid, J W (1994), *Valuing Biodiversity for Use in Pharmaceutical Research*, Resources for the Future, Washington DC

Sittenfeld, A and Gamez, R (1993), 'Bioprospecting by INBio', in Reid, W V, Laird, S A, Meyer, C A, Gamez, R, Sittenfeld, A, Janzen, D H, Gollin, M A and Juma, C (eds), *Biodiversity Prospecting: Using Genetic Resources for Sustainable Development*, WRI, Washington DC

Smith, J E (1996), *Biotechnology*, 3rd edn, CUP, Cambridge

Smith, R, Geier, M, Reno, J, Sarasohn-Kahn, J (eds) (1997), *Medical and Healthcare Marketplace Guide*, 12th edn, IDD Enterprises, Philadelphia PA

Soejarto, D D (1994), 'Biodiversity Prospecting and Benefit-Sharing', Perspective from the Field Proceedings of the Symposium on Intellectual Property Rights, Naturally Derived Bioactive Compounds and Resource Conservation, San Jose, 20–22 October

Steinmetz, M (1998), 'Venturing into Drug Discovery', *Nature Biotechnology*, Supplement, vol 16

Stockholm Environment Institute and International Academy of the Environment (1994), *Co-ordinated Arrangements for the Conservation and Sustainable Use of Genetic Resources, Material and Technology Transfer and Benefit Sharing*, SEI, Stockholm and IAE, Geneva

Straus, J (1995), 'Patenting Human Genes in Europe', *International Review of Industrial Property and Copyright Law*, vol 26, no 6

Svarstad, H (1995), 'Biodiversity Prospecting: Biopiracy or Equitable Sharing of Benefits?', Working Paper 1995/7, Centre for Development and the Environment (SUM), University of Oslo

Swanson, T M and Luxmore, R A (1996), *Industrial Reliance Upon Biodiversity: A Darwin Initiative Project*, World Conservation Monitoring Centre, Cambridge, UK

Tempesta, M S and King, S (1994), *Tropical Plants as a Source of New Pharmaceuticals*, Pharmaceutical Manufacturing International, Sterling, London, 47–50

Thayer, A M (1998), 'Pharmaceuticals: Redesigning R&D' and 'Double-Digit Growth for Drug Producers', *Chemical and Engineering News*, 23 February

Thomas, R T (1996), 'Qualitative and Quantitative Assessment of the Role of Natural Products in the Development of New Leads by European Pharmaceutical and Agrochemical Industries', a study on biodiversity prospecting undertaken by Biotics Ltd for the Institute for Prospective Technological Studies, Guildford, Surrey

Tufts Center for the Study of Drug Development (1998), 'Total Drug Development Time from Synthesis to Approval' as in *PhRMA Facts*, 1998, Washington DC

Ubillas, R, et al (1994), 'SP-303: An Antiviral Oligomeric Proanthocyanidin from the Latex of *Croton lechleri* (Sangre de Drago)', *Phytomedicine*, vol 1, 77–106

Vogel, J H (1996), *The Successful Use of Economic Instruments to Foster Sustainable Use of Biodiversity: Six Case Studies from Latin America and the Caribbean*, White Paper produced for the Bolivia Summit on Sustainable Development, Santa Cruz, December

Waterman, P (1997), unpublished paper on the investment and capacities needed to establish natural product screening facilities, given at UNIDO seminar in Davao, Philippines, July 1997

Webster, A J and Etzkowitz, H (1991), *Academic–Industry Relations: The Second Academic Revolution: A Framework Paper for a Proposed Workshop on Academic–Industry Relations*, Science Policy Support Group, London

WHO (1998), *The World Drug Situation*, Geneva, Switzerland

Wolf, R (1994), 'Yellowstone Discovery: Should US Get Profits?', *San Jose Mercury News*, Business Monday, 25 July

Wolff, M (1995) (ed), *Burger's Medicinal Chemistry and Drug Discovery*, 5th edn, vol I: Principles and Practice, Wiley-Interscience, John Wiley, New York, Chichester

World Business Council for Sustainable Development and IUCN (1997), *Business and Biodiversity: A Guide for the Private Sector*, WBCSD and IUCN, Gland, Switzerland

Wrigely, S, Hayes, M, Thomas, R and Chrystal, E (1997), *Phytochemical Diversity: A Source of New Industrial Products*, Royal Society of Chemistry Information Services, London

Chapter 4: The Botanical Medicine Industry

ABDA (ed) (1998), *Die Apotheke Zahlen, Daten, Fakten, 1997*, Bundesvereinigung Deutscher Apothekerverbande, Germany

Adams, C (1998), 'Fiji Loses Its Wonder Drug to Western Stress-Busters', *Guardian*, 8 October

Akerele, O, Heywood, V and Synge, H (1991), *Conservation of Medicinal Plants*, Columbia University Press, New York

Antoniak, M (1997), 'America's Top Selling Herbs', *Vitamin Retailer*, East Brunswick NJ, February, 38-41

Anuradha, R V (1998), 'Sharing with the Kanis: A Case Study from Kerala, India', *Benefit Sharing Case Studies*, 4th meeting of the Conference of the Parties to the Convention on Biological Diversity, Bratislava, May

Awang, D (1998), 'The Health of the Canadian Herbal Medicinal Market: An Editorial', *Herbalgram*, no 44, Fall

BAH (Bundesfachverband der Arzneimittel-Hersteller eV) (ed) (1992-8), *Der Selbstmedikationsmerkt in der Bundesrepublik Deutschland in Zahlen, 1991–7*, Annual Reports, Germany

Balick, M, Elisabetsky, E and Laird, S A (1996), *Medicinal Resources of the Tropical Forest: Biodiversity and Its Importance to Human Health*, Columbia University Press, New York

Bingham, R (1998), 'Health Business Partners', personal communication, 23 September, in Brevoort, 'The Booming US Botanical Market: A New Overview', *Herbalgram*, no 44 Fall

Bisset, N G (1994), *Herbal Drugs and Phytopharmaceuticals*, Medpharm and CRC Press, Stuttgart and Boca Raton

Brevoort, P (1996), 'The US Botanical Market: An Overview', *Herbalgram*, no 36, 49-57

Brevoort, P (1998), 'The Booming US Botanical Market: A New Overview', *Herbalgram*, no 44 Fall

British Medical Journal (1994), vol 39, July

Brody, J E (1999), 'Americans Gamble on Herbs as Medicine: With Few Regulations No Guarantee of Quality', *The New York Times*, 9 February

Canedy, D (1998), 'Big Manufacturers Enter Herbal-Supplement Market', *New York Times*, 23 July

Cech, R (1997), 'The UpS At Risk List', *UpS Newsletter*, Fall

Chi, Judy (1998), 'Au Naturel: Capitalizing on their Names, Major OTC Firms Sally Forth into Herbals', *Drug Topics*, 17 August, 60-62

Commission on Dietary Supplement Labels, The (1997), *Commission on Dietary Supplement Labels: Draft for Public Comment: June*, Office of Disease Prevention and Health Promotion, Washington DC, 17 pp

Consumer Guide (1997), *Natural Health*, September-October, 116-17

Correa, T (1997), 'Making Room for "Alternatives": Natural Medications Go Mainstream', *Fresno Bee*, 11 December, C1

Cunningham, M C, Cunningham, A B and Schippmann, U (1997), *Trade in* Prunus africana *and the Implementation of CITES*, German Federal Agency for Nature Conservation, Bonn

Cunningham, A B and Mbenkum, F T (1993), *Sustainability of Harvesting* Prunus africana *Bark in Cameroon: A Medicinal Plant in International Trade*, People and Plants Working Papers no 2, UNESCO/WWF/RBG, Paris, Kew, London

Daily Camera Newspaper (1997), 'Business Plus', January

Dansby, R (1997), 'Sweet Science: Over-Expression of Monellin in Yeast', *Nature Biotechnology*, vol 15, May

Datamonitor (1997), *US Nutraceuticals Overview*, New York

Day, K (1997), 'Drug Stores Find Profits in Alternative Remedies', *Daily Camera Newspaper*, Boulder CO, 21 January, 16

Dennis, F R (1997), 'The Trade in Medicinal Plants in the UK', unpublished report for WWF UK

Downes, D R and Laird, S A (1999), 'Innovative Mechanisms for Sharing Benefits of Biodiversity and Related Knowledge: Case Studies on Geographic Indications, Trademarks and Databases', paper prepared for UNCTAD Biotrade Initiative, Geneva, March

Eisenberg, D M et al (1998), 'Trends in Alternative Medicine Use in the United States, 1990–1997', *The Journal of the American Medical Association (JAMA)*, 280, 11 November, 1569–75

Foster, S (1995), *Forest Pharmacy: Medicinal Plants in American Forests*, Forest History Society, Durham NC

Fuller, D (1991), *Medicine from the Wild: An Overview of the US Native Medicinal Plant Trade and Its Conservation Implications*, TRAFFIC USA, Washington DC

Gallup (1995) (1996) (1997), Study of Attitudes Toward and Usage of Herbal Supplements, USA

Goldstein, L (1998), 'A New Recipe For Well-Being: Vitamin, Herb Sales Part of Booming Billion Dollar Health Craze', *Eastern Pennsylvania Business Journal*, 9 February, vol 9, no 6, 16

Gruenwald, J (1994), 'The European Phytomedicines Market: Figures, Trends, Analyses', *Herbalgram*, no 34,60-65

Gruenwald, J (1997), 'Europe Leads in Herbal Remedies', *NBJ*, January/February, vol 12

Gruenwald, J (1998), 'Herbal Healing', *Time Magazine*, 23 November

Gruenwald, J (1999), 'The International Herbal Medicine Market', *Nutraceuticals World*, January/February

Gruenwald, J and Buettel, K (1996), 'The European Phytotherapeutics Market', *Drugs Made in Germany*, vol 39, no 1, 6-11

Gruenwald, J and Goldberg, A (1997), 'The Herbal Remedies Market in the US: Market Development, Consumers, Legislation and Organizations', *Drugs Made in Germany*, vol 40, no 3

Harris/Celestial Seasonings (1998), Press Release, 14 April

Hauser (1997), *Annual Report*, Boulder CO

Herb Research Foundation (1997), *Herb Industry Statistics Fact Sheet*, Boulder CO

Herb Research Foundation (1998), *Herb Safety and Drug Interactions*, Boulder CO

Herbalgram (1998), no 43, American Botanical Council, Texas

Herbs for Health (1998), 'Rule-Playing: FDA Proposal Under Attack', *Herbs for Health*, vol 3, no 5, November/December

Hollis, S and Brummit, R K (1992), *World Geographical Scheme for Recording Plant Distributions, Plant Taxonomic Database Standards* no 2, version 1.0, January, Carnegie Mellon University, Hunt Institute for Botanical Documentation, Pittsburgh, 104 pp

Indena SpA (1996), 'From Act to Actions When Dealing with Botanicals', *Botanicals Update*, Fall

Institut fuer Demoskopie Allensbach (1997), *Wichtigste Erkenntnisse aus der Studie Naturheilmittel*, Allensbacher Archiv, IfD-Umfrage 6039, January

International Research Institute (IRI), Mission Hills, CA, July 1998

IUCN Medicinal Plant Specialist Group (1997), *Medicinal Plant Conservation Newsletter*, vol 3, Bonn

Iwu, M M (1996), 'Production of Phytomedicines and Cosmetics from Indigenous Genetic Resources: From Lab to Market', BDCP workshop on Commercial Production of Indigenous Plants as Phytomedicines and Cosmetics (proceedings in press: Iwu, M M, Sokomba, E and Akubue, P I (eds) Nigeria)

Johnston, B A (1997), 'One-Third of Nation's Adults Use Herbal Remedies', *Herbalgram*, November, no 40, 49

Kalorama Information Market Intelligence Reports (1996), *The Homeopathic Products Market*, Packaged Facts, May, FIND/SVP Publishers, New York

Kalorama Information Market Intelligence Reports (1997a), *The US Herbal Supplements Market*, Packaged Facts, November, FIND/SVP Publishers, New York Kalorama Information Market Intelligence Reports (1997b), *The US Market for Vitamins, Supplements and Minerals*, Packaged Facts, January, FIND/SVP Publishers, New York

Kalorama Information Market Intelligence Reports (1998), *The Market for Nutraceuticals*, April, FIND/SVP Publishers, New York NY Keating, B (1996), 'Herbal Tea Poised to Enter New Era', *Teatrade*, Premier Issue, 52

Kilham, C (1996), *Kava: Medicine Hunting in Paradise*, Park Street Press, Rochester VT

Kilham, C (1998), 'Kava: A Review', proceedings of Medicines from the Earth: Phytomedicines: Their Expanding Role, 30 May–1 June, Gaia Herbal Research Institute, North Carolina

King, S R et al (in press), 'Issues in the Commercialization of Medicinal Plants', in *From Plants in the South to Medicines in the North: Perspectives on Bioprospecting*, SUM, Oslo, Norway

Laird, S A (1994), 'Sustainable Sourcing of Raw Materials: Weighing the Benefits', paper prepared for WWF People and Plants Program, Gland, Switzerland

Laird, S A (1995), *The Natural Management of Tropical Forests for Timber and Non-Timber Products*, Oxford

Forestry Institute, Occasional Papers no 49

Laird, S A and Lisinge, E E (1998), 'Benefit-Sharing Case Studies from Cameroon: *Ancistrocladus korupensis* and *Prunus africana*', 4th meeting of the Conference of the Parties to the Convention on Biological Diversity, May, Bratislava

Laird, S A and Wynberg, R (1996), 'Biodiversity Prospecting in South Africa: Towards the Development of Equitable Partnerships, 1997', in J Mugabe, C Barber, G Henne, L Glowka and A La Vina (eds), *Managing Access to Genetic Resources: Towards Strategies for Benefit-Sharing*, ACTS, Nairobi and WRI, Washington DC

Landes, P (1997), 'Market Report', *Herbalgram*, no 40, Summer, 53

Lange, D (1997), 'Trade in Plant Material for Medicinal and Other Purposes: A German Case Study', *TRAFFIC Bulletin*, vol 17, no 1, 21-32

Lange, D (1998a), *Europe's Medicinal and Aromatic Plants: Their Use, Trade and Conservation*, TRAFFIC International, Cambridge

Lange, D (1998b), *Training Unit for CITES-Listed Plant Species Used for Medicinal Purposes*, version 1.1, 15 January, Federal Agency for Nature Conservation, Germany

Lange, D (1999), 'Medicinal Plant Trade in Europe: Conservation and Supply', proceedings of the First International Symposium on the Conservation of Medicinal Plants in Trade in Europe, 22-3 June 1998, Royal Botanic Gardens, Kew, London

Lange, D and Schippmann, U (1997), *Trade Survey of Medicinal Plants in Germany*, German Federal Agency for Nature Conservation, Bonn, Germany

Lebot, V, Merlin, M and Lindstrom, L (1997), *Kava: The Pacific Elixir: The Definitive Guide to Its Ethnobotany, History and Chemistry*, Healing Arts Press, Rochester VT

Leung, A (1997a), 'Use and Acceptance of Herbs in Consumer Products: Part 1', *DCI Magazine*, New York, February, 40-47

Leung, PhD, Albert, Y (1997b), 'Use of Herbs in Consumer Products: Part 2', *DCI Magazine*, New York, May, 34-41

Lewington, A (1993), *A Review of the Importation of Medicinal Plants and Plant Extracts into Europe*, TRAFFIC International, Cambridge

Mabberley, D J (1987), *The Plant Book: A Portable Dictionary of the Higher Plants*, CUP, Cambridge

Martin, M (1998), 'How to Sell a Wonder Herb', *Down to Earth*, Centre for Science and Environment (CSE), New Delhi, vol 7, no 12, 15 November

McAlpine Thorpe and Warrier Ltd (1996), *Future World Trends in the Supply, Utilisation and Marketing of Endangered Medicinal Plants*, London, November

McCaleb, R (1996), 'The Top Ten Herbs for Healthcare', Herb Research Foundation, Boulder CO, July, 1-4

McCaleb, R and Blumenthal, M (1998), 'President's Commission on Dietary Supplement Labels Issues Final Report: Botanicals are a Key Issue', *Herbalgram*, January, no 41, 24-6, 57 and 64

McCaleb, R S (1997), 'Medicinal Plants for Healing the Planet: Biodiversity and Environmental Health Care' in F Grifo and J Rosenthal (eds), *Biodiversity and Human Health*, Island Press, Washington DC

McCann, B (1997), 'Integrating Botanicals With Drugs: A Confounding Job', *Drug Topics*, Montvale NJ, 8 December, 74

McPhee, M (1999), 'Dangerous Herbal Supplement is Billed as Miracle Cure-All', *Daily News*, New York, 1 March

McNamara, S H (1995), 'Dietary Supplements of Botanicals and Other Substances: A New Era of Regulation', *Food and Drug Law Journal*, Washington DC, vol 50, 341-8

McSweeney, D (1995), 'Using Information under DSHEA', *Vitamin Retailer*, East Brunswick NJ, May, 27-8

Mullins, R (1998), 'Praise God – and Pass the Regulations', *Milwaukee Business Journal*, 27 March, p 23

Natural Health (1997), September-October

Natural Pharmacy (1996), October–November

Neergaard, Lauran (1999), 'Crackdown on "Functional Food" Health Claims Urged', Associated Press, 25 March

Nepstad, D C and Schwartzman, S (eds) (1992), 'Non-Timber Products from Tropical Forests: Evaluation of a Conservation and Development Strategy', *Advances in Economic Botany*, vol 9, Botanical Garden Press, New York

Neurath, P (1997), 'The Boom in Self-Medication Trend Creating Good Times for State's Distributors, Retailers and Supplement Manufacturers', *Puget Sound Business Journal*, vol 18, no 2, Seattle WA, 23 May, 3

New England Journal of Medicine (1993), vol 328, 28 January, 242-52

New Hope Communications (1995), 'The Medicinal Herb Market: An Analysis of Sales, Trends and Research', *Natural Foods Merchandiser*, June

New Hope Communications (1998), 'Taking the Mystery out of Chinese Patent Medicine', www.hartman-newhope.com

NBJ (1996), 'Raw Materials I', October, vol 1, no 3

NBJ (1997), 'Retail Trends', vol 2, October/November, no 9

NBJ (1998a), 'Natural Personal Care', January, vol 3, no 1

NBJ (1998b), 'Sales and Marketing Strategy and Spending', March, vol 3, no 3

NBJ (1998c), 'Raw Materials II', June, vol 3, no 6

NBJ (1998d), 'Industry Overview', September, vol 3, no 9

Palevich, A (1991), 'Agronomy Applied to Medicinal Plant Conservation', in O Akerele, V Heywood and H Synge (1991), *Conservation of Medicinal Plants*, Columbia University Press, New York

Peters, C (1994), *Sustainable Harvest of Non-Timber Forest Resources in Tropical Moist Forest: An Ecological Primer*, Biodiversity Support Program, Washington DC

Peteru, C (1997), 'Indigenous Innovations and Practices: A Case Study of Kava in the South Pacific', presentation to the Workshop on Traditional Knowledge and Biodiversity, Madrid, 24-8 November

PHAGRO (1997), *Geschaftsbericht 1996/7*, Bundesverband des pharmazeutischen Grosshandels eV (ed) Germany

Pilarski, M (1997), 'Is "Sustainable Wildcrafting" an Oxymoron?', *UpS Newsletter*, Fall

Portyansky, P E (1998), 'Alternative Medicine', *Drug Topics*, Montvale, NJ, 6 April, 44-50

Prevention Magazine (1997), 'Survey of Use of Herbs in America', Rodale Press, PA

Quality Botanicals International (199) www.4qbi.com

Richman, A and Witkowski, J P (1996), 'A Wonderful Year for Herbs', *Whole Foods*, South Plainfield NJ, October, 52-60

Richman, A and Witkowski, J P (1997), 'Echinacea #1 Natural Food Trade', *Whole Foods*, South Plainfield, NJ, October

Richman, A and Witkowski, J P (1998), 'Herb Sales Still Strong', *Whole*

Foods, South Plainfield NJ, October, 19-26

Robbins, C S (1998), 'Medicinal Plant Conservation: A Priority at TRAFFIC', *Herbalgram*, vol 44, Fall

Rocky Mountain Herbalists Coalition (1995), 'Deep Green Herbal Activism Comes of Age', Spring, Lyons CO

Schwabe, U (1997), *Arzneiverodnungs-Report '97*, Gustav Fischer, Stuttgart

Schwabe, U and Pfaffrath, D (1995), *Arzneiverordnungs-Report '95*, Gustav Fischer, Stuttgart

Sears, C (1995), 'The Easy Way to Sell Drugs', *New Scientist*, 4 November

Sheldon, J W, Balick, M and Laird, S A (1997), 'Medicinal Plants: Can Utilisation and Conservation Coexist?', *Advances in Economic Botany*, vol 12, New York Botanical Garden Scientific Publications Department, New York

Smith, G (1998), *An Overview of the Herbal Supplement Industry, Trends and Companies*, a market report prepared for Conservation International, Washington DC, 9 September

Spence Information Services (SPINS) (1998), 185 Berry St, Suite 5405, San Francisco, CA 94107

Srivastava, J, Lambert, J and Vietmeyer, N (1996), *Medicinal Plants: An Expanding Role in Development*, World Bank Technical Paper no 320, Washington DC

Steinhoff, B (1998), 'A Potential Scientific Basis for a Rational Assessment of Herbal Medicinal Products in Europe under Specific Aspects of the Regulatory Situation', ESCOP Monographs, paper given at Royal Botanic Gardens, Kew Symposium, June

Steinhoff, B (1998), *Regulatory Situation of Herbal Medicines: A World-Wide Review*, WHO, Geneva

Sunderland, T et al (1997), *The Ethnobotany, Ecology and Natural Distribution of Yohimbe* (Pausinystalia johimbe (K Schum) Pierre ex Bielle): *An Evaluation of the Sustainability of Current Bark Harvesting Practices and Recommendations for Domestication and Management*, report prepared for the International Council for Research in Agroforestry (ICRAF), Yaoundé, Cameroon

Taylor, D (1996), 'Herbal Medicine at a Crossroads', *Environmental Health-Perspectives*, Journal of the National Institute of Environmental Health Sciences/ National Institutes of Health, Washington DC, September, 924-8

UNCTAD COMTRADE database (1996–8), New York Statistical Division, United Nations, New York

VDR eG (Vereinigung Deutscher Reformhauser eG) (1997), *Annual Report*, Germany

Welt Amsonntag (1997), 'Natural Medicines More and More Popular', no 12, 23 March, 40

Whole Foods (1997), 'US Nutrition Industry Seen at $17.2 Billion in Sales', South Plainfield NJ, February, 12

WHO, IUCN and WWF (1993), *Guidelines on the Conservation of Medicinal Plants*, IUCN, Gland, Switzerland

Wood, L (1997), 'Today's Proactive Consumer and Herbal Supplements', *Herbalgram*, Austin TX, November, no 40, pp 50-51

Wood, L (1997), 'The Herbal Tea Drinker', *Health Foods Business*, Hackensack, NJ, September

Yuan, R and Gruenwald, J (1997), 'Germany Moves to the Forefront of the European Herbal Medicine Industry', *Genetic Engineering News*, 15 April, 14

Yuquan, W (1998), *China Medipharm Insights*, 8 April

Zhan, S (1997), 'Global Trends and Developments: This Bud's for You', *World Trade*, December Freedom Magazines, Irvine CA, 12

Chapter 5: The Development of Major Crops by the Seed Industry

AgBiotech Bulletin (1998), vol 6, issue 8/9, September

Agra Europe (1998a), *AgraFile: Grain and Oilseeds*, 147, June

Agra Europe (1998b), *Fruit and Vegetable Markets*, 93, June

Agrobiodiversity and Farmers' Rights (1996), M S Swaminathan Research Foundation, Madras, Proceedings no 14, May

Ajai, O (1997) 'Access to Genetic Resources and Biotechnology Regulations in Nigeria', *Review of European Community and International Environmental Law (RECIEL)*, vol 6, no 1, Blackwell, Oxford, 42

Anderson, B (1988), 'Market Access for Transgenic Plants in Japan, SARAS Column', *AgBiotech News Bulletin*, November

Anderson, C (1998), 'Potato Blight', *Washington (AP)*, 24 May

Anderson, W (1988), 'Regulatory Column: Labelling of GMOs in Europe', *Agbiotech Bulletin*, September

Anon (1997), *UK Protocol for Distinctness, Uniformity and Stability Tests for Wheat, Barley and Oats*, September

Anon (1997), United Kingdom National List Trials Protocol 1996/97: Cereals *Protocol for Conducting Trials of Cereals Varieties to Establish Value for Cultivation and Use for National List Purposes*, January

APHIS (1998), *User's Guide for Introducing Genetically Engineered Plants and Micro-Organisms* and website Animal and Plants Health Inspection Service, www.aphis.nsda.gov

Ascencio, A (1997), 'The Transboundary Movement of Living Modified Organisms: Issues Relating to Liability and Compensation', *Review of European Community and International Environmental Law (RECIEL)*, vol 6, no 3, 293

Asgrow, (1998) *Asgrow Vegetable Seeds: We Never Forget What Our Seed Grows Up To Be*, Asgrow Vegetable Seeds, USA

ASSINSEL (1998), *ASSINSEL Position on Access to Plant Genetic Resources for Food and Agriculture and the Equitable Sharing of Benefits Arising from their Use* Statement adopted by the General Assembly, Monte Carlo, 5 June, International Association of Plant Breeders (ASSINSEL), Nyon, Switzerland

ASTA (1996), 'Strengthening the Germplasm Base of USA Hybrids by Enhancing Corn Germplasm', memo from American Seed Trade Association Corn and Sorghum Basic Research Committee to the US Senate Agriculture, Rural Development and Related Agencies Subcommittee, 17 May

Bagnara, D (1992), 'Developing Countries Contribute Germplasm to Italian Plant Breeding Programs through the CGIAR Centers', *Diversity*, vol 8, no 1, 6ff

Bell, J (1997), 'Will the US Breadbasket Last?', *Seedling*, vol 14, no 4, 8–18

Biotechnology and Development Monitor 35, June 1998 Focus: *Impacts of Modern Biotechnologies on Agriculture*, Amsterdam

Boef, W de, Amanor, K, Wellard, K and Bebbington, A (1993), *Cultivating Knowledge: Genetic Diversity, Farmer Experimentation and Crop Research*, Intermediate Technology Publications, London

Byerlee, D and Moya, P (1993), *Impacts of International Wheat Breeding*

Research in the Developing World, 1966–1990, CIMMYT, Mexico DF

Carney, D (1998), *Changing Public and Private Roles in Agricultural Service Provision*, ODI Natural Resources Group, London

Christopher, W (1994), Letter to Senator George Mitchell, 16 August 1994, urging the US Senate to ratify the CBD

CIMMYT (1996), *Understanding Global Trends in the Use of Wheat Diversity and International Flows of Wheat Genetic Resources*, World Wheat Facts and Trends Series, Mexico, DF, www.cgiar.org/cimmyt

CIMMYT, (1998a), *South Africa and CIMMYT: a Twenty-year Partnership*, www.cgiar.org/cimmyt

CIMMYT, (1998b), *'Working' the Seed: Farmers' Management of Genetic Resources in Two Indigenous Communities*, www.cgiar.org/cimmyt

CIMMYT, (1998c), *Farmer Management and Genetic Resource Conservation in Oaxaca, Mexico*, www.cgiar.org/cimmyt

Copeland, L O and McDonald, M B (1995), *Principles of Seed Science and Technology*, Ch 13, Chapman and Hall, New York

CPRO-DLO (1997), *Annual Report CPRO-DLO 1997*, Agricultural Research Department, Centre for Plant Breeding and Reproduction Research, Wageningen

Dutfield, G (1999), *Intellectual Property Rights, Trade and Biodiversity: The Case of Seeds and Plant Varieties*, IUCN, Gland, Switzerland (forthcoming)

Duvick, D N (1996), 'Seed Company Perspectives', 253–61

Economic Intelligence Unit Ltd (1998), 'Grains', *EU World Commodity Forecasts: Food, Feedstuffs and Beverages*, 2nd quarter, 16ff

Economist, The (1998a), 'In Defence of the Demon Seed', 13 June, 13–14

Economist, The (1998b), 'Food Fights: GMOs', 13 June, 113–14

Eggers, B (1997), 'International Biosafety: Novel Regulations for a Novel Technology', *RECIEL*, vol 6, no 1, 68

EIU, 1998 *World Commodity Forecasts: Food, Feedstuffs and Beverages*, 2nd quarter

Evans, J (1992), *Plantation Forestry in the Tropics*, 2nd edn, Clarendon Press, Oxford

FAO (1994a), *Seed Marketing*, FAO Agricultural Services Bulletin 114, FAO, Rome

FAO (1994b), *FAO Seed Review: 1989–90*, FAO, Rome

FAO (1996c), *World List of Seed Sources*, FAO Plant Production and Protection Division, Rome

FAO (1998), 'Basic Facts of World Cereal Situation' in *Food Outlook*, no 2, April

Farrington, J (ed) (1989) *Agricultural Biotechnology: Prospects for the Third World*, ODI, London

FIS (1998a), *Consolidation of FIS Objectives and Motions (Adopted by the General Assembly, the Council and the Sections since 1967)* International Seed Trade Federation web page

FIS/ASSINSEL (1998), *Recommendations by the Seed Industry of Developing Countries on the Revision of the International Undertaking*, Joint Statement from the General Assembly, Monte Carlo, 5 June, International Seed Trade Federation (FIS) and the International Association of Plant Breeders (ASSINSEL), Nyon, Switzerland

Frison, E A, Horry, J-P and de Waele, D (eds) (1996), *New Frontiers in Resistance Breeding for Nematode, Fusarium and Sigatoka* INIBAP, IPGRI, CIRAD, MARDI, Rome

Frison, E A, Orjeda, G and Sharrock, S L (eds) (1997), *ProMusa: a Global Programme for Musa Improvement*, proceedings of a meeting held in Gosier, Guadeloupe, 5 and 9 March, IPGRI, INIBAP, The World Bank, Rome

Future Food Security and Plant Genetic Resources (1992), SAREC Documentation: Report on a Consultation on a Global System for the Security and Sustainable Use of Plant Genetic Resources, 11–13 January 1992, Stockholm

GRAIN (1998), 'Japan: Genetech's Late Bloomer', *Seedling*, vol 15, no1, 2–11 GRAI, Barcelona

GRAIN (1998) 'Genetech Preys on the Paddy Field', *Seedling*, vol 15, no 2, 10–20 GRAI, Barcelona

Grain and Oilseeds Monthly (1988), June, no 147

Hawtin, G and Reeves, T (1997), *Intellectual Property Rights and Access to Genetic Resources in the Consultative Group on International Agricultural Research (CGIAR)*, paper presented at the workshop 'Intellectual Property Rights III – Global Genetic Resources: Access and Property Rights', 4–6 June 1997, Washington DC In 1987, an Inter-Centre Working Group on Genetic Resources (ICWG-GR), comprising the heads of the genetic resources programmes of the various Centres, was established to develop common policies

Heffer, P (1998), speech during consultation on SINGER at the FAO, Rome, 11–13 November, ASSINSEL, Nyon, Switzerland

INTERAGRES (1990), 'The Germplasm Contribution of the CGIAR Centres to the Italian Plant Breeding Programme', International Agricultural Research European Service (INTERAGRES), Rome

IPGRI, (1993), *Diversity for Development*, the strategy of the IPGRI, IPGRI, Rome

IPGRI (1997a), *Ethics and Equity in Conservation and Use of Genetic Resources for Sustainable Food Security*, proceedings of a workshop to develop guidelines for the CGIAR, 21–5 April, Foz do Iguaçu, Brazil, IPGRI, Rome

IPGRI (1997b), *Geneflow: a Publication about the Earth's Plant Genetic Resources*, IPGRI, Rome

Jackson, J H, Davey, W J and Sykes, A O (1995), *Legal Problems of International Economic Relations: Cases, Materials and Text* (3rd edn), West Pub Co, St Paul MN

Juma, C (1989), *The Gene Hunters: Biotechnology and the Scramble for Seeds*, ACTS Research Series no 1, Zed Books, London

ten Kate, K and Collis, A (1998), *Benefit-Sharing Case Study: The Genetic Resources Recognition Fund of the University of California, Davis*, Submission to the Executive Secretary of the Convention on Biological Diversity by the Royal Botanic Gardens, Kew, London

ten Kate, K and Diaz, L (1997), 'The Undertaking Revisited: a Commentary on the Revision of the International Undertaking on Plant Genetic Resources for Food and Agriculture', in *RECIEL*, vol 6, issue 3, 1997

ten Kate, K and Wells, A (1998), 'The Access and Benefit-Sharing Policies of the US National Cancer Institute: A Comparative Account of the Discovery and Development of the Drugs Calanolide and Topotecan', in *Case Studies on Benefit Sharing Arrangements*, Conference of the Parties to the Convention on Biological Diversity, 4th meeting, Bratislava, May

Kelly, A F and George, R Q T (eds), (1998), *Encyclopaedia of Seed Production of World Crops*, Wiley, Chichester

Khush, G S and Virmani, S S (1985), 'Breeding Rice for Disease

Resistance', in G F Russell (ed) *Progress in Plant Breeding*, vol 1, 239ff

Khush, Gurdev S, (1995), 'Modern Varieties – Their Real Contribution to Food Supply and Equity', *GeoJournal* vol 35, no 3, 275–84, Kluwer Academic Press

Levidow, L, Carr, S and Wield, D (1998), 'European Regulation: Harmony – or Cacophony?', *BINAS News*, vol 4, no 1

Lotschert, W and Beese, G (1992), *Collins' Guide to Tropical Plants*, Collins, London

Mabberley, D J (1990), *The Plant-Book: A Portable Dictionary of the Higher Plants*, CUP, Cambridge

MAFF (1995), *Guide to National Listings of Varieties of Agricultural and Vegetable Crops in the UK*, PB 2153, MAFF/Plant Variety Rights Office, London?

MAFF (1996), *Plant Genetic Resources: Characterisation and Evaluation*, MAFF, 22–4 October, Japan MAFF, Tokyo

MAFF (1998), *Genetic Modification of Crops and Food* (November fact sheet), Joint Food Safety and Standards Group, MAFF, London

MAFF (1998), *Regulatory Appraisal (Draft) Labelling of Modified Soya and Maize Genetically*, November, MAFF, London

MAFF (1998), *The UK Plant Breeders' Rights Handbook*, PB3760, MAFF/Plant Variety Rights Office, Cambridge

Maxted, N, Hawkes, J G, Guarino, L and Sawkins, M (1997), 'Towards the Selection of Taxa for Plant Genetic Conservation', *Genetic Resources and Crop Evolution*, vol 44, 337-48

Mazzucato, V and Ly, S (1994), *An Economic Analysis of Research and Technology Transfer of Millet, Sorghum and Cowpeas in Niger*, Michigan State University, Development Working Paper no 40

Meyerson, A R (1997), 'Breeding Seeds of Discontent: Cotton Growers Say Strain Cuts Yields', *New York Times*, 19 November

Monsanto (1998), 'Monsanto releases seed piracy case settlement details', press release by Karen K Marshall and Jennifer O'Brien, Monsanto Company, 800 North Lindbergh Boulevard, St Louis, Missouri 63167, web: www.monsanto.com

Morrison, A B (1996), *Fundamentals in American Law*, OUP, Oxford and New York

Novartis (1998a), *Novartis in Brief*, Novartis Corporate Communications, Basle

Novartis (1998b), *The World of Novartis Seeds*, Novartis Seeds AG, Basle

Ochave, J M A (1997), 'Barking Up the Wrong Tree: Intellectual Property Law and Genetic Resources', *BINAS News*, vol 3, nos 3 and 4

OECD (1992), *The Tomato Market in OECD Countries*, OECD, Paris

OECD (1995), 'Schemes for the Varietal Certification of Seed Moving in International Trade', Annex I: Control Plot and Field Inspection Manual AGR/CA/S(95)22, OECD, Paris

OECD (1996) Up-to-Date Version of the Seed Schemes as of June 1996: Annex I (Addendum), AGR/CA/S(95)15, OECD, Paris

OECD (1997), *List of Cultivars Eligible for Certification*, OECD, Paris

OECD (1998), 'Schemes for the Varietal Certification of Seed Moving in International Trade: List of Countries Participating in the OECD Codes and Schemes' (or having sent an application for admission) as at 25 March 1998, AGR/CA/S(98)1, OECD, Paris

Pollak, L (1996), 'The US Germplasm Enhancement of Maize (GEM) Project, Status Report', USDA, May, Washington DC

Prescott-Allen, C and Prescott-Allen, R (1987), *The First Resource: Wild Species in the North American Economy*, Yale University Press, with support of WWF and Philip Morris Inc, New Haven and London

Prescott-Allen, C and Prescott-Allen, R (1988), *Genes from the Wild: Using Wild Genetic Resources For Food and Raw Materials*, Earthscan, London

Rabobank, (1994), *The World Seed Market: Developments and Strategy*, Agricultural Economic Institute (LEI-DLO)/Rabobank/Ministry of Agriculture, Nature Management and Fisheries, Netherlands

RAFI (nd) *Conserving Indigenous Knowledge: Integrating Two Systems of Innovation*, an independent study by RAFI, commissioned by UNDP, RAFI, Ottawa

RAFI (1994), *Declaring the Benefits*, Occasional Paper Series, vol 1, no 3, October, RAFI, Ottawa

RAFI (1998), 'Plant Breeders' Wrongs: 147 Reasons to Cancel the WTO's Requirements for Intellectual Property on Plant Varieties; the Biopiracy and Plant Patent Scandal of the Century' *Rural News*, 15 September, www.rafi.org

RAFI (1998), 'Sprouting Up: Monsanto Rounds Up Seed-Saving Farmers', *Seedling*, vol 15, no 1, 28, GRAI, Barcelona

Reichman (1993), 'The TRIPS Component of the GATT's Uruguay Round: Competitive Prospects for Intellectual Property Owners in an Integrated World Market, 4 Fordham Intellectual Property', *Media and Entertainment Law Journal*, vol 171,

Reid, W V, Barber, C V and La Vina, A (1995), 'Translating Genetic Resource Rights into Sustainable Development: Gene Co-operatives, the Biotrade and Lessons from the Philippines', *Plant Genetic Resources Newsletter*, no 102, 1ff

Sands, P (1994), *Principles of International Environmental Law*, vol 1, Manchester University Press, Manchester and New York

Scrinis, G (1998), 'Colonizing the Seed', in *Arena Magazine*, no 36, August–September

Shand, H (1998), *Human Nature: Agricultural Biodiversity and Farm-Based Security*, RAFI, Ottawa

Shiva, V (1998a), *Monocultures, Monopolies, Myths and the Masculinisation of Agriculture*

Shiva, V (1998b), 'Statement to the Workshop on "Women's Knowledge, Biotechnology and International Trade – Fostering a New Dialogue into the Millenium"', International Conference on Women in Agriculture, Washington, 28 June–2 July

Smale, M (1996), *Understanding Global Trends in the Use of Wheat Diversity and International Flows of Wheat Genetic Resources*, Economics Working Paper 96–02, CIMMYT, Mexico, DF

Smale, M (1997), 'The Green Revolution and Wheat Genetic Diversity: Some Unfounded Assumptions', *World Development*, vol 25, no 8, 1257–69

Smartt, J and Simmonds, N W (1995), *Evolution of Crop Plants: Second Edition*, Longman, Harlow

Starnes, R (1998), 'Altered Potato Fights Diarrhoea', *Calgary Herald*, 23 May

Swanson, T (ed) (1995) *Intellectual Property Rights and Biodiversity Conservation: an Interdisciplinary Analysis of the Values of Medicinal Plants*, CUP, Cambridge

Swanson, T M and Luxmoore, R A (1996), *Industrial Reliance upon Biodiversity: a Darwin Initiative Project*, DOE

Tanksley, S D and McCouch, S R (1997), 'Seed Banks and Molecular Maps: Unlocking Genetic Potential from the Wild', *Science* (reprint series), 22 August, vol 277, 1063-6

Thrupp, L A (1997), *Linking Biodiversity and Agriculture: Challenges and Opportunities for Sustainable Food Security*, WRI, Washington, 6–7

Times, The (1988), 'Plants Take to the Hard Stuff', 2 February

Tribe, D (1991), 'Report to the Crawford Fund for International Agricultural Research, Australia', cited in RAFI, qv, 1994

Tribe, D (1994), *Feeding and Greening the World: The Role of International Agricultural Research*, CAB International, Wallingford

Tripp, R (1997), 'The Structure of National Seed Systems' in R Tripp (ed) *New Seed and Old Laws: Regulatory Reform and the Diversification of National Seed Systems*, Intermediate Technology Publications, for ODI, London

Turner, R (1998), 'Agricultural Biotechnology in the UK: Its Role in the Technology Foresight Programme', in *The UK Biotechnology Handbook 97/98*, Biocommerce Data, Slough

USDA, Agricultural Research Service (1996), *Germplasm Enhancement of Maize Newsletter*, August, vol 3, no 2,

USDA, Foreign Agricultural Service (1998), *World Agricultural Production*, Circular Series, May

USDA(1998), *World Agricultural Production*, USDA Foreign Agricultural Service WAP 05–98, May, Washington DC

van Dorp, M and Rulkens, T (1993), 'Farmers' Crop-Selection Criteria and Genebank Collections in Indonesia' in W de Boef et al, *Cultivating Knowledge: Genetic Diversity, Farmer Experimentation and Crop Research*, Intermediate Technology Publications, London

van Wijk, J (1995), 'Plant Breeders' Rights Create Winners and Losers' *Biotechnology and Development Monitor*, no 23, June 1995, 15–19

Vaughan, D A and Sitch, L A (1991), 'Gene Flow from the Jungle to Farmers: Wild-Rice Genetic Resources and their Uses', *Bioscience*, vol 41, no 1

Vellve, R (1992), *Saving the Seed*, Earthscan, London

Wijk, van J (1997), 'The Impact of Intellectual Property on Seed Supply', in Tripp (1997), 185-97

World Conservation Monitoring Centre (WCMC) (1992), *Global Biodiversity: Status of the Earth's Living Resources*, Chapman and Hall, London

Chapter 6: Horticulture

Agra Europe (1998), *Fruit and Vegetable Markets Monthly*, no 93, London, June

CAB International (1998), *Horticultural Abstracts*, CABI, vol 68, no 12, Oxon, December

Desmond, R (1982), *The India Museum, 1801–1879*, HMSO, Alden Press, Oxford

Desmond, R (1998), *The History of the Royal Botanic Gardens, Kew*, Harvill Press and Royal Botanic Gardens, Kew, London

Eastwood, A, Bytebier, B, Tye, H, Tye, A, Robertson, A and Maunder, M (1998), 'The Conservation Status of Saintpaulia', *Curtis's Botanical Magazine*, February, vol 15, pt 1

Field, H and Semple, R H (1878), *Memoirs of the Botanic Gardens at Chelsea*, Gilbert and Rivington, London

FIS/ASSINSEL (1998), *World Seed Statistics*, available from the FIS Secretariat website: www.worldseedorg/~assinsel/stat

Fletcher, H and Brown, W (1970), *Royal Botanic Garden Edinburgh, 1670–1970*, HMSO, Edinburgh, 1970

Floriculture Crops Summary (1998), available from the National Agricultural Statistics Service of the USDA (see below)

Flower Council of Holland, (1998), *Facts and Figures about Dutch Horticulture*, Leiden

Fresh Produce Desk Book (1996), Fresh Produce Journal, Lockwood Press, London

HDC (1997), *How It Works*, HDC, Dronfield, UK

HDC (1997), *Project Report Index: Bulbs and Outdoor Flowers*, HDC, Dronfield, UK

HDC (1998a), *Profit From Research: Nursery Stock, Bedding Plants, Pot Plants*, September, HDC, Dronfield, UK

HDC (1998b), *Project Report Index: Hardy Nursery Stock*, HDC, Dronfield, UK

HDC (1998c), *Project Report Index: Soft and Stone Fruit*, HDC, Dronfield, UK

HDC (1998d), *Project Report Index: Protected Crops*, HDC, Dronfield, UK

HDC (1998e), *Project Report Index: Field Vegetables*, HDC, Dronfield, UK

Heinrichs, F and Siegmund, I (eds) (1998), *Yearbook of the International Horticulture Statistics*, vol 46, AIPH – Union Fleurs, Institut fur Gartenbauokonomie der Universitat Hannover, Hanover

Horticultural Directory, The (1997), *Sources of Supply for Garden Centres, Commercial Growers and Amenity Buyers*, Nurseryman and Garden Centre, Nexus Media, Swanley, Kent

Horticulture Week (1997), *Buyers' Guide 1998*, Haymarket Business Publications, London

Institute of Horticulture/ RHS(1998), *Come Into Horticulture*, RHS, London, June

Lacey, S (1997), 'Plant-Gathering Can Have all the Drama of a *Boy's Own* Adventure' in *Daily Telegraph Weekend Section*, 15 November

Lyte, C (1983), *The Plant Hunters*, Orbis Publishing, London

MAFF (1998), *Basic Horticulture Statistics for the United Kingdom, Calendar and Crop Years, 1987–97*, PB 3759, MAFF

Marshall, N T (1993), *The Gardener's Guide to Plant Conservation*, WWF and The Garden Club of America, Baltimore

McCracken, D and E (1988), *The Way to Kirstenbosch*, National Botanic Gardens, Cape Town

Pathfast, *International Floriculture Address Book 1996/97*, Pathfast Publishing, Frinton on Sea

Pathfast, *International Floriculture Quarterly Report*, Pathfast Publishing, Frinton on Sea

Pattison, G and Cook, L (1998), 'Introduction to the NCCPG: Conservation in Action', *Curtis's Botanical Magazine*, May, vol 15, pt 2, 86-91

Pavord, A (1998), 'The Tulip', in W Benton (ed) *Encyclopaedia Britannica*, Bloomsbury Publishing, London

Pertwee, J (1998), *International Floriculture Trade Statistics 1997*, Pathfast Publishing, Frinton on Sea

Posey, D A and Dutfield, G (1996), *Beyond Intellectual Property*, IDRC, Canada

Rabobank, (1994), *The World Seed Market: Developments and Strategy*, Agricultural Economic Institute (LEI-DLO)/Rabobank/Ministry of Agriculture, Nature Management and Fisheries, Netherlands

Stearn, W T (1961), *Botanical Gardens and Botanical Literature in the Eighteenth Century*, British Museum (Natural History), London

USDA (1996–8), *National Agricultural Statistics Service*, produced by USDA and available from USDA – NASS website: www.usda.gov

Xue Dayuan (1998), 'Access to Genetic Resources and Traditional Knowledge for Biodiversity Conservation in

China', paper delivered at the South and South East Asia Regional Workshop on Access to Genetic Resources and Traditional Knowledge, Madras, 22–5 February, IUCN – World Conservation Union and the M S Swaminathan Research Foundation

Chapter 7: Crop Protection

Adirukmi-Noor-Saleh (1992), 'Natural Products From Several Important Plant Families', in Ho, Y W et al (eds), *Proceedings of the National IPRA*, University of Pertanian, Malaysia, National IRPA Seminar, Kuala Lumpur, 6–11 January

Agnello, A M and Bradley, J R (1991), 'Improved Safety Through Reduction in Use of Existing Chemicals', in E Hodgson and R J Kuhr (eds), *Development of Safer Insecticides*, 509-27

Agouron Pharmaceuticals (1998), 'Alanex and Zeneca Form Strategic Collaboration to Discover Novel Agrochemicals', *PR Newswire*, 18 June

Agri Marketing (1996), 'US Farmer Spending on Crop Protection', Research Reports, 15

AGROW (1998a), 'Global Agrochemical Market up to 1.3% in 1997', no 299, 27 February, 16

AGROW (1998b), 'European Agrochemical Growth in 1996', no 300, 13 March, 11

Ahmed, S et al (1984), 'Some Promising Plant Species for Use as Pest Control Agents Under Traditional Farming Systems', in *Natural Pesticides from the Neem Tree and Other Tropical Plants*, Proceedings from the Second International Neem Conference, Federal Republic of Germany, 565–80

Alkofahi, A (1989), 'Search for New Pesticides from Higher Plants', in J T Arnason et al (eds), *Insecticides of Plant Origin*, American Chemical Society, Washington, 25–43

Arima, K, Imanaka, H, Kousaka, M, Fukuda, A and Tamura, G (1964), 'Pyrrolnitrin: a New Antibiotic Substance Produced by Pseudomonas', *Agri-Biological Chemistry*, vol 28, 575-6

Arima, K, Imanaka, H, Kousaka, M, Fukuda, A and Tamura, G (1965), 'Studies on Pyrrolnitrin, a New Antibiotic Isolation and Properties of Pyrrolnitrin', *Journal of Antibiotics*, vol 28, 201-10

Arnason, J T (1989) 'Naturally Occurring and Synthetic Thiophenes as Photoactivated Insecticides', in J T Arnason et al (eds), *Insecticides of Plant Origin*, American Chemical Society, Washington, 164–71

Balkenhohl, F, Bussche-Hünnefeld, C von dem, Lansky, A and Zechel, C (1996), 'Combinatorial Synthesis of Small Organic Molecules', *Agnew: Chem Int Ed Engl*, Weinheim, Germany, vol 35, 2288-337

Bateman, R (1997a), 'Mycoinsecticide Formulation Development Work in the United Kingdom', *Abstracts of Presentations Made by LUBILOSA Scientists*, Society for Invertebrate Pathology, 30th Annual Meeting, Banff, Canada, August

Bateman, R, (1997b) 'The Development of a Mycoinsecticide for the Control of Locusts and Grasshoppers', *Outlook on Agriculture*, vol 26, no 1, 13–18

Bates, J A R (1989), *Pesticides and International Organisations: An Overview*, World Directory of Pesticide Control Organisation, Royal Society of Chemistry, Essex

Beil, G (1997), 'What Happened to the Seed Business?', *Seed and Crop Digest*, June/July, vol 42

Bell, E A et al (1991), 'Natural Products from Plants for the Control of Insect Pests', in E Hodgson and R J Kuhr (eds), *Development of Safer Insecticides*, 337–50

Benner, J P (1996), 'Pesticides from Nature', in L G Copping, *Crop Protection Agents from Nature: Natural Products and Analogues*, Royal Society of Chemistry, Essex, 217–28

Biological Control Products and LUBILOSA (1998), *Green Muscleâ, Handbook for Central and Southern Africa*

Brent, K J (1996), 'Pathways to Success in Fungicidal Research and Technology', in H Lyr et al (eds), *Modern Fungicides and Antifungal Compounds*, Intercept, UK, 3–15

British Agricultural Association (1998), 'The World Market', *Annual Review*, 29–31

British Agrochemicals Association (BAA) (1997), *Annual Review and Handbook*, 27–31

Buchel, K H (1990), *Chemistry of Pesticides*, Wiley, New York

Canadian Seed Trade Association Brochure

Chiu, S-F (1989), 'Recent Advances in Research on Botanical Insecticides in China', in J T Arnason et al (eds), *Insecticides of Plant Origin*, American Chemical Society, Washington DC, 69–77

Cline, M N and Re, D B (1997), 'Plant Biotechnology: a Progress Report and Look Ahead', *Feedstuffs*, 11 August, 17–19

Clough, J M and Godfrey, C R A (1994), 'The Strobilurin Fungicides' in D H Hutson and J Miyamoto (eds), *Fungicidal Activity*, John Wiley

Clough, J M et al (1994), 'Role of Natural Products in Pesticide Discovery', in P A Hedin et al (eds), *Natural and Engineered Pest Management Agents*, American Chemical Society Symposium Series 551, 37–52

Copping, L G (ed) (1996), *Crop Protection Agents from Nature: Natural Products and Analogues*, Critical Reports on Applied Chemistry, vol 35, Royal Society of Chemistry, Essex

Cremlyn, R J (1990), *Agrochemicals: Preparation and Mode of Action*, John Wiley

Cutler, H G (1988), 'Natural Products and their Potential in Agriculture: a Personal Overview', in H G Cutler (ed) *Biologically Active Natural Products*, ACS Symposium Series 380, Washington DC, 1–23

Dale, P J and Irwin, J A (1995), 'The Release of Transgenic Plants from Containment and the Move Towards their Widespread Use in Agriculture', *Euphytica*, 85, 425–31

Dale, P J (1995), 'R&D Regulation and Field Trialing of Transgenic Crops', *Tibtech*, September, vol 13, 398-401

DETR/ACRE (1997), *Newsletter*, June, issue 7

Dinham, B (ed) (1996), *Growing Food Security: Challenging the Link Between Pesticides and Access to Food*, Pesticides Trust/PAN, London

Duke, S, Menn, J and Plimmer, J (eds) (1993), *Pest Control With Enhanced Environmental Safety*, American Chemical Society, Washington DC

Duke, S O and Abbas, H K (1995), 'Natural Products with Potential Use as Herbicides', in S O Duke (ed), *Allelopathy: Organisms, Processes and Applications*, American Institute of Biological Sciences, Botanical Society of America Section, American Chemical Society, Washington DC, 117–26

Duke, S O et al (1996), 'Phytotoxins of Microbial Origin with Potential for Use as Herbicides', in L G Copping (ed) *Natural Products and their Potential in Agriculture*, Critical Reviews in Applied Chemistry Series, Royal Society of Chemistry, Essex, 82–113

Duke, S O (ed) (1996), *Herbicide-Resistant Crops: Agricultural,*

Environmental, Economic, Regulatory and Technical Aspects, CRC Press, Boca Raton FL

Duke, S O, Dayan, F and Rimando, A (1998), *Natural Products as Tools for Weed Management*, Reprint of the Proceedings of Special Lecture Meeting on 'Recent Topics of Weed Science and Weed Technology', USDA, Mississippi

Duke, S O et al (1999), 'Strategies for the Discovery of Bioactive Phytochemicals', in W R Bidlak et al (eds), *Phytochemicals as Bioactive Agents*, Technomic Publishing, Lancaster PA (forthcoming)

Duvick, N D (1996), 'Seed Company Perspective', in S O Duke (ed) *Herbicide-Resistant Crops*, CRC Press, Boca Raton FL, 253–61

Dyer, W E (1996), 'Techniques for Producing Herbicide-Resistant Crops', in S O Duke (ed) *Herbicide-Resistant Crops*, CRC Press, Boca Raton FL, 37–51

Engel, J F et al (1991), 'Challenges: The Industrial Viewpoint', in E Hodgson and R J Kuhr (eds), *Development of Safer Insecticides*, 551–73

Ernst and Young (1997), *European Biotech 97: A New Economy*, Industry Annual Report, Ernst and Young Report on the European Biotechnology Industry, Thought Leadership Series, London, 4

Eubios Ethics Institute News – Extracts 1991–4

European Chemical News, (1997), 9–15 June

European Crop Protection Association, *Approval of Crop Protection Products within the EU*, ECPA, 1160 Brussels, Belgium

Ferreira, J F S and Duke, S O (1997), 'Approaches to Maximising Biosynthesis of Medicinal Plant Secondary Metabolites', in *AgBiotech – News and Information*, vol 9, no 12, 309N–319N

Fischer, D C et al, 'Inducers of Plant Resistance to Insects', in E Hodgson and R J Kuhr (eds), *Development of Safer Insecticides*, 257–9

Fuchs, R A and Schröder, R (1990), 'Agents for Control of Animal Pests', in K H Büchel (ed) *Chemistry of Pesticides*, Wiley, New York, Chichester

Gehmann, K, Nyfeler, R, Leadbeater, A, Nevill, D and Sozzi, D (1990), 'CGA 173506: A New Phenylpyrrole Fungicide for Broad-Spectrum Disease Control, Brighton Crop Protection Conference', *Pests and Diseases*, vol 2, 399-406

Gerth, K, Trowitzsch, W, Wray, V, Hofle, G, Irschik, H and Reichembach, H (1982), 'Pyrrolnitrin from Myxococcus Fulvus (myxobacteriales)', *Journal of Antibiotics*, vol 35, 1101-13

Georgis, R (1997), 'Commercial Prospects of Microbial Insecticides in Agriculture', *Microbial Insecticides: Novelty or Necessity?*, BCPC Symposium Proceedings, no 68, Farnham, Surrey

Gordee, R and Matthews, T R (1969), 'Systemic Antifungal Activity of Pyrrolnitrin', *Applied Microbiology*, vol 17, 690-94

Gray, M (1996), 'The International Crime of Ecocide', *California Western International Law Journal*, vol 26, 215

Gressel, J (1996), 'The Potential Roles for Herbicide-Resistant Crops in World Agriculture', in S O Duke (ed) *Herbicide-Resistant Crops*, CRC Press, Boca Raton FL, 231–50

Grosjean, O (1998), 'BASF Agriculture Research and Development: Philosophy and Practice', BASF (AG) Germany – Agricultural Centre, Limburgerhof D-67114 Germany

GTZ Locust Projects, <www.gtz.de/locust/englisch/>

Guerrant, E O and McMahon, I (1995), 'Saving Seeds for the Future', *Plant Talk*, Botanical Information Co, Richmond, Surrey, UK, 21–32

Hajduch, M and Libantova, J (1996), 'Review of Aspects of Biosafety of Transgenisis', *Eubios Journal of Asian and International Bioethics*, vol 6], 134–5

Hammer, P, Hill, D S, Lam, S, van Pée K-H and Ligon, J (1997), *Applied Environmental Microbiology*, vol 63, 2147–57

Harada, H (1996), 'Advancement of Plant Breeding Techniques: Scientific, Social and Global Impact', *Eubios Journal of Asian and International Bioethics*, vol 6, 131–4

Hassanali, A and Lwande, W (1989), 'Antipest Secondary Metabolites from African Plants', in J T Arnason et al (eds), *Insecticides of Plant Origin*, American Chemical Society, Washington, 78-93

Hedin, P A (1991), 'Use of Natural Products in Pest Control', in P A Hedin (ed) *Naturally Occurring Pest Bioregulators*, American Chemical Society, Washington, 2–9

Hedin, P A, Menn, J J and Hollingworth, R M (1994), 'Development of Natural Products and their Derivatives for Pest Control in the 21st Century', in P A Hedin et al (eds), *Natural and Engineered Pest Management Agents*, American Chemical Society Symposium Series, 551, American Chemical Society, Washington, 2–9

Hess, F D (1996), 'Herbicide-Resistant Crops: Perspectives from a Herbicide Manufacturer', in S O Duke (ed) *Herbicide-Resistant Crops*, CRC Press, Boca Raton FL, 263–70

Hodgson, E and Kuhr, R J (1991), 'Introduction', in E Hodgson and R J Kuhr (eds), *Development of Safer Insecticides*, 1–18

Hosada, H, et al (1996), 'Agrochemical Registration in Japan', Brighton Crop Protection Conference: Pests and Disease, Brighton, Sussex

Ikan, R (1991), *Natural Products: A Laboratory Guide*, Academic Press, San Diego

Iles, M and Christophers, A (eds) (1994), IPM Working for Development, NRI, Chatham, Kent

'Intercept', *Biofungicide*, Soil Technologies Corporation, Iowa

Jacobson, M (1984), 'Neem Research in the US Department of Agriculture: an Update', *Natural Pesticides from the Neem Tree and Other Tropical Plants*, Proceedings from the Second International Neem Conference, Federal Republic of Germany, 31–42

Jacobson, M (1989), 'Botanical Pesticides', in J T Arnason et al (eds), *Insecticides of Plant Origin*, American Chemical Society, Washington, 1–19

Jenkins, N E G, Heviefo, J U, Langewald, A J, Cherry and Lomer, C J (1998), 'Development of Mass Production Technology for Aerial Conidia of Mitosporic Fungi for Use as Mycopesticides', *Biocontrol Information and News Service*,19, 21N–31N

Johnen, B G and Urech, P A (1997), 'The Myths and Facts about Crop Protection Products and Food Quality', paper given at Crop Protection and Food Quality: Meeting Customers' Needs, University of Kent, UK, 17–19 September

Johnson, D (1997), *CMR Special Report*, 14 April

Jotwani, M G and Srivastava, K P (1984), 'A Review of Neem Research in India in Relation to Insects', in *Natural Pesticides from the Neem Tree and Other Tropical Plants*, Proceedings from the Second International Neem Conference, Federal Republic of Germany, 43–56

Kathirithamby, J (1998), 'Relating to Research on Oil Palm Insect Pests and their Natural Enemies', Agreement between the Chancellor, Masters and Scholars of the University of Oxford and the Papua New Guinea Oil Palm Research Association, 18 March

Karthirithamby, J, Simpson, S, Solulu, T and Caudwell, R (1998), 'Strepsiptera Parasites: Novel Biocontrol Tools for Oil Palm Integrated Pest Management in Papua New Guinea', *International Journal of Insect Pests Management*, vol 44, no 2 (forthcoming)

Kawanishi, C Y and Held, G A (1991), 'Viruses and Bacteria as Sources of Insecticides', in E Hodgson and R J Kuhr (eds), *Development of Safer Insecticides*, 351–83

Kirner, S, Hammer, P, Hill, D S, Altmann, A, Fischer, I, Weislo, L, Lanahan, M, van Pée, K-H and Ligon, J (1998), *Journal of Bacteriology*, vol 180, 1939–43

Koller, W (1996), 'Recent Developments in DMI Resistance', in H Lyr et al (eds), *Modern Fungicides and Antifungal Compounds*, Intercept, UK, 301–3

Kooyman, C and O M Abdalla (1998), 'Application of Metarhizium Flavoviridae (Deuteromycotina: Hyphomycetes) Spores Against the Tree Locust, Anacridium melanorhodon (Orthoptera: Acrididae) in Sudan', *Biocontrol Science and Technology* 8, 215–19

Kooyman, C R P, Bateman, J, Langewald, C J, Lomer, Z, Ouambama and Thomas, M B (1997), 'Operational-Scale Application of Entomopathogenic Fungi for the Control of Sahelian Grasshoppers', Proceedings of The Royal Society, London, B 264, 541– 6

Kuesgen, K (1997), 'Registration Improvements Urgently Needed', *ECPA Annual Report 1996–7*, November, 5

Leadbeater, N J, Nyfeler, R and Elmsheuser, H (1994), 'The Phenylpyrroles: The History of their Development at CIBA, 1994 Brighton Crop Protection Conference', *Pests and Diseases*, no 57, Seed Treatment: Progress and Prospects, 129-84

Legg, M (1998), 'Opportunities for Microbial Natural Products in the Agrochemical Industry', Abstract from 139th meeting of Society for General Microbiology, Bradford, UK

Lewis, W J, Lenteren, J C, Phatak, S C and Tumlinson, J H (1997), 'A Total Systems Approach to Sustainable Pest Management', *Proc-National Academy of Science, USA*, 11 November, vol 94, no 23, 122243–8

Ligon, J, Hill, D, Hammer, P, Torkewitz, N, Hofmann, D, Kempf, H J and van Pée, K-H (1998), 'Natural Products with Antimicrobial Activity from Pseudomonas Biocontrol Bacteria', paper given at the 9th IUPAC Conference, Pesticide Chemistry, London, 2–7 August

Lomer, C and H de Groote (1997), 'LUBILOSA Participatory Trials', *LUBILOSA 3, The Newsletter of Phase 3 of the LUBILOSA Programme*, issue no 3, October 1997

LUBILOSA (April 1997), 'The Efficacy Issue', *LUBILOSA 3, The Newsletter of Phase 3 of the LUBILOSA Programme*, issue no 2, April 1997

LUBILOSA (December 1997), 'The Two Technologies', *LUBILOSA 3, The Newsletter of Phase 3 of the LUBILOSA Programme*, issue no 4, December 1997

LUBILOSA, (1998a) 'Welcome', *LUBILOSA 3, The Newsletter of Phase 3 of the LUBILOSA Programme*, issue no 5, March 1998

LUBILOSA, (1998b) 'Intellectual Property Rights', *LUBILOSA 3, The Newsletter of Phase 3 of the LUBILOSA Programme*, issue no 5, March 1998

LUBILOSA, 'R&D to Commercialisation' (1998c), *LUBILOSA 3, The Newsletter of Phase 3 of the LUBILOSA Programme*, issue no 5, March 1998

MAFF/PSD (1997), *The Work of the Pesticides Safety Directorate*, MAFF/PSD, UK

MAFF/PSD/Health and Safety Executive (1998), *Pesticides Safety Directorate: The Registration Handbook*, January, PSD/Health and Safety Executive, UK

Magasin Agricole (1995), 'French Vine Insecticides Market in 1995', 17 November

Merck Index (1999), 12th edn

Miyakado, M, Nakayama, I and Ohno, N (1989), 'Insecticidal Unsaturated Isobutylamides', in J T Arnason et al (eds), *Insecticides of Plant Origin*, American Chemical Society, Washington, 173–85

Miyakado, M, Watanabe, K and Miyamoto, J (1997), 'Natural Products as Leads in Structural Modification Studies Yielding New Agrochemicals', in *Phytochemicals for Pest Control*, ACS Symposium Series 658, 168–82

Morhy, L (1995), 'Proteins from Amazonia: Studies and Perspective for their Research', in *Chemistry of the Amazon: Biodiversity, Natural Products and Environmental Issues*, Seidl et al, 93–8

NABC (1998), *Resource Management in Challenged Environments*, Report 9

Nair, M G (1994), 'Natural Products as Sources of Potential Agrochemicals', in Hedin, P A et al (eds), *Natural and Engineered Pest Management Agents*, American Chemical Society Symposium Series 551, 145–59

Novartis (1997), 'Novartis and Chiron Sign Agreement on Combinatorial Chemistry for Crop Protection Research', press release, 9 September

Novartis Crop Protection (1998), *Celest/Maxim: A Healthy Crop at the End of the Winter*, Novartis Canada Inc, info.canada@cp.novartis.com

OECD (1994), 'Data Requirements for Pesticide Registration in OECD Member Countries: Survey Results', *Environmental Monograph*, no 77, OECD, Paris

OECD (1996), 'Data Requirements for Registration of Biopesticides in OECD Member Countries: Survey Results', *Environmental Monograph*, no 106, OECD, Paris, 37

Oerke, E-C (1996), 'The Impact of Diseases and Disease Control on Crop Production', in H Lyr et al (eds), *Modern Fungicides and Antifungal Compounds*, Intercept, UK, 17–24

Perrin, R M (1997), 'Crop Protection: Taking Stock for the New Millennium', *Crop Protection*, vol 16, no 5, 449–56

Persley, G J (ed) (1996), 'Biotechnology and Integrated Pest Management', *CABI*, Biotechnology in Agriculture, no 15 ISBN 0851989306, Wallingford, Oxon, UK

Pesticide Referee Group (1998), Report to the FAO, seventh meeting, Rome, 2-6 March

Plant Genetic Engineering News – Extracts since January 1994

Pletsch M, Sant'Ana, E G and Charlwood, B V (1995), 'Secondary Compound Accumulation in Plants: The Application of Plant Biotechnology to Plant Improvement: A Proposed Strategy for Natural Product Research in Brazil', in *Chemistry of the Amazon: Biodiversity, Natural Products and Environmental Issues*, 51–64

Price, R E, Bateman, R P, Brown, H D, Butler E T and Muller, E J (1997), 'Aerial Spray Trials Against Brown Locust (*Locustana pardalina*, Walker) nymphs in South Africa Using Oil-Based Formulations of Metarhizium flavoviridae', *Crop Protection* 16, 345–51

Prior, C and Greathead, 'Biological Control of Locusts: the Potential for the Exploitation of Pathogens' *FAO Plant Protection Bulletin*, no 37, 37–48

Prior, C (1992), 'Discovery and Characterisation of Fungal Pathogens for Locust and Grasshopper Control,' in

C Lomer and C Prior (eds), *Biological Control of Locusts and Grasshoppers*, CAB International, 159–80
RAFI (1997), *The Life Industry 1997: The Global Enterprises that Dominate Commercial Agriculture, Food and Health*, RAFI Communiqué, November/December
Rejesus, R M, Smale, M and van Ginkel, M (1996), 'Wheat Breeders' Perspectives on Genetic Diversity and Germplasm Use: Findings from an International Survey', *Plant Varieties and Seeds*, vol 9, no 3, 129–47
Rembold, H (1989), 'Azadirachtins – Their Structure and Mode of Action', in J T Arnason et al (eds), *Insecticides of Plant Origin*, American Chemical Society, Washington, 150–63
Sandman, G et al (1996), 'Steps Towards Genetic Engineering of Crops Resistant Against Bleaching Herbicides', in S O Duke (ed) *Herbicide-Resistant Crops*, CRC Press, Boca Raton FL, 189–99
Sands, P (1994), *Principles of International Environmental Law*, vol 1, Manchester University Press, Manchester and New York
Saxena, R C (1989), 'Insecticides from Neem', in *Insecticides and Plant Origin*, American Chemical Society Symposium Series 387, 110–12
Schmutterer, H (1984), 'Neem Research in the Federal Republic of Germany Since the First International Neem Conference', in *Natural Pesticides from the Neem Tree and Other Tropical Plants*, Proceedings from the Second International Neem Conference, Federal Republic of Germany, 21–30
Schmutterer, H (1995), *Neem Tree and Other Sources of Natural Products for Integrated Pest Management*, VCH, Weinheim
Schoonhoven, L M (1984), 'Second International Neem Conference: Afterword', in *Natural Pesticides from the Neem Tree and Other Tropical Plants*, Proceedings from the Second International Neem Conference, Federal Republic of Germany, 581–8
Schulten, G G M (1997), *The FAO Code of Conduct for the Import and Release of Exotic Biological Control Agents*, Bulletin OEPP/EPPO Bulletin 27, Plant Protection Service, FAO, Rome, 29–36
Seddon, B, Edwards, S G and Rutland, L (1996), 'Developments of Bacillus Species as Antifungal Agents in Crop Protection', in H Lyr, et al (eds), *Modern Fungicides and Antifungal Compounds*, Intercept, UK, 555–64
Seminis Inc (1998), 'Seminis-Zeneca Collaboration Produces Successful Biotech Tomato Product in the UK', *PR Newswire*, 3 June
Smartt, J and Simmonds, N (eds) (1995), *Evolution of Crop Plants*, 2nd edn
Society of Agricultural Chemical Industry, Japan (1984) Agricultural Chemical Laws and Regulations: Japan (I)
Solulu, T M, Simpson, S J and Kathirithamby, J (1998), 'The Effect of Strepsipteran Parasitism on a Tettigoniid Pest of Oil Palm in Papua New Guinea', *Physiological Entomology*, vol 23, no 3
Tauer, L W (1996), 'Farmer and Public Perspectives of Herbicide-Resistant Crops', in S O Duke (ed), *Herbicide-Resistant Crops*, CRC Press, Boca Raton FL, 271–9
Umino, S, et al (1987), 'Antifungal Composition Employing Pyrrolnitrin in Combination with an Imidazole Compound', US Patent 4, 636, 520
UK Biotechnology Handbook 97/98, The (1998), BioIndustry Association and BioCommerce Data
United States Environmental Agency, Office of Pesticides Programs, Registration Division (1992), *General Information on Applying for Registration of Pesticides in the United States*, 2nd edn
Urech, P A (1996), 'Is More Legislation and Regulation Needed to Control Crop Protection Products in Europe?', *Brighton Crop Protection Conference: Pests and Diseases*, vol 2, European Crop Protection Association, Brussels, 549–58
Urech, P A, Watelet, N and Gardiner, G R (1997) 'Crop Protection Between the Needs of Market and Regulation', *Brighton Crop Protection Conference: Weeds*, vol 5B-3, European Crop Protection Association, Brussels, 427–34
Uttley (1997), 'Coming to Market', European Chemical News, 9–15 June 1997, 18–20
Vilich, V (1996), 'Aspects of Diversity in Cereal Crop Stands – Plant Health and Plant Resistance' in H Lyr et al (eds), *Modern Fungicides and Antifungal Compounds*, Intercept, UK, 535–8
Walsh, K (1997a), 'Pesticides Demand Thrives on Higher Acreage', *Chemical Week*, 9 April, 29
Walsh, K (1997b), 'Pyrethroid Market Under Pressure', *Chemical Week*, 5 November, 50
Wilcut, J W (1996), 'The Niche for Herbicide-Resistant Crops in US Agriculture', in S O Duke (ed) *Herbicide-Resistant Crops*, CRC Press, Boca Raton FL, USA, 213–27
Wink, M and Latz-Bruning, B (1995), 'Allelopathic Properties of Alaloids and Other Natural Products: Possible Modes of Action', in S O Duke and HK Abbas, *Allelopathy: Organisms, Processes and Applications*, 117–26
Wood, H A (1997), 'Risks and Safety of Insecticides: Chemicals vs Natural and Recombinant Viral Pesticides', *Biosafety Journal*, vol 3, paper 7
Wymer P (1988), 'Genetically Modified Food: Ambrosia or Anathema', *Chemistry and Industry*, 1 June, 422–6
Zarcone, C (1994), 'Minimal Growth', *Chemical Marketing Reporter*, 25 April
Zeneca Agrochemicals (1996) *Azoxystrobin – A New Broad Spectrum Fungicide*, Zeneca

Chapter 8: Biotechnology in Fields other than Healthcare and Agriculture

Adams, A (1997), 'Let a Thousand Flowers Bloom', *New Scientist*, 20–27 December, 26–7
Amann, R I, Ludwig, W and Schleifer, K (1995), 'Phylogenetic Identification and In Situ Detection of Individual Microbial Cells Without Cultivation', *Microbiol Review* 59, 143–69
Anon (1998), 'Fat-Free Fried Food Could Net 1$bn/a', *Chemistry and Industry*, 2 March, p 155
Anon (1998), 'Lichen Gets Uranium Licked', *Chemistry and Industry*, 16 February, 117
'Biocatalysis: Enzymes and Micro-organisms in Organic Synthesis', database available from Synopsis Scientific Systems, Leeds, UK
British Agrochemicals Association (1998), *Annual Report 1998*, Peterborough
Black, H (1998), 'Bean Bag', *New Scientist*, 21 February, 14
Black, H (1998), 'Plant Power: a New Technique Turns Farm Waste into High Energy Gas', *New Scientist*, 22 August, 7
Bull, A T, Goodfellow, M and Slater, J H (1992) 'Biodiversity as a Source of Innovation in Biotechnology', *Annual Review of Microbiology*, 46, 219–52
Burgiel, S (1998), 'The "Biosafety Protocol": Fourth Session of the Ad Hoc Working Group', *BINAS News*, vol 4, 1

Cambio Ltd (1996/7), 'Thermostable Enzymes', 1996/7 Catalogue, Cambridge, UK, 79–92

Coghlan, A (1998), 'From a Watery Grave: Bacteria that Feast on Whale Bones Reveal the Secret of Cool Washes', *New Scientist*, 14 March, 24

Current Opinion in Microbiology (1998), vol 1, no 3, June 1998

DTI (undated), 'Waste Treatment, Status Report', BMB (Biotechnology Means Business) Initiative, HMSO, London, 134–7

Dordick, J S, et al (1998), 'Biocatalytic Plastics', *Chemistry and Industry*, 5 January, 17–20

Economist, The (1997), 'Hello, Dolly: Cloning and its Temptations', 1–7 March

Novo Nordisk (1989), 'Enzymes at Work', Denmark, May

Novo Nordisk (1995), 'Enzymes at Work', Denmark, September

Ernst and Young (1995a), *European Biotech 95: Gathering Momentum*, Industry Annual Report, Second Annual Ernst and Young Report on the European Biotechnology Industry, Thought Leadership Series, London

Ernst and Young (1995b), *Biotech 96 Pursuing Stability*, Tenth Industry Annual Report on the Biotechnology Industry, California

Ernst and Young (1998), *European Life Sciences 98: Continental Shift*, Fifth Annual Ernst and Young Report on the European Entrepreneurial Life Sciences Industry, Thought Leadership Series, London

Ernst and Young (1998), *New Directions 98*, Twelfth Biotechnology Industry Annual Report, California

Evans, C, et al (1998), 'Bioremediation by Fungi', *Chemistry and Industry*, 16 February

FAO Yearbook (1995), 'Forest Products 1991–5', FAO Forestry Series no 30, FAO Statistics Series no 137, FAO, Rome

FAO (1996a), *Global Plan of Action : For the Conservation and Sustainable Utilization of Plant Genetic Resources for Food and Agriculture*, FAO, Rome

FAO (1996b), *Report on the State of the World's Plant Genetic Resources for Food and Agriculture*, FAO, Rome

Griffiths, M H (1997), 'Bioremediation for a Better Environment', paper given at 8th European Congress on Biotechnology, Budapest, 18-22 August

Griffiths, M and Hesselink, P (1996), 'Environmental Biotechnology for the Millennium', OECD Quarterly Journal, *STI Review*, no *19, OECD 1996, 93 ff*

Hawksworth, D L (1991a), in *Global Biodiversity Assessment*, UNEP, CUP, vol 3, 81

Hawksworth, D L (1991b) 'The Fungal Dimension of Biodiversity: Magnitude, Significance and Conservation', *Mycological Research*, 95, 641–55

Hongu, T and Phillips, G (1990), 'Biomimetic Chemistry and Fibres', in E Horwood, *New Fibres*

Hugenholtz, P, Brett, M and Pace, N (1998), 'Impact of Culture-Independent Studies on the Emerging Phylogenetic View of Bacterial Diversity', in *Journal of Bacteriology*, September, vol 180, no 18, 4765–74, American Society for Microbiology, USA

IDRC (1994), *People, Plants and Patents*: The Impact of IP on Biodiversity, Conservation, Trade and Rural Society, Crucible Group, IDRC, Ottawa, Canada

Johnson, D (1997), 'Ag Biotech Putting Roots in Crop Protection', *Chemical Marketing Reporter (CMR) Special Report*, 14 April

ten Kate, K, Touche, L and Collis, A (1998), *Benefit-Sharing Case Study: Yellowstone National Park and the Diversa Corporation*, Submission to the Executive Secretary of the Convention on Biological Diversity by the Royal Botanic Gardens, Kew, London

Kendall, A, McDonald, A and Williams, A (1997), 'The Power of Biomass', *Chemistry and Industry*, 5 May, 342–5

Kleiner, K (1998), 'Yellowstone's Bugs Land Up in Court', *New Scientist*, 14 March

Lawton, G (1997), 'Biofirms Get Geyser Fever', *Chemistry and Industry*, 15 December, 983

Malongoy, K J (ed) (1997), *Biosafety Needs and Priority Actions for West and Central Africa*, International Academy of the Environment, Geneva, Switzerland

Malongoy, K J (ed) (1997), *Transboundary Movement of Living Modified Organisms Resulting from Modern Biotechnology: Issues and Opportunities for Policy-Makers*, International Academy of the Environment, Geneva, Switzerland

Messner, K (1997), 'Biopulping', in B Palfreyman and F Taylor (eds), *Forest Products Biotechnology*, 63–79

Morita, R Y (1975), 'Psychrophilic Bacteria', *Bacteriological Reviews*, 39, 144–67

Nature Biotechnology (1998), 'Building a Biotechnology Company from the Ground Up', Bioentrepreneurship Supplement, May, vol 16

Nader, W F and Rojas, M (1996), 'Gene Prospecting for Sustainable Use of the Biodiversity in Costa Rica', *Genetic Engineering News*, New York, 1 April, 35

Nader, W F and Mateo, N (1998), 'Biodiversity: Resource for New Products, Development and Self-Reliance', in W Barthlott and M Winiger (eds), *Biodiversity: A Challenge for Development and Policy*, Springer-Verlag, Heidelberg, New York and Tokyo

Nisbet, L J and Fox, F M (1991) 'The Importance of Microbial Biodiversity to Biotechnology', in *The Biodiversity of Microorganisms and Invertebrates: Its Role in Sustainable Agriculture* (ed D L Hawksworth), CABI, Wallingford (UK), 229–44

Obst, J R (1997), 'Special (Secondary) Metabolites from Wood', in B Palfreyman and F Taylor (eds), *Forest Products Biotechnology*, 151–65

OECD (1994), *Biotechnology for a Clean Environment: Prevention, Detection, Remediation*, Executive Summary, Ad Hoc Task Force chaired by Mike Griffith, OECD, Paris

OECD (1998), *Biotechnology for Clean Industrial Products and Processes: Towards Industrial Sustainability*, Ad Hoc Task Force chaired by Alan Bull, OECD, Paris

Pain, S (1998), 'Extreme Worms', *New Scientist*, 25 July, 48–50

Pearce, F (1998), 'Metal Munching Poplars Could Clear Toxic Dumps', *New Scientist*, 3 October, 11

Podila, G K and Karnosky, D F (1996), 'Fibre Farms of the Future: Genetically Engineered Trees', *Chemistry and Industry*, 16 December, 976–81

Ross-Murphy, S (1998), 'Starch Qualities', *Chemistry and Industry*, 7 September, 693

Saddler, J N and Gregg, D J (1997), 'Ethanol Production from Forest Products Waste', in B Palfreyman and F Taylor (eds), *Forest Products Biotechnology*

Schlegel, H G and Jannasch, H W (1992), 'Prokaryotes and their Habitats', in Balows, A, Truper, H, Dworkin, M, Harder, W and Schleifer (eds), *The Prokaryotes*, vol I, Springer-Verlag, New York, 75–125

Seguin et al (1997), 'Transgenic Trees', in B Palfreyman and F Taylor (eds), *Forest Products Biotechnology*, 151–65

Seife, C (1998), 'Designer Enzymes Enjoy Life in the Hot Seat', *New Scientist*, 7 March, 10

Smith, B L (1998), 'Studying Shells: A Growth Industry', *Chemistry and Industry*, 17 August, 649–53

Stackebrandt, E (1994), 'The Uncertainties of Microbial Diversity', in Kirsop, B and Hawksworth, D (eds), *The Biodiversity of Microorganisms and the Role of Microbial Resource Centres*, WFCC, Germany

Sugawara, H and Kirsop, B (1994), 'The WFCC Data Center on Microorganisms and Global Statistics on Microbial Resource Centres', in Kirsop, B and Hawksworth, D (eds), *The Biodiversity of Microorganisms and the Role of Microbial Resource Centres*, WFCC, Germany

Talbot et al (1996), *Fungal Genet Biol.*; 20(4): 254-267, 1996

Tamayo, G, Nader, W F and Sittenfeld, A (1997), 'Biodiversity for the Bioindustries', in J A Callow, B V Ford-Lloyd and H J Newbury (eds), *Biotechnology and Plant Genetic Resources*, CABI, Wallingford, UK, 255–80

UNEP (1995), *Global Biodiversity Assessment*, CUP, Cambridge

UNEP (1996), CBD Secretariat, SBSTTA 2/15

Viikari, L, et al (1997), 'Enzymes in Pulp Bleaching', in B Palfreyman and F Taylor (eds), *Forest Products Biotechnology*, 83–97

Waage, J (1997), 'Global Developments in Biological Control and the Implications for Europe', *Bulletin OEPP/EPPO Bulletin*, vol 27, no 1, 5–13

Wilson, E O (1988) 'The Current State of Biological Diversity', in E O Wilson and F M Peter (eds), *Biodiversity*, National Academic Press, Washington DC, 3–18

World Conservation Monitoring Centre (ed) (1992) *Global Biodiversity: Status of the Earth's Living Resources*, Chapman and Hall, London

WFCC (1994), *The Biodiversity of Microorganisms and the Role of Microbial Resource Centres*, WFCC, UK

WFCC (1996), 'Information Document on Access to *Ex Situ* Microbial Genetic Resources Within the Framework of the Convention on Biological Diversity', WFCC Biodiversity Committee (E-mail: bio@biostrat.demon.co.uk), 20 June

WFCC (1998), *The Economic Value of Microbial Genetic Resources*, Proceedings of the Workshop held at the Eighth Symposium on Microbial Ecology (ISME-8), WFCC, Brazil

Ward, M (1997), 'Ancient Inks Will Clean Up Desk Top Computers', *New Scientist*, 20–27 December

Wells, W (1997), 'Extreme Chemistry', *Biotech Today*, December

Wood, P J and Burley, J (1991), *A Tree for all Reasons: The Introduction and Evaluation of Multipurpose Trees for Agroforestry*, International Centre for Research in Agroforestry, Nairobi

Zhang, J H, Dawes, G and Stemmer, W P C (1997), 'Directed Evolution of a Fucosidase from a Galactosidase by DNA Shuffling and Screening', *Proceedings of the National Academy of Science*, USA 94, 4504-9

Chapter 9: The Natural Personal Care and Cosmetics Industry

Anderson, J (1996), 'Utilization of Natural Products in the Cosmetics Industry', paper presented at The Royal Society of Chemistry Conference: Phytochemical Diversity: A Source of New Industrial Products Conference, The University of Sussex, April 1996

Aveda Corporation (1994), 'The Source Book', Minneapolis MN

Body Shop, The (1996), 'Ethical Guidelines/Criteria for Buyers', West Sussex

Body Shop, The (1997), 'The Body Shop Community Trade Programme – Trading with Communities in Need', West Sussex

Boom, B (1989), 'Use of Plant Resources by the Chacobo', in DA Posey and W Balee (eds) 'Resource Management in Amazonia: Indigenous and Folk Strategies', *Advances in Economic Botany*, **7**, Botanical Garden Press, New York

Brown, A (1996), 'Bioprospecting for Leads', *DDT*, vol 1, no 17, July 1996

Brun, E, Blinder B, and Ryan, T (1998), 'Cosmetics and Botanicals as We Approach the Year 2000', *Drug and Cosmetic Industry (DCI)*, February 1998, Advanstar Communications, New York, 16–18

Bucalo, AJ (1998), 'Overview of the US Cosmetics and Toiletries Market', *Drug and Cosmetic Industry (DCI)*, June 1998, Advanstar Communications, New York, 26–31

Chittum, S (1996), 'Organic Shampoo? Check the Label', *New York Post*, 20 November

Clay, J (1999), *Greening the Amazon: Communities and Corporations in Search of Sustainable Business Practices*, (forthcoming)

Corry, S (1993), 'Harvest Moonshine Taking You for a Ride: a Critique of the "Rainforest Harvest" – its Theory and Practice', Survival International, London

Duke, JA (1989), *Handbook of Nuts*, CRC Press, Boca Raton, Florida

Dunn, CA (1998a), 'The Skin Care Market', *Happi Magazine*, May 1998, Rodman Publishing, 5 pages

Dunn, CA (1998b), 'What's New in Cosmetic R&D', *Happi Magazine*, January 1998, Rodman Publishing, 6 pages

EcoMaya and Conservation International (1998), 'Cohune Oil', Conservation Enterprise Product Fact Sheet, Conservation International, Washington DC

Egginton, C (1997), 'State of the Industry: Latin American Cosmetics and Toiletries Market', *Drug and Cosmetic Industry (DCI)*, September 1997, Advanstar Communications, New York, 20–30

Fost, D (1997), 'Expanding Asian Markets Offer New Horizons for Manufacturers of Performance Chemicals', *Drug and Cosmetic Industry (DCI)*, September 1997, Advanstar Communications, New York, 33–8

Gain, B (1996), 'Natural Products Gain Favor', *Chemical Week*, 11 December, New York, 35–7

Goldemberg, RL (1998), 'Natural Ingredients', *Drug and Cosmetic Industry (DCI)*, February 1998, Advanstar Communications, New York, 26–8

Hampton, A (1994), *Natural Ingredients Dictionary, Plus 10 Synthetic Cosmetic Ingredients to Avoid*, Organica Press, Tampa, Florida

Happi Magazine, (1997), 'The 1997 International Top 30', *Happi Magazine*, August, Rodman Publishing

Happi Magazine (1998), 'Special Report: The HAPPI International Top 30', *Happi Magazine*, August, Rodman Publishing

Hecht, S and Posey, S A (1989), 'Preliminary Results on Soil Management Techniques of the Kayapo Indians' in D A Posey and W Balee (eds) 'Resource Management in Amazonia: Indigenous and Folk Strategies', *Advances in Economic Botany*, 7, Botanical Garden Press, New York

IKW (Industrieverband Koerperpflege-und Waschmittel eV) (1998), 'Annual Report 1997/98', Frankfurt

INIREB INFORMA, Instituto nacional de investigaciones sobre recursos bióticos (1979) 'Communicado no 35', Guatemala, 1–5

Iwu, M M (1996), 'Production of Phytomedicines and Cosmetics from Indigenous Genetic Resources: from Lab to Market', paper presented at BDCP workshop on Commercial Production of Indigenous Plants as Phytomedicines and Cosmetics

Kalorama Information Market Intelligence Reports; Packaged Facts (1996a), *The International Skincare Market*, February, FIND/SVP, New York

Kalorama Information Market Intelligence Reports; Packaged Facts (1996b), *The International Suncare Market*, May, FIND/SVP, New York, NY

Kalorama Information Market Intelligence Reports; Packaged Facts (1996c), *The Market for Natural Personal Care Products*, April, FIND/SVP, New York,

Kalorama Information Market Intelligence Reports (1997a), *The Market For Cosmetics and Toiletry Products (Southeast Asia)* FIND/SVP, New York, NY

Kalorama Information Market Intelligence Reports; Packaged Facts (1997b), *The US Market for Color Cosmetics*, November, FIND/SVP, New York

Kalorama Information Market Intelligence Reports; Packaged Facts (1998), *The US Cosmeceuticals Market*, February, FIND/SVP, New York

Kline Reports (1997) *Cosmetics and Toiletries USA 1997*, Kline and Co, New Jersey

Milliken, W, Miller, R P, Pollard, S R and Wandelli, E C (1992) *Ethnobotany of the Waimiri Atroari Indians of Brazil*, Royal Botanic Gardens, Kew, London

Natural Pharmacy (1997) 'Filling More Than Prescriptions ... More Than Skin Deep', *Natural Pharmacy*, vol 1, no 2, February

Natural Pharmacy (1997) 'Are You on Board the Botanical Bandwagon?', *Natural Pharmacy*, vol 1, no 1a, January

NBJ (1998), *NBJ*, vol 3, no 1 January

Orejuela, J E (1992), 'Traditional Productive Systems of the Awa (Cuaiquer) Indians of Southwestern Colombia and Neighboring Ecuador', in K H Redford and C Padoch (eds) *Conservation of Neotropical Forests: Working from Traditional Resource Use*, Colombia University Press, New York

Ouellette, J (1997a), 'Cosmeceuticals', *Chemical Market Reporter*, 12 May (Special Report), Schnell Publishing, New York, SR 26

Ouellette, J (1997b), 'A Matter of Preservation', *Chemical Market Reporter*, 12 May (Special Report), Schnell, New York, SR 24–6

Phyto-Lierac (Ales Group) (1997), 'Annual Report', Paris, France

Richman, A and Witkowski, J P (1996), 'A Wonderful Year For Herbs', *Whole Foods*, October, South Plainfield NJ, 54

Tom's of Maine (1997), 'Annual Report', Maine, USA

Vam Arkum (1997), 'Milder and Milder with Surfactants', *Chemical Market Reporter*, 12 May (Special Report), Schnell, New York

Wilck, J (1997), 'Baby Boomers and Natural Ingredients Lift Personal Care', *Chemical Market Reporter*, 12 May (Special Report), Schnell, New York, SR 3–6

Winter, R (1994), *A Consumer's Dictionary of Cosmetics Ingredients*, 4th edn, Crown Trade Paperbacks, New York

Women's Wear Daily (1997), 'The Beauty Top 75: A Who's Who of Cosmetics – 1997', September

Yves Rocher (1997), 'Annual Report', France

Note

1 Agra Europe publishes updated information on legislation for fruits and vegetables (see Agra Europe 1998b, pp 349–41). There is also a publication called *EU Food Law Monthly*. Try the web page <www.agra-food-news.com>

Research Methodology and Team

Introduction

Research and methodology

This book is based on research conducted over a two-year period, including 193 interviews focused on access, benefit-sharing and the CBD, and based on the questions described below. In addition, members of the research team spoke to at least 100 individuals about the markets and scientific, technological, and regulatory trends in the industry sectors covered. Each of the seven chapters on individual sectors of industry (Chapter 3 on Pharmaceuticals; Chapter 4 on Botanical Medicines; Chapter 5 on Crop Breeding; Chapter 6 on Horticulture; Chapter 7 on Crop Protection; Chapter 8 on Biotechnology; and Chapter 9 on Personal Care and Cosmetics) and on companies' opinions of the CBD were based on information given to us during these interviews, and the results of research on markets and scientific and regulatory trends.

The interviewees were drawn mostly from companies and trade associations, but 31 individuals engaged in product discovery from government agencies, research institutes, genebanks, universities and botanic gardens development (representing 16 per cent of the total) were also interviewed. The vast majority of the interviews were conducted by telephone, and generally ranged in length from 45 to 90 minutes. A few interviews were conducted face to face during meetings. Interviewees were not asked identical questions, but the questions asked of each interviewee were similar, and followed the core set of questions summarised below. Interviewees answered our questionnaire on behalf of their organisations, but also spoke from their personal experience. Perspectives often differed significantly from colleague to colleague, as interviews conducted with two people from the same organisation sometimes revealed. Most of the interviewees were engaged in research and development involving genetic resources, but others worked in management and the legal, public affairs and marketing departments of their organisations.

Questions

The questionnaire for each industry sector was divided into four themes: an overview of the company's activities; the manner in which the company sources materials for product discovery and development; the kind of partnerships and benefit-sharing arrangements that this involves; and the interviewee's opinions on the CBD, experiences with access legislation and collecting permits and recommendations to governments and to other companies. Not all interviewees were asked every question, and the precise questions and the terminology were tailored to the industry sector, and to the nature of the institution being interviewed. For example, while pharmaceutical companies refer to 'natural products', this terminology is not appropriate in the biotechnology or horticulture industries, where the terms 'genetic resources' and 'germplasm' are more commonly used. Another example of a reason for tailoring the questions asked to individual interviewees was that some companies acquire genetic resources (or their derivatives) and sell finished products, while some organisations are 'intermediaries', both acquiring and passing on genetic resources and their derivatives to subsequent recipients for use in research and development.

The term 'company' is used in the questions that follow, since the majority of interviewees (83 per cent) were from companies, but 'organisation' was used when interviewing research institutes, university departments, etc. Despite the distinctions discussed, the questions put to each interviewee were broadly similar and fell into the following categories:

Overview of the company's activities:

- Position and role of interviewee in the company.
- Size of the company (turnover and employees).
- Categories/kinds of products the company produces.
- Stages of discovery and development in which the company is involved.
- Whether the company uses natural products/ 'genetic resources' in its discovery and development activities.

- The proportion of the company's research that focuses on the use of genetic resources, compared with alternative approaches such as synthetic chemistry.
- Whether the company's interest in/demand for access to natural products is changing; whether the interviewee believes the company's demand for access to genetic resources would be greater or less in ten years' time, and why.
- The proportion of the company's research with natural products that involves the search for and introduction of novel organisms, compared with the manipulation of known resources.

Sourcing material for product discovery and development:

- Whether the company has an in-house library or collection of genetic resources.
- Whether that library or collection is used for in-house work only, or access to it licensed as a commercial product.
- What kind of materials the library/collection contains (eg dried plant samples, living plants, culture collections, extracts, compounds, other – if so, what).
- What kind of genetic resources the company's research focuses on (eg plants, insects, human genetic material, other animals, fungi, streptomycetes, actinomycetes, other microorganisms, marine organisms, other – if so, what.
- Who provides the company with the samples/genetic resources used in its research. Whether company staff conduct field collections. Whether the company acquires samples from other departments or companies within the same group and/or from search and discovery companies. Whether the company acquires samples from external collectors or from *ex situ* collections such as universities, culture collections, botanical gardens, seedbanks or other genebanks, or other organisations.
- The proportion of these samples/genetic resources obtained from foreign countries.
- Whether the company would be prepared to pay more for access to genetic resources for research.
- The methods used by the company to prioritise the selection of samples (eg Random/blind; Chemotaxonomic; Ecological/Biorational; Ethnobotanical; Other).
- The relative importance to the company of sourcing samples from different environments such as soil, forest/field, marine ecosystems, extreme ecosystems, other.
- Whether the company acquires raw materials (eg dried plants, soil samples) or value-added products from collaborators. Depending on the industry sector in question, the value-added products included extracts (organic or aqueous), materials provided with ethnobotanical information, results of screens provided with materials, identified, bioactive compound (with known structure and activity), elite strains licensed in as parents, isolated genes, animal model data supplied with identified bioactive compound, clinical data supplied with identified bioactive compound, or a final commercial product.
- Whether the company obtained any of these value-added materials from foreign countries.
- Whether the company needs to return to the source for more material for development or scale-up for production.
- The company's criteria and priorities for selecting a country or institution from which to obtain samples (eg biological diversity, quality of sample, confidence in resupply, cost per sample, simplicity of process for obtaining permits, calibre of scientists/technology of in-country collaborators, IPRs in provider country, other).
- Whether the company uses traditional knowledge/ethnobotanical data in product discovery and development. If so, how the data are sourced (eg by recording ethnobotanical information when collecting in the field, by searching literature or databases), and how the data are used (eg to identify materials for screening/use in product discovery and development; to eliminate materials from screening/discovery and development; to demonstrate safety and/or efficacy; other use).
- Whether the company supplies the genetic resources, or products derived from them, to other companies by allowing access to libraries of samples/extracts/compounds to others, licensing compounds or other value-added derivatives on to other companies, entering into joint ventures, or through mergers and acquisitions.
- The kind of customers and clients of the company.

Partnership/benefit-sharing:

- Whether the company pays organisations who provide it with genetic resources.
- If so, what kind of monetary benefits the company offers to its collaborators (including payment per sample, research grant to conduct agreed research on genetic resources, milestone payments, royalties, stakes in equity, joint ventures, other).
- The importance to the company of the following criteria for defining payments to providers of

genetic resources/value-added products: current market rates for royalties, etc; degree of derivation of the final product from the genetic resource supplied; the role of partners in the inventive step/discovery and development of the resource; the provision of ethnobotanical data; the likely market share of the product; other.

- Whether the company offers any kind of non-monetary benefits, such as information or technology.
- If so, which kinds of non-monetary benefit the company offers to its collaborators (including information, research results, technology transfer – including equipment and know-how – training, capacity building, other).
- The market rates for payment to collaborators for supplying products such as the different kinds of raw material and value-added products outlined in 'sourcing', above.
- Whether the company has any collaborations with providers of genetic resources that involve some kind of value-added joint research.
- What, if any, activities the company is prepared to outlicense, and which ones it will only conduct in-house.

Opinions of the CBD, experiences with access and recommendations:

- Whether the interviewee had heard of the CBD.
- If yes, whether the interviewee would characterise the importance of the CBD in that industry as negligible, moderate or significant.
- Whether the interviewee perceived the effect of the CBD as positive, negative or indifferent.
- In what ways the CBD affects the interviewee's company.
- Which regulations the company follows if it collects genetic resources, and its experience in applying for permits.
- Whether the company is concerned to ensure that any supplier of genetic resources has obtained them legally.
- How the company is sure that organisations or individuals supplying it with materials pass the company watertight legal title, so that it is entitled to commercialise products derived from them.
- Whether the company ensures that the government of the country from where the materials came authorised the organisation supplying the company with materials to do so.
- Whether the company has a policy on the CBD or on the acquisition of genetic resources.
- Whether recent policy and legal developments on access and benefit-sharing have influenced the manner in which the company sources genetic resources for its discovery and development, and if so, how.
- What recommendations the interviewee would make to the Conference of the Parties or to countries deciding how to regulate access to their genetic resources.
- What recommendations the interviewee would make to other companies and collectors of genetic resources.

Sector interviews

The figures below relate to the interviews conducted on access, benefit-sharing and the CBD, and not to the additional interviews related to market data, and trends in science, technology and regulations. The geographical indications refer to the location where the interviewee was working, and not necessarily to the headquarters of the company. **It should be noted that the listing of companies and other organisations interviewed in this section in no way implies their endorsement of the information and opinions contained in this book. It is simply a record of those companies and organisations interviewed that agreed to be acknowledged in this way. The authors are grateful to all those interviewed for their cooperation.**

Chapter 3: Pharmaceuticals

Total number of interviews – 26: 24 companies; 2 government institutes, including:

AMRAD Discovery Technologies Pty Ltd (Australia), Andes Pharmaceuticals (USA), Boehringer Ingelheim Pharma KG (Germany), Biotics (UK), Biodiversity (UK), Bioresources International (Ghana and USA), Bristol-Myers Squibb (USA), Glaxo Wellcome PLC (UK), Magainin Pharmaceuticals (USA), Marine Biotechnology Institute Co Ltd (Japan), MDS Panlabs (USA) (subsidiary of MDS, Canada), Monsanto (USA), Merck & Co (USA), Millennium Pharmaceuticals (USA), National Cancer Institute (USA), Novo Nordisk (Denmark), Pfizer (USA), Pharmacognetics (USA), Sankyo Co Ltd (Japan), Scotia Pharmaceuticals Ltd (UK), Shaman Pharmaceuticals (USA), SmithKline Beecham (UK), T-Cell Science (USA), Tularik (USA), Xenova (UK).

Annual revenues/sales of companies interviewed	
<US$100 m	12
<US$500 m	6
> US$1 bn	8

Geographical range of organisations interviewed	
Europe	9
North America	13
Asia	2
Australia	1
Africa	1

Chapter 4: Botanical medicines

Total number of interviews – 21 representatives from companies, including:
Axxon Biopharm Inc (USA and Nigeria), Botanical Liaisons (USA), Celestial Seasonings (USA), East West Herbs Ltd and East West Biotech Ltd (UK), ExtractsPlus (USA), Hauser Inc (USA), Herbal Apothecary (UK), Kneipp Werke (Germany), Pure World Botanicals (USA), Martin Bauer (Germany), Medical Supplies (Sabah) Sdn Bhd (Malaysia), Nature's Herbs (a division of TwinLabs, USA), Nature's Way (USA), Phytopharm plc (UK), Quality Botanicals Inc (USA), Qualiphar (Belgium), Salus-Haus (Germany), Schwabe Group (Germany), Serturner Arzneimittel (a division of Lichtwer Pharmaceuticals, Germany), Shaman Botanicals (USA), Trout Lake Farm Co (USA).

Annual revenues/sales of companies interviewed	
<US$100 m	14
<US$500 m	7
> US$1 bn	0
Geographical range of organisations interviewed	
Europe	9
North America	10
Asia	1
Australia	0
Africa	1
South America	0

Chapter 5: Crop development

Recognising that, in a study of the small scale of this one, it would be impossible to cover the full range of agricultural crops, each with its unique approach to breeding and demand for access to genetic resources, we decided to select four 'indicator' crops: wheat, rice, sorghum and tomato. These cover a good range of factors that might influence breeding programmes and the demand for acces to genetic resources, such as whether the crop was hybrid or inbred; historical genetic diversity in breeding stock; length of breeding history; centre of origin; major centres of consumption; and public and private breeding efforts. We divided our interviews between plant breeders working mainly on these four crops (although they frequently referred to their companies' work with many other crops), and aimed for a geographical spread and allocation of interviews between public and private breeders that reflected the location of the main breeding efforts for each crop.

Total number of interviews – 32: 20 representatives from companies; 12 from universities and research institutions, including:
Advanta Seeds (UK), Agricultural Research Institute of Chile (INIA) (Chile), AgriPro Biosciences (USA), CIMMYT (Mexico), Cornell University (USA), DEKALB (USA), ICRISAT (India), Indo-American Hybrid Seeds Ltd (India), KSU Agricultural Research Centre (USA), Lochow-Petkus (GMBH) (Germany), Mahyco Life Sciences (India), National Agricultural Research Centre (Japan), National Institute of Agrobiological Resources (Japan), NC Plus Hybrids (USA), Nickerson Seeds (UK), Novartis Seeds (Switzerland), Novartis Seeds BV (Netherlands), Nunhems Zaden (Netherlands), Pannar Pty Ltd (RSA), Pioneer Hi-Bred (USA), R AGT (France), Seminis Vegetable Seeds (USA), University of California, Davis (USA), University of Alberta (Canada), University of Florida (USA).

Size of organisation/ number of employees	
1–10	1
11–100	7
100–1,000	14
>1,000	9
Geographical range of organisations interviewed	
Europe	11
North America	15
Asia	2
Australia	0
Africa	2
South America	2

Chapter 6: Horticulture

For reasons similar to those described above for agriculture, it is impossible to do justice to the full range of product categories and organisations producing them in horticulture. To complement the chapter on the development of major crops, we decided to focus mainly on ornamental horticulture, although several interviews were conducted with breeders of fruit and vegetables. Interviews focused mainly on herbaceous ornamental horticulture. Some interviewees also worked on woody ornamental horticulture, cut flowers, foliage plants, bulbs, and fruit and vegetables.

Total number of interviews – 30: 21 representatives from companies and trade associations; 9 from universities and research institutions, including:
ADAS (UK), British Bedding Plant Association (UK), CIOPORA (Netherlands), Crug Farm Plants (UK), David Austin Roses (UK), Eden Botanical Institute (UK), Elidia (Limagrain) (France), Elsoms Seeds Ltd (UK), FHIA (Honduras), Flora Nova (UK), Horticulture Research International (UK), Merensky Technological Services (UK), National Botanic Gardens, Wales (UK), National Institute of Agricultural Botany (UK), Nir Nursery (Israel), Overman Services Ltd (UK), Pacific Seeds (Thailand), Plant Finder (UK), Samuel Yates Seeds (Netherlands), Sande BV (Netherlands), The University Botanical Gardens, Jerusalem (Israel).

Size of organisation/ number of employees	
1–10	5
11–100	13
100–1,000	6
>1,000	2
Geographical range of organisations interviewed	
Europe	20
North America	2
Asia	0
Australia	1
Africa	1
South America	1

Chapter 7: Crop protection

Total number of interviews – 31: 28 representatives from companies; 3 from universities and research institutions, including:

Agra Quest Inc (USA), AgrEvo (Schering Agrochemicals) (UK), Agrisense UK (UK), American Cyanamid (USA), BASF AG (Germany), Bayer AG (Germany), Bayer PLC (UK), Biobest bvba (Belgium), Biotechnology South West Ltd (UK), CSIRO Entomology (Australia), Dow Agro (USA), DuPont (USA), Ecosafe Systems Pty Ltd (India), FMC Europe NV (Belgium), FMC US (USA), INBio (Costa Rica), Institute of Arable Crop Research (UK), Intrachem Bio (International) (Switzerland), Jeloise Company (Ghana), Kemira (Finland), Koppert Biological Systems (Netherlands), Monsanto (USA), Murkumbi Bioagro Pvt Ltd (India), Novartis Crop Protection AG (Switzerland), Zeneca Agrochemicals (UK).

Size of organisation/ number of employees	
1–10	0
11–100	9
100–1,000	3
>1,000	16

Geographical range of organisations interviewed	
Europe	19
North America	7
Asia	2
Australia	1
Africa	1
South America	1

Chapter 8: Biotechnology

It is important to note that the interviews in this chapter addressed applications of biotechnology other than healthcare and agriculture.

Total number of interviews – 31: 28 representatives from companies; 3 from universities and research institutions, including:

University of Aberdeen, Department of Molecular and Cell Biology (UK), Biotal Ltd (UK), Brewing Research International (UK), British Textile Technology Group (UK), CABI Bioscience (UK), Celsis International PLC (UK), Clair-Tech (Netherlands), Cleveland Biotechnology Ltd (UK), DSMZ (German Culture Collection of Microorganisms and Cell Culture) (Germany), Diversa (USA), Ensynthase Engineering Ltd (UK), Genecor International BV (USA and Netherlands), Gist-brocades (Netherlands), Iogen (Canada), Institute of Virology and Environmental Microbiology, (UK), University of Kent, Department of Biosciences (UK), KWS (Planta) (Germany), Kyowa Hakko Kogyo (Japan), Marine Biotechnology Institute (Japan), Micro-Bio Limited (UK), Montana Biotech (USA), New England Biolabs Inc (USA), Shell International Renewables Ltd (UK), Slater UK (UK).

Size of organisation/ number of employees	
1–10	5
11–100	13
100–1,000	6
>1,000	5

Geographical range of organisations interviewed	
Europe	19
North America	8
Asia	2
Australia	0
Africa	0
South America	0

Chapter 9: Personal care and cosmetics

Total number of interviews – 22 representatives from companies, including:

Ales Group (France), Aroma Vera (USA), Aveda Corporation (USA), AE Hobbs (UK), Aubrey Organics (USA), The Body Shop (UK), Croda Inc (UK), H2O+ (USA), Jason's Naturals (USA), Kneipp Werke (Germany), Liz Claiborn (USA), Pt Martina Berto (Indonesia), Neal's Yard (UK), NuSkin (USA), Raintree Nutrition Inc (USA), Tom's of Maine (USA), Weleda (Germany/Switzerland), Yves Rocher (France).

Annual revenues/sales of companies interviewed	
<US$100 m	11
<US$500 m	7
> US$1 bn	4

Geographical range of organisations interviewed	
Europe	8
North America	13
Asia	1
Australia	0
Africa	0
South America	0

The team

The authors

Kerry ten Kate is Head of the Convention on Biological Diversity Unit at the Royal Botanic Gardens, Kew. Formerly a practising barrister, and member of the Secretariat of the United Nations Conference on Environment and Development (UNCED) during the preparations for the 1992 Rio Earth Summit, she is now responsible for coordinating the design and implementation of Kew's policies in response to the CBD. She has advised developing and developed country governments, companies, the Secretariat of the Convention on Biological Diversity, multilateral aid organisations, and research institutions such as botanic gardens on their strategies on access to genetic resources and the sharing of benefits, and has published on this subject and other sustainable development issues. She holds an MA in law from Oxford University.

In the course of this study, she conducted research on pharmaceuticals, biotechnology, crop development, crop protection, and horticulture. She wrote the chapters on biotechnology, crop development, crop protection, horticulture and on the CBD, national access legislation and material transfer agreements, and co-authored with Sarah Laird the chapters on pharmaceuticals, industry and the CBD, the introduction, conclusion and recommendations. With Sarah Laird and Laura Touche, she coordinated the study itself.

Sarah Laird is an independent consultant with a focus on the commercial and cultural context of biodiversity and forest conservation. She has conducted research and provided advice on access and benefit-sharing issues for a range of NGOs, governments, research institutes, and community groups, most recently in Cameroon and Malaysia. Projects undertaken of late include those for the WWF People and Plants Programme, the World Resources Institute, WWF Cameroon, and the United Nations Environment Programme. Publications in this field include co-authorship of *Biodiversity Prospecting* (1993), *Biodiversity Prospecting in South Africa: Towards the Development of Equitable Partnerships* (1996), and *Benefit Sharing Case Studies from Cameroon* (1998). She received an MSc in Forestry from Oxford University.

In the course of this study, Sarah conducted and oversaw research into the pharmaceutical, botanical medicine, and personal care and cosmetic industries. This included conducting the majority of interviews for each sector. She authored the chapters on the botanical medicine and personal care and cosmetic sectors, and co-authored, with Kerry ten Kate, chapters on the pharmaceutical industry, industry and the CBD, the introduction, conclusions and recommendations. With Kerry ten Kate and Laura Touche she designed and coordinated the project upon which this book is based.

The project coordinators

The study was administered by the Royal Botanic Gardens, Kew, and coordinated by Kerry ten Kate, Sarah Laird and Laura Touche.

Laura Touche was the Senior Advisor in the CBD Unit at the Royal Botanic Gardens, Kew. An American, with degrees from the Woodrow Wilson School of Princeton University and Harvard Law School, Laura was a practising attorney before joining Kew. At Kew, she developed access and benefit-sharing agreements, liaised on these issues between Kew and its partners in the UK and overseas, and conducted research on legal and policy issues related to access and benefit-sharing.

In addition to her help in administering this project as a whole, she contributed to the research and analysis of the data gathered and helped to edit the entire volume. Without her enthusiasm, enquiring mind, eye for detail and meticulous management and editing, this project could not have happened. Her tragic death in February 1999 at the age of 31 from a complication following the delivery of healthy twin sons is a terrible loss to the conservation movement and to her family, friends and colleagues.

The research team

Harry Barton: Currently Director of the Council for the Protection of Rural England, Kent, Harry was formerly Programmes Manager at the RBG, Kew Foundation & Friends. He has a BSc in Geography from Durham University and an MSc in Rural Resources and Environmental Policy (Wye). His work on this project centred on the chapter on major crops, for which he conducted research on IR64 for the preparation of the case study, and on national genebanks and general reseach on the horticulture chapter.

Steve Brewer: Now an independent consultant to the biotechnology, agriculture and pharmaceutical industries, Steve was formerly a director in product discovery at Monsanto, where he was responsible for developing their natural products acquisitions programme. He has a PhD in Biochemistry. For this project he conducted interviews in the crop protection and biotechnology sectors, and provided background reports on scientific, technological and regulatory trends and on markets in crop protection and biotechnology.

Amanda Collis: A programme manager for the Engineering and Biological Systems Directorate at the Biotechnology and Biological Sciences Research Council, UK. Her contribution to the IPTS survey included questionnaire development, interviewing biotech industry representatives and interview data analysis. She has a PhD in molecular biology, and her academic background is in plant and fungal molecular biology and biotechnology.

Mike Griffiths: Now managing his own environmental biotechnology consultancy, Mike has a DPhil in Biochemistry from Oxford University. He spent most of his working life in biotechnology R & D and new business development at Shell, both in the UK and Brazil, particularly in the areas of forestry, alternative energy sources such as bioethanol, and the production of industrial enzymes. He has participated in a number of major OECD studies on environmental biotechnology. Mike conducted a number of the interviews for the biotechnology chapter, assisted in developing the analytical approach used for biotech industries and participated in the expert review of this chapter.

Dagmar Lange: An independent consultant with a focus on commercial and conservation aspects in the medicinal plants trade, Dagmar has a PhD in Botany and a background in taxonomy. She has conducted research on the medicinal plants trade in Europe (principally Germany) and India. She recently developed a training unit for CITES-listed medicinal plants. Projects undertaken include those for the German Federal Agency for Nature Conservation, WWF Germany, and TRAFFIC Europe. Her publications in this field include *Europe's Medicinal and Aromatic Plants: Their Use, Trade and Conservation* (1998), and co-authorship of *Trade Survey of Medicinal Plants in Germany* (1997), and *Checklist of Medicinal and Aromatic Plants and their Trade Names covered by CITES and EU Regulation 2307/97* (1999). For this project Dagmar conducted interviews with the German botanical medicine companies included in this study, as well as a German pharmaceutical company. She also contributed valuable market data to the chapter on botanical medicines, and participated in the expert review of this chapter.

Lucia Liscio: Lucia studied at the University of Toronto and the London School of Economics before working in investment banking for two years. She has decided to change career to law, and is finishing her LLB studies. Lucia helped with the quantitative analysis of interviewees' responses on the development of major crops.

Ben Lyte: Currently based at the Conservation Projects Development Unit at RBG, Kew, Ben concentrates on researching threatened plant species. He has practical horticultural experience and an HND in Horticulture from the Scottish Agricultural College. On this project Ben conducted many of the horticulture interviews.

Craig Metrick: While pursuing a Master's Degree at George Mason University, Fairfax,VA, Craig has undertaken environmental and research consulting projects for, among others, the Biodiversity Action Network. His academic research analyses the Mesoamerican Biological Corridor as a financial mechanism for biodiversity conservation. Craig compiled and organised the contacts and websites sections of this study.

Fiona Mucklow: Formerly associated with FIELD (Foundation of International Environmental Law and Development), where she worked on biodiversity and climate change projects, and investigated EU Directives relevant to environmental issues, Fiona has worked on projects involving the Rainforest Foundation UK; the Lauterpacht Research Centre of International Law (University of Cambridge); as well as with the Legal Affairs Division of the World Trade Organisation. She has an LLM from the London School of Economics, specialising in Public International Law, and a BSc in Geography. In September 1999, she is to commence as a trainee solicitor with Clyde & Co (London). For this project Fiona conducted interviews in the major crops and crop protection sector. She researched and prepared background reports on regulations in the seeds and crop protection sectors, and prepared the Latin American ICBG case study.

Kristina Plenderleith: Kristina undertakes research on issues relating to traditional agricultural systems, agricultural sustainability, and farmers' rights. With Darrell Posey and Graham Dutfield she set up the Working Group on Traditional Resource Rights, a forum for the rights of indigenous peoples and local communities. She is a geographer with an MSc in Forestry and its Relation to Land Use from Oxford University. Kristina's contribution to this volume was to conduct interviews with agricultural seed companies and the horticulture industry; to provide reports on breeding trends and markets in the seed industry; and background research on the horticulture industry.

Markus Radscheit: Currently Technical Manager at the University Botanic Garden, Bonn, Marcus was recently a background researcher for the Eden Project, Cornwall, and also worked at the Conservation Projects Development Unit office at Kew. He trained in Horticulture at Bonn University, RHS Wisley, Kew and Writtle College. For this project he conducted over 20 interviews with commercial plant breeders for the horticulture chapter.

Helene Weitzner: Spent six years at the Rainforest Alliance, primarily as Chief Development Officer, and also helping to coordinate the 1992 Rainforest Alliance – New York Botanical Garden 1992 symposium 'Tropical Forest Medical Resources and the Conservation of Biodiversity'. She has an MSc in Environmental Conservation from New York University, and recently has been working as a project analyst for a Zimbabwean venture capital firm in Harare. Helene conducted marketing research for this project, primarily in the area of personal care and cosmetic products, and also for botanical medicines and pharmaceuticals.

Adrian Wells: Currently undertaking a Masters degree in International Environmental, Trade and IP Law at the University of London, Adrian worked at the CBD Unit, RBG Kew, for a year, during which time he assisted with interviews and market research for the crop protection chapter. He researched the case studies of crop-protection products and LUBILOSA.

China Williams: Formerly a practising barrister and now working in the CBD Unit, RBG Kew, China is currently completing a Masters degree in International Environmental Law at SOAS, University of London. Her work on this project included research for the introduction and conclusion, the horticulture chapter, and on biotechnology markets and regulations. She also prepared the NEB case study in the biotechnology chapter.

Administrative assistance

Helen Armitage: Helen is a writer and book editor. She was in-house senior and executive editor for a variety of major publishing houses for 12 years before going freelance in 1989. Since then she has worked as a copywriter for Saatchi and Saatchi, researched, edited and ghostwritten fiction and non-fiction, and taught writing courses at the European Business School. For this project she corresponded with interviewees over quotations and worked on the bibliography.

Amanda Collis: see above.

Stuart Huntley: Completed his BA LLB in 1998 at the University of Cape Town, South Africa, where he worked as a part-time researcher for the Institute of Marine Law. Stuart came to the UK to travel, watch the Rugby World Cup, and gain a further understanding of International law. He worked on the bibliography and methodology chapters while also conducting further research on various organisations interviewed.

Index

Italics refer to figures; **bold** refers to tables.

Milton Keynes UK
Ingram Content Group UK Ltd.
UKHW051856071024
449327UK00025B/1986